Lecture Notes in Computer Science 7808

Commenced Publication in 1973
Founding and Former Series Editors:
Gerhard Goos, Juris Hartmanis, and Jan van Leeuwen

Yoshiharu Ishikawa   Jianzhong Li
Wei Wang   Rui Zhang   Wenjie Zhang (Eds.)

Web Technologies and Applications

15th Asia-Pacific Web Conference, APWeb 2013
Sydney, Australia, April 4-6, 2013
Proceedings

 Springer

Volume Editors

Yoshiharu Ishikawa
Nagoya University
Graduate School of Information Science
Nagoya 464-8601, Japan
E-mail: ishikawa@is.nagoya-u.ac.jp

Jianzhong Li
Harbin Institute of Technology
Department of Computer Science and Technology
Harbin 150006, China
E-mail: ljz@mail.banner.com

Wei Wang
Wenjie Zhang
University of New South Wales
School of Computer Science and Engineering
Sydney, NSW 2052, Australia
E-mail: {weiw, zhangw}@cse.unsw.edu.au

Rui Zhang
University of Melbourne
Department of Computing and Information Systems
Melbourne, VIC 3052, Australia
E-mail: rui@csse.unimelb.edu.au

ISSN 0302-9743   e-ISSN 1611-3349
ISBN 978-3-642-37400-5   e-ISBN 978-3-642-37401-2
DOI 10.1007/978-3-642-37401-2
Springer Heidelberg Dordrecht London New York

Library of Congress Control Number: 2013934117

CR Subject Classification (1998): H.2.8, H.2, H.3, H.5, H.4, J.1, K.4, I.2

LNCS Sublibrary: SL 3 – Information Systems and Application, incl. Internet/Web
and HCI

Typesetting: Camera-ready by author, data conversion by Scientific Publishing Services, Chennai, India

Printed on acid-free paper

Springer is part of Springer Science+Business Media (www.springer.com)

Message from the General Chairs

Welcome to APWeb 2013, the 15th Edition of the Asia Pacific Web Conference. APWeb is a leading international conference on research, development, and applications of Web technologies, database systems, information management, and software engineering, with a focus on the Asia-Pacific region. Previous APWeb conferences were held in Kunming (2012), Beijing (2011), Busan (2010), Suzhou (2009), Shenyang (2008), Huangshan (2007), Harbin (2006), Shanghai (2005), Hangzhou (2004), Xi'an (2003), Changsha (2001), Xi'an (2000), Hong Kong (1999), and Beijing (1998).

The APWeb 2013 conference was, for the first time, held in Sydney, Australia — a city blessed with a temperate climate, a beautiful harbor, and natural attractions surrounding it. These proceedings collect the technical papers selected for presentation at the conference, during April 4–6, 2013.

The APWeb 2013 program featured a main conference, a special track, and four satellite workshops. The main conference had three keynotes by eminent researchers H.V. Jagadish from the University of Michigan, USA, Mark Sanderson from RMIT University, Australia, and Dan Suciu from the University of Washington, USA. Three tutorials were offered by Haixun Wang, Microsoft Research Asia, China, Yuqing Wu, Indiana University, USA, George Fletcher, Eindhoven University of Technology, The Netherlands, and Lei Chen, Hong Kong University of Science and Technology, Hong Kong, China. The conference received 165 paper submissions from North America, South America, Europe, Asia, and Oceania. Each submitted paper underwent a rigorous review by at least three independent referees, with detailed review reports. Finally, 39 full research papers and 22 short research papers were accepted, from Australia, Bangladesh, Canada, China, India, Ireland, Italy, Japan, New Zealand, Saudia Arabia, Sweden, Norway, UK, and USA. The special track of "Distributed Processing of Graph, XML and RDF Data: Theory and Practice" was organized by Alfredo Cuzzocrea. The conference had four workshops

- The Second International Workshop on Data Management for Emerging Network Infrastructure (DaMEN 2013)
- International Workshop on Location-Based Data Management (LBDM 2013)
- International Workshop on Management of Spatial Temporal Data (MSTD 2013)
- International Workshop on Social Media Analytics and Recommendation Technologies (SMART 2013)

We were extremely excited with our strong Program Committee, comprising outstanding researchers in the APWeb research areas. We would like to extend our sincere gratitude to the Program Committee members and external reviewers. Last but not least, we would like to thank the sponsors, for their strong support of this conference, making it a big success. Special thanks go to the Chinese University of Hong Kong, the University of New South Wales, Macquarie University, and the University of Sydney.

Finally, we wish to thank the APWeb Steering Committee, led by Xuemin Lin, for offering us the opportunity to organize APWeb 2013 in Sydney. We also wish to thank the host organization, the University of New South Wales, and Local Arrangements Committee and volunteers for their assistance in organizing this conference.

February 2013   Vijay Varadharajan
                                                                       Jeffrey Xu Yu

Conference Organization

Conference Co-chairs

Vijay Varadharajan   Macquarie University, Australia
Jeffrey Xu Yu   Chinese University of Hong Kong, China

Program Committee Co-chairs

Yoshiharu Ishikawa   Nagoya University, Japan
Jianzhong Li   Harbin Institute of Technology, China
Wei Wang   University of New South Wales, Australia

Local Organization Co-chairs

Muhammad Aamir Cheema   University of New South Wales, Australia
Ying Zhang   University of New South Wales, Australia

Workshop Co-chairs

James Bailey   University of Melbourne, Australia
Xiaochun Yang   Northeastern University, China

Tutorial/Panel Co-chairs

Sanjay Chawla   University of Sydney, Australia
Xiaofeng Meng   Renmin University of China, China

Industrial Co-chairs

Marek Kowalkiewicz   SAP Research in Brisbane, Australia
Mukesh Mohania   IBM Research, India

Publication Co-chairs

Rui Zhang   University of Melbourne, Australia
Wenjie Zhang   University of New South Wales, Australia

Publicity Co-chairs

Alfredo Cuzzocrea	University of Calabria, Italy
Jiaheng Lu	Renmin University of China, China

Demo Co-chairs

Wook-Shin Han	Kyungpook National University, Korea
Helen Huang	University of Queensland, Australia

APWeb Steering Committee Liaison

Xuemin Lin	University of New South Wales, Australia

WISE Society Liaison

Yanchun Zhang	Victoria University, Australia

WAIM Steering Committee Liaison

Qing Li	City University of Hong Kong, China

Webmasters

Yu Zheng	East China Normal University, China
Chen Chen	University of New South Wales, Australia

Program Committee

Toshiyuki Amagasa	University of Tsukuba
Djamal Benslimane	University of Lyon
Jae-Woo Chang	Chonbuk National University
Haiming Chen	Chinese Academy of Sciences
Jinchuan Chen	Renmin University of China
David Cheung	The University of Hong Kong
Bin Cui	Beijing University
Alfredo Cuzzocrea	ICAR-CNR & University of Calabria
Ting Deng	Beihang University
Jianlin Feng	Sun Yat-Sen University
Yaokai Feng	Kyushu University

Sergio Flesca	University of Calabria
Hong Gao	Harbin Institute of Technology
Yunjun Gao	Zhejiang University
Stephane Grumbach	INRIA
Giovanna Guerrini	University of Genoa
Mohand-Said Hacid	University of Lyon 1
Qi He	IBM
Jun Hong	Queen's University Belfast
Michael Houle	National Institute of Informatics
Bin Hu	Lanzhou University
Zi Huang	University of Queensland
Jeong-Hyon Hwang	State University of New York at Albany
Seung-won Hwang	POSTECH
Mizuho Iwaihara	Waseda University
Adam Jatowt	Kyoto University
Cheqing Jin	East China Normal University
Anastasios Kementsietsidis	IBM T.J. Watson Research Center
Jin-Ho Kim	Kangwon National University
Markus Kirchberg	Institute for Infocomm Research
Manolis Koubarakis	University of Athens
Byung Lee	Vermont University
Chiang Lee	National Cheng Kung University
Jae-Gil Lee	KAIST
SangKeun Lee	Korea University
Carson Leung	University of Manitoba
Jianxin Li	Swinburne University
Xue Li	Queensland University
Yingshu Li	Georgia State University
Yinsheng Li	Fudan University
Zhanhuai Li	Northwestern Polytechnical University
Xiang Lian	University of Texas - Pan American
Guimei Liu	National University of Singapore
Mengchi Liu	Carleton University
Chengfei Liu	Swinburne University of Technology
Bo Luo	University of Kansas
Jiangang Ma	University of Adelaide
Qiang Ma	Kyoto University
Shuai Ma	Beihang University
Zakaria Maamar	Zayed University
Sanjay Madria	University of Missouri-Rolla
Weiyi Meng	State University of New York at Binghamton
Yang-Sae Moon	Kangwon National University
Michael Mrissa	University of Lyon
Akiyo Nadamoto	Konan University
Shinsuke Nakajima	Kyoto Sangyo University

Miyuki Nakano	University of Tokyo
Werner Nutt	Free University of Bozen-Bolzano
Satoshi Oyama	Hokkaido University
Helen Paik	University of New South Wales
Chaoyi Pang	CSIRO
Apostolos Papadopoulos	Aristotle University
Eric Pardede	Latrobe University
Sanghyun Park	Yonsei University
Zhiyong Peng	Wuhan University
Tieyun Qian	Wuhan University
Weining Qian	East China Normal University
Joao Rocha-Junior	Univ. Estadual de Feira de Santana
KeunHo Ryu	Chungbuk National University
Markus Schneider	University of Florida
Marc Scholl	Universität Konstanz
Aviv Segev	KAIST
Bin Shao	Microsoft Research Asia
Derong Shen	Northeastern University
Heng Shen	Queensland University
Jialie Shen	Singapore Management University
Timothy Shih	National Taipei University of Education
Lidan Shou	Zhejiang University
Shaoxu Song	Tsinghua University
Konstantinos Stefanidis	Norwegian University of Science and Technology
Kazutoshi Sumiya	University of Hyogo
Aixin Sun	Nanyang Technological University
Claudia Szabo	University of Adelaide
Changjie Tang	Sichuan University
Nan Tang	University of Edinburgh
David Taniar	Monash University
Alex Thomo	University of Victoria
Chaokun Wang	Tsinghua University
Daling Wang	Northeastern University
Fan Wang	Microsoft
Guoren Wang	Northeastern University
Hongzhi Wang	Harbin Institute of Technology
Hua Wang	University of Southern Queensland
Jianyong Wang	Tsinghua University
X. Wang	Fudan University
Xiaoling Wang	East China Normal University
Jef Wijsen	University of Mons-Hainaut
Jianliang Xu	Hong Kong Baptist University
Xiaochun Yang	Northeastern University
Jian Yin	Sun Yat-Sen University
Haruo Yokota	Tokyo Institute of Technology

Jian Yu	Swinburne University of Technology
Ming Zhang	Beijing University
Xiao Zhang	Renmin University of China
Baihua Zheng	Singapore Management University
Rui Zhou	Swinburne University
Shuigeng Zhou	Fudan University
Xiaofang Zhou	University of Queensland
Xuan Zhou	Renmin University
Zhaonian Zou	Harbin Institute of Technology

External Reviewers

Xuefei Li	Mahmoud Barhamgi
Hongyun Cai	Xian Li
Jingkuan Song	Yu Jiang
Yang Yang	Saurav Acharya
Xiaofeng Zhu	Syed K. Tanbeer
Scott Bourne	Hongda Ren
Yasser Salem	Wei Shen
Shi Feng	Zhenhua Song
Jianwei Zhang	Jianhua Yin
Kenta Oku	Liu Chen
Sukhwan Jung	Wei Song

An Overview of Probabilistic Databases

Dan Suciu

University of Washington
suciu@cs.washington.edu
http://homes.cs.washington.edu/ suciu/

A major challenge in modern data management is how to cope with uncertainty in the data. Uncertainty may exists because the data was extracted automatically from text, or was derived from the physical world such as RFID data, or was obtained by integrating several data sets using fuzzy matches, or may be the result of complex stochastic models. In a *probabilistic database* uncertainty is modeled using probabilities, and data management techniques are extended to cope with probabilistic data.

The main challenge is query evaluation. For each answer to the query, its degree of uncertainty is the probability that its lineage formula is true. Thus, query evaluation reduces to the problem of computing the probability of a Boolean formula. This problem generalizes model counting, which has been extensively studied in the AI and model checking literature. Today's state of the art methods for computing the exact probability are extensions of Davis Putnam's (DP) procedure [3, 2, 1, 4]. In probabilistic databases we can take a new approach, because here we can fix the query, and consider only the database as variable input (called *data complexity* [7]). An interesting dichotomy theorem holds: for every query, either its complexity is in PTIME or is #P-hard. A new probabilistic inference algorithm was needed in order to compute *all* PTIME queries, which uses the inclusion/exclusion principle [6]. This technique is missing from today's extensions of DP, yet necessary: without it one can show that probabilistic inference for certain simple PTIME queries requires exponential time [5].

References

1. Bacchus, F., Dalmao, S., Pitassi, T.: Algorithms and complexity results for #sat and bayesian inference. In: FOCS, pp. 340–351 (2003)
2. Birnbaum, E., Lozinskii, E.L.: The good old davis-putnam procedure helps counting models. J. Artif. Int. Res. 10(1), 457–477 (1999)
3. Davis, M., Logemann, G., Loveland, D.: A machine program for theorem-proving. Commun. ACM 5(7), 394–397 (1962)
4. Gomes, C.P., Sabharwal, A., Selman, B.: Model counting. In: Handbook of Satisfiability, pp. 633–654 (2009)
5. Jha, A.K., Suciu, D.: Knowledge compilation meets database theory: compiling queries to decision diagrams. In: ICDT, pp. 162–173 (2011)
6. Suciu, D., Olteanu, D., Ré, C., Koch, C.: Probabilistic Databases. In: Synthesis Lectures on Data Management. Morgan & Claypool Publishers (2011)
7. Vardi, M.Y.: The complexity of relational query languages (extended abstract). In: STOC, pp. 137–146 (1982)

An Overview of Probabilistic Databases

Dan Suciu

University of Washington
suciu@cs.washington.edu

References

Challenges with Big Data on the Web

H.V. Jagadish*

University of Michigan
jag@umich.edu

The promise of data-driven decision-making is now being recognized broadly, and there is growing enthusiasm for the notion of "Big Data.ÕÕ This is true of Big Data in the enterprise, but this is even more true of Big Data on the web. While the promise of Big Data is real – for example, it is estimated that Google alone contributed 54 billion dollars to the US economy in 2009 – there is currently a wide gap between its potential and its realization.

Heterogeneity, scale, timeliness, complexity, and privacy problems with Big Data impede progress at all phases of the pipeline that can create value from data. The problems start right away during data acquisition, when the data tsunami requires us to make decisions, currently in an ad hoc manner, about what data to keep and what to discard, and how to store what we keep reliably with the right metadata. Much data today is not natively in structured format; for example, tweets and blogs are weakly structured pieces of text, while images and video are structured for storage and display, but not for semantic content and search: transforming such content into a structured format for later analysis is a major challenge. The value of data explodes when it can be linked with other data, thus data integration is a major creator of value. Since most data is directly generated in digital format today, we have the opportunity and the challenge both to influence the creation to facilitate later linkage and to automatically link previously created data. Data analysis, organization, retrieval, and modeling are other foundational challenges. Finally, presentation of the results and its interpretation by non-technical domain experts is crucial to extracting actionable knowledge.

A recent white paper[CCC12] mapped out the many challenges in this space. In this talk, drawing upon this white paper, I will present these challenges, particularly as they relate to the web. I will draw upon examples from database usability to show how size and complexity of Big Data can create difficulties for a user, and mention some directions of work in this regard. In particular, I will highlight how Big Data issues arise in surprising contexts, such as in browsing[SIGMOD12].

References

[CCC12] Jagadish, H.V., et al: Challenges and Opportunities with Big Data, http://cra.org/ccc/docs/init/bigdatawhitepaper.pdf
[SIGMOD12] Singh, M., Nandi, A., Jagadish, H.V.: Skimmer: rapid scrolling of relational query results. In: SIGMOD Conference, pp. 181–192 (2012)

---

* Supported in part by NSF under grant IIS-1017296.

Twenty Years of Web Search – Where to Next?

Mark Sanderson

School of Computer Science and Information Technology
RMIT University
GPO Box 2476, Melbourne 3001
Victoria, Australia

Abstract. This year, (2013) marks the 20th anniversary of the first public web search engine JumpStation launched in late 1993. For those who were around in those early days, it was becoming clear that an information provision and an information access revolution was on its way; though very few, if any would have predicted the state of the information society we have today. It is perhaps worth reflecting on what has been achieved in the field of information retrieval since these systems were first created, and consider what remains to be accomplished. It is perhaps easy to see the success of systems like Google and ask what else is there to achieve? However, in some ways, Google has it easy. In this talk, I will explain why Web search can be viewed as a relatively easy task and why other forms of search are much harder to perform accurately.

Search engines require a great deal of tuning, currently achieved empirically. The tuning carried out depends greatly on the types of queries submitted to a search engine and the types of document collections the queries will search over. It should be possible to study the population of queries and documents and predictively configure a search engine. However, there is little understanding in either the research or practitioner communities on how query and collection properties map to search engine configurations. I will present the some of the early work we have conducted at RMIT to start charting the problems in this particular space.

Another crucial challenge for search engine companies is how to ensure that users are delivered the best quality content. There is a growth in systems that recommend content based not only on queries, but also on user context. The problem is that the quality of these systems is highly variable; one way of tackling this problem is gathering context from a wider range of places. I will present some of the possible new approaches to providing that context to search engines. Here diverse social media, and advances in location technologies will be emphasized.

Finally, I will describe what I see as one of the more important challenges that face the whole of the information community, namely the penetration of computer systems to virtually every person on the planet and the challenges that such an expansion presents.

Table of Contents

Tutorials

Distributed Processing:

Graphs

Web Search and Web Mining

XML, RDF Data and Query Processing

Social Networks

Probabilistic Queries

Multimedia and Visualization

Spatial-Temporal Databases

Data Mining and Knowledge Discovery

Privacy and Security

Performance

Query Processing and Optimization

The Second International Workshop on Data Management for Emerging Network Infrastructure

International Workshop on Social Media Analytics and Recommendation Technologies

International Workshop on Management of Spatial Temporal Data

Understanding Short Texts

Haixun Wang

Microsoft Research Asia, China
haixunw@microsoft.com

Abstract. Many applications handle short texts, and enableing machines to understand short texts is a big challenge. For example, in Ads selection, it is is difficult to evaluate the semantic similarity between a search query and an ad. Clearly, edit distance based string similarity does not work. Moreover, statistical methods that find latent topic models from text also fall short because ads and search queries are insufficient to provide enough statistical signals.

In this tutorial, I will talk about a knowledge empowered approach for text understanding. When the input is sparse, noisy, and ambiguous, knowledge is needed to fill the gap in understanding. I will introduce the Probase project at Microsoft Research Asia, whose goal is to enable machines to understand human communications. Probase is a universal, probabilistic taxonomy more comprehensive than any current taxonomy. It contains more than 2 million concepts, harnessed automatically from a corpus of 1.68 billion web pages and two years worth of search-log data. It enables probabilistic interpretations of search queries, document titles, ad keywords, etc. The probabilistic nature also enables it to incorporate heterogeneous information naturally. I will explain how the core taxonomy, which contains hypernym-hyponym relationships, is constructed and how it models knowledge's inherent uncertainty, ambiguity, and inconsistency.

1 Speakers

Haixun Wang is a senior researcher at Microsoft Research Asia in Beijing, China, where he manages the Data Management, Analytics, and Services group. Before joining Microsoft, he had been a research staff member at IBM T. J. Watson Research Center for 9 years. He was Technical Assistant to Stuart Feldman (Vice President of Computer Science of IBM Research) from 2006 to 2007, and Technical Assistant to Mark Wegman (Head of Computer Science of IBM Research) from 2007 to 2009. Haixun Wang has published more than 120 research papers in referred international journals and conference proceedings. He is on the editorial board of Distributed and Parallel Databases (DAPD), IEEE Transactions of Knowledge and Data Engineering (TKDE), Knowledge and Information System (KAIS), Journal of Computer Science and Technology (JCST). He is PC co-Chair of WWW 2013 (P&E), ICDE 2013 (Industry), CIKM 2012, ICMLA 2011, WAIM 2011. Haixun Wang got the ER 2008 Conference best paper award (DKE 25 year award), and ICDM 2009 Best Student Paper run-up award.

Y. Ishikawa et al. (Eds.): APWeb 2013, LNCS 7808, p. 1, 2013.

Managing the Wisdom
of Crowds on Social Media Services

Lei Chen

Hong Kong University of Science and Technology, Hong Kong
leichen@cse.ust.hk

Abstract. Recently, the "Wisdom of Crowds" has attracted a huge amount of interests from both research and industrial communities. For the current stage, most of focus has been put on several specific crowdsourcing marketplaces like Amazon MTurk or CrowdFlower, on which "requesters" publish tasks and "workers" select tasks according to their own benefits. However, users on social media services can also serve as candidate "workers" for crowdsourcing tasks, and it is possible for the "requesters" to actively manage the quality and cost of such crowds.

In this tutorial, we will first review the basic concept of crowdsourcing and its applications. Then, we will discuss the current popular crowdsourcing platforms and several interesting crowdsourcing related algorithms. Finally, we will discuss the benefits of using social media services as the crowdsourcing platform and propose several research challenges on managing the wisdom of crowds on social media.

1 Speakers

Lei Chen received the BS degree in Computer Science and Engineering from Tianjin University, Tianjin, China, in 1994, the MA degree from Asian Institute of Technology, Bangkok, Thailand, in 1997, and the PhD degree in computer science from the University of Waterloo, Waterloo, Ontario, Canada, in 2005. He is currently an Associate Professor in the Department of Computer Science and Engineering, Hong Kong University of Science and Technology. His research interests include crowd sourcing on social media, social media analysis, probabilistic and uncertain databases, and privacy-preserved data publishing. So far, he has published nearly 200 conference and journal papers. He got the Best Paper Awards in DASFAA 2009 and 2010. He is PC Track Chairs for VLDB 2014, ICDE 2012, CIKM 2012, SIGMM 2011. He has served as PC members for SIGMOD, VLDB, ICDE, SIGMM, and WWW. Currently, he serves as an Associate Editor for IEEE Transaction on Data and Knowledge Engineering and Distribute and Parallel Databases. He is a member of the ACM and the chairman of ACM Hong Kong Chapter.

Y. Ishikawa et al. (Eds.): APWeb 2013, LNCS 7808, p. 2, 2013.

Search on Graphs: Theory Meets Engineering

Yuqing Wu[1] and George H.L. Fletcher[2]

[1] Indiana University, Bloomington, USA
`yuqwu@cs.indiana.edu`
[2] Eindhoven University of Technology, The Netherlands
`g.h.l.fletcher@tue.nl`

Abstract. The last decade has witnessed an explosion of the availability of and interest in graph structured data. The desire to search and reason over these increasingly massive data collections pushes the boundaries of search languages, from pure keyword search to structure-aware searches in the graph. These phenomena have inspired a rich body of research on query languages, data management and query evaluation techniques for graph data, both from the theoretical and engineering angles. In this tutorial, we present an overview of the progress on graph search queries, focusing specifically on how the theoretical and engineering perspectives meet and together advanced the field.

1 Tutorial Overview

Exploratory keyword-style search has been heavily studied in the past decade, both in the context of structured [18] and semi-structured [9] data. Given the ubiquity of massive (loosely structured) graph data in domains such as the web, social networks, biological networks, and linked open data (to name a few), there recently has been a surge of interest and advances on the problem of search in graphs (e.g., [1, 3, 12, 15, 17, 19, 20]). As graph exploration leads to deeper domain understanding, user queries begin to shift from unstructured searching to richer structure-based exploration of the graph. Consequently, there has been a flurry of language proposals specifically targeting this style of structure-aware querying in graphs (e.g., [2, 3, 5, 13, 16]).

In this tutorial, we survey this growing body of work, with an eye towards both bringing participants up to speed in this field of rapid progress and delimiting the boundaries of the state-of-the-art. A particular focus will be on recent results in the theory of graph languages on the design and structural characterization of simple yet powerful algebraic languages for graph search, which bridge structure-oblivious and structure-aware graph exploration [4–6]. At the heart of these results is the methodology of coupling the expressive power of a given query language with an appropriate structural notion on data instances. Here, the idea is to characterize language equivalence of data objects in instances (i.e., the inability of queries in the language to distinguish the objects) purely in terms of the structure of the instance (i.e., equivalence under notions such as homomorphism or bisimilarity). Recently, first steps towards graph

Y. Ishikawa et al. (Eds.): APWeb 2013, LNCS 7808, pp. 3–6, 2013.

indexing have shown promise in transferring this theoretical framework into practice [7, 14]. The basic intuition behind this approach is that data should be organized to optimally reflect the type and style of queries being asked, and that the optimality of this organization can be formally established, as above. Recent results have established the practical feasibility of computing and maintaining these structural organizations on massive graphs [8, 10, 11].

2   Tutorial Outline

The tutorial will be presented as follows:

Part 1: Searching the Graph
We start the tutorial with an overview of variants of graph data and the evolution of the search queries on graph, leading into the discussion of the challenges posed by the massive size and complex nature of graph data and the flexible nature of graph queries and their evaluation.

Part 2: Bridging Theory and Engineering
We next examine the lines of work in the theoretical study of query languages and the engineering efforts in developing novel techniques for managing graph data and evaluating various types of search queries on such data. We then present our methodology of coupling the expressive power of a given query language with an appropriate structural notion on data instances, as a tool for reasoning about and guiding engineering efforts.

Part 3: Indexing Graph Data – A Case Study
We follow this discussion with a presentation of the design of indices for semi-structured and graph data, as a case study, to illustrate our methodology.

Part 4: Looking Forward
We close the tutorial with indications of ongoing and future research directions.

3   Speakers

Yuqing Wu and George Fletcher, together with a group of collaborators in the USA, the Netherlands, and Belgium, have been conducting research in this area in recent years and have published several papers in both the theory and engineering branches of database research.

Yuqing Wu, Indiana University, Bloomington, USA
Prof. Wu is an Associate Professor at School of Informatics and Computing, Indiana University, Bloomington, USA. Prof. Wu received her Ph.D. degree from University of Michigan, Ann Arbor, in 2004. Her research area is in data management, especially semi-structure and non-structured data, with an emphasis on query language, query processing and query optimization.

George Fletcher, Eindhoven University of Technology, The Netherlands
Dr. Fletcher is an Assistant Professor in the Databases and Hypermedia group at Eindhoven University of Technology, The Netherlands. Dr. Fletcher was awarded a doctorate in computer science from Indiana University, Bloomington (2007), with a dissertation on the topic of query learning for data integration. His current research focuses on the study of database query languages for data integration and web data.

References

1. Delbru, R., Campinas, S., Tummarello, G.: Searching web data: An entity retrieval and high-performance indexing model. J. Web Sem. 10, 33–58 (2012)
2. Fazzinga, B., Gianforme, G., Gottlob, G., Lukasiewicz, T.: Semantic web search based on ontological conjunctive queries. J. Web Sem. 9(4), 453–473 (2011)
3. Fletcher, G.H.L., Van den Bussche, J., Van Gucht, D., Vansummeren, S.: Towards a theory of search queries. ACM Trans. Database Syst. 35(4), 28 (2010)
4. Fletcher, G.H.L., Gyssens, M., Leinders, D., Van den Bussche, J., Van Gucht, D., Vansummeren, S.: Similarity and bisimilarity notions appropriate for characterizing indistinguishability in fragments of the calculus of relations. CoRR, abs/1210.2688 (2012)
5. Fletcher, G.H.L., Gyssens, M., Leinders, D., Van den Bussche, J., Van Gucht, D., Vansummeren, S., Wu, Y.: Relative expressive power of navigational querying on graphs. In: Proc. ICDT, Uppsala, Sweden, pp. 197–207 (2011)
6. Fletcher, G.H.L., Gyssens, M., Leinders, D., Van den Bussche, J., Van Gucht, D., Vansummeren, S., Wu, Y.: The impact of transitive closure on the boolean expressiveness of navigational query languages on graphs. In: Lukasiewicz, T., Sali, A. (eds.) FoIKS 2012. LNCS, vol. 7153, pp. 124–143. Springer, Heidelberg (2012)
7. Fletcher, G.H.L., Hidders, J., Vansummeren, S., Picalausa, F., Luo, Y., De Bra, P.: On guarded simulations and acyclic first-order languages. In: Proc. DBPL, Seattle, WA, USA (2011)
8. Hellings, J., Fletcher, G.H.L., Haverkort, H.: Efficient external-memory bisimulation on DAGs. In: Proc. ACM SIGMOD, Scottsdale, AZ, USA, pp. 553–564 (2012)
9. Liu, Z., Chen, Y.: Processing keyword search on XML: a survey. World Wide Web 14(5-6), 671–707 (2011)
10. Luo, Y., de Lange, Y., Fletcher, G.H.L., De Bra, P., Hidders, J., Wu, Y.: Bisimulation reduction of big graphs on MapReduce (manuscript in preparation, 2013)
11. Luo, Y., Fletcher, G.H.L., Hidders, J., Wu, Y., De Bra, P.: I/O-efficient algorithms for localized bisimulation partition construction and maintenance on massive graphs. CoRR, abs/1210.0748 (2012)
12. Mass, Y., Sagiv, Y.: Language models for keyword search over data graphs. In: Proc. ACM WSDM, Seattle, Washington, USA (2012)
13. Pérez, J., Arenas, M., Gutierrez, C.: nSPARQL: A navigational language for RDF. J. Web Sem. 8(4), 255–270 (2010)
14. Picalausa, F., Luo, Y., Fletcher, G.H.L., Hidders, J., Vansummeren, S.: A structural approach to indexing triples. In: Simperl, E., Cimiano, P., Polleres, A., Corcho, O., Presutti, V. (eds.) ESWC 2012. LNCS, vol. 7295, pp. 406–421. Springer, Heidelberg (2012)
15. Tran, T., Herzig, D.M., Ladwig, G.: SemSearchPro - using semantics throughout the search process. J. Web Sem. 9(4), 349–364 (2011)

16. Tran, T., Wang, H., Rudolph, S., Cimiano, P.: Top-k exploration of query candidates for efficient keyword search on graph-shaped (RDF) data. In: Proc. IEEE ICDE, Shanghai, pp. 405–416 (2009)
17. Wu, Y., Van Gucht, D., Gyssens, M., Paredaens, J.: A study of a positive fragment of path queries: Expressiveness, normal form and minimization. Comput. J. 54(7), 1091–1118 (2011)
18. Yu, J.X., Qin, L., Chang, L.: Keyword search in relational databases: A survey. IEEE Data Eng. Bull. 33(1), 67–78 (2010)
19. Zhou, M., Pan, Y., Wu, Y.: Conkar: constraint keyword-based association discovery. In: Proc. ACM CIKM, Glasgow, UK, pp. 2553–2556 (2011)
20. Zhou, M., Pan, Y., Wu, Y.: Efficient association discovery with keyword-based constraints on large graph data. In: Proc. ACM CIKM, Glasgow, UK, pp. 2441–2444 (2011)

A Simple XSLT Processor for Distributed XML

Hiroki Mizumoto and Nobutaka Suzuki

University of Tsukuba
1-2, Kasuga, Tsukuba Ibaraki 305-8550, Japan
{s0911654@u,nsuzuki@slis}.tsukuba.ac.jp

Abstract. Recently, the sizes of XML documents have rapidly been increasing, and due to geographical and administrative reasons XML documents are tend to be partitioned into fragments and managed separately in plural sites. Such a form of XML documents is called *distributed XML*. In this paper, we propose a method for performing XSLT transformation efficiently for distributed XML documents. Our method assumes that the expressive power of XSLT is restricted to unranked top-down tree transducer, and all the sites storing an XML fragment perform an XSLT transformation in parallel. We implemented our method in Ruby and made evaluation experiments. This result suggests that our method is more efficient than a centralized approach.

1 Introduction

XML has been a de facto standard format in the Web, and the sizes of XML documents have rapidly been increasing. Recently, due to geographical and administrative reasons an XML document is partitioned into fragments and managed separately in plural sites. Such a form of XML documents is called *distributed XML*[5,2,1]. For example, Figures 1 and 2 show a distributed XML document of an auction site. In this example, one XML document is partitioned into four fragments F_0, F_1, F_2, and F_3, and fragment F_1 is stored in site S_1, fragment F_2 is stored in site S_2, and so on. We say that S_1 and S_3 are the *child sites* of S_0 (S_0 is the *parent site* of S_1 and S_3).

In this paper, we consider XSLT transformation for distributed XML documents. An usual centralized approach for performing an XSLT transformation of a distributed XML document is to send all the fragments to a specific site, then merge all the fragments into one XML document, and perform an XSLT transformation to the merged document. However, this approach is inefficient due to the following reasons. First, in this approach an XSLT transformation processing is not load-balanced. Second, an XSLT transformation becomes inefficient if the size of the target XML document is large[12]. This implies that the centralized approach is inefficient even if the size of each XML fragment is small, whenever the merged document is large.

In this paper, we propose a method for performing XSLT transformation efficiently for distributed XML documents. In our method, all the sites having an XML fragment perform an XSLT transformation in parallel. This avoids centralized XSLT transformation for distributed XML documents and leads to an

Y. Ishikawa et al. (Eds.): APWeb 2013, LNCS 7808, pp. 7–18, 2013.

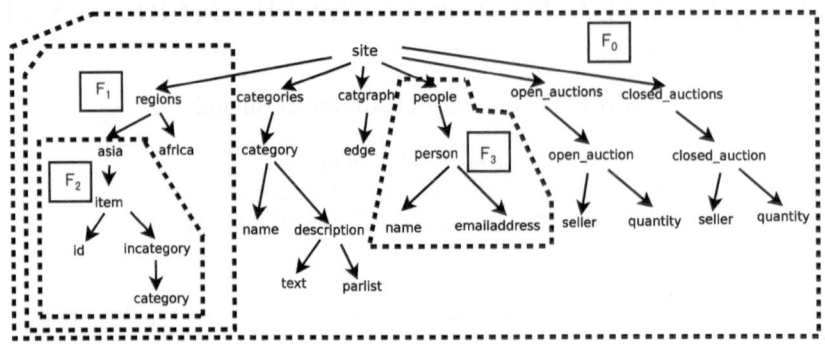

Fig. 1. Auction XML document

efficient XSLT processing. For distributed XML, however, we have to consider handling modes of XSLT templates carefully, since multiple templates may be defined to a single element, e.g., in Fig. 3 two templates are applicable to a and the other two templates are applicable to b. This implies that we cannot determine the template that should be applied to a fragment F until the transformation of the "parent" fragment of F is completed. For example, in Fig. 1, in order to transform F_2 we need the mode applied to F_2 that is obtained after the transformation of F_1 is completed. Thus, a child site have to wait until the transformation of its parent site is completed, and thus fragments stored in distributed sites cannot be transformed in parallel. To handle this problem, our method takes the following approach.

1. For a site S having fragment F, S applies all the templates applicable to F simultaneously using native threads.
2. When the parent site S' of S completes the transformation of its fragment, the mode m that should be applied to F is identified. S' sends the mode m to S.
3. S chooses the result transformed in mode m, and returns it as the result.

Since XSLT is Turing-complete[6], it is impossible to plan a complete strategy of XSLT transformation for distributed XML. Therefore, in this paper we restrict the expressive power of XSLT to unranked top-down tree transducer[8]. We implemented our method in Ruby and made evaluation experiments. The result suggests that our method is more efficient than the centralized approach.

Related Work

A distribution design of XML documents is firstly proposed in [3]. There have been several studies on evaluations of XPath and other languages for distributed XML. [4,5] propose an efficient XPath evaluation algorithms for distributed XML. [7] proposes a method for evaluating XQ, a subset of XPath, for vertically partitioned XML documents. [10] considers a regular path query evaluation in an distributed environment. [11] extensively studies the complexities of regular

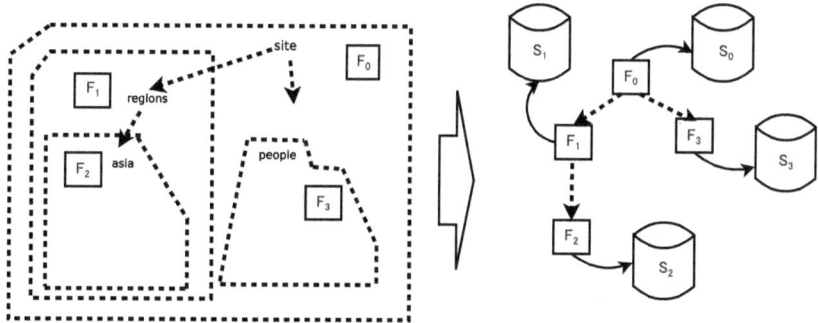

Fig. 2. Fragments of an XML document

path query and structural recursion over distributed semistructured data. Besides query languages, [2,1] study on the complexities of schema design problems for distributed XML. To the best of the authors' knowledge, there is no study on XSLT evaluation for distributed XML.

2   Definitions

Trees, Hedges, and Tree Transducer

Since our method is based on unranked top-down tree transducer, we first show related definitions. Let Σ be a set of labels. By $\mathcal{T}_\Sigma$ we mean the set of unranked Σ-trees. A tree whose root is labeled with $a \in \Sigma$ and has n subtrees $t_1, \cdots, t_n$ is denoted by $a(t_1 \cdots t_n)$. In the following, we always mean Σ-tree whenever we say tree. A *hedge* is a finite sequence of trees. The set of hedges is denoted by $\mathcal{H}_\Sigma$. In the following, we use $t, t_1, t_2, \cdots$ to denote trees and $h, h_1, h_2, \cdots$ to denote hedges. For a set Q, by $\mathcal{H}_\Sigma(Q)$ ($\mathcal{T}_\Sigma(Q)$) we mean the set of Σ-hedges (resp., Σ-trees) such that leaf nodes can be labeled with elements from Q.

A *tree transducer* is a quadruple (Q, Σ, q_0, R), where Q is a finite set of *states*, Σ is the set of labels, $q_0 \in Q$ is the initial state, and R is a finite set of rules of the form $(q, a) \to h$, where $a \in \Sigma, q \in Q$ and $h \in \mathcal{H}_\Sigma(Q)$. If $q = q_0$, then h is restricted to $\mathcal{T}_\Sigma(Q) \setminus Q$. A state corresponds to a mode of XSLT.

The translation defined by a tree transducer $Tr = (Q, \Sigma, q_0, R)$ on a tree t in state q, denoted by $Tr^q(t)$, is inductively defined as follows.

R1: If $t = \epsilon$, then $Tr^q(t) := \epsilon$.

R2: If $t = a(t_1 \cdots t_n)$ and there is a rule $(q, a) \to h \in R$, then $Tr^q(t)$ is obtained from h by replacing every node u in h labeled with $p \in Q$ by the hedge $Tr^p(t_1) \cdots Tr^p(t_n)$.

R3: If there is no rule $(q, a) \to h$ in R, then $Tr^p(t) := \epsilon$.

The transformation of t by Tr, denoted by $Tr(t)$, is defined as $Tr^{q_0}(t)$.

```
<xsl:template match="a" mode="p">
  <x>
    <z/>
  </x>
</xsl:template>

<xsl:template match="a" mode="q">
  <z>
    <xsl:apply-templates mode="q" />
  </z>
</xsl:template>

<xsl:template match="b" mode="p">
  <x>
    <xsl:apply-templates mode="p" />
    <xsl:apply-templates mode="q" />
  </x>
</xsl:template>

<xsl:template match="b" mode="q">
  <y>
    <xsl:apply-templates mode="p" />
  </y>
</xsl:template>
```

Fig. 3. An example XSLT script

Example 1. Let $Tr = (Q, \Sigma, p, R)$ be a tree transducer, where

$$Q = \{p, q\},$$
$$\Sigma = \{a, b, x, y, z\},$$
$$R = \{(p, a) \to x(z), (q, a) \to z(q), (p, b) \to x(p\,q), (q, b) \to y(p)\}.$$

T_r corresponds to the XSLT script shown in Fig. 3. For example, consider the rule $(p, a) \to x(z)$ in R. This corresponds to the first template in Fig. 3. The mode p in the left-hand side of the rule corresponds to the mode of the template, and label a in the left-hand side of the rule corresponds to the match attribute. The tree t in Fig. 4 is transformed to $Tr(t)$ in Fig. 5.

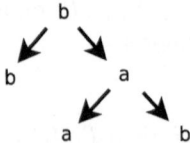

Fig. 4. Input tree t

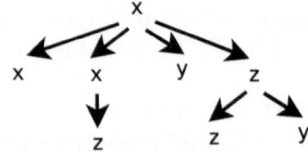

Fig. 5. Output tree $Tr(t)$

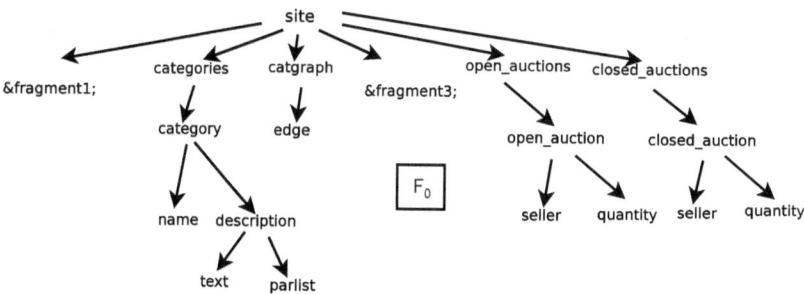

Fig. 6. Fragment F_0

Distributed XML

In this paper, we consider settings in which an XML tree t is partitioned into a set $\mathcal{F}_t$ of disjoint subtrees of t, where each subtree is called *fragment*. We assume that each fragment $F_i \in \mathcal{F}_t$ is stored in a distinct site. We allow arbitrary "nesting" of fragments. Thus, fragments can appear at any level of the tree. For example, the XML tree $t \in \mathcal{T}_\Sigma$ in Fig. 1 is partitioned into four fragments, $\mathcal{F}_t = \{F_0, F_1, F_2, F_3\}$. For a tree t, the fragment containing the root node of t is called *root fragment*. In Fig. 2, the root fragment is F_0.

For two fragments F_i and F_j, we say that F_j is a *child fragment* of F_i if the root node of F_j corresponds to a leaf node v of F_i. In order to represent a connection between F_i and F_j, we use an EntityReference node at the position of v, which refers the root node of F_j. For example, in Fig. 6 EntityReference nodes "&fragment1" and "&fragment3" are inserted into F_0 at the positions of "regions" and "people", respectively.

Each fragment is stored in a *site*. The site having the root fragment is called *root site* and the other sites are called *slave sites*. For example, in Fig. 2 S_0 is the root site and S_1, S_2, S_3 are slave sites. We assume that no two fragments are stored in the same site. If the fragment in site S has a child fragment stored in S', then S' is a *child* site of S (S is the *parent* site of S'). For example, in Fig. 2 S_0 has two child sites S_1 and S_3.

3   Transformation Method

3.1   Outline

In our transformation method, all the sites S transform the fragment stored in S in parallel, in order to avoid transformation processes being centralized on a specific site. To realize this strategy, we have a problem to solve. For a fragment F, the template that should be applied to F cannot be determined until the transformation of the parent fragment of F is completed. For example, consider the right child a of the root in Fig.4. Since this node is labeled with a, we have two rules that can be applied to the node, $(p, a) \to x(z)$ and $(q, a) \to z(q)$, and we cannot determine which of the rules should be applied to this node until the transformation of the root node is completed. Thus, for a fragment F and the rules $r_1, \cdots, r_n$ that can be applied to F, our method proceeds the transformation of F as follows.

1. Let S be the site storing F. S applies $r_1, \cdots, r_n$ to F in parallel, using native threads. Thus S will obtain n transformation results.
2. When the parent site S' of S completes the transformation of the parent fragment of F, we can identify the mode(s) that should be applied to F. S' sends the mode(s) to S.
3. S receives the mode(s) that should be applied to F from S'. S choose the appropriate transformation result(s) of the n transformation results and send the result(s) to the root site.

3.2   Our Transformation Method

We now present the details of our method. We use two XSLT possessors Master-XSLT and Slave-XSLT. First, Master-XSLT is used in the root site and has the following functions.

- Transforming the root fragment.
- Sending appropriate modes to its child sites according to the transformation result of the root fragment.
- Merging (a) the transformed root fragment and (b) the transformed fragments received from the slave sites.

Second, Slave-XSLT is used in a slave site and has the following functions.

- Transforming a fragment F stored in the slave site.
- Applying rules to F in parallel, using native threads of Ruby.
- Sending appropriate modes to its child sites according to the transformation result of F.
- Sending the transformed result of F to the root site.

Let us first present Master-XSLT. This procedure first sends a tree transducer (XSLT styleseet) to each slave site (lines 1 to 3), then transforms the root fragment by procedure Transform shown later (line 5), and receives the transformed fragments from the slave sites (line 6), and finally merges the fragments (line 7).

Master-XSLT
Input: A tree transducer $T_r = (Q, \Sigma, q_0, R)$ and the root fragment F.
Output: Tree $T \in \mathcal{T}_\Sigma$.

1. **for** each slave site S **do**
2.   Send tree transducer T_r to S.
3. **end**
4. $v \leftarrow$ the root node of F;
5. $F' \leftarrow \text{Transform}(T_r, F, v, q_0)$;

6. Wait until transformed fragment F_i is received from each slave site S_i.
   Let $F_1, \cdots, F_k$ be the received fragments.
7. Merge F' and $F_1, \cdots, F_k$ into T.
8. Return T;

We next present procedure Transform used in line 5 above. This procedure transforms a given fragment recursively. Lines 4 to 21 transform the subfragment rooted at v' of F according to T_r. Line 4 to 13 apply rule R2 and line 21 applies rule R3 of the definition of tree transducer. In lines 1 to 3, when the procedure encounters an EntityReference node v', then the procedure identifies the mode(s) q that should be applied to fragment $F(v')$ and thus sends the mode(s) to site $S(v')$, where $F(v')$ denotes the fragment that v' referrers to and $S(v')$ denotes the site storing $F(v')$.

Procedure Transform
Input: A tree transducer $T_r = (Q, \Sigma, q_0, R)$, a fragment F, the context node v' of F, and a mode q.
Output: A transformed fragment of F in mode q.

1. **if** v' is an EntityReference node **then**
2.   By referring to the parent node of v', identify the mode q that should be applied to $F(v')$.
3.   Send the mode q to $S(v')$.
4. **else if** $(q, v') \to h \in R$ for some h **then**
5. $Q' \leftarrow \{q' \mid q'$ is a leaf node of $h\} \cap Q$;
6.   **if** v' has a child node $v_1, \cdots, v_k$ **then**
7.   **for** each $q' \in Q'$ **do**
8.   **for** each child node $v_i \in \{v_1, \cdots, v_k\}$ of v' **do**
9. $F_i \leftarrow$ the subtree rooted at v_i of F;
10. $T_i \leftarrow \text{Transform}(T_r, F_i, v_i, q')$;
11.   **end**
12.   Replace node q' in h with hedge $T_1 \cdots T_k$.
13.   **end**
14.   **else**
15.   **for** each $q' \in Q'$ **do**
16. $q' \leftarrow \epsilon$;
17.   **end**
18.   **end**
19.   Replace the subtree rooted at v' of F with h.
20. **else**
21.   Replace the subtree rooted at v' of F with ϵ.
22. **end**
23. Return F;

Finally, we present Slave-XSLT. This procedure runs in each slave site S and transforms the fragment F stored in S. This procedure starts when it receives a tree transducer from the root site, and then transforms F by using the rules that can be applied to F in parallel, using threads (lines 3 to 8). Thus more than one transformation results may be obtained. Then the procedure waits until the mode(s) p that should be applied to F is received from the parent site (line 9), and sends the fragment(s) transformed in mode(s) p to the root site (line 11). Note that if F contains EntityReference nodes v, we have to tell the appropriate modes of v to child sites $S(v)$. This is done in lines 12 to 15.

Procedure Slave-XSLT
Input: A fragment F stored in its own site.
Output: none

1. Wait until tree transducer $T_r = (Q, \Sigma, q_0, R)$ is received from the root site.
2. $v_r \leftarrow$ the root node of F;
3. $Modes \leftarrow \{q \mid (q, v_r) \to h \in R$ for some $h\}$;
4. **for** each $q \in Modes$ **do**
5.   Thread **start**
6. $T_q \leftarrow$ Transform(T_r, F, v_r, q);
7.   Thread **end**
8. **end**
9. Wait until mode(s) p is received from the parent site.
10. **if** $T_p \neq null$ **then**
11.   Send T_p to the root site.
12.   **for** each EntityReference node v in F **do**
13.   Identify the mode q' of v in T_p.
14.   Send q' to $S(v)$.
15.   **end**
16. **else**
17.   Send ϵ to the root site.
18. **end**

For example, let us consider the distributed XML document consisting of two fragments m and f in Fig. 7. Let T_r be the tree transducer defined in Example 1. Moreover, let S_r be the root site storing m and S_s be the slave site storing f. Master-XSLT running in S_r sends T_r to site $S(\&fragment) = S_s$ and starts the transformation of m with the initial mode p (Figure 8 shows the the transformation result). Slave-XSLT running in S_s receives T_r and starts to transform f. In this case, two rules $(p, a) \to x(z)$ and $(q, a) \to z(q)$ can be applied to the root node a of f. Thus, Slave-XSLT applies the two rules to f in parallel, and the obtained results are shown in Fig. 9. When Master-XSLT encounters the EntityReference node "$\&fragment$" of m, the procedure sends the appropriate mode(s) to $S(\&fragment)$. In this case, both modes p and q are send to $S(\&fragment)$. After Slave-XSLT completes the transformation of f, the procedure sends the transformed fragments. In this case, the two fragments shown in Fig. 9 are send to the root site S_r. Finally, S_r merges the three fragments and we obtain the output tree in Fig.10.

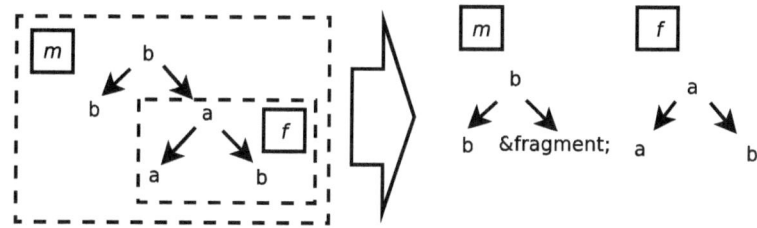

Fig. 7. A distributed XML document consisting of fragments m and f

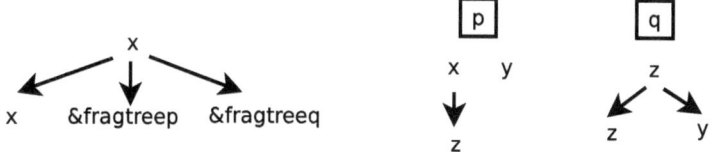

Fig. 8. Transformed fragment of m **Fig. 9.** Transformed fragments of f

4   Evaluation Experiment

In this section, we present experimental results on our method. We implemented our method in Ruby 1.93. We used 4 Linux machines, distributed over a local LAN (100base-TX). Each machine has a 2.4GHz Intel Xeon CPU and 4GB of memory. We used five XML documents generated by XMark[9], each of the documents were divided into four fragments (Table 1). Each XML document is divided as follows. As shown in Fig. 1, the root fragment F_0 has "site" as the root node, fragment F_1 has "regions" as the root node, fragment F_2 has "asia" as the root node, and fragment F_3 has "people" as the root node. Fragment F_i is stored in site S_i for $0 \le i \le 3$. We measure the execution time of (a) a centralized method and (b) our transformation method. In approach (a), fragments F_1, $F2$, and F_3 are first sent to S_0, then F_0 to F_3 are merged into one document T and an XSLT transformation is performed on T in site S_0.

We used the following three XSLT stylesheets. These are synthetic stylesheets generated by our Ruby program.

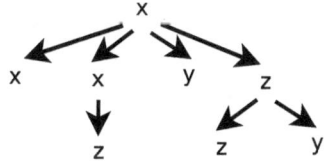

Fig. 10. Transformation result

Table 1. Sizes of distributed XML documents

	Fragment size			
Total size	F_0	F_1	F_2	F_3
50MB	23MB	25MB	5.5MB	2.6MB
100MB	46MB	50MB	11MB	5.1MB
150MB	69MB	75MB	17MB	7.6MB
200MB	92MB	100MB	22MB	11MB
250MB	115MB	125MB	28MB	13MB

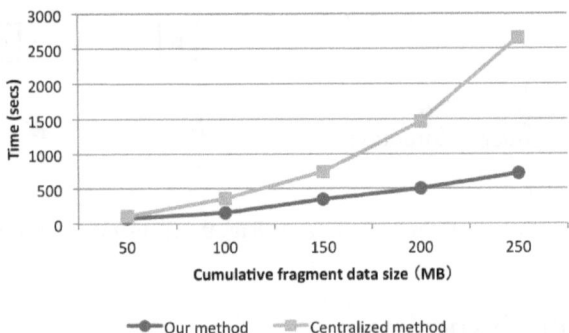

Fig. 11. Experimental result with Sheet-1

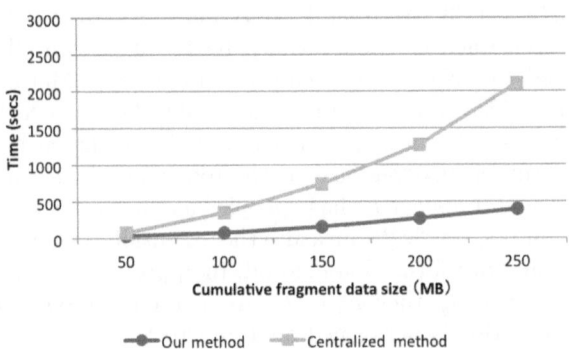

Fig. 12. Experimental result with Sheet-2

Sheet-1: This stylesheet has, for *every* element m of an XML document, at least one template applicable to m. *Two modes* are used in this stylesheet. This stylesheet consists of 137 templates.

Sheet-2: This stylesheet has, for *every* element m of an XML document, a template applicable to m. A *single mode* is used in this stylesheet. This stylesheet consists of 68 templates.

Sheet-3: This stylesheet has, for *some* element m of an XML document, at least one template applicable to m. *Two modes* are used in this stylesheet. This stylesheet consists of 94 templates.

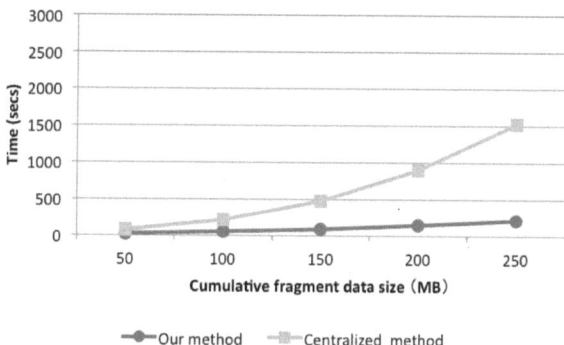

Fig. 13. Experimental with Sheet-3

The results are shown in Figs. 11 to 13. As shown in these figures, our method is about four times faster than the centralized method, regardless the used stylesheets. This suggests that our method works well for distributed XML documents.

5   Conclusion

In this paper, we proposed a method for performing XSLT transformation for distributed XML documents. The experimental results suggest that our method work well for distributed XML documents.

However, we have a lot of future work to do. First, in this paper the expressive power of XSLT is restricted to unranked top-down tree transducer. We have investigated XSLT elements and functions, and we have found that about half of the elements/functions can easily be incorporated into our method (Type A of Table 2). The elements/functions f of this type can be calculated within the fragment in which f is used, e.g., `xslt:text`. On the other hand, it seems difficult to incorporate the rest elements/functions into our method (Type B of Table 2). An example element of this type is `xslt:for-each`, which accesses several fragments beyond the fragment in which the `xslt:for-each` element is used. Thus, we have to handle XSLT elements/functions of Type B carefully in order to extend the expressive power of our method. Another future work relates to experimentation. In our experimentation we use only three synthetic XSLT stylesheets. Thus we need to make more experiments using real-world XSLT stylesheets.

Table 2. XSLT 1.0 elements and functions

	elements	function
Type A	18	18
Type B	17	16
Total	35	34

Acknowledgement. This research is partially supported by Research Center for Knowledge Communities, University of Tsukuba.

References

1. Abiteboul, S., Gottlob, G., Manna, M.: Distributed XML design. J. Comput. Syst. Sci. 77(6), 936–964 (2011)
2. Abiteboul, S., Gottlob, G., Manna, M.: Distributed XML design. In: Proc. PODS 2009 Proceedings of the Twenty-Eighth ACM SIGMOD-SIGACT-SIGART Symposium on Principles of Database Systems, pp. 247–258 (2009)
3. Bremer, J.M., Gertz, M.: On distributing XML repositories. In: Proc. of WebDB, pp. 73–78 (2003)
4. Buneman, P., Cong, G., Fan, W., Kementsietsidis, A.: Using partial evaluation in distributed query evaluation. In: Proc. VLDB 2006 (2006)
5. Cong, G., Fan, W., Anastasios: Distributed Query Evaluation with Performance Guarantees. In: Proc. SIGMOD 2007 Proceedings of the 2007 ACM SIGMOD International Conference on Management of Data, pp. 509–520 (2007)
6. Kepser, S.: A simple proof for the Turing-completeness of XSLT and XQuery. In: Extreme Markup Languages (2004)
7. Kling, P., Özsu, M.T., Daudjee, K.: Generating efficient execution plans for vertically partitioned XML databases. Proc. VLDB Endow. 4(1), 1–11 (2010)
8. Martens, W., Neven, F.: Typechecking Top-Down Uniform Unranked Tree Transducers. In: Calvanese, D., Lenzerini, M., Motwani, R. (eds.) ICDT 2003. LNCS, vol. 2572, pp. 64–78. Springer, Heidelberg (2002)
9. Schmidt, A., Waas, F., Kersten, M., Carey, M.J., Manolescu, I., Busse, R.: Xmark: a benchmark for XML data management. In: Proceedings of the 28th International Conference on Very Large Data Bases, Proc. VLDB 2002, pp. 974–985 (2002)
10. Stefanescu, D.C., Thomo, A., Thomo, L.: Distributed evaluation of generalized path queries. In: Proc. the 2005 ACM Symposium on Applied Computing, SAC 2005, pp. 610–616 (2005)
11. Suciu, D.: Distributed query evaluation on semistructured data. ACM Trans. Database Syst. 27(1), 1–62 (2002)
12. Zavoral, F., Dvorakovam, J.: Perfomance of XSLT processors on large data sets. In: Proc. Applications of Digital Information and Web Technologies, pp. 110–115 (2009)

Ontology Usage Network Analysis Framework

Jamshaid Ashraf and Omar Khadeer Hussain

School of Information Systems, CBS, Curtin University, Perth, WA 6845
jamshaid.ashraf@gmail.com, O.Hussain@cbs.curtin.edu.au

Abstract. Recently, there is tremendous growth in the use of ontologies to publish semantically rich structured data on the Web. In order to understand the adoption and uptake of ontologies in real world setting, it is important to analyse the ontology usage. In this paper, we propose Ontology Usage Network Analysis Framework (OUN-AF) which models the ontology usage by different data publishers in the form of affiliation network. Metrics are defined to measure the ontology usage, their co-usability and the cohesive subgroups emerging from the dataset.

1 Introduction

A decade-long effort by the Semantic Web community regarding knowledge representation and assimilation has resulted in the development of methodologies, tools and technologies to develop and manage ontologies [1]. As a result, numerous domain ontologies have been developed to describe the information pertaining to different domains such as Healthcare and Life Science (HCLS), entertainment, financial services and eCommerce. Consequently, we now have billions of triples [2] on the Web in different domains, annotated by using different domain-specific ontologies [3] to provide structured and semantically rich data on the Web.

In our previous work [4], Ontology Usage Analysis Framework (OUSAF) is presented in order to analyse the use of ontologies on the web. The OUSAF framework is comprised of four phases namely *identification, investigation, representation* and *utilization*. The identification phase, which is the focus of this paper, is responsible for identifying different ontologies that are being used in a particular application area or in a given dataset for further analysis. A few of the common requirements which form the selection criteria for the identification of ontologies in this scenario are:

1. What are the widely used ontologies in the given application?
2. What ontologies are more interlinked with other ontologies to describe domain-specific entities?
3. What ontologies are used more frequently and what is their usage percentage based on the given dataset?
4. Which ontology clusters form cohesive groups?

In order to establish a better understanding of ontology usage and to identify the ontologies, based on the abovementioned criteria, this paper proposes the

Y. Ishikawa et al. (Eds.): APWeb 2013, LNCS 7808, pp. 19–30, 2013.

Ontology Usage Network Analysis Framework (OUN-AF) that models the use of ontologies by different data sources using an Ontology Usage Network (OUN). OUN represents the relationships between ontologies and data sources based on the actual usage data available on the Web in the form of a graph-based relationship structure. This structure is then analysed using metrics to study the structural, typological and functional characteristics of OUN by applying Social Network Analysis (SNA) techniques.

2   Background

2.1   Related Work

In [5], the authors have analyzed the social and structural relationships available on the Semantic Web, focusing mainly on FOAF and DC vocabularies. The study was performed on approximately 1.5 million FOAF documents to analyze instance data available on the Web and their usefulness in understanding social structures and networks. They identified the graphical patterns emerging in social networks and represented this using the FOAF vocabulary and the degree distribution of the network. As this research provides a detailed analysis of Semantic Web data by focusing on a specific vocabulary, it does not provide a framework or methodology to make it applicable to different vocabularies.

Other work in the literature has analyzed a different number of ontologies using SNA. For example, [6] covered five ontologies, [7] analyzed 250 ontologies and [8] used approximately 3,000 vocabularies. In all these work, ontologies were investigated to measure their structural properties, the distribution of different measures and terminological knowledge encoded in ontologies, but none includes how they are being used on the Web. The use of SNA and its techniques to analyse the use of ontologies and measure the relationships based on usage has only been applied marginally. In the identification phase of the OUSAF framework, the ontologies and their use by different data sources are represented in a way that allows the "affiliation" between ontologies and different data publishers to be measured.

2.2   Affiliation Network

Networks, in general sense, record relationships between entities. However, in affiliation network relationship reflect the affiliation between entities and the events in which entities participate. Affiliation Networks are essentially two-mode networks comprised of two disjoint sets of nodes with links which always are between a node of one set and a node of the other. Affiliation networks are represented through an affiliation matrix, $A = \{a_{ij}\}$. Matrix A is a two-mode sociomatrix in which rows represent actors and columns represent events. Generally, affiliation network A is defined as: A is a bipartite graph $A = (U, V, E)$ where U (often known as actors) and V (often known as events) are disjoint set of nodes and $E \cup (UXV)$ is the set of edges. With $p = |V|$ and $q = |U|$,

A is represented by an incident matrix with p lines and q columns. Formally, $A = \{a_{ij}\}$ records the affiliation of each actor with each event in an affiliation matrix such that:

$$a_{ij} = \begin{cases} 1, & \text{if actor } i \text{ is affiliated with event } j \\ 0, & \text{otherwise} \end{cases} \tag{1}$$

The value of 1 is put in the (i,j)th cell if ith actor (ith row) is affiliated with jth event (jth column) and an entry of 0 if ith actor is not affiliated with jth event as shown in Fig. 1.

	paper 1	paper 2	paper 3	paper 4
Author 1	1	0	1	0
Author 2	1	1	0	1
Author 3	0	1	0	1

(a)

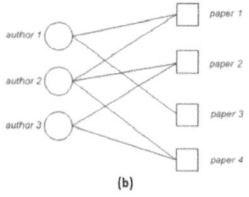

(b)

Fig. 1. (a) Affiliation matrix of author-paper affiliation network, and (b) an example of an author-paper affiliation network

Often, two-mode networks are transformed into a one-mode network by a procedure called projection which generates a one-mode network by selecting nodes of one set (e.g. authors in Fig. 1) and linking two nodes from the set if both are connected to the same node of the other set.

3 Ontology Usage Network Analysis Framework (OUN-AF)

The objective of the ontology identification is to identify the use of different ontologies by different data publishers to discover hidden usage patterns. Therefore, in order to mine such analysis, the Ontology Usage Network Analysis Framework (OUN-AF) is proposed, as shown in Fig. 2. OUN-AF comprises three phases, namely *Input* phase, *Computation* phase and *Analysis* phase.

3.1 Input Phase

The input phase is responsible for managing the dataset. The two key components in this phase are a crawler and an RDF triple store. In order to point the crawler to relevant and interesting data sources, the bootstrapping process first builds the seed URIs as multiple starting points. A list of seed URIs is obtained by accessing semantic search engines which return the URIs (URLs) of the data sources (web sites) with structured data. The crawler collects the Semantic Web data (RDF data) and after preprocessing it, loads it to the RDF triple store (database). From a data management point of view, since RDF data comprises

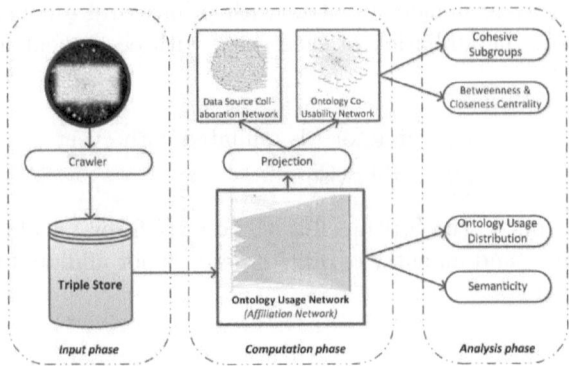

Fig. 2. Ontology Usage Network Analysis Framework (OUN-AF) and its phases

triples (statements) which do not provide a default mechanism to group or associate certain sets of triples to a context, the Named Graph approach is used which enables the provision of contextualization by introducing an additional URI (context) to a set of related triples.

3.2 Computation Phase

The computation phase provides the computational architecture to transform the RDF data to a model so that further analysis can be performed. The OUN-AF transforms the data into two formats: two-mode affiliation network (i.e. *Ontology Usage Network*) and the subsequent one-mode networks (i.e. *Ontology Co-Usability* and *Data-Source Collaboration* networks).

Ontology Usage Network (OUN) : OUN is an affiliation network represented as a bipartite graph. OUN comprises *ontology* and *datasource* sets of nodes with an edge between the ontology node and data-source node if the ontology has been used by the data source. To formally define the Ontology Usage Network, first the two sets of nodes, namely <u>*ontology set*</u> and <u>*data-source set*</u> are defined and then the OUN definition is presented.

- **An ontology set** is defined as the set O which represents the nodes of the first mode of the affiliation network. An ontology set O contains the list of ontologies used on the Web-of-Data such that there is a triple $t < s, p, o >$ anywhere in the dataset (specifically, otherwise in general, on the Web-of-Data) where $o \in O$ is the URI of either p or o.
- **A data-source set** is defined as the set D which has the list of hostnames on the Web-of-Data such that there exist a triple $t < s, p, o >$ in the dataset (specifically, otherwise in general, on the Web-of-Data), where s is the hostname and either p or o is a member of O.
- **The Ontology Usage Network (OUN)** is a bipartite graph, denoted as $OUN(O, D)$ that represents the affiliation network, with a set of ontologies

O on one side and a set of data- sources D on the other and edge (o, d) represents the fact that o is "used" by d.

In order to infer the connectedness present within one set of nodes based on their co-participation in the other set of nodes, OUN is transformed from a two-mode network to a one-mode network using the process of projection. Projection is used to generate two one-mode graphs; one for nodes in the ontology set known as the *Ontology Co-Usability* network, and second for the data-source set known as the *Data-Source Collaboration* network as shown in Fig. 2.

The Ontology Co-Usability network is an ontology-to-ontology network, in which two nodes are connected if both of the ontologies are being used by the same data source. This means that the *Ontology Co-Usability* network represents the connectedness of an ontology with other ontologies, based on their co-membership in the data source.

The Data-Source Collaboration network is a data-source-to-data-source network in which two nodes are connected if both of them have used the same ontology to describe their data. The Data-Source Collaboration network represents the similarity of data-sources in terms of their need to semantically describe the information on the Web.

3.3   Analysis Phase

The analysis phase is the third and last phase of OUN-AF. The objective of this phase is to mine the hidden relationships explicitly or implicitly present in the two-mode network (i.e. OUN) and one-mode networks (Ontology Co-Usability and Data-Source Collaboration). This is done by developing metrics for analyse the two- and one-mode network as described in next section.

4   Metrics for Ontology Identification

The metrics used for the identification phase are presented below.

4.1   Ontology Usage Distribution (OUD)

The metric Ontology Usage Distribution, OUD_k, is used to identify which fraction of the ontologies in the network have a degree k. OUD measures the degree distribution of ontologies in an affiliation network. In affiliation network, the degree of a node is the number of ties it has with the number of nodes of the other set. The degree centrality $C_D(o_i)$ of an ontology o_i is measured as:

$$C_D(o_i) = d(o_i) = \sum_{j=1}^{n_1} A_{ij} \qquad (2)$$

Where $i = 1, \ldots n_1$, $n_1 = |O|$, $d(o_i)$ is the degree of i_{th} ontology o, and A is the affiliation matrix representing OUN.

4.2   Semanticity

Semanticity distribution identifies the fraction of the data sources in the network which have a degree k. Semanticity measures the participation of different ontologies in a given data source. The more ontologies are being used by a data source, the higher semanticity value it has. Semanticity is measured by calculating the degree centrality and degree distribution on the second set of nodes in an affiliation network, which is the set representing the data sources present in the dataset. The degree centrality $C_D(ds_i)$ of a data source ds_i is measured as:

$$C_D(ds_i) = d(ds_i) = \sum_{j=1}^{n_2} A_{ij} \tag{3}$$

Where $i = 1, \ldots n_2$, $n_2 = |D|$, $d(ds_i)$ is the degree of i_{th} data source ds, and A is the affiliation matrix representing OUN

4.3   Betweenness and Closeness Centrality

Like Ontology Co-Usability, both these centrality measures are computed on the projected one-mode network.

Betweenness Centrality is the number of shortest paths between any two nodes that passes through the given node. The betweenness centrality $C_B(v_i)$ of a node v_i is measured as :

$$C_B(v_i) = \sum_{v_s \neq v_i \neq v_t \in V} \frac{\sigma_{st}(v_i)}{\sigma_{st}} \tag{4}$$

where σ_{st} is the number of shortest paths between vertices v_s and v_t.

And $\sigma_{st}(v_i)$ is the number of shortest paths between v_s and v_t that pass through v_i.

Closeness centrality is a measure of the overall position of a node (actor) in the network, giving an idea about how long it takes to reach other nodes in the network from a given starting node. Closeness measures reachability, that is, how fast a given node (actor) can reach everyone in the network [9] and is defined as:

$$c_v = \frac{n-1}{\sum_{u \in V} d(v, u)} \tag{5}$$

where $d(v, u)$ denotes the length of a shortest-path between v and u. The closeness centrality of v is the inverse of the average (shortest path) distance from v to any other vertex in the graph. The higher the c_v, the shorter the average distance from v to other vertices, and v is more important by this measure.

4.4   Cohesive Subgroups

Generally speaking, cohesive subgroups refer to the areas of the network in which actors are more closely related to each other than actors outside the group. Cohesive subgroups are often measures using n-clique, in which it is not required that each member of the clique has a direct tie with the others, but instead that it has to be no more than distance n from each other. Formally, a clique is the maximum number of actors who have all possible ties present between them.

5 Analysis of Ontology Usage Network

The detail of dataset which is used to populate OUN and the analysis performed using metrics defined above, are discussed as follows.

5.1 Dataset and Its Characteristics

In order to build a dataset which has a fair representation of the Semantic Web data described using domain ontologies, semantic search engines such as Sindice and Swoogle are used to build the seed URLs. These seed URLs are then used to crawl the structured data published on the Web using ontologies. The dataset built for the identification phase comprises 22.3 millions of triples, collected from 211 data sources[1]. In this dataset, 44 namespaces are used to describe entities semantically.

The resulting Ontology Usage Network is comprise of 1390 edges linking 44 ontologies to 211 data sources. In terms of generic OUN properties, the *density* of the network is 0.149 and the *average degree* is 10.90. The average degree shows that the network is neither too sparse nor too dense which is a common pattern in information networks. Details on the other properties and metrics are given in the following subsections.

5.2 Analysing Ontology Usage Distribution (OUD)

Through OUD, we would like to determine how the use of an ontology is distributed over the data sources in the dataset. Using Eq. 2, Fig. 3(a) shows the percentage of the ontologies being used by a number of different data sources. The relative frequency of OUD on the dataset shows that there is both extreme and average ontology usage by data sources. It also shows that 13.6% of ontologies are not used by any of the data sources and approximately half of the ontologies are exclusively used by the data sources. The second row of the Fig. 3(a) shows that 47.7% of ontologies (21 ontologies) are being used by a data source that has not used any other ontology. This means that there are several ontologies in the dataset which either conceptualize a very specialized domain, restricting their reusability, or are of a proprietary nature. From the third row of Fig. 3(a) onwards, there is an increase in the reusability factor of ontologies. This is because an increasingly large number of data sources are using them. The last row shows that 4.5% (2 ontologies) of ontologies are being used equally by

---

[1] https://docs.google.com/spreadsheet/ccc?key=OAqjAK1TTtaSZdGpIMkVQUTRNenlrTGctR2J1bkl6WEE

208 data sources. Through this analysis, we can see that there are less ontologies which are not being used at all and there are a few which have almost optimal utilization.

Fig. 3(b) shows the degree distribution of ontology usage in a number of data sources. The value of degree is shown on the x-axis and the number of ontologies with that degree is shown on the y-axis. It can be seen that there are a large number of ontologies with a small degree value and only a few ontologies have a larger degree value.

# of data sources	# ontologies	% ontologies
0	6	13.6
1	21	47.7
2	1	2.3
3	1	2.3
4	1	2.3
11	1	2.3
15	1	2.3
16	1	2.3
18	1	2.3
38	1	2.3
75	1	2.3
115	1	2.3
126	1	2.3
141	1	2.3
164	1	2.3
190	1	2.3
208	2	4.5

(a)

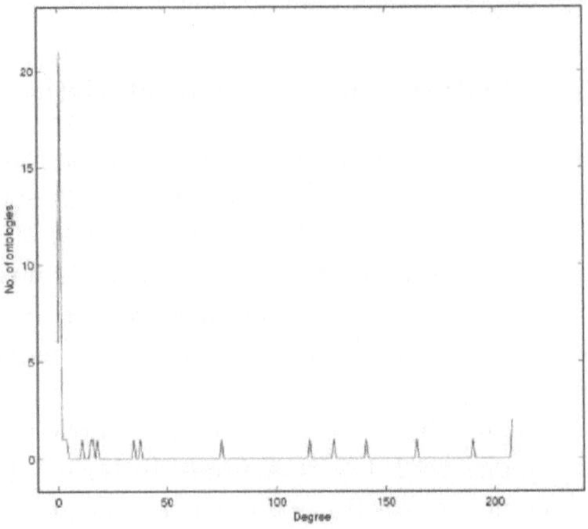

(b)

Fig. 3. (a) Distribution of Ontology Usage in data sources, and (b) Degree distribution of ontology usage (data sources per ontology)

5.3   Analysing Semanticity

Semanticity measures the richness of a data source in terms of ontology usage. Using Eq. 3, in the OUN, it is found that on average, 6.6 ontologies per data source are used in the dataset which, in our view, is an encouragingly high semanticity value, particularly bearing in mind that there are several ontologies with very low ontology usage degree values such as 0 and 1, as described in the previous section on Ontology Usage Distribution section. After determining the average semanticity of the data sources, we then look at their degree distribution. Figure 4(a) shows the relative frequency of ontologies being used by a number of data sources. The degree distribution of ontology usage per data source is different from ontology usage distribution.

At the lowest, in the network, two ontologies www.oettl.it and www.openlinksw. com are used by one data source, which shows the lowest semanticity value and 14 is the maximum semanticity value which is also used by one data source. When the data sources' degree distribution is plotted, it follows the Gaussian distribution, as shown in Fig. 4(b). Gaussian distribution [10], which is essentially a bell

# ontologies	# of data source	% data source
2	1	0.5
3	4	1.9
4	3	1.4
5	38	18.0
6	59	28.0
7	51	24.2
8	39	18.5
9	13	6.2
10	2	0.9
14	1	0.4

(a)

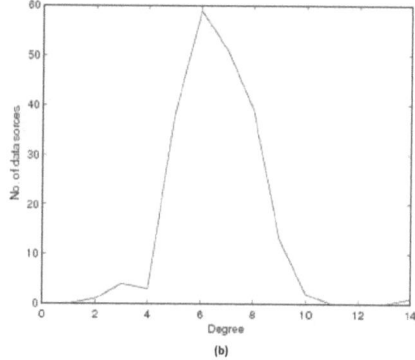

(b)

Fig. 4. (a) Distribution of Semanticity (Ontology used per data source), and (b)Degree distribution of semanticity (Ontologies per Data source)

shaped curve, is normally concentrated in the centre and decreases on either side. This signifies that degree has less tendency to produce extreme values compared to power law distribution.

5.4 Analysing Betweenness and Closeness

Betweenness centrality measures the number of shortest paths going through a certain node. However, on the other hand, the closeness centrality of a node in the network is the inverse of the average shortest path distance from the node to any other node in the network [11].

In the context of the Ontology Co-usage network, the interpretation of betweenness and closeness measures are different from other collaboration networks such as co-authorship collaboration networks [12]. In betweenness, which is measured using Eq. 4, the nodes of larger values are considered to be the hub of the network, controlling the communication flow (or becoming the major facilitator) between nodes with a geodesic path passing through these hub. In the Ontology Co-usage network, ontologies are linked based on the fact that these are being co-used by the data sources, therefore we believe the ontologies with maximum betweenness centrality act as a semantic gateway and become a major motivational factor for the usage of other ontologies.

Likewise, in closeness centrality which is measured using Eq. 5, the larger the value, the shorter the average distance from the node to any other node, and thus the node (with a larger value) is positioned in the best location to spread information quickly [13]. This centrality measure in the ontology co-usage graph enables the establishment of correspondence between ontologies which have concepts related to each other, supplementing each others conceptual model to form an exploded domain. The utilization of ontology indexing based on closeness centrality is very similar to the features discussed in [14] in supporting the application specific use of ontologies such as: (i) the ontologies closer to each other in their usage are better candidates for vocabulary alignment;

(ii) ontologies closer to each other have more entities which correspond to entities of other ontologies; and (iii) closely related ontologies tend to facilitate query answering on the Semantic Web.

The betweenness and closeness centrality of ontology co-usage nodes is shown in Fig. 5(a) and 5(b), respectively and node size reflects the centrality value.

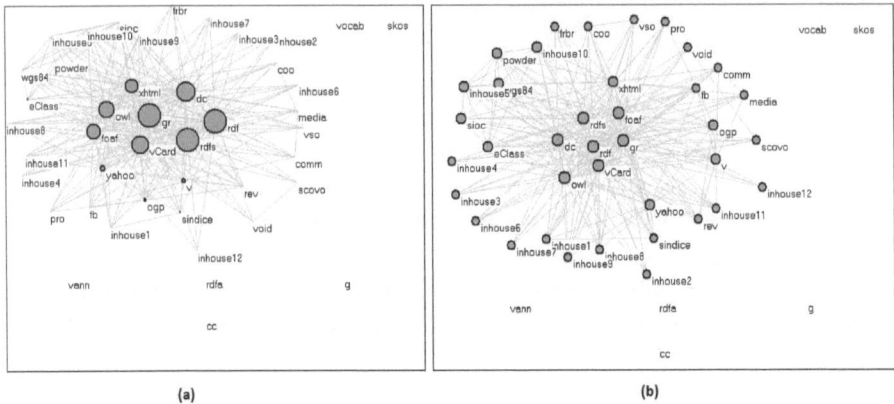

(a)   (b)

Fig. 5. (a) Betweenness centrality of Ontology Co-Usage network, and (b) Closeness centrality of Ontology Co-Usage network

As we can see in Fig. 5(a), the Ontology Co-Usage network has very few nodes with a higher betweenness value which means that the ontologies represented by the green nodes (which are not many) are the ones falling in between the geodesic path of many other nodes and acting as the gateway (or hub) in the communication between other ontologies in the graph. These are the nodes, namely rdf, rdfs, gr, vCard and foaf which, in our interpretation, are acting as the semantic gateway by becoming the reason for the adoption of other ontologies on the Web. However, on the other hand, closeness centrality is approximately distributed evenly in the network. Thus, it is safe to assume that almost every node is reachable to other nodes except those which are not connected.

5.5   Analysing Cohesive Subgroups

In a collaboration graph such as the Ontology Co-Usage network, a connected component is a maximal set of ontologies that are mutually reachable (and connected) through a chain of (co-usage) links. A cohesive sub-group analysis to identify connected components of the Ontology Co-Usage network shows that the network is widely connected. The connected component is 86.36% (See Figure 6), in which only six are not connected in the network (this means 0-core), while others are connected with varying k-core values) making it a giant network since it encompasses the majority of the nodes. This means that 86.36% of the ontologies are reachable to each other by following the links (domain names

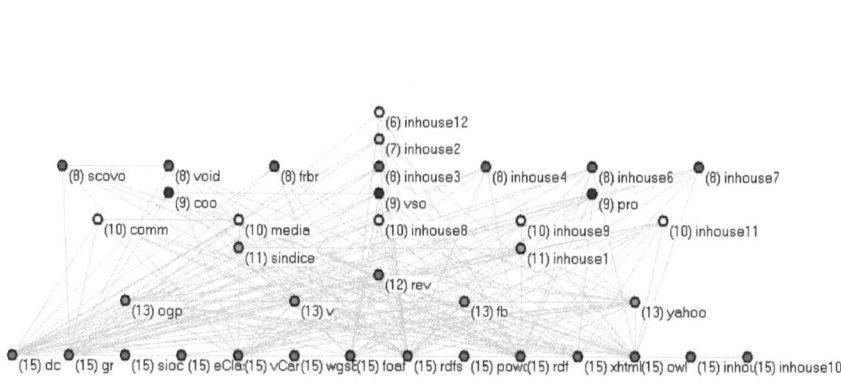

Fig. 6. Stacking of k-cores of Ontology Co-Usage network

URIs) of the data sources included in the dataset (or on the Web to generalize it). Note that the size of the cohesive sub-group, in terms of percentage, closely matches the findings of [15] for the classical Web which was 91%. Within the giant connected component, to know the sub-component based on the equal distribution of the concentration of links around a set of nodes, k-core is computed. k-core is the maximum sub-graph in which each node has at least degree k within the sub-graph. Fig. 6 stacks the k-core components, based on ascending k values from highest to lowest. From Figure 6, it is easy to see which ontologies are highly linked, based on ontology usage patterns invariance across data sources.

6   Conclusion and Future Work

In this paper, the use ontologies by different data sources are analysed by representing the ontology usage related data in the form of affiliation network. Using different metrics the network is analysed to understand the usage patterns in real world settings. In future work, we would like to enrich the dataset by crawling more data sources and automating the analyses process.

References

1. Hitzler, P., van Harmelen, F.: A Reasonable Semantic Web. Semantic Web Journal 1(1-2), 39–44 (2010)
2. Bizer, C., Jentzsch, A., Cyganiak, R.: State of the Linked Open Data (LOD) Cloud. Technical Report April 5, 2011 (March 2011),
   http://www4.wiwiss.fu-berlin.de/lodcloud/state/
3. Ashraf, J., Cyganiak, R., O'Riain, S., Hadzic, M.: Open eBusiness Ontology Usage: Investigating Community Implementation of GoodRelations. In: Proceedings of Linked Data on the Web Workshop (LDOW) at WWW 2011, Hyderabad, India, March 29. CEUR Workshop Proceedings, vol. 813, pp. 1–11. CEUR-WS.org (2011)

4. Ashraf, J.: A Framework for Ontology Usage Analysis. In: Simperl, E., Cimiano, P., Polleres, A., Corcho, O., Presutti, V. (eds.) ESWC 2012. LNCS, vol. 7295, pp. 813–817. Springer, Heidelberg (2012)
5. Ding, L., Zhou, L., Finin, T., Joshi, A.: How the Semantic Web is Being Used: An Analysis of FOAF Documents. In: Proceedings of the 38th Annual Hawaii International Conference on System Sciences, vol. 4, pp. 113–120. IEEE Computer Society, Washington, DC (2005)
6. Zhang, H.: The scale-free nature of semantic web ontology. In: Proceeding of the 17th International Conference on World Wide Web, pp. 1047–1048. ACM (2008)
7. Theoharis, Y., Tzitzikas, Y., Kotzinos, D., Christophides, V.: On Graph Features of Semantic Web Schemas. IEEE Transactions on Knowledge and Data Engineering 20(5), 692–702 (2008)
8. Cheng, G., Qu, Y.: Term Dependence on the Semantic Web. In: Sheth, A.P., Staab, S., Dean, M., Paolucci, M., Maynard, D., Finin, T., Thirunarayan, K. (eds.) ISWC 2008. LNCS, vol. 5318, pp. 665–680. Springer, Heidelberg (2008)
9. Oliveira, M., Gama, J.: An overview of social network analysis. Wiley Interdisciplinary Reviews: Data Mining and Knowledge Discovery 2(2), 99–115 (2012)
10. Weisstein, E.: Normal distribution (2005) From MathWorld,A Wolfram Web Resource, http://mathworld.wolfram.com/NormalDistribution.html
11. Newman, M.: The mathematics of networks. The New Palgrave Encyclopedia of Economics 2 (2008)
12. Newman, M.E.J.: Coauthorship networks and patterns of scientific collaboration. Proceedings of the National Academy of Sciences of the United States of America 101(suppl. 1), 5200–5205 (2004)
13. Okamoto, K., Chen, W., Li, X.-Y.: Ranking of closeness centrality for large-scale social networks. In: Preparata, F.P., Wu, X., Yin, J. (eds.) FAW 2008. LNCS, vol. 5059, pp. 186–195. Springer, Heidelberg (2008)
14. David, J., Euzenat, J.: Comparison between Ontology Distances (Preliminary Results). In: Sheth, A.P., Staab, S., Dean, M., Paolucci, M., Maynard, D., Finin, T., Thirunarayan, K. (eds.) ISWC 2008. LNCS, vol. 5318, pp. 245–260. Springer, Heidelberg (2008)
15. Broder, A., Kumar, R., Maghoul, F., Raghavan, P., Rajagopalan, S., Stata, R., Tomkins, A., Wiener, J.: Graph structure in the web. Computer Networks 33(1), 309–320 (2000)

Energy Efficiency in W-Grid Data-Centric Sensor Networks via Workload Balancing

Alfredo Cuzzocrea[1], Gianluca Moro[2], and Claudio Sartori[3]

[1] ICAR-CNR and University of Calabria, Italy
cuzzocrea@si.deis.unical.it
[2] DISI Department – Cesena Branch, University of Bologna, Italy
[3] DISI Department – Bologna Branch, University of Bologna, Italy
{gianluca.moro,claudio.sartori}@unibo.it

Abstract. *Wireless sensor networks* are usually composed by small units able to sense and transmit to a *sink* elementary data which are then processed by an external machine. However, recent improvements in the memory and computational power of sensors, together with the reduction of *energy consumption*, are rapidly changing the potential of such systems, moving the attention towards *data-centric sensor networks*. In this kind of networks, nodes are smart enough either to store some data and to perform basic processing allowing the network itself to supply higher level information closer to the network user expectations. In other words, sensors no longer transmit each elementary data sensed, rather they cooperate in order to assemble them in more complex and synthetic information, which will be locally stored and transmitted according to queries and/or events defined by users and external applications. Recently, we proposed W-Grid, a *fully decentralized cross-layer infrastructure for self-organizing data-centric sensor networks* where wireless communication occur through multi-hop routing among devices. In this paper, we show that network traffic, and thus the energy consumption, can be *balanced* among sensors by assigning multiple *virtual coordinates* to nodes trough a *fully decentralized workload balancing algorithm*, which extends W-Grid.

1 Introduction

Wireless sensor networks consist of large number of distributed nodes that organize themselves into a *multi-hop* (wireless) network. Sensor nodes are generally powered by batteries with an inherently limited lifetime, hence *energy consumption* issues play a leading role. In more detail, we focus the attention on so-called *data-centric sensor networks*, where nodes are smart enough either to store some data and to perform basic processing allowing the network itself to supply higher level information closer to the network user expectations.

Energy saving in wireless sensor networks involves both MAC layer and network layer. In particular, here we state that the routing protocol must be lightweight, must not require too much computation, such as complex evaluations of possible paths, and must not need a wide knowledge of the network

Y. Ishikawa et al. (Eds.): APWeb 2013, LNCS 7808, pp. 31–42, 2013.

organization. In previous work, we showed that W-Grid [16,15] routing scheme satisfies both previously described requirements. In fact, W-Grid routing protocol needs information about only *one-hop away devices* and the choice of the next hop requires bit-a-bit comparison of simple binary strings. We also proved that W-Grid allows sensors to generate a *decentralized* wireless network in which devices, thanks to uni-cast multi-hop transmissions (i.e., no broadcast propagations), can communicate each other independently of their location in the network. Our solution enhances typical sensor networks, in such a way that queries can be posed to the network anytime through any sensor by overcoming limitations of a dedicated *sink* node, which collects raw data in a fixed position and under time constraints (e.g., sensor synchronization). Thus, the resulting sensor network is capable of executing, in principle, any distributed algorithm for processing raw data they generate, in such a way to supply external users with the high level information needed or something almost close to their expectations. Also, the resulting network, which is W-Grid-aware, is able to *efficiently index and query multi-dimensional data* without reliance either on *Global Positioning System* (GPS) or flooding/broadcasting operations [16,15].

As mentioned, energy consumption plays a central role in wireless sensor networks. Indeed, in ad-hoc and sensor networks, most of the energy consumption is due to radio transmissions and hence protocol design for these networks must be lightweight and directed towards reducing communication cost. Strictly adhering to this paradigm, in this paper, we show that network traffic, and thus the energy consumption, can be *balanced* among sensors by assigning multiple *virtual coordinates* to nodes trough a *fully decentralized workload balancing algorithm*, which extends W-Grid. The energy consumption balancing, besides improving the entire network efficiency by lowering collisions and resource contentions among nodes, also prolongs the network life avoiding premature partitions caused by non-uniform battery discharging.

As a non-secondary contribution, we evaluate network routing performances, which have direct influence on energy consumption, as the quantity of *network knowledge* at nodes (namely routing table size) vary. An extensive number of experiments have been conducted in order to proof the improvement in energy consumption allowed by workload balancing in W-Grid.

The remaining part of the paper is organized as follows. Section 2 describes related work. Section 3 briefly presents W-Grid. Section 4 contains the results of our comprehensive experimental campaign. Finally, Section 5 concludes the paper with some considerations and introducing future work.

2   Related Work

Existing routing protocols have been developed according to different approaches. Basically, routing is necessary whenever a data is sensed (we also say generated) and stored in the system or whenever a query is submitted to the system.

As stated before, we do not consider sensor network systems which store sensor data externally at a remote *base station*, but rather we focus on advances wireless sensor networks in which data or events are kept at sensors, by means of representing them in terms of relations in a virtual distributed database and, for efficiency purposes, indexed by suitable attributes. For instance in [12,10,21,20], data generated at a node is assumed to be stored at the same node, and queries are either *flooded* throughout the network. In [19], a *Geographic Hash Tables* (GHT) approach is proposed, where data are hashed by name to a location within the network, enabling highly efficient rendezvous. GHTs are built upon the GPSR protocol [11] thus leveraging some interesting properties of that protocol, such as the ability of routing to sensors nearest to a given location, together with some of well-known limits, such as the risk of dead ends. Dead end problems, especially under low density environments or scenarios with obstacles and holes, are caused by the inherent *greedy nature* of routing algorithms that can lead to situation in which a packet gets stuck at a local optimal sensors that appears closer to the destination than any of its known neighbors. In order to solve this flaw, correction methods such as *perimeter routing*, which tries to exploit the right hand rule, have been implemented. However, packet losses still remain and, in addition to this, using perimeter routing causes loss of efficiency both in terms of average path length and energy consumption. Besides, another limitation of GHT-based routing is that it needs sensors to know their physical position, which causes additional localization costs to the whole system. In [9], Greenstein *et al.* have designed a *spatially-distributed index*, called DIFS, to facilitate *range search* over attributes. In [12], Li *et al.* have built a *distributed index*, called DIM, for *multidimensional range queries* over attributes but they require nodes to be aware of their physical location and of network perimeter. Moreover, they exploit GPSR for routing. Our solution extending W-Grid also behaves like a distributed index, but its indexing feature is cross-layered with routing, meaning that no physical position nor any external routing protocol is necessary, routing information is given by index itself.

Another research area that is directly related to our work is represented by the problem of *effectively and efficiently managing multidimensional data over sensor networks*, as W-Grid may represent a very efficient indexing layer for techniques and algorithms supporting this critical task. The problem of performing *multidimensional and OLAP analysis of data streams*, like the ones originated by sensor networks, has received great attention recently (e.g., [2]). Due to computational overheads introduced by these time-consuming tasks, several solutions have been proposed in literature, such as *data compression* (e.g., [4]), usage of *high-performance Computational Grids* (e.g., [6,5]), extensions to *uncertain and imprecise domains* (e.g., [3]) that occur very frequently in sensor environments, and so forth. This problem also exposes very interesting and challenging correlations with cross-disciplinary scientific areas, such as *mobile computing* (e.g., [7]), thus opening the door to novel research perspectives still poorly explored, such as *data management issues in mobile sensor networks* (e.g., [8,14,13]).

3   W-Grid Overview

W-Grid[1] can be viewed as a *binary tree index* cross-layering both routing and data management features over a wireless sensor network. Two main phases are performed in W-Grid: (1) implicitly generate *coordinates* and *relations* among nodes that allow efficient message routing; (2) determine a *data indexing space partition* by means of so-generated coordinates in order to efficiently support multidimensional data management. Also, each node can have one or more *virtual coordinates* on which an *order relation* is defined and through which the routing occurs. At the same time, each virtual coordinate represents a portion of the data indexing space for which a device is assigned with the (data) management tasks. By assigning *multiple* coordinates at nodes, we aim at reducing query path length and latency. This is obtained via bounding the probability that two nodes physically close have very different virtual coordinates, which may happen whenever a multidimensional space is translated into a one-dimensional space. In the next Sections, we provide a formal description on the W-Grid main features.

3.1   W-Grid Static Network Properties

The target sensor network is represented as a graph S:

$$S = (D, L) \tag{1}$$

such that D is the set of participating devices and L is the set of physical connectivity between couples of devices, which is defined as follows:

$$L = \{(d_i, d_j) : two-way \; connection \; between \; d_i \; and \; d_j\} \tag{2}$$

Each device is assigned one or more (virtual) coordinate(s). We define C as the set of existing coordinates. Each coordinate c_i is represented as a string of bits starting with $\star$. According to the *regular expression* formalism, coordinates are defined as follows:

$$C = \{c: \; c = \star(0 \mid 1)^*\} \tag{3}$$

For instance, $\star01001$ is a valid W-Grid coordinate. Given a coordinate i and a bit b, their concatenation will be indicated as $c_i b$. For instance, let: $c_i = \star0100, b = 0$; then: $c_i b = \star01000$. Given a bit b, we denote its *complementary* as follows: $\bar{b}$. For instance, $\bar{1} = 0$.

Some functions are defined on top of C. We provide their description.

Given a coordinate c, $length(c)$ returns the number of bits in c. ($\star$ excluded). For instance, $length(\star01001) = 5$. $length(c)$ is formally defined as follows:

$$length(c) : C \to \mathbb{N} \tag{4}$$

---

[1] From now on, we will refer at devices with the terms "sensors" or "nodes" indistinctly.

Given a coordinate c and a positive integer $k \leq length(c)$, $bit(c, k)$ returns the k-th bit of c. Position 0 is out of the domain since it is assigned to $\star$. $bit(c, k)$ is formally defined as follows:

$$bit(c, k) : (C, \mathbb{N} - \{0\}) \rightarrow \{0, 1\} \tag{5}$$

Given a coordinate c and a positive integer $k \leq length(c)$, $pref(c, k)$ returns the first k bits of c. For instance, $pref(\star 01001, 3) = \star 010$. $pref(c, k)$ is formally defined as follows:

$$pref(c, k) : (\mathbb{C}, \mathbb{N}) \rightarrow C \tag{6}$$

Given a coordinate c, we define the complementary of c, denoted by $\bar{c}$, as follows:

$$\bar{c} = pref(c, length(c) - 1)\overline{bit(c, length(c))} \tag{7}$$

For instance, $\overline{\star 01001} = \star 01000$.

Given a coordinate c, we define the *father* of c, denoted by $father(c)$, as follows:

$$father(c) : (C - \{\star\}) \rightarrow C \tag{8}$$

$$father(c) = pref(c, length(c) - 1) \tag{9}$$

Given a coordinate c, we define the *left child* of c, denoted by $lChild(c)$, as follows:

$$lChild(c), rChild(c) : (C) \rightarrow C \tag{10}$$

$$lChild(c) = c0 \tag{11}$$

Given a coordinate c, we define the *right child* of c, denoted by $rChild(c)$, as follows:

$$rChild(c) = c1 \tag{12}$$

To give examples, given a coordinate $c_i = \star 011$, then: $father(\star 011) = \star 01$, $lChild(\star 011) = \star 0110$, $rChild(\star 011) = \star 0111$.

Given a coordinates in the domain C and devices in the domain D, we introduce the *mapping function* M that maps each coordinate c to the device d holding it, as follows:

$$M : C \rightarrow D \tag{13}$$

A W-Grid *network* W is represented in terms of a graph, as follows:

$$W = (C, P) \tag{14}$$

such that P is the set of *parent-ships* between pair of coordinates, defined as follows:

$$P = \{(c_i, c_j) : c_j = c_i(0 \mid 1)\} \tag{15}$$

For instance, $p_i = (\star 010, \star 0101)$.

Given a parent-ship $p = (c_i, c_j)$, we define the complementary of parent-ship $p = (c_i, c_j)$, denoted by $\bar{p}$, as follows:

$$\bar{p} = (c_i, \overline{c_j}) \tag{16}$$

For instance, if $p = (\star 010, \star 0101)$, then: $\bar{p} = (\star 010, \star 0100)$.

Finally, given a graph W, W is a *valid* W-Grid network if all the following properties are satisfied: (*i*) $\forall p = (c_i, c_j) \in P, (M(c_i) = M(c_j)) \vee ((M(c_i), M(c_j)) \in L)$; (*ii*) $\forall p = (c_i, c_j) \in P : M(c_i) \neq M(c_j) \Rightarrow \exists\, \bar{p} = (c_i, \overline{c_j}) \in P : M(c_i) = M(\overline{c_j})$.

3.2 W-Grid Dynamic Rules

W-Grid network is generated according to this few simple rules, which we describe next.

1. The first node that joins the networks (hence it initiates a coordinate space) gets the coordinate $\star$. We say that a node that holds a W-Grid coordinate is *active*. Given a device d, the function *last* returns the last coordinate received by d, and it is defined as follows:

$$last(d) : (D) \rightarrow C \tag{17}$$

If d is *not active*, *last* returns $\{\emptyset\}$. Let n_1 denote the first node that joined the network, then $last(n_1) = \star$.

2. $\forall\, l = (d_i, d_j) \in L : last(d_i) \neq \{\emptyset\}$, two parent-ships are generated: (*i*) $p = (last(d_i), c') : M(c') = d_j$; (*ii*) $\bar{p}$, where $c' = lChild(last(d_i)) \mid rChild(last(d_i))$. Namely, c' corresponds to the *non-deterministic choice* of one of the children of c.

According to a pre-defined *coordinate selection strategy*, nodes progressively get new coordinates from each of its physical neighbors, in order to establish parent-ships with them. Coordinate getting is also called "split", and it is actually related to data management tasks of W-Grid. Actors of the split procedure are the so-called *joining node* and *giving node*, respectively. We say that a coordinate c_i is split by concatenating a bit to it, and, after, one of the new coordinates is assigned to the joining node, while the other one is kept by the giving node. Obviously, an already split coordinate c_i cannot be split anymore since this would generate *duplicates*. Besides, in order to guarantee uniqueness of coordinates even in case of simultaneous requests, each joining node must be acknowledged by the giving node. Thus, if two nodes ask for the same coordinate to split, only one request will succeed, while the other one will be temporarily rejected and postponed. Coordinate discovering is gradually performed by implicitly overhearing of neighbor sensor transmissions.

3.3 W-Grid Routing Algorithm

The routing of any message/query is based on the concept of *distance* among coordinates. Given two coordinates c_i and c_j, the distance between c_i and c_j, denoted by $d(c_i, cj)$, is measured in terms of logical hops and corresponds to the sum of the number of bits of c_i and c_j which are *not* part of their common prefix. To give and For instance: $d(*0011, *011) = 5$.

Given a target binary string c_t, each device of the network is able of constantly getting closer to it by choosing the neighbor that exposes the *shortest* distance to c_t. It is important to notice that each device needs neither global nor partial knowledge about network topology to route messages, as its routing table is limited to information about its direct neighbors' coordinates. This means ensuring *scalability* with respect to network size, which is a nice amenity supported by W-Grid.

3.4   W-Grid Data Management

W-Grid distributes data (i.e., tuples of attributes) gathered by sensors among them in a *data-centric manner*. Values of tuples are linearized into binary strings (see [18]) and stored at nodes whose W-Grid coordinates have the *longest* common prefix with so-generated strings. Thus, a W-Grid network acts directly as a distributed database and coordinates c are used as items stored in a data repository. This means that each coordinate represents a portion (region) of the global data space. In order to balance data load at each region, a maximum number of data items is pre-fixed, namely the *bucket size* b. Whenever a coordinate c turns overloaded with respect to b, c splits into two coordinates c' and $\overline{c'}$. Split operation on coordinates is slightly different from the split operation on nodes described in Section 3.2. The operations supporting the split of c into c' and $\overline{c'}$ are described next: (i) both newly generated coordinates c' and $\overline{c'}$ are (temporarily) kept at the node $M(c)$; (ii) data held by c are distributed between c' and $\overline{c'}$; (iii) parent-ships $p' = (c, c')$ and $\overline{p'} = (c, \overline{c'})$ are generated; (iv) coordinate $\overline{c'}$ is labeled as "givable", meaning that, in the case of a node asking $M(c)$ for a coordinate, $\overline{c'}$ is immediately returned without performing any other split.

If the first split has not solved the overload in one between c or $\overline{c'}$, the split operation is performed again by repeating the previously described steps. In order to further improve the data distribution balancing, the *Storage Load Balancing algorithm* (SLOB) [16] is executed.

In addition to the data management solutions described above, in W-Grid sensed data are replicated in *each* of the existing spaces, thus allowing us to improve data availability as to cope with sensor failures and periodically turn-off a partition of them without compromising the network functionalities (the latter for energy efficiency purposes).

4   Experimental Assessment and Analysis

An extensive number of experiments have been conducted in order to assess the effectiveness and the efficiency of the W-Grid load balancing solutions presented in this paper. Again, we stress the concept that the proposed experimental analysis is focused to *intelligent sensor networks* in which data collected by sensors are not simply downloaded at a fixed sink or in which a sink node is in charge of periodically performing queries but, contrary to this, networks where each active sensor can be queried *at any time*.

Effectiveness and efficiency of W-Grid load balancing solutions have been evaluated in terms of the following experimental parameters: *query average path length, data and traffic load at sensors*, and *energy consumption* of the resulting sensor network. The functional parameters chosen are, instead, the following: *device density, number of coordinates per node*. In the following, we first describe the simulation model of our experimental campaign (Section 4.1). Then, we provide experimental results for the following experimental aspect assessing the W-Grid's performance: *query efficiency* (Section 4.2), *network load* (Section 4.3) and *energy consumption* (Section 4.4).

4.1   Simulation Model

We performed our experiments on top of a Java-based in-laboratory simulation environment. We adopted the following simulation model. Simulation area size is $800 \times 800m$, where nodes are spread in a *uniformly-random manner*. Each sensor has its own identifier (ID), a radio range of $100m$, and it adhere to an ideal transmission mode. As a result, the average number of one-hop neighbors varies from 11 to 4. Due to space limitation, here we show the results for the range $11 - 8$. For each scenario, we ran five simulations and, for each simulation, we submitted $10,000$ queries to the system.

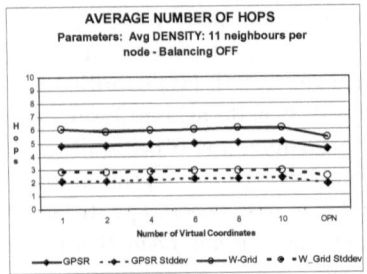
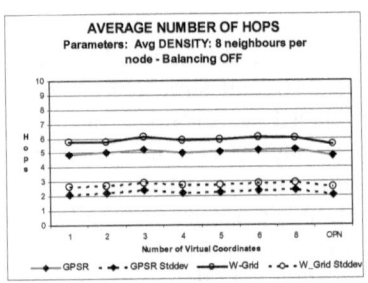

Fig. 1. Query efficiency due to W-Grid without load balancing solutions, in comparison to GPSR, for the configurations having 11 (left) and 8 (right) neighbors per node

We let the simulator perform the following tasks: (*i*) uniformly random placement of n sensors in a user-defined area; (*ii*) gradual generation of W-Grid coordinates at sensors that exploit implicit overhearing; (*iii*) random generation of data sensing and query in a user-defined ratio $q : i$.

At each simulator run, we observed the following experimental parameters: (*i*) average path length of queries (compared with the one due to GPSR [11]); (*ii*) loads at sensors, modeled in terms of the relative difference between network traffic and managed data space at sensors; (*iii*) network energy consumption in each different scenario.

4.2   Query Efficiency

In the first experiment, we tested the ratio of succeeded queries due to W-Grid with respect to the ones accomplished by GPSR. Indeed, for the sake of clarity, here we highlight that the latter comparison is unfair as GPSR assumes to know sensor physical positions over the network. This means that the probability of committing a query is always very high, especially in quite dense application scenarios like the one we tested. However, our main experimental goal was to test our solution against the most efficient one available in literature, even if the comparison one was advantaged. Figure 1 and Figure 2 show the average number of hops needed to succeed queries with respect to the number of virtual coordinates for W-Grid and GPSR in the two distinctive scenarios of not applying (Figure 1) and applying (Figure 2) the load balancing solutions. As Figure 1 and Figure 2 show, it clearly follows that W-Grid behavior is absolutely efficient and not too much distant from the one by GPSR, which reasonably represents the favorite case.

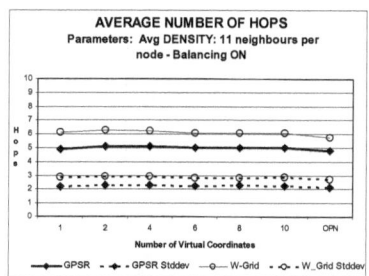 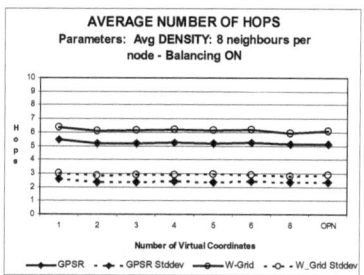

Fig. 2. Query efficiency due to W-Grid with load balancing solutions, in comparison to GPSR, for the configurations having 11 (left) and 8 (right) neighbors per node

4.3   Network Load

Network load trends of W-Grid has also been observed as well. Network workload variations are the result of the interaction between the following two factors: (i) data load balancing among nodes and (ii) increase of transmission costs due to path lengthening. As the number of coordinates grows, these two factors expose better trends, as confirmed by Figure 3 and Figure 4, where we depict the average and standard deviation evolutions of network traffic distribution among nodes with respect to the number of coordinates for W-Grid and GPSR in the two distinctive scenarios of not applying (Figure 3) and applying (Figure 4) the load balancing solutions. Similarly, for cross-comparison purposes, we studied the variation of the standard deviation of data space at sensors (Figure 5) (clearly, with load balancing solutions). Retrieved results clearly confirm the benefits deriving from W-Grid.

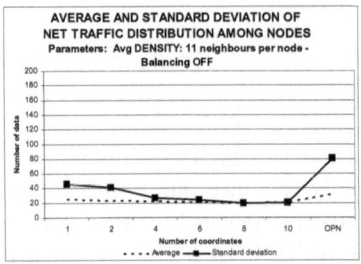 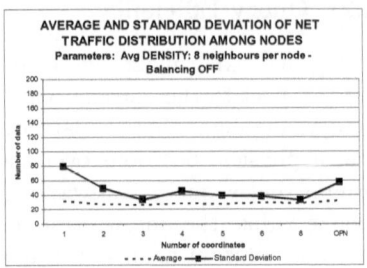

Fig. 3. Network load due to W-Grid without load balancing solutions for the configurations having 11 (left) and 8 (right) neighbors per node

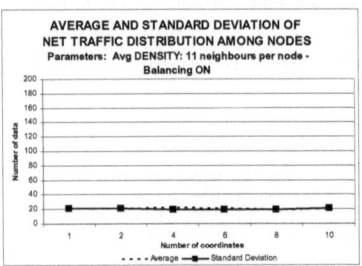 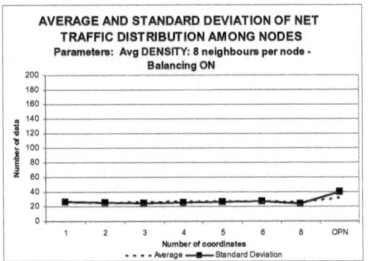

Fig. 4. Network load due to W-Grid with load balancing solutions for the configurations having 11 (left) and 8 (right) neighbors per node

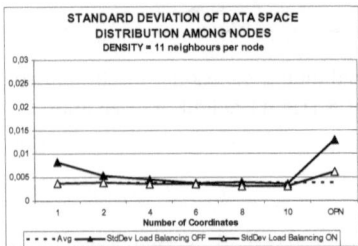 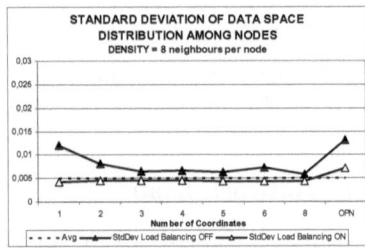

Fig. 5. Variation of data space at sensors due to W-Grid with load balancing solutions for the configurations having 11 (left) and 8 (right) neighbors per node

4.4   Energy Consumption

Finally, we focused the attention on energy consumption due to W-Grid for each application scenario considered in the experimental campaign (i.e., 11, 8 and 4 neighbors per node). Energy consumption was measured according to [17]. Figure 6 shows the retrieved observations for the two different cases of applying (left) and not applying (right) load balancing solutions. Even in this study, W-Grid observations expose a very positive trend.

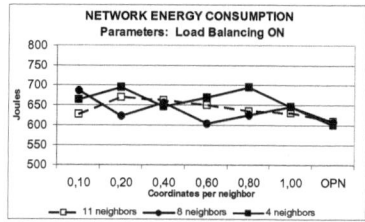 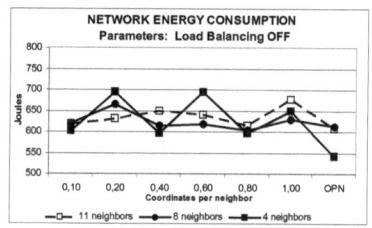

Fig. 6. Energy consumption due to W-Grid for different configurations with (left) and without (right) load balancing solutions

5   Conclusions and Future Work

In this paper, we proposed load balancing extensions of W-Grid and assessed them by means of a comprehensive experimental campaign that has focused the attention on several interesting experimental parameters, ranging from query efficiency to energy consumption. Experiments have been conducted on top of a Java-based simulation environment. Retrieved results have clearly confirmed the benefits due to the usage of W-Grid as cross-layer infrastructure for self-organizing data-centric sensor networks.

Future work is mainly oriented towards three directions. First, we are considering the presence of *outliers* in the data layer of such networks. *"How their presence impacts on the performance of W-Grid?"* is a challenging question that demands for solution. Second, we are considering the integration of W-Grid's architecture with modern *Cloud computing platforms*. Finally we are addressing in W-Grid the issue of network security with malicious nodes by extending decentralized approach based on distributed supervised data mining [1].

References

1. Cerroni, W., Moro, G., Pirini, T., Ramilli, M.: Peer-to-peer data mining classifiers for decentralized detection of network attacks. In: Wang, H., Zhang, R. (eds.) ADC 2013, Adelaide, South Australia. CRPIT, pp. 1–8. ACS (2013)
2. Cuzzocrea, A.: CAMS: OLAPing multidimensional data streams efficiently. In: Pedersen, T.B., Mohania, M.K., Tjoa, A.M. (eds.) DaWaK 2009. LNCS, vol. 5691, pp. 48–62. Springer, Heidelberg (2009)
3. Cuzzocrea, A.: Retrieving accurate estimates to OLAP queries over uncertain and imprecise multidimensional data streams. In: Bayard Cushing, J., French, J., Bowers, S. (eds.) SSDBM 2011. LNCS, vol. 6809, pp. 575–576. Springer, Heidelberg (2011)
4. Cuzzocrea, A., Chakravarthy, S.: Event-based lossy compression for effective and efficient OLAP over data streams. Data Knowl. Eng. 69(7), 678–708 (2010)
5. Cuzzocrea, A., Furfaro, F., Greco, S., Masciari, E., Mazzeo, G.M., Saccà, D.: A distributed system for answering range queries on sensor network data. In: PerCom Workshops, pp. 369–373 (2005)

6. Cuzzocrea, A., Furfaro, F., Mazzeo, G.M., Saccá, D.: A grid framework for approximate aggregate query answering on summarized sensor network readings. In: Meersman, R., Tari, Z., Corsaro, A. (eds.) OTM Workshops 2004. LNCS, vol. 3292, pp. 144–153. Springer, Heidelberg (2004)

7. Cuzzocrea, A., Furfaro, F., Saccà, D.: Enabling OLAP in mobile environments via intelligent data cube compression techniques. J. Intell. Inf. Syst. 33(2), 95–143 (2009)

8. El-Moukaddem, F., Torng, E., Xing, G.: Mobile relay configuration in data-intensive wireless sensor networks. IEEE Trans. Mob. Comput. 12(2), 261–273 (2013)

9. Greenstein, B., Estrin, D., Govindan, R., Ratnasamy, S., Shenker, S.: Difs: A distributed index for features in sensor networks. In: Proceedings of First IEEE WSNA, pp. 163–173. IEEE Computer Society (2003)

10. Intanagonwiwat, C., Govindan, R., Estrin, D., Heidemann, J., Silva, F.: Directed diffusion for wireless sensor networking. IEEE/ACM Trans. Netw. 11(1), 2–16 (2003)

11. Karp, B., Kung, H.: GPRS: greedy perimeter stateless routing for wireless networks. In: MobiCom 2000, pp. 243–254. ACM Press (2000)

12. Li, X., Kim, Y., Govindan, R., Hong, W.: Multi-dimensional range queries in sensor networks. In: SenSys 2003, pp. 63–75. ACM Press, New York (2003)

13. Li, Z., Liu, Y., Li, M., Wang, J., Cao, Z.: Exploiting ubiquitous data collection for mobile users in wireless sensor networks. IEEE Trans. Parallel Distrib. Syst. 24(2), 312–326 (2013)

14. Liu, B., Dousse, O., Nain, P., Towsley, D.: Dynamic coverage of mobile sensor networks. IEEE Trans. Parallel Distrib. Syst. 24(2), 301–311 (2013)

15. Monti, G., Moro, G., Sartori, C.: W^R-Grid: A scalable cross-layer infrastructure for routing, multi-dimensional data management and replication in wireless sensor networks. In: Min, G., Di Martino, B., Yang, L.T., Guo, M., Rünger, G. (eds.) ISPA 2006 Ws. LNCS, vol. 4331, pp. 377–386. Springer, Heidelberg (2006)

16. Moro, G., Monti, G.: W-Grid: a self-organizing infrastructure for multi-dimensional querying and routing in wireless ad-hoc networks. In: IEEE P2P 2006 (2006)

17. Moro, G., Monti, G.: W-grid: A scalable and efficient self-organizing infrastructure for multi-dimensional data management, querying and routing in wireless data-centric sensor networks. Journal of Network and Computer Applications 35(4), 1218–1234 (2012), http://dx.doi.org/10.1016/j.jnca.2011.05.002

18. Ouksel, A.M., Moro, G.: G-Grid: A class of scalable and self-organizing data structures for multi-dimensional querying and content routing in P2P networks. In: Moro, G., Sartori, C., Singh, M.P. (eds.) AP2PC 2003. LNCS (LNAI), vol. 2872, pp. 123–137. Springer, Heidelberg (2004)

19. Ratnasamy, S., Karp, B., Shenker, S., Estrin, D., Govindan, R., Yin, L., Yu, F.: Data-centric storage in sensornets with ght, a geographic hash table. Mob. Netw. Appl. 8(4), 427–442 (2003)

20. Xiao, L., Ouksel, A.: Tolerance of localization imprecision in efficiently managing mobile sensor databases. In: ACM MobiDE 2005, pp. 25–32. ACM Press, New York (2005)

21. Ye, F., Luo, H., Cheng, J., Lu, S., Zhang, L.: A two-tier data dissemination model for large-scale wireless sensor networks. In: MobiCom 2002, pp. 148–159. ACM Press, New York (2002)

Update Semantics for Interoperability among XML, RDF and RDB*

A Case Study of Semantic Presence in CISCO's Unified Presence Systems

Muhammad Intizar Ali[1], Nuno Lopes[1], Owen Friel[2], and Alessandra Mileo[1]

[1] DERI, National University of Ireland, Galway, Ireland
{ali.intizar,nuno.lopes,alessandra.mileo}@deri.org
[2] Cisco Systems, Galway, Ireland
ofriel@cisco.com

Abstract. XSPARQL is a transformation and querying language that provides an integrated access over heterogeneous data sources on the fly. It is an extension of XQuery which supports a subset of SPARQL and SQL to provide unified access over XML, RDF and RDB formats. In practical applications, data integration does not only require the integrated access over distributed heterogeneous data sources, but also the update of underlying data.

XSPARQL in its present state is only a querying and transformation language, hence lacking the update facility. In this paper, we propose an extension of the XSPARQL language with update facility. We present the syntax and semantics for this extension, and we use the real world scenario of semantic presence in CISCO's Unified Presence Systems to demonstrate the requirement of update facility. Preliminary evaluation of the *XSPARQL Update Facility* is also presented.

Keywords: XSPARQL, SPARQL , XQuery, XMPP, Updates, DML.

1 Introduction

Relational databases (RDB) are the most common way of storing and managing data in majority of the enterprise environments [1]. Particularly in the enterprises where integrity and confidentiality of the data are of the utmost importance, relational data model is still the most preferred option. However, with the growing popularity of the Web and other Web related technologies (e.g. Semantic Web), various data models have been introduced. Extensible Markup Language (XML) is a very popular data model to store and specify the semi-structured data over the Web [7]. It is also considered as a de-facto data model for data exchange. XPath and XQuery are designed to access and query data in XML

---

* This work has been funded by Science Foundation Ireland, Grant No. SFI/08/CE/l1380 (Lion2) and CISCO Systems Galway, Ireland.

Y. Ishikawa et al. (Eds.): APWeb 2013, LNCS 7808, pp. 43–50, 2013.

format[10,5]. Semantic Web is built upon data represented in Resource Description Format (RDF) [4]. RDF is a standard data format for establishing semantic interpretability and machine readability among the various data sources scattered over the Web. SPARQL is a query language designed to query linked data in RDF data format [17]. All of the above mentioned data models were designed to perform a specific task and their importance in their own domains can not be denied. Several query languages including SQL, XQuery and SPARQL have been designed to access and query RDB, XML and RDF data respectively. It is inevitable for modern data integration applications to avoid such heterogeneity of data formats and query languages. Therefore, often these applications require integrated access over distributed heterogeneous data sources.

XSPARQL is a W3C Member Submission (http://www.w3.org/Submission/2009/01/) which combines XQuery, SPARQL and SQL to provide an integrated access over XML, RDF and RDB data sources [2]. However, in practice the integrated applications not only require the access of integrated data, but in many scenarios an update in the existing data is also desired. XSPARQL in its present state only provides Data Query Language (DQL) while lacking Data Manipulation Language (DML). In this paper we define syntax and semantics of XSPARQL Update Facility which extends the capabilities of the XSPARQL language to perform update operations over RDB, XML and RDF data on the fly while keeping the data sources in their original data formats. XSPARQL Update Facility is fully compatible with the update semantics defined individually for SQL, XQuery and SPARQL [16,18,12]. Our main motivation for this extension was the real world scenario of semantic presence in the CISCO Unified Presence (CUP) systems, based on processing and integrating information retrieved from XMPP(http://xmpp.org) messages and update the underlying data in RDB and RDF data sources to keep track of up-to-date semantic presence history of registered users. In this scenario we require not only the simultaneous access to heterogeneous data represented in RDB, XML and RDF, but also the update operations are desired to update the underlying RDB and RDF data on the fly [8].

Our main contributions in this paper can be summarised as follows: (i) we extend XSPARQL with updates, providing what we call *XSPARQL Update Facility*, (ii) we define formal semantics of the XSPRQL Update Facility, (iii) we demonstrate the need for XSPARQL Update Facility by presenting a use case scenario for semantic presence in the CUP systems, and (iv) we successfully implement and evaluate XSPARQL Update Facility for dealing with group chats history in the same scenario.

2   Semantic Presence in Cisco's Unified Presence System

Cisco Unified Presence (CUP) systems is a standards-based enterprise platform that aims to provide an effective way of communication among people in and across the organisation using instant messaging (IM) over XMPP standard protocol. A general architecture to collect several presence information from various heterogeneous and dynamic sources has been introduced in [14],which provides some insights on how richer semantic presence services and applications can be deployed. As a first step towards the enhancement of CUP systems with richer semantic presence, a mapping of the core XMPP messages into RDF is formally defined in [8]. Figure 1 depicts a simplified high-level

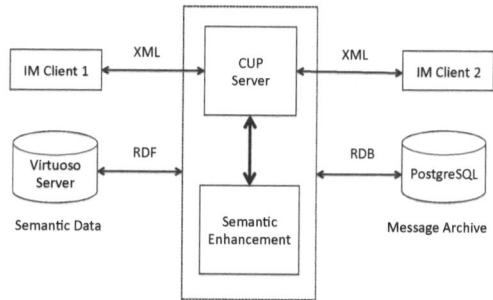

Fig. 1. Semantic Presence in CUP Systems

version of the architecture of semantic presence in the CUP systems focusing only on heterogeneous data formats that need to be manipulated during communication. All XMPP messages are mapped into RDF using various ontologies e.g SIOCC (http://rdfs.org/sioc/chat#), Nepomuk (http://www.semanticdesktop.org/ontologies/2007/03/22/nmo#) and SIOC (http://rdfs.org/sioc/ns#). CUP server stores history of all the group chat messages and other communications in an external PostgreSQL (http://www.postgresql.org/) database, while semantic enhancement component stores the semantic information in a graph database using Virtuoso Server (http://openlinksw.com/virtuoso). Listing 1 shows a sample XMPP message generated by IM client while requesting for the creation of a chat room. Listing 2 depicts mapping of XMPP messages into RDF using SIOC and SIOCC ontologies. We use the XSPARQL Update Facility to interpret the XMPP message and generate/update RDB and RDF data on the fly. This enables a semantically enriched CUP system to update/generate data from one data model to any other data model. Later, in section 4 we evaluate our XSPARQL Update facility by using the use case of semantic presence in CUP systems.

```
#intizar@deri.org seeks to enter
#in the DERICollab room

<presence
    from=
"intizar@deri.org/intizar-notebook"
        to=
"DERICollab@conference.corp.com/
intizar">
<x xmlns="http://jabber.org/protocol/
muc"/>
</presence>
```

Listing 1. XMPP request message for chat room creation

```
@base <http://lion2pcisco.com
/example#>.
:DERICollab      a
sioct:ChatChannel ;
    sioc:has_owner
:intizar@corp.com       .
:DERICollabChatSession      a
siocc:ChatSession ;
    sioc:has_container
:DERICollab;
  sioc:has_publishing_source
:intizarNotebook
    siocc:has_nickname   "ali"
```

Listing 2. Mapping XMPP Message of Listing 1 into RDF [8]

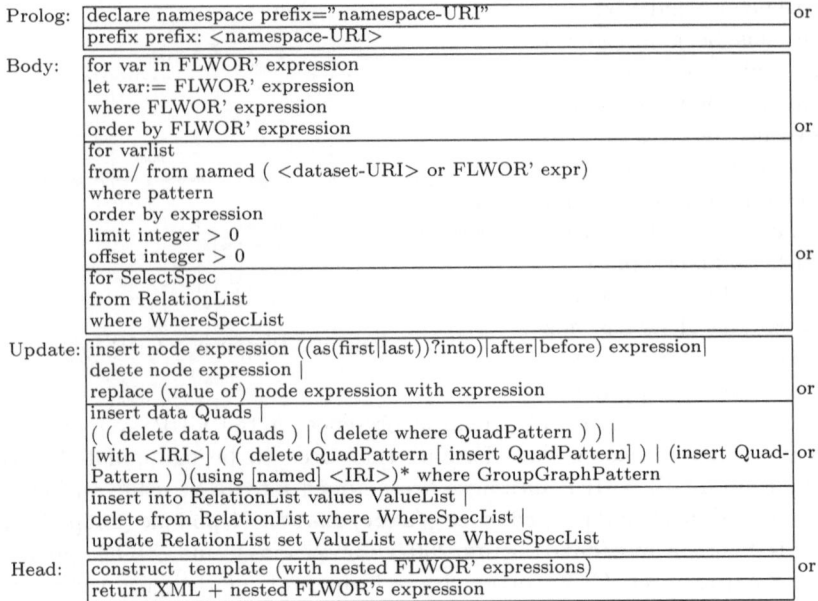

| Prolog: | declare namespace prefix="namespace-URI" | or |
| | prefix prefix: \<namespace-URI\> | |

Body:	for var in FLWOR' expression	
	let var:= FLWOR' expression	
	where FLWOR' expression	
	order by FLWOR' expression	or
	for varlist	
	from/ from named (\<dataset-URI\> or FLWOR' expr)	
	where pattern	
	order by expression	
	limit integer > 0	
	offset integer > 0	or
	for SelectSpec	
	from RelationList	
	where WhereSpecList	

Update:	insert node expression ((as(first|last))?into)|after|before) expression|	
	delete node expression |	
	replace (value of) node expression with expression	or
	insert data Quads |	
	((delete data Quads) | (delete where QuadPattern)) |	
	[with \<IRI\>] ((delete QuadPattern [insert QuadPattern]) | (insert Quad-	or
	Pattern))(using [named] \<IRI\>)* where GroupGraphPattern	
	insert into RelationList values ValueList |	
	delete from RelationList where WhereSpecList |	
	update RelationList set ValueList where WhereSpecList	

| Head: | construct template (with nested FLWOR' expressions) | or |
| | return XML + nested FLWOR's expression | |

Fig. 2. Schematic View of XSPARQL with Updates

3   XSPARQL Update Facility

In this section we specify the XSPARQL Update Facility by giving a detailed description of its syntax and semantics.

3.1   Syntax

XSPARQL is built on XQuery semantics, XSPARQL Update Facility is also extended using formal semantics of XQuery Update Facility. It merges the subset of SPARQL, XQuery and SQL update clauses. We limit ourself to three major update operations of *INSERT:* to insert new records in the data set, *DELETE:* to delete already existing records from data set and *UPDATE:* to replace the existing records or their values with the new records or values. Figure 2 presents a schematic view of the XSPARQL Update Facility which allows its users to select data from one data source of any of its data format and update the results into another data source of different format within a single XSPARQL query while preserving the bindings of the variables defined within query. Contrary to the select queries, the update queries have no return type. On successful execution of the update queries no results are returned, only a response is generated which can be either *successful* upon successful execution of the query or an appropriate *error* is raised if for some reasons the XSPARQL query processor is unable to update the records successfully. However if a valid updating or delete query with or without *where* clause does not match any relevant tuple in the data source, the XSPARQL query processor will still response with successful execution, but there will be no effect on the existing data source. The basic building block of XQuery is the expression, an XQuery expression takes zero or more XDM instances and returns an XDM instance. Following new expressions are introduced in XSPARQL Update Facility

- A new category of expression called *updating expression* is introduced to make persistent changes in the existing data.
- A basic updating expression can be any of the *insert*, *delete* or *update*.
- *for, let, where* or *order by* clause can not contain an updating expression.
- *return* clause can contain an updating expression which will be evaluated for each tuple generated by its *for* clause.

```
RevalidationDecl::=  "declare" "revalidation" ("strict" | "lax" | "skip" )
XSPARQLExpr    ::=  (FLWORExpr | SPARQLForClause | SQLForClause) |
                    (InsertClause | DeleteClause | UpdateClause)
                    (ReturnClause | ConstructClause)
InsertClause   ::=  XQueryInsert | SPARQLInsert | SQLInsert
DeleteClause   ::=  XQueryDelete | SPARQLDelete | SQLDelete
UpdateClause   ::=  XQueryUpdate | SPARQLUpdate | SQLUpdate
...
```

Fig. 3. XSPARQL Update Facility Grammar Overview

Figure 3 presents an overview of the basic syntax rules for XSPARQL Update Facility, for complete grammar of XSPARQL Update Facility, we refer our reader to the latest version of the XSPARQL available at `http://sourceforge.net/projects/xsparql`.

3.2 Semantics

Similar to the semantics of the XSPARQL language [6], for the semantics of updates in XSPARQL we rely on the semantics of the original languages: SQL, XQuery, and SPARQL. Notably, the semantics for SPARQL updates are presented only in the upcoming W3C recommendation: SPARQL 1.1 [12]. We start by presenting a brief overview of the semantics of update languages for the different data models: relational databases, XML, and RDF.

RDB: updates in the SQL language rely on a procedural language that specify the operations to be performed, respectively: $ins_{rdb}(r,t)$, $del_{rdb}(r,C)$, and $mod_{rdb}(r,C,C')$, where r is a relation name, t is a relational tuple, and C,C' are a set of conditions of the form $A - c$ or $A \neq c$ (A is an attribute name and c is a constant). Following [1], $ins_{rdb}(r,t)$ inserts the relational tuple t into the relation r, $del_{rdb}(r,C)$ deletes the relational tuples from relation r that match the condition, and $mod_{rdb}(r,C,C')$ updates the tuples in relation r that match conditions C to the values specified by C'.

XML: For XML data, the XQuery language is adapted such that an expression can return a sequence of nodes (as per the XQuery semantics specification [9]) or *pending update list* [18], which consists of a collection of *update* *primitives*, representing state changes to XML nodes. For this paper we are focusing on *insert*, *delete*, and *replace* operations, whose semantics are procedurally defined in [18] and in this paper we represent these procedural semantics by the functions $ins_{xml}(SourceExpr, InTargetChoice, TargetExpr)$, $del_{xml}(TargetExpr)$, and $mod_{xml}(TargetExpr, ExprSingle)$.

RDF: SPARQL Update's semantics are similarly defined in terms of changes to the underlying triple store [12], where the result of an update operation is a new triple store. We rely on the semantics functions ins_{rdf}, del_{rdf}, and mod_{rdf} (shorthand for the functions *OpInsertData*, *OpDeleteData*, and *OpDeleteInsert*, respectively). The functions signatures are as per the SPARQL specification: $ins_{rdf}(GS, QuadPattern)$, $del_{rdf}(GS, QuadPattern)$, and $mod_{rdf}(GS, DS, QuadPattern_{del}, QuadPattern_{ins}, P)$ presented in [12]. The functions ins_{rdf} and del_{rdf} insert and delete triples matching *QuadPattern* from the triple store *DS*, respectively. The mod_{rdf} evaluates the graph pattern *P* against the dataset *DS* and apply these bindings to *QuadPattern$_{del}$* in order to determine the triples to be removed from the graph store *GS* and to *QuadPattern$_{ins}$* for the inserted triples.

4   Implementation and Evaluation

Implementation: We use java platform for the implementation of the XS-PARQL Update Facility. XSPARQL grammar rules are defined using ANTLR (http://www.antlr.org). XSPARQL uses Saxon (http://www.saxonica.com) as a query processor for XQuery, Jena ARQ (http://jena.apache.org) for SPARQL queries and for SQL queries we provide option to connect to multiple relational database management systems including PostgreSQL and MySQL (http://www.mysql.com).

Evaluation of XSPARQL Update Facility: In order to demonstrate and evaluate performance and practicality of XSPARQL Update Facility, we consider a generic use case of group chat of the semantic presence in CUP systems as described in Section 2. However XSPARQL Update Facility can be used by any application which requires access and updates in distributed heterogeneous data stored in any of the RDB, XML or RDF data models.

```
for $x in //*[name='presence']
if $x/status[@code=201] then
insert data
{Graph <SemanticPresence>
    {
        :DERICollab      a      sioct:ChatChannel ;
        sioc:has_owner      :intizarali@corp.com .
        :DERICollabChatSession      a      siocc:ChatSession;
        sioc:has_container      :DERICollab;
        siocc:has_publishing_source      :intizaraliNotebook
        siocc:has_nickname      "ali"
    }}
```

Listing 3. A Sample XSPARQL Update Query

A User Creates a Chat Group: Consider a scenario where a user creates a persistent chat room over a specific topic as shown in Listing 1 in the Section 2. When an IM client sends a request to create a chat room an XMPP message will be generated by IM client and sent to the CUP Server. CUP Server will process the message. The semantic presence component of CUP systems wants to update the semantic data stored in Virtuoso Server. Listing 3 shows an XSPARQL query that interprets XMPP messages using XQuery and if certain condition is met, the XSPARQL update query will insert the relevant information into RDF data store using SPARQL update query.

5   Related Work

Several efforts are made to integrate distributed XML, RDF and RDB data on the fly. These approaches can be divided into two types, (i) *Transformation Based:* Data stored in various format is transformed into one format and can be queried using a single query language [11]. The W3C GRDDL working group addresses the issues of extracting RDF from XML documents. (ii) *Query Rewriting Based:* Query languages are used to transform or query data from one format to another format. SPARQL queries are embedded into XQuery/XSLT to pose against pure XML data [13]. XSPARQL was initially designed to integrate data from XML and RDF and later extended to RDB as well. DeXIN is another approach to integrate XML, RDF and RDB data on the fly with more focus on distributed computing of heterogeneous data sources scattered over the Web [3]. Realising the importance of the updates, the W3C has recommendations for XQuery and SPARQL updates [18,12]. In [15], SPARQL is used to update the relational data. However, to the best of our knowledge their is no work available which can provide simultaneous access over the distributed heterogeneous data source with updates.

6   Conclusion and Future Work

In this paper we have extended XSPARQL to provide the Update Facility which further strengthens the capabilities of XSPARQL by enabling simultaneous access and update of heterogeneous data sources. We have defined syntax and semantics of XSPARQL Update Facility, which merges update operations of SQL, XQuery and SPARQL. Using XSPARQL Update Facility user not only can query, integrate and transform heterogeneous data on the fly, but can also update the already stored data in the data sets or in-memory data generated as a result of a query before an update operation. XSPARQL Update Facility will lead to the wide adoption of XSPARQL in many applications (e.g. similar to semantic presence in CUP systems) which require integrated access over distributed heterogeneous data sources with updates.

In the future, we plan to further investigate query optimisation techniques that would make it possible to evaluate even complex XSPARQL queries in tractable time, that would make XSPARQL usable for large-scale applications. In this paper, we elaborated the practicality of XSPARQL Update Facility using real world scenario, however in future, we plan to define detailed formal semantics of the language, evaluation of static and dynamic rules and experimental evaluation considering various parameters including scalability and query executaon time. We also plan to extend XSPARQL Update Facility to include Data Definition Language (DDL) for its subsequent languages.

References

1. Abiteboul, S., Hull, R., Vianu, V.: Foundations of Databases. Addison-Wesley (1995)
2. Akhtar, W., Kopecký, J., Krennwallner, T., Polleres, A.: XSPARQL: Traveling between the XML and RDF Worlds – and Avoiding the XSLT Pilgrimage. In: Bechhofer, S., Hauswirth, M., Hoffmann, J., Koubarakis, M. (eds.) ESWC 2008. LNCS, vol. 5021, pp. 432–447. Springer, Heidelberg (2008)
3. Ali, M.I., Pichler, R., Truong, H.-L., Dustdar, S.: DeXIN: An Extensible Framework for Distributed XQuery over Heterogeneous Data Sources. In: Filipe, J., Cordeiro, J. (eds.) ICEIS 2009. LNBIP, vol. 24, pp. 172–183. Springer, Heidelberg (2009)
4. Beckett, D., McBride, B.: RDF/XML Syntax Specification. W3C Proposed Recommendation (February 2004) (revised)
5. Berglund, A., Boag, S., Chamberlin, D., Fernández, M.F., Kay, M., Robie, J., Siméon, J.: XML Path Language (XPath) 2.0. W3C Recommendation (December 2010)
6. Bischof, S., Decker, S., Krennwallner, T., Lopes, N., Polleres, A.: Mapping between RDF and XML with XSPARQL. Journal on Data Semantics 1, 147–185 (2012)
7. Bray, T., Paoli, J., Maler, E., Yergeau, F., Sperberg-McQueen, C.M.: Extensible Markup Language (XML) 1.0 (5th edn.). W3C Recommendation (November 2008)
8. Dabrowski, M., Scerri, S., Rivera, I., Leggieri, M.: Dx– Initial Mappings for the Semantic Presence Based Ontology Definition (November 2012), http://www.deri.ie/publications/technical-reports/
9. Draper, D., Fankhauser, P., Fernández, M., Malhotra, A., Rose, K., Rys, M., Siméon, J., Wadler, P.: XQuery 1.0 and XPath 2.0 Formal Semantics. W3C Recommendation (January 2007)
10. Fernández, M.F., Florescu, D., Boag, S., Robie, J., Chamberlin, D., Siméon, J.: XQuery1.0: An XML query language. W3C Proposed Recommendation (April 2009)
11. Gandon, F.: GRDDL Use Cases: Scenarios of extracting RDF data from XML documents. W3C Proposed Recommendation (April 2007)
12. Gearon, P., Passant, A., Polleres, A.: SPARQL 1.1 Update. W3C Working Draft (January 2012)
13. Groppe, S., Groppe, J., Linnemann, V., Kukulenz, D., Hoeller, N., Reinke, C.: Embedding SPARQL into XQuery/XSLT. In: Proc. of SAC (2008)
14. Hauswirth, M., Euzenat, J., Friel, O., Griffin, K., Hession, P., Jennings, B., Groza, T., Handschuh, S., Zarko, I.P., Polleres, A., Zimmermann, A.: Towards Consolidated Presence. In: Proc. of CollaborateCom 2010, pp. 1–10 (2010)
15. Hert, M., Reif, G., Gall, H.: Updating relational data via SPARQL/update. In: Proc. of EDBT/ICDT Workshops (2010)
16. Negri, M., Pelagatti, G., Sbattella, L.: Formal Semantics of SQL Queries. ACM Trans. Database Syst. 16(3), 513–534 (1991)
17. Prud'hommeaux, E., Seaborne, A.: SPARQL Query Language for RDF. W3C Proposed Recommendation (January 2008)
18. Robie, J., Chamberlin, D., Dyck, M., Florescu, D., Melton, J., Siméon, J.: XQuery Update Facility 1.0. W3C Recommendation (March 2011)

GPU-Accelerated Bidirected De Bruijn Graph Construction for Genome Assembly

Mian Lu[1], Qiong Luo[2], Bingqiang Wang[3], Junkai Wu[2], and Jiuxin Zhao[2]

[1] A*STAR Institute of High Performance Computing, Singapore
lum@ihpc.a-star.edu.sg
[2] Hong Kong University of Science and Technology
{luo,jwuac,zhaojx}@cse.ust.hk
[3] BGI-Shenzhen, China
wangbingqiang@genomics.org.cn

Abstract. De Bruijn graph construction is a basic component in de novo genome assembly for short reads generated from the second-generation sequencing machines. As this component processes a large amount of data and performs intensive computation, we propose to use the GPU (Graphics Processing Unit) for acceleration. Specifically, we propose a staged algorithm to utilize the GPU for computation over large data sets that do not fit into the GPU memory. We also pipeline the I/O, GPU, and CPU processing to further improve the overall performance. Our preliminary results show that our GPU-accelerated graph construction on an NVIDIA S1070 server achieves a speedup of around two times over previous performance results on a 1024-node IBM Blue Gene/L.

1 Introduction

Genome assembly refers to the process of reconstructing a genome sequence from a large number of sequence fragments (known as *reads*). These reads are generated by sequencing machines through randomly sampling the original sequence. The De Bruijn graph based genome assembly algorithms have been shown effective for assembling a large number of short reads and have been adopted in state-of-the-art assemblers [1–4]. In this paper, we focus on the bidirected De Bruijn graph construction, which is the first as well as one of the most expensive steps in genome assembly. We propose to utilize the GPU (Graphics Processing Units) to accelerate this process, in particular, the construction of a bidirected De Bruijn graph from a large set of short reads.

The bidirected De Bruijn graph construction is expensive in both memory consumption and running time. A previous study [2] has shown that the graph construction for human genome took around 8 hours on a 16-core machine with 2.3GHz AMD quadcore CPUs and consumed 140GB main memory. With limited GPU memory (up to 6 GB per GPU in the market), the first challenge in GPU-based De Bruijin graph construction is to develop algorithms that can handle data larger than the GPU memory. Second, given the superb computation power of the GPU, the overall performance is likely to be dominated by disk

Y. Ishikawa et al. (Eds.): APWeb 2013, LNCS 7808, pp. 51–62, 2013.

I/O and CPU processing. Therefore, we study how to utilize the hierarchy of disk, main memory, and GPU memory to pipeline processing and to involve the CPU for co-processing. These issues are essential for the feasibility and overall performance on practical applications; unfortunately, they are seldom studied in current GPGPU research.

Specifically, we address the GPU memory limit by developing a staged algorithm for GPU-based graph construction. We divide the reads into chunks and load the data chunk by chunk from disk through the main memory to the GPU memory. To estimate the chunk size that can fit into the GPU memory, we develop a memory cost model for each processing step. We further utilize the CPU and main memory to perform result merging each time the GPU finishes processing a chunk. Finally, we pipeline the disk I/O, GPU processing, and CPU merging (*DGC*) to improve the overall performance. We expect this *pipelined-DGC* processing model to be useful for a wide range of GPGPU applications that handle large data.

We have implemented the GPU-based bidirected De Bruijn graph construction and evaluated it on an NVIDIA Tesla S1070 GPU device with 4 GB memory. Our initial results show that this implementation doubles the performance reported on a 1024-node IBM Blue Gene/L and is orders of magnitude faster than state-of-the-art CPU-based sequential implementations.

The remainder of this paper is organized as follows. In Section 2, we briefly introduce the graph construction algorithm and related work. We present our design of the staged algorithm in Section 3. We describe the details of the pipelined-DGC model in Section 4. The experimental results are reported in Section 5. We conclude in Section 6.

2   Preliminary

2.1   Genome Assembly

The second generation of sequencing produces very short reads at a high throughput. Popular algorithms to reconstruct the original sequence for such a large number of short reads are based on the *De Bruijn [5]* or *bidirected De Bruijn* graph [6]. The graph is constructed through generating *k-mers* from reads as graph nodes. For example, suppose a short read is ACCTGC and $k = 4$, then this read can generate three 4-mers, which are ACCT, CCTG, and CTGC. The major difference between the two graph models is that for each *k-mer*, its reverse complement is represented by a separate node (De Bruijn) or the same node (bidirected De Bruijn).

At the beginning of assembly, each l-length read generates $(l - k + 1)$ *k-mers*. Then the De Bruijn graph is constructed using information about overlap between *k-mers*. Next, the graph is simplified and corrected by some heuristic algorithms. After the simplification and correction, several long *contigs* are generated. Finally, if reads are generated through paired-end sequencing, these *contigs* are joined to produce *scaffolds* using relevant information. Another alternative of joining contigs is through Eulerian paths. The detailed algorithms of these steps can be found in previous work [2, 4]. In this work, we focus on the De Bruijn graph construction.

2.2 Bidirected De Bruijn Graph

The double-stranded structure of DNA sequences can be naturally mapped to the bidirected De Bruijn graph [6]. For a birdirected graph, each edge has independent directions at two ends. A valid path from node v_1 to v_k is represented as a sequence $v_1 e_1 v_2 e_2 ... e_{k-1} v_k$, where e_i is the edge connecting two node v_i and v_{i+1}, and e_{i-1} and e_i have different directions at the node v_i for $1 < i < k$. Figure 1 shows an example of bidirected graph. A De Bruijn graph is a directed graph, where each node represents an ordered sequence, and all sequences have the same length. An edge from node A representing $a_1 a_2 ... a_n$ to node B representing $b_1 b_2 ... b_n$ exists when $a_2 a_3 ... a_n$ is identical to $b_1 b_2 ... b_{n-1}$.

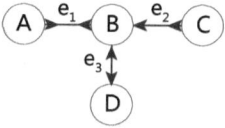

Fig. 1. An example of bidirected graph

For genome assembly, each node in a bidirected De Bruijn graph represents a *k-mer* containing k bases. Given the double-stranded structure of DNA sequences, each node also implicitly represents its reverse complementary *k-mer*, i.e., the base on each position in one sequence is the complement (A to T, C to G) of another sequence in the reverse order. For a given node v representing two reverse complementary *k-mers*, we denote the lexicographically greater one as v^+ to be the canonical form of these two *k-mers* as well as the node representative, and v^- the reverse complement of v^+. Note that, when k is an odd number, v^+ and v^- will never be identical.

Suppose an edge exists between nodes A and B. There are four cases when A overlaps B.

- Case 1. A^+ overlaps B^+ $\Rightarrow$ A ▷—▷B
- Case 2. A^- overlaps B^- $\Rightarrow$ A ◁—◁B
- Case 3. A^- overlaps B^+ $\Rightarrow$ A ◁—▷B
- Case 4. A^+ overlaps B^- $\Rightarrow$ A ▷—◁B

Note that duplicate edges are expected due to the high coverage of input reads. However, only distinct edges and their frequency counts (known as *multiplicities*) need to be maintained. The multiplicity is useful when removing errors from graphs. After constructing all distinct edges and recording their multiplicities, we generate an ordered pair $<c_X, c_Y>$ corresponding to an edge representing X overlaps Y. c_X and c_Y are the label to output when walking from node X to Y and from node Y to X, respectively. The label is generated as follows corresponding to the four cases of overlap. $X^+[1]$ and $X^-[1]$ denote the first character of the canonical form and its reverse complement of node X, respectively.

- Case 1. A^+ overlaps $B^+ \Rightarrow <A^+[1], B^-[1]>$
- Case 2. A^- overlaps $B^- \Rightarrow <A^-[1], B^+[1]>$
- Case 3. A^- overlaps $B^+ \Rightarrow <A^-[1], B^-[1]>$
- Case 4. A^+ overlaps $B^- \Rightarrow <A^+[1], B^+[1]>$

Distinct canonical *k-mers* are collected as representatives for graph nodes, and adjacency information is built for the graph representation for further manipulation.

2.3   Related Work

There are several genome assembly tools for short reads on the CPU. Velvet [4] is a pioneer software for short reads assembly, which is still one of the most popular short reads assemblers today. SOAPdenovo [2] has been used successfully for human genome projects. There are other similar assemblers, such as Euler-SR [7], Shorty [8], Edena [9], ALLPATHS [10], SSAKE [11], SHARCGS [12], ABySS [3], and YAGA [1]. These assemblers are different in implementation details, but are all based on the (bidirected) De Bruijn graph model.

Among parallel short reads assemblers, ABySS [3] was implemented using MPI. SOAPdenovo [2] has parallelized several key time-consuming steps. Jackson et al. [13] have reported their bidirected De Bruijn graph construction on an IBM Blue Gene/L. Later, the authors implemented the complete genome assembly procedure on the Blue Gene/L machine and reported the performance scalability [1]. Kundeti et al. [14] have further improved the parallel graph construction algorithm presented by Jackson et al. [13] to avoid a large amount of message passing. However, to the best of our knowledge, the only work on GPU-accelerated genome assembly is GPU-Euler [15], which implements a similar construction algorithm. The major issue of their work is that it cannot handle large genome data, e.g., human genome, which cannot fit into the GPU memory.

3   Design and Implementation

In this section, we first present the in-memory implementation of GPU-based bidirected De Bruijn graph construction assuming sufficient memory. Then, we propose a staged algorithm for out-of-core processing.

3.1   In-Memory Implementation

We consider the graph construction as a component in de novo assembly, where the program input is a plain text file storing short reads and the output is the adjacency list representation of the bidirected De Bruijn graph stored in the main memory. Overall, there are three steps for the graph construction.

Step 1. Encoding. This step loads short reads from the disk to the main memory and uses two bits per character to represent the reads.

Step 2. Canonical Edge Generation. We adopt an edge-oriented building method to avoid validating edges [1]. For an l-length short read, we generate $(k+1)$-mers, and the number of $(k+1)$-mers is $(l-k)$. Each $(k+1)$-mer corresponds to an edge connecting two overlapping k-mers. The two k-mers represent two nodes connected by the edge. We do not extract the two k-mers in this step since there are many redundant edges and nodes at this time. A 64-bit data type, such as *long*, is sufficient to hold a $(k+1)$-mer since $(k+1)$ is usually set to less than 32 in practice. Similarly, we do not extract edge directions and output labels in this step due to the redundancy.

In the GPU-based implementation, generating edges is done through a map primitive efficiently. Each thread takes charge of one encoded read and generates corresponding $(l-k)$ edges. In our implementation, the encoding and edge generation are done in one kernel program. We use the fast *shared memory* on the GPU to hold encoded reads, and bitwise operations to generate all $(k+1)$-mers. Duplicate elimination can be implemented through either sorting, or hashing. In our evaluation, the two methods have a similar performance. Considering the merging step in out-of-core processing, we have adopted the sorting-based approach to perform the duplicate elimination.

Step 3. Adjacency List Representation After edges are generated, we can extract two corresponding k-mers, edge directions, and output labels from each $(k+1)$-mer. This step can be implemented as a map on GPUs. To keep the adjacency information, we use a pair $(k$-mer, edge_id$)$ to represent a node. An additional duplicate elimination step on the field k-mer is performed for all pairs to obtain distinct nodes as well as associated edges.

3.2   The Staged Graph Construction Algorithm

The in-memory implementation assumes unlimited GPU memory for the graph construction. In practice, a staged graph construction algorithm is necessary for the GPU-based implementation to handle data that cannot fit into the GPU memory. In common cases, the most memory consumption occurs in Step 2, when edges are generated from encoded reads before the duplicates are eliminated. Suppose there are n short reads of length l each, for a user-defined k, the total size of $(k+1)$-mers is $8 \times n \times (l-k)$ bytes, e.g., several hundreds of gigabytes for the human genome. Step 3 is executed using a similar staged algorithm when all distinct edges are produced. Therefore, we focus on introducing the staged algorithm for Step 1 and 2.

The basic idea in our staged algorithm is to load and process the input reads through multiple passes. In each pass, only a subset of input reads, denoted as a *chunk* of reads, is loaded from the disk to the memory for processing. Figure 2 shows the workflow of chunk-based processing. The chunk size is set so

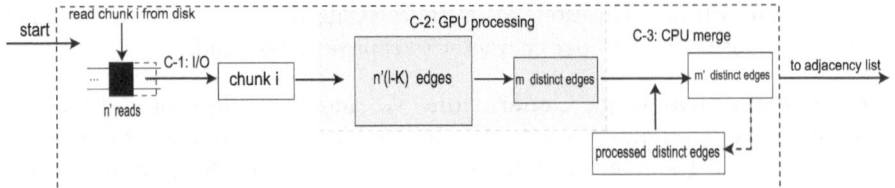

Fig. 2. The workflow of one chunk based processing of GPU-accelerated bidirected De Bruijn graph construction. The color coding of black, grey, and white indicates the data is stored in the disk, GPU memory, and CPU main memory, respectively

that the memory required for the chunk-based processing does not exceed the GPU memory size. Specifically, there are three processing steps for one chunk:

1. **C-1: I/O.** n' $(n' \leq n)$ short reads are loaded from the disk to the main memory as a chunk of data.
2. **C-2: GPU processing.** We transfer the data chunk to the GPU memory. The GPU performs the encoding and generates $n' \times (l-k)$ edges. Duplicates are eliminated through a sorting-based method for these $n' \times (l-k)$ edges. After the duplicate elimination, there are m distinct edges for this chunk.
3. **C-3: CPU merge.** These m distinct edges are copied from the GPU memory to the main memory, and merged with the distinct edges generated from previously processed data chunks. Since both the newly generated edges and the existing edges are ordered, the merge step is efficient. The m' distinct edges after merging will be used for the next chunk. The multiplicity information are also updated in this step. Due to the high coverage of input reads, the number of distinct edges is around tens of times smaller than that of all generated edges. Therefore we assume that the main memory is sufficient to hold these distinct edges.

3.3   The Memory Cost Model

Given the GPU memory size M_g we estimate the memory consumption for each step to decide the suitable chunk size. Suppose each chunk contains n' reads, and each read is l-length. Then the memory size of input reads is $n' \times l$ bytes (one byte per character). After encoding, the memory size for encoded reads is $\frac{n' \times l}{4}$ bytes. Thus the total GPU memory consumption for this encoding step is: $(n' \times l + \frac{n' \times l}{4})$ bytes. Next we use encoded reads in the memory for generating edges. The number of edges generated for one chunk is $n' \times (l-k)$, and each edge is represented using a 64-bit word type. Thus the memory size required for all generated *(k+1)-mers* in one chunk is $8 \times n' \times (l-k)$ bytes. Therefore the total GPU memory consumption of this edge generation step for one chunk is: $\left(\frac{n' \times l}{4} + 8 \times n' \times (l-k)\right)$ bytes. We perform the sorting-based duplicate elimination algorithm on $n' \times (l-k)$ edges. The memory consumption of the GPU-based radix sort is an output buffer with the same size of the input array. Thus the total memory consumption for the radix sort can be estimated as $(2 \times 8 \times n' \times (l-k))$

bytes. Finally, we remove duplicates from these sorted edges. The memory consumption for this operation is at most $(2 \times 8 \times n' \times (l - k))$ bytes. Therefore, the maximum value of n' can be estimated as follows:

$$min \left(\frac{4M_g}{5l}, \frac{M_g}{8.25l - 8k}, \frac{M_g}{16 \times (l - k)} \right)$$

4   The Pipelined Processing Model

4.1   The DGC Model

Our staged algorithm can be generalized as a processing model, which can be adopted in many GPGPU applications to handle the data set that cannot entirely fit into the GPU memory or of which some steps are not suitable to be processed on the GPU. We call it the DGC (Disk-GPU-CPU) model.

The DGC model is defined as follows. There are three components: I/O, the computation on GPUs, and the processing on CPUs. Suppose the total size of input data is D, the chunk size d, there are $\lceil \frac{D}{d} \rceil$ passes. For each pass: (1) A chunk of data is loaded from the disk to the main memory. (2) The in-memory chunk is transferred from the main memory to the GPU memory, and processed by the GPU. (3) The result is transferred back to the main memory, and a merge (or other post-processing) step is performed on the CPU.

In the DGC model, at least three memory buffers are required: b_i is the CPU memory buffer to hold the data from the disk, b_g is the GPU memory buffer used to store the input data, and b_c is the CPU memory buffer used to hold the computation result transferred from the GPU memory. To simplify the presentation, we assume the computation result on the GPU is also stored in the input buffer, that is b_g.

4.2   The Pipelined-DGC Model

The DGC model can be improved using pipelining. Since b_g is the GPU memory buffer independently accessed from b_i and b_c, the DGC model can be pipelined without additional memory allocated for data exchange. Without loss of generality, we assume that during processing, b_g and b_c will not be released by the GPU and CPU programs, respectively.

We maintain three threads to independently take charge of the I/O, GPU, and CPU processing. The I/O thread will be blocked when b_i is full and also when the memory copy to b_g is ongoing. The GPU thread may be blocked in two cases. First, the data is not ready, i.e., b_g is not full. Second, the processing on the GPU has been done, but the data is being copied to b_c. The CPU thread may be blocked only when the data is not ready in b_c.

At the beginning, we load the first data chunk to b_i. When b_i is full, a memory copy from b_i to b_g is executed. The I/O is blocked when performing the memory copy. As long as the data has been uploaded to b_g, b_i becomes available to load the next chunk of data until it is full again. However, the second memory copy from b_i to b_g (and all following memory copies) should wait until the GPU processing for the previous chunk is done to release the use of b_g. Similarly, we pipeline the GPU and CPU processing. A memory copy from the GPU to CPU is performed when the computation on the GPU is finished. The GPU processing is blocked when coping the data to b_c. After the memory copy, the CPU can perform the merge step for the data in b_c, and the the next chunk of data can be copied from b_i to b_g if the data is ready in b_i, otherwise the GPU thread is blocked to wait for the data in b_i becoming ready. We can see that with pipelining, the I/O, GPU processing, and CPU merge can overlap in time.

5   Experiments

5.1   Experimental Setup

Hardware Configuration. We conduct our experiments on a server machine with two Intel Xeon E5520 CPUs (16 threads in total) and an NVIDIA Tesla S1070 GPU. The NVIDIA Tesla S1070 has four GPU devices. In our current implementation, we only utilize one GPU device, and the CPU merge step also only uses one core on the CPU. This computation capability is sufficient for the overall performance as the I/O is the bottleneck. The main memory size of the server is 16 GB.

Data Sets. We use a small and a medium sized data set for the evaluation. As the previous research [13], we randomly sample known chromosomes to simulate short reads. Two data sets are sampled from Arabidopsis chromosome 1 (denoted as *Arab.*) and human chromosome 1 (denoted as *Human*), which can be accessed from the NCBI genome databases [16]. The details of the two data sets are shown in Table 1. Additionally, we fix the *k-mer* length k to 21, which is a common value for most genome assemblers.

Table 1. Data sets

	Arabidopsis chromosome 1	Human chromosome 1
#nucleotide	30 million	247 million
Read length	36	36
Coverage	17	36
#read	12 million	230 million
File size (FASTA)	800 MB	13 GB

5.2   Performance Results

Time Breakdown without Pipelining. We first study the performance bottleneck of our GPU-based graph construction without the pipeline optimization. Figure 3 shows the time breakdown of our GPU-accelerated bidirected De Bruijn graph construction. In this figure, the I/O, the encoding and edge generation on the GPU, the duplicate elimination for each chunk on the GPU, and the merge

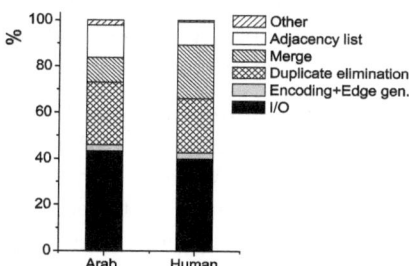

Fig. 3. The time breakdown of GPU-based bidirected De Bruijn graph construction without pipelining

step on the CPU together take around 90% of the execution time. The DGC model is applied to these four components. Particularly, among these four components, I/O takes around 50%, and both the GPU processing (including encoding and generating edges) and CPU processing take around 25% of the elapsed time. Since the efficiency of the pipeline is limited by the most time-consuming component, we expect that the overall performance of these four components can be improved by around two times through pipelining.

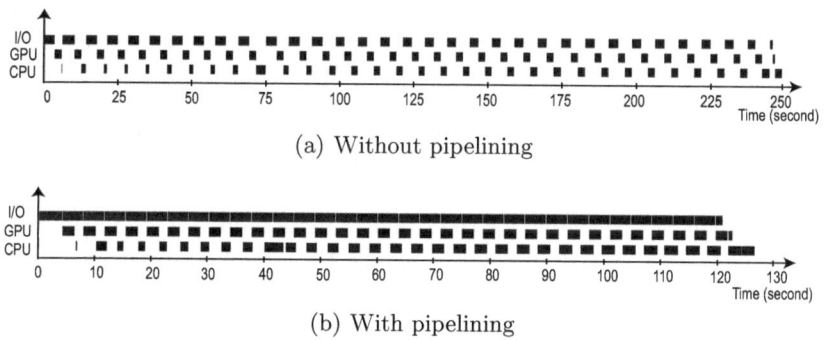

(a) Without pipelining

(b) With pipelining

Fig. 4. Executions of the staged algorithm without and with pipelining for GPU-based bidirected De Bruijn graph construction. The solid black rectangles represent the busy time of a component.

The Pipelining Optimization. We focus on the four steps (I/O, Encoding+Edge gen., Duplicate elimination, Merge) which as shown in Figure 3 that consume around 90% of the overall elapsed time. We take the larger data set human chromosome 1 to demonstrate the result. There are 33 passes in the staged algorithm for this medium size data set. Figure 4(a) shows the execution of the staged algorithm without pipelining. The I/O, GPU and CPU processing are done sequentially in each pass. As a result, the disk is idle around half of the overall time, and the utilization of the GPU and CPU is only around 25%. Figure 4(b) shows the execution status of the staged algorithm when pipelining optimization is adopted. With pipelining, I/O is nearly non-blocking.

Even though the GPU and CPU processing cannot completely overlap, their utilization is improved considerably. As a result, the overall execution time for these four steps are reduced from around 250 seconds to 128 seconds.

Comparison to Existing Implementations. There are four CPU-based implementations for comparison: Velvet [4], SOAPdenovo [2], ParBidirected [14], and Jackson's implementation [13]. We only compare the performance of graph construction. We perform the evaluation on the same machine for all these software except the Jackson's implementation. Note that SOAPdenovo, ParBidirected, and Jackson's implementation are parallel implementations.

Table 2. Comparison of the elapsed time in seconds

	Arab.	Human 1
GPU-accelerated	18	177
Velvet [4]	86	-
SOAPdenovo [2]	78	1,245
ParBidirected [14]	1,740	32,400
Jackson2008 [13]	15	327

Table 2 shows the comparison result for the running time of four different implementations. The memory required by preprocessing in Velvet for the Human data set exceeds our main memory limit and takes excessively long time, thus we do not report the performance number for the Human data set in the table. SOAPdenovo has parallelized the hash table building. In the evaluation, it consumes a similar main memory size (around 8 GB for the Human data set) to our implementation. However, the GPU-accelerated graph construction is around 4-7x faster than SOAPdenovo for the similar functionality. Since ParBidirected adopts a more conservative method to handle the out-of-core processing, intermediate results need to be written into the disk in most steps, which results in a slow execution time, but the memory consumption is stable and very low. In our experiments, we have already modified the default buffer size used in the external sorting to improve the performance. The published performance results from Jackson's implementation [13] is based on a 1024-node IBM Blue Gene/L. The input and output are the same as our program. However, they have adopted a node-oriented graph building approach, and the message passing is very expensive. In summary, compared with existing implementations, our GPU-based graph construction is significantly faster. Specifically, compared with the massively parallel implementation on the IBM Blue Gene/L, our implementation is still around two times faster.

6   Conclusion

In this paper, we have presented the design and implementation of our GPU-accelerated bidirected De Bruijn graph construction, which is a first step to build a complete GPU-accelerated genome assembler. We have addressed the GPU memory limit issue through a staged algorithm, and the overall performance can be further improved by around two times through pipelining I/O, GPU and CPU processing. Furthermore, such optimized processing flow can be generalized as a pipelined-DGC model, which can be applied to other GPGPU applications to handle large data sets. Compared with existing implementations on CPUs, the performance of our implementation is up to two orders of magnitude faster. In particular, our GPU-accelerated graph construction is around 2X faster than the published performance numbers for a parallel implementation on a 1024-node IBM Blue Gene/L.

Acknowledgement. This work was supported by grants 617509 and 616012 from the Research Grants Council of Hong Kong, and grant MRA11EG01 from Microsoft China.

References

1. Jackson, B., Regennitter, M., Yang, X., Schnable, P., Aluru, S.: Parallel de novo assembly of large genomes from high-throughput short reads. In: IPDPS 2010: Proceedings of the 2010 IEEE International Symposium on Parallel&Distributed Processing, pp. 1–10 (April 2010)
2. Li, R., Zhu, H., Ruan, J., Qian, W., Fang, X., Shi, Z., Li, Y., Li, S., Shan, G., Kristiansen, K., Li, S., Yang, H., Wang, J., Wang, J.: De novo assembly of human genomes with massively parallel short read sequencing. Genome Research 20(2), 265–272 (2010)
3. Simpson, J.T., Wong, K., Jackman, S.D., Schein, J.E., Jones, S.J., Birol, I.: Abyss: a parallel assembler for short read sequence data. Genome Research 19(6), 1117–1123 (2009)
4. Zerbino, D.R., Birney, E.: Velvet: algorithms for de novo short read assembly using de bruijn graphs. Genome Research 18(5), 821–829 (2008)
5. Pevzner, P.A., Tang, H.: Fragment assembly with double-barreled data. Bioinformatics 17(suppl. 1), S225–S233 (2001)
6. Medvedev, P., Georgiou, K., Myers, G., Brudno, M.: Computability of models for sequence assembly. In: Giancarlo, R., Hannenhalli, S. (eds.) WABI 2007. LNCS (LNBI), vol. 4645, pp. 289–301. Springer, Heidelberg (2007)
7. Chaisson, M.J., Pevzner, P.A.: Short read fragment assembly of bacterial genomes. Genome Research 18(2), 324–330 (2008)
8. Hossain, M.S.S., Azimi, N., Skiena, S.: Crystallizing short-read assemblies around seeds. BMC Bioinformatics 10(suppl. 1) (2009)
9. Hernandez, D., François, P., Farinelli, L., Østerås, M., Schrenzel, J.: De novo bacterial genome sequencing: Millions of very short reads assembled on a desktop computer. Genome Research 18(5), 802–809 (2008)
10. Butler, J., MacCallum, I., Kleber, M., Shlyakhter, I.A., Belmonte, M.K., Lander, E.S., Nusbaum, C., Jaffe, D.B.: Allpaths: De novo assembly of whole-genome shotgun microreads. Genome Research 18(5), 810–820 (2008)
11. Warren, R.L., Sutton, G.G., Jones, S.J., Holt, R.A.: Assembling millions of short dna sequences using ssake. Bioinformatics 23(4), 500–501 (2007)

12. Dohm, J.C., Lottaz, C., Borodina, T., Himmelbauer, H.: Sharcgs, a fast and highly accurate short-read assembly algorithm for de novo genomic sequencing. Genome Research 17(11), 1697–1706 (2007)
13. Jackson, B.G., Aluru, S.: Parallel construction of bidirected string graphs for genome assembly. In: International Conference on Parallel Processing, pp. 346–353 (2008)
14. Kundeti, V., Rajasekaran, S., Dinh, H.: Efficient parallel and out of core algorithms for constructing large bi-directed de bruijn graphs. CoRR abs/1003.1940 (2010)
15. Mahmood, S.F., Rangwala, H.: Gpu-euler: Sequence assembly using gpgpu. In: Proceedings of the 2011 IEEE International Conference on High Performance Computing and Communications, HPCC 2011, pp. 153–160. IEEE Computer Society (2011)
16. National Center for Biotechnology Information, http://www.ncbi.nlm.nih.gov/

K Hops Frequent Subgraphs Mining
for Large Attribute Graph

Haiwei Zhang, Simeng Jin, Xiangyu Hu, Ying Zhang,
Yanlong Wen*, and Xiaojie Yuan

Department of Computer Science and Information Security, Nankai University.
94, Weijin Road, Tianjin, China
{zhanghaiwei,jinsimeng,huxiangyu,zhangying,wenyanlong,
yuanxiaojie}@dbis.nankai.edu.cn
http://dbis.nankai.edu.cn

Abstract. Attribute Graphs are widely used to describe complex data
in many applications such as bio-informatics and social network. With
the rapid growth of scale for graph data, traditional solutions for mining
frequent subgraphs cannot performed well in large attribute graph because of time-consuming candidates generation and isomorphism testing.
In this paper, we investigate the problem for k hops subgraph mining in
large attribute graph. The attribute graph is transformed into labeled
graphs by projection for each attribute. K hops frequent subgraph mining algorithm FSGen consists of three procedures is performed. Firstly,
frequent vertices and edges will be extended to frequent subgraphs from
root vertices. Secondly, frequent edges joining frequent vertices will be
added into extended subgraphs. Thirdly, if necessary, isomorphism testing will be used to summarize frequent subgraphs based on Graph Edit
Distance. Then, frequent labeled subgraphs will be merged into attribute
subgraphs by integration according to designated attributes. The complexity of our mechanism is approximately $O(2^n)$, which is more efficient
than existing algorithms. Real data sets are applied in experiments to
demonstrate the efficiency and effectiveness of our technique.

Keywords: Attribute graph, k hops frequent subgraph, mining, projection, integration.

1 Introduction

With the rapidly increasing amount of data in Web and related applications,
more and more data can be represented by graphs. Most mechanism and algorithms can only work on labeled graphs, e.g. a vertex has only one label, while
vertices in many real graph data have more than one label, and this kind of
graphs are named multi-labeled graphs [1], such as graphs describing social network and proteins. Attribute graph is characterized as a kind of multi-labeled

---

* Corresponding author.

Y. Ishikawa et al. (Eds.): APWeb 2013, LNCS 7808, pp. 63–74, 2013.

graph. There are a set of attributes associated with each vertex in attribute graph, and each attribute has an exacted name.

Mining frequent subgraphs from graph databases is of great importance in graph data mining. Frequent subgraphs are useful in data analysis and data mining tasks, such as similarity search, graph classification, clustering and indexing [2]. In similarity search and classification, frequent subgraphs are usually used as feature leading to exact results and high scalability. In graph clustering, frequent subgraphs can be used as a solution for traditional algorithms such as k-means. Similarity searching is an important issue in many applications of graph data. The task is considered NP-complete and is inefficient for large graph.

Existing research on frequent subgraphs mining is conducted mainly on two types of graph database. The first one is transaction graph databases [3] that consist of a set of relatively smaller graphs. Transaction graph databases are prevalently used in scientific domains such as chemistry, bioinformatics, etc. The Second one is a graph with a large number of vertices such as social networks [4]. This paper focuses on the latter type of graph database with several attributes in each vertex.

Mining frequent subgraphs is a iterative processing consists of two main steps [5]: candidates generation and subgraphs counting. In the first step, candidates can be mined extending. In the second step, subgraph isomorphism will be investigated for counting. Subgraph isomorphism is an NP-complete problem [6] and it is time-consuming for large graphs.

Considering the increasing importance of frequent subgraphs mining, this paper will present a mechanism for mining frequent subgraphs from a large attribute graph. Attribute graph will be transformed into labeled graphs by projection firstly. Then algorithm FSGen will be performed to mining k hops frequent subgraphs by extending, joining and isomorphism testing. Finally, frequent labeled subgraphs will be merged into attribute subgraphs by integration according to designated attributes. There is no candidates generation in algorithm FSGen, and isomorphism testing is optional. Thus, the time occupation of our mechanism for mining frequent labeled subgraphs is better than existing works.

The rest of this paper is organized as follows. Related works are summarized in section 2 on subgraph mining algorithms for graph databases consist of multiple small labeled graphs. Section 3 is dedicated to basic concepts and general definitions of frequent subgraphs mining. We present the k hops frequent subgraph mining algorithm FSGen consisting of extending, joining and isomorphism testing procedures in section 4. Section 5 shows experimental studies for our technique and section 6 concludes our work.

2 Related Works

Frequent subgraph mining has been widely studied for graph databases consist of multiple small labeled graphs. To the best of our knowledge, no mechanisms can be directly performed in frequent subgraphs mining from graph databases consist of one large attribute graph.

One popular approach for frequent subgraphs mining is raised in Apriori algorithm. AGM [7] defines induced subgraph $G' = (V', E')$ which satisfies $V' \subseteq V$ and $E' \subseteq E$ where E' includes all edges connecting vertices in V'. In the algorithm, all frequent induced subgraphs were mined according to designed support by using vertices for candidates. FSG [8] algorithm stores labels of all (k-1)-subgraphs (subgraphs including k-1 edges), then the common structure between two k-graphs can be found by computing intersection of the label sets of their subgraphs. As a result, a large amount of common subgraphs are generated as candidates, which reduces efficiency of the algorithm.

The other popular approach for frequent subgraphs mining is based on extending. Extending-based approaches usually use depth first search to find frequent subgraphs. Initially, frequent subgraphs are frequent vertices. Then, a new vertex extending from each subgraphs by another edge is added to the subgraph in order to generate a new subgraph. Isomorphism testing is used to determine whether the new subgraph is frequent. All subgraphs can be mined by repeating these steps. But this case creates many redundant candidates. So all existing methods based on extending attempt to reduce the generation of redundant candidates. gSpan [9] defines DFS (Depth First Search) Lexicographic Order and DFS code of a graph. GraphGen [6] improves gSpan by mining frequent subtree instead of frequent subgraph in order to reduce the complexity of time. When preforming a depth first search in a graph, a DFS tree is constructed. One graph can have several DFS trees and respective DFS codes. Because the minimum DFS code of a graph is unique, a DFS tree can be denoted by this minimum DFS code. Isomorphism testing can be performed based on minimum DFS codes of each subgraph. gSpan and GraphGen can only extend vertices in the right path of DFS tree. Candidates can only generate from finite vertices or edges for reducing redundant subgraphs. FFSM [10] designs canonical adjacent matrix (CAM) as formal data model for graphs and guarantee all frequent subgraphs are enumerated unambiguously. FFSM uses two operations: FFSM-Join and FFSM-Extension to generate candidates efficiently and completely avoids subgraph isomorphism testing by maintaining an embedding set for each frequent subgraph. Indeed, in FFSM, edges extending and isomorphism testing has been transformed into operations on CAMs. And the complexity of FFSM concentrates on operations of CAMs. CAMs will be larger together with the increasing scale of graphs and mining frequent subgraphs in large graph is inefficient.

We can summary from existing works that candidates generation and isomorphism testing is essential stages for mining subgraphs. Since the complexity of isomorphism testing cannot be actually reduced (including substitution of isomorphism testing stage in FFSM), all researches focus on reducing creation of candidates. We will present a new algorithm FSGen for k hops frequent subgraphs mining from a large labeled graph. Under the restriction of k, candidates generation and isomorphism testing can be avoided by the mechanism.

3   Preliminaries

For simplicity of presentation, we restrict our discussion to undirected, multi-attributes graphs. And graph database consists of only one large attribute graph. However, our approach can be applied to directed graphs and labeled graphs with minor changes.

A graph G is defined as a binary (V, E), where V is the set of vertices, E is the set of edges. A labeled graph G_L is defined as a triple (V, E, L), where V and E is the same as G, L is a labeling function that assigns a label to each vertex and edge. $L(v) = l$ denotes that the label of vertex v is l. An attribute graph G_A is extending the function L to multiple attributes, as definition 1 interprets.

Definition 1. Attribute Graph. An Attribute Graph G_A is a triple (V, E, F_A) where V is a set of vertices and $E \subseteq V \times V$ is a set of edges. A denotes the set of attributes. $F_A = \bigcup_{j=1}^{n} \bigcup_{i=1}^{m} f_{a_i}(v_j)$ (Assuming there are n vertices in V and m attributes in each vertex.) is a set of functions , where $f_{a_i}(v_j)$ ($1 \leq i \leq m$, $1 \leq j \leq n$) denotes to the value of attribute a_i of vertex v_j ($a_i \in A$, $v_j \in V$).

Fig. 1 shows examples of graph, labeled graph and attribute graph. In the attribute graph, each vertex has three attributes: *id*, *name* and *major*.

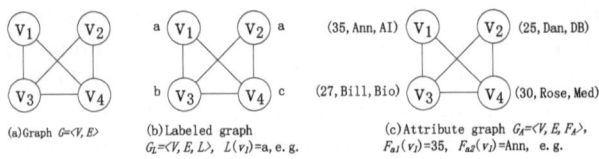

(a) Graph $G = \langle V, E \rangle$ (b) Labeled graph $G_L = \langle V, E, L \rangle$, $L(v_1) = a$, e. g.   (c) Attribute graph $G_A = \langle V, E, F_A \rangle$, $F_{a1}(v_1) = 35$, $F_{a2}(v_1) = $Ann, e. g.

Fig. 1. Graph, Labeled Graph and Attribute Graph

Definition 2. K Hops Neighbor. $\forall v \in V$, $N(v, k)$ denotes to a set of vertices that are at most k hops from v. Each vertex $u \in N(v, k)$ is a k hops neighbor of vertex v. For example, in Fig. 1(c) $N(v_1, 1) = \{v_3, v_4\}$.

In this paper, attribute graph will be transformed into labeled graphs by projection according to attributes. Thus, following definitions in this section are designed for labeled graphs.

Definition 3. Support. The support of a subgraph G_S in labeled graph G_L is denoted by $Sup(G_S)$ and is given as the total quantity of its occurrence in G_L. $Sup(v)$ denotes the support of vertex v by counting the occurrences of $L(v)$. $Sup(e)$ denotes the support of edge e by counting the occurrences of e whose labels of begin and end vertices are the same respectively.

Definition 4. Frequent. Given a minimum support *MinSup* and a labeled graph G_L, a vertex $v \in V$ is a frequent vertex iff. $Sup(v) \geq MinSup$. An edge $e \in E$ is a frequent edge iff. $Sup(e) \geq MinSup$. A subgraph $g \in G_L$ is frequent iff. $Sup(g) \geq MinSup$.

Definition 5. K Hops Frequent Subgraph. A k hops frequent subgraph $G_f = (V_f, E_f, L_f)$ of labeled graph G_L satisfies: (1)$V_f \subseteq V$, $E_f \subseteq E$, (2)quantity of G_f in G is not less than a given $MinSup$, (3)minimum hops (distance) between all vertices and a designated root vertex are not more than a given constant k, where the root vertex is one of frequent vertices in G_L. For example, in Fig. 1(b), assuming $k=1$ and $MinSup=2$, 1 hop frequent subgraphs in the labeled graph are edges (a, b) and (a, c).

4   K Hops Frequent Subgraph Mining

In this section, we discuss algorithms for mining k hops frequent subgraphs in a large attribute graph. Attribute graph is transformed into labeled graphs by projection firstly. Then, algorithm FSGen for k hops frequent subgraph mining in a large labeled graph will be proposed. Finally, integration will be described in order to merge frequent labeled subgraphs into frequent attribute subgraphs by designated attributes.

4.1   Projection

Attribute graph will be transformed into labeled graphs by projection before frequent subgraphs mining. Projection can simplify the complexity of attribute graph. And mechanisms for processing labeled graphs can be used to work on attribute graph after projection.

Definition 6. Projection. Given attribute graph $G_A=(V, E, F_A)$, a labeled graph $G_p = (V_p, E_p, L_p)$ is a projection of G_A, iff. $V_p \subseteq V$, $E_p \subseteq E$, and $\forall v \in V_p$, $\forall a \in A$ then $L(v) \in f_a(v)$.

Fig. 2 shows labeled graphs transformed from the attribute graph shown in Fig. 1(c) by projection. For each attribute, a projection g_{pi} can be used as a labeled graph for frequent subgraph mining algorithm presented in next subsection. And integration will be performed to combine mining results as frequent attribute subgraphs.

For each projection of an attribute, a linear scanning will be performed to get each value of the attribute as label of related vertex in labeled graph. Then a double iteration will be executed to get edges of labeled graph according to the attribute. Assuming there are m attributes in each vertex of n vertices in the attribute graph G_A. The complexity of time in projection stage is $O(n^2)$ as the number of attributes m is a constant.

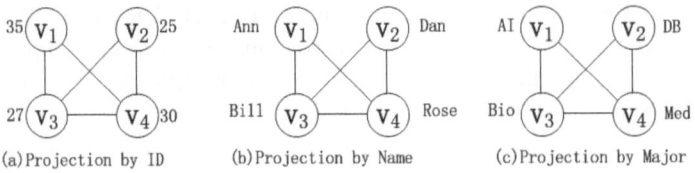

Fig. 2. Labeled graphs of attribute graph in Fig. 1(c) by projection

4.2   K Hops Frequent Subgraph Mining Algorithm

In this subsection, k hops frequent subgraph mining algorithm FSGen will be presented. In FSGen, frequent subgraphs are extended from root vertices until k steps. In each extending step, the subgraphs extended from previous steps are frequent. Thus, the ultimate extending structures after k steps are frequent. This idea is based on property 1.

Property 1. Assuming a set of frequent vertices $V_R=\{v_1, v_2, ..., v_n\}(V_R \subseteq V)$ where $L(v_1) = L(v_2) = ... = L(v_n)$ and $Sup(v_i) \geq MinSup(1 \leq i \leq n)$. If $\exists u_{hj} \in N(v_i, h)(1 \leq h \leq k)$ and occurrences of $L(u_{hj})$ is no less than given $MinSup$, then vertex u_{hj} is frequent and subgraph $g_i = (V_i, E_i, L_i)$ is frequent where $V_i = \{v_1\} \bigcup_{h=1,j}^{k} u_{hj}$, $E_i = \{\bigcup_j (v_1, u_{1j})\} \bigcup_{h=1,j}^{k-1} (u_{hj}, u_{(h+1)j})$.

Proof: The occurrence of $L(u_{hj})$ is no less than given $MinSup$, so in all $N(v_i, h)$, $Sup(u_{hj}) \geq MinSup$, and vertex u_{hj} is frequent. $\forall e \in E_i$, the start and end vertices of e is frequent, so edge e is frequent according to definition 3 and 4. In labeled graph g_i, $\forall v \in V_i$ and $e \in E_i, Sup(v_i) \geq MinSup$ and $Sup(e_i) \geq MinSup$, then all components of g_i are frequent. As a result, g_i is frequent.

V_R is considered as the set of root vertices. An iterated processing will be invoked to get root vertices. If a frequent vertex is not contained in any frequent subgraph, it will be extracted as new root vertex until all frequent vertices are contained in frequent subgraphs. Frequency of vertices in a labeled graph will be calculated by a polynomial time scanning, and the worst complexity of time is $O(n^2)$.

In order to brief mining processing, we attempt to avoid overlaps among frequent subgraphs mined by each stage in the algorithm. Overlap is a repeated structure in subgraph, such as paths, subgraphs and so on. We discuss the probability of overlap basing on property 2.

Property 2. Assuming frequent vertices v, u, w, if $\exists u, w \in N(v, 1)$ and $L(u) = L(w)$, then the frequent subgraphs extended from v and u are probably overlap.

Proof: Given a constant k as the number of hops, assuming $\exists u_1 \in N(u, 1)$, $u_2 \in N(u, 2), ..., u_{(k-1)} \in N(u, k-1)$ and $\exists w_1 \in N(w, 1), w_2 \in N(w, 2), ..., w_{(k-1)} \in N(w, k-1)$, where $L(u_i) = L(w_i)$, $1 \leq i \leq k-1$. Subgraphs consist of paths $u, u_1, u_2, ..., u_{(k-1)}$ and $w, w_1, w_2, ..., w_{(k-1)}$ are overlap.

Based on above properties, our algorithm for mining frequent subgraph has following features: (1) Vertices and edges in each extending step are frequent in all $N(v_r, h)(1 \leq h \leq k)$, where $v_r \in V_R$ is a root vertex. (2) The same edge (labels of start and end vertices are the same respectively) occurs once in each extending step.

K hops frequent subgraph mining algorithm FSGen consists of three stages: extending, joining and isomorphism testing. Frequent subgraphs are initially denoted as frequent vertices, and then these vertices will be extended by other frequent vertices and edges until k hops in extending stage. Frequent subgraphs obtained from extending stage are constructed as tree model. So in joining stage, frequent edges in labeled graph will be added into extended k hops frequent subgraphs where they do not occur. After above two stages, k hops frequent subgraphs can be enumerated. If we want to summarize all k hops frequent subgraphs, isomorphism testing will be performed and this stage is optional according to actual requirements. In order to complete our algorithm, isomorphism testing will be presented along with the other two stages in this subsection. Algorithm 1 shows FSGen performed on labeled graphs by projection. The algorithm consists of three main procedures: Extending, Joining and Isomorphism_Testing.

Algorithm 1. FSGen

Input: (1) k, (2) labeled graph G_p by projection, (3) frequent vertices set V_f, (4) frequent edges set E_f

Output: k hops frequent subgraphs of G_p

1: Designing V_{mf} denotes the set of frequent vertices as root vertices set
2: int $h = 0$;
3: Designing G_f denotes to set of k hops frequent subgraphs and $G_f = \Phi$;
4: **for all** v_i in V_{mf} and v_i does not belong any extended frequent subgraph **do**
5:     Designing g as a k hops frequent subgraph in G_f and $g.V = v_i$, $g.E = \Phi$;
6:     Extending($v_i, g, k, ++h$)
7:     Joining(g)
8:     add g to G_f
9: **end for**
10: Isomorphism_Testing(G_f)
11: **return** G_f

As each G_p is a labeled graph, an inverted table will be built as index for each label and sorted by frequency. Then root vertices can be chosen by the inverted table. In extending stage, frequent vertices that do not occur in any extended frequent subgraphs can be used as new root vertices in the iteration. Extending is a recursive procedure as Algorithm 2 shows and the complexity of time is $O(2^n)$. We design h as the current number of hops. In each steps of extending, a frequent vertex and a related frequent edge will be added into frequent subgraph extended in previous steps as line 6 to line 8 shows. In order to avoid overlap, we use a list of vertices for $N(v, 1)$ which include non-repeated neighbors of vertex v. While h is equal to k, the procedure will return and y will be extended in vertices set and edges set.

Algorithm 2. Extending

Input: (1) k, (2)current number of hops h, (3) frequent vertex v, (4) a reference of frequent subgraph g

Output: reference g as a frequent subgraph

1: **if** $h<k$ **then**
2:     add v to $g.V$
3:     Designing list L of frequent vertices in $N(v,1)$, and L includes different vertex only once
4:     **for all** v_{ni} in L **do**
5:        **if** v_{ni} is frequent and edge (v,v_{ni}) is frequent **then**
6:           add v_{ni} to $g.V$
7:           add (v,v_{ni}) to $g.E$
8:           Extending(v_{ni}, g, k, $++h$)
9:        **end if**
10:    **end for**
11: **end if**

Joining procedure is shown in Algorithm 3 and double iteration will be performed in the procedure. Assuming n is the number of frequent vertices in the labeled graph which is a projection of attribute graph G_A and m is the number of edges in the edges set of frequent subgraph G_f. Time occupation is $O(n^2*m)$. Since m is approximately a constant. The complexity of time in Joining procedure is $O(n^2)$.

Algorithm 3. Joining

Input: (1) k, (2) k hops frequent subgraph g, (3) frequent edges set E_f

Output: Joined k hops frequent subgraphs g

1: **for all** vertex v_i in $g.V$ **do**
2:     **for all** vertex v_j in $g.V$ **do**
3:        **if** edge $(v_i,v_j) \in E_f$ and $(v_i,v_j) \notin g.E$ **then**
4:           add (v_i,v_j) to $g.E$
5:        **end if**
6:     **end for**
7: **end for**
8: **return** g

Algorithm 4 shows Isomorphism_Testing procedure. All k hops frequent subgraphs in labeled graph can be enumerated after executing Extending and Joining procedure. If we want to summarize the set of k hops frequent subgraphs, repeated subgraphs will be reduced, and Isomorphism_Testing procedure will be performed. In existing works, isomorphism testing is considered as NP-complete problem. In our mechanism, the procedure will be replaced by a approximated solution named Graph Edit Distance(GED) [11]. GED can be computed in polynomial time $O(m^3)$ (m denotes to the larger number of vertices in two graphs) by the idea of dynamic programming. If GED between two frequent subgraphs is no more than a given threshold of similarity, these two frequent subgraphs are considered similar subgraph and one of them will be removed from G_f. The total complexity of time in isomorphism testing is $O(m^3 \cdot n^2)$ where m is the number of vertices in each frequent labeled subgraph and n denotes the number of all frequent labeled subgraphs. By the restriction of k, the number of vertices in a frequent subgraph can be considered as a constant, so the complexity of time in Isomorphism_Testing procedure is $O(n^2)$.

Algorithm 4. Isomorphism_Testing

Input: (1)set of k hops frequent subgraphs G_f, (2)similar threshold s
Output: reduced G_f as summary
1: **for all** g_i in G_f **do**
2:    **for all** g_j in G_f **do**
3:        **if** Graph_Edit_Distance(g_i, g_j)$\leq s$ **then**
4:            remove g_j from G_f
5:        **end if**
6:    **end for**
7: **end for**
8: **return** G_f

K hops frequent subgraphs mining algorithm FSGen consists of three procedures. The complexities of time in each procedure are: $O(2^n)$, $O(n^2)$ and $O(n^2)$. Therefore, the total complexity of the algorithm FSGen is $O(2^n + n^2 + n^2)$, which can be merged into $O(2^n)$ no matter what including the stage of isomorphism testing.

4.3 Integration

Attribute graph is transformed into labeled graphs by projection. Algorithms are performed on labeled graphs for mining k hops frequent subgraphs. Each frequent subgraph is also a labeled subgraph. In order to merge frequent labeled subgraphs into attribute subgraphs, integration will be used. Integration described in definition 7 is the opposite operation of projection.

Definition 7. Integration. Given n labeled graphs $g_{p1}, g_{p2}, ..., g_{pn}$, where $g_{pi} = (V_{pi}, E_{pi}, L_{pi})$. Integration is a aggregated attribute graph $G_A = (V, E, F_A)$, where $V = \bigcap_{i=1}^{n} V_{pi}$, $E = \bigcap_{i=1}^{n} E_{pi}$ and $F_A - \bigcup_{i=1}^{n} L_{pt}$.

According to the definition of integration, many operations of intersection computing will be carried out. Common solution as definition 7 shows is time-consuming. On the other hand, there may be no frequent subgraphs mined from labeled subgraphs for attributes whose values are unique. While integrating all frequent labeled subgraphs, no frequent attribute subgraphs will be obtained. Thus, the mining results are not significant. In order to solve problems presented above, we adopt a flexible method for integration. We add a flag for each attribute of a vertex, if value of the attribute is frequent, the flag will be set to 1, and otherwise set to 0. Then we can designate attributes and its related frequent labeled subgraphs for integration. In each frequent labeled subgraph, scanning the set of vertices, if flags of the values in designated attributes of a vertex are equal to 1, the vertex will be selected as a frequent vertex and added to vertices set of frequent attribute subgraph. This processing is a linear scanning and the complexity of time is $O(m)$, where m is the number of vertices in the scanned frequent subgraph. For all frequent subgraphs, the total complexity of time for integration is $O(m \cdot n)$ where n is the number of frequent labeled subgraphs mined by FSGen.

5   Experimental Studies

The mechanism presented in this paper is written in C++ using Microsoft Visual Studio.net 2008. All experiments are carried out on a PC with Intel core i7 CPU 2.67GHz, 4GB memory and Microsoft Windows 7.

The experiment consists of two parts to demonstrate the effectiveness of our work. Firstly, we perform our mechanism for mining frequent subgraphs in attribute graph with different k, different size of data and different support. And then testing time occupation in each case. Secondly, we perform algorithm FSGen for mining frequent subgraphs in labeled graphs together with other traditional algorithms including gSpan, FFSM and GraphGen, which must be rewritten for large labeled graph instead of multiple small graphs, to compare the time occupations of these algorithms. We obtain large attribute graph data by crawling from YouTube API and arrange each item to a vertex of graph including 9 attributes. We choose three attributes: *category*, *length* and *views* for frequent subgraphs mining.

5.1   Mining in Attribute Graph

K is the number of hops denoting distance from root vertex to another in each frequent subgraph. In this experiment, we firstly discuss the impact of k in time occupation while running the k hops frequent subgraph mining algorithm FS-Gen on labeled graphs transformed from the attribute graph based on YouTube dataset. And then testing the effectiveness with different size of data and different support.

Fig. 3 shows time occupation according to different k. We can see from the figure that the time occupation is increasing along with the growth of k while the number of vertices in attribute graph is 500K. The same tendency occurs in Fig. 4 while the number of hops is set to 3, with the growing amount of vertices, time occupation of our work will increase. The time occupation is also related to the support as Fig. 5 shows. The values of x axis is denoted by the ratio between support (given by definition 3) and total number of vertices in the labeled graph. More time will be occupied while the support is low because low support will bring more subgraphs.

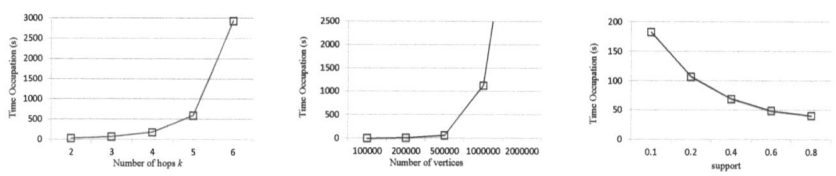

Fig. 3. Effectiveness of k **Fig. 4.** Effectiveness of sizes   **Fig. 5.** Effectiveness of supports

5.2 Mining in Labeled Graph

In this experiment, we compare algorithm FSGen presented in this paper with gSpan, FFSM and GraphGen. All the algorithms comparing to FSGen can be used to mine frequent subgraphs in a large labeled graph with few modification. As a result, these algorithms can be used to mining frequent subgraphs from labeled graph transformed from attribute graph by projection. The same processing of projection and integration will be performed while executing gSpan, FSSM, GraphGen and FSGen. We only discuss the effectiveness of mining frequent subgraphs from large labeled graph for these algorithms.

Designing the size of attribute graph (assuming 500K vertices in the experiment) and the number of hops (assuming $k = 3$). We can see from Fig. 6, FSGen is more efficient than any other algorithms. With the size of graph data growing, this advantage becomes more and more apparent. And the same tendency will occur when we compare the time occupation with given supports as Fig. 7 shows.

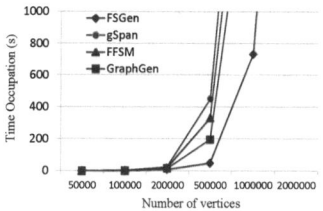
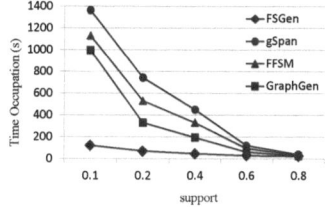

Fig. 6. Effectiveness of sizes   **Fig. 7.** Effectiveness of supports

6 Conclusion

We introduced a new mechanism for mining k hops frequent subgraphs in large attribute graph. In our work, large attribute graph was transformed into labeled graphs by projection, and then algorithm FSGen for mining k hops frequent subgraphs in large labeled graph was presented. Finally, integration was performed to merge frequent labeled subgraphs into frequent attribute subgraphs. Experimental studies show that our mechanism can efficiently mining k hops frequent attribute subgraphs. Comparing with existing algorithms for mining frequent subgraphs in large labeled graph, algorithm FSGen presented in this paper is more efficient than gSpan, FFSM and GraphGen on real datasets.

Acknowledgments. This work is partly supported by the National Natural Science Foundation of China under Grant Nos. 61170184, 61070089; the National High Technology Development 863 Programs of China under Grant No. 2013AA013204; the Ph.D. Programs Foundation of Ministry of Education of China under Grand No. 20120031120038.

References

1. Yang, J., Zhang, S., Jin, W.: DELTA: Indexing and Querying Multi-labeled Graphs. In: The 20th ACM International Conference on Information and Knowledge Management (CIKM), pp. 1765–1774 (2011)
2. Keyvanpour, M.R., Azizani, F.: Classification and Analysis of Frequent Subgraphs Mining Algorithms. Journal of Software 7(1), 220–227 (2012)
3. James, C., Yiping, K., Ada, W.C.F., Jeffrey, X.Y.: Fast graph query processing with a low-cost index. The International Journal on Very Large Data Bases 20(4), 521–539 (2011)
4. Koren, Y., North, S.C., Volinsky, C.: Measuring and extracting proximity in networks. In: The 12th ACM SIGKDD International Conference on Knowledge Discovery and Data Mining (KDD), pp. 245–255 (2006)
5. Meinl, T., Borgelt, C., Berthold, M.R.: Discriminative Closed Fragment Mining and Perfect Extensions in MoFa. In: The 2nd Starting AI Researchers' Symposium (STAIRS), pp. 3–14 (2004)
6. Li, X.T., Li, J.Z., Gao, H.: An Efficient Frequent Subgraph Mining Algorithm. Journal of Software 18(10), 2469–2480 (2007)
7. Inokuchi, A., Washio, T., Motoda, H.: Complete Mining of Frequent Patterns From Graphs: Mining graph data. Machine Learning 50(3), 321–354 (2003)
8. Kuramochi, M., Karypis, G.: Frequent Subgraph Discovery. In: The 1st IEEE International Conference on Data Mining (ICDM), pp. 313–320 (2001)
9. Yan, X., Han, J.: GSpan: Graph-Based Substructure Pattern Mining. In: The 2nd IEEE International Conference on Data Mining (ICDM), pp. 721–724 (2002)
10. Huan, J., Wang, W., Prins, J.: Efficient mining of frequent subgraph in the presence of isomorphism. In: The 3rd IEEE International Conference on Data Mining (ICDM), pp. 549–552 (2003)
11. Riesen, K., Bunke, H.: Approximate graph edit distance computation by means of bipartite graph matching. Image and Vision Computing 27(4), 950–959 (2009)

Privacy Preserving Graph Publication in a Distributed Environment

Mingxuan Yuan[1,2], Lei Chen[2], Philip S. Yu[3], and Hong Mei[4]

[1] Huawei Noah Ark Lab, Hong Kong
[2] The Hong Kong University of Science & Technology
[3] University of Illinois at Chicago
[4] Peking University

Abstract. Recently, many works studied how to publish privacy preserving social networks for "safely" data mining or analysis. These works all assume that there exists a single publisher who holds the complete graph. While, in real life, people join different social networks for different purposes. As a result, there are a group of publishers and each of them holds only a subgraph. Since no one has the complete graph, it is a challenging problem to generate the published graph in a distributed environment without releasing any publisher's local content. In this paper, we propose a SMC (Secure Multi-Party Computation) based protocol to publish a privacy preserving graph in a distributed environment. Our scheme can publish a privacy preserving graph without leaking the local content information and meanwhile achieve the maximum graph utility. We show the effectiveness of the protocol on a real social network under different distributed storage cases.

1 Introduction

Recently, privacy preserving graph publication has become a hot research topic and many graph protecting models have been proposed [9,7,3,12,11,13,14,1]. All these models assume there is a trustable centralized data source, which has a complete original graph G, and can directly generate the privacy preserving graph G' from G.

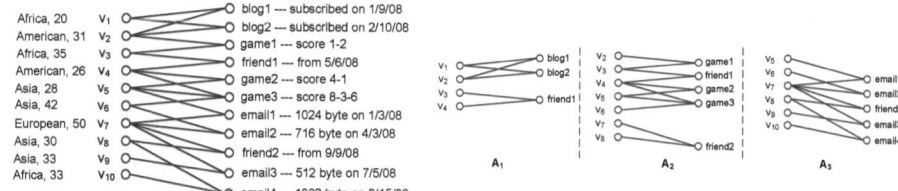

Fig. 1. The complete graph   **Fig. 2.** The distributed storage

However, in reality, people join different social networks due to different interests or purposes. For example, a person uses Facebook to share his information with his classmates and coworkers. At the same time, he may also build his blog on a blog server to share his interests with others. As a result, his connection information is stored in two social networks. A consequence of this joining preference is that each social network website only holds partial information (a subgraph) of the complete social

Y. Ishikawa et al. (Eds.): APWeb 2013, LNCS 7808, pp. 75–87, 2013.

network G. We call such a social network website as a data agent. Consider a social graph as shown in Figure 1 where each person has two labels and people are involved in different interactions, the graph can be stored on three data agents, A_1, A_2 and A_3 separately as shown in Figure 2.

Although people join different networks, it is often necessary to obtain a privacy preserving graph generated from the complete graph for criminal investigation [8] or mining useful patterns/influential persons [4,8,6]. This requests that the distributed agents should cooperatively generate a privacy preserving graph G'. There are three potential approaches to do this (Figure 3).

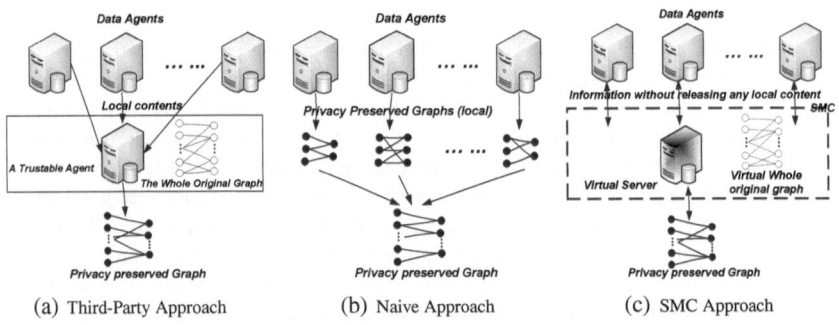

(a) Third-Party Approach   (b) Naive Approach   (c) SMC Approach

Fig. 3. Architectures for privacy preserving data publishing

A Third-Party approach (Figure 3(a)) is to let all the data agents send their local contents to a trustable third party agent, where the published graph by integrating all the data. However, since the data of each site is its most valuable asset, no one is willing to share its data with others, finding such a trustable third party data agent is not feasible in the real world. The naive approach (Figure 3(b)) is to let each data agent generate a privacy preserving graph based on its own content and *securely combine* them into a large privacy preserving graph [1]. Secure combination means no local content is released during the process. However, this approach encounters two problems: (1) It results in the low quality of the published data since the graph is constructed only based on local information instead of the complete graph; (2) A privacy preserving graph generated only by local information may violate the privacy requirement when globally more connection information is provided considering the connections between nodes [9,13,14,2,1]. The secure combination needs to delete some connections to ensure the final published graph satisfies the privacy requirement. Therefore, the published graph will have incomplete information. Another approach (Figure 3(c)) is that each data agent participates in a protocol to produce a privacy preserving graph. The privacy preserving graph is generated just as there virtually exists an agent who has the integrated data. During the computation, except the content derived from the final published graph, the protocol controls no additional local content of a data agent is released. This solution is

---

[1] The overlapping between subgraphs should be considered. [5] showed it is not safe even when each publisher publishes privacy preserved data independently. Here we use Fig. 3(b) to show the basic workflow of the naive approach.

known as the famous Secure Multi-Party Computation (SMC). SMC deals with a problem where a set of parties with private data wish to jointly compute a function of their data. A lot of SMC protocols have been proposed for different computation problems, but not for privacy preserving graph publishing in a distributed environment. SMC has two requirements: 1) Correctness requirement: the computation is performed in a distributed environment just the same as doing it on an agent who holds the integrated data; 2) Security requirement: each data agent should not know the local information of other agents even with the intermediate results passing through each other.

In this paper, we follow the SMC approach and design a secure protocol SP to allow the data agents to cooperately generate a graph G' based on a recently proposed protection model [1], called S-Clustering. Through clustering methods, S-Clustering publishes a graph G' that only contains super nodes (clusters) where each super node represents multiple nodes in G. We call a published graph which satisfies S-Clustering as the S-Clustering graph. We refer the algorithm which generates the S-Clustering graph in a centralized environment as the centralized algorithm. The distributed version of the centralized algorithm, SP, should work the same as running the centralized algorithm on the complete original graph. For any data agent, SP protects its local information when generating the published graph. We propose novel solutions in SP based on random lock, permutation, and Millionaire Protocol [10].

2   Problem Description

2.1   Graph, Storage Model and Problem Definition

An online social network with rich interactions can be represented as a bipartite graph $G(V, I, E)$ [1], where each node in V represents an entity in the network and each node in I stands for an interaction between a subset of entities in V. An edge $e(u, i) \in E$ ($u \in V$, $i \in I$) means entity u participants in the interaction i. Each entity has an identity and a group of attributes. Without loss of generality, each entity's identifier can be represented as a unique id within $[1, |V|]$. The commutative encryption scheme [8] can assign each name a unique number id without releasing the real name values in a distributed environment. In the rest part of this paper, we refer entity u as the same as the entity with id u. Each interaction also has an identity and a set of properties as shown in Figure 1. An interaction can involve more than two entities, such as the interaction "game3". Two entities can also be involved in different interactions at the same time. For example, in Figure 1, v_1 and v_2 participate in "blog1" and "blog2" simultaneously.

In a distributed environment, the graph G is distributively stored on l different data agents. That is, each agent A_i holds a portion of the graph $G_i(V_i, I_i, F_i)$ such that $V = \cup_{i=1}^{l} V_i$, $I = \cup_{i=1}^{l} I_i$ and $E = \cup_{i=1}^{l} E_i$. The interactions that an entity participants may be stored on different agents. The G_i held by different data agents may overlap, such that the intersection between two graphs stored on different agents might not be an empty set. [2] That is, $\cap_{i=1}^{l} V_i \neq \phi$ and $\cap_{i=1}^{l} I_i \neq \phi$.

---

[2] When V_is do not have overlap ($\cap_{i=1}^{l} V_i = \phi$), G is a disconnected graph since each data agent only holds a subgraph that is isolated from other parts.

In the rest of this paper, we use **node** to specifically represent an entity in V and use **interaction** to represent an item in I. When we say two nodes u, v have a **connection**, it means u and v are involved in a same interaction i in I (In the graph, we have two edges $e(u, i)$ and $e(v, i)$).

We propose a protocol which securely generates a S-Clustering graph [1] for interaction based graphs in a distributed environment. S-Clustering [1] assumes that an attacker can know the id, label and any connection information of a node. An attacker uses the information he knows about some users to analyze the published graph in order to learn more about these users. Given a constant k, S-Clustering publishes a clustered graph (S-Clustering graph) which guarantees the following three *Privacy objectives*:

Objective 1: For any node u, an attacker has at most $\frac{1}{k}$ probability to recognize it.

Objective 2: For any interaction i, an attacker has at most $\frac{1}{k}$ probability to know a node u involves in it.

Objective 3: For any two nodes u and v, an attacker has at most $\frac{1}{k}$ probability to know they have a connection (u and v participate in a same interaction).

We assume that all participants joined in the computation, including all data agents and necessary extra servers are semi-honest. The local information beyonds the *Privacy objectives* of S-Clustering is protected. The problem we solve in this paper is:

1. *Input*: A graph G distributed stored on l data agents and a privacy parameter k.
2. *Output*: A S-Clustering graph G' of G under k.
3. *Constraints*: (a) Correctness: G' is computed the same as generating it on an agent who holds G; (b) Security: The three *Privacy objectives* of S-Clustering are guaranteed even when a participant gets the intermediate results during the computation.

2.2   S-Clustering Model

Given a graph $G(V, I, E)$, to satisfy the three privacy objectives, a S-Clustering graph $G'(CV, I, CE)$ is published, where CV is a super node (cluster) set in which each super node represents a group of entities in V. We use c to denote a cluster in CV and $|c|$ to denote the cluster size, which is the number of entities that c represents. For each interaction i in I, if c contains an entity which participates in i, there is an edge $e(c, i)$ in G'. Figure 4 is a clustered graph of Figure 1.

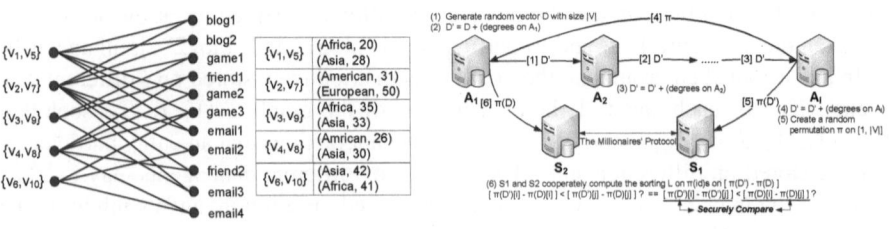

Fig. 4. A clustered graph   **Fig. 5.** Stage 1: SSSP

To guarantee the three privacy objectives, the S-Clustering graph G' must satisfy:

- Each cluster represents at least k entities. This guarantees the Objective (1).
- The *Clustering Safety Condition (CSC)* [1]:

Definition 1. *A division (clustering) of nodes V into clusters satisfies the **Clustering Safety Condition (CSC)** if for any node $v \in V$, v participates in interactions with at most one node in any cluster $S \subset V$, that is:*

- $\forall e(v,i), e(w,i), e(v,j), e(z,j) \in E : w \in S \land z \in S \Rightarrow z = w;$
- $\forall e(v,i), e(w,i) \in E : v \in S \land w \in S \Rightarrow v = w;$

This condition guarantees Objectives (2) and (3).

The CSC requires that any two nodes in a cluster cannot be connected or connect to the same node. Figure 4 is a clustered graph which satisfies CSC. For any two nodes that appear in the same cluster of Figure 4, they do not have a connection or connected to the same third node. For any two nodes u and v, if they satisfy (or do not satisfy) the CSC, we denote this as $CSC(u,v)$ (or $\neg CSC(u,v)$). Similarly, for a cluster c and a node u, if u satisfies (or does not satisfy) CSC with all the nodes in c, we denote this as $CSC(c,u)$ (or $\neg CSC(c,u)$).

3 Secure Protocol (SP)

3.1 Overview

We proposed SP to generate a S-Clustering graph in a distributed environment as shown in Algorithm 1. We call a computation that satisfies the Security requirement as a Secure Computation. Thus, SP contains four stages such that each stage conducts one Secure Computation in lines 1, 7, 15, and 16, respectively. Stage 1 sorts nodes without releasing any degree information. Stage 2 clusters nodes by correctly checking

---

Algorithm 1. Clustering with CSC in distributed environment

```
1   Securely Sort(V) by degree;                                    /* Stage1 */
2   CV=null;                                                        /* Begin Stage2 */
3   for v ∈ V do
4       flag = true;
5       for cluster c ∈ CV do
6           if SIZE(c)<k then
7               Distributed computing r = CSC(c, v);
8               if r then
9                   Insert(c,v) without generating/updating edges between clusters;
10                  flag = false;
11                  break;

12      if flag then
13          Create a New Cluster c and add c to CV;
14          Insert(c, v);                                          /* End Stage2 */

15  Compute interactions from distributed agents;                 /* Stage3 */
16  Publish a graph with node attributes only links to clusters;  /* Stage4 */
```

$CSC(c, v)$ without releasing any connection information. Stage 3 and Stage 4 generate the interactions on clusters and attributes for clusters respectively without disclosing any interaction-node mapping and node-attribute mapping. To summarize, we must avoid the releasing of the following information during computation: (1) Degree information (from Stage 1); (2) Specific node-attribute mapping (from Stage 4); (3) Specific connection information, including node-interaction mapping and connection between nodes (from Stage 2 & 3); (4) Any $\neg CSC$ result and the computation order of nodes (from Stage 1 & 2).

Next, we introduce our design of SP for each stage in detail. Due to space limitation, we ignore all the proofs. Before presenting SP, we assume each node/interaction has a weight that represents how many duplicated copies of this node/interaction have been stored in the system. We call these weights *duplicated weights*. Details of the method to generate duplicated weights is shown in Appendix A.

3.2   Stage 1: Secure Sorting Sub-Protocol (SSSP)

Protocol Design. In this stage, we sort the nodes without revealing any degree information and the computation order of nodes (the sorting order of nodes on real ids) information. The basic idea is to do the sorting on permuted ids with their corresponding "locked" degree values. We use the random number to lock the real degree of each node. That is, we make one agent hold a key (a random number) and another agent hold the locked degree value (real degree plus this random number). Then, we sort all nodes on permuted ids through the cooperation of these two agents. During the sorting, the agent who has the locked degree values cannot learn any key value and the agent who holds the keys cannot learn any locked value. The Secure Sorting Sub-Protocol (SSSP) works as follows (Figure 5):

1. A_1 generates a random real number vector D with size $|V|$. A_1 constructs a variable vector $D' = D$. Then, for each node u stored on A_1, A_1 adds $\sum_{e(u,i)} \frac{1}{w_i}$ to $D'[u]$ (w_i is the duplicated weight of the interaction i). Finally, A_1 sends D' to A_2;
2. When A_2 receives D', for each node u stored on A_2, we add $\sum_{e(u,i)} \frac{1}{w_i}$ to $D'[u]$. Then, A_2 sends D' to A_3;
3. Each A_i ($i > 1$) does the same operation as A_2 and sends D' to A_{i+1} if A_{i+1} exists;
4. A_l generates a random permutation function π, passes π to A_1 and sends $\pi(D')$ to S_1
5. A_1 sends $\pi(D)$ to S_2.
6. S_1 and S_2 cooperately sort all the nodes. During the sorting, each time when S_1 needs to compare two values $\pi(D')[i] - \pi(D)[i]$ and $\pi(D')[j] - \pi(D)[j]$, S_1 and S_2 uses the Millionaires' Protocol [10][3] to securely get the result by comparing the value $(\pi(D')[i] - \pi(D')[j])$ on S_1 and $(\pi(D)[i] - \pi(D)[j])$ on S_2.

Theorem 1. $SSSP$ sorts all nodes on $\pi(id)s$ under the Security requirement.

---

[3] The Millionaires' Problem is: suppose there are two numbers a and b hold by two people, they want to know the inequality $a > b$ or $a < b$ without revealing the actual values of a and b. The Millionaires' Protocol [10] can securely compare a and b based on the techniques such as Homomorphic Encryption. When we want to compare two degree values d_1 and d_2 in case that S_x knows r_1, r_2 and S_y knows $(r_1 + d_1), (r_2 + d_2)$, we can use the Millionaires' Protocol to compare the value $((r_1 + d_1) - (r_2 + d_2))$ on S_x and the value $(r_1 - r_2)$ on S_y.

3.3   Stage 2: Secure Clustering Sub-Protocol (SCSP)

Protocol Design. In Stage 2, we need to cluster nodes based on the sorted order L without revealing any connection information.

Since S_1 holds L after Stage 1, in $SCSP$, we continue to make S_1 generate clusters on $\pi(id)$s. Before S_1 can do clustering on $\pi(id)$s, the connection information between nodes should be mapped on $\pi(id)$s firstly. So, $SCSP$ contains the following three steps:

1. *Generate a noise matrix MR on A_1 and a noised adjacent matrix MR' of G on A_l. For any two nodes u and v, $MR'[u][v] - MR[u][v] > 0$ is equivalent to u, v has a connection;*
2. *Create a matrix NMR on A_1 and NMR' on A_l based on MR and MR', respectively. For any two nodes u and v, $NMR'[u][v] - NMR[u][v]$ is a $[0, 1]$ matrix where $NMR'[u][v] - NMR[u][v] = 1$ indicates that u, v have a connection;*
3. *Do clustering on $\pi(id)$s with $\pi(NMR')$, $\pi(NMR)$ through the cooperations among S_1, S_2 and S_3.*

The first two steps map the connection information on $\pi(id)$s (The $[0, 1]$ character of $NMR' - NMR$ will be used for securely clustering). The last step does the clustering on the connection information between $\pi(id)$s. During the computation, to satisfy the Security requirement, we make the agent who holds MR, NMR or $\pi(NMR)$ cannot get the noised content MR', NMR' or $\pi(NMR')$ and vice versa.

The first step computes MR and MR' securely as follows:

1. *A_1 generates a random real number matrix MR with size $|V| \times |V|$. A_1 constructs a variable matrix $MR' = MR$. Then, a self-edge is added for each node (The self-edge will be used to determine the CSC). For each u, A_1 sets $MR'[u][u] = MR'[u][u] + a_1$ (a_1 is a random positive number). For any interaction i and any two nodes u and v that participate in i, A_1 sets $MR'[u][v] = MR'[u][v] + a_2$ (a_2 is a positive random number), and then A_1 sends MR' to A_2;*
2. *After A_2 getting MR', for any interaction i and any two nodes u and v that participate in i, A_2 sets $MR'[u][v] = MR'[u][v] + a_3$ (a_3 is a positive random number), and then A_2 sends MR' to A_3;*
3. *Each A_i ($i > 1$) does the same operation as A_2 and sends MR' to A_{i+1} if A_{i+1} exists;*

After the first step, A_l gets a noised matrix MR'. For any nodes u and v, $MR'[u][v] - MR[u][v] > 0$ is equivalent to u and v have a connection. The second step securely computes the matrices NMR and NMR' as shown in Figure 6:

1. *A_l generates a random permutation function π_2 on $|V| \times |V|$ numbers and sends π_2 to A_1. A_l sorts all the numbers in MR' row by row to obtain a vector L_2' whose length is $|V| \times |V|$;*
2. *A_1 sorts MR row by row to obtain a vector L_2;*
3. *A_1 sends $\pi_2(L_2)$ to S_3 and A_l sends $\pi_2(L_2')$ to S_3;*
4. *S_3 computes two random number vectors LN_2 and LN_2' which satisfy*

$$LN_2'[i] - LN_2[i] = \begin{cases} 0 & \pi_2(L_2')[i] - \pi_2(L_2)[i] = 0 \\ 1 & \pi_2(L_2')[i] - \pi_2(L_2)[i] > 0 \end{cases}$$

Then, S_3 passes LN_2 to A_1 and LN_2' to A_l;
5. *A_1 computes $\pi_2^{-1}(LN_2)$ and converts the results to a $|V| \times |V|$ matrix NMR. A_1 passes $\pi(NMR)$ to S_2; Similarly, A_l generates the corresponding matrix NMR' based on $\pi_2^{-1}(LN_2')$ and passes $\pi(NMR')$ to S_1;*

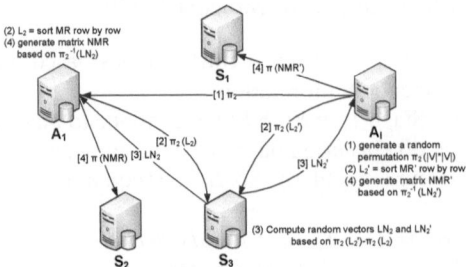

Fig. 6. Stage 2: SCSP 2

The third step is to do the clustering on $\pi(id)$s. Before introducing how S_1 conducts the clustering, we prove a property we use in the next computation. For a node u, we call E_u ($|E_u| = |V|$) the *connection vector* of u. E_u is a $[0,1]$ vector and $E_u[u] = 1$. If nodes u and v have a connection, $E_u[i] = 1$, otherwise $E_u[i] = 0$. For example, node v_1's connection vector in Figure 1 $E_1 = [1,1,0,0,0,0,0,0,0,0]$. For a cluster of nodes C, we call $E_C = (\sum_{\forall u \in C} E_u)$ as C's *connection vector*. The connection vector has the following property:

Theorem 2. *For any cluster of nodes C, if there exists a t where $E_C[t] > 1$, then C does not satisfy CSC; otherwise, C satisfies the CSC.*

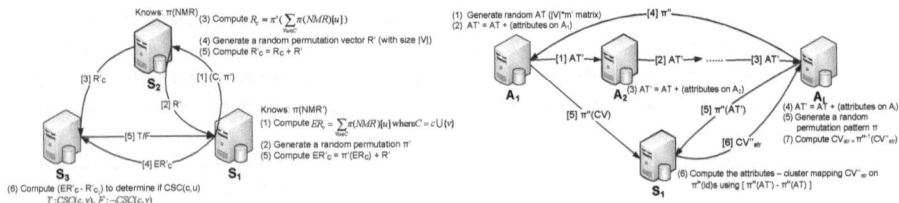

Fig. 7. Stage 2: SCSP 3   **Fig. 8.** Stage 4: SLSP

For a cluster of nodes C on $\pi(id)$s, it is obvious $E_C = \sum_{\forall u \in C}(\pi(NMR')[u] - \pi(NMR)[u])$. Based on the property of connection vector, when S_1 needs to check whether v can join a cluster c based on CSC, the protocol works as follows (Figure 7):

1. S_1 computes a connection vector $ER_C = \sum_{\forall u \in C} \pi(NMR')[u]$. Where $C = c \cup \{v\}$. The size of this vector is $|V|$.;
2. S_1 generates a random permutation pattern π' and sends (C, π') to S_2;
3. S_2 computes a vector $R_C = \pi'(\sum_{\forall u \in C} \pi(NMR)[u])$. A_2 continues to generate a random vector R' with size $|V|$ and sends R' to S_1. S_2 sends $R'_C = R_C + R'$ to S_3;
4. S_1 computes $ER'_C = \pi'(ER_C) + R'$ and passes ER'_C to S_3;
5. S_3 computes $E_C = ER'_C - R'_C$. If E_C contains a number bigger than 1, it returns "cannot cluster" to S_1. Otherwise, it returns "can cluster" to S_1.

S_1 does the clustering on L and uses the above method to test CSC. S_1 finally gets the cluster set CV' on $\pi(id)$s and passes CV' to A_1. A_1 computes the cluster set on real ids $CV = \pi^{-1}(CV')$.

Theorem 3. *SCSP clusters all nodes without violating the Security requirement.*

3.4   Stage 3: Secure Edge Generation Sub-Protocol (SESP)

Protocol Design. In Stage 3, we generate the interactions among clusters. After A_1 gets the cluster set CV on real ids, A_1 reports CV to all A_is one by one. Each A_i generates the interactions between clusters and sends the results to A_{i+1}. Finally A_l gets a clustered graph with correct interactions. The above process only passes the connection information through interactions without the attribute information. Each A_i directly sends the attribute of each interaction to A_l since all the interactions will be clearly published. Since the interactions between super nodes reported by each A_i is a subset of the final result, the middle results are already included in the published graph. Therefore the Security requirement is satisfied.

3.5   Stage 4: Secure Label Generation Sub-Protocol (SLSP)

Protocol Design. In Stage 4, we generate the node attributes for each cluster without disclosing any specific node-attribute mapping. Suppose there are m' attributes for each node. The Secure Label Generation Protocol (SLGP) works as shown in Figure 8. This is similar to Stage 1. Due to page limitation, we ignore the detailed description.

Theorem 4. *SLSP assigns node attributes to each cluster without violating the Security requirement.*

Based on the above step by step illustration of Algorithm, we conclude that the whole SP satisfies the Correctness and Security requirements. The intermediate results of one stage do not influence other stages since computation contents are different. Since each stage satisfies the Security requirement, the whole SP also satisfies the Security requirement.

4   Experiment

To demonstrate the effectiveness of SP protocol, we implement a Relaxed Secure Protocol (RSP) which is based on the naive approach (Fig. 3(b)). The brief introduction to the design of RSP is shown in Appendix B. We compare the graph generated We compare the graph generated by RSP and SP.

4.1   Criteria

We focus on two aspects to show the benefit of designing a SMC protocol: the information loss and the utilities. We estimate the information loss of RSP comparing with SP (SP does not delete any interaction). Assume del is the number of interactions that are deleted by RSP and the original complete graph contains $|I|$ interactions, we use the ratio of deleted interactions ($\frac{100del}{|I|}\%$) to represent the information loss.

Utility is used to estimate the quality of the published graph. For the clustering-based protection models [7,12,1], the utility testing is operated by drawing sample graphs from the published one, measuring the utility of each sample and aggregating utilities across samples. We test the following measures:

1. *Degree distribution*: Suppose the sorted degree sequence of the original graph G is D_G and the one of a sampled graph G_c is D_{G_c}. The different between the degree distributions of G and G_c is represented as: $ED_{DD}(D_G, D_{G_c}) = \sqrt{\frac{1}{|V|} \sum_{i=1}^{|V|} (D_G[i] - D_{G_c}[i])^2}$. We compute the average ED_{DD} of 30 sampled graphs for each protocol. Suppose the SP's result is $ED_{DD,SP}$ and RSP's result is $ED_{DD,RSP}$, we use $\frac{ED_{DD,RSP} - ED_{DD,SP}}{ED_{DD,SP}} \times 100\%$ to compare these two protocols.

2. *A group of randomly selected queries*: We test three aggregate queries as the same as Paper [1]. We compute the average query error between the sampling graphs and the original graph. Suppose SP's result is $error_{SP}$ and RSP's result is $error_{RSP}$, we use $\frac{error_{RSP} - error_{SP}}{error_{SP}} \times 100\%$ to compare them. The three aggregate queries include: (1) Pair Queries: how many nodes with certain attribute interact with nodes with another attribute; (2) Trio Queries: how many two hop neighbors. (3) Triangle Queries: how many triangles given three attributes. We select a random set of queries of each type to do the testing. For each type, we select 20 queries and compute the average query error between the sampling graphs and original graph.

4.2   Data Set and Distributed Storage

Our experiment is operated on a real life social network we crawled from ArXiv (arXiv.org). ArXiv(arXiv.org) is an e-print service system in Physics, Computer Science etc. We extract the co-author graphs in Computer Science. Each node denotes an author, and each interaction means a unique coauthor paper. The ArXiv provides 37 categories in Computer Science to search the papers. Since most people work in multiple sub-fields and each paper also belongs to multiple sub-fields, the authors of papers in different categories are overlapped and finally form a large graph. We set each author's research field as his/her node label. We divide the 37 categories to 6 sets and store the crawled graph of each category set on one data agent. This can be seen as different data agents maintain the relationship of people in different fields. The integrated graph contains 28868 nodes and 23290 interactions. On average, each interaction is involved by 2.4 nodes.

We consider the following two distributed storage cases:

1. Real Distributed Case: The 6 subgraphs we crawled in different research fields can be seen as 6 data agents maintain the relationship of people in different research fields. The number of nodes stored in the 6 data agents are 6371, 6370, 7705, 5923, 4676 and 7876 respectively. There are $r = 34\%$ nodes stored multiple times on different agents. We call r the node overlapping ratio.

2. Simulated Distributed Case: We manually divide the integrated graph G to 6 parts that have similar node numbers with different node overlapping ratios (rs).

4.3   Result

Result on Real Distributed Case

We compare the performance of SP and RSP under different privacy parameter ks. Figure 9(a) shows the result of information loss. RSP deletes 13% to 17% interactions. This is a fairly significant information loss. The published graph by RSP fails

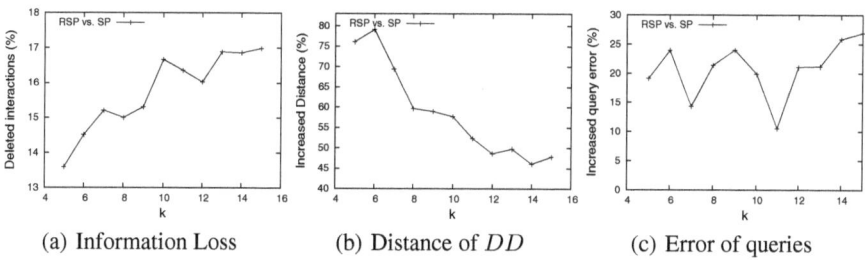

(a) Information Loss   (b) Distance of DD (c) Error of queries

Fig. 9. Real Distributed Case

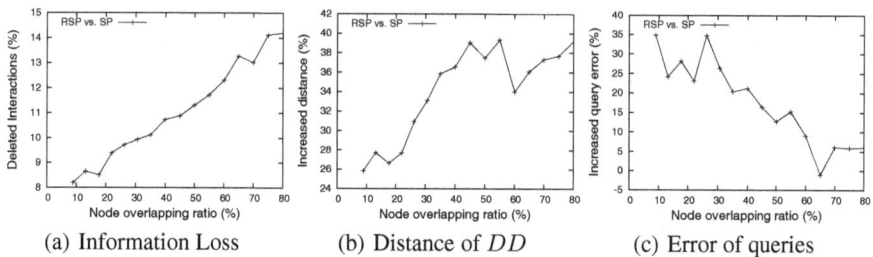

(a) Information Loss   (b) Distance of DD (c) Error of queries

Fig. 10. Simulated Distributed Case

to correctly represent the original graph. Figure 9(b) shows the comparison of degree sequence distance. RSP performs 45% to 80% worse than SP. The published graph by RSP is much worse than SP when estimating by the degree distribution. Figure 9(c) shows the results of queries. In most cases, the RSP performs 10% to 25% worse than SP. In most cases, SP gets a much better graph than RSP.

Result on Simulated Distributed Case

For the Simulated Distributed Case, we set $k = 5$ and compare the performance of SP and RSP with different node overlapping ratio rs. Figure 10(a) shows the number of deleted interactions. Larger r means more overlapping between data agents and more chances to delete an interaction. In most cases, RSP deletes 10% to 20% interactions. Figure 10(b) shows the comparison of degree sequence distance. RSP performs 25% to 45% worse than SP. Figure 9(c) shows the results of queries. In most cases, the RSP performs 5% to 35% worse than SP. SP generates a graph with higher utilities than RSP.

From the testing results, we can find SP exhibits benefits on both information completeness and the utilities of the published graphs. Firstly, the RSP deletes roughly 13%-19% interactions, publishing a graph which loses such a large portion of information is not acceptable. While SP guarantees no information loss in the published graph. Secondly, the utilities of the graph generated by SP are much better than RSP. It is necessary to design a SMC protocol such as SP for the privacy preserving graph publication problem. Actually, since SP has the same effect as running the state-of-art centralized graph construction algorithm on the complete original graph, the published graph generated by SP can be seen as the best.

5 Related Work

Two models are proposed to publish a privacy preserving graph: the edge editing based model [9,13,14,2,11] and the clustering based model [7,12,1]. The edge editing based model is to add/delete edges to make the published graph satisfy certain properties according to the privacy requirements. For example, Liu[9] defined and implemented the k-Degree anonymous model on network structure, with which there exists at least other $k-1$ nodes having the same degree as any node in a published graph. The clustering based model is to cluster "similar" nodes into super nodes. Since a clustered graph only contains super nodes, the probability of re-identifying any user can be bounded to $\frac{1}{k}$ by making each cluster's size at least k.

Frikken [4] designed a protocol which allows a group of agents to generate an integrated graph. However, the graph generated by this protocol cannot provide the protections against the complex attacks proposed in recent works [9,7,3,12,11,13,1,14]. Kerschbaum [8] designed a protocol to generate an anonymized graph from multiple data agents. Each edge in the anonymized graph has a digital signature which can help trace where this edge comes from. While, simple removing the identifiers of nodes in a graph cannot resist an attack which aims to re-identify the nodes/links [7]. Therefore, it is essential to investigate protocols that support the stronger protection models [9,7,3,12,11,13,1,14] in a distributed environment. Our work generates the published graph which follows the recently proposed *S-Clustering* protection model. This model protects nodes and links against the strongest attack which may use any information around nodes. Moreover, our work supports a more flexible graph model while the protocols of Keith [4] and Florian [8] only support the homogeneous graphs.

6 Conclusion

In this paper, we target on the secure multi-party privacy preserving social network publication problem. We design a SMC protocol SP for the latest clustering based graph protection model. SP can securely generate a published graph with the same quality as the one generated in the centralized environment. As far as our knowledge, this is the first work on SMC privacy preserving graph publication against the "structure attack". In the future, one interesting direction is to study how to enhance the current solution to handle the malicious or accessory agents. How to design SMC protocols for the editing based graph protection models will be another interesting direction.

Acknowledgments. This work is supported in part by Hong Kong RGC grants N_HKU-ST612/09, National Grand Fundamental Research 973 Program of China under Grant 2012CB316200 and NSFC 619308.

References

1. Bhagat, S., Cormode, G., Krishnamurthy, B., Srivastava, D.: Class-based graph anonymization for social network data. Proc. VLDB Endow. 2, 766–777 (2009)
2. Cheng, J., Fu, A.W.-C., Liu, J.: K-isomorphism: privacy preserving network publication against structural attacks. In: SIGMOD 2010, pp. 459–470. ACM, New York (2010)

3. Cormode, G., Srivastava, D., Yu, T., Zhang, Q.: Anonymizing bipartite graph data using safe groupings. Proc. VLDB Endow. 1, 833–844 (2008)
4. Frikken, K.B., Golle, P.: Private social network analysis: how to assemble pieces of a graph privately. In: WPES 2006, pp. 89–98. ACM, New York (2006)
5. Ganta, S.R., Kasiviswanathan, S., Smith, A.: Composition attacks and auxiliary information in data privacy. CoRR (2008)
6. Garg, S., Gupta, T., Carlsson, N., Mahanti, A.: Evolution of an online social aggregation network: an empirical study. In: IMC 2009, pp. 315–321. ACM, New York (2009)
7. Hay, M., Miklau, G., Jensen, D., Towsley, D., Weis, P.: Resisting structural re-identification in anonymized social networks. Proc. VLDB Endow. 1, 102–114 (2008)
8. Kerschbaum, F., Schaad, A.: Privacy-preserving social network analysis for criminal investigations. In: WPES 2008, pp. 9–14. ACM, New York (2008)
9. Liu, K., Terzi, E.: Towards identity anonymization on graphs. In: SIGMOD 2008, pp. 93–106 (2008)
10. Yao, A.C.: Protocols for secure computations. In: SFCS 1982, pp. 160–164. IEEE Computer Society, Washington, DC (1982)
11. Ying, X., Wu, X.: Randomizing social networks: a spectrum preserving approach. In: SDM 2008 (2008)
12. Zheleva, E., Getoor, L.: Preserving the privacy of sensitive relationships in graph data. In: Bonchi, F., Malin, B., Saygın, Y. (eds.) PinKDD 2007. LNCS, vol. 4890, pp. 153–171. Springer, Heidelberg (2008)
13. Zhou, B., Pei, J.: Preserving privacy in social networks against neighborhood attacks. In: ICDE 2008, pp. 506–515 (2008)
14. Zou, L., Chen, L., Özsu, M.T.: k-automorphism: a general framework for privacy preserving network publication. Proc. VLDB Endow. 2, 946–957 (2009)

A   Method for Duplicated Weights Computation

Since each node and interaction has a unique id, A_1 generates two random vectors RV_{dup} and RI_{dup}. RV_{dup} has size $|V|$ and RI_{dup} has size $|I|$. Then A_1 generates two variable vectors V_{dup} and I_{dup} with $V_{dup} = RV_{dup}$ and $I_{dup} = RI_{dup}$. $V_{dup}[i] - RV_{dup}[i]$ will be used to represent the duplicated weight of node i. $I_{dup}[i] - RI_{dup}[i]$ will be used to represent the duplicated weight of interaction i. Each A_i does the following operations: (1) For each node i stored by A_i, $V_{dup}[i] = V_{dup}[i] + 1$; (2) For each interaction i stored by A_i, $I_{dup}[i] = I_{dup}[i] + 1$; (3) Passes V_{dup} and I_{dup} to A_{i+1} if A_{i+1} exists. After doing this, A_l sends the V_{dup} and I_{dup} to A_1. Finally, A_1 gets the duplicated weights of all nodes and interactions by computing $V_{dup} - RV_{dup}$ and $I_{dup} - RI_{dup}$. A_1 passes the result to each A_i and A_i stores the duplicated weights for the nodes and interactions in G_i.

B   Relaxed Secure Protocol (RSP)

We implement RSP by clustering nodes on local data agents and deleting some cross agent interactions. By doing this, the Security can be guaranteed. A *cross agent interaction* is an interaction whose participants contain at least two nodes which have duplicates in the system. We only delete the cross agent interactions which break or have the potential risk to break the CSC. A_1 generates a clustered graph and passes it to the next agent A_2. Each A_i generates a larger clustered graph by adding its local content into the received clustered graph. The cross agent interactions, which have potential to break the CSC, are deleted.

Correlation Mining in Graph Databases
with a New Measure

Md. Samiullah[1], Chowdhury Farhan Ahmed[1], Manziba Akanda Nishi[1],
Anna Fariha[1], S M Abdullah[2], and Md. Rafiqul Islam[3]

[1]Department of Computer Science and Engineering, University of Dhaka, Bangladesh
sami_cse_du1@yahoo.com, farhan@khu.ac.kr, nishifriend786@yahoo.com,
purpleblueanna@gmail.com
[2]Department of Computer Science and Engineering,
United International University, Bangladesh
smab@cse.uiu.ac.bd
[3] School of Computing and Mathematics, Charles Sturt University, Australia
mislam@csu.edu.au

Abstract. Correlation mining is recognized as one of the most important data mining tasks for its capability to identify underlying dependencies between objects. Nowadays, data mining techniques are increasingly applied to such non-traditional domains, where existing approaches to obtain knowledge from large volume of data cannot be used, as they are not capable to model the requirement of the domains. In particular, the graph modeling based data mining techniques are advantageous in modeling various real life complex scenarios. However, existing graph based data mining techniques cannot efficiently capture actual correlations and behave like a searching algorithm based on user provided query. Eventually, for extracting some very useful knowledge from large amount of spurious patterns, correlation measures are used. Hence, we have focused on correlation mining in graph databases and this paper proposed a new graph correlation measure, *gConfidence*, to efficiently extract useful graph patterns along with a method *CGM* (*C*orrelated *G*raph *M*ining), to find the underlying correlations among graphs in graph databases using the proposed measure. Finally, extensive performance analysis of our scheme proved two times improvement on speed and efficiency in mining correlation compared to existing algorithms.

Keywords: Correlation mining, knowledge discovery, correlated graph patterns, graph mining, graph correlation.

1 Introduction

Data mining extracts useful implicit patterns within data along with important correlations/affinities between the patterns to discover knowledge from databases. Moreover, in data mining, correlation analysis is a special measure which finds the underlying dependencies between objects. The frequent itemset can be mined but correlation calculation among items can define the dependencies and mutual correlation among items within an itemset. The hidden information within databases,

Y. Ishikawa et al. (Eds.): APWeb 2013, LNCS 7808, pp. 88–95, 2013.

and mainly the interesting association relationships among sets of objects, may disclose useful patterns for decision support, financial forecast, marketing policies, even medical diagnosis and many other applications.

Nowadays, data mining techniques are applied to non-traditional domains e.g. the most complicated real life scenarios where objects are interacting with their surrounding other objects. To model such scenarios graph can be used, where vertices of the graph will correspond to entities and edges will correspond to relations among entities. Because of combinatorially explosive search for subgraphs which includes subgraph isomorphism testing, the graph structured data mining is difficult. Moreover, in mining graph data, the emphasis is on frequent labels and common topologies unlike traditional data. Here mining can be divided into level-by-level *generate-and-test* method and *pattern growth-based approach*. AGM[1] and FSG[2] are of the former type, where as gSpan[3] and graph pattern [4] are of later type, which require no candidate generation. In order to discover correlations, several measures are used [5] and to mine correlation in graph databases existing works such as [6] mainly focus on structural similarity search. However, graphs those are structurally dissimilar but always appear together within the database, may be more interesting , such as, the chemical properties of isomers [7].

For capturing effective correlations, authors of [7] proposed CGS algorithm which mines graph correlation by adopting $Pearson's\ correlation\ coefficient$ by taking into account the occurrence distributions of graphs. However, CGS works for searching correlation of a specific query graph with the database. Therefore, it has some limitations in describing inherent correlation among graphs and the domain knowledge is obligatory in using CGS, otherwise lots of queries would be meaningless. These facts motivated us in developing a new measure which can prune a large number of un-correlated graphs and designing an algorithm to efficiently mine graph correlation. Our contributions are: a new graph correlation measure $gConfidence$ to mine inherent correlation in graph databases, pruning a large number of candidates using the downward closure property of the measure and an algorithm CGM (Correlated Graph Mining) to efficiently mine correlation by constructing a hierarchical reduced search space.

Rest of the paper is organized as follows: Section 2 contains our proposed scheme and Section 3 focuses on the performance analysis of our proposed algorithm. Finally, we concluded our work in Section 4.

2   Our Proposed Approach

Correlated graph mining is one of the most important graph mining tasks. So, we have proposed a new measure, $gConfidence$ and a new method, $CGM(Correlated Graph Mining)$, to search correlation among graphs within a graph database. We have mined correlated graphs by constructing a hierarchical search space using our proposed measure and algorithm.

In graph domains, transactions can be represented by $G = \{V(G), E(G), L(V(G)), L(E(G))\}$ that is the set of vertices, set of edges and set of labels for vertices and edges respectively. Size of a graph varies in various algorithms and can be the number of nodes or edges or disjoint paths. To tackle graph isomorphism problem, various labeling or coding is introduced in many algorithms. For a graph database, GD, with G_s, G_b and G_h are sub-graphs of any graph $G \in GD$, the support of G_s is the ratio among the frequency of G_s's supergraph and total number of transaction graphs. Moreover, confidence of G_b with respect to G_h is the ratio between the joint-occurrence frequency of G_b along with G_h and frequency of the super graph for G_b. The canonical labeling for the normal form representation, X of a graph G can be defined as[3], the minimum DFS code among all possible DFS codes of G . The proposed measure $gConfidence$, is defined as follows where the correlation is defined based on the co-occurrence probability of edges of the graphs within the database.

Definition 1. *(gConfidence) Given a graph database GD and one of its transaction graph G, then we have defined the gConfidence of G_s, a subgraph of G as*

$$gConfidence(G_s) = \frac{\{No.\,of\,graphs\,G \mid G_s \subseteq G \in GD\}}{max(\{No.\,of\,graphs\,G_i \subseteq G \in GD \mid \forall G_i \subseteq G_s\})} \tag{1}$$

"$max(\{No.\,of\,graphs\,G_i \subseteq G \in GD \mid \forall G_i \subseteq G_s\})$" is the maximum support of any sub-graph of G_s. This prescribed clarity merely entails that $gConfidence$ is the smallest correlation of any graph, G_s and fall within the range $[0, 1]$. The measure $gConficence$, has numerous crucial properties those made it efficient in mining graph correlation more effectively. These are the *downward closure property, null invariance, lower bound* marker of correlation and *co − occurrence* observer.

Lemma 1. *Given a graph database GD and one of its transaction graph G, then we can define the gConfidence of $G_s \subseteq G$ derived in Equation 1 as*

$$gConfidence(G_s) = \frac{\{No.\,of\,Graphs\,G \mid G_s \subseteq G \in GD\}}{max(\{\forall e_i \in E(G_s),\ No.\,of\,graphs\,G_j \mid e_i \in E(G_j),\,G_j \in GD\})} \tag{2}$$

Consider a scenario shown in Figure 1, where two frequent closed graphs found from a set of graphs representing a group of people. Each graph in the set represents friend circle of an individual where nodes represent individuals and edges represent interaction among individual pairs. The circles around closed graphs represent the interaction of a group of people all together. Labels of edges and circles represent the frequencies of the edges and closed graphs respectively. However, in mining the most correlated group, frequency of closed-frequent graphs can't help due a tie i.e. 30 for each. As a consequence, Group G_2 will be suggested as most correlated by our proposed measure. Since, maximum interaction of any pair in G_1 is 100 and in G_2 is 60. Therefore, $gConfidence(G_1) = \frac{30}{100} = 0.3 < gConfidence(G_2) = \frac{30}{60} = 0.5$.

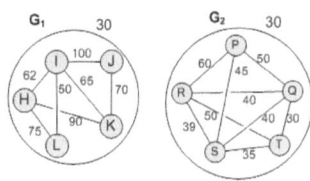

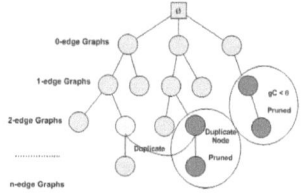

Fig. 1. Motivating Scenario   **Fig. 2.** gConfidence Tree : A search space

Definition 2. *(Correlated Graph Mining) Given a graph database $GD =$ $\{G_1, G_2, ..., G_N\}$, a user specified minimum support threshold, σ and a user specified minimum correlation threshold, θ. We have to search for interesting graph/subgraphs that is, we have to search for the set of graphs $G_I = \{\forall G_i \mid supp(G_i) \geq \sigma; gConfidence(G_i) \geq \theta\}$. It means the problem is to search for graphs having support count greater than or equal to σ and gConfidence value greater than or equal to θ.*

We have created a hierarchical tree like structure for efficiently searching the correlation among graphs within graph databases. The tree is defined as follows:

Definition 3. *(gConfidence Tree) A tree, where each node represents a graph or subgraph by storing corresponding DFS code and represents correlation by storing gConfidence value "gC". Moreover, the relation between parent node and child node complies with the relation that a parent is one edge shorter in size than its child and a child is one edge larger than its parent and no child has "gC" greater than its parent. The relation between siblings is consistent with the DFS lexicographic order. That is, the pre-order search of gConfidence tree follows the DFS lexicographic order.*

In Figure 2, we have shown a *gConfidence* code tree where the $(n+1)$-th level of the tree has nodes which contain DFS codes of n-edge graphs and a value "*gC*" which represents the inherent correlation among the nodes, edges and subgraphs of that particular graph. Any node in the *gConfidence tree* contains a valid DFS code. Certainly some of the nodes contain a minimum DFS code while others do not. And there could be some nodes having "*gC*" values smaller than the minimum correlation threshold. The value for "*gC*" also maintains a parent and child relationship that is $gC(\alpha) \geq gC(\beta)$ where $\alpha = (a_0, a_1, ..., a_m)$ and $\beta = (a_0, a_1, ..., a_m, b)$ that is α is β's parent.

Now we will describe our algorithm for mining graph correlation using our proposed measure *gConfidence* within graph databases. Since correlation is searched based on user specified minimum correlation threshold then the search space can be pruned based on two values, one for minimum support threshold and another one is minimum confidence threshold. Proposed *CGM* is an "edge" based correlation mining algorithm and we also used the concept of "Projected Database" to reduce the costly searching operation for counting occurrences of any graph/subgraph.

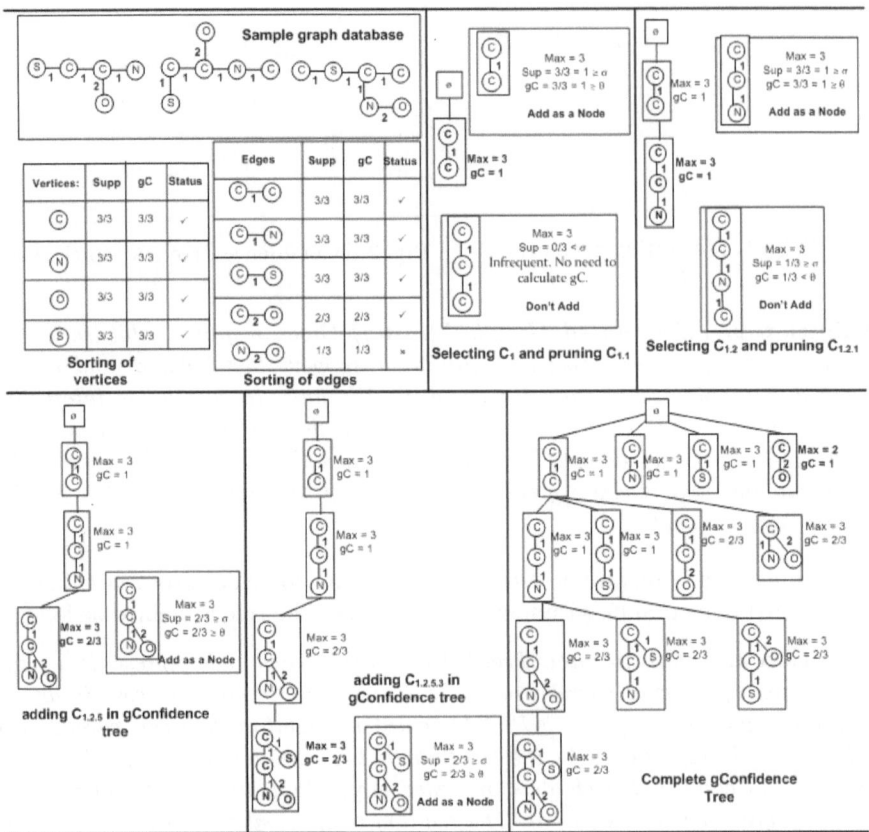

Fig. 3. gConfidence mining Illustration

For illustrating the working procedure of our CGM algorithm, we can consider the graph database for the chemical dataset shown in Figure 3. We have assumed the support threshold, $\sigma = \frac{1}{3}$ and correlation threshold, $\theta = \frac{2}{3}$. According to our algorithm we have calculated *support* and *gConfidence* of each edge and vertex and then selected the frequent and correlated vertices and edges. Now, we have to construct single edge graph for each frequent-correlated edge and the edge set is also used to construct *GLOBAL EDGE MATRIX*. This matrix is sorted based on *support* count and DFS code and can be used in a chronological order for constructing potential children (smallest DFS code oriented child first) and also helps in counting maximum elementary edge support count of a graph.

We have created a Null-rooted *gConfidence* tree and then started mining for the first 1-edge graph C_1. Since it satisfies both the threshold values, we have added it in the search space and start mining the correlation of its potential children recursively. Its first child $C_{1.1}$ is not frequent, hence we have pruned it. The second child $C_{1.2}$ is frequent and correlated, hence added to the search space. However, recursive calculation of its children shows that $C_{1.2.1}$ and $C_{1.2.4}$ are frequent but not correlated and $C_{1.2.2}$ and $C_{1.2.3}$ are not frequent.

As a consequence, the fifth child of $C_{1.2}$ that is $C_{1.2.5}$ is found correlated and added to $gConfidence$ tree. The eighth child of $C_{1.2.5}$ is found correlated and added in the search space but first seven children of $C_{1.2.5}$ are not frequent hence will not be correlated and been pruned from the search space. Therefore, we have to further search for the correlation of all possible children of $C_{1.2.5.8}$. Since no children of $C_{1.2.5.8}$ found frequent-correlated, we can backtrack to $C_{1.2.5}$. But already we have checked all possible children of $C_{1.2.5}$ hence we can trace back again to $C_{1.2}$ for checking correlation of its remaining children.

In this way we have calculated the correlation for the graph database of Figure 3 and found a complete $gConfidence$ tree, where nodes of the tree contain correlated graphs along with the amount of correlation. All the above discussed steps are illustrated in Figure 3.

3   Experimental Results

We have performed a comprehensive performance study in our experiments on both synthetic and real world datasets. We have used our own synthetic graph generator to construct synthetic dataset. The real life datasets we have tested are cancer dataset, found from [8], namely MOLT-4 and NCI-H23. On an average, the real datasets contain about 40K graphs with average 25 nodes and 30 edges along with an average of 17 distinct labels for vertices and edges. The synthetic dataset can be identified by four properties, $\mid D \mid$ representing numbers of graphs, $\mid N \mid$ indicating number of distinct labels for vertices and edges, $\mid T \mid$ and $\mid V \mid$ for average graph size wrt. edges and vertices respectively.

In both types of data (real and synthetic) CGM algorithm is proved to be sound and efficient as well as found scalable and faster enough that it can mine correlation among various graphs with any size and any level of complexity in comparison to CGS[7] and gSpan[3]. All experiments of CGM, CGS and gSpan have been performed on a 2.1 GHz Intel(R) Core(TM) Duo PC with 1GB RAM, running Windows 7 operating system, using C/C++ programming language. We have kept the support threshold fixed at 5% and varied the correlation threshold from 45% to 85% unless stated otherwise. In some cases, we have varied the size of database by adding or removing graph transactions randomly from the actual database.

In Figure 4 and Figure 5, we have shown the performance of our proposed CGM algorithm for Processing time vs. Graph Density with varying Correlation Threshold, when run on the MOLT-4 graph database and a synthetic database characterized by $D200kN30T80V50$ respectively. Since we are assessing the performance of our algorithm against graph density, we have varied the size of the graph database. It can be noticed that maximum 150 seconds were needed and minimum required time was 25 seconds in mining real life database, where most of the time processing completes within 1000 seconds for synthetic dataset with any confidence threshold within range.

Figure 6 and Figure 7 contain the performance analysis of our proposed algorithm for Scalability wrt Time with varying Data Size, when run on the NCI-H23

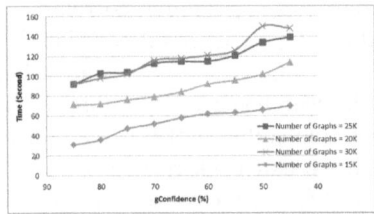

Fig. 4. Processing time wrt Graph Density on (MOLT-4)

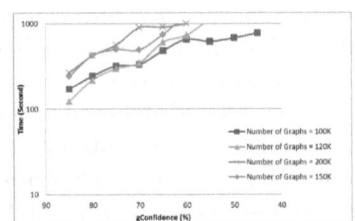

Fig. 5. Processing time wrt Graph Density on (D200kN30T80v50)

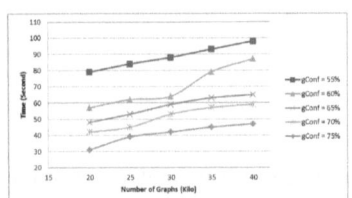

Fig. 6. Scalability wrt processing time on (NCI-H23)

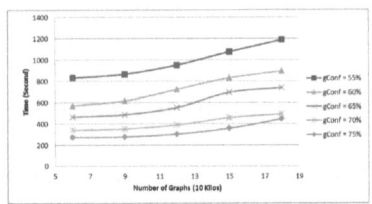

Fig. 7. Scalability of CGM wrt processing time for Synthetic Data(D200kN20T40v30)

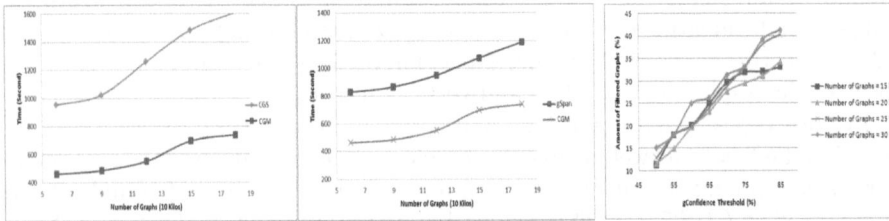

Fig. 8. CGM vs CGS on (D200kN20T40v30)

Fig. 9. CGM vs gSpan on (D200kN20T40v30)

Fig. 10. Performance in Filtering(%) on (MOLT-4)

graph database and a synthtic database characterized by $D200kN20T40v30$ respectively. We have found that 50 to 100 seconds are required in mining NCI-H23 database and 300 to 1200 seconds are required for the synthetic dataset with any confidence threshold within range.

To compare the performance of CGS with CGM, once again we have used the denser synthetic dataset used in the scalability assessment earlier. However, the support threshold is considered 5% and confidence threshold 50%. The comparison is shown in the Figure 8, which illustrates a significant performance gain. We have also provided the performance of our proposed algorithm in filtering less significant graphs. By comparing it with the well known frequent subgraph mining algorithm gSpan, using denser synthetic dataset, used in the scalability assessment, we found our algorithm efficient enough in comparison with existing algorithms. Here, we have considered the support threshold 5% and confidence threshold 50%. The comparison can be found in Figure 9. Figure 10 contains the percentage of graphs, those are un-correlated, filtered by CGM with respect to the graphs selected by gSpan. Figure 10, shows that the filtering percentage of CGM can be from 10% up to 40%, on a real life data set MOLT-4, for various $gConfidence$ threshold.

4   Conclusions

Mining frequent patterns or sub-patterns with larger support threshold could miss some interesting patterns. At the same time if the threshold considered is small enough to capture such rare but interesting items could generate lots of spurious patterns. Therefore, association and correlation analysis is used in association of frequent pattern mining for mining frequent-interesting patterns from a collection of itemsets, but correlation searching is a challenging task. In a graph database, correlation searching is more challenging due to the fact that searching frequent subgraphs faces the graph isomorphism problem. Therefore, a new measure $gConfidence$ and an algorithm CGM are proposed for correlation mining in graph databases which can capture more interesting inherent correlation among graphs. $gConfidence$ has downward closure property, which helps in pruning descendants of non-correlated candidates. Proposed method constructs a tree-like search space named $gConfidence$ tree to efficiently mine the correlation. We have performed extensive performance analysis of CGM and found it efficient enough which outperforms existing works in correlation search based on speed. The proposed algorithm can be applied in both traditional and non-traditional domains such as bio-informatics, computer vision, various networks, machine learning, chemical domain and various other real life domains.

References

1. Inokuchi, A., Washio, T., Motoda, H.: An apriori-based algorithm for mining frequent substructures from graph data. In: Zighed, D.A., Komorowski, J., Żytkow, J.M. (eds.) PKDD 2000. LNCS (LNAI), vol. 1910, pp. 13–23. Springer, Heidelberg (2000)
2. Kuramochi, M., Karypis, G.: Frequent subgraph discovery. In: ICDM, pp. 313–320. IEEE Computer Society (2001)
3. Yan, X., Han, J.: gSpan: Graph-based substructure pattern mining. In: ICDM, pp. 721–724. IEEE Computer Society (2002)
4. Li, J., Liu, Y., Gao, H.: Efficient algorithms for summarizing graph patterns. IEEE Trans. Knowl. Data Eng. 23(9), 1388–1405 (2011)
5. Tan, P.N., Kumar, V., Srivastava, J.: Selecting the right interestingness measure for association patterns. In: KDD, pp. 32–41. ACM (2002)
6. Yan, X., Zhu, F., Yu, P.S., Han, J.: Feature-based similarity search in graph structures. ACM Trans. Database Syst. 31(4), 1418–1453 (2006)
7. Ke, Y., Cheng, J., Ng, W.: Efficient correlation search from graph databases. IEEE Trans. Knowl. Data Eng. 20(12), 1601–1615 (2008)
8. Pubchem web site for information on biological activities of small molecules (2011), http://pubchem.ncbi.nlm.nih.gov

Improved Parallel Processing
of Massive De Bruijn Graph for Genome Assembly

Li Zeng[1], Jiefeng Cheng[1,*], Jintao Meng[1,2,3], Bingqiang Wang[4],
and Shengzhong Feng[1]

[1] Shenzhen Institutes of Advanced Technology, CAS, Shenzhen, 518055, P.R. China
[2] Institute of Computing Technology, CAS, Beijing, 100190, P.R. China
[3] Graduate University of Chinese Academy of Sciences, Beijing, 100049, China
{li.zeng,jf.cheng,jt.meng,sz.feng}@siat.ac.cn
[4] Beijing Genomics Institute, Shenzhen, 518083, China
wangbingqiang@genomics.cn

Abstract. De Bruijn graph is a vastly used technique for developing genome
assembly software nowadays. The scale of this kind of graph can reach billions of
vertices and edges which poses great challenges to the genome assembly task. It is
of great importance to study scalable genome assembly algorithms in order to
cope with this situation. Despite some recent works which begin to address the
scalability problem with parallel assembly algorithms, massive De Bruijn graph
processing is still very time consuming which needs optimized operations. In this
paper, we aim to significantly improve the efficiency of massive De Bruijn graph
processing. Specifically, the time consuming and memory intensive processing are
the De Bruijn graph construction phase and the simplification phase. We observe
that the existing list ranking approach repeatedly performs parallel global sorting
over all De Bruijn graph vertices, which results in a huge amount of
communications between computing nodes. Therefore, we propose to use depth-
first traversal over the underlying De Bruijn graph once to achieve the same
objective as the existing list ranking approach. The new method is fast, effective
and can be executed in parallel. It has a computing complexity of $O(g/p)$ and
communication complexity of $O(g)$, which is smaller than the existing list
ranking approach, here g is the length of genome reference, p is the number of
processors. Our experimental results using error-free data show that, when the
number of processors scales from 8 to 128, our algorithm has a speedup of 10
times on processing simulated data of Yeast and C.elegans.

Keywords: Parallelized Graph Algorithm, De Bruijn Graph, Genome
Assembler.

1   Introduction

Whole Genome sequencing is one of the most fundamental problems in Biology.
Since 1977, when Sanger sequencing method appeared, tens of thousands of genome

---

* Corresponding author.

Y. Ishikawa et al. (Eds.): APWeb 2013, LNCS 7808, pp. 96–107, 2013.

sequence has been sequenced. Due to the diversity of species, there are still a large number of species that need to be sequenced. Moreover, modern medical research shows that most of the diseases and gene have connections. Genome sequencing results can lead to reveal the mystery of genetic variation, and is now widely used in genetic diagnosis, gene therapy and drug design, etc.[4].

Whole Genome sequencing has two main steps. The first step is to obtain the sequence of short reads. It needs amplifying the DNA molecule (multiples of amplification is referred to as coverage) first, and then these DNA molecules are randomly broken into many small fragments (each small fragment is referred to as a read), the short read is then determined by sequencing devices, which is illustrated in Figure 1. The second step is to reconstruct genome sequence from these short reads, which is usually referred as *genome assembly*. Genome assembly problem is proved to be is NP-hard [1], reduced from the Shortest Common Superstring(SCS) problem.

break the DNA into short fragments randomly

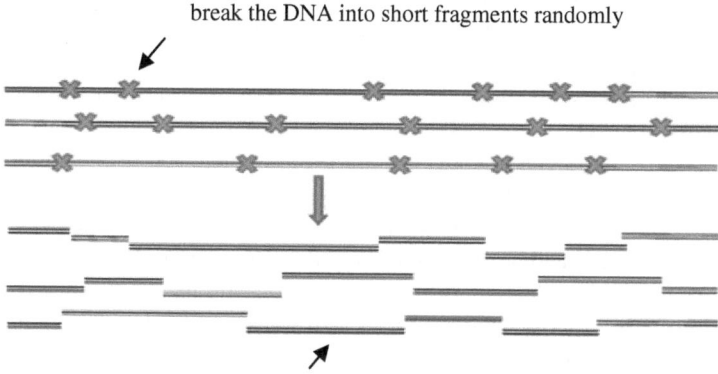

read, sequenced by sequencing instrument

Fig. 1. DNA and its reads

There are two types of sequence assembly algorithms. one is called overlap graph model [2]. On an overlap graph, each vertex corresponds to a read. There is a directed edge between two reads if overlap exists between them and the overlapping length exceeds a certain threshold. Therefore, sequence assembly problem is converted into finding a Hamilton path through each node in the overlap graph, which is NP-hard [2]. Another model is based on De Bruijn graph [3]. In a De Bruijn graph, each read is cut into small fragments of length k, called *k*-mer. Each *k*-mer contributes a node in the graph. If there are *k*-1 characters overlapping between two adjacent *k*-mers, then a directed edge exists between the two corresponding nodes. In this way, each read can produce a corresponding path in the graph. So the sequence assembly problem has been transformed into finding the shortest tour of the De Bruijn graph that include all the read path [4]. The problem is extremely difficult because repeat fragments of various lengths exist in the original sequence. Also, sequencing machines will introduce errors into these reads, where the error rate is about 1%~3% in modern sequencing machines. These two issues make the sequence assembly more complicated.

Many optimization criteria and heuristics have been presented to deal with assembly problem. Early, the length of fragments obtained by the Sanger sequencing method can be achieved 200bp (base pairs). Overlap graph model algorithm is more effective, but the cost of sequencing is relatively high. For example, the Human Genome Project, by the completion of the first-generation sequencing technology, already spent $3 billion in 3 years. More recently, genome sequencing entered in a era of large-scale applications due to the second-generation sequencing technology (also known as the next-generation sequencing technology), such as Solexa, 454, SOLID [5]. There are three prominent features of the second-generation sequencing technology, namely, high throughput, short sequence and high coverage. For high-throughput sequencing, a sequencing machine can simultaneously sequence millions of reads, which greatly reduces the whole cost. For short sequence, it stands for that the sequence length is between 25-80 bp in general. In this regard, genome assembly has to greatly increase DNA coverage [5] to ensure the integrity of the information. With the increasing of the coverage, the total number of reads also increases greatly. For overlap graph, then the size of the graph will significantly increase accordingly. But for De Bruijn graph, the size of the graph is only linear to the length of the reference genome. Therefore, in the face of the second generation of gene sequencing technology, De Bruijn graph model has a great advantage and become much popular. The proposed algorithms based n De Bruijn graph includes Euler[3], ALLPATHS[6], Velvet[7], IDBA[8], SOAPdenovo[9], ABySS[10] and YAGA[11]. Euler, ALLPATHS, Velvet and IDBA algorithms are serial algorithms, only suitable for small data sets. SOAPdenovo is multi-threaded SMP mainframe-based algorithm, which can process a large data set. But it also spends more than 40 hours on human genome [9] and the required infrastructure is very expensive.

The main phases of genome assembly based on De Bruijn graph is as follows: First, De Bruijn graph construction. Second, errors correction. To remove the known erroneous structures in the De Bruijn graph. Third, graph simplification. It is to compress a single chain in the De Bruijn graph into a single compact node, which also generates preliminary contig information. Fourth, generating scaffold. It is to merge contigs from the previous step with pair-end information obtained by sequencing, then to derive longer contigs (known as scaffold). Fifth, to output the obtained scaffold. Generally, the first phase has the largest memory consumption, and the second phase and the third phase has the largest time expense. Initially, the graph after constructed is extremely sparse, chain nodes ratio is more than 99% [12].

YAGA[11] is a distributed algorithm with well parallelism and scalability which represent state of the art. In this paper, we improve the parallel processing of De Bruijn graph by constructing a De Bruijn graph in distributed manner for the construction phase and then a parallel graph traversal to obtain all chains for the simplification phase. After assuming that the error has been removed, our work focuses on improving the efficiency of parallel graph simplification. The main contribution of this paper is to formulate the graph simplification problem into graph traversal problem. In YAGA algorithm, Graph simplification is approached as a list

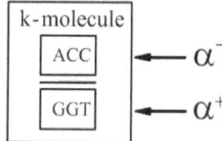

Fig. 2. k-mer and *k*-molecule

ranking problem with a total of four times global sorting and a multi-round recursion. Our algorithm directly traverses from end node to access all chains and visit each node only once. Our algorithm has a computing complexity of $O(g/p)$ and communication complexity of $O(g)$, which is smaller than YAGA algorithm, here g is the length of genome reference, p is the number of processors.

Section 2 gives the definition of De Bruijn and the detailed description on graph construction. Section 3 describes parallel graph simplification. We show experimental results in section 4. Section 5 concludes this paper.

2   Parallel De Bruijn Graph Construction

We first explain the definition of De Bruijn graph. Then, in order to achieve efficient parallel De Bruijn graph construction, we discuss a compact representation of De Bruijn graph. Finally, a parallel De Bruijn graph construction is given, which distributes the storage of the graph across multiple computational nodes.

2.1   Defining De Bruijn Graph

Let s be a DNA sequence of length n, which is a sequence consists of the four bases, namely A, C, T and G. A substring derived from s with length k, represented as $s[j]s[j + 1] \ldots s[j + k - 1], n - k + 1 \geq j \geq 0$, is called as a k-mer. The set of all k-mers derived from s is referred to as the k-spectrum of s. For a k-mer α, $\bar{\alpha}$ represents the reverse sequence of α, s complementary sequence (for example $\alpha = AAGTA$, then $\bar{\alpha} = TACTT$). $\bar{\alpha}$ can be obtained by first complementing each character in alpha and then reversing and the resulted sequence. The complementary rules are $\sum: \{A \to T, T \to A, C \to G, G \to C\}$.

A k-molecule represents a pair of k-mers $\{\alpha, \bar{\alpha}\}$, and α and $\bar{\alpha}$ is reverse complementary. The notation $>$ is the ordering relation between the two k-mers with length k, such that $\alpha > \bar{\alpha}$ indicates that α is larger than $\bar{\alpha}$ lexicographically. We designate the lexicographically larger one as the positive k-mer , denoted α^+, and the other one as negative k-mer , denoted α^- (as shown in Figure 2), here there is $\alpha > \bar{\alpha}$. When the context is clear, α^+ stands for a k-molecule $\{\alpha^+, \alpha^-\}$, that is $\alpha^+ = \{\alpha^+, \alpha^-\} = \{\alpha, \alpha^-\}$. Fig.2 depicts the relationship between a k-mer and its corresponding k-molecule with an example.

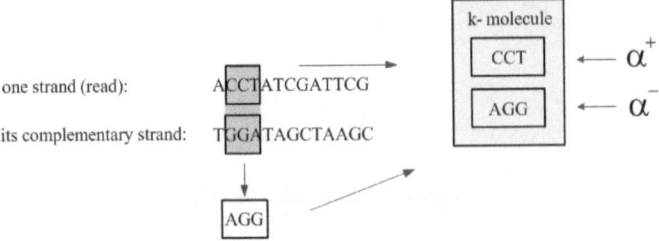

Fig. 3. *k*-molecule from read

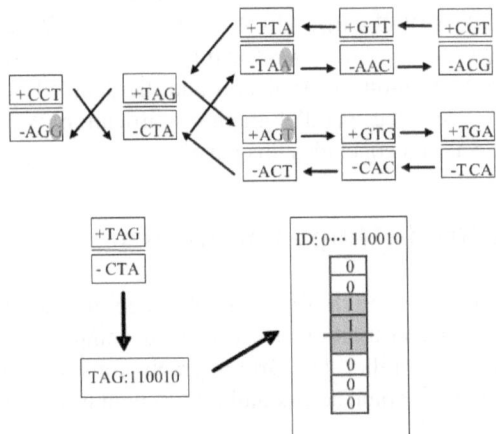

Fig. 4. The structure of De Bruijn graph node

2.2   Structure of De Bruijn Graph

The node structure in our De Bruijn graph is different from the existing one of YAGA algorithm [11]. Specifically, we design a compacted representation of De Bruijn graph, where an ID field and an arc field are devised for every node. And this node structure not only uniquely identifies a node by the ID field, it also encodes all incident edges for that node in the arc field. Therefore, our De Bruijn graph representation does not store edges explicitly.

Let's consider a De Bruijn graph node, which is a *k*-molecule (as shown in Figure 3). We compute a 64 bit unsigned integer based on the sequence of all characters in the positive *k*-mer using the rule {A: 00, C: 01 G: 10 T: 11} (shown in Figure 4). In turn, the 64 bit unsigned integer forms the ID field (or simply ID) of the node. Since we have a maximum value for *k* as 31, therefore, all characters in the positive *k*-mer can at most encode the low 2k bits of the ID. And the remaining high bits are filled with 0.

All edges are stored in the arc field of the node. Because DNA consists of four bases {A, C, G, T}, a node can connect to eight adjacent nodes at most, corresponding to eight edges. Since the two adjacent nodes have *k*-1 characters overlapped in their

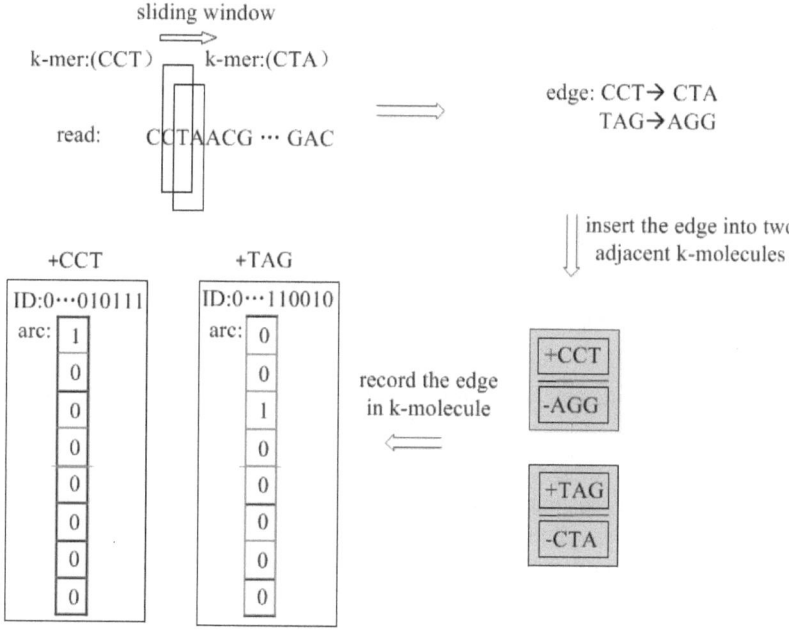

Fig. 5. Obtaining all *k*-mers

k-mers, positive or negative, only one different character can be found over the two adjacent nodes. So we can simply use this character to represent the corresponding edge. We use an 8-bit char type variable to store these edges. The bit with value of 1 indicates that the existence of this edge, otherwise, it means there is no corresponding edge. The rule is: the top 4 bits represent the edges of positive *k*-mer, 0 to 3 corresponding to the A C G T; the after 4 bits represent the edges of negative *k*-mer, 4 to 7 corresponding to A C G T. For example, the node TAG owns three edges connected to three nodes (TTA AGT and TAG), as defined above, we use the positive *k*-mer TAG to represent the node. We set the third bit of arc field to 1. It means that an edge exist from the positive *k*-mer(TAG) to the negative *k*-mer(AGG) of node CCT, for the number of the bit is the third and in the top 4. The edges record of node TAG is shown in Figure 4. The bits of the shaded part in Figure 4 with value of 1 represent the existence of these edges. Accordingly, only 9 bytes is used to store one node in our De Bruijn graph.

2.3   Parallel De Bruijn Graph Construction Algorithm

The parallel De Bruijn graph construction algorithm has three main steps. The first step is to read all short sequences in parallel; the second step is to obtain all *k*-mers and the edges between any two consecutive *k*-mers, and to obtain all *k*-molecules based on those *k*-mers, repeated edges in a *k*-molecule are merged. The third step to

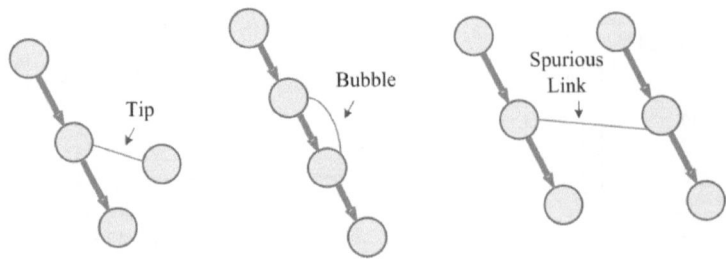

Fig. 6. Three types of erroneous connections

distribute those k-molecules in a number of computational nodes. The details are given below.

Step 1: Read short sequences in parallel.

Each processor is assigned with different blocks of short sequences according to the rank of processor and the file size of short sequences. Assume that the length of the file is L and there are n processors, then the i-th processor get short sequences from the $((i-1)*L)/n$ to $(i*L)/n$.

Step 2: Parallel segmentation of k-mers (see Figure 5).

Each processor cuts all short reads into k-mers by a sliding window with length k from the first k-mer to the last k-mer. Then, each k-mer is handled as follows:

a)   By the rules of complementation, we get all reverse complementary k-mers of the existing k-mers. Moreover, we obtain all k-molecule $\{\alpha^+, \alpha^-\}$ based on those k-mers. The lexicographically larger k-mer is encoded by the rule {A: 00, C: 01 G: 10 T: 11} as the ID of the k-molecule.

b)   Record edges between adjacent k-molecules. Form the compact De Bruijn graph node representation for all k-molecules.

Step 3: Distribute all k-molecules.

Each processor stores a partition of the De Bruijn graph. All processors are assigned with a certain portion of De Bruijn graph nodes with a hash function over the 64-bit node representation. Specifically, the location of a De Bruijn graph is computed by taking the ID of the k-molecule mod of the number of processors. So, all k-molecule are mapped to different processors by an all-to-all communication operation. After this step, any processor can computed a node directly by the same hash function. This is to boost the performance of the following graph simplification phase.

3   Parallel Simplification of De Bruijn Graph

In this section, we first explain the problem of De Bruijn simplification. Then, we discuss our parallel De Bruijn graph simplification algorithm based on concurrent graph traversal starting from multiple sources in the graph.

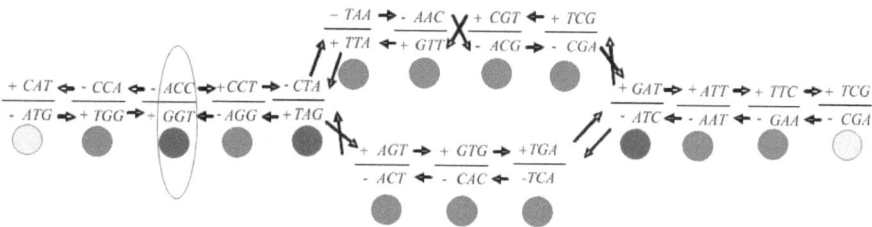

Fig. 7. De Bruijn graph in our algorithm

3.1   The De Bruijn Graph Simplification Problem

There can be a lot of sequencing errors introduced by gene sequencing devices. Therefore, the De Bruijn graph obtained by the construction algorithm can have a large number of erroneous paths, which give rise to three types of *erroneous connections* in the graph, namely, *tip*, *bubble* and *spurious link*, which are depicted in Figure 6. Therefore, after the De Bruijn graph construction, the immediate processing is to remove those erroneous connections using the knowledge from Biology study [12]. After this, the De Bruijn graph is error-free and very sparse such that it contains a large number of chains (as shown in Figure 7) [13]. Therefore, an important processing is to simplify the De Bruijn graph by finding all chains in the graph. After simplification, the size of the graph can be reduced sharply, almost about 90% off and the graph can be processed for latter tasks such as processing branches and repeats in a single machine easily.

Since our focus in this paper is an efficient and effective parallel De Bruijn graph simplification algorithm, we assume the processing to remove erroneous connections is completed. So, in the following of this paper, our discussion is on the De Bruijn graph without erroneous connections, that is, the error-free data.   First, we give a few notations and a detailed description of the problem. We call a node with a degree of 1 as the *1 degree node*. A node with degree of 2, by an incoming edge and an outgoing edge, is called as the *chain node*. The chain node can lead to a chain. all nodes with a degree of 2, by two incoming edges or two outgoing edges, as well as all nodes with the degree larger than 3 is *bifurcation node*. A *chain* is a simple path in the graph containing chain nodes only. Nodes at both ends of the chain are called as *end node*. Although an end node is not part of the chain, the chain can be obtained by traversal from it.

In Figure 7, the green nodes are chain nodes, which are the target of simplification. The red node is bifurcation node and the yellow is 1 degree node, both of them are end node. Although the degree of the circled node in Figure 7 is 2, it cannot be traversed for it has two incoming edges or two outgoing edges.

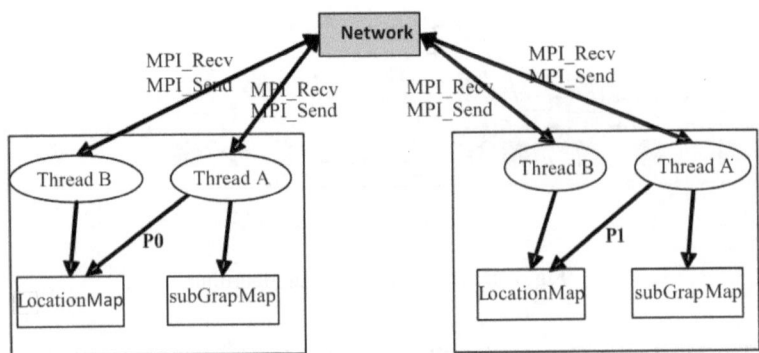

Fig. 8. The merging thread and the communication thread

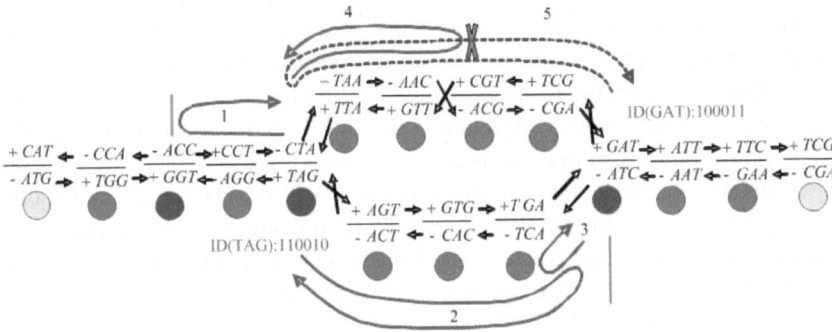

Fig. 9. Traversing De Bruijn graph

3.2   Parallel De Bruijn Graph Simplification Algorithm

Overview. Our algorithm directly traverses from an end node of a chain to the other end node to simplify the chain.

Therefore, we use two threads for each processor, which collectively performs the concurrent graph traversal to identify all chains. As shown in Figure 8, thread A is responsible for the simplification task as to merge all chain nodes identified over a chain, on the other hand, thread B is dedicated for inter-process communications. The communication is for resolving local requests and dealing with other processor's requests on traversing end nodes.

Data Structure. Each processor has two Maps. One is locationMap, which store nodes before simplification. Another is subGraphMap which stores nodes after simplification. During initialization, the subGraphMap is empty and the subGraphMap is inserted with nodes during the simplification by the process of identifying chains.

Algorithm. Based on the above data structure, we outline the algorithm as below:

Step 1: Initialization: De Bruijn graph is distributed on all processor and each processor has a locationMap and a subGraphMap.

Step 2: Traversal of the graph (as shown in Figure 9): Every processor starting to traverse from an end node in local locationMap. Since a chain has two ends nodes, traversal can be divided into the following three cases: 1) a processor visits a same chain twice from the two end nodes; 2) two different processors visit the same chain twice, each one from a different end node; 3) two processors visit the same chain simultaneously by starting from the two end node at almost the same time. In order to reduce repeated access to the same chain, we add a endNodeID variable to store the ID of end node in the direction of the end. According to the ID number, the merging thread can determine whether to continue or exit.

Step 3: Simplification: Delete all chain nodes and append their information to the corresponding end node, and then insert them into subGraphMap.

Step 4: Select another end node to repeat the processing from Step 2 until no end node can be found.

3.3   Analysis

YAGA algorithm transforms the problem of simplifying all chains into list ranking problem which requires four times of global sorting over all graph nodes. In this procedure, a large number of nodes are moved four times. The time and communication cost is very large. The computational complexity of four times global sorting is $O((g/p)\ln(g/p))$, and the communication complexity is $O(g)$. In addition to the cost of four times of global sort, the cost of other involved processing is also very large.

In our algorithm, each processor directly traverses from the local end node to get all chains. A chain only holds two end nodes, so each chain node is accessed twice at most. Also our algorithm successfully prevents a chain from being visited twice. So, in the worst case, all nodes are calculated twice, the time complexity is $O(g/p)$. And each node is requested twice at most, the communication complexity is $O(g)$.

4   Performance Evaluation

Our assembly software is written in C++ and MPI. The experimental platforms includes 10 servers interconnected by InfiniBand network, which is configured as 16-core, 32G shared memory. We choose Yeast and C.elegans genomes, using Perl scripts automatically generated yeast test data with 17, 007, 362 reads and C.elegans test data, which consists of 140, 396, 108 reads. The length of reads ranges from 36bp to 50bp with error rate of 0, and the coverage for both data is 50X. Our goal is to demonstrate the scalability of our assembler to graph construction and graph compaction on a large distributed De Bruijn graph.

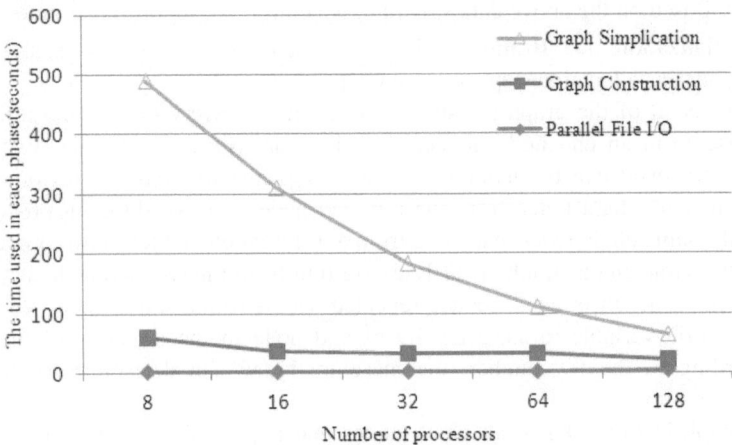

Fig. 10. The time used in three phases on Yeast dataset

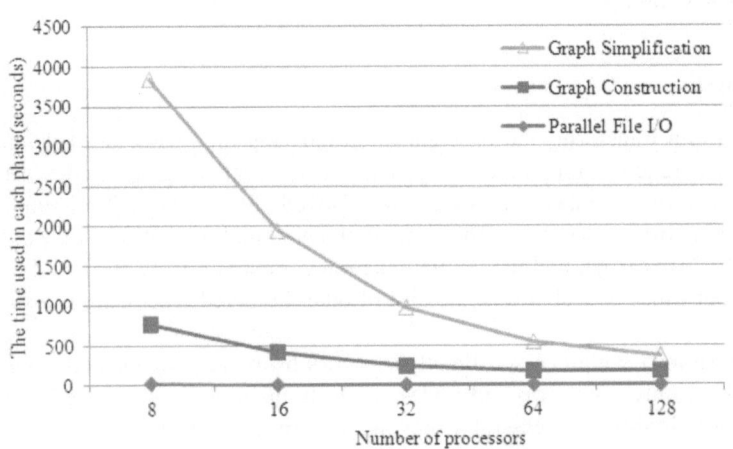

Fig. 11. The time used in three phases on C.elegans dataset

Firstly, we tested assembler on the Yeast data set. The runtime is displayed in Figure 10 and the time is divided into three phrases. First, parallel read file from a distributed file system. Second, graph construction. Third, graph compaction. The most part time on the third phase is network transmission cost. But, in our algorithms most nodes only be moved once, that is an efficient optimization. Figure 10 indicates that the algorithm has good scalability. When the number of processors is increased from 8 to 128, the total time is reduced from 490s to 63s that means the speedup is about 8 times.

The C.elegans dataset is 10 times larger than Yeast dataset. The running time is shown in Figure 11. When the number of processors is increased from 8 to 128, the total time is reduced from 3844s to 375s with the speedup being 10 times.

5   Conclusion

We propose to use depth-first traversal over the underlying De Bruijn graph once to simplify it. The new method is fast, effective and can be executed in parallel. By testing the two data sets, the experimental results show that the algorithm has great scalability. After the completion of the graph simplification, the scale of the graph is sharply reduced, thus the graph simplification is of great value. The latter stage includes resolving branches and repeats using pair-end information in a single machine. We will study these problems in our future work.

6   Conclusion

This work is supported by NSFC of China (Grant No. 61103049) and Shenzhen Internet Industry Development Fund (Grant No. JC201005270342A). The authors would like to thank the anonymous reviewers for their helpful comments.

References

[1]   Kundeti, V.K., Rajasekaran, S., Dinh, H., et al.: Efficient parallel and out of core algorithms for constructing large bi-directed de Bruijn graphs. BMC Bioinformatics 11(560) (2010)

[2]   Kececioglu, J.D., Myers, E.W.: Combinatorial algorithms for DNA sequence assembly. Algorithmica 13(1), 7–51 (1995)

[3]   Pevzner, P.A., Tang, H., Waterman, M.S.: An Eulerian path approach to DNA fragment assembly. Proceedings of the National Academy of Sciences of the United States of America 98(17), 9748–9753 (2001)

[4]   Medvedev, P., Georgiou, K., Myers, G., Brudno, M.: Computability of models for sequence assembly. In: Giancarlo, R., Hannenhalli, S. (eds.) WABI 2007. LNCS (LNBI), vol. 4645, pp. 289–301. Springer, Heidelberg (2007)

[5]   Jackson, B.G., Aluru, S.: Parallel Construction of Bidirected String Graphs for Genome Assembly, 346–353 (2008)

[6]   Butler, J., Maccallum, I., Kleber, M., et al.: ALLPATHS: de novo assembly of whole-genome shotgun microreads. Genome Res. 18(5), 810–820 (2008)

[7]   Zerbino, D.R., Birney, E.: Velvet: algorithms for de novo short read assembly using de Bruijn graphs. Genome Res. 18(5), 821–829 (2008)

[8]   Peng, Y., Leung, H.C.M., Yiu, S.M., Chin, F.Y.L.: IDBA–A Practical Iterative de Bruijn Graph De Novo Assembler. In: Berger, B. (ed.) RECOMB 2010. LNCS, vol. 6044, pp. 426–440. Springer, Heidelberg (2010)

[9]   Li, R., Zhu, H., Ruan, J., et al.: De novo assembly of human genomes with massively parallel short read sequencing. Genome Res. 20(2), 265–272 (2010)

[10] Simpson, J.T., Wong, K., Jackman, S.D., et al.: ABySS: a parallel assembler for short read sequence data. Genome Res. 19(6), 1117–1123 (2009)

[11] Jackson, B., Regennitter, M., Yang, X., et al.: Parallel de novo assembly of large genomes from high-throughput short reads. IEEE (2010)

B3Clustering: Identifying Protein Complexes from Protein-Protein Interaction Network

Eunjung Chin and Jia Zhu

School of ITEE, The University of Queensland, Australia
{eunjung,jiazhu}@itee.uq.edu.au

Abstract. Cluster analysis is one of most important challenges for data mining in the modern Biology. The advance of experimental technologies have produced large amount of binary protein-protein interaction data, but it is hard to find protein complexes in vitro. We introduce new algorithm called *B3Clustering* which detects densely connected subgraphs from the complicated and noisy graph.

B3Clustering finds clusters by adjusting the density of subgraphs to be flexible according to its size, because the more vertices the cluster has, the less dense it becomes. B3Clustering bisects the paths with distance of 3 into two groups to select vertices from each group. We experiment B3Clustering and two other clustering methods in three different PPI networks. Then, we compare the resultant clusters from each method with benchmark complexes called CYC2008. The experimental result supports the efficiency and robustness of B3Clustering for protein complex prediction in PPI networks.

1 Introduction

Protein complexes play essential roles in biological process and they are also important to discover drugs in pharmaceutical process. Recent progress in experimental technologies has led to an unprecedented growth in binary protein-protein interaction (PPI) data. Despite this accumulation of PPI data, considerably small amount of protein complexes are identified in vitro. Thus, computational approaches are used to overcome the technological limit for detecting protein complexes. It becomes one of the most important challenges for modern biology to identify protein complexes from binary PPI data. Proteins are represented as vertices and their interactions are represented as edges in the PPI network. Since protein complexes are shown as densely connected subgraphs in PPI network, clustering methods are applied to find them from the network.

For example, some used clique finding algorithms to predict protein complexes from PPI network (Spirin and Mirny [23] and Liu et al.[18]). They devised their own methods to merge overlapping cliques.

Besides, Van Dongen (2000)[27] introduced MCL (Markov Clustering) as graph partitioning method by simulating random walks. It used two operators called expansion and inflation, which boosts strong connections and demotes weak connections. Brohee et al. (2006)[6] shows the robustness of MCL with comparison to three other clustering algorithms (RNSC[12], SPC[5], MCODE[4]) for detecting protein complexes.

One of the recent emerging techniques is to use protein core attachments methods, which identify protein-complex cores and add attachments into these cores to form protein complexes (CORE[15] and COACH[28]). Li et al.(2010)[17] shows that COACH

Y. Ishikawa et al. (Eds.): APWeb 2013, LNCS 7808, pp. 108–119, 2013.

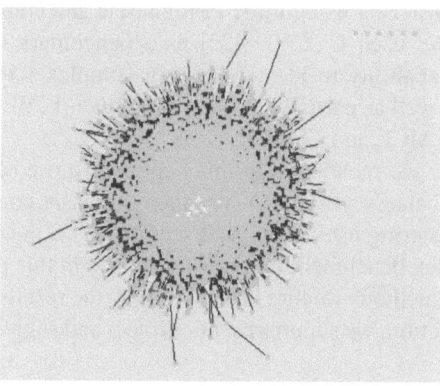

Fig. 1. The graph comes from the Biogrid database using Cytoscape

performs better than seven other clustering algorithms (MCODE[4], RNSC[12], MCL[27], DPClus[3], CFinder[1], DECAFF[16], and CORE[15]) when they are applied to two PPI datasets (DIP[30] and Krogan et al.[13] data).

PPI data has increased by about forty times for the last six years from 50,726(July 2006) to 2,013,176(April 2012) in Biogrid database[25] for Sacchromyces Cerevisiae called baker's yeast, while number of yeast proteins has grown by just 1.2 times from 5,300 to 6,226 for the same period. It means that PPI networks become more complicated and tightly interconnected. For example, DIP and Krogan data have average degrees of 6.98 and 7.86, respectively, but Biogrid 3.1.85 has an average degree of 62.22 for Sacchromyces Cerevisiae.

Even though the clustering methods which we mentioned before prove their competency in the small size of PPI data (DIP and Krogan data), they show poor performance in the large size of tightly interconnected PPI network (Biogrid 3.1.85). The existing clustering methods are not suitable to detect protein complexes from the tightly interconnected network such as Biogrid 3.1.85. Normally, they has been experimented with the yeast datasets produced before 2006.

Since Biogrid 3.1.85 is densely connected and hard to partition as shown in Fig.1, the existing algorithms do not perform very well. MCL has low F-measure of 0.0261 by producing 52 clusters for Biogrid 3.1.85 and one of the clusters has included 5,786 proteins out of 6,194 total proteins in PPI network. COACH has predicted a large number of clusters as many as 3,138 and it also shows low F-measure of 0.055 for Biogrid 3.1.85 Scchromyces Cerevisiae.

To detect clusters in such a tightly interconnected network, we introduce new algorithm called *B3Clustering* which finds densely connected subgraphs from tightly interconnected PPI network. B3Clustering tries to get protein complexes from each pair of proteins by bisecting vertices of paths with distance of 3 into two groups and calculating the probability of interaction between them. It chooses vertices with the probability of interaction and check how much two groups have common vertices. If there remain vertices in each group and two groups overlap more than the given threshold, then we identify the union of two groups as a complex.

We evaluated B3Clustering using three PPI datasets and compared the results with MCL and COACH. We used CYC2008[22] as a benchmark of protein complexes. B3Clustering shows capability to identify protein complexes in the tightly interconnected PPI network as well as partially missing PPI network. We found that it achieved higher F-measure than MCL and COACH.

All the proteins that we mentioned in this paper are those of Sacchromyces Cerevisiae, so to speak, a baker's yeast. We introduce the basic concepts and explain the methods which B3Clustering uses in the next section. In section 3, we give full details of PPI data and complex benchmark data which we use in this paper. In section 4, we explain the evaluation methods we hire here. We show the results of our experiments in section 5. In the last section, we summarize our project and suggest future improvement of B3Clustering.

2   Method

Protein-protein interaction(PPI) data come in the form of connections between proteins, which is easily described as a graph model. Proteins are represented as vertices and their interactions are represented as edges in the graph. Since protein complexes are shown as densely connected subgraphs in PPI network, clustering methods are applied to find them from the PPI network. We introduce B3Clustering to identify clusters in the graph and we apply it to detect protein complexes in the PPI network.

B3Clustering proceeds clustering in three steps. Firstly, it selects vertices and we call it vertex selection step. Secondly, it decides if the vertices can form clusters or not and it is called cluster selection step. Lastly, it merges clusters, which is called trimming step. We introduce terminology we use before we explain three steps.

2.1   Terminology

For graph $G = (V, E)$, $V(G)$ and $E(G)$ denote the sets of vertices and edges of G, respectively. We assume that graph G is an undirected and simple graph. One of important objects in B3Clustering is a path with distance of 3. A path in a graph is a sequence of vertices which are connected to the next vertex in the sequence. We use paths with distinct vertices in B3Clustering.

Definition 1   (3-Path Reachable). *Let a graph G on a vertex set $V(G)$ be an undirected and simple graph with an edge set $E(G)$. Given two vertices $u, v \in V(G)$, we say that u is 3-path reachable to v if there exists a path with two distinct internal vertices and three different edges between u and v. We also say that the path is 3-path reachable from u to v.*

That is, u is 3-path reachable to v if there exists a path whose distance is 3.

Definition 2   (3-Path Set). $Path^3(u, v)$ *is defined as the set of paths 3-path reachable from u to v.*

$$Path^3(u,v) = \{ \{u,i,j,v\} \mid (u,i),(i,j),(j,v) \in E(G), u \neq i \neq j \neq v, \text{ for } i,j \in V(G) \}$$

We assume that u, i, j, v are member of $V(G)$ and they should be different from each other. $Path^3(u, v)$ could be empty set if u is not 3-path reachable to v. Otherwise, it will have paths with distance of 3 as members.

Definition 3 (Bi-sets). $D_u(u, v)$ *is defined as the set of* u, v, *and vertices which are linked to* u *in paths* $\in Path^3(u, v)$. $D_v(u, v)$ *is defined as the set of* u, v, *and vertices which are linked to* v *in paths* $\in Path^3(u, v)$. $D_u(u, v)$ *and* $D_v(u, v)$ *are called* Bi-sets *of* $Path^3(u, v)$.

$$D_u(u,v) = \{ u,v,i \mid (u,i) \in E(G), \{u,v,i\} \subset z, \text{for any } z \subset Path^3(u,v) \}$$

$$D_v(u,v) = \{ u,v,j \mid (j,v) \in E(G), \{u,v,j\} \subset z, \text{for any } z \subset Path^3(u,v) \}$$

We extracts two sets of vertices from 3-path set and call them $D_u(u, v)$ and $D_v(u, v)$. Each set includes u and v. Additionally, $D_u(u, v)$ includes the set of vertices which are linked to u in paths $\in Path^3(u, v)$ and $D_v(u, v)$ includes the set of vertices which are linked to v in paths $\in Path^3(u, v)$.

2.2 Vertex Selection Step

B3Clustering detects clusters from every pair of vertices in PPI network. It has three steps to proceed clustering. The first step is vertex selection. We assume that a vertex

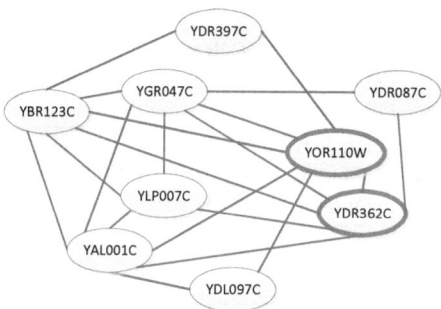

Fig. 2. Sample subgraph from krogan data

pair (u, v) is given. We can get $D_u(u, v)$ and $D_v(u, v)$ from Definition 3. Then, we calculate the probability to interact with members of other set for each vertex. We define the probability of interaction $p_u(i)$ and $p_v(j)$ as follows.

$$p_u(i) = \frac{\sum_{j \in D_v(u,v)} w(i,j)}{|D_v(u,v)|}$$

$$p_v(j) = \frac{\sum_{i \in D_u(u,v)} w(i,j)}{|D_u(u,v)|}$$

$w(i,j)$ is an edge weight between i and j. If there exists an edge between i and j, then $w(i,j)$ should be 1. If i is the same vertex with j, then $w(i,j)$ should 1. Otherwise, $w(i,j)$ should be 0.

$$w(i,j) = \begin{cases} 1, & \text{if } (i,j) \in E(G) \text{ or } i = j; \\ 0, & \text{otherwise.} \end{cases}$$

We calculate $p_u(i)$ for every vertex $i \in D_u(u,v)$ and we calculate $p_v(j)$ for every vertex $j \in D_v(u,v)$. If $p_u(i)$ is higher than σ, then i remains in $D_u(u,v)$. Otherwise, we remove i from $D_u(u,v)$. Likewise, if $p_v(j)$ is higher than σ, then v remains in $D_v(u,v)$. Otherwise, we remove j from $D_v(u,v)$.

σ is the threshold which decides lowest margin for probability of interaction. As the protein complex consist of more proteins, they show lower connectivity, Therefore, we set σ a lower value when there are more paths with distance of 3, and vice versa.

$$\sigma = \alpha \times 2^{\frac{|4 \times Path^3(u,v)|}{m}}$$

where $m = max_{i,j \in V(G)} |Path^3(u,v)|$ and α is given. We set $\alpha = 0.7$ in this paper.

Let's see Fig.2 as a example. If we start from $(YOR110W, YDR362C)$, then we get bi-sets as shown Table 1. From now on, we call $D_{YOR110W}(YOR110W, YDR362C)$ $D1$ and $D_{YDR362C}(YOR110W, YDR362C)$ $D2$ in short.

Table 1. Bi-sets of YOR110W and YDR362C before removing vertices

D1	YOR110W,	YDR362C,	YAL001C,
	YBR123C,	YGR047C,	YPL007C,
	YDR087C		
D2	YOR110W,	YDR362C,	YAL001C,
	YBR123C,	YGR047C,	YDR397C,
	YDL097C		

If we calculate the probability of interaction for each vertex in $D1$ and $D2$, each vertex has a value in Table 2.

If We set that σ is equal to 0.5, then YDR087C is removed from D1 and YDL397C and YDL097C are removed from D2. Finally, we get the D1 and D2 as shown in Table 3.

2.3   Cluster Selection Step

After we get D1 and D2 after removing vertices with lower probability of interaction value than σ, we decide whether the union of D1 and D2 can be a cluster or not. We use similarity value $s(D1,D2)$ between D1 and D2 and define $s(D1,D2)$ as follows.

$$s(D1,D2) = \frac{|D1 \cap D2|}{|D1 \cup D2|}$$

If $s(D1,D2)$ is higher than γ, $D1 \cup D2$ is selected as a cluster. We set that γ is equal to 0.6 in this paper.

Table 2. Probability of interaction

$P_{YOR110W}$	$P_{YDR362C}$
p1(YOR110W)=1	p2(YOR110W)=0.714
p1(YDR362C)=0.714	p2(YDR362C)=1
p1(YAL001C)=0.857	p2(YAL001C)=0.857
p1(YBR123C)=0.857	p2(YBR123C)=0.857
p1(YGR047C)=0.714	p2(YGR047C)=1
p1(YPL007C)=0.571	p2(YDR397C)=0.281
p1(YDR087C)=0.281	p2(YDL097C)=0.281

Table 3. Bi-sets of YOR110W and YDR362C after removing vertices

D1	YOR110W, YDR362C, YAL001C, YBR123C, YGR047C, YPL007C
D2	YOR110W, YDR362C, YAL001C, YBR123C, YGR047C

Let's look at example in Table 3. Then the union of $D1$ and $D2$ has 6 members. Since $D1$ and $D2$ has 5 common vertices, $s(YOR110W, YDR362C) = 0.833$. Hence, $D1 \cup D2$ is considered as a cluster because it has higher similarity value than γ.

2.4 Merging Step

Since B3Clustering calculates similarity value from every pair of vertices, it can produce a large number of overlapping clusters. Therefore, we need to merge similar clusters together. We calculate the overlapping ratio between clusters and merge similar clusters. We define overlapping ratio $l(C1, C2)$ between cluster C1 and cluster C2 as follows.

$$l(C1, C2) = \frac{|C1 \cap C2|}{|C1 \cup C2|}$$

where θ is a marginal overlapping threshold. We set $\theta = 0.9$ in this paper.

After merging step, we get final clusters.

3 Evaluation

3.1 Data

We use three PPI datasets for *Saccharomyces cerevisiae*. One is a dataset from Krogan et al.(2006)[25] and another comes from Dip database[30]. The last dataset is biogrid 3.1.85 downloaded from BIOGRID[25] database [1].

---

[1] http://www.thebiogrid.org/

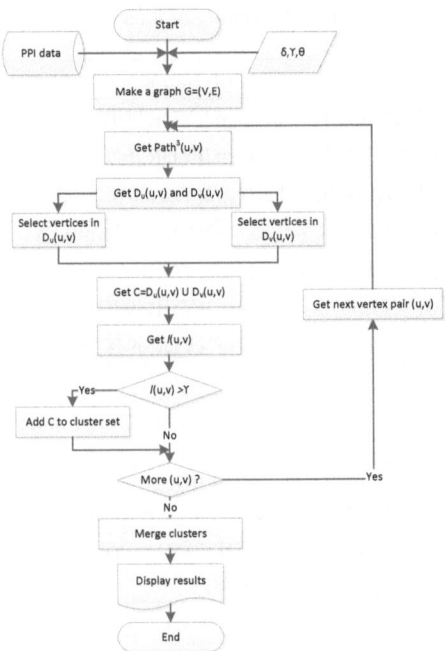

Fig. 3. Overall algorithm of B3Clustering

Krogan and dip datasets are used by Li et al.(2010) to evaluate the performance of several clustering algorithms[17]. As shown in Table 4, Krogan and dip dataset has similar number of average degree, but biogrid has much higher average degree than them. Biogrid has about ten times higher average degree then dip dataset.

PPI data have a high rate of false positives, which has been estimated to be about 50% [24]. The noise of the data disturbs clustering methods to detect protein complexes from PPI data.

We use CYC2008 complexes as a benchmark set. CYC2008 is published by Pu et al.(2008)[22] and it can be downloaded from (http://wodaklab.org/cyc2008/). CYC2008 provides a comprehensive catelogue of manually curated 408 protein complexes in Sac-charoyces cerevisiae.

CYC2008 complexes consists of 1627 proteins, which is about one quarter of proteins in Biogrid 3.1.85. Among the 408 protein complexes, 259 protein complexes are composed of less than four proteins with 97% of connectivity. As the protein complex consist of more proteins, they have lower connectivity.

There are 172 protein complexes consisting of two proteins out of 408 total protein complexes in CYC2008. Two-member complexes exist as a form of edge in a graph so that they should be chosen from edges in PPI networks. For example, we try to detect 172 protein complexes from 193,679 edges in Biogrid 3.1.85.

Table 4. Features of PPI datasets

Dataset	# Nodes	# Edges	Average degree
Biogrid(3.1.85)	6194	193976	62.22
Dip	4930	17201	6.98
Krogan et al.	3181	14076	7.86

3.2 F-Measure

Precision, recall, and F-measure have been used to evaluate the performance in information retrieval and data mining. We calculate affinity score used by Li et al.[17] to get F-measure. The neighbourhood affinity score NA(p,b) is defined as follows.

$$NA(p,b) = \frac{|V_p \cap V_b|^2}{|V_p| \cdot |V_b|}$$

where $P = (V_p, E_p)$ is a predicted complex and $B = (V_b, E_b)$ is a benchmark complex.

Precision. Precision is calculated as follows.

$$Precision = \frac{N_{cp}}{|P|}$$

where $N_{cp} = |\{p|p \in P, NA(p,b) \geq \omega, for \exists b \in B\}|$.

Recall. Recall is calculated as follows.

$$Recall = \frac{N_{cb}}{|B|}$$

where $N_{cb} = |\{b|b \in B, NA(p,b) \geq \omega, for \exists p \in P\}|$.

F-measure. F-measure is the geometric mean of Precision and recall as follows.

$$F\text{-}measure = \sqrt{Precision \times Recall}$$

4   Result

We experiment B3Clustering with three datasets. Three databases are Biogrid 3.1.85, Dip, and Krogan as describes in section 3. We experiment MCL and COACH with the same datasets to compare the performance of B3Clustering. We set σ to be 0.7, γ to be 0.6, and θ to be 0.9.

Table 5. Result from Krogan et al. dataset

Algorithms	# Clusters	N_{cp}	N_{cb}
MCL	624	135	159
COACH	570	269	155
B3Clustering	629	398	144

Table 6. Result of Dip dataset

Algorithms	# Clusters	N_{cp}	N_{cb}
MCL	835	171	186
COACH	746	311	215
B3Clustering	786	306	211

From the results, B3Clustering has the highest F-measure in Krogan data and Biogrid data, but it has lower F-measure than COACH in Dip data. COACH and B3Clustering score F-measure twice higher than MCL in Krogan and Dip datasets as shown in Fig. 4. MCL supposes to have the lowest F-measure because it partitions PPI graph. On the contrary, COACH and B3Clustering consider some of vertices and overlap clusters instead of partitioning.

For Biogrid 3.1.85 data, MCL has a F-measure of 0.0261 by producing 52 clusters, one of which includes 5,786 proteins out of 6,194 total proteins. COACH predicts a large number of clusters as many as 3,138 and only 105 are similar to protein complexes in CYC2008. Since COACH has low precision for Biogrid 3.1.85, it shows low F-measure of 0.055 as well.

B3Clustering has the similar F-measure to COACH in Krogan and Dip data, but it has six times higher F-measure than COACH in Biogrid 3.1.85.

The experimental results show that B3Clustering has the capability to detect clusters from the noisy graph as well as tightly interconnected graph. They support the efficiency and robustness of B3Clustering for protein complex prediction in PPI networks.

5   Related Works

There exist several clustering algorithms applied to biological networks in order to find protein complexes. One of traditional clustering approaches are maximal clique finding algorithm, which induces complete subgraphs that is not a subset of any larger subgraph. Clique finding algorithms have been used to predict protein complexes from PPI network (Spirin and Mirny [23] and Liu et al.[18]). They devised their own methods to merge overlapping cliques. Liu et al.[18] proposed an algorithm called CMC (Clistering-based on Maximal Cliques), which used a variant of Czekanowski-Dice distance to rank cliques and merged highly overlapped cliques. If the interaction networks

Table 7. Result of Biogrid 3.1.85

Algorithms	# Clusters	N_{cp}	N_{cb}
MCL	52	2	2
COACH	3,138	105	63
B3Clustering	467	117	52

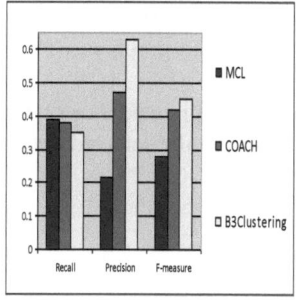 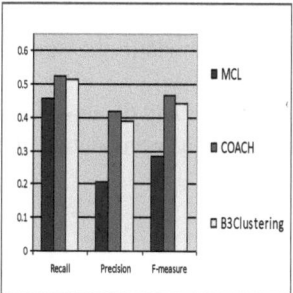 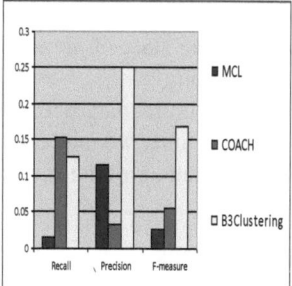

Fig. 4. Result from Krogan et al., Dip and Biogird 3.1.85 datasets

are accuate and complete, then maximal clique finding algorithm [26] can be ideal for detecting protein complexes from the networks. However, since biological data is highly noisy and incomplete, protein complexes are not formed as shapes of complete subgraphs in the biological networks. To reduce the impact of the noise in the network, Pei et al.[21] proposed using quasi-cliques instead of maximal cliques. Other approaches employ vertex similarity measures to improve confidence of noisy networks [14].

Besides, Van Dongen (2000)[27] introduced MCL (Markov Clustering) as graph partitioning method by simulating random walks. It used two operators called expansion and inflation, which boosts strong connections and demotes weak connections. Iterative expansion and inflation separate the graphs into many subgraphs. Brohée et al. (2006)[6] shows the robustness of MCL with comparison to three other clustering algorithms (RNSC[12], SPC[5], MCODE[4]) for detecting protein complexes. Each clustering algorithm was applied to binary PPI data in order to test the ability to extract complexes from the networks and the clusters were compared with the annotated MIPS complexes [20]

MCODE [4] uses a vertex-weighting method based on the clustering coefficient which assesses the local neighbourhood density of a vertex. MCODE is the overlapping clustering approach which allows to overlap clusters.

One of the recent emerging techniques is to use protein core attachments methods, which identify protein-complex cores and add attachments into these cores to form protein complexes (CORE[15] and COACH[28]). Li et al.(2010)[17] shows that COACH performs better than seven other clustering algorithms (MCODE[4], RNSC[12], MCL[27], DPClus[3], CFinder[1], DECAFF[16], and CORE[15]) when they are applied to two PPI datasets (DIP[30] and Krogan et al.[13] data).

6 Conclusion

As the advances of technologies has produced a wealth of data, data becomes complicated and erroneous. The noise of the data makes it hard to find desirable results from the overflow of data. We introduce B3Clustering to detect clusters in the complicated and noisy graph. B3Clustering controls the connectivity of clusters according to the size of clusters by two thresholds. Since protein complexes has lower connectivity with larger size, it is reasonable to control threshold according to the size of protein complexes. As we can see in the experimental results, B3Clustering performs better than the existing state-of-art clustering methods in the densely connected graph as well as noisy graph.

In the future, data size will be bigger and data may be more complicated. For saving operation time, we may consider B3Clustering detect clusters from the selected vertex pairs instead of all the vertex pairs.

References

1. Adamcsek, B., Palla, G., Farkas, I.J., Derenyi, I., Vicsek, T.: CFinder: locating cliques and overlapping modules in biological networks. Bioinformatics 22(8), 1021–1023 (2006)
2. Aloy, P., et al.: Structure-based assembly of protein complexes in yeast. Science 303(5666), 2026–2029 (2004)
3. Altaf-Ul-Amin, M., Shinbo, Y., Mihara, K., Kurokawa, K., Kanaya, S.: Development and implementation of an algorithm for detection of protein complexes in large interaction networks. BMC Bioinformatics 7, 207 (2006)
4. Bader, G., Hogue, C.: An automated method for finding molecular complexes in large protein interaction networks. MBC Bioinformatics 4, 2 (2003)
5. Blatt, M., Wiseman, S., Domany, E.: Superparamagnetic clustering of data. Phys. Rev. Lett. 76(18), 3251–3254 (1996)
6. Brohëe, S., van Helen, J.: Evaluation of clustering algorithms for protein-protein interaction networks. BMC Bioinformatics 7, 488 (2006)
7. Cho, Y., Hwang, W., Ramanthan, M., Zhang, A.: Semantic integration to identify overlapping functional modules inprotein interaction networks. BMC Bioinfotmatics 8, 265 (2007)
8. Dwight, S.S., et al.: Saccharomyces Genome Database provides secondary gene annotation using the Gene Ontology. Nucleic Acids Research 30(1), 69–72 (2002)
9. Friedel, C.C., Krumsiek, J., Zimmer, R.: Boostrapping the interactome: Unsupervised Identification of Protein Complexes in Yeast. In: Vingron, M., Wong, L. (eds.) RECOMB 2008. LNCS (LNBI), vol. 4955, pp. 3–16. Springer, Heidelberg (2008)
10. Gavin, A., Aloy, P., Grandi, P., Krause, R., Boesche, M., Marzioch, M., Rau, C., Jensen, L.J., Bastuck, S., Dumpelfeld, B., et al.: Proteome survey reveals modularity of the yeast cell machinery. Nature 440(7084), 631–636 (2006)
11. Gentleman, R., Huber, W.: Making the most of high-throughput protein-interaction data. Genome Biology 8(10), 112 (2007)
12. King, A., Przulj, N., Jurisica, I.: Protein complexes prediction via cost-based clustering. Bioinformatics 20(17), 3013–3020 (2004)
13. Krogan, N., Cagney, G., Yu, H., Zhong, G., Guo, X., et al.: Global landscape of protein complexes in the yeast Saccharomyces cerevisiae. Nature 440(7082), 637–643 (2006)
14. Leight, E., Holme, P., Newman, J.: Vertex similarity in networks. Physical Review E 73, 026120 (2006)

15. Leung, H.C., Yiu, S.M., Xiang, Q., Chin, F.Y.: Predicting Protein Complexes from PPI Data: A Core-Attachment Approach. Journal of Computational Biology 16(2), 133–144 (2009)
16. Li, X., Foo, C., Ng, S.K.: Discovering protein complexes in dense reliable neighborhoods of protein interaction networks. In: Comput. Sys. Bioinformatics Conf., pp. 157–168 (2007)
17. Li, X., Wu, M., Kwoh, C.K., Ng, S.K.: Computational approaches for detecting protein complexes from protein interaction networks: a survey. BMC Bioinformatics 11(suppl. 1), S3 (2010)
18. Liu, G.M., Chua, H.N., Wong, L.: Complex discovery from weighted PPI networks. Bioinformatics 25(15), 1891–1897 (2009)
19. Mete, M., Tang, F., Xu, X.D., Yuruk, N.: A structural approach for finding functional modules from large biological networks. BMC Bioinformatics 9(suppl. 9), S19 (2008)
20. Mewes, H.W., et al.: MIPS: Analysis and annotation of proteins from whole genomes. Nucleic Acids Res. 32(Database Issue), D41–D44 (2004)
21. Pei, J., Jiang, D., Zhang, A.: On mining cross-grph quasi-cliques. In: Proceedings of the ACM SIGKDD International Conference on Knowledge Discovery and Data Mining (2005)
22. Pu, S., Wong, J., Turner, B., Cho, E., Wodak, S.J.: Up-to-date catalogues of yeast protein complexes. Nucleic Acids Res. 37(3), 825–831 (2009)
23. Spirin, V., Mirny, L.: Protein complexes and functional modules in molecular networks. PNAS 100(21), 12123–12128 (2003)
24. Sprinzak, E., Sattah, S., Magalit, H.: How reliable are experimental protein-protein interaction data. Journal of Molecular Biology 327(5), 919–923 (2003)
25. Stark, C., Breitkreutz, B.J., Reguly, T., Boucher, L., Breitkreutz, A., Tyers, M.: Biogrid: A general Repository for Interaction Datasets. Nucleic Acids Res. 34, D535–D539 (2006)
26. Tomita, E., Tanaka, A., Takahashi, H.: The worst-case time complexity for generating all maximal cliques and computational experiments. Theoretical Computer Sciene 363, 28–42 (2006)
27. Van Dongen, S.: Graph Clustering by Flow Stimulation. University of Utrecht (2000)
28. Wu, D.D., Hu, X.: An efficient approach to detect a protein community from a seed. In: 2005 IEEE Symposium on Computational Intelligence in Bioinformatics and Computational Biology (CIBCB 2005), pp. 135–141. IEEE, La Jolla (2005)
29. Wu, M., Li, X., Kwoh, C.K., Ng, S.K.: A core-attachment based method to detect protein complexes in PPI networks. BMC Bioinformatics 10, 169 (2009)
30. Xenarios, I., Salwinski, L., Duan, X., Higney, P., Kim, S., Eisenberg, D.: DIP, the Database of Interacting Proteins: a research tool for studying cellular networks of ptoein interactions. Nucleic Acids Research 30, 303–305 (2002)

Detecting Event Rumors on Sina Weibo Automatically

Shengyun Sun[1], Hongyan Liu[2,*], Jun He[1,*], and Xiaoyong Du[1]

[1] Key Labs of Data Engineering and Knowledge Engineering, Ministry of Education, China
School of Information, Renmin University of China, 100872, China
{sun_shengyun,hejun,duyong}@ruc.edu.cn
[2] Department of Management Science and Engineering, Tsinghua University, 100084, China
hyliu@tsinghua.edu.cn

Abstract. Sina Weibo has become one of the most popular social networks in China. In the meantime, it also becomes a good place to spread various spams. Unlike previous studies on detecting spams such as ads, pornographic messages and phishing, we focus on identifying event rumors (rumors about social events), which are more harmful than other kinds of spams especially in China. To detect event rumors from enormous posts, we studied the characteristics of event rumors and extracted features which can distinguish rumors from ordinary posts. The experiments conducted on real dataset show that the new features are effective to improve the rumor classifier. Further analysis of the event rumors reveals that they can be classified into 4 different types. We propose an approach for detecting one major type, text-picture unmatched event rumors. The experiment demonstrates that this approach is well-performed.

Keywords: rumor, Sina Weibo, rumor detection, social network.

1   Introduction

Over the past several years, online social networks such as Facebook, Twitter, Renren, and Weibo have become more and more popular. Among all these sites, Sina Weibo is one of the leading micro-blogging service providers with eight times more users than Twitter [1]. It is designed as a micro-blog platform for users to communicate with their friends and keep track of hot trends. It allows users to publish micro-blogs including short text messages with no more than 140 Chinese characters, attached videos, audios and pictures to show their opinions and interests. All these micro-blogs published by users will appear in their followers' pages. Also users can make comments on or retweet others' micro-blogs to express their view.

Weibo is a kind of convenient sites for users to show themselves and communicate with others. It is put into service in 2009 by Sina Corporation. A recent official report shows that Sina Weibo has more than 300 million registered users by the end of February 29, 2012 and posts 100 million micro-blogs per day. Unfortunately, the wealth of information also attract interests of spammers who attempt to send kinds of spams

---

* Corresponding authors.

Y. Ishikawa et al. (Eds.): APWeb 2013, LNCS 7808, pp. 120–131, 2013.
© Springer-Verlag Berlin Heidelberg 2013

such as rumors, ads, pornographic messages or links to phishing or malicious web sites. Some of these malicious spams may cause significant economic loss to our society and even threaten national security and harmony.

In this paper, we focus on a kind of spam, and we call it ***event rumor,*** which is more harmful than other kinds of spams especially in China. Event rumors are spams that make up untruthful social events and they're harmful to society security and harmony in potential. Figure 1 shows an instance of event rumor which says that the picture is not a scene of a movie or a war, but shows that an officer was on his way to investigate people's life under the protection of several soldiers. It was posted by a Sina weibo user (for the purpose of privacy protection we hide the user's name) at 08:31 on May 11.

Fig. 1. Instance of event rumor on Sina Weibo   **Fig. 2.** Instance of Sina Weibo Rumor-Busting

To detect spams, Sina Weibo has tried several ways such as adding a "report as a spam" button to each micro-blog as shown in the left bottom of Figure 1. Any users considering a micro-blog to be suspicious can report it to Sina official Weibo account (with the translated English user name of "Weibo Rumor-Busting" [1]) who specializes in identifying fake micro-blogs manually. Figure 2 shows a micro-blog published by the Sina official account, which points out that the message showed in Figure 1 is a rumor and gives detailed and authentic reasons. Though artificial method is accurate, it costs a lot of human efforts and financial resources. What's more, there is a certain delay by this approach. It cannot identify a spam quickly once it appears. The rumor may have been spread widely and produced bad effects on society before it is identified manually. So an automatic approach to detecting event rumors is essential. To the best of our knowledge, [1] is the first paper that detect rumor automatically on Sina Weibo, but it aims to detect general rumors not event rumors. And the accuracy has big room to be improved.

In this paper we study how to detect the special kind of rumor, event rumor, because it is more harmful to our society than others. We consider the problem of rumor detection as a classification problem firstly. To distinguish rumors from ordinary posts, we design special features about event rumors from Sina Weibo and build classifiers based on these features. Then we divide event rumors into 4 types and propose a method to identifying one major type among them. The experimental results indicate that our approaches are effective.

The rest of the paper is organized as follows. In section 2 we give a review of related work. In section 3 we describe our two approaches to identifying event rumors and provide a detailed description of the new features. In section 4 we describe how to collect data and present the experimental results. Finally, we conclude this paper in section 5.

2   Related Work

The previous research about spam detection mainly focuses on detecting spams related to ads, pornography, viruses, phishing and so on. These works can be divided into two types. One is detecting spammers who publish social spams [5-10], while the other is detecting spams directly [1-4].

Lee et al. [6] proposed a honeypot-based approach to uncover social spammers in online social systems including MySpace and Twitter. They deployed social honeypots to harvest deceptive spam profiles from social networking communities firstly. Then they analyzed the properties of these spam profiles and built spam classifiers to filter out spammers. Benevento et al. [11] also used honeypots to collect spam profiles. Benevento et al. [5] extracted attributes related to content of tweets and attributes about user behavior and built a SVM classifier to detect spammers of Twitter. This approach is well-performed in filtering out spammers who focus on posting ads, especially link deception. Stringhini et al. [7] did detailed analysis of spammer on Facebook, MySpace and Twitter. They created honey profiles accepting all requests but sending none request to make friends in order to collect data about spamming activity. Spam bots were divided into four categories Displayer, Bragger, Poster and Whisperer. Random forest algorithm was used to identify spammers.

Wang [3] extracted three graph-based features and three content-based features, and formulated the problem of spam bots detection as classification problem. And then he applied different classification algorithms such as decision tree, neural network, support vector machines, and k-nearest neighbor. Finally they found that Bayes classifier is best well-performed. The three graph-based features were extracted from Twitter's social network graph which represents the following relationship among users. The content-based features contained the number of duplicate tweets, the number of HTTP links, and the number of replies or mentions. Similar to this work [3], in Wang's another work [4], a new content-based feature named trending topics representing the number of tweets that contains the hash tag # in a user's 20 most recent tweets was added to features set. Besides the features studied in previous works, Yang, et al. [1] extracted two new features, client-based feature and location-based feature. The former refers to the client program that user has used to post a micro-blog while the later refers to the actual place where the event mentioned in the micro-blog has happened. They trained a Support Vector Machine classifier to measure the impact of proposed features on the classification performance. A noteworthy point is that the dataset they used is from Sina Weibo. To the best of our knowledge, this is the first paper that studies rumor detection of Sina Weibo. The difference between this work and ours lies in that we focus on event rumors. Gao, et al. [2] studied the wall posts in Facebook, which are usually used to spread malicious content by spammers. They built wall post similarity graph, clustered similar wall posts into a group and then identified spam clusters. The difference from previous works is using clustering method.

As mentioned above, majority of the previous works are about detecting spams such as ads, viruses, phishing, pornography and so on. Unlike previous studies, we focus on identifying event rumors (rumors about social events), which are more harmful to national security and society harmony than other kinds of spams especially in China.

3   Event Rumors Detection

3.1   Problem Statement

Driven by different purposes, spammers spread various kinds of spams such as adver-
tise, pornography, viruses, phishing websites and rumors on Sina Weibo. Most of the
previous works focus on detecting spams like ads and pornography as mentioned above.
In this paper, we focus on the detection of event rumor which is more harmful than
other kinds of spams in China. Other rumors like gossips about stars are not covered by
our study. Event rumors are messages which describe fake social events. Usually, an
event is described by attributes such as time, location, character and content.

In the following we list the notations to be used in the subsequent sections.

- $U = \{u_i\}$: set of users
- $T = \{t_i\}$: set of micro-blogs
- $Followers(u_i) = \{u_j | u_j \in U\ and\ u_j\ is\ u_i's\ follower\}$: a set of u_i's followers
- $Friends(u_i) = \{u_j | u_j \in U\ and\ u_i\ is\ u_j's\ follower\}$: a set of u_i's friends
- $t_i.postingtime$: the posting time of micro-blog t_i
- $t_i.text$: the text content of micro-blog t_i
- $t_i.picture$: the picture attached to micro-blog t_i

3.2   Detecting Event Rumors

In this paper, we consider event rumors detection as a classification problem. To identify the event rumors on Sina Weibo automatically, we need to train classifiers and the
key point is to extract powerful features. By analyzing micro-blogs' characters we
extract 15 features, which are divided into three types: *content-based features,
user-based features* and *multimedia-based features*. Some of these features have been
studied in previous works [1,3-7], and others are proposed in this paper. Here we
introduce them briefly.

Content-Based Features.   Content-based features are 8 attributes related to the message content including the number of comments, the number of retweets, the number
of "@" tags, the number of URLs, the number of duplications, the client program
type, the number of verbs used to describe event, and the message contains strong
negative words or not. The former 6 features have been studied already while the
latter two are new features proposed by us.

The number of comments and the number of retweets. Event rumors are usually trending topics that attract more comments and retweets than common micro-blogs.

The number of "@" tags.   "@" is considered as a mention tag in Sina Weibo similar
to that in Twitter. In Sina Weibo, users use "@"+username anywhere in the micro-
blog messages as a reply or mention of another user. A user can mention anyone in a
micro-blog as long as the user follows him or her. If a user is mentioned in a micro-
blog by the mention tag, he or she will see the notification message. The reply and
mention function is often utilized by the rumor accounts to draw users' attention to
spread rumors.

The number of URLs. Since Sina Weibo only allows users to post a short message within 140 Chinese characters, the URLs are always shown in a shorten format [3]. Shorten URLs can hide the URLs' source and users can't get direct information from the URLs. So URL is an attribute that is often utilized by rumor publisher.

The number of duplications. To spread rumors widely, users may post the same micro-blog several times. So the number of duplications is also a noteworthy feature. Given two micro-blogs named t_i and t_j, we measure the similarity based on Jaccard coefficient as shown in Equation 1.

$$similarity(t_i, t_j) = \frac{|t_i.text \cap t_j.text|}{|t_i.text \cup t_j.text|} \tag{1}$$

Where $t_i.text$ is represented by a set of keywords in the text content of t_i. Keywords are the words of the micro-blog text after removing stopping words. When the value of similarity is bigger than 0.8, we consider the two micro-blogs are duplicates.

Client. It refers to the client program used to post micro-blogs by users. Clients are divided into two types [1]: non-mobile client and mobile client. The non-mobile client program includes Sina Weibo web-app, times posting tools and embedded Sina Weibo's third party applications, while the mobile client program includes mobile-based client and Tablet Personal Computer client. Due to the convenience of non-mobile clients, they are often utilized to post and spread event rumors by users.

The number of event verbs. In order to discriminate event rumors from other micro-blogs, we regard the number of *event verbs* as one of the features. We create an event verb dataset using the following 2-step method. First, we extract all the verbs from news of Sina and Fenghuang. Second, we filter out the common verbs usually used in daily life. Figure 3 compares the number of event verbs in event rumor and non-rumor micro-blogs. In this figure, each point represents a micro-blog message.

The message contains strong negative words or not. We find that most of event rumors are negative social events and contain strong negative sentiment words and opinion words. Figure 4 shows the comparison result. As we expect, event rumors are more likely to contain strong negative words than normal micro-bogs.

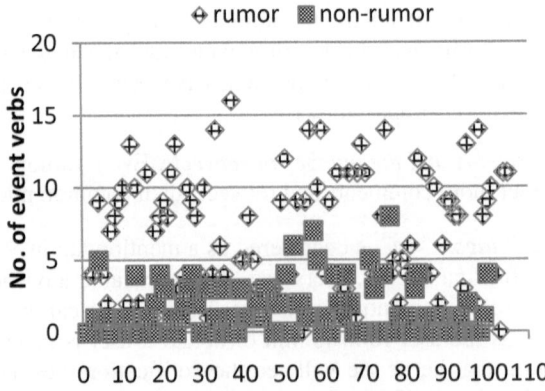

Fig. 3. The number of event verbs used in each micro-blog

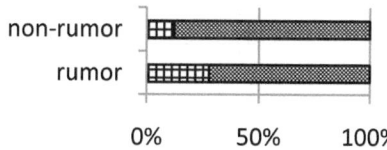

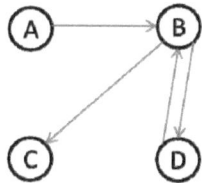

Fig. 4. The distribution of micro-blogs that contains strong negative words

Fig. 5. An example social network graph on Sina Weibo

User-Based Features. User-based features are 6 attributes about users, including the number of user's followers, the number of user's friends, reputation of user, whether the user has VIP authentication, *ratio of micro-blogs containing event verbs*, and *ratio of micro-blogs containing strong negative words*. The former 4 features have been studied already, while the later 2 features are new features proposed by us. Sina Weibo provides a platform to users to create social connections by following others and allowing others to follow them. Figure 5 shows a simple social network graph and the arrows depict the following relationship between users. We can see that user A and user D are the followers of user B, while user C and user D are friends of user B. User B and user D are mutual followers.

The number of followers. The micro-blogs posted by a user will be seen by all his or her followers. The more followers the user has, the wider the rumors spread.

The number of friends. Anyone can follow a user to be this user's follower without seeking permission. As Sina Weibo will send a private message to you when someone follows your account actively, rumor publishers use the following function to attract attention of users' attention. If you follow other accounts, you can also mention their accounts by using a "@" tag in your micro-blogs, then they will receive your micro-blogs even though they are not your followers.

Reputation. If a user has a small number of followers but a lot of friends, the possibility that he or she posts rumors is higher than others. The reputation of a user is defined as follows by Equation 2.

$$reputation(u_i) = \frac{|Followers(u_i)|}{|Followers(u_i)| + |Friends(u_i)|} \qquad (2)$$

Where $|Followers(u_i)|$ means the number of followers and $|Friends(u_i)|$ means the number of friends. The value of reputation is between 0 and 1. Value 1 means a high reputation while 0 means a low reputation.

VIP authentication. If user's identity is verified by Sina Weibo officially, then a VIP tag will occur after the user name. The verified users include famous actors, exporters, organizations, famous corporations and so on. Usually, VIP users are less likely to post rumors.

Ratio of micro-blogs containing event verbs. This attribute means the proportion of micro-blogs that contain event verbs over all micro-blogs posted by the user. The higher this value, the more likely the users post event rumors.

Ratio of micro-blogs containing strong negative words. This attribute means the proportion of micro-blogs that contain strong negative words over all micro-blogs posted by the user. The higher this value, the more likely the users post event rumors.

Multimedia-Based Features. Multimedia-based features are attributes about picture. Sina Weibo allows the micro-blogs containing multimedia like pictures, videos and audios. The most powerful feature we extracted from picture is time span.

Timespan. To make the rumors look like normal micro-blogs, users usually attach corresponding pictures to the micro-blogs. These pictures often come from Internet and were published before. We call them outdated pictures. Users utilize the outdated pictures and new messages to make up a new micro-blog related to event rumors. The timespan function is defined by Equation 3.

$$timespan(t_i.\,text, t_i.\,picture) = \begin{cases} 0, & t_i.\,picture = \emptyset \\ 1, & time(t_i.\,text) - time(t_i.\,picture) > tw \\ 2, & time(t_i.\,text) - time(t_i.\,picture) \le tw \end{cases} \quad (3)$$

Where $time(\,)$ is the time function, $time(t.\,text) = t.\,postingtime$, and tw is a timespan threshold. In our experiment we set the threshold value as a week.

The time of picture is calculated by the following steps. First, we use the picture-to-action function provided by Baidu[1] search engine which is used to identify similar pictures to find out the attached picture on the Internet. If we submit a picture shown in Figure 6 which is an attached picture of the rumor shown in Figure 1 as a query, the search engine will output lots of result records and each record includes the corresponding picture, title, description, URL, and time as shown in Figure 7. Second, according to the time order, we identify the oldest picture and consider this picture's time as the attached picture's time. If a micro-blog doesn't contain any picture, then we consider the value of timespan to be 0. If the timespan between the text and the picture is bigger than the threshold, then the value is 1. Otherwise the value is 2.

---

[1] http://shitu.baidu.com/

大洋新闻-中国驻伊拉克大使馆加紧寻找临时馆舍
新华社记者 聂晓阳 摄...
http://news.dayoo.com 2009-11-22

400x307 32k jpg

400x307

Fig. 6. An attached picture   **Fig. 7.** An result record

Figure 8 depicts the distribution of timespan of micro-blogs and Figure 9 depicts the distribution of timespan of micro-bogs that contain pictures. The two distributions indicate that rumors are more likely to contain an outdated picture than normal micro-blogs indeed.

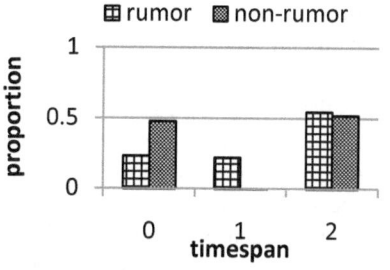

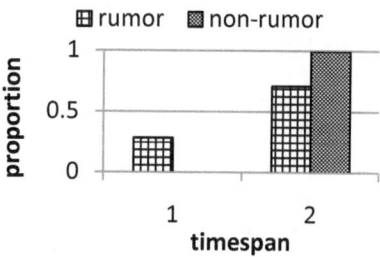

Fig. 8. The distribution of times pan of micro-blogs on Sina Weibo

Fig. 9. The distribution of timespan of micro-blogs containing pictures on Sina Weibo

Based on these features, classifiers are built and used to predict whether a micro-blog is an event rumor or not.

3.3   Detecting Text-Picture Unmatched Rumors

Further analysis of sample event rumors reveals that event rumors can be classified into different types. In the following discussion, we provide our analysis of several common types of event rumors and propose an effective method to one major type, detect text-picture unmatched rumors.

Based on our observation, most event rumors belong to one of the following types:

- Complete fiction: The social event itself is a pure fabrication.
- Time-sensitive: First, time-sensitive event rumors were true messages in the past, but the user published it again later and made it like just happened, so these messages are out of date and become time-sensitive event rumors.
- Fabricated details: As mentioned in section 3.1, an event contains the following attributes: time, location, character and content. This kind of rumor is made up by modifying the time, location or characters of another real event happened before.

The difference between fabricated details type and time-sensitive type is that the latter was true but is out of date while the former is fake.

- Text-picture unmatched rumors: As shown in Figure 1, to increase event rumors' credibility, users often attach a picture to the text, but there is no relationship between the picture and the text.

From Figure 8 we can see that about 80% event rumors contain a picture and a majority of them are text-picture unmatched rumors according to our observation. Therefore, we focus on study how to detect this kind of event rumors. Figure 1 shows an instance of text-picture unmatched rumor, whose text describes an officer's investigation of people's life under the protection of soldiers, but the picture actually describes the team leader of Chinese Embassy in the Republic of Iraq visiting temporary embassy accompanied by security personnel in fact. The rumor account utilizes the unrelated picture and new text to make up a new micro-bog and we call it *text-picture* unmatched rumor. We propose a 5-step method to identify this kind of event rumors.

First, we create a list, including 60 news websites consisting of almost all the major domestic media, and foreign media[2]. Second, we submit the picture attached to a micro-blog as a query to search engine to look for similar pictures. If the output result is nothing, then we consider that the picture matches the text, and the micro-blog is non-rumor. Otherwise, we order the result records according to their website's reliability and their posting time descending. The websites in the list created in the first step are credible and their credibility is the same while others are not credible. Third, we crawl the main content of the top ranked website. Finally, we use Jaccard coefficient to calculate the similarity between micro-blog's text and the crawled content after removing stopping words. If they are similar, we consider the micro-blog is non-rumor. Otherwise, they are text-picture unmatched rumors.

4 Experiments

4.1 Collection of Dataset

In order to evaluate the proposed approach for detecting event rumors, we need a labeled dataset of micro-blogs. To the best of our knowledge, no such collection is publicly available, so we need to build one. Labeling manually costs lots of human labor. Besides, the result may not be accurate by the errors in human judgment. Fortunately, Sina Weibo has an official account named "Weibo Rumor-Busting" for publishing micro-blogs relevant to rumor topics. All the announced rumors are verified by authoritative sources. Based on this official rumor busting service, we build a high quality dataset.

The micro-blogs posted by "Weibo Rumor-Busting" are messages explaining which micro-blog is rumor and why it is rumor, but they are not the original rumors themselves, so we need to construct queries manually to find out the original rumors

---

[2] http://news.hao123.com/wangzhi

by the search function provided by Sina Weibo. We select the top one micro-blog from the search results. We collect 104 event rumors matching the keywords of micro-blogs published by "Weibo Rumor-Busting" from May 2011 to December 2011. Also we extract the profiles (number of followers, number of followings and VIP authentication) of users who published the event rumors correspondingly and all their micro-blogs. The profiles of the followers and followings and their micro-blogs are also included in our dataset. Finally we create a high quality dataset that consists of 1943 users and 26972 micro-blogs, which includes 104 event rumors and 26868 non-rumors.

4.2   Evaluation

To assess the effectiveness of our methods, we use the standard information retrieval metrics of precision, recall and F1 [12]. The precision is the ratio of the number of rumors classified correctly to the total number of micro-blogs predicted as rumors. The recall is the ratio of the number of rumors correctly classified to the total number of true rumors. The F1 is the harmonic mean between both precision and recall, and is defined as $F1 = \frac{2 \times precision \times recall}{precision + recall}$. We train 4 different classifiers used in previous works [1, 3-5] including Naïve Bayes, Bayesian Network, Neural Networks and Decision Tree. 10-fold cross validation strategy is used to measure the impact of those features introduced in Section 3.2 on the classification performance.

Table 1. The comparison of different classifiers using different set of features

Classifier	Precision		Recall		F-measure	
	Without	With	Without	With	Without	With
Naïve Bayes	0.375	0.176	0.606	0.827	0.463	0.29
Bayesian Network	0.5	0.850	0.058	0.654	0.103	0.739
Neural Network	0.5	0.783	0.163	0.692	0.246	0.735
Decision Tree	0.553	0.750	0.202	0.663	0.296	0.704

The experimental results are shown in Table 1. The experimental results indicate that before adding the 5 new features the precision, recall and F-measure are 0.555, 0.202 and 0.269 respectively. There are two reasons why the performance of these old features in our experiment is worse than that in previous works. One is that the datasets are different, and the other is that the purposes are different. Previous studies focus on identifying spams such as ads, while we focus on detecting event rumors. After introducing the 5 new features to the classification process, the precision, recall and F1 are improved to 0.85, 0.654 and 0.739 respectively. As shown in Figure 10, the precision, recall and f1-measure have increased 0.475, 0.048 and 0.276 separately, demonstrating the effectiveness of our proposed new features.

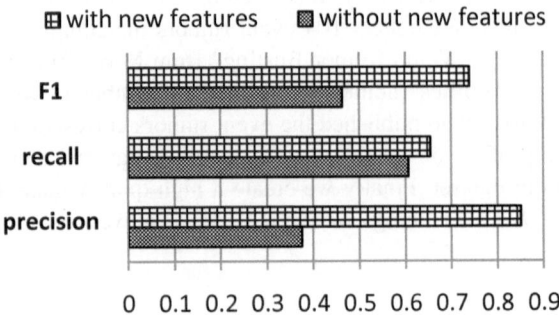

Fig. 10. The comparison between effect of feature set without and with new features

To evaluate the effectiveness of our approach to detecting text-picture unmatched event rumors, we create a dataset including 40 messages, while 17 of them are text-picture unmatched event rumors and 23 non-rumors. All these micro-blog messages contain a picture. Our experiment result shows that the approach is well-performed and the precision, recall and F-measure are 0.833, 0.823 and 0.857 respectively. The disadvantage of this method is that the effectiveness of the picture search engine affects the experimental results to some extent.

5   Conclusion

In this paper we focus on detecting event rumors on Sina Weibo, which are more harmful than other kinds of spams especially in China. To distinguish event rumors from normal messages, we propose five new features and use classification technique to predict which one is event rumor. Further, we divide event rumors into four types and propose a method to identifying text-picture unmatched event rumors with the help of picture search engine. The experiments show the proposed approaches are effective. In the future we will focus on study how to detect other three forms of event rumors.

Acknowledgement. This work was supported by the HGJ PROJECT 2010ZX01042-002-002-03 and National Natural Science Foundation of China under Grant No. 71272029, 70890083, 71110107027 and 61033010.

References

1. Yang, F., Liu, Y., Yu, X., Yang, M.: Automatic detection of rumor on Sina Weibo. In: Proceedings of the ACM SIGKDD Workshop on Mining Data Semantics, Beijing, China, pp. 1–7. ACM (2012)
2. Gao, H.Y., Hu, J., Wilson, C., Li, Z.C., Chen, Y., Zhao, B.Y.: Detecting and characterizing social spam campaigns. In: Proceedings of the 10th ACM SIGCOMM Conference on Internet Measurement, Melbourne, Australia, pp. 35–47. ACM (2010)

3. Wang, A.H.: Don't follow me: Spam detection in Twitter. In: Proceedings of the 2010 International Conference on Security and Cryptography (SECRYPT) (2010)
4. Wang, A.H.: Detecting spam bots in online social networking sites: A machine learning approach. In: Foresti, S., Jajodia, S. (eds.) Data and Applications Security and Privacy XXIV. LNCS, vol. 6166, pp. 335–342. Springer, Heidelberg (2010)
5. Benevenuto, F., Magno, G., Rodrigues, T., Almeida, V.: Detecting Spammers on Twitter. In: Collaboration, Electronic messaging, Anti-Abuse and Spam Conference, Redmond, Washington, US (2010)
6. Lee, K., Caverlee, J., Webb, S.: Uncovering social spammers: social honeypots + machine learning. In: Proceedings of the 33rd International ACM SIGIR Conference on Research and Development in Information Retrieval, Geneva, Switzerland, pp. 435–442. ACM (2010)
7. Stringhini, G., Kruegel, C., Vigna, G.: Detecting spammers on social networks. In: Proceedings of the 26th Annual Computer Security Applications Conference, Austin, Texas, pp. 1–9. ACM (2010)
8. Yang, C., Harkreader, R.C., Gu, G.: Die free or live hard? Empirical evaluation and new design for fighting evolving twitter spammers. In: Sommer, R., Balzarotti, D., Maier, G. (eds.) RAID 2011. LNCS, vol. 6961, pp. 318–337. Springer, Heidelberg (2011)
9. Benevenuto, F., Rodrigues, T., Almeida, V., Almeida, J., Goncalves, M.: Detecting spammers and content promoters in online video social networks. In: Proceedings of the 32nd International ACM SIGIR Conference on Research and Development in Information Retrieval, Boston, MA, USA, pp. 620–627. ACM (2009)
10. Benevenuto, F., Rodrigues, T., Almeida, V., Almeida, J., Zhang, C., Ross, K.: Identifying video spammers in online social networks. In: Proceedings of the 4th International Workshop on Adversarial Information Retrieval on the Web, Beijing, China, pp. 45–52. ACM (2008)
11. Webb, S., Caverlee, J., Pu, C.: Social Honeypots: Making Friends with a Spammer near You. In: Collaboration, Electronic messaging, Anti-Abuse and Spam Conference, Mountain View, California, US (2008)
12. Yang, Y.: An Evaluation of Statistical Approaches to Text Categorization. Journal of Information Retrieval 1(1/2), 67–88 (1999)

Uncertain Subgraph Query Processing over Uncertain Graphs*

Wenjing Ruan[1], Chaokun Wang[1,2,4], Lu Han[1], Zhuo Peng[1], and Yiyuan Bai[1,3]

[1] School of Software, Tsinghua University, Beijing 100084, China
[2] Tsinghua National Laboratory for Information Science and Technology
[3] Department of Computer Science and Technology, Tsinghua University
[4] Key Laboratory of Intelligent Information Processing, Institute of Computing Technology, Chinese Academy of Sciences
{ruanwj11,han109,pengz09,baiyy10}@mails.tsinghua.edu.cn,
chaokun@tsinghua.edu.cn

Abstract. In some real-world applications, data cannot be measured accurately. Uncertain graphs emerge when this kind of data is modeled by graph data structures. When the graph database is uncertain, our query is highly possibly uncertain too. All of the prior works do not consider the uncertainty of the query. In this paper, we propose a new algorithm called Mutual-Match to solve the problem of uncertain subgraph query under the situation where both the graph dataset and the query are uncertain. Considering that the subgraph isomorphism verification is an NP-hard problem, and it will be more complex on uncertain graph data, the Mutual-Match algorithm uses MapReduce processes to find out all the subgraph matches and can adapt to dynamic graphs well. Experimental results on both real-world and synthetic datasets show the correctness and effectiveness of our proposed methods.

Keywords: Uncertain subgraph query, uncertain graphs, mutual-match, dynamic graphs, mapreduce.

1 Introduction

With the rapid development of web technology, graph has considerable importance in modeling of data in diverse scenarios, e.g., social networks, mobile networks and sensor networks. If the graph-structured data is uncertain, it is called an uncertain graph. For example, each edge of the graph has a probability of its existence in this paper. That is to say the uncertainty is embodied in the relationships between nodes of the graph. A typical example of the uncertain graph is protein-protein interaction networks (PPI networks for short) [5].

---

* This work was supported by the National Natural Science Foundation of China (No. 61170064, No. 61133002), the National High Technology Research and Development Program of China (No. 2013AA013204) and the ICT Key Laboratory of Intelligent Information Processing, CAS.

Y. Ishikawa et al. (Eds.): APWeb 2013, LNCS 7808, pp. 132–139, 2013.

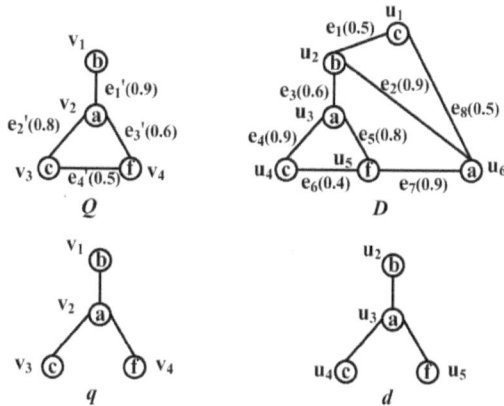

Fig. 1. An Example of Matching Pairs

In PPI networks, nodes denote protein molecules while edges denote the interaction relationships between protein molecules. Limited by the measuring methods, we cannot accurately detect whether there is an interaction between two protein molecules. Therefore, an existence probability is set for each interaction.

Recently, uncertain graphs have become an increasingly attractive focus in many fields. There are a number of algorithms proposed to solve mining and query problems in uncertain graph database. Representative mining techniques in this area include finding top-k maximal cliques [11] and mining frequent subgraph patterns [10]. Various query methods have been proposed to solve the essential problems of uncertain graph database, e.g., KNN query [4], the shortest path query [9], reliable subgraph query [3], and so on. All of the prior works haven't considered the uncertainty of the query graph, which is used to search matches in the uncertain graph database.

In the research field of uncertain graph management, the problem of *uncertain subgraph query* emerges when both the dataset and the query are uncertain. As shown in Figure 1, Q is an uncertain query graph, and D is an uncertain data graph. The uncertain subgraph query returns all the pairs of isomorphic existences between Q and D.

There are two challenges to deal with the above-mentioned problem. Firstly, the subgraph isomorphism verification is an NP-hard problem, and it will be more complex when the dataset and the query are both uncertain. Secondly, uncertain subgraph query problem has high complexity, especially when the size of the uncertain dataset is large.

To address the above challenges, a novel algorithm called *Mutual-Match* is proposed in this paper. The intuitive idea of Mutual-Match is using distributed computing methods to greatly improve the query performance. Thanks to the MapReduce model [1], Mutual-Match can effectively find out all matching pairs, each of which is composed by two isomorphic subgraphs derived, respectively, from the query graph and the data graph.

To the best of our knowledge, this paper is the first one addressing the problem of uncertain subgraph query. The main contributions of the paper are summarized as follows.

1. The problem of uncertain subgraph query over uncertain graph is formally presented.
2. An algorithm called Mutual-Match is proposed to process the uncertain subgraph query over uncertain graphs with the MapReduce model. It can also execute on dynamic uncertain graphs.
3. Extensive experiments are conducted to show the correctness and effectiveness of our proposed method.

The rest of the paper is organized as follows. Section 2 gives the formal definition of the problem of uncertain subgraph query processing. We present a brute-force algorithm, and then propose an index structure and the Mutual-Match algorithm in Section 3. Moreover, we conduct an empirical study on both real-world and synthetic datasets in Section 4. Finally, we conclude in Section 5.

2   Problem Definition

Definition 1. *An uncertain graph is defined as $G(V, E, \Sigma, \theta, \tau)$, where (1) V is a set of nodes; (2) $E \subseteq V \times V$ is a set of edges; (3) Σ is a set of labels; (4) $\theta : V \to \Sigma$ is a function that sets a label to each node of V; and (5) $\tau : E \to [0,1]$ is a probability function for edges.*

Given an uncertain graph $G(V, E, \Sigma, \theta, \tau)$, we can find out a certain graph $G^*(V, E, \Sigma, \theta, \tau)$ which is generated from G when all the edges' probabilities of G are 1. We call G^* the underlining graph of G.

For a clear formalization of the problem of uncertain subgraph query over uncertain graphs, we use the possible world model. Each possible world of G corresponds to a subgraph of G^* and has an existence probability that is computed according to the probability of every edge in G. In this paper, we assume that all the probabilities of edges distribute independently.

Definition 2. *Given an uncertain query graph Q and an uncertain data graph D, uncertain subgraph query is to find a set of matches. Each match which is denoted as a pair $m(q, d)$ should satisfy the following two requirements.*
1. q and d are certain graphs from the possible worlds of Q and D, respectively.
2. q and d are isomorphic.

As shown in Figure 1, if we limit that the edge number of isomorphic subgraph is not less than 3, there are 5 matching pairs when searching uncertain subgraph matches of Q in D. One of the pairs is $m(q, d)$.

Definition 3. *For n given uncertain query graph Q, an uncertain data graph D and one of their matches $m(q, d)$, the matching probability of $m(q, d)$ is defined as*

$$Pro(q, d) = Pro(q) \times Pro(d), \tag{1}$$

where $Pro(q)$ and $Pro(d)$ are the existence probabilities of the corresponding possible worlds of Q and D, respectively.

3   Algorithms

In this section, we first present the brute-force algorithm of uncertain subgraph query and then propose an effective index and the mutual-match algorithm which transforms the complex uncertain subgraph query on a large uncertain database to a series of MapReduce operations with the index structure.

3.1   Brute-Force Algorithm

For a given uncertain query Q and an uncertain data graph D, the brute-force algorithm first transforms them into possible worlds $W_Q = \{g'_1, g'_2, \cdots, g'_s\}$ and $W_D = \{g_1, g_2, \cdots, g_t\}$, respectively. Let $CS = \{q_1, q_2, \cdots, q_m\}$ be a subset of W_Q, which contains all connected components of W_Q. Each element of CS or W_D is a certain graph with an existence probability, so that the brute-force algorithm evaluates the uncertain subgraph query by processing certain subgraph queries on W_D m times. We build an index structure on all elements of W_D with gIndex [7,8], which performs the best for most queries for sparse datasets in recent researches about indexes [2]. And then, search subgraph matches against each graph in CS using the gIndex.

This brute-force algorithm has to deal with exponential times of subgraph queries within an extremely large search space, if the query and data graphs are large. Clearly, it can only be applied to small datasets.

3.2   Mutual-Match Algorithm

With the Mutual-Match algorithm, we need not know the possible worlds of uncertain data and query graphs before the query processing. In order to get a matching pair $m(q, d)$ from the uncertain query graph Q and the uncertain data graph D, the Mutual-Match algorithm first splits both Q and D into some small components which are denoted as bi-edge graphs. Then, with the help of the edge join operation (EJoin) which joins components together, q and d are gradually constructed respectively.

Definition 4. *Given an uncertain graph G, a subgraph of G is called a bi-edge graph, denoted as $\lambda(e_l(v_l, v_m), e_r(v_m, v_r))$, if it has three nodes v_l, v_m, v_r and two corresponding edges e_l, e_r. In a bi-edge graph, v_m is called the middle node. $Bie(G)$ is the set containing all bi-edge graphs of G.*

Definition 5. *Given two uncertain graphs G_1 and G_2, and e is one of their common edges. The operator $EJoin(G_1, G_2, e)$ joins two graphs by merging the common edge e.*

Bi-Edge Index Structure. Before the query processing, we build an index structure of the uncertain graph database based on the bi-edge graphs. Given an uncertain data graph D, we generate $Bie(D)$ by decomposing D into bi-edge graphs. If $\lambda(e_l(v_l, v_m), e_r(v_m, v_r)) \in Bie(D)$, the label of λ is "$label_l/label_m/label_r$", where $label_m$ is from v_m and $label_l < label_r$ ($<$ is the partial order). The value of λ is saved as $(dbi\_id, e_l, e_r, v_m, v_l, v_r)$, where dbi_id is the unique identifier of λ. We index bi-edge graphs by their labels. This bi-edge graph based index is implemented by a hash chain H. The ids of bi-edge graphs which have the same keyword are saved in a list denoted by $blist(label)$.

We generate $Bie(Q)$ and store each bi-edge graph in $Bie(Q)$ with the same method as the data graph. And then, we search the database index for every label of bi-edge graphs of $Bie(Q)$ and get the input file for MapReduce jobs. The format of the input file is $(e_l, e_r, qbi\_id, v_m, posv, pose)$, where e_l, e_r, v_m are from $\lambda(\lambda \in Bie(D))$ and qbi_id is the unique identifier of $\lambda'(\lambda' \in Bie(Q))$. $posv$ and $pose$ are vectors that save the matched nodes and edges, respectively. Every record of the input file is an initial matching pair.

Algorithm Description. The Mutual-Match algorithm decomposes the query and data graphs into bi-edge graphs and then joins different bi-edge graphs according to the matching relation of these two graphs by EJoin. It coincides with the idea of MapReduce. The map and reduce phase can process this kind of query very well.

A MapReduce job chain assembles these initial matching pairs to find out all the query answers. The mapper of the Job_1 maps each line of the input file into two key-value pairs. Every edge of a bi-edge graph of the data graph acts as the key of a pair. Pairs that have the same key may be joined together with the common edge so as to form a bigger matching pair.

The reducer constructs matching pairs by the EJoin operation. It finds out the pairs with the same keyword, and then finds the joinable pairs, according to some join rules. These join rules contain: (1) common edge check, (2) join manner check, (3) identity check. The first two rules can be obtained from the process that decomposes the query graph into bi-edge graphs, and the last one will be obtained through the EJoin process. The common edge check aims to find out whether the two bi-edge graphs of the query graph, matched by bi-edge graphs from the data graph respectively, have a common edge. If there is a common edge, the matched bi-edge graphs from the data graph are joinable. Apart from this, there are two kinds of join manners about different matching pairs. As shown in Figure 2, when the two bi-edge graphs are joined as the above manner, they have the same middle node. This is a *middle-join*. On the contrary, when joined as the below manner, their middle nodes are different, and this is a *non-middle-join*. The EJoin operation requires the two bi-edge graphs from data graph have the same join manner with the two matched bi-edge graphs from the query graph. If the two key-value pairs are joinable, nodes from Q and D must satisfy the one-to-one correspondence in their two $posv$ vectors. The identity check will do the job to ensure that nodes of the two $posv$ vectors satisfy the one-to-one correspondence.

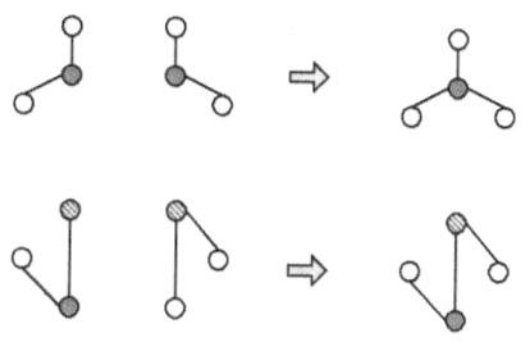

Fig. 2. Two Different Join Manners. The above is a middle-join, and the below is a non-middle-join.

3.3   Improvement

Our method can adapt to dynamic uncertain graphs well. That is the Mutual-Match algorithm can be improved to address the problem of uncertain subgraph query over dynamic uncertain graphs.

When new edges are inserted or existing edges are removed and original graph D becomes the new graph D', we don't need to rebuild the bi-edge index for D' considering all edges of D'. Instead, we only take the new edges and the removed edges into consideration.

Suppose we have run the Mutual Match Algorithm for the data graph D and the query graph Q, and get the uncertain subgraph query result file RF which contains all matching pairs of D and Q. When D changes, for each removed edge e, we find matching pairs which contain e and remove these pairs from RF. For each new edge e, we build new bi-edge index related with e. Then we get the new input file F'. We process F' in Mapper to produce key-value pairs P' and choose records in RF whose keys are the same with the keys in P'. Then we add these records to the file F'. Finally, we construct new matching pairs of D' and Q by EJoin operation in the MapReduce job chain ($Job_2 \rightarrow Job_3 \rightarrow \dots \rightarrow Job_n$) with F' as the input file.

4   Experiments

Our experiments are conducted on two different environments. The index construction, MapReduce input file construction and the brute-force algorithm are performed on a 2.4GHZ, 2GB-memory, Intel(R) Core(TM) i5 CPU PC, running Ubuntu 10.10 operation system. The other is a MapReduce platform with 5 multithreaded cores. All the algorithms are implemented with Java.

4.1   Results on the Real-World Dataset

We use a real-world dataset of CAL road network in our experiments. It contains 21,692 edges and 21,047 nodes. Every edge has a weight that indicates the uncertainty. We distribute labels to all of the nodes randomly.

Figure 3 shows the results on the real dataset. Firstly, we compare the time costs of the index construction of both Mutual-Match and brute-force algorithms. As shown in Figure 3(a), the time cost of the index construction of the brute-force

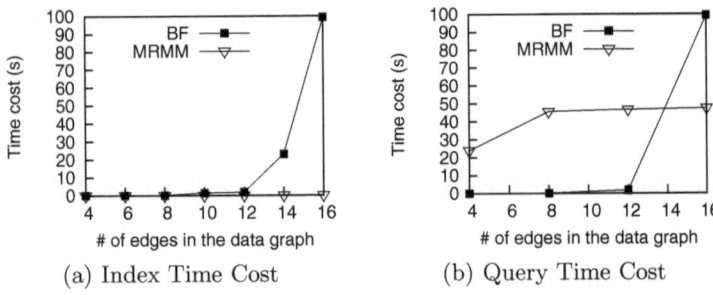

(a) Index Time Cost   (b) Query Time Cost

Fig. 3. Results on the Real-world Dataset

algorithm increases rapidly. Meantime, that of the Mutual-Match algorithm is satisfactory and nearly constant. For example, the Mutual-Match method can index a 16-edge uncertain graph within just 2 ms.

The query efficiencies of both methods are showed as Figure 3(b). The query time of Mutual-Match is computed by the sum of time costs to get MapReduce input file and MapReduce jobs. It shows that the MapReduce-Based Mutual-Match has a poor performance when the uncertain data size is too small. MapReduce is not designed for small data size. With the increasing of the data size, Mutual-Match is much more effective than brute-force.

4.2   Results on the Synthetic Dataset

The synthetic dataset is generated using the gengraph_win [6]. It is a node-labeled, undirected, uncertain data graph which contains 50,000 edges, and the average degree of the nodes is 3. As same as the real-world dataset experiments, labels of every node are distributed randomly. The probabilities of every edge that describe the uncertainty are distributed as Gaussian distribution.

Figure 4 shows the efficiency of our method on the synthetic dynamic graph dataset. We use MRMM to stand for the algorithm that computes matches

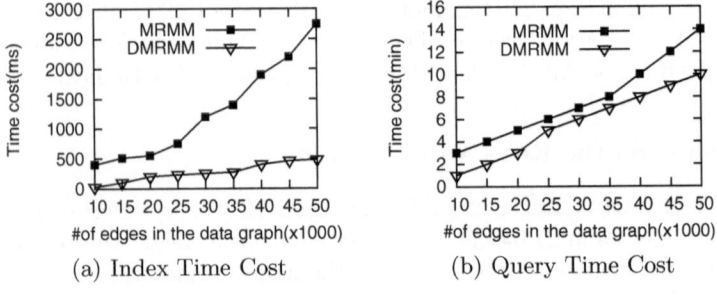

(a) Index Time Cost   (b) Query Time Cost

Fig. 4. Results on the Synthetic Dynamic Dataset

from the beginning and DMRMM to stand for the algorithm that only takes changed edges into consideration. Figure 4(a) shows that the time cost of the index building of the DMRMM method is much smaller than that of the MRMM method for DMRMM does not compute all edges of the new graph. Figure 4(b) shows the efficiency of query processing in the MapReduce jobs. Clearly, the DMRMM method is faster than the MRMM method.

5   Conclusions

Uncertain subgraph query is an essential operation on uncertain graph database. It is complicated due to the NP-hard isomorphism verification and uncertainty. MapReduce provides a good solution to this problem. In this paper, we have explored a Mutual-Match algorithm and an applicative index method to conduct the uncertain subgraph query problem within MapReduce framework. Our MapReduce-Based Mutual-Match algorithm has effective performances on uncertain subgraph matching processing over static uncertain graphs and dynamic uncertain graphs.

References

1. Dean, J., Ghemawat, S.: Mapreduce: Simplified data processing on large clusters. Communications of the ACM 51(1), 107–113 (2008)
2. Han, W., Lee, J., Pham, M., Yu, J.: iGraph: A framework for comparisons of disk-based graph indexing techniques. Proceedings of the VLDB Endowment 3(1-2), 449–459 (2010)
3. Hintsanen, P., Toivonen, H.: Finding reliable subgraphs from large probabilistic graphs. Data Mining and Knowledge Discovery 17(1), 3–23 (2008)
4. Potamias, M., Bonchi, F., Gionis, A., Kollios, G.: K-nearest neighbors in uncertain graphs. Proceedings of the VLDB Endowment 3(1-2), 997–1008 (2010)
5. Saito, R., Suzuki, H., Hayashizaki, Y.: Interaction generality, a measurement to assess the reliability of a protein–protein interaction. Nucleic Acids Research 30(5), 1163 (2002)
6. Viger, F., Latapy, M.: Efficient and simple generation of random simple connected graphs with prescribed degree sequence. In: Wang, L. (ed.) COCOON 2005. LNCS, vol. 3595, pp. 440–449. Springer, Heidelberg (2005)
7. Yan, X., Yu, P., Han, J.: Graph indexing: A frequent structure-based approach. In: SIGMOD, pp. 335–346. ACM (2004)
8. Yan, X., Yu, P., Han, J.: Graph indexing based on discriminative frequent structure analysis. ACM Transactions on Database Systems (TODS) 30(4), 960–993 (2005)
9. Yuan, Y., Chen, L., Wang, G.: Efficiently answering probability threshold-based shortest path queries over uncertain graphs. In: Kitagawa, H., Ishikawa, Y., Li, Q., Watanabe, C. (eds.) DASFAA 2010, Part I. LNCS, vol. 5981, pp. 155–170. Springer, Heidelberg (2010)
10. Zou, Z., Gao, H., Li, J.: Discovering frequent subgraphs over uncertain graph databases under probabilistic semantics. In: SIGKDD, pp. 633–642. ACM (2010)
11. Zou, Z., Li, J., Gao, H., Zhang, S.: Finding top-k maximal cliques in an uncertain graph. In: ICDE, pp. 649–652. IEEE (2010)

Improving Keyphrase Extraction
from Web News by Exploiting Comments Information

Zhunchen Luo, Jintao Tang, and Ting Wang

College of Computer, National University of Defense Technology,
410073 Changsha, Hunan, China
{zhunchenluo,tangjintao,tingwang}@nudt.edu.cn

Abstract. Automatic keyphrase extraction from web news is a fundamental task for news documents retrieval, summarization, topic detection and tracking, etc. Most existing work generally treats each web news as an isolated document. With the rapidly increasing popularity of Web 2.0 technologies, many web news sites provide various social tools for people to post comments. These comments are highly related to the web news and can be considered as valuable background information which can potentially help improve keyphrase extraction. In this paper we propose a novel method to integrate the comment posts into the task of extracting keyphrases from a web news document. Since comments are typically more casual, conversational, and full of jargon, we introduce several strategies to select useful comments for improving this task. The experimental results show that using comments information can significantly improve keyphrase extraction from web news, especially our comments selection method, using machine learning technology, yields the best result.

Keywords: Keyphrase Extraction, Comments, Web News, Machine Learning.

1 Introduction

As thousands of web news are posted on the Internet everyday, it is a challenge to retrieve, summarize, classify or mine this enormous repository effectively. Keyphrases can be seen as condensed represents of web news documents which can be used to help these applications. However only a small minorities of documents have author-assigned keyphrases and manually assigning keyphrases for each document is very laborious. Therefore it is highly desirable to extract keyphrases automatically. In this paper we consider this task in web news scenario.

Most existing work on keyphrase extraction from a web news document only considers the internal information, such as the phrase's Tf-idf, position and other syntactic information, which neglects the external information of the document. Some researches also utilize background information. For example, Wan *and* Xiao proposed a method, which selected similar documents for a given document from a large documents corpus, to improve the performance of keyphrase extraction [14]. Mihalcea *and* Csomai used Wikipedia to provide background knowledge for this task [11]. However both of these work needs to retrieve topic-related documents as external information, which is a non-trivial problem. Fortunately, nowadays many web news sites provide various social tools for community interaction, including the possibility to comment on the web

Y. Ishikawa et al. (Eds.): APWeb 2013, LNCS 7808, pp. 140–150, 2013.

news. These comments are highly related to the web news documents and naturally bound with them. The motivation of our work is that comments are valuable resource providing background information for each web news document, especially some useful comments contain rich information which is the focus of the readers. Hence, in this paper we consider exploiting comments information to improve keyphrase extraction from the web news documents.

Unfortunately, although the main entry and its comments are certainly related and at least partially address similar topics, they are markedly different in several ways [16]. First of all, their vocabulary is noticeably different. Comments are more casual, conversational, and full of jargon. They are less carefully edited and therefore contain more misspellings and typographical errors. There is more diversity among comments than within the single-author post, both in style of writing and in what commenters like to talk about. Simply using all the comments as external information does not improve the performance of keyphrase extraction (see section 5.2). Therefore, we propose several methods to select useful comments, based on some criterions such as textual similarity and helpfulness of the comment. The experimental results show our comment selection approaches can improve keyphrase extraction for web news. Especially our machine leaning approach which integrates many factors of comments can significantly improve this task.

The rest of this paper is organized as follows. Section 2 introduces the related work. Section 3 describes the framework of our approach to extract keyphrases from a web news document. In section 4 we propose the strategies for selecting useful comments. Section 5 presents and discusses the evaluation results. Lastly we conclude the paper.

2   Related Work

Most existing work only uses the internal information of a document for keyphrase extraction. The simplest approach is to use a frequency criterion (or Tf-idf model [12]) to select keyphrases in a document. This method was generally found to give poor performance [3]. However, Hasan recently conducted a systematic evaluation and analysed some algorithms on a variety of standard evaluation datasets, his results indicated that Tf-idf remained offering very robust performance across different datasets [4]. Another important clue for keyphrase extraction is the location of phrase in the document. Kea [2] and GenEx [13] used this clue for their studies. Hulth added more linguistic knowledge, such as syntactic features, to enrich term representation, which were effective for keyphrase extraction [5]. Mihalcea *and* Tarau firstly proposed a graph-based ranking method for keyprase extraction [10] and currently graph-based ranking methods have become the most widely used approaches [3,8,9]. However, all the approaches above never consider the external information when extracting keyphrases.

Another important information, which can be used for keyphrase extraction, is external resource. Wikipedia[1] has been intensively used to provide background knowledge. Mihalcea *and* Csomai used the link structure of Wikipedia to derive a novel word feature for keyphrase extraction [11]. Grineva *et al.* utilized article titles and the link structure of Wikipedia to construct a semantic graph to extract keyphrases [3]. Xu *et al.*

---

[1] http://www.wikipedia.org/

extracted the in-link, out-link, category, and inbox information of the documents related the article in Wikipedia for this task [15]. All approaches using Wikipedia significantly improve the performance of keyphrase extraction. Another external resource is neighbor documents. Wan *and* Xiao proposed an idea that a user would better understand a topic expressed in a document if the user reads more documents about the same topic [14]. He used a textual similarity measure (e.g., cosine similarity) to find documents topically closed to the document from a large documents corpus. These neighbor documents are beneficial to evaluate and extract keyphrases from the document. However using the external resource introduced above needs to retrieve topic-related documents, which is a non-trivial problem. Unlike these resources, we exploit comments of web news as external information. This information is topic-related to the original document. Moreover, the characteristic of naturally bound with the original news makes the comments easier to obtain.

3 Framework

A web news document has two parts: an article and comments. The article contains the text of event written by the author. Comments can be seen as the discussion of the article including the attitude of readers. Most of comments are topic-related to the article. Especially some comments contain rich information which is focused by the readers. We call the comments, which provide additional information for readers to better understand the content of the article, **useful comments**. Intuitively, these useful comments are expected to be beneficial to evaluate and extract the salient keyphrases from the article. However traditional keyphrase extraction methods only consider the information of the article and ignore its comments information. This study we propose an approach exploiting comments information to improve keyphrase extraction for the web news documents.

Figure 1 gives the framework of our approach for keyphrase extraction. First we segment a web news document into the article part and the comments part based on its natural structure. Next, since not all comments are helpful for keyphrase extraction (see section 5.2), we use a comments selector, based on some strategies, to select useful comments from the comments part. Then we integrate the selected comments with the article part to form a new document. At last we use a keyphrase extractor to extract phrases from the new document. The extracted phrases, occurred in the article, are taken as keyphrases. Our approach can be easily incorporated with any state-of-the-art keyphrase extractor to extract keyphrases from web news.

Since some comments contain noise information such as advertisement, irrelevant texts and texts only containing meaningless word emoticons. For example, a comment *"Who cares!"* just reflects a reader's attitude to the news without any useful information for other readers to understand the article. Hence, the key point of our approach is to find the useful comments for keyphrase extraction. We propose three strategies to select useful comments, which are expected to find the article's keyphrases with higher accuracy. The first is selecting comments which is similar to the original article, called **Similar Comments Selector**. The idea of this approach is the same as Wan *and* Xiao [14]. Many web news sites provide ratings setting about the comments for other readers

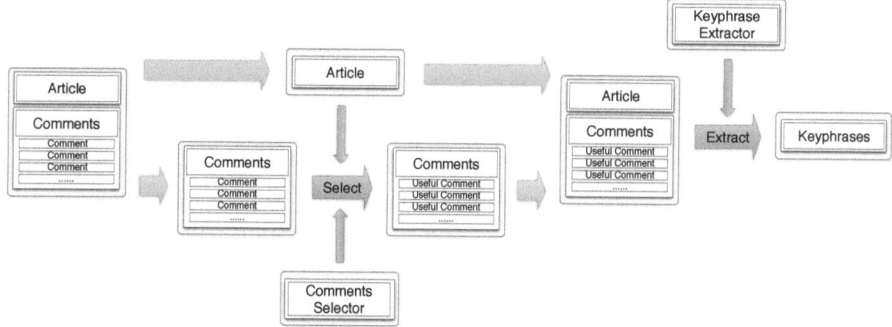

Fig. 1. Framework of Keyphrase Extraction from a Web News Document

(e.g, thumb up or thumb down votes). These meta ratings information can help filter useful comments more efficiently. We call the approach based on ratings information **Helpfulness Comments Selector**. Furthermore, since existing technologies of natural language processing and machine learning provide strong supports to mine the potential knowledge, we use these technologies to select useful comments, called **KeyInfo Comments Selector**. Next section we will introduce the three comments selectors in detail.

4   Selecting Useful Comments

4.1   The Similar Comments Selector

Wan *and* Xiao proposed a method to select similar documents for a given document to improve the performance of keyphrase extraction [14]. Their assumption is that multiple documents within an appropriate cluster context usually have mutual influences and contain useful clues, which can help to extract keyphrases from each other. For example, two documents about the same topic "earthquake" would share a few common phrases, e.g. "earthquake", "victim", and they can provide additional knowledge for each other to better evaluate and extract salient keyphrases from each other.

The similar comments selector adopts the same idea to find useful comments. Firstly it uses the widely used cosine measure to evaluate document similarity between the web news article and its comments. The term weight is computed by Tf-idf. Then the selector chooses top N comments with the highest similarity values as the useful comments for the web news keyphrase extraction.

4.2   The Helpfulness Comments Selector

Due to the lack of editorial and quality control, comments on web news vary greatly. The low-quality comments may hurt the performance of keyphrase extraction.

Many web news sites also allow readers to vote for the helpfulness of each comment and provide a function to assess the quality. The helpfulness function h is usually defined as follow:

$$h = \frac{Num_{thumbup}}{Num_{thumbup} + Num_{thumbdown}} \tag{1}$$

where $Num_{thumbup}$ is the number of people, who consider the comment is helpful. $Num_{thumbdown}$ is the number of people, who consider the comment is unhelpful. This function actually provides a direct and convenient way to estimate the utility value of a given comment. Therefore, we use the helpfulness of comments based on this function to select useful comments. The top N comments with highest scores voted by at least 5 users are selected for the following keyphrase extraction step.

4.3   The KeyInfo Comments Selector

A simple idea is that given an article, if the comments include more keyphrases, they are more likely to be useful comments than the other comments which have less or no keyphrase, since these comments contain more information which the readers focus on. Intuitively, integrating these useful comments into original article is expected to improve the performance of keyphrase extraction. However, we can not obtain the accurate number of keyphrases in each comment, since it is the final purpose of the task. Fortunately, some factors of comments are related to the probability of comments containing more keyphrases such as the textual similarity and helpfulness of comments. Moreover, existing natural language processing technologies can help us find many factors related to the probability of comments containing keyphrases, and using machine learning technologies, based on these factors, can also estimate the probability accurately. Therefore, we propose another comments selector called KeyInfo comments selector using natural language processing and machine learning technologies.

The principle of the KeyInfo comments selector is that if a shorter comment contains more keyphrases, it is more likely to improve the performance of keyphrases extraction from the web news when integrating it into the original article. We define a function k called keyInfo function as follow:

$$k = \frac{Num_{keyphrase}}{Num_{word}} \tag{2}$$

where $Num_{keyphrase}$ is the number of keyphrases in the comment and Num_{word} is the number of words in the comment. We choose the top N comments with highest score of k as the useful comments for keyphrase extraction.

The key point of this approach is to predict the value of $Num_{keyphrase}$. We take this prediction as a regression-learning problem. Firstly, some web news and their comments are collected as training data. These news documents have been originally annotated keyphrases. Then using these gold standard keyphrases, each comment's $Num_{keyphrase}$ can be easily obtained. Some features, which can potentially help to predict the $Num_{keyphrase}$, are designed for each comment. Next each comment is represented by a feature vector. Then a classical regression tool-simple linear regression

(SLR) implemented in WEKA[2] is used to train a prediction model, called KeyInfo model. Lastly, give a new comment represented as a feature vector, this model can predict its $Num_{keyphrase}$ value.

To the features for each comment, textual similarity and helpfulness are not enough. Since the most similar comments are similar to the original entire news article text, the probability of containing keyphrases in each part of the article text, however, is different. For example, the headline of article, the beginning and the end part of article may have higher probability than another parts of text. If a comment is more similar to these key parts of the article, it is more likely to be useful comment for keyphrase extraction. Additionally, for the comments with high score of helpfulness, the phenomenon of "rich-get-richer" might hurt the performance of this approach. The reason is that, how fast a comment accumulate votes depends on the number of votes they already have [7] and this led to a bias to the comments with more user voting. Therefore, we develop more features, based on natural language processing technologies, for the KeyInfo model. These features are designed to maintain the advantages and overcome the disadvantages of two approaches introduced above. Table 1 shows all the features for the KeyInfo Model. All developed features are classified into five categories: **Structural**, **Similar**, **Lexical**, **Syntactic**, **Semantic**, and **Meta-data**. We use *Stanford Log-linear Part-Of-Speech Tagger*[3] to tag the comment and use *General Inquirer*[4] for sentiment analysis.

5 Experiment

We empirically evaluate our approach for keyphrase extraction from a web news document, especially comparing the performance of various strategies to select useful comments for this task. In addition we conduct in-depth analysis of the differences among useful comments sets chosen by different comments selectors.

5.1 Experimental Setup

The AOL Corpus. We construct our own keyphrase extraction data set by collecting 60 web news posted on the AOL websites[5]. All these news are world news posted from November 16th 2010 to November 26th 2010. Every news document contains at least 20 comments. The main reason we choose AOL web news for the experiment is that every web news document has its own tags. We take the tagged phrases that occur in the article as gold standard keyphrases. Table 2 provides an overview of the AOL corpus.

Evaluation Metric. For the evaluation of keyphrase extraction results, the automatic extracted keyphrases are compared with the gold standard keyphrases. The precision $p = count_{correct}/count_{extrator}$, recall $r = count_{correct}/count_{gold}$, and F1 score $f = 2pr/(p + r)$

---

[2] http://www.cs.waikato.ac.nz/ml/weka/
[3] http://nlp.stanford.edu/software/tagger.shtml
[4] http://www.wjh.harvard.edu/~inquirer/
[5] http://www.aol.com/

Table 1. Features of the KeyInfo Model

Feature category	Feature name	Description
Structural	NumTokens	The number of tokens in the comment
	NumSentences	The number of sentences in the comment
	NumAskMarks	The number of ask marks in the comment
	NumExclamationMarks	The number of exclamation marks in the comment
Similar	ArticleSimilar	The similarity between the comment and the article
	TitleSimilar	The similarity between the comment and the title
	HelpfulnessCommentsSimilar	The similarity between the comment and other comments with the highest score of helpfulness (the number of other comments is 5)
Syntactic	RatioNouns	The percentage of nouns in the comment
	RatioVerb	The percentage of verb in the comment
	RatioAdj	The percentage of adjective in the comment
	RatioAdv	The percentage of adverb in the comment
Semantic	NumPosWord	The number of positive words in the comment
	NumNegWord	The number of negative words in the comment
MetaData	NumThumbUps	The number of thumb up votes in the comment
	NumThumbDowns	The number of thumb down votes in the comment
	HelpfulNess	The helpfulness score of the comment
	TimeLines	The number of comments having been posted before this comment

Table 2. AOL Corpus Statistics

	Ave	Max	Min
#Tokens in Article	570.77	1558	162
#Comments	103.98	316	22
#Gold Standard Keyphrases	7.6	17	3

are used as evaluation metrics, where $count_{correct}$ is the total number of correct keyphrases extracted by the extractor, $count_{extrator}$ is the total number of automatic extracted keyphrases, and $count_{gold}$ is the total number of gold keyphrases.

The Keyphrase Extractors. Most state-of-the-art keyphrase extractors can be used in our approach. Hasan conducted a systematic evaluation and analysed many unsupervised keyphrase extraction methods on a variety of standard evaluation datasets. He found that Tf-idf and SingleRank extractors give very robust performance across different datasets [4]. Therefore, we use these two keyphrase extractors in our experiment. A publicly available implementation of these two extractors can be downloaded[6].

Tf-idf Keyphrase Extractor assigns a score to every term in a document based on its frequency (term frequency, tf) and the number of other documents containing this term (inverse document frequency, idf). Given a document the extractor first computes the Tf-idf score of each candidate word, then extracts all the longest n-grams consisting of candidate words and scores each n-gram by summing the Tf-idf scores of its constituent unigrams, and finally output the top N n-grams as keyphrases.

SingleRank Keyphrase Extractor is expanded from **TextRank Keyphrase Extractor** [10,14]. Both of these two extractors take a text represented by a graph, in which each vertex corresponds to a word type and the edge connects two vertices if the two words co-occur within a window of W words in the associated text. The goal is to compute the score of each vertex, which reflects its importance. The two extractors use the word types that correspond to the highest-scored vertices to form keyphrases for the text. Some graph-based ranking algorithms such as Google's PageRank [1] and HITS [6], which compute the importance of vertices, can be easily integrated into the TextRank model and SingleRank model. There are some differences between these two extractors. First, each edge has a weight equal to the number of times the two corresponding word types co-occur within a window of W words in a SingleRank graph, while each edge has the same weight in a TextRank graph. Second, SingleRank only uses high score sequence nouns and adjectives words to form keyphrases, but any high score sequence words can form keyphrases in TextRank.

5.2 Evaluation of the Proposed Approaches

First we investigate all the comments information can help keyphrase extraction. We integrate all comments into their related articles in our AOL corpus to form a new dataset. We extract 20 keyphrases in each news document for evaluation using **Tf-idf Keyphrase Extractor** and **SingleRank Keyphrase Extractor** respectively. Table 3 shows the result. We can see that, compared the approaches extracted keyphrases from original article (**Base_Tf-idf** and **Base_SingleRank**), the approaches adding all comments into the article (**All_Tf-idf** and **All_SingleRank**) decrease the performance of extracting keyphrases. We conducted a paired t-test between the results of these methods and found statistically significant differences (at $p = 0.05$). All these means simply using all comments can not help for keyphrase extraction.

Table 3. Performance of All Comments Testing

	Precision	Recall	F1
Base_Tf-idf	0.139	0.366	0.202
All_Tf-idf	0.153	0.259	0.192
Base_SingleRank	0.120	0.316	0.174
All_SingleRank	0.096	0.163	0.121

---

[6] http://www.utdallas.edu/~ksh071000/code.html

Next we investigate some useful comments can help keyphrase extraction. We ranked all the comments based on keyInfo function k. We use the gold keyphrases in each article to calculate the accurate $Num_{keyphrase}$ value for each comment. For each web news document, we choose the top 15 comments with high value of k and integrate them into the original article (called **Oracle_Tf-idf** and **Oracle_SingleRank**), which achieve the best result of keyphrase extraction in our experiment (see Table 4). Moreover, they are oracle tests and the results are significantly different (at $p = 0.05$). It means there is a subset of useful comments, which can help for keyphrase extraction, in comments part for each web news (see Figure 1).

Table 4. Performance of Useful Comments Testing

	Precision	Recall	F1
Base_Tf-idf	0.139	0.366	0.202
Oracle_Tf-idf	0.165	0.432	0.239
Base_SingleRank	0.120	0.316	0.174
Oracle_SingleRank	0.146	0.382	0.211

At last we investigate the performance of our three comments selectors, choosing top 15 high score comments, for keyphrase extraction. The **Similar_Tf-idf (SingleRank)** and **Helpfulness_Tf-idf (SingleRank)** methods used the measures of comments' similar and helpfulness score to choose the comments. To **KeyInfo_Tf-idf (SingleRank)** approach, we used 5 new AOL web news comments excluding AOL corpus to train a regression model, which can predict the $Num_{keyphrase}$ of each new comment. The 5 new documents contain 524 comments and each document has gold standard keyphrases. We extract all features in Table 1 to represent each comment as a feature vector for training. The trained model can be used to predict the KeyInfo score of a new comment. Table 5 gives the comparison results of various selecting methods for keyphrase extraction. We can see that the performance of most methods integrating comments outperform the **Base_Tf-idf(SingleRank)** for keyphrase extraction. It shows comments information indeed can help for keyphrase extraction. We conducted a paired t-test between the results and found all methods significantly improve the keyphrase extraction (at $p = 0.05$) except **Helpfulness_Tf-idf**. Especially the **KeyInfo_Tf-idf (SingleRank)** achieves the best result. It shows similar and helpfulness information are helpful for selecting useful comments. Moreover, using more knowledge, extracted based on natural language processing and machine learning technologies, can collect more useful comments. All these shows exploiting comments information are effective for keyphrase extraction.

We are interested in the detail of comments selected by these three selectors. We compare three selected comment sets with the comments selected based on the accurate $Num_{keyphrase}$ value (see Oracle Test). All of comparisons use the top 15 high score comments for each document. Table 6 gives the result. We can see that the KeyInfo selector collect the largest most useful comments. It demonstrates our machine learning approach is effective to find more comments with high function K value for keyphrase extraction.

Table 5. Performance of Extracting 20 Keyphrases

	Precision	Recall	F1
Base_Tf-idf	0.139	0.366	0.202
Similar_Tf-idf	0.151	0.393	0.218
Helpfulness_Tf-idf	0.144	0.379	0.209
KeyInfo_Tf-idf	0.157	0.406	0.226
Base_SingleRank	0.120	0.316	0.174
Similar_SingleRank	0.134	0.324	0.190
Helpfulness_SingleRank	0.133	0.332	0.190
KeyInfo_SingleRank	0.144	0.331	0.201

Table 6. Useful Comments Sets Comparing Evaluation

	Number
$Set_{Oracle} \cap Set_{KeyInfo}$	437
$Set_{Oracle} \cap Set_{Helpfulness}$	331
$Set_{Oracle} \cap Set_{Similar}$	300

6   Conclusion

In this paper we propose a novel approach to use comments information for keyphrase extraction from web news documents. Since comments are full of noisy and low-quality data, simply integrating comment into an original article does not improve the performance of this task. Therefore we give three strategies to select useful comments. The experimental results show that our comments selection approaches can improve keyphrase extraction from web news. Especially our machine leaning approach which considers many factors of comments can significantly improve this task.

Acknowledgements. This research is supported by the National Natural Science Foundation of China (Grant No. 61170156 and 61202337) and a CSC scholarship.

References

1. Brin, S., Page, L.: The anatomy of a large-scale hypertextual web search engine. In: Proceedings of the Seventh International Conference on World Wide Web 7, WWW7, pp. 107–117. Elsevier Science Publishers B. V., Amsterdam (1998)
2. Frank, E., Paynter, G.W., Witten, I.H., Gutwin, C., Nevill-Manning, C.G.: Domain-specific keyphrase extraction. In: Proceedings of the Sixteenth International Joint Conference on Artificial Intelligence, IJCAI 1999, pp. 668–673. Morgan Kaufmann Publishers Inc., San Francisco (1999)
3. Grineva, M., Grinev, M., Lizorkin, D.: Extracting key terms from noisy and multitheme documents. In: Proceedings of the 18th International Conference on World Wide Web, WWW 2009, pp. 661–670. ACM, New York (2009)
4. Hasan, K.S., Ng, V.: Conundrums in unsupervised keyphrase extraction: making sense of the state-of-the-art. In: Proceedings of the 23rd International Conference on Computational Linguistics: Posters, COLING 2010, pp. 365–373. Association for Computational Linguistics, Stroudsburg (2010)

5. Hulth, A.: Improved automatic keyword extraction given more linguistic knowledge. In: Proceedings of the 2003 Conference on Empirical Methods in Natural Language Processing, EMNLP 2003, pp. 216–223. Association for Computational Linguistics, Stroudsburg (2003)
6. Kleinberg, J.M.: Authoritative sources in a hyperlinked environment. J. ACM 46(5), 604–632 (1999)
7. Liu, J., Cao, Y., Lin, C., Huang, Y., Zhou, M.: Low-quality product review detection in opinion summarization. In: Proceedings of the 2007 Joint Conference on Empirical Methods in Natural Language Processing and Computational Natural Language Learning (EMNLP-CoNLL), pp. 334–342 (2007)
8. Liu, Z., Huang, W., Zheng, Y., Sun, M.: Automatic keyphrase extraction via topic decomposition. In: Proceedings of the 2010 Conference on Empirical Methods in Natural Language Processing, EMNLP 2010, pp. 366–376. Association for Computational Linguistics, Stroudsburg (2010)
9. Liu, Z., Li, P., Zheng, Y., Sun, M.: Clustering to find exemplar terms for keyphrase extraction. In: Proceedings of the 2009 Conference on Empirical Methods in Natural Language Processing, EMNLP 2009, vol. 1, pp. 257–266. Association for Computational Linguistics, Stroudsburg (2009)
10. Mihalcea, R., Tarau, P.: Textrank: Bringing order into texts. In: Proceedings of EMNLP, Barcelona, Spain, vol. 4, pp. 404–411 (2004)
11. Mihalcea, R., Csomai, A.: Wikify!: linking documents to encyclopedic knowledge. In: Proceedings of the Sixteenth ACM Conference on Information and Knowledge Management, CIKM 2007, pp. 233–242. ACM, New York (2007)
12. Salton, G., Buckley, C.: Term-weighting approaches in automatic text retrieval. Inf. Process. Manage. 24(5), 513–523 (1988)
13. Turney, P.D.: Learning to extract keyphrases from text. CoRR cs.LG/0212013 (2002)
14. Wan, X., Xiao, J.: Single document keyphrase extraction using neighborhood knowledge. In: Proceedings of the 23rd National Conference on Artificial Intelligence, AAAI 2008, vol. 2, pp. 855–860. AAAI Press (2008)
15. Xu, S., Yang, S., Lau, F.: Keyword extraction and headline generation using novel word features. In: Proc. of the Twenty-Fourth AAAI Conference on Artificial Intelligence (2010)
16. Yano, T., Cohen, W., Smith, N.: Predicting response to political blog posts with topic models. In: Proceedings of Human Language Technologies: The 2009 Annual Conference of the North American Chapter of the Association for Computational Linguistics, pp. 477–485. Association for Computational Linguistics (2009)

A Two-Layer Multi-dimensional Trustworthiness Metric for Web Service Composition

Han Jiao[1], Jixue Liu[1], Jiuyong Li[1], and Chengfei Liu[2]

[1] School of Information Technologies and Math Sciences,
University of South Australia, SA 5095, Australia
[2] Faculty of Information and Communication Technologies
Swinburne University of Technology, Melbourne, VIC3122, Australia
{han.jiao,jixue.liu,jiuyong.li}@unisa.edu.au, cliu@siwn.edu.au

Abstract. Web service composition is a promising area to revolutionize the collaboration among heterogeneous applications. Trust is recognized as an important topic in this research area. In this paper, we propose a two-layer multi-dimensional metric to represent the trustworthiness of web service composition programs. We explored several dimensions for both composition level and service level and finalize the metric, as an indicator, to represent the trustworthiness of the automatic service composition. We also propose a method to transfer the multi-dimensional metric into a single value for comparison purpose. A simulation experiment shows that the proposed metric shows better results than the empirical judgement based on past experiences.

Keywords: trust, trustworthiness, web service, composition, metric.

1 Introduction

The Web originates for enabling users the information sharing power. As evolving, it has been enriched to provide more flexible ways to link different sources, one of which is to use web services, a set of related functionalities that can be programmatically accessed through the web [16], usually via XML-based protocols. Web service composition, aiming at composing distributed web services to carry out complex business processes, is considered to be a promising area to revolutionize the collaboration among heterogeneous applications[3]. Due to the constantly changing environment and the large scalability of the Web, the research focus on web service composition is moving from static to dynamic technology, known as Automatic Service Composition (ASC) [15]. The software programs that provide ASC functionalities are termed ASC agents. An ASC agent has to compose several web services to deliver results, which brings them a few particular characteristics that are different from single-functionality web services.

Researchers has realized the significance of trust in the web service environment. Methods, models and frameworks are proposed to help service consumers identify trustworthy web services [17]. However, determining the trustworthiness

Y. Ishikawa et al. (Eds.): APWeb 2013, LNCS 7808, pp. 151–162, 2013.

for ASC agents is left as an open issue. We adopt the major consideration in [6] by defining the trustworthiness of ASC agents to be a two-layer structure, namely composition-layer and service-layer. This caters to the unique characteristics owned by ASC agents that their deliveries depend on not only their own activities, but also the single-functionality web services they compose.

The previous work did not give a complete investigation on how to use QoS factors to represent the trustworthiness of an ASC agent. The conducted experiment cannot reflect the effectiveness of the proposed model. In this paper, we will thoroughly explore the relationships between trustworthiness and QoS factors. A proper set of indicators, based on QoS factors, will be constructed that can better present ASC's trustworthiness. We will also use experiment to prove that our proposed metric shows better result than result of the forecast simply using historical data. The contributions of this paper are:

- We explore the relationships between trustworthiness QoS factors for ASC agents. And compare them with their counterparts in normal web services.
- We design a metric of Trustworthiness Status Indicators (TSIs) that can be used to evaluate the trustworthiness of ASC agents. The metric is open and customizable. Users can config it to cater for their own preferences.
- We provide a mathematic method to transfer the multi-dimensional TSI metric into single-valued format to ease comparison and evaluate the method using experiments.

The rest of the paper is structured as follows. In Section 2, we explore the relationship between the trustworthiness of ASC agent and QoS factors. In Section 3, we present trustworthiness metric in detail. A method to integrate TSIs for trustworthiness comparison is introduced in Section 4. A simulation experiment is presented in Section 5. Section 6 gives the recent related works and Section 7 summarizes the paper and looks at the future work.

2   Trust-Related Concepts and QoS Factors

In this section, we clarify the concept of trust and trustworthiness. We will also discuss the commonly agreed dimensions of trustworthiness and what QoS factors can be used to represent those trustworthiness dimensions.

2.1   Trust and Trustworthiness

Trust and trustworthiness are different. Trust is the willingness of a trustor to believe that the trustee will behave as expected. This willingness makes the trustor confident to be vulnerable to some trustee's behaviors. Different from trust, trustworthiness is a trustee's attribute to represent the extent that it can be trusted [5]. In other words, a trustor gives its trust to a trustee based on the trustworthiness of the trustee. Strictly speaking, trust is a mental activity. People are not necessarily to give their trust based on trustworthiness. In reality, there are people that irrationally trust others without any reasons. However, rational people give their attitude based on gathered information on trustworthiness.

In this paper, the ASC requesters are trustors and the ASC agents are the trustees. Instead of directly modeling trust, we design a metric to represent the trustworthiness of ASC agents, which has impact on the trust from requesters.

2.2 Trustworthiness Dimensions and QoS Factors

In [11,9], three dimensions of trustworthiness were studied. They are ability, benevolence and integrity. This taxonomy is agreed by most researcher since proposed, for example in [4,5,12]. Since the process of automatic service composition is entirely computer-controlled and a computer software does not have emotions to act altruistically or kindheartedly, it makes no senses to measure the dimension of benevolence in this paper. The concept of trustworthiness only covers the first two dimensions: ability and integrity.

As we mentioned, ASC agents own the speciality that their deliverables are based on their composition ability as well as the single-functionality web service they use. When we investigate the trustworthiness of an ASC agent, we have different concerns on these two sources. At composition layer, people implicitly make an assumption that they believe the ASC agent has the ability to complete the required tasks. What they concern is that whether their actual performance can be really as good as what they claim to be. In other words, people concerns the deviation between the actual action with the advertised one. At single service layer, people know that even the ASC agent cannot control the single services, because they are provided by third parties. So people concern whether an ASC agent have enough qualified resources to compose. The selected QoS factor at service layer should be able to reflect the general conditions of the resource pool maintained by the ASC agents.

Based on the above discussion, we design the following trustworthiness metric consisting of 4 aspect at each layer. Each aspect is represented by a Trustworthiness Status Indicator (TSI) to distinguish it from the basic corresponding QoS. At composition layer, we have: 1)Functional Effectiveness Deviation, 2)Execution Efficiency Deviation, 3)Availability Deviation, and 4)Reputation. At service layer, another four TSIs are taken into account: 1)Average Execution Duration, 2)Average Availability, 3)Average Reliability, and 4)Service Monitoring. Please note that at composition layer, we concern "deviations" between the promises and the actual performance. At service layer, we concern the general quality of the maintained resource pool. In the next section, we present all the TSIs in detail.

3   Trustworthiness Indicators

3.1   Composition-Layer TSIs

Functional Effectiveness Deviation (T_{fed})**.** This TSI represents the status of functional effectiveness. To define the functional effectiveness deviation, we first need to define functional error rate. Functional error rate is the proportion of requests answered with functional errors over the total number of requests. Given an ASC agent c and a certain period du, it is calculated as $\mu_*(c, du) = \frac{N_{err}}{N_{total}}$, where N_{err} is the total number of requests answered in error and N_{total} is the total number of composition requests. The functional effectiveness deviation $T_{fed}(c, du)$ represents the difference between the actual functional error rate $\mu_a(c, du)$ and the claimed one $\mu_c(c, du)$ within that duration. It is calculated as $T_{fed}(c, du) = max\{0, \mu_a(c, du) - \mu_c(c, du)\}$. In practice, $\mu_a(c, du)$ can be monitored by consumers or a third party and $\mu_c(c, du)$ is claimed by the ASC agent itself. The lower the $T_{fed}(c, du)$ is, the more trustworthy the ASC agent is. In general, this TSI reflects how trustworthy the ASC agent is in the aspect of functional capability.

Execution Efficiency Deviation (T_{eed}). This TSI indicates the status of the execution efficiency of an ASC agent. For an ASC agent c within duration du, the execution efficiency deviation $T_{eed}(c, du)$ is the difference between the claimed on-time response rate $\eta_c(c, du)$ and the actual on-time response rate $\eta_a(c, du)$. On-time response rate η is the proportion of on-time response times over the total times an ASC agent should response, i.e., $\eta = \frac{N_{ontime}}{N_{total}}$. Therefore, $T_{eed}(c, du)$ is calculated as $T_{eed}(c, du) = max\{0, \eta_c(c, du) - \eta_a(c, du)\}$. In general, execution efficiency deviation stands for how trustworthy the ASC agent is in the aspect of on-time response.

Availability Deviation (T_{ad}). In [14,17], availability is defined to be the proportion of the available duration in the total duration. Based on those definitions, we define availability deviation $T_{ad}(c, du)$ to be the difference between the claimed availability and the actual availability for ASC agent c in duration du. It is difficult to make real-time availability monitoring of an ASC agent. One possible solution is to change this "real-time" continuous monitoring to randomly discrete monitoring. For every time the ASC agent is called by consumers, its availability information is recorded. After a certain period of time du, the actual availability of an ASC agent c can be represented by the proportion of the available times in the total accessing times, i.e., $\rho = \frac{N_{available}}{N_{total}}$. Then, we use the following formula to calculate availability deviation: $T_{ad}(c, du) = max\{0, \rho_c(c, du) - \rho_a(c, du)\}$. Same as the above two TSIs, the lower the availability deviation is, the more trustworthy the ASC agent is.

Public Reputation (T_{pr}). The public reputation of an ASC agent is a measure of its trustworthiness from public opinions. There is no doubt that in an open environment, different people's opinions will impact each other. Different consumers may have different opinions on the same ASC agent. Our public reputation is defined as the average ranking scored by consumers. This can be represented by the following formula: $T_{pr}(c, du) = \frac{\sum_{i=1}^{n} \gamma_i}{n}$, where γ_i is the score of ranking given by consumer i in the last duration du and n is total number of consumers who give scores in that duration. To simplify the calculation in our framework, we define the range of ranking score is $[0,1]$. Therefore, the range of $T_{pr}(c, du)$ is also $[0,1]$. Different from the above three TSIs, the higher the public reputation is, the more trustworthy the ASC agent is.

The above four TSIs are composition-layer TSIs that reflect the trustworthiness of an ASC agent itself. In the next sub-section, we will introduce several service-layer TSIs which reflects the status of candidate elementary services maintained by an ASC agent. Their properties also impacts the final deliverables of ASC agents and therefore, impact the trust consumers give.

3.2 Service-Layer TSIs

Average Execution Duration (T_{aed}). Given an ASC agent c, the average execution duration $T_{aed}(c, du)$ is the average processing time of the elementary services that ASC agent c calls during the last duration du, i.e., $T_{aed}(c, du) = \frac{\sum_{i=1}^{n} Norm(t_i)}{n}$, where t_i is the execution duration of the ith call to an elementary service and n is the total number of calls in duration du. To simplify the future calculation, we would like to normalize t_i into range $[0,1]$ by giving a value of maximum tolerable time t_{max}, which is the largest execution time that consumers would like to tolerate, and a minimum satisfactory time t_{min}, which is the execution time that consumers are 100% satisfied with (consumers will not get more satisfaction even the web service is faster than that). These two reference values can be determined by consumer surveys. Here they are treated as environmental constants. If the average execution duration exceeds t_{max}, the score from this aspect deserves 1 point. If it is lower than t_{min}, the score is 0. If it is in between, we use the following formula to calculate t_i: $Norm(t_i) = \frac{t_i - t_{min}}{t_{max} - t_{min}}$ and the final value is ranging from $[0,1]$. Therefore, T_{aed} will be in "the-less-the better" family. In general, this TSI represents how fast generally the elementary services can complete their tasks. The less it is, the better the overall quality of elementary services is.

Average Service Availability (T_{asa}). Average service availability $T_{asa}(c, du)$ is the average value of the availability of each elementary service maintained in ASC agent c in the last duration du, i.e., $T_{asa}(c, du) = \frac{\sum_{i=1}^{n} Avai_i}{n}$, where $Avai_i$ is the availability of the ith elementary service and n is the total number of elementary service. The definition of the availability of a single elementary service can be referred to [14]. The higher the average service availability is, the better the quality of elementary services is. Generally, average service availability reflects the overall quality of elementary services in the aspect of on-line available duration.

Average Service Reliability (T_{asr}). Reliability is defined as the probability that a request is correctly responded within the maximum expected time frame indicated in the web service description. Therefore, the reliability of service s in duration du can be calculated as $Rel(s, du) = \frac{n_{resp}}{n_{total}}$, where n_{resp} is the number of correct responses and n_{total} is the number in total. Our average service reliability $T_{asr}(c, du)$ is just the average value of all the maintained elementary services in ASC agent c, i.e., $T_{asr}(c, du) = \frac{\sum_{i=1}^{n} Rel(s_i, du)}{n}$, where n is the total number the elementary services in the service pool of ASC agent c. Same as average service availability, the higher the average service reliability is, the better the quality of elementary services is. Average service reliability stands for the overall quality in the aspect of correctly on-time responding.

Monitoring Strength (T_{ms}). Monitoring strength stands for the strength that elementary services are monitored by ASC agents or a third party. It is reasonable to think that if elementary services are monitored regularly, their services will be of consistent quality. Two factors are considered to have impact on the monitoring strength: monitoring coverage mc and monitoring frequency mf. Monitoring coverage is the proportion of the monitored services over all the services. Its range is $[0,1]$. Monitoring frequency stands for how often the monitoring session is. To simplify the calculation, we normalize monitoring frequency by giving satisfactory monitoring frequency smf as a reference so that the range of mf becomes $[0,1]$. The satisfactory monitoring frequency is a frequency in which if elementary services are monitored, consumers will consider the monitoring to be satisfactory. This reference value can be decided by conducting consumer survey. Once it is decided, it keeps constant and seldom changes. For a certain ASC agent c in duration du, we use the following formula to calculate Monitoring Strength: $T_{ms}(c, du) = \frac{\sum_{i=1}^{n} mc_i}{n} \times Min\{\frac{mf}{smf}, 1\}$, where mc_i is the monitoring coverage of the ith monitoring process. The higher the monitoring strength is, the better the quality that it represents is.

In summary, we proposed several composition-layer and service-layer TSIs, which form a comprehensive metric, reflecting the trustworthiness status of an ASC agent. As mentioned before, it is impossible to give complete metric that shows every aspects of a composer, since the properties are countless and different contexts may concern different things. Our metric is extensible so that consumers can add or remove TSIs as they preferred, as far as the new added ones can be scaled into the range $[0,1]$. In the next part, we will integrate all the TSIs together to form a single value through which the trustworthiness of ASC agents can be comprehensively distinguished.

4   Trustworthiness Comparison

The trustworthiness of ASC agent c in the last duration du can be represented by the follow vector: $(T_{fed}(c, du), T_{eed}(c, du), T_{ad}(c, du), T_{rp}(c, du), T_{aed}(c, du), T_{asa}(c, du), T_{asr}(c, du), T_{ms}(c, du))$. To simplify the representation, the eight TSIs are numbered from 1 to 8 in the order as shown. Then, the above vector can be re-written as $\{T_j(c, du)|j = 1, 2, ..., 8\}$. Given an ASC agent c_i, the jth TSI $TSI_{i,j}$ can be represented as $TSI_{i,j} = T_j(c_i, du)$. We adopt simple additive weighting technique to help transfer the multi-dimensional TSI metric into a single value. This technique has two phases: normalization and weighting.

Table 1. Settings for ASC agents

ASC agent	Avai.	Exec.Drtn	Func.Err.Rate	Moni.Cov.	Moni.Freq.
Agent01	75%	40ms	15%	30%	1/4
Agent02	77%	36ms	14%	40%	1/4
Agent03	79%	32ms	13%	40%	1/4
Agent04	81%	28ms	12%	50%	1/2
Agent05	83%	24ms	11%	60%	1/2
Agent06	85%	20ms	10%	70%	1/2
Agent07	87%	16ms	09%	80%	2/3
Agent08	89%	12ms	08%	80%	2/3
Agent09	91%	8ms	07%	90%	4/5
Agent10	93%	4ms	06%	100%	1

4.1   Normalization Phase

The eight TSIs can be grouped into two, as four of them (No. 4,6,7,8) are "the bigger the better" and the rest are entirely opposite. The normalization processes for these two types of TSIs are different. Given n ASC agents in total, the ith TSI of the jth ASC agent can be normalized by the following formula:

$$v_{i,j}^{N} = \begin{cases} \dfrac{TSI_i^{max}-TSI_{i,j}}{TSI_i^{max}-TSI_i^{min}} & \text{if j=1,2,3,5} \\ & \text{and } TSI_i^{max} \neq TSI_i^{min}; \\[2mm] \dfrac{TSI_{i,j}-TSI_i^{min}}{TSI_i^{max}-TSI_i^{min}} & \text{if j=4,6,7,8} \\ & \text{and } TSI_i^{max} \neq TSI_i^{min}; \\[2mm] 0 & \text{if } TSI_i^{max} = TSI_i^{min} \end{cases} \tag{1}$$

In the above formula, TSI_i^{max} and TSI_i^{min} are respectively the highest and lowest values of the ith TSI of all the ASC agents, i.e., $TSI_i^{max} = Max(TSI_{i,j}), 1 \leq j \leq n$ and $TSI_i^{min} = Min(TSI_{i,j}), 1 \leq j \leq n$. By applying the above formula, we obtain a $8 \times n$ matrix V^N,

$$\begin{bmatrix} v_{1,1} & v_{1,2} & \cdots & v_{1,8} \\ v_{2,1} & v_{2,2} & \cdots & v_{2,8} \\ v_{3,1} & v_{3,2} & \cdots & v_{3,8} \\ \cdots & \cdots & \cdots & \cdots \\ v_{n,1} & v_{n,2} & \cdots & v_{n,8} \end{bmatrix} \tag{2}$$

where each column stands for one TSI dimension and each row represents an ASC agent.

4.2   Weighting Phase

Even though we can normalize the TSI values of all the ASC agents into one matrix, we still cannot judge which ASC agent is more trustworthy, since one

ASC agent may be better in some TSIs and the other may be better in other TSIs. This problem is related to how important each TSI is in the context that the comparison occurs. There are no fixed rules saying what TSI is the most important one and what is the second one. The significance of each TSI depends on user preference.

In our trustworthiness comparison method, users are given an 1×8 matrix $W = (w_j | 1 \leq j \leq 8, 0 \leq w_j \leq 1, \sum_{j=1}^{8} w_j = 1)$ to allocate weight values for each TSI, where element w_j is the weight value for the jth TSI we described before. Given an $8 \times n$ normalized TSI matrix $V^N = (v_{i,j}^N | 1 \leq i \leq n, 1 \leq j \leq 8)$, by applying the following formula, an $1 \times n$ matrix T will be achieved, in which each value T_i stands for trustworthiness of an ASC agent.

$$T = V^N \times W = \begin{bmatrix} v_{1,1} & v_{1,2} & \cdots & v_{1,8} \\ v_{2,1} & v_{2,2} & \cdots & v_{2,8} \\ v_{3,1} & v_{3,2} & \cdots & v_{3,8} \\ \cdots & \cdots & \cdots & \cdots \\ v_{n,1} & v_{n,2} & \cdots & v_{n,8} \end{bmatrix} \times \begin{bmatrix} w_1 \\ w_2 \\ w_3 \\ \cdots \\ w_n \end{bmatrix} \qquad (3)$$

The above formula results in n T_j, each of which is the final normalized trustworthiness for the jth ASC agent. That value is ranging in [0,1]. The higher the value is, the more trustworthy the ASC agent is. Please note that if users only want to calculate an absolute trustworthiness value for an ASC agent (no comparison), they can directly multiple weight values with the result calculated from the formulas given in TSI description. In that case, normalization is not necessary.

As we mentioned, it is very difficult to know which TSI is more important. However, there are some general rules that can be applied to any contexts. The first rule is effort-oriented division, which means that we put more weight to where the major computation jobs are done. If you know most of your work will be done by elementary services and the composer is just for integrating the results, you should put more weight on service-layer TSIs, and vice versa. The second rule is difficulty-driven. We put more weight to the TSIs that are hard to achieve. For example, if you have a massive computation task in which computation logic is very simple but the amount is quite large, rather than putting too much weight on the TSIs reflecting the ability factors, you should give more attention on the quality factors such as execution efficiency. If the computation logic is very complex, you should focus on ability factors such as Functional Effectiveness Deviation. The third rule is concern-driven. If you are quite keen in certain aspects, e.g., in some time-crucial cases where even one microsecond determines everything, you should definitely put the highest weight to the corresponding factors. Again, our metric is open and it accepts any new TSIs as its member, as far as they can be scaled to [0,1]. Users are allowed to create their own TSIs based on their contexts.

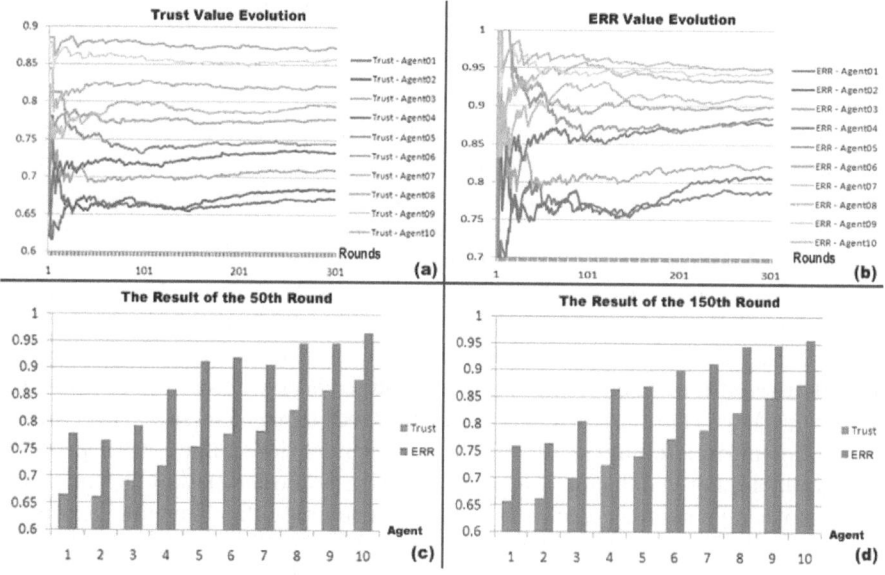

Fig. 1. Trustworthiness Evolution

5   Experiments and Simulation

In this section, we present a simulation experiment to illustrate the effectiveness of our proposed TSI metric. We want to test that given several ASC agents with the same claimed capabilities status and different actual capabilities, the higher their calculated trustworthiness values are, the better the actual Eligible Return Rate (ERR) they give. Here ERR stands for the proportion of the eligible returned results over all the results. The eligible returns are the on-time return with valid composition results. An ASC agent giving an eligible return means it successfully delivered its job without any flaws. We use this indicator to represent the actual level of ASC agents. The whole experiment is conducted under Microsoft Office Excel 2007 Professional with Visual Basic for Application (VBA) as the programming language.

Suppose that there is a business process that needs to composing three types of elementary web services. For each types of function, there are 500 elementary web services available on the web. Their availability follows the normal distribution with mean value of 90% and variance of 10%. Same probability distributions are also followed by their reliability with mean value of 85% and variance of 10% and by actual execution duration with mean value of 20 milliseconds and variance of 5 milliseconds. ASC agents regard web services that can return results in 10ms as perfect and the web services that exceed 35ms to return as non-tolerable.

There are 10 available ASC agents which claim to offer the corresponding composition service. They claim to provide results with 95% score for TSI 1, 2 and 3 and 100% for TSI 8. However, their actual capabilities are shown in the above table. Except the MC and MF data, the rest data in the table are just the average values. The performance of ASC agents follows a normal distribution with the mean values given in the table and 1% variance for availability, 5 milliseconds variance for execution duration and 0.5% for error rate. 200ms is the longest duration consumers would like to wait for one composition job. Each ASC agent maintain their own elementary service pool that consists of 60 web services in total with 20 for each function. Those services are randomly selected from the web. The ideal monitoring strength in our experiment is suppose to check service pool every 10 compositions with coverage 100%.

We simulate the composition processes for each ASC agents for 300 rounds. In each round, for a certain ASC agent, we first check whether it is available by comparing a even-distribution random number with its availability that follows normal distribution. Then we use same method to check whether the composition has errors. After that, the ASC agent has to select three elementary web service with different functionalities. If a selected web service cannot work properly (not available, not reliable or execution duration is too long), ASC agent has to select another elementary service with the same functionality. The final execution duration of ASC agent is calculated as $D_{asc} = \sum_{i=1}^{n} T_i^{ws} + (n-2) \times T_{asc}$, where n is the total times that the WSC interact with different web services, T_i^{ws} is the execution duration of ith web service and T_{asc} is the duration. If the final total execution duration is greater than 200ms, the composition is regarded as not efficient. Until now, we finish calculation of the composition-layer TSIs. The calculation of the first three service-layer TSIs is done based on the current services maintained in each ASC agent. Then based on the monitoring frequency, we decide whether a monitoring process must be started after the current round. If a monitoring process has to be conducted, we will randomly select several web services for each function type according to monitoring coverage and check their availability, reliability and execution duration. The web services that give ineligible results will be removed. In those cases, new services whose quality is better than the removed ones will be added in from the web. A web service that was once selected by a certain ASC agent is not allowed to be added again. After each round, a trustworthiness value (equal weight for TSIs) and a ERR value of each ASC agent are calculated based on the previous composition records. The results are shown in Fig.1.

Fig.1(a) and (b) give the general evolution curves of trustworthiness values and eligible returning rate (ERR) for each ASC agent. In general, as we expected, the better actual capability an ASC agent has, the better its ERR and trustworthiness are. This shows that our calculated trustworthiness result effectively reflect the actual performance. We note that at the first several rounds, the trustworthiness curves and ERR curves fluctuate drastically and can not reflect the actual capability status. This is because the performance of the ASC agents and elementary services are of random nature (normal distributions in the settings). As the experiment goes, our calculated trustworthiness values become stable much faster than ERR. It tells that our selected TSIs and trustworthiness calculation method have better sensitivity of reflecting the actual performance than ERR which purely reflects capability by following previous historical records. This is more clear in Fig.1(c) and Fig.1(d). (c) give the trustworthiness and ERR status after 50 rounds. At that time ERR was still fluctuating, while trustworthiness values had already correctly shown the actual capability comparison. In Fig.1(d) after 150 rounds when ERR gives general capability status, our trustworthiness values are able to present clear and precise differences among the ASC agents.

6 Related Works

Research on trust in the online environment has been studied mainly in the information systems domain and computer science domain.

In the information systems domain, people concerns the fundamental conceptualization and the implication of each concepts. The seminal work that clarify the subtilty includes [9,4,12,5]. There is a branch of trust research which focuses on analyzing the institutional trust elements and their functions [18,10,13]. Recently, some brain visualization technologies are used to analyze trust reactions [8,2]. Basically, the trust research in information systems focuses on exploring the nature and roles of trust and use different methods to analyze the impact of people's trustworthy and untrustworthy behaviors. Their research is valuable in guiding people and organizations to form trust relationships.

Researchers from computer science paid more attention on trust modeling and trust framework construction. There have been some very good survey papers in this field [17,7,1]. Details cannot be listed due to the limit of the required page size.

7 Conclusions and Future Work

In this paper, we clarified several trust-related concepts and explore the dimensions, manifestations and formation of trustworthiness. Based on the special characteristics of automatic web service composition, we proposed a combined composition-layer and service-layer trust status indicator metric to calculate the trustworthiness of ASC agents. For comparison purposes, we use simple additive weight method to transfer the multi-dimensional TSI metric into a single trustworthiness value. We use experimental simulation to prove that our proposed metric and the comparison method can effectively differentiate the trustworthiness of ASC agents. They also present a better performance than the method that purely use historical data to judge trustworthiness.

Our next step will focus on implementing the proposed structure into a software framework and doing experiments with real data to determine the reasonable weights for TSIs under different situations. Exploring new service-layer and composition-layer trustworthiness factors to enrich the framework is also a job on our schedule.

References

1. Artz, D., Gil, Y.: A survey of trust in computer science and the Semantic Web. Journal of Web Semantics: Science, Services and Agents on the World Wide Web 5(2), 58–71 (2007)
2. Dimoka, A.: What Does the Brain Tell Us about Trust and Distrust? Evidence From A Functional Neuroimaging Study. MIS Quarterly 34(2), 373–396 (2010)
3. Dustdar, S., Schreiner, W.: A Survey on Web Service Composition. International Journal of Web and Grid Services 1(1), 1–30 (2005)
4. Gefen, D.: Reflections on the Dimensions of Trust and Trustworthiness among Online Consumers. SIGMIS Database 33(3), 38–53 (2002)
5. Gefen, D., Benbasat, I., Pavlou, P.: A Research Agenda for Trust in Online Environments. Journal of Management Information Systems 24(4), 275–286 (2008)
6. Jiao, H., Liu, J., Li, J., Liu, C.: Trust for Web Service Composers. In: The 12th Global Information Technology Management Association World Conference, GITMA, pp. 180–184 (2011)
7. Jøsang, A., Ismail, R., Boyd, C.A.: A survey of trust and reputation systems for online service provision. Decision Support Systems 43(2), 618–644 (2007)
8. King-Casas, B., Tomlin, D., Anen, C., Camerer, C.F., Quartz, S.R., Montague, P.R.: Getting to Know You: Reputation and Trust in a Two-Person Economic Exchange. Science 308(5718), 78–83 (2005)
9. Mayer, R.C., Davis, J.H., Schoorman, F.D.: An Integrative Model of Organizational Trust. The Academy of Management Review 20(3), 709–734 (1995)
10. McKnight, D.H., Chervany, N.L.: What Trust Means in E-Commerce Customer Relationships: An Interdisciplinary Conceptual Typology. International Journal on Electronic Commerce 6(2), 35–59 (2001)
11. McKnight, D.H., Choudhury, V., Kacmar, C.: Developing and Validating Trust Measures for e-Commerce: An Integrative Typology. Info. Sys. Research 13(3), 334–359 (2002)
12. Pavlou, P.A., Dimoka, A.: The Nature and Role of Feedback Text Comments in Online Marketplaces: Implications for Trust Building, Price Premiums, and Seller Differentiation. Information Systems Research 17(4), 392–414 (2006)
13. Pavlou, P.A., Tan, Y.-H., Gefen, D.: The Transitional Role of Institutional Trust in Online Interorganizational Relationships. In: 36th Hawaii International Conference on Sytem Sciences, HICSS 2003, p. 215a (2003)
14. Ran, S.: A Model for Web Service Discovery with QoS. ACM SIGecom Exchanges 4(1), 1–10 (2003)
15. Rao, J., Su, X.: A Survey of Automated Web Service Composition Methods. In: Cardoso, J., Sheth, A.P. (eds.) SWSWPC 2004. LNCS, vol. 3387, pp. 43–54. Springer, Heidelberg (2005)
16. Tsur, S., Abiteboul, S., Agrawal, R., Dayal, U., Klein, J., Weikum, G.: Are Web Services the Next Revolution in e-Commerce. Paper Presented at the The 27th International Conference on Very Large Data Bases, VLDB 2001 (2001)
17. Wang, Y., Vassileva, J.: A Review on Trust and Reputation for Web Service Selection. In: 27th International Conference on Distributed Computing Systems Workshops, ICDCSW 2007, pp. 25–33 (2007)
18. Zucker, L.G.: Production of trust: Institutional sources of economic structure, 1840-1920. Research in Organizational Behavior 8(1), 53–111 (1986)

An Integrated Approach
for Large-Scale Relation Extraction from the Web

Naimdjon Takhirov[1], Fabien Duchateau[2], Trond Aalberg[1], and Ingeborg Sølvberg[1]

[1] Norwegian University of Science and Technology, 7491 Trondheim, Norway
{takhirov,trondaal,ingeborg}@idi.ntnu.no
[2] Université Lyon 1, LIRIS, UMR5205, Lyon, France
fduchate@liris.cnrs.fr

Abstract. Deriving knowledge from information stored in unstructured documents is a major challenge. More specifically, binary relationships representing facts between entities can be extracted to populate semantic triple stores or large knowledge bases. The main constraint of all knowledge extraction approaches is to find a trade-off between quality and scalability. Thus, we propose in this paper SPIDER, a novel integrated system for extracting binary relationships at large scale. Through series of experiments, we show the benefit of our approach, which in general, outperforms existing systems both in terms of quality (precision and the number of discovered facts) and scalability.

Keywords: Relation Extraction, Knowledge Bases, Web Mining.

1 Introduction

Information available on the Web has the potential to be a great source of structured knowledge. However this potential is far from being realized. The main benefit of obtaining exploitable facts such as relationships between entities from natural language texts is that machines can automatically interpret them. The automatic processing enables advanced applications such as semantic search, question answering and various other applications and services. The **Linked Open Data** (LOD) aims at making the vision of creating a large structured database a reality. In this domain, the building of semantic knowledge bases such as DBpedia, MusicBrainz or Geonames is (semi-)automatically performed by adding new facts which are usually represented by triples. However, most of these triples express a simple relationship between an entity and one of its properties, such as the *birthplace of a person* or the *author of a book*. By mining structured and unstructured documents from the Web, one can provide more complex relationships such as *parodies*. A different vision known as the **web of concepts** shares similar objectives with LOD [12]. As a consequence, knowledge harvesting [9,10,14] and more generally open-domain information extraction [5] are emerging fields with the goal of acquiring knowledge from textual content.

In this paper, we propose a relation extraction approach named **SPIDER**[1]. It aims at addressing the previously mentioned issues by integrating the most relevant techniques

---

[1] **S**emantic and **P**rovenance-based **I**ntegration for **D**etecting and **E**xtracting **R**elations.

Y. Ishikawa et al. (Eds.): APWeb 2013, LNCS 7808, pp. 163–175, 2013.

to generate trustworthy patterns and relationships. SPIDER is based on a perpetual process of generating patterns and examples either in supervised or unsupervised mode. The main intuition is that generic patterns which are derived from similar sentences and which are discovered in trustworthy documents are useful for detecting relationships in other documents. In summary, our contributions in this paper are:

- Extracting relationships from a Web-scale source is a major bottleneck. To the best of our knowledge, SPIDER is the first approach which does not require several days to perform a single iteration.
- Contrary to most knowledge extraction tools, we tackle the problem of uniquely identifying entities to extend their list of spelling forms and to facilitate the matching to LOD.
- SPIDER includes a flexible pattern definition scheme. This scheme is used to merge similar patterns for efficiency purposes. In addition, we introduce the notion of confidence score that controls the ranking of patterns. The confidence score evolves over time as the system runs continuously.
- Experiments confirm the benefits of our approach w.r.t. similar tools (*ReadTheWeb*, *Prospera*) in terms of quality and performance.

2 Overview

2.1 Problem Definition

The overall goal of SPIDER is to continuously generate relationships and patterns. Let us first define a **relationship**. It is a triplet $<e_1, \tau, e_2>$ where e_1 and e_2 represent **entities** and τ stands for a **type of relationship**. An entity might be represented with different labels in natural language text, and we note $l_e \in \mathcal{L}$ one of the mentions for the entity e. An example of a relationship is *createdBy* between the work entity *The Lord of the Rings* and the person entity *J.R.R. Tolkien*. Note that both the entities and the type of relationship are uniquely identified using an URI. An **example** denotes a **pair of entities** (e_1, e_2) which satisfies a type of relationship.

The patterns are extracted from **a collection of documents** $\mathcal{D} = \{d_1, d_2, ..., d_n\}$. Although not limited to, these documents are webpages in our context. Each document is composed of **sentences**, which may contain mentions of the two entities. In that case, the sentence is extracted as a **candidate pattern**. We note $\mathcal{CP}$ the set of candidate patterns given a collection of documents and a set of initial examples. Each candidate pattern is defined as a tuple $cp = \{t_b, e_1, t_m, e_2, t_a\}$ with t_b, t_m and t_a respectively standing for the *text before*, the *text in the middle* and the *text after* the entities. A sentence *"Bored of the Rings is a parody of Lord of the Rings"* is transformed to a corresponding candidate pattern *{"", e_1, "is a parody of", e_2, ""}*.

From the set of candidate patterns $\mathcal{CP}$, we derive a set $\mathcal{P}$ of **generic patterns** by applying a strategy s. A **strategy** is defined as a sequence of **operations** $s = < o_1, o_2, ..., o_k >$, each operation aiming at generalizing the candidate patterns. Namely, this generalization implies the detection of frequent terms and the POS-tagging of the other terms of the candidate patterns. Thus, a strategy is a function such that $s(\mathcal{CP}) \rightarrow \mathcal{P}$. All generated patterns are associated to a specific type of relationship τ. For instance, a generic pattern for the *parody* type is illustrated with *"{e₁} is/VBZ a/DT {parody, illusion, spoof} of/IN {e₂}"* [2].

Finally, a pattern and an example both have a **confidence score** noted $conf_p$ and $conf_e$ respectively. This score is based on the support, the provenance, the number of occurrences, the number of strategies and the iteration. A **pattern similarity** metric indicates the proximity between two (candidate) patterns. The notion of confidence score, as well as the one for operation, strategy, pattern similarity and (candidate) pattern, are further detailed in the next sections.

2.2 Workflow

Given two labels, SPIDER generates patterns and derives to a relationship. This pattern generation capability is guaranteed by the following two processes:

- **Pattern Generation** is in charge of detecting candidate patterns by using examples or provided entities and of generalizing these candidate patterns to obtain patterns for a given type of relationship (see Figure 1).
- **Example Generation** exploits the previously generated patterns in order to discover new examples which satisfy the type of relationship.

The **knowledge base** stores all generated examples and patterns. These examples can be used to maintain the system continuously running, but they can be exploited from a user perspective too.

3 Pattern Generation

The pattern generation process either requires a few examples for a given type of relationship so that patterns for this type of relationship can be automatically generated (supervised), or it directly tries to guess the type of relationship for two given labels (unsupervised). The process is similar in both modes and it is composed of three main steps: extension of entities, extraction of candidate patterns from the collection of documents and their refinement into patterns.

3.1 Extending Entities

In a document, entities are not uniquely identified by a label but they have alternative labels or spelling forms. Therefore, extending these entities with their alternative labels is a crucial step and it requires the correct identification of the entity. For instance, the entity *"Lord of the Rings"* can be labeled *"LOTR"* or *"The Lord of the Rings"*. To avoid missing potentially interesting relationships, we search for these alternative forms of spelling in the documents. Given an entity e represented by a label l, the goal is to discover its set of alternative labels $\mathcal{L}_e = \{l, l^1, l^2, ..., l^n\}$. The idea is to match the entity against LOD semantic knowledge bases to obtain this list of alternative labels. Namely,

---

[2] VBZ=Verb, 3rd person sing. present, DT=Determiner, IN=Preposition or subordinating conjunction.

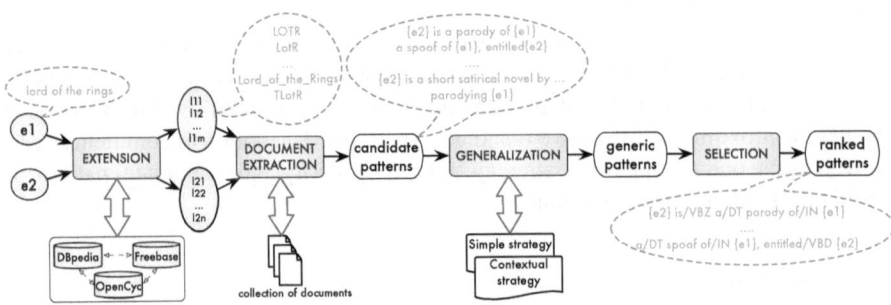

Fig. 1. The Pattern Generation Process

we build various queries by decomposing the initial label and we query in the aliases attributes of knowledge bases (i.e., *common.topic.alias* for Freebase, *wikiPageRedirects* for DBpedia, etc.). In most cases, several candidate entities are returned and the tool tries to automatically select the correct one.

The process of automatically selecting the correct entity is achieved as follows. First, an AND query is constructed with the two labels. Clusters of documents are built representing documents belonging to a set of specific type of entities. The n number of words around labels are extracted and stemming performed on words. Our assumption is based on the fact that documents mentioning the same entities tend to have similar words. Therefore, a graph of semantically related words is built. The most important documents in the cluster are then compared against the abstract of the automatically selected entities. Next, we extract frequent terms from the most important documents in the result set and use these frequent words as extensions. Note that if disambiguation is not possible, we discard the example and we do not use it for subsequent pattern generation. The result of the extension process is a list of alternative labels as illustrated in Figure 1.

The main issue in this step deals with the absence of the two labels in any knowledge base, which means that the entities cannot be extended. The number of retrieved documents in that case could not be sufficient to extract good candidate patterns. The first solution consists of analyzing these retrieved documents to detect potential alternative spellings by applying metrics such as tf-idf and Named Entity Recognition techniques. Another possibility is to relax the similarity constraint when searching a label in a knowledge base. In other words, a strict equality measure would not be applied between the label and the candidate spelling forms from a knowledge base. Rather n-grams or Levenshtein similarity metrics with a high threshold would be a better choice.

3.2 Extraction of Candidate Patterns

As depicted in Figure 1, the outcome of document extraction process is candidate patterns. Given the lists of extended labels for both entities, our tool associates all alternative labels of the first entity to all labels of the second entity (Cartesian product) to build different queries. The documents resulting from these queries are ranked according to their relevance score. The candidate patterns are extracted by parsing these documents and locating the sentences with co-occurrence of both entities (defined by a maximum number of words between them, currently 15 words). Note that we include in the candidate patterns the text before and after the entities to obtain full sentences. The final step aims at refining the candidate patterns to obtain patterns.

3.3 Selection of Patterns

The last issue deals with the selection or ranking of the generic patterns. Thus, a **confidence score** noted $conf_p$ is computed for each pattern p with Formula 1. Our intuition is to exploit all information which allowed the discovery of the patterns and to compare a pattern with the ones of the same type of relationship.

$$conf(p) = \left(\frac{\alpha sup_p + \beta occ_p + \gamma prov_p}{\alpha + \beta + \gamma} \right) \tag{1}$$

The **support** sup_p is defined as the ratio between the number of examples ex_p that this pattern is able to discover and the total number of examples ex_τ discovered by all patterns of the same type of relationship τ. Note that the support cannot be computed at the first iteration.

Similarly, the **occurrency** occ_p stands for the number of candidate patterns which led to the generation of the pattern p. It is normalized by the total number of candidate patterns used to generalize all patterns of the same type of relationship τ.

$$supp_p = \frac{ex_p}{ex_\tau} \qquad\qquad occ_p = \frac{occ_p}{occ_\tau}$$

The **provenance** $prov_p$ refers to the relevance of the documents from which the candidate patterns which generalize a given pattern have been extracted. The relevance is evaluated given three metrics: the relevance score namely applies tf-idf on the content of the document and its values are bounded by a maximal value which depends on the query. PageRank[3] is widely known due to the Google search engine. The PageRank scores of our collection are in the range $[0.15, 10]$. Finally, SpamScore indicates the probability that a document is a spam or not [8]. The idea is to average the scores returned by these three metrics for all documents from which patterns have been derived. We note $\mathcal{D}_p$ this set of documents for the pattern p, and d_p^i a document of this set. The following formula computes a score in the range $[0, 1]$ to evaluate the average relevance of this set of documents, and thus the provenance of the pattern:

$$prov_p = \frac{\sum^{d_p^i} \frac{1}{3} \left(\frac{relevance(d_p^i)}{max(relevance(\mathcal{D}_p))} + \frac{spamscore(d_p^i)}{100} + \frac{prank(d_p^i)}{10} \right)}{|\mathcal{D}_p|}$$

4   Relationship and Example Discovery

In the previous step, we have generated patterns for a given type of relationship. SPIDER aims at discovering relationships between entities and discovering new examples.

---

[3] http://j.mp/Clueweb09-Pagerank

4.1   Challenges for Exploiting Patterns

Using the generated patterns to discover knowledge from plain texts involves the addressing of the following two challenges: pattern similarity and NER.

Pattern Similarity. When analyzing sentences in a document, SPIDER needs to evaluate the similarity between a sentence and a pattern. Thus, we have designed a pattern similarity metric. The intuition which underlies our metric is twofold: (i) the presence of frequent terms in the sentence is crucial while there is more flexibility for less important terms and (ii) the position of the words should be taken into account. First, the sentence is transformed (cleaning, POS-tagging and replacing the mentions of the entities) so that both the sentence s and the pattern p are composed of POS-tagged terms. The pattern may also include a list of frequent terms, which is noted $\mathcal{FT}_p$. The idea is to compute Δ, the minimal total distance to transform the sentence into the pattern. Thus, our metric is an adaptation of the Levenshtein distance [7]. However, we do not compute a number of operations (delete, transform or add a term) between two words or characters, but rather we evaluate the semantic distance between two words. Given the ith word w_p^i associated to the tag t_p^i in the pattern p and the word w_s^j with the tag t_s^j in the sentence s, we compute their semantic distance $semdist(w_p^i, w_s^j)$ using the following formula:

$$semdist(w_p^i, w_s^j) = \begin{cases} 0.0 & \text{if } w_s^j \in \mathcal{FT}_p \\ resnik(w_p^i, w_s^j) & \text{if } w_s^j \notin \mathcal{FT}_p \text{ and } t_p^i = t_s^j \\ 1.0 & \text{otherwise} \end{cases} \qquad (2)$$

Namely, the distance is equal to 0 if the word in the sentence is a frequent term. POS-tagged terms whose tags are identical (e.g., two verbs) have a similarity obtained by applying the Resnik distance in Wordnet [13]. Else, words with different tags have the maximal distance. The minimal total distance $\Delta(p, s)$ between a pattern p and a sentence s is computed by the matrix algorithm of the Levenshtein distance[4] using the semantic distance for all pairs of words. The distance is normalized in the range $[0, 1]$ with Formula 3 which assesses the similarity $pattsim$.

$$pattsim(p, s) = \frac{1}{1 + \Delta(p, s)} \qquad (3)$$

Our metric is flexible because extra or missing words in a sentence do not significantly affect the similarity value. Similarly, less important words are mainly compared on their nature (POS-tag). Finally, we can select the sentences which are modeled by a pattern according to a threshold.

NER. When a sentence in a document corresponds to a pattern, our approach needs to identify and extract entities contained in the sentence. Thus, the NER issue is crucial as it determines (part of) the output. Indeed, it is necessary to correctly identify the (labels of) entities in the sentence based on a pattern. To solve this issue, we rely on the formalism of our patterns: since they have been POS-tagged, the tags serve as a delimiter and may constrain the candidate entities.

---

[4] http://j.mp/Levenschtein

4.2   Discovering the Type of Relationship

Given two labels (representing an entity), the goal is to determine the possible type(s) of relationships between these two entities. The two entities are first extended to obtain their alternative labels (see Section 3.1). A set of documents is analyzed to extract all sentences which contain one label of each entity, and these sentences are then compared to all patterns stored in SPIDER's knowledge base (see Section 4.1). If a sentence is similar to a trustable pattern (with a sufficient confidence score), then the type of relationship corresponding to this pattern is proposed to the user. Note that if there is no trustable pattern in the knowledge base, the user has the possibility to provide training data to the system.

4.3   Discovering New Examples

The discovery of new examples is at the basis of the *never-ending* feature which enables the feeding of the knowledge base with additional training data for generating new patterns. SPIDER selects in the knowledge base the patterns of a given type of relationship. It retrieves a set of documents by querying each de-tagged pattern (or only their frequent terms). We compute the similarity between a pattern and the sentences of the documents which include a frequent term of this pattern. If the sentence is modeled by the pattern, then we apply the NER techniques for discovering the two entities. Note that each discovered example has a confidence score, which is computed with the same formula as in Section 3.3 for the confidence of a pattern, except that sup_e replaces sup_p and it indicates the number of patterns which discovered this example. Examples with a very low confidence are discarded while others are stored in the knowledge base.

5   Scalability

In order to efficiently process documents, we distribute the jobs into several machines. MapReduce inspired techniques have been popular to tackle such tasks. Therefore, the collection is indexed with Hadoop enabling efficient indexing and searching. To compute statistics, SPIDER makes use of Pig[5] which is a high level platform for analyzing a large collection of data.

Additionally, we propose a document partition to incrementally provide results on a subset of the collection. The general idea is that highly ranked documents should be a better source for obtaining patterns than those with lower PageRank, SpamScore and relevance score values. The size of a partition depends on the quality of the obtained patterns and examples. SPIDER is able to automatically tune the ideal size of a partition. Initially, the documents are sorted by their PageRank, SpamScore and the relevance score as described above. The top k documents are selected for analysis from the head of the ranked list of documents. For the i-th round, the cursor is moved to the range $[k, i \times k]$ and the documents in that range are picked out for analysis. Furthermore, when there are too few patterns discovered for two given labels out of these initial set of documents, the partition size is subsequently adjusted to a higher number.

---

[5] http://pig.apache.org

The combination of scores as well as the partitioning mechanism makes obtaining the URLs of the documents and their content for a given query fast, i.e., it only requires a few seconds. This efficiency is demonstrated in Section 7.2.

6   Related Work

Relationship extraction has been widely studied in the last decade. **Supervised systems** for relationship extraction are mainly based on kernel methods, but they suffer from the processing costs and the difficulty for annotating new training data. One of the earliest semi-supervised system, **DIPRE**, uses a few pairs of entities for a given type of relationship as initial seeds [3]. It searches the Web for these pairs to extract patterns representing a relationship, and use the patterns to discover new pairs of entities. These new entities are integrated in the loop to generate more patterns, and then find new pairs of entities. **Snowball** [1] enhances DIPRE in two different directions. First, a verification step is performed so that generated pairs of examples are checked with MITRE named entity tagger. Secondly, the patterns are more flexible because they are represented as vectors of weighted terms, thus enabling the clustering of similar patterns.

Espresso [11] is a weakly-supervised relation extraction system that also makes use of generic patterns. The system is efficient in terms of reliability and precision. However, experiments were performed on smaller datasets and it is not known how the system performs at Web-scale.

TextRunner brought further perspective in the field of Open Information Extraction, for which the types of relationships are not predefined [2]. A self-supervised learner is in charge of labeling the training data as positive or negative, and a classifier is then trained. Relations are extracted with the classifier while a redundancy-based probabilistic model assigns confidence scores to each new examples. The system was further developed into a **ReVerb** framework [6] which improves the precision-recall curve of the TextRunner.

ReadTheWeb / NELL [4] is another project that aims at continuously extracting categories (e.g., the type of an entity) and relationships from web documents and improving the extraction step by means of **learning** techniques. Four components including a classifier and two learners are in charge of deriving the facts with a confidence score. According to the content of the online knowledge base, more iterations provide high confidence scores (almost 100%) for irrelevant relationships. Contrary to NELL, we do not assume that one document returning a high confidence score for a given relationship is sufficient for approving this relationship. In addition, NELL is mainly dedicated to the discovery of categories (95% of the discovered facts) rather than relationships between entities.

A recent work reconciles three main issues in terms of precision, recall and performance [10]. Indeed, **Prospera** utilizes both pattern analysis with n-gram item sets to ensure a good recall and rule-based reasoning to guarantee an acceptable precision. The performance aspect is handled by partitioning and parallelizing the tasks in a distributed architecture. A restriction of this work deals with the pattern which only covers the middle text between the two entities. This limitation affects the recall, as shown with the example *"Lord Of The Rings, which Tolkien has written"*.

Contrary to the oldest systems which include hard representations of patterns, Prospera and SPIDER includes a more **flexible definition of the patterns**, so that similar patterns can be merged. In addition, the patterns are at the sentence level, which means that the texts before, after and between the entities are considered. Our **confidence score** for a pattern or an example takes into account crucial criteria such as provenance. In addition, the support and the occurrence scores are correlated within the same type of relationship, thus the confidence in a pattern or an example may decrease over time. To the best of our knowledge, none of these approaches deal with the issue of **identifying the alternative labels** of an entity. In the future, we plan to demonstrate the impact of these alternative labels on the recall. Finally, our approach is **scalable** with document partitioning based on smart sorting using the SpamScore, PageRank and relevance score of the documents.

7   Experimental Results

Our collection of documents is the English portion of the ClueWeb09 dataset (around 500 million documents). For the components, we have used the contextual strategy, with the Maxent POS-tagger provided with StandfordNLP[6]. The NER component is based on the OpenNLP toolkit[7]. Evaluation is performed using well-known metrics such as precision (number of correct discovered results divided by the total number of discovered results). The recall can only be estimated since we cannot manually parse the 500 million documents to check if some results have been forgotten. However, we show that the number of correct results increases over time.

7.1   Quality Results

In this section, we present our results in terms of quality of label extension, relationship discovery and comparison with state of the art baseline knowledge extraction tools. The evaluation of relationship discovery depends on the quality of generated patterns and hence we present this evaluation rather than the pattern generation itself.

Relationship Discovery. We evaluate the quality obtained by SPIDER when running the first use case (Section 4.2). Given two labels (representing entities), we **search for the correct type of relationship** which links them. To fulfill this goal, we have manually designed a set of 200 relationships, available at this URL[8]. Note that the type of relationship associated to each example is the most expected one, but several types of relationship are possible for the same example. Table 1 provides a sample of examples (e.g., *Obama, Hawai*) and some candidate types of relationship discovered by SPIDER (e.g., *birthplace*). A **bolded** type of relationship indicates that it is correct for this example. A second remark about our set of relationships deals with the complexity of some relationships (e.g., *<cockatoo, tail, yellow>*). The last column shows the initial confidence score computed for the candidate relationship. The quality is measured

---

[6] http://nlp.stanford.edu/software/CRF-NER.shtml
[7] http://opennlp.apache.org/
[8] http://j.mp/apweb2013

Table 1. Examples of Discovered Types of Relationship and Confidence Scores

Example	Discovered type of relationship	Confidence score
	birthplace	0.42
Obama, Hawai	senator	0.31
	president-elect	0.18
	amazon	0.32
cockatoo, yellow	parrot	0.31
	tail	0.16
	plant	0.51
eucalyptus, myrtaceae	**family**	0.43
	specie	0.27
	inventor	0.60
Bartolomeo Cristofori,	instrument	0.43
piano	**maker**	0.19

in terms of precision at different top-k. Indeed, SPIDER outputs a ranked list of relationship types according to their confidence scores. In addition, our approach is able to run with or without training data. Thus, we have tested the system when a few training data have been provided. Using 1 training example means that the system has randomly selected 1 correct example for bootstrapping the system. Experiments with the training data are based on cross-validation and 5 runs reduce the impact of randomness. The manual validation of the discovered relationships has been performed by 8 people. This manual validation includes around 3000 invalid relationships and 600 correct ones, and it facilitates the automatic computation of precision. In addition, we are able to estimate the recall, i.e. to evaluate the number of correct types discovered during a run w.r.t. all validated types. This is an estimation because there may exist more correct types of relationship than the ones which have been validated. Besides, a discovered type may have a different spelling from a validated type while both have the same meaning, thus decreasing the recall.

Figures 2(a) and 2(b) respectively depict the **average precision and the average recall** for the 200 relationships by top-k and by number of provided training data. We first notice that SPIDER achieves low quality without training data (precision from 40% at top-1 to 30% at top-10). The estimated recall values are also quite low at top-1 because there is an average of 3 correct types of relationships for each example. The top-3 results are interesting with 5 training data: the precision is acceptable (more than 80%) while the recall value (32%) indicates that one type of relationship out of three is correctly identified. Since our dataset contains complex relationships, this configuration is promising for bootstrapping the system. Precision strongly decreases at top-5 and top-10, mainly because each example roughly includes 3 types. However, the top-5 and top-10 recall values indicate that we discover more correct examples. Finally, we notice that providing a few training data (5 examples) enables at least a 10% improvement both for precision and recall. This remark is important since our approach aims at running perpetually by reusing previously discovered examples and patterns.

The quality results are subject to the complexity of the set of relationships, since we have selected some complex ones to discover, such as < *"cockatoo"*, *"tail"*, *"yellow"*>. Other problems of disambiguation occurred, for instance the example *"Chelsea"*, *"London"* mainly returns types of relationships about accommodations because Chelsea is identified as a district of London and not as the football team.

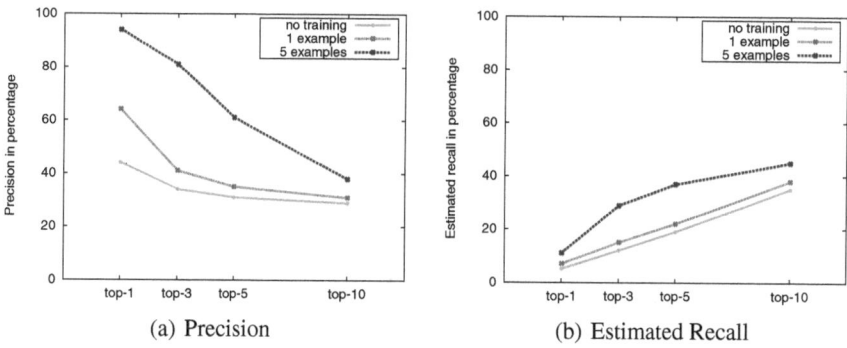

(a) Precision   (b) Estimated Recall

Fig. 2. Quality results according to top-k and Training Data

Baseline Comparison. A final experiment aims at comparing our system with two other approaches, ReadTheWeb (NELL) [4] and Prospera [10], both described in Section 6. These two approaches have been chosen as baseline because the dataset along with the results are available online. An evaluation of these tools is described online[9]. Since the seed examples are available, we have used them as training data. Table 2 summarizes the **comparison between the three systems** in terms of estimated precision, as explained in the experiments reported in [4,10]. Similarly to Prospera and ReadTheWeb, our precision is an estimation due to the amount of relationships to validate. Namely, 1000 random types have been validated for each relationship. The average precision of the three systems is the same (around 0.91). However, the total number of facts discovered by SPIDER (71,921) is 36 times higher than ReadTheWeb (2,112) and 1.3 times higher than Prospera (57,070), outperforming both baselines.

Prospera provides slightly better quality results than our approach on *AthletePlaysForTeam* relation. However, several factors have an influence on the precision results between Prospera, ReadTheWeb and SPIDER. First, Prospera is able to use seeds and counter seeds while we only rely on positive examples. On the other side, Prospera includes a rule-based reasoner combined with the YAGO ontology. Although SPIDER does not support this feature, the combination of POS-tagged patterns and NER techniques achieves outstanding precision values.

7.2   Performance

Since knowledge extraction systems deal with large collections of documents, they need to be scalable. Figure 3(a) depicts the performance of SPIDER for retrieving and preprocessing (i.e., clean up the header, remove html tags) the documents. The total time (sum of retrieval and preprocessing) is also indicated. Although there is no caching, the total time is not significant for collecting and preprocessing one million documents (around 40 seconds). Note that in real cases, a conjunctive query composed of two labels rarely returns more than 20,000 documents. The peak for retrieval at 600,000 documents is due to an overhead processing from the thread manager. Increasing the

---

[9] http://www.mpi-inf.mpg.de/yago-naga/prospera/

Table 2. Estimated Precision (with Number of Discovered Facts) values obtained by ReadTheWeb (RTW), Prospera and SPIDER

Relation	RTW	Prospera	SPIDER
AthletePlaysForTeam	**1.00** (456)	0.82 (14,685)	0.80 (15,234)
TeamWonTrophy	0.68 (397)	0.94 (98)	**0.96** (92)
CoachCoachesTeam	**1.00** (329)	0.88 (1,013)	0.90 (1,629)
AthleteWonTrophy	n/a	0.92 (10)	**0.94** (124)
AthletePlaysInLeague	n/a	0.94 (3,920)	**0.95** (4,211)
CoachCoachesInLeague	n/a	**0.99** (676)	0.89 (741)
TeamPlaysAgainstTeam	0.99 (1,068)	0.89 (15,170)	0.93 (15,729)
TeamPlaysInLeague	n/a	0.89 (1,920)	**0.95** (2,409)
TeamMate	n/a	**0.86** (19,578)	0.84 (31,752)

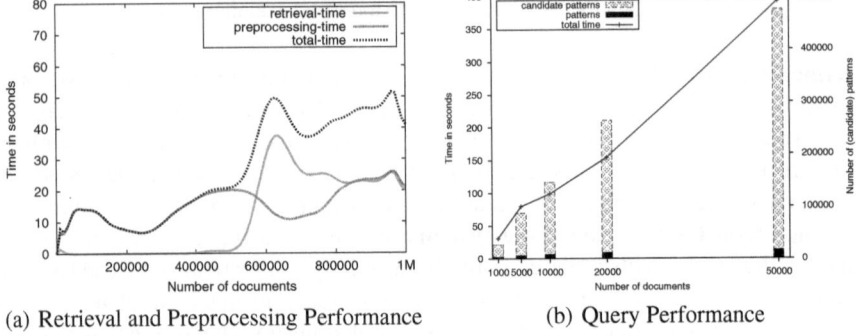

(a) Retrieval and Preprocessing Performance   (b) Query Performance

Fig. 3. Performance results

number of threads above 400 leads to higher thread switching latency while decreasing this number only reports the peak earlier during the process. This issue could be simply solved by dispatching this task on different servers.

ReadTheWeb and Prospera expose their knowledge base but not their tools. Thus, it is not possible to evaluate the three systems on the same hardware and the following comparison is based on the performance described in the original research papers. It mainly aims at showing the significant improvement of SPIDER over ReadTheWeb and Prospera, for a better average quality. To produce the results shown in Table 2, Prospera needed more than 2 days using 10 servers with 48 GB RAM [10]. In a similar fashion, ReadTheWeb has generated an average of 3618 facts per day during the course of 67 days using the Yahoo $M45$ supercomputing cluster [4]. On the contrary, our approach performed the same experiment in a few hours. The generation of facts for each type of relationship took between 20 to 60 minutes with four servers equipped with 24 GB RAM. Although these values are mainly indicative, SPIDER is more efficient than the two other systems when dealing with dynamic and large-scale environments.

8 Conclusion

In this paper, we have presented SPIDER, an approach to **automatic extraction of binary relationships from large text corpora**. The main advantage of our system is to guarantee both **a better quality** and **a strong improvement in terms of performance** over similar approaches, thus providing new opportunities for discovering relationships at large scale. Finally, we have demonstrated **the feasibility of SPIDER** at Web-scale.

References

1. Agichtein, E., Gravano, L.: Snowball: Extracting relations from large plain-text collections. In: Proc. of DL, pp. 85–94. ACM (2000)
2. Banko, M., Cafarella, M.J., Soderland, S., Broadhead, M., Etzioni, O.: Open information extraction from the web. In: Proc. of IJCAI, pp. 2670–2676. Morgan Kaufmann (2007)
3. Brin, S.: Extracting patterns and relations from the world wide web. In: Atzeni, P., Mendelzon, A.O., Mecca, G. (eds.) WebDB 1998. LNCS, vol. 1590, pp. 172–183. Springer, Heidelberg (1999)
4. Carlson, A., Betteridge, J., Kisiel, B., Settles, B., Hruschka Jr., E., Mitchell, T.M.: Toward an architecture for never-ending language learning. In: Proc. of AAAI. AAAI Press (2010)
5. Etzioni, O., Banko, M., Soderland, S., Weld, D.S.: Open information extraction from the web. Communication of ACM 51, 68–74 (2008)
6. Fader, A., Soderland, S., Etzioni, O.: Identifying relations for open information extraction. In: Proc. of EMNLP, pp. 1535–1545. ACL (2011)
7. Levenshtein, V.: Binary Codes Capable of Correcting Deletions, Insertions and Reversals. Journal of Soviet Physics Doklady 10, 707 (1966)
8. Lynam, T.R., Cormack, G.V., Cheriton, D.R.: On-line spam filter fusion. In: Proc. of SIGIR, pp. 123–130. ACM (2006)
9. Mausam, Schmitz, M., Soderland, S., Bart, R., Etzioni, O.: Open language learning for information extraction. In: Proc. of EMNLP, pp. 523–534. ACL (2012)
10. Nakashole, N., Theobald, M., Weikum, G.: Scalable knowledge harvesting with high precision and high recall. In: Proc. of WSDM, pp. 227–236. ACM (2011)
11. Pantel, P., Pennacchiotti, M.: Espresso: leveraging generic patterns for automatically harvesting semantic relations. In: Proc. of ACL, pp. 113–120. ACL (2006)
12. Parameswaran, A., Garcia-Molina, H., Rajaraman, A.: Towards the web of concepts: extracting concepts from large datasets. VLDB Endowment 3, 566–577 (2010)
13. Resnik, P.: Semantic similarity in a taxonomy: An information-based measure and its application to problems of ambiguity in natural language. Journal of Artificial Intelligence Research 11, 95–130 (1999)
14. Takhirov, N., Duchateau, F., Aalberg, T.: An evidence-based verification approach to extract entities and relations for knowledge base population. In: Cudré-Mauroux, P., Heflin, J., Sirin, E., Tudorache, T., Euzenat, J., Hauswirth, M., Parreira, J.X., Hendler, J., Schreiber, G., Bernstein, A., Blomqvist, E. (eds.) ISWC 2012, Part I. LNCS, vol. 7649, pp. 575–590. Springer, Heidelberg (2012)

Multi-QoS Effective Prediction
in Web Service Selection

Zhongjun Liang[1,*], Hua Zou[1],
Jing Guo[2], Fangchun Yang[1], and Rongheng Lin[1]

[1] State Key Laboratory of Networking and Switching Technology
[2] School of Computer Science
Beijing University of Posts and Telecommunications, Beijing, China
{kkliangjun,guo.apple8}@gmail.com, {zouhua,fcyang,rhlin}@bupt.edu.cn

Abstract. The rising development of service-oriented architecture makes Web service selection a hot research topic. However, there still remains challenges to design accurate personalized QoS prediction approaches for Web service selection, as existing algorithms are all focused on predicting individual QoS, without considering the relationship between them. In this paper, we propose a novel Multi-QoS Effective Prediction (MQEP for short) problem, which aims to make effective Multi-QoS prediction based on Multi-QoS attributes and their relationships. To address this problem, we design a novel prediction framework Multi-QoS Effective Prediction Approach (MQEPA for short). MQEPA first takes use of Gaussian method to normalize the QoS attribute values, then exploits Non-negative Matrix Factorization to extract the feature of Web services from Multi-QoS attributes, and last predicts the Multi-QoS of unused services via Multi-output Support Vector Regression algorithm. Comprehensive empirical studies demonstrate the utility of the proposed method.

Keywords: Web service, Multi-QoS, QoS Prediction.

1 Introduction

In recent years, Web services are designed as computational components to build service-oriented distributed systems [1]. The rising development of service-oriented architecture makes more and more alternative Web services offered by different providers with equivalent or similar functions. So how to select Web services from a large number of functionally-equivalent candidates become an important issue.

In an intuitive way, Web service can be selected from all candidates by their QoS values. However, it is hard to obtain all their QoS observed by a target user, as some Web service invocations may be charged. Hence QoS prediction [2-5] become a challenging research topic. For example, Jieming Zhu et al [2] proposes a Web service positioning framework for response time prediction. Zibin Zheng

---

* Corresponding author.

Y. Ishikawa et al. (Eds.): APWeb 2013, LNCS 7808, pp. 176–183, 2013.

and Michael R. Lyu [3] propose a collaborative reliability prediction approach. Yilei Zhang et al [4] propose a neighborhood-based approach for collaborative and personalized quality prediction of cloud components.

However, there remains challenges to design accurate personalized QoS prediction approaches for Web service selection.

- How to predict Multi-QoS simultaneously? In Web service selection, the user wants to obtain Multi-QoS at same time. Existing algorithms are all focused on predicting individual QoS one by one, which cannot scale to Multi-QoS prediction problem. So how to effective predict the Multi-QoS is an interesting problem.

- How to improve the accuration of Multi-QoS prediction? Existing algorithms make QoS prediction without considering the relationship between QoS attributes. In fact, there are certain relationships among the QoS attributes of a same Web service. For instance, the response time and the throughput obtained by users will become worse when the network performance is poor. In addition, probable QoS prediction should be made based on many aspects, and multi-dimensional QoS information is helped for rational decision-making. So introducing Multi-QoS attributes and their relationships for Multi-QoS prediction deserves further research.

In this paper, we propose a novel Multi-QoS Effective Prediction problem, which aims to make effective Multi-QoS prediction based on Multi-QoS attributes and their relationships. To address this problem, we design a novel prediction framework Multi-QoS Effective Prediction Approach. It first take use of Gaussian method to normalize the QoS attribute values, then exploits Non-negative Matrix Factorization to extract the latent feature of users and services from Multi-QoS attributes, and last predicts the Multi-QoS value of unused services via Multi-output Support Vector Regression algorithm. Comprehensive empirical studies demonstrate the utility of the proposed method.

The contribution is as follows.

First, we propose a novel Multi-QoS Effective Prediction problem, which aims to make effective Multi-QoS predict based on Multi-QoS attributes and their relationships.

Second, we propose a novel prediction framework Multi-QoS Effective Prediction Approach to address this MQEP problem. Unlike existing algorithms that all focused on predicting QoS independently one by one, and taking no account of the relationship between them, MQEPA taking use of these Multi-QoS attributes and their relationships to predict QoS.

Last, we conduct several experiments on real-world Web service dataset, comprehensive empirical studies demonstrate the utility of the proposed method.

2   Problem Statement

In this section, we formally describe the Multi-QoS effective prediction problem as follow.

Given a set of users and a set of Web services, based on the existing Multi-QoS values from different users, predict the missing Multi-QoS values of Web service when invoked by a target user at same time. For example, a bipartite graph $G = (U \cup S, E)$ represents the user-Web services invocation (shown as Fig.1), where U denotes the users set, and S denotes the Web services set, E denotes the invocations set between U and S. If user u_i has invoked Web service s_i, where $u_i \in U$, $s_j \in S$, and $e_{ij} \in E$, the weight $Q_{ij} = \{q_{ij}^1, q_{ij}^2, ..., q_{ij}^m\}$ on edge e_{ij} corresponds to the Multi-QoS values (e.g., throughput, response-time in this example) of that invocation. Our task is to effectively predict the Multi-QoS values of potential invocations (the broken lines) for a targeted user.

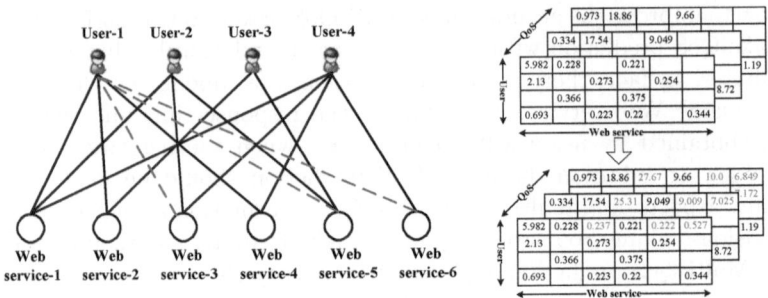

Fig. 1. Bipartite Graph of User-Web service Invocation

In fact, the Multi-QoS effective prediction problem can be considered as how to effective predict the missing entries in the user-Web service matrix based on the existing entries, where the existing entries are Multi-QoS values between U and S, and can be observed in Fig.1.

3 Multi-QoS Effective Prediction Approach

The personalized QoS experience, obtained from a user after the invocation, is co-determined by the feature of corresponding invoked Web service and the user. Therefore, for a target user, he always obtains different QoS experience after invoking the different Web services, as the feature of these Web services is different. As a result, we predict the Multi-QoS based on the extracted feature of Web services and historical QoS experience of a target user.

In this paper, we propose a novel three-step prediction framework Multi-QoS Effective Prediction Approach (shown as Figure.2).

- First, preprocess step, which aims to normalize the QoS attribute values by Gaussian method. This step can map the QoS values to the interval [0,1] for feature extraction.
- Second, extraction step, which aims to extract the feature of Web services from existing Multi-QoS values. This step can handle the acquired sparse Multi-QoS information.

– Third, prediction step, which aims to predict the missing Multi-QoS values via Multi-output Support Vector Regression algorithm. This step finally solves our problem based on the extraction feature of Web services and the targeted user's historical QoS experience.

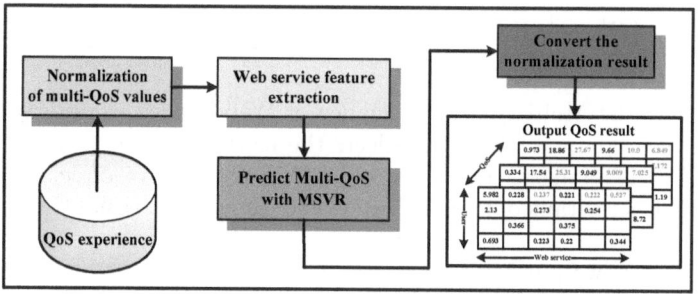

Fig. 2. The Whole Framework of Proposed Approach

3.1   The QoS Normalization

Since different QoS attributes have different range of values, we employ Gaussian method[6] to normalize the QoS attribute values for feature extraction. It can not only map the QoS values to the interval [0,1], but also can better avoid the influence of abnormal values (such as high value or low value) compared with other normalization methods (such as Extremum Regularization Method). The detail can be described as follow.

– First, exploit formulation (1) to normalize the Multi-QoS attribute values in each dimension. It has been proved that 99% data can be directly mapped to the [0,1] interval[6].

$$Q_{ij}^k = 0.5 + \frac{q_{ij}^k - \tilde{q}_{ij}^k}{2 \times 3\sigma} \tag{1}$$

Where Q_{ij}^k is normalized value of k-th QoS attribute, $\tilde{q}_{ij}^k$ denote the average value of k-th QoS attribute obtained from all users, σ is the standard deviation of the k-th QoS attribute.

– Second, assign the Q_{ij}^k with 0 or 1 based on the distance with 0 or 1, if it is out of the [0,1] interval.

3.2   Feature Extraction Based on Multi-QoS Attributes

To better reflect the feature of Web service, we extract it based on Multi-QoS attributes. Due to the existing QoS matrix is always sparse, we employ matrix factorization to extract the features of Web services. The main idea of this method is to derive a high-quality low-dimensional matrix to approximate the historical QoS data, which represent the feature of Web service and user respectively.

Consider Multi-QoS matrix $Q = \{Q^1, Q^2, ..., Q^m\}(Q \in R^{l \times m \times n})$ is consisting of l users, n Web services and m QoS attributes. Our method is to factorize Q into two matrices $C \in R^{l \times m \times r}$ and $W \in R^{r \times n}$, which satisfy the following equation:

$$Q_{ij} = \{Q_{ij}^1, Q_{ij}^2, ..., Q_{ij}^m\} \approx \{\sum_{k=1}^{r} c_{ik}^1 w_{jk}, \sum_{k=1}^{r} c_{ik}^2 w_{jk}, ..., \sum_{k=1}^{r} c_{ik}^m w_{jk}\} \qquad (2)$$

Where Q_{ij} is the Multi-QoS performance of Web service s_j observed by user u_i, vector $w_j = [w_{j1}, w_{j2}, ..., w_{jr}]$ reflects the feature of Web service s_j, matrix $c_i = [c_i^1, c_i^2, ..., c_i^m]$ reflects the feature of user u_i, vector $c_i^d = [c_{i1}^d, c_{i2}^d, ..., c_{ir}^d]$ is its sub feature, r is the rank of matrix C and W, it generally satisfies $(l \times m + n) \times r < m \times l \times n$.

In this paper, we put Multi-QoS attribute values together and use NMF algorithm[7] to learn the feature of Web services. The detail can be described as follow:

- First, we put normalized Multi-QoS attribute values together to form a super matrix $Q^* = [Q^1, Q^2, ..., Q^m]^T$.

- Second, we decompose the Q^* to extract the matrix C and W with following objective function min $f(C, W)$.

$$\min f(C, W) = \sum_{h=1}^{l \times m} \sum_{j=1}^{n} I_{ij}[Q_{hj}^* \times log(CW)_{hj} - (CW)_{hj}] \qquad (3)$$

where I_{ij} is the indicator function that is equal to 1, if user u_i invoked service s_j and 0 otherwise. Then, an incremental gradient descent method is employed to extract the feature of Web service by following iterative process.

$$C_{ha} = C_{ha} \sum_{j} W_{aj} \frac{Q_{hj}^*}{(CW)_{hj}} \qquad (4)$$

$$C_{ha} = \frac{C_{ha}}{\sum_{k} C_{ka}} \qquad (5)$$

$$W_{aj} = W_{aj} \sum_{h} C_{ha} \frac{Q_{hj}^*}{(CW)_{hj}} \qquad (6)$$

3.3   Predict QoS with Multi-output Support Vector Regression

Based on the extracted feature of Web services and user's historical QoS experience, we predict the missing Multi-QoS values for a targeted user. Consider the relationship between QoS attributes, we employ the Multi-output Support

Vector Regression (MSVR for short) algorithm[8] to predict the missing Multi-QoS values at same time by following steps:

- First, we select the training data according to the invoked experience of the targeted user. Suppose a targeted user u_e has invoked d Web services, then the training data can be represented by $T = \{(x_1, y_1), ..., (x_d, y_d)\}$, where $x_d \in W$ is the obtained feature of Web service s_d, y_d represents QoS performance of Web service s_d observed by u_e.

- Second, we employ MSVR to build a function f, which represents the relation between the feature of Web service and Multi-QoS values obtained from a target user. Consider R^r is the feature space of Web service, R^m is the QoS space. A function f can be described as formulation (7).

$$\begin{cases} f : R^r \to R^m \\ x_k \to W^T \varphi(x_k) + b \end{cases} \tag{7}$$

where $W^T \varphi(x_k) + b$ reflects a map from the feature space of Web service to the QoS space, which is learning from the training data by MSVR.

- Last, we predict the miss Multi-QoS values for a targeted user, based on the feature of unused Web service. let x_u is the feature of unused Web service s_u, the predicted Multi-QoS Q_{eu} for a targeted user u_e can be calculated as following:

$$Q_{eu} = W^T \varphi(x_u) + b \tag{8}$$

4   Experiment

In this section, we conduct several experiments to show the prediction quality of our proposed approach on real-world Web service dataset[9]. In the experiments, we randomly remove 90% entries from the original QoS matrix, only use the remaining 10% entries to predict the removed entries of a targeted user (which is chosen by random). The parameter settings are $r = 4, \varepsilon = 0.02, C = 678, \sigma = 0.6$.

Figure.3 and Figure.4 show the predictions of response-time and throughput for different users. The experimental results show that our approach have good performance on predicting the QoS value of different Web services.

4.1   Performance Comparison

In order to evaluate the performance of our approach, we compare our prediction precision with following methods on Mean Absolute Error (MAE) and Root Mean Squared Error (RMSE).

- UserMean : predict the missing QoS values based on the mean QoS value of each user.
- UPCC: predict the missing QoS values based on similar users[10].
- IPCC: predict the missing QoS values based on similar items (item refers to Web service in this paper)[11].

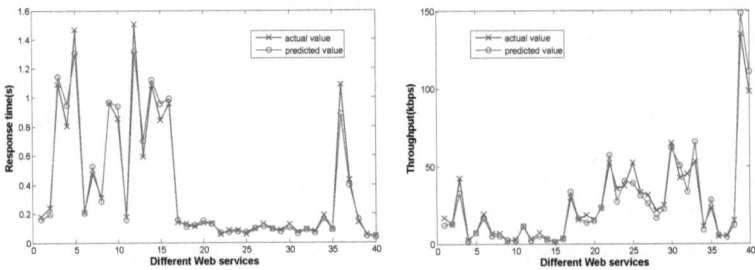

Fig. 3. Multi-QoS Prediction for Random Chosen User-1

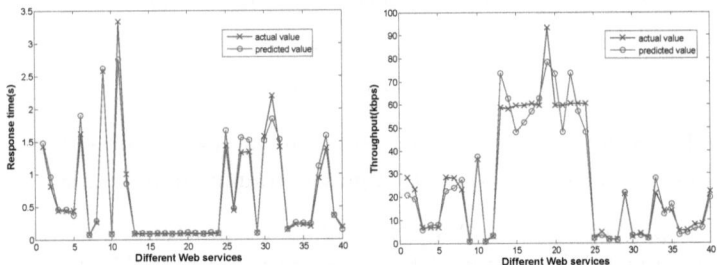

Fig. 4. Multi-QoS Prediction for Random Chosen User-2

Table 1. Accuracy Comparison with Different Users

Different User	RandomUser-1				RandomUser-2			
QoS	Response time		Throughput		Response time		Throughput	
Metrics	MAE	RMSE	MAE	RMSE	MAE	RMSE	MAE	RMSE
UserMean	0.5783	0.8173	21.2979	27.6767	0.3792	0.4279	18.2875	28.7697
IPCC	0.4466	0.6941	21.6252	26.9089	0.2730	0.3890	15.9715	19.1147
UPCC	0.3069	0.5800	5.8532	10.2054	0.3872	0.5604	8.2635	12.5219
MQEPA	**0.0809**	**0.1444**	**4.1930**	**6.1993**	**0.0396**	**0.0653**	**3.8689**	**5.6997**

Table.1 shows the experimental results of different approaches, and the result of MQEPA is specially highlighted. From Table.1, we can observe that our approach can obtain smaller MAE and RMSE, shows that MQEPA has better performance on prediction accuracy on both response-time and throughput for different users. This indicates that our approach suits better for Web service selection.

5   Conclusion

In this paper, we propose a novel Multi-QoS Effective Prediction problem, which aims to make effective Multi-QoS prediction based on Multi-QoS attributes and their relationships. To address this problem, we propose a novel prediction framework Multi-QoS Effective Prediction Approach. Comprehensive empirical studies demonstrate the utility of the proposed method.

Acknowledgments. This research is supported by the 973 Program of China No. 2009CB320406; 863 Program of China No.2011AA01A102; the CPSF No.2011M500226; the RFDP No.20110005130001.

References

1. Zhang, L., Zhang, J., Cai, H.: Services Computing: Core Enabling Technology of the Modern Services Industry. Tsinghua University Press (2007)
2. Zhu, J., Kang, Y., Zheng, Z., Lyu, M.R.: WSP: A Network Coordinate Based Web Service Positioning Framework for Response Time Prediction. In: Proc. of the IEEE ICWS 2012, pp. 90–97 (2012)
3. Zheng, Z., Lyu, M.R.: Collaborative reliability prediction of service-oriented systems. In: Proc. of 32nd International Conference on Software Engineering, pp. 35–44 (2010)
4. Zhang, Y., Zheng, Z., Lyu, M.R.: Exploring Latent Features for Memory-Based QoS Prediction in Cloud Computing. In: Proc. of the 30th IEEE Symposium on Reliable Distributed Systems, pp. 1–10 (2011)
5. Jiang, Y., Liu, J., Tang, M., Liu, X.F.: An effective web service recommendation method based on personalized collaborative filtering. In: Proc. of the IEEE ICWS 2011, pp. 211–218 (2011)
6. Ortega, M., Rui, Y., Chakrabarti, K., Mehrotra, S., Huang, T.S.: Supporting similarity queries in MARS. In: Proc. of the ACM Multimedia, pp. 403–413 (1997)
7. Lee, D.D., Seung, H.S.: Learning the parts of objects by non-negative matrix factorization. Nature 401(6755), 788–791 (1999)
8. Yu, S.P., Yu, K., Volker, T.: Multi-Output Regularized Feature Projection. IEEE Transactions on Knowledge and Data Engineering, 1600–1613 (2006)
9. Zheng, Z., Zhang, Y., Lyu, M.: Distributed QoS Evaluation for Real-World Web Services. In: Proc. of ICWS 2010, pp. 83–90 (2010)
10. Shao, L., Zhang, J., Wei, Y., Zhao, J., Xie, B., Mei, H.: Personalized qos prediction for web services via collaborative filtering. In: Proc. of ICWS 2007, pp. 439–446 (2007)
11. Resnick, P., Iacovou, N., Suchak, M., Bergstrom, P., Riedl, J.: GroupLens: an open architecture for collaborative filtering of netnews. In: Proc. of CSCW 1994, pp. 175–186 (1994)

Accelerating Topic Model Training on a Single Machine

Mian Lu[2], Ge Bai[2], Qiong Luo[2], Jie Tang[3], and Jiuxin Zhao[2]

[1] A*STAR Institute of High Performance Computing, Singapore
lum@ihpc.a-star.edu.sg
[2] Hong Kong University of Science and Technology
{luo,gbai,zhaojx}@cse.ust.hk
[3] Tsinghua University
jietang@tsinghua.edu.cn

Abstract. We present the design and implementation of GLDA, a library that utilizes the GPU (Graphics Processing Unit) to perform Gibbs sampling of Latent Dirichlet Allocation (LDA) on a single machine. LDA is an effective topic model used in many applications, e.g., classification, feature selection, and information retrieval. However, training an LDA model on large data sets takes hours, even days, due to the heavy computation and intensive memory access. Therefore, we explore the use of the GPU to accelerate LDA training on a single machine. Specifically, we propose three memory-efficient techniques to handle large data sets on the GPU: (1) generating document-topic counts as needed instead of storing all of them, (2) adopting a compact storage scheme for sparse matrices, and (3) partitioning word tokens. Through these techniques, the LDA training which would take 10 GB memory originally, can be performed on a commodity GPU card with only 1 GB GPU memory. Furthermore, our GLDA achieves a speedup of 15X over the original CPU-based LDA for large data sets.

1 Introduction

Statistical topic models have recently been successfully applied to text mining tasks such as classification, topic modeling, and recommendation. Latent Dirichlet Allocation (LDA) [1], one of the recent major developments in statistical topical modeling, immediately attracted a considerable amount of interest from both research and industry. However, due to its high computational cost, training an LDA model may take multiple days [2]. Such a long running time hinders the use of LDA in applications that require online performance. Therefore, there is a clear need for methods and techniques to efficiently learn a topic model.

In this paper, we present a solution to the efficiency problem of training the LDA model: acceleration with graphics processing units (GPUs). GPUs have recently become a promising high-performance parallel architecture for a wide range of applications [3]. The current generation GPU provides a $10\times$ higher computation capability and a $10\times$ higher memory bandwidth than a commodity multi-core CPU. Earlier studies of using GPUs to accelerate the LDA have shown significant speedups compared with CPU counterparts [4,5]. However, due the limited size of GPU memory, these methods are not scalable to large data sets. The GPU memory size is relatively small compared with the main memory size, e.g., up to 6GB for GPUs in the market. This limits the

Y. Ishikawa et al. (Eds.): APWeb 2013, LNCS 7808, pp. 184–195, 2013.

use of GPU-based LDA for practical applications since the memory size required of many real-world data sets clearly exceeds the capability of the GPU memory. Moreover, some parallel implementations of LDA, such as AD-LDA [2], cannot be directly adapted on the GPU either since the overall memory consumption would easily exceed the available GPU memory size due to a full copy of word-topic counts maintained for each processor.

To address this issue, we study how to utilize limited hardware resources for large-scale GPU-based LDA training. Compared with the existing work on GPU-base LDA, our implementation is able to scale up to millions of documents. This scalability is achieved through the following three techniques:

1. *On-the-fly generation of document-topic counts.* We do not precompute and store the global document-topic matrix. Instead, the counts for specific documents are generated only when necessary.
2. *Sparse storage scheme.* We investigate a more compact storage scheme to store the document-topic and word-topic count matrices since they are usually sparse.
3. *Word token partitioning.* Word tokens in the data set are divided into a few disjoint partitions based on several criteria. Then for each partition, only a part of document-topic and word-topic matrices are needed in each iteration.

Based on these techniques, we have successfully trained LDA models on data sets originally requiring up to 10 GB memory on a GPU with only 1 GB memory. Furthermore, our GPU-based LDA training achieved significant performance speedups. These results show that our approach is practical and cost-effective. Also, our techniques can be extended to solutions that involve multiple GPUs.

The remainder of this paper is organized as follows. In Section 2, we briefly introduce the LDA algorithm and review GPGPU. We present the design and implementation of our GLDA in detail in Section 3. In Section 4, we experimentally evaluate our GLDA on four real-world data sets. We conclude in Section 5.

2 Preliminary

2.1 Gibbs Sampling of Latent Dirchlet Allocation

We develop GLDA based on Gibbs sampling of LDA [6] due to its simplicity and high effectiveness for real-world text processing. In the LDA model, D documents are modeled as a mixture over K latent topics. Given a training data set, we use a set x to represent the words in documents, where x_{ij} is the jth word in the ith document. Furthermore, a corresponding topic assignments set z is maintained, where z_{ij} is the assigned topic for the word x_{ij}. The total number of words in the data set is N. Two matrices, n_{jk} and n_{wk} are used in LDA. Specifically, n_{jk} is the document-topic count – the number of words in document j assigned to topic k, and n_{wk} is the word-topic count – the number of occurrences of word w assigned to topic k. The sizes of matrices n_{jk} and n_{wk} are $D \times K$ and $W \times K$, respectively, where W is the number of words in the vocabulary for the given data set.

The original Gibbs sampling is inherently sequential, as each iteration in the training process depends on the results from the previous iteration. To implement it in parallel, the previous study has proposed a parallel approximate LDA algorithm [2], which has been shown similar accuracy to the sequential algorithm. Therefore, in this work, we adopt a similar parallel approximate LDA algorithm for our GPU implementation.

2.2 Graphics Processing Units (GPUs)

GPUs are widely available as commodity components in modern machines. We adopt the massive thread parallelism programming model from NVIDIA CUDA to design our algorithm. The GPU is modeled as an architecture including multiple SIMD multi-processors. All processors on the GPU share the *device memory*, which has both a high bandwidth and a high access latency. Each multi-processor has a fast on-chip *local memory* (called *shared memory* in NVIDIA's term) shared by all the scalar processors in the multi-processor. The size of this local memory is small and the access latency is low. The threads on each multi-processor are organized into *thread blocks*. *Threads* in a thread block share the computation resources on a multi-processor. Thread blocks are dynamically scheduled on the multi-processors by the runtime system. Moreover, when the addresses of the memory accesses of the multiple threads in a thread block are consecutive, these memory accesses are grouped into one access. This feature is called *coalesced access*.

2.3 Parallel and Distributed LDAs

Two parallel LDA algorithms, named AD-LDA and HD-LDA, are proposed by Newman et al. [2]. The speedup of AD-LDA on a 16-processor computer was up to 8 times. Chen et al. [7] have implemented AD-LDA using MPI and MapReduce for large-scale LDA applications. Their implementations achieved a speedup of $10\times$ when 32 machines were used. Another asynchronous distributed LDA algorithm was proposed by Asuncion et al. [8], which had a speedup of $15\text{-}25\times$ when there were 32 processors.

Among existing studies on GPU-based LDAs, Masada et al. [4] have accelerated collapsed variational Bayesian inference for LDA using GPUs and achieved a speedup of up to 7 times compared with the standard LDA on the CPU. Moreover, Yan et al. [5] have implemented both collapsed Gibbs sampling (CGS) and variational Bayesian (CVB) methods of LDA on the GPU. Compared with the sequential implementation on the CPU, the speedups for CGS and CVB are around 26x and 196x, respectively. However, none of these GPU-based LDAs has effectively addressed the scalability issue. In contrast, our GLDA can train very large data sets on the GPU efficiently.

3   GLDA Design and Implementation

3.1   Parallel Gibbs Sampling of LDA on GPUs

A traditional parallel approach [2] for LDA (AD-LDA) is that both D documents and document-topic counts n_{jk} are distributed evenly over p processors. However, each processor needs to maintain a local copy of all word-topic counts. Then each processor performs LDA training based on its own local documents, document-topic, and word-topic counts independently. At the end of each iteration, a reduction operation is performed to update all copies of local n_{wk} counts. The major difficulty in applying this approach to GPUs is that multiple copies of the word-topic count matrix may not fit into the limited GPU memory. Therefore, Yan et al. [5] have proposed a data partitioning method to avoid access conflicts while maintaining only one copy of n_{wk} matrix. The major issue of their implementation is it may cause workload imbalance since the partitioning scheme is based on both documents and vocabulary, and different partitions are processed in parallel. The approximate algorithm they have proposed may not work well for non-uniform data distributions. Moreover, it still cannot handle the data set when either n_{jk} or n_{wk} matrix cannot be entirely stored in the GPU memory. Instead, we have adopted a different algorithm without such 2-dimension partitioning and solved the scalability issue.

Algorithm 1. GLDA algorithm for one iteration.

Input:
x^p: word tokens assigned to the pth processor
Output: n_{jk}, n_{wk}, z_{ij}

1   **for** *all processors in parallel* **do**
2   **foreach** $x_{ij} \in x^p$ **do**
3   Sample z_{ij} with global counts n_{jk} and n_{wk}
4   /* Global synchronization */
5   Update n_{jk}
6   Atomic update n_{wk}

Our parallel algorithm is based on the following fact: With atomic increment and decrement operations, concurrent execution of these operations is guaranteed to produce a correct result. Therefore, in our implementation, we also maintain only one copy of n_{wk} matrix, and adopt two modifications. (1) We use atomic increment and decrement operations, which are supported by recent generation GPUs, in count updates. (2) We serialize the computation and update on the n_{wk} matrix. With these modifications, Algorithm 1 outlines our parallel LDA algorithm for one iteration. Compared with AD-LDA, which does not perform global updates until the end of each iteration, our algorithm performs global updates after each p words are sampled in p processors in parallel and guarantees the updated results are correct. Since the multi-processors on the GPU are tightly coupled, the communication overhead is insignificant in this implementation. Specifically, we map each thread block in CUDA as a processor, and the inherent data parallelism of the sampling algorithm is implemented using multiple threads within each thread block.

3.2 Memory Consumption of Original LDA

We estimate the total memory consumption of the original LDA algorithm in the following four components: the memory used by the word token set $x = \{x_{ij}\}$ and topic assignment set $z = \{z_{ij}\}$ are both $sizeof(int) \times N$ bytes. The document-topic and word-topic count matrix consume $sizeof(int) \times D \times K$ bytes and $sizeof(int) \times W \times K$ bytes, respectively. Therefore, the estimated total memory size is at least M bytes:

$$M = sizeof(int) \times (2 \times N + (D + W) \times K)$$

Where N, D, and W are fixed for a given data set, and the number of topics K is assigned by users. A large K is usually required for large data sets, for example, a typical value is $\sqrt{D}$. Figure 1 shows the estimated memory consumption for NYTimes and PubMed data sets that are used in our evaluations (see Table 1 for experimental setup). Obviously, current generations of GPUs cannot support such large amounts of required memory space. A straightforward solution is to send the data set x_{ij} and topic assignment set z_{ij} to the GPU on-the-fly. However, this method does not solve the problem when either n_{jk} or n_{wk} matrix cannot be entirely stored in the GPU memory. Therefore, we study more advanced techniques to enable our GLDA to handle large data sets.

(a) NYTimes   (b) PubMed

Fig. 1. Estimated memory consumptions for NYTimes and PubMed data sets (Table 1) with number of topics varied

3.3   Discarding Global Document-Topic Matrix

We have observed that to generate the n_{jk} counts for a given document j, only the word tokens in that document are necessary to scan. Moreover, the sequential sampling of words in a single document can share an array with K elements storing the document-topic counts corresponding to that document. Therefore, we consider discarding the storage for the global document-topic counts. We only produce the necessary n_{jk} counts for specific documents when these documents are processed (the technique is denoted as no_n_{jk}). Since we process p words in p different documents in parallel, the estimated memory size required for this method is $M_{no\_n_{jk}}$ bytes.

$$M_{no\_n_{jk}} = sizeof(int) \times (2 \times N + W \times K + p \times K)$$

Moreover, the additional cost introduced is equivalent to one scan on the topic assignment set z for each iteration. Due to the high memory bandwidth of the GPU, the generation of document-topic counts can be implemented efficiently through the coalesced access pattern.

Algorithm 2. Generating document-topic counts for a specific document through the coalesced access.

Input:
z: the topic assignment array for a given document in the device memory
l: the array with K elements in the local memory
n: the number of elements in z
tx: the thread index in a thread block
nt: the number of threads in a thread block
Output:
n_{jk}: the document-topic count array with length K in the device memory for the given document
1   l is initialized to 0;
2   **for** $i = tx; i < n; i = i + nt$ **do**
3 | atomicInc($l[z[i]]$);

4   /*synchronization within the thread block*/
5   **for** $k = tx; k < K; k = k + nt$ **do**
6 | $n_{jk}[k] = l[k]$;

We use a thread block to generate the document-topic counts for a given document. Algorithm 2 demonstrates the GPU code for a thread to produce document-topic counts based on the coalesced access. We have adopted a temporary array stored in the local memory to improve the memory access efficiency for counting as well as to facilitate the coalesced access for both reads and writes on the device memory. Without storing all n_{jk} counts, the memory saving is significant with a low computation overhead.

In contrast, the n_{wk} counts are stored. This design choice is made because re-calculating n_{wk} is expensive - as words are distributed across documents, generating n_{wk} for each iteration would require approximately N scans of the original data set. Fortunately, the number of topics and the size of vocabulary are both much smaller than the data set size; therefore, it is feasible to store the n_{wk} counts even for large data sets. In the following, we further discuss how to store these matrices efficiently and how to partition them if they do not fit into memory.

3.4 Sparse Storage Scheme for Count Matrices

We observe that a considerable portion of the values in the n_{wk} and n_{jk} matrices become zeros after a burn-in period. For example, after about 100 iterations of training on the NIPS data set, which is used in our evaluation, the percentages of non-zero values are around 45% and 10% for n_{jk} and n_{wk} counts, respectively. This observation has motivated us to study the sparse matrix storage scheme for these two counts.

We only store the topic-count pair for non-zero elements for the n_{jk} or n_{wk} count matrix. An array of K topic counts is generated from the sparse storage decompressed for a given document or word token. The decompression can also be implemented efficiently through the coalesced access as well as the local memory usage, which is similar to Algorithm 2. Suppose we use a data structure including two elements to store a topic-count pair, namely id and $count$. Then to generate an array including all topic counts for a specific document or word, the value of $count$ will be written to the idth element in the array for specific pairs. Since p word tokens in p different documents are processed in parallel, we need to decompress those corresponding n_{jk} and n_{wk} counts. The estimated memory required is M_{sparse} bytes.

$$M_{sparse} = sizeof(int) \times (2 \times N + 2 \times p \times K +$$
$$2 \times (D \times K \times nz_{jk}\% + W \times K \times nz_{wk}\%))$$

Where $nz_{jk}\%$ and $nz_{wk}\%$ refer to the percentage of non-zero elements for n_{jk} and n_{wk} counts, respectively. Note that, when $nz_{jk}\%$ or $nz_{wk}\%$ is over 50%, this storage scheme will consume more memory than the original storage format for the n_{jk} or n_{wk} matrix, respectively. Therefore, to decide whether adopting the sparse matrix storage, we should first estimate the memory space saving.

Although the introduced computation overhead is inexpensive, there is a critical problem for the GPU-based implementation. Since the percentage of non-zero elements is not guaranteed to monotonically decrease with iterations, a new topic-count pair may be inserted when an original zero element becomes non-zero. Thus an additional checking step is necessary after each word sampling process to examine whether new pairs should be inserted. Unfortunately, our evaluated GPUs do not support dynamical memory allocation within the GPU kernel code, and the kernel code cannot invoke the CPU code at runtime. To perform the checking step and also the GPU memory allocation, some information must be copied from the device memory to the main memory. The number of such memory copies is around N/p. The overhead of such a large number of memory copies may be expensive. Our evaluation results confirm this concern. With the newer generation GPUs, this concern may be partially or fully addressed.

3.5   Data Set Partitioning

Either no_n_{jk} or sparse storage scheme only partially addresses the memory limitation issue. To perform LDA for an arbitrary size of data set on the GPU, we consider a partitioning scheme for word tokens. The basic idea is that the GPU only holds the necessary information for a subset of word tokens.

Specifically, we divided the set of documents and the vocabulary into n and m disjoint subsets by ranges of document IDs or word IDs, respectively. Then there are totally $(n \times m)$ partitions and corresponding $(n \times m)$ computation phases in each iteration. For the $(i \times m + j)$th phase, where $i \in \{0, 1, ..., (n-1)\}$ and $j \in \{0, 1, ..., (m-1)\}$, only the word tokens belonging to the ith document subset and jth vocabulary subset are processed on the GPU, and corresponding document-topic and word-topic counts are generated on the GPU. The topic assignments set z_{ij} is processed exactly the same as the word token set x_{ij}. Through this partitioning scheme, the GPU can process an arbitrary size of data set since we can always decompose a large data set to several smaller partitions that can fit into the GPU memory.

The algorithms of generating necessary n_{jk} and n_{wk} counts are similar to Algorithm 2. The computation overhead depends on the order of partition processing. Since generating n_{wk} counts is more expensive, we process the partitions in a vocabulary-oriented style. Specifically, for the given jth vocabulary subset, the partitions $\{i \times m + j | 0 \leq i < n\}$ are processed sequentially. This way, each n partitions can share n_{wk} partitioning counts for the same word token subset. Therefore the overhead is $(1 + m)$ more scans on the original data set. We denote this global partitioning scheme as *doc-voc-part*. Partitioning schemes with $m = 1$ and $n = 1$ are denoted as *doc-part* and *voc-part*, respectively. *doc-part* and *voc-part* are suitable when the word-topic and document-topic matrix can be entirely stored in the GPU memory, respectively. For a uniform data distribution with range partitioning, the memory size requirement is estimated as M_{part} bytes:

$$M_{part} = sizeof(int) \times (\frac{2 \times N}{n \times m} + \frac{D \times K}{n} + \frac{W \times K}{m})$$

Note that the workload difference between different partitions will not hurt the overall performance significantly since these partitions are processed in different computation phases sequentially rather than in parallel on the GPU. Moreover, if the entire word token set, the topic assignment set, and all n_{jk} and n_{wk} counts can be stored in the main memory, we can directly copy the required words and counts from the main memory to the GPU memory and avoid the additional computation for the corresponding n_{jk} and n_{wk} counts generation in each iteration. This can further reduce the overhead but requires large main memory. In our implementation, we do not store n_{jk} and n_{wk} counts in the main memory thus our implementation can also be used on a PC with limited main memory.

3.6 Discussion

The three techniques we have proposed can also be combined, e.g., either no_n_{jk} or sparse storage scheme can be used together with the word token partitioning. To choose appropriate techniques, we give the following suggestions based on our experimental evaluation results: *doc-part* should be first considered if the data set and topic assignments cannot be entirely stored in the GPU memory. Then if all n_{jk} counts cannot be held in the GPU memory, no_n_{jk} or *doc-part* should be considered. Finally, the *voc-part* should be further adopted if the GPU memory is insufficient to store all n_{wk} counts, and the sparse matrix storage may be used for n_{wk}.

4 Empirical Evaluation

4.1 Experimental Setup

We conducted the experiments on a Windows XP PC with 2.4 GHz Intel Core2 Duo Quad processor, 2GB main memory and an NVIDAI GeForce GTX 280 GPU (G280). G280 has 240 1.3 GHz scalar processors, and 1GB device memory. Our GLDA is developed using NVIDIA CUDA. The CPU counterparts (CPU LDA) are based on a widely

Table 1. Statistics of four real-world data sets

	KOS	NIPS	NYTimes	PubMed
D	3,430	1,500	300,000	1,000,000
W	6,906	12,419	102,660	116,324
N	0.47×10^6	1.9×10^6	100×10^6	70×10^6

used open-source package GibbsLDA++[1]. Note that CPU LDA is a sequential program using only one core. Due to the unavailability of the code from the previous work on GPU-based LDA [5], we could not compare our GLDA with it. However, the speedup reported by their paper is similar to ours when the data can be entirely stored on the GPU.

Four real-world data sets are used in our evaluations [2], which are illustrated in Table 1. We use a subset of the original PubMed data set since the main memory cannot hold the original PubMed data set. The two small data sets KOS and NIPS are used to measure the accuracy of GLDA, and the two large data sets NYTimes and PubMed are used to evaluate the efficiency. Moreover, we also have a randomly sampled subset with a size half of the original NY Times dataset (denoted as *Sub-NYTimes*) to study the overhead of different techniques when all data can be stored in the GPU memory. For each data set, we randomly split it to two subsets. 90% of the original documents are used for training, and the remaining 10% are used for test. Throughout our evaluations, the hyperparameters α and β are set to $50/K$ and 0.1, respectively [6].

We use the *perplexity* [9] to evaluate the accuracy of GLDA. Specifically, after obtaining the LDA model from 1000 iteration training, we calculate the perplexity of the test data.

4.2 Perplexity Performance

Figure 2 shows the perplexity results for the KOS and NIPS data sets with number of thread blocks varied when $K=64$ and $K=128$. Note that when there is only one thread block, the implementation is equivalent to the standard sequential LDA. Therefore, perplexity with multiple thread blocks is expected to be similar to that with one thread block. This figure shows that there is no significant difference for the perplexity results between the standard LDA model and our parallel LDA algorithm.

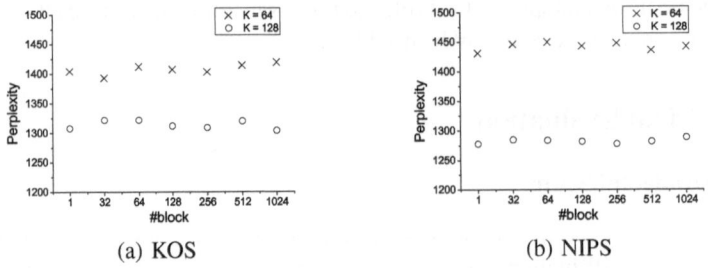

(a) KOS   (b) NIPS

Fig. 2. Perplexity of KOS and NIPS data sets with number of thread blocks varied

4.3 Memory Consumption and Efficiency

We first study the performance impact of the thread parallelism when all data is stored in the GPU using the data set Sub-NYTimes. We first set the number of threads in

---

[1] http://gibbslda.sourceforge.net/
[2] http://archive.ics.uci.edu/ml/datasets/Bag+of+Words

each thread block to 32 when varying the number of blocks. Then the number of thread block is fixed to 512 and the number of threads in each thread block is varied. Figure 3 shows the elapsed time of one iteration for the GLDA with the number of thread blocks and number of threads in each thread block varied. Compared with the CPU-LDA's performance of 10 minutes per iteration, this figure shows significant speedups when either number of thread blocks or threads per block is increased. However, it nearly maintains a constant or slightly reduced when the GPU resource is fully utilized.

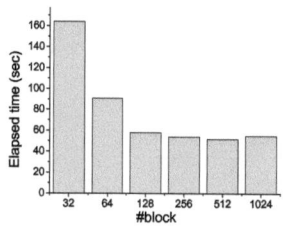 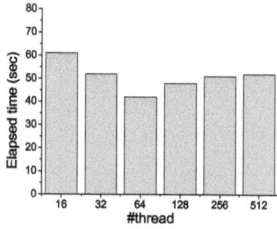

(a) Number of blocks varied.   (b) Number of threads varied.

Fig. 3. Elapsed time of one iteration on the GPU for Sub-NYTimes with the number of blocks and threads per block varied, K=256

Next, we study the overhead of techniques used to address the scalability issue. We first measure the performance when only a single method is adopted. Figure 4(a) shows the elapsed time of GLDA adopting different techniques. *opt* refers to the implementation without any additional techniques and is optimized through the thread parallelism. There are four partitions for both *doc-part* and *voc-part*. *doc-voc-part* has divided both documents and vocabulary into two subsets, thus also resulting four partitions. The figure shows that the sparse storage scheme has a relatively large overhead compared with the other methods. Through our detailed study, we find the additional overhead is mainly from the large number of memory copies between the main memory and GPU memory. Moreover, *voc-part* is slightly more expensive than *doc-part* and no_n_{jk} since it needs to scan the original data set for each partition. For data sets used in our evaluations, the sparse storage scheme is always more expensive than other techniques, thus we focus on the other two techniques in the following evaluations. We further study the performance overhead when the technique no_n_{jk} and partitioning are used together. Figure 4(b) shows the elapsed time when these two techniques are combined. The combined approach is slightly more expensive than only one technique adopted since the overhead from two techniques are both kept for the combined method.

To investigate the memory space saving from different techniques, Figure 5 shows the GPU memory consumptions corresponding to Figure 4. It demonstrates that the memory requirement is reduced by around 14%-60% through various techniques. Since the number of topics is not very large and the partitioning scheme can also reduce the memory size consumed for the word token set and topic assignment set, partitioning is more effective than the other two techniques. Moreover, Figure 5(b) shows that the combined technique can further reduce the memory consumption compared with a single technique adopted.

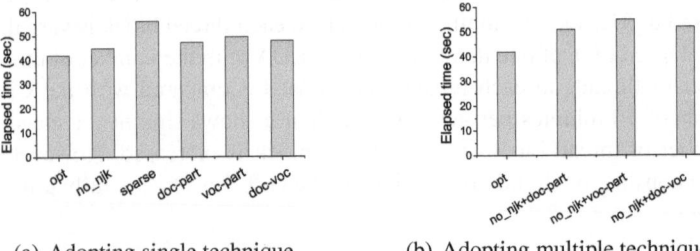

(a) Adopting single technique   (b) Adopting multiple techniques

Fig. 4. Elapsed time of one iteration on the GPU for Sub-NYTimes when adopting different techniques, K=256

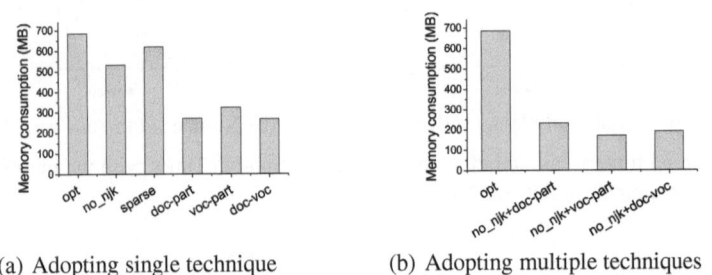

(a) Adopting single technique   (b) Adopting multiple techniques

Fig. 5. Overall memory consumption on the GPU for Sub-NYTimes of adopting different techniques, K=256

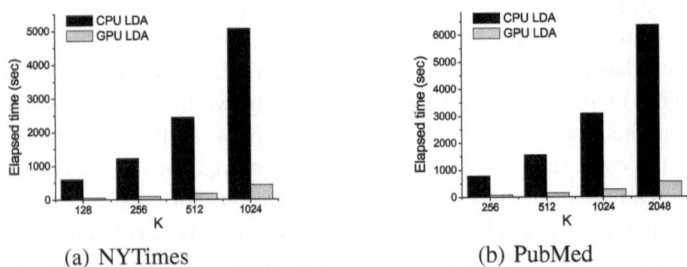

(a) NYTimes   (b) PubMed

Fig. 6. Performance comparisons for one iteration of NYTimes and PubMed data sets with number of topics varied

Finally, we show the overall performance comparisons based on the NYTimes and PubMed data sets with reasonable numbers of topics. The CPU counterparts are implemented using the same set of techniques since the main memory is also insufficient to handle such data set for the original CPU implementation. The estimated memory consumptions using a traditional LDA algorithm for these two data sets are presented in Figure 1, and the actual GPU memory consumption is around 700MB for each data sets. The technique selection is based on the discussion in the implementation section. Specifically, the evaluation with NYTimes has adopted the techniques *doc-part*, while *doc-voc-part* and no_j_{jk} are both adopted for the PubMed data set. Figure 6 shows that our GLDA implementations are around 10-15x faster than their CPU counterparts. Such speedup is significant especially for large data sets. For example, suppose 1000 iterations are required for the PubMed with 2048 topics, GLDA could accelerate the computation from originally more than two months to only around 5 days.

5   Conclusion

We implemented GLDA, a GPU-based LDA library that features high speed and scalability. On a single GPU with 1 GB GPU memory, we have evaluated the performance using the data sets originally requiring up to 10 GB memory successfully. Our experimental studies demonstrate that GLDA can handle such large data sets and provide a performance speedup of up to 15X on a G280 over a popular open-source LDA library on a single PC.

Acknowledgement. This work was supported by grant 617509 from the Research Grants Council of Hong Kong.

References

1. Blei, D.M., Ng, A.Y., Jordan, M.I.: Latent dirichlet allocation. Journal of Machine Learning Research 3, 993–1022 (2003)
2. Newman, D., Asuncion, A., Smyth, P., Welling, M.: Distributed inference for latent dirichlet allocation. In: NIPS (2007)
3. Owens, J.D., Luebke, D., Govindaraju, N.K., Harris, M., Kruger, J., Lefohn, A.E., Purcell, T.J.: A survey of general-purpose computation on graphics hardware. In: Eurographics 2005, State of the Art Reports (2005)
4. Masada, T., Hamada, T., Shibata, Y., Oguri, K.: Accelerating collapsed variational bayesian inference for latent dirichlet allocation with nvidia CUDA compatible devices. In: Chien, B.-C., Hong, T.-P., Chen, S.-M., Ali, M. (eds.) IEA/AIE 2009. LNCS, vol. 5579, pp. 491–500. Springer, Heidelberg (2009)
5. Yan, F., Xu, N., Qi, Y.: Parallel inference for latent dirichlet allocation on graphics processing units. In: NIPS 2009, pp. 2134–2142 (2009)
6. Griffiths, T.L., Steyvers, M.: Finding scientific topics. Proceedings of the National Academy of Sciences, PNAS 2004 (2004)
7. Chen, W.Y., Chu, J.C., Luan, J., Bai, H., Wang, Y., Chang, E.Y.: Collaborative filtering for orkut communities: discovery of user latent behavior. In: WWW 2009 (2009)
8. Asuncion, A., Smyth, P., Welling, M.: Asynchronous distributed learning of topic models. In: NIPS (2008)
9. Azzopardi, L., Girolami, M., van Risjbergen, K.: Investigating the relationship between language model perplexity and ir precision-recall measures. In: SIGIR 2003, pp. 369–370 (2003)

Collusion Detection in Online Rating Systems

Mohammad Allahbakhsh[1], Aleksandar Ignjatovic[1], Boualem Benatallah[1],
Seyed-Mehdi-Reza Beheshti[1], Elisa Bertino[2], and Norman Foo[1]

[1] The University of New South Wales, Sydney, NSW 2052, Australia
{mallahbakhsh,ignjat,boualem,sbeheshti,norman}@cse.unsw.edu.au
[2] Department of Computer Science and CERIAS, Purdue University, Indiana, USA
bertino@cs.purdue.edu

Abstract. Online rating systems are subject to unfair evaluations. Users may try to individually or collaboratively promote or demote a product. Collaborative unfair rating, i.e., *collusion*, is more damaging than individual unfair rating. Detecting massive collusive attacks as well as honest looking intelligent attacks is still a real challenge for collusion detection systems. In this paper, we study impact of collusion in online rating systems and asses their susceptibility to collusion attacks. The proposed model uses frequent itemset mining technique to detect candidate collusion groups and sub-groups. Then, several indicators are used for identifying collusion groups and to estimate how damaging such colluding groups might be. The model has been implemented and we present results of experimental evaluation of our methodology.

1 Introduction

Today, in e-commerce systems, buyers usually rely on the feedback posted by others via *online rating systems* to decide on purchasing a product [13]. The higher the number of positive feedback on a product, the higher the possibility that one buys such a product. This fact motivates people to promote products of their interest or demote the products in which they are not interested by posting *unfair rating scores* [7, 6]. Unfair rating scores are scores which are cast regardless of the quality of a product and usually are given based on personal vested interests of the users. For example, providers may try to submit supporting feedback to increase the rating of their product in order to increase their income [6]. The providers also may attack their competitors by giving low scores on their competitor's products. Also, sometimes sellers in eBay boost their reputations unfairly by buying or selling feedback [7]. Unfair rating scores may be given individually or collaboratively [18]. Collaborative unfair ratings which are also called *collusion* [17, 18] by their nature are more sophisticated and harder to detect than individual unfair ratings [17]. For that reason, in this work we focus on studying and identifying collusion.

Although collusion is widely studied in online collaborative systems [4, 8, 10, 13, 12, 9, 19], such systems still face some serious challenges. One major challenge arises when a group of raters try to completely take control of a product i.e., when the number of unfair reviewers is significantly higher than the number of honest users; the existing models usually can not detect such groups. Also, the existing models do not perform

Y. Ishikawa et al. (Eds.): APWeb 2013, LNCS 7808, pp. 196–207, 2013.

well against intelligent attacks, in which group members try to give an appearance of honest users. For example, typically they will not deviate from the majority's ranking on most of the cast feedback and target only a small fraction of the products. Such attacks are hard to identify using the existing methods [19].

Moreover, there are cases in which a large group of people have rated the same group of products, and thus could be considered as a potential collusion group. Such a group might be subsequently deemed as non collusive. However, in such cases there may exist some smaller collusive sub-groups inside the large group which are collusive, but when put along with others they might be undetected. Detection of such sub-groups is not addressed in the existing collusion detection models.

In this paper we propose a framework for detecting and representing collusion in online rating systems. Besides indicators used in the already existing work, we also define two novel indicators. One indicator which we call *Suspiciousness* of a reviewer, which is a metric to estimate to what extent ratings posted by such reviewer correspond to majority consensus. Such indicator is calculated using two distance functions: the L^p *norm* and the *Uniform distance* (see Section 4.1 for more details). The other indicator which we call *spamicity* estimates the likelihood that a particular rating score given to a product by a reviewer is unfair.

We also propose a graph data model for representing rating activities in online rating system. This model allows representing products, reviewers and the rating scores reviewers have cast on products and identifying collusive groups. We propose a new notion and representation for collusion groups called *biclique*. A biclique is a group of users and a group of products such that every reviewer in such a group has rated every product in the corresponding group of products. We use bicliques to detect collusion.

Moreover, we propose an algorithm which employs two clustering algorithms built upon frequent itemset mining (FIM) technique [1] for finding bicliques and sub-bicliques. Then the algorithm uses the proposed collusion indicators to find any possible collusion in online rating systems. We have implemented our model and tested it and the evaluation results show the efficiency of our model.

The rest of the paper is organized as follows. In section 2 we propose our data model. In Section 3 we propose our example scenario and data preprocessing details. Our collusion detection method is proposed in Section 4 and the corresponding algorithm appears in Section 5. In section 6 we propose implementation details and also evaluate results. We discuss related work in section 7 and conclude in section 8.

2   Representing Online Rating Entities

We define a graph data model (i.e. ORM: Online Rating Model) for organizing a set of entities (e.g. reviewers and products) and relationships among them in an online rating system. ORM can be used to distinguish between fair and unfair ratings in an online rating system and helps to calculate a more realistic rating score for every product. In ORM, we assume that interactions between reviewers and products are represented by a directed graph $G = (V, E)$ where V is a set of nodes representing entities and E is a set of directed edges representing relationships between nodes. Each edge is labeled by a triplet of numbers, as explained in the followings.

An entity is an object that exists independently and has a unique identity. ORM consists of three types of entities: *products*, *reviewers* and *bicliques*. We use the concept of *folder* used in our previous work [3] to represent bicliques.

Product. A product is an item which has been rated by system users in terms of quality or any other possible aspects. Products are described by a set of attributes such as the unique indicator (i.e. ID), title, and category (e.g. book, cd, track, etc). We assume that there are N_p products $P = \{p_j | 1 \leq j \leq N_p\}$ in the system to be rated.

Reviewer. A reviewer is a person who has rated at least one product. Reviewers are described by a set of attributes and are identified by their unique identifier (i.e. ID). We assume that there are N_u reviewers $U = \{u_i | 1 \leq i \leq N_u\}$ rating products.

Rating Relationship. A relationship is a directed link between a pair of entities, which is associated with a predicate defined on the attributes of entities that characterizes the relationship. We assume that no reviewer can rate a product more than once. So, in ORM, there is at most one relation between every pair of products and reviewers.

When a reviewer rates a product, a rating relationship is established between corresponding reviewer and product. We define $R = \{e_{ij} \mid u_i \in U \wedge p_j \in P\}$ as the set of all rating relationships between reviewers and products i.e. $e_{ij} \in R$ is the rating score which the u_i has given to p_j. A rating relationship is weighted with the values of the following three attributes:

1. The **value** is the evaluation of the the reviewer from the quality of the product and is in the range $[1, M], M > 1$. M is a system dependent constant. for example in Amazon, rating scores reside between 1 and 5. We denote the value field of the rating by $e_{ij}.v$.
2. The **time** shows the time in which the rating has been posted to the system. The time field of the rating is an integer number and is denoted by $e_{ij}.t$.
3. As we mentioned earlier we assume that every reviewer can have at most one relation to each product (i.e., we do not allow multi-graphs). However, in real world, e.g. in Amazon rating system, one reviewer can rate a product several times. Some collusion detection models like [12] just eliminate duplicate rating scores. We rather use them for the purpose of detecting unfair ratings. **Spamicity** shows what fraction of all scores cast for this particular product are cast by the reviewer in the considered relationship with this product. We denote the spamicity of a rating relation by $e_{ij}.spam$.

2.1   Biclique

A biclique is a sub-graph of ORM containing a set of reviewers R, a set of products P and their corresponding relationships Rel. All reviewers in R of a biclique have rated all products in the corresponding P, i.e., there is a rating relationship between every $r \in R$ and every $p \in P$. A biclique is denoted by $BC = \{R, P, Rel\}$. For example, $BC = \{\{r_1, r_2, r_3\}, \{p_1, p_2, p_3\}, \{r_1 \rightarrow p_1, ...\}\}$ means that reviewers u_1, u_2, and u_3 all have voted on p_1, p_2 and p_3. We will use terms 'biclique' and 'group' interchangeably throughout this paper.

3   Example Scenario

Amazon is one of the well-known online markets. Providers or sellers put products on the Amazon online market. Buyers go to Amazon and buy products if they find them of an acceptable quality, price, etc. Users also can share their experiences with others as reviews or rating scores they cast on products. Rating scores are numbers in range $[1, 5]$. Amazon generates an overall rating score for every product based on the rating scores cast by users. Some evidence like [6] show that the Amazon rating system has widely been subject to collusion and unfair ratings.

We use the log of Amazon online rating system[1] which was collected by Leskovec et. al. for analyzing dynamics of viral Marketing [11], referred in the following as AM-ZLog. This log contains more than 7 million ratings cast on the quality of more than 500 thousands of products collected in the summer of 2006.

We preprocess AMZLog to fit our graph data model. We delete inactive reviewers and unpopular products and only keep reviewers who reviewed at least 10 products and the products on which at least 10 ratings are cast. We also change the date format from 'yyyy-mm-dd' to an integer number reflecting the the number of days between the date on which the first rating score in the system has been posted and the day in which this rating score has been cast.

We use redundant votes from one rater on same product to calculate spamicity of their relation. Let $E(j)$ be the set of all ratings given to product p_j and $E(i, j)$ be the set of all ratings given by reviewer u_i to the product p_j. We calculate spamicity degree of the relationship between the reviewer r_i and the product p_j as follows:

$$e_{ij}.spam = \begin{cases} 0 & \text{if } (|E(i,j)| \leq 2) \\ \frac{|E(i,j)|}{|E(j)|} & \text{otherwise} \end{cases}$$

In the above equation allowing casting two votes instead of one accommodates for the situations where a genuine "mind change" has taken place.

4   Collusion Detection Method

To find collusive bicliques, first we have to identify all collaborative rating groups as collusion biclique candidates, then check them to find real collusion groups. We employs FIM technique [1] to find bicliques. More details on our biclique detection algorithm can be found here [2].

4.1   Collusion Indicators

It is almost impossible to identify collusion bicliques just by analyzing the behavior of individual members or even only based on the posted rating scores [12]. Rather, several indicators must be checked to realize in what extent a group is a collusion group [13]. In the followings, we propose some indicators and show how they indicate to a possible collusive activity.

---

[1] http://snap.stanford.edu/data/amazon-meta.html

Group Rating Value Similarity (GVS). A group of reviewers can be considered as formers of a collusive biclique (denote by g), if they have posted similar rating scores for similar products. For example, all of them promoted a set of products and demoted another set of products by similar rating scores. To find this similarity we calculate the *Pairwise Rating Value Similarity* between every pair of reviewers in the group. Pairwise rating value similarity denoted by $VS(i,j)$ shows to what extent u_i and u_j have cast similar rating scores to every product in $g.P$. We use cosine similarity model, a well-known model for similarity detection [15], for finding similarities between group members. Since $VS(i,j)$ is the cosine of the angle between two vectors containing ratings of two reviewers, it is a value in range $[0, 1]$. The value 1 means completely same and 0 means completely different. Suppose that u_i and u_j are two members of group g i.e. u_i and $u_j \in g.R$. We calculate similarity between them as follows:

$$VS(i,j) = \frac{\sum_{k \in g.R} v_{ik} \times v_{jk}}{\sqrt{\sum_{k \in g.R}(v_{ik})^2} \times \sqrt{\sum_{k \in g.R}(v_{jk})^2}} \tag{1}$$

We then calculate an overall degree of similarity for every group to show how all members are similar in terms of values they have posted as rating scores and call it *Group rating Value Similarity* (GVS). GVS for every group is the minimum amount of pairwise similarity between group members. The GVS of group g is calculated as follows.

$$GVS(g) = \min(VS(i,j)), \text{ for all } u_i \text{ and } u_j \in g.R \tag{2}$$

The bigger the GVS, the more similar the group members are in terms of their posted rating values.

Group Rating Time Similarity (GTS). Another indicator for finding collusive groups is that they usually cast their ratings in a short period of time. In other words, when a group of users aim to promote or demote a product they should do it in a short period of time in order to gain benefit, get paid, etc. We use this fact as an indicator to find collusive groups and call it *Group rating Time Similarity (GTS)*. To calculate GTS, we find the the Time Window (TW) in which the rating scores have been posted on each product $p \in g.P$. The beginning of the TW is the time of casting the first rating score on the products and its end is the time of casting the last rating score on the product by members of the group. We use the size of the TW in comparison to a constant $MaxTW$ to show how big such a TW is. The parameter $MaxTW$ is the maximum size of a TW which might be suspected to be collusive. So, for every product in the group we have:

$$TW(j) = \begin{cases} 0 & \text{if } (MaxT(j) - MinT(j) > MaxTW) \\ 1 - \frac{MaxT(j) - MinT(j)}{MaxTW} & \text{Otherwise} \end{cases} \tag{3}$$

Where $MinT = \min(e_{ij}.t)$, $MaxT = \max(e_{ij}.t)$, $j \in g.P$ for all $i \in g.R$

Now, we choose the largest $TW(j)$ as the degree of time similarity between the ratings posted by group members on the target products. We say

$$GTS(g) = \max(TW(j)), \text{ for all } p_j \in g.P \tag{4}$$

The bigger the GTS, the more similar the group members are in terms of the time of posting their rating scores.

Group Ratings Spamicity (GRS). As we described in Sections 2 and 3, the spamicity of a rating relationship shows the suspiciousness of the rating score because of high number of ratings posted by a same user to a same product. We define the *Group Rating Spamicity* (GRS) as follows:

$$GRS(g) = \frac{\sum_{e \in g.Rel} e.v \times e.spam}{\sum_{e \in g.Rel} e.v} \tag{5}$$

Group Members Suspiciousness (GMS). We identify suspicious users in four steps.

Step 1: Suppose that $E(j)$ is the set of all ratings posted on the product p_j. We find the median of the members of $E(j)$ and denote it by m_j. We then calculate the average distance of all ratings posted on p_j from the median m_j and denote it by d_j. We calculate the average distance using equation (6).

$$d_j = \sqrt{\frac{\sum_{i \in E(j)} (e_{ij}.v - m_j)^2}{||E(j)||}} \tag{6}$$

Now, we calculate a credibility degree for every $e_{ij} \in E(j)$. We denote this credibility degree with φ_{ij} and use it to marginalize outliers. The φ_{ij} is calculated as follows.

$$\varphi_{ij} = \begin{cases} 1 & \text{if } (m_j - d_j) \leq r_{ij} \leq (m_j + d_j) \\ 0 & \text{otherwise} \end{cases} \tag{7}$$

Equation (7) implies that only the ratings which fall in range $(m_j \pm d_j)$ are considered as credible.

Step 2: In this step, we calculate the averages of all credible ratings on the product p_j and assume that it is a dependable guess for the real rating of the product. We show it by g_j and calculate it as follows.

$$g_j = \frac{\sum_{i \in E(j)} (e_{ij}.v \times \varphi_{ij})}{\sum_{i \in E(j)} \varphi_{ij}} \tag{8}$$

Step 3: In the third step, using the ratings we guessed for every product (Equation (8)), we calculate two error rates for every reviewer. The first error rate is the L^P *error rate* (here L^2) which is $L^2 - norm$ of all differences between the ratings cast by the reviewer on the quality of p_j and the g_j. We denote the L^p error rate of reviewer u_i by $LP(i)$. Suppose that J_i is the set of indices of all products have been rated by u_i. The $LP(i)$ is calculated as follows.

$$LP(i) = \sqrt{\sum_{j \in J_i} \left(\left| e_{ij}.v - g_j \right| \right)^2} \tag{9}$$

The second error rate we calculate for every reviewer is the uniform norm of differences between $e_{ij}.v$ and g_j for all products have been rated by u_i. We call this error rate *uniform error rate* of u_i, denote it by $UN(i)$ and calculate it as follows.

$$UN(i) = \max \left(|e_{ij}.v - g_j| \right) \quad , \quad j \in J_i \tag{10}$$

Algorithm 1. Collusion Detection Algorithm

Input: ORM Graph, δ, $MaxTW$
Output: C as the set of all collusive Bicliques
  Find all bicliques and store them in $TempC$
  for all $tc \in TempC$ **do**
    Calculate all indicators for tc
  end for
  repeat
    $tc = $ first member of $TempC$
    if $tc.POC > \delta$ **then**
      Move tc to C
    else
      if $tc.DI < \delta$ **then**
        Remove tc from $TempC$
      else
        $SC = $ Sub-Bicliques of tc
        Calculate Indicators for all members of SC
        $TempC = TempC \cup SC$
        Remove tc from $TempC$
      end if
    end if
  until $(TempC.size > 0)$
  return C

Step 4: The suspicious reviewers are the people who have have large $LP(i)$ or while they have normal $LP(i)$ they have high $UN(i)$ values. To identify suspicious reviewers, we identify the normal range of error rates for all reviewers. Based on the calculated range, the outliers are considered as suspicious reviewers. Suppose that $\widehat{LP}$ is the median of all $LP(i)$ and $\widehat{UN}$ is the median of all $UN(i)$. Also, we assume that $\overline{LP}$ and $\overline{UN}$ are the standard distance of all $LP(i)$ and $UN(i)$ respectively, calculated similar to the method being used in Equation (6). The list of suspicious reviewers is denoted by S and built using Equation (11).

$$S = \left\{ u | \left(u \in U \right) \wedge \left(\left(LP(u) < \widehat{LP} - \overline{LP} \right) \vee \left(LP(u) > \widehat{LP} + \overline{LP} \right) \right. \right.$$
$$\left. \left. \vee \left(UN(u) < \widehat{UN} - \overline{UN} \right) \vee \left(UN(u) > \widehat{UN} + \overline{UN} \right) \right) \right\}. \tag{11}$$

Now, we define GMS of a group as follows.

$$GMS(g) = \frac{|S(g)|}{|g.R|} \text{ where } S(g) = (g.R) \cap S \tag{12}$$

4.2   How Destructive a Biclique Might Be (The Damaging Impact)

The damaging impact of a collusion group is its ability to impact normal behavior of the system and final rating scores of products. There are two important parameters reflecting damaging impact of a group: (i) size of the group and (ii) the number of products have been attacked by the group. The bigger these two parameters are, the more defective such a collusive group is.

Group Size (GS). Size of a collusive group (GS) is proportional to the number of reviewers who have collaborated in the group i.e., $g.R.size$ and is calculated as follows.

$$GS(g) = \frac{|g.R|}{\max(|g.R|)} \text{ where } g \text{ is a group} \tag{13}$$

GS is a parameter between $(0, 1]$ and showing how large is the number of members of a group in comparison with other groups.

Group Target Products Size (GPS). The size of the target of a group (GPS) is proportional to the number of products which have been attacked by members of a collusion group. The bigger the GTS, the more defective the group will be. GTS is a number in range $(0, 1]$ and is calculated as follows.

$$GPS(g) = \frac{|g.P|}{\max(|g.P|)} \text{ where } g \text{ is a group} \tag{14}$$

4.3   Finding Collusive Sub-bicliques

When finding bicliques, we try to maximize size of the group, i.e. $g.R.size$, to find the largest possible collusive bicliques. It is possible that a large group is identified as an honest group due to large and diverse number of users who have just rated same subset of products. But possibly, there might exist some smaller collusive sub-groups inside. To find such sub-groups, we use again the idea of finding candidate bicliques using FIM technique. But in this stage, we only investigate bicliques which have damaging impact greater than a threshold called δ. δ is the collusion threshold and is specified when collusion cliques are found using our proposed algorithm. Due to space limitation, both biclique and sub-biclique detection algorithms are available in [2].

5   Collusion Detection Algorithm

It is quite impossible to certainly find out if a group is collusive or not [13]. Therefore, we define a metric called *Probability of Collusion (POC)* to show to what extent a group seems to be collusive. POC is an aggregation of four collusion indicators. Since in different environments, the importance of these indicators may be different, our model enables user to assign weight to every indicator to have adequate impact on POC. Suppose that W_{GVS}, W_{GTS}, W_{GRS} and W_{GMS} are corresponding weights for GVS, GTS, GRS and GMS respectively so that $W_{GVS} + W_{GTS} + W_{GRS} + W_{GMS} = 1$. The POC is calculated as follows.

$$POC(g) = GVS \times W_{GVS} + GTS \times W_{GTS} + GRS \times W_{GRS} + GMS \times W_{GMS}$$

Moreover, for every group we calculate its damaging impact (DI) as follows.

$$DI(g) = \frac{GPS + GS}{2}$$

Finally we propose Algorithm 1 for collusion detection. This Algorithm uses POC, DI and ORM graph along with constants δ and $MaxtTW$ to detect collusion in online rating systems.

6   Implementation and Evaluation

We have implemented a framework for querying and analyzing collusion in online rating systems. Our framework comprises several parts which enable user to easily customize collusion detection process and query and analyze collusion with the simple query language provided for this purpose. More details of the framework architecture and implementation can be found in [2] and [3].

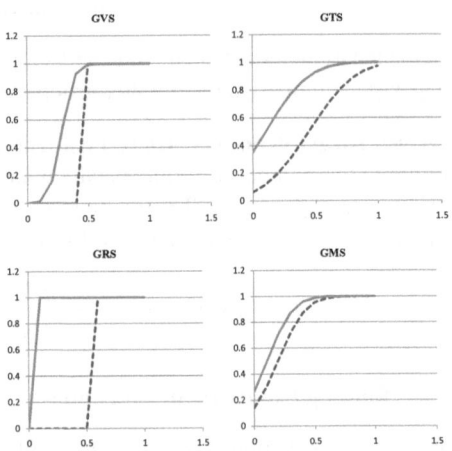

Fig. 1. Cumulative distribution of values of collusion indicators for collusive(red/dashed) and non-collusive (blue/solid) selected groups

To evaluate our model, we randomly chose 20 groups identified within AMZLog. Then we asked domain experts to manually check whether these groups are collusive or not. Experts using their experience and information provided in Amazon web site such as rating scores, corresponding written reviews,etc analyzed these 20 selected groups. 7 groups out of 20 were identified as collusive and 13 as honest. We used this dataset to evaluate our model.

Statistical Evaluation. For each group in dataset, we calculate all four collusion indicators. Then we calculate the cumulative distribution of every indicator for both collusive and non-collusive groups. The results of these calculations are shown in Figure 1. In every section of the chart, the red/dashed line shows the cumulative distribution of the corresponding indicator of collusive groups while the blue/solid line reflects the corresponding value for non-collusive groups. Also, the vertical axis is the cumulative value and horizontal axis is the percentage.

For every indicator in the chart, the cumulative distribution charts of the non-collusive groups are on the left side while collusive charts are more closer to one. It simply means that in average, for every indicator, values calculated for collusive groups are larger than values calculated for non-collusive groups. Therefor, these indicators and the way they are calculated are truly reflecting the behavior of the group, and can be used as indicators to collusive behavior.

Evaluating Quality of the Results. To evaluate quality of the results, we use the well-known measures of precision and recall [16]. Precision measures the quality of the results and is defined by the ratio of the relevant results to the total number of retrieved results. Recall measures coverage of the relevant results and is defined by the ratio of relevant results retrieved to the total number of existing relevant results in database.

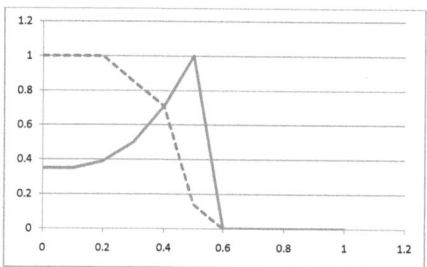

Fig. 2. How precision (blue/solid) and recall (red/dashed) change with different threshold values.

Table 1. A Sample Set of Identified bicliques

Group ID	Probability of Collusiveness	Damaging Impact
...	...	...
2	0.598	0.599
4	0.37	0.49
6	0.45	0.42
9	0.056	0.173
11	0.121	0.41
...	...	...

An effective model should achieve a high precision and a high recall; but its is not possible in real world, because these metrics are inversely related [16]. It means that cost of improvement in precision is reduction of recall and vice versa. We calculate precision and recall with different thresholds to show how value of δ impacts the quality and accuracy of the results. Figure 2 shows the results of running our algorithm with different threshold values. We do not specify particular value for precision and recall. We only say that if the user wants to achieve the highest possible values for both precision and recall metrics, Figure 2 obviously shows that the optimal value for threshold (δ) is 0.4. In this case 71% of the bicliques are retrieved and 71% of retrieved results are really collusive. One can change δ to change quality or coverage of data. Also, for more accuracy one can run the model with different groups of randomly chosen bicliques.

Our model also calculates a damaging impact (DI) factor for every group to show how damaging the group is. DI helps users to identify groups that have a high potential damaging power. These groups are mainly groups with high number of targeted products. Table 1 shows a sample set of bicliques and their DI and POC. The group 2 has high values for POC and DI, so it is a damaging group. On the other hand, group 9 has small values for POC and DI, and is not a really defective group. Looking at group 11 reveals that although the POC of the group is small (0.121), but it still can be damaging because its DI is 0.41. DI is also useful when comparing groups, e.g, comparing 4 and 6. Although, the POC of group 6 is 0.08 higher than POC of group 4, the DI of group 4 is 0.07 higher than DI of group 6. Therefore we can classify them similar rather than deeming group 4 more damaging than group 6. Without having DI, the damaging potential of groups like 11 or 4 may be underestimated and it can lead to incorrect calculation of rating scores.

7 Related Work

Collusion detection has been widely studied in P2P systems [4, 8, 14]. For instance, EigenTrust [8] tries to build a robust reputation score for P2P collaborators but a research [14] shows that it is still prone to collusion. A comprehensive survey on collusion detection in P2P systems can be found here [4]. Models proposed for detecting colluders in P2P systems are not directly applicable to online rating systems, because in P2P systems models are mostly built on relations and communications between people; but in online rating systems there is no direct relation between raters. Reputation management systems are also targeted by collusion. Very similar to rating systems, colluders in reputation management systems try to manipulate reputation scores by collusion. Many efforts are put on detecting collusion using majority rules, weight of the voter and temporal analysis of the behavior of the users [17] but none of these models is enough strong to detect all sorts of collusion [17, 18]. Yang et. al. [19] try to identify collusion by employing both majority rule and temporal behavior analysis. Their model is not still tested thoroughly and just is applied to a specific dataset and a particular type of attack.

The most similar work to ours is [13]. In this work Mukherjee et.al. propose a model for spotting fake review groups in online rating systems. The model analyzes textual feedbacks cast on products in Amazon online market to find collusion groups. They use 8 indicators to identify colluders and propose an algorithm for ranking collusion groups based on their degree of spamicity. However, our model is different from this model in terms of proposed indicators, analyzing personal behavior of the raters and dealing with redundant rating scores. Also a recent survey [5] shows that buyers rely more on scores and ratings when intend to buy something rather than reading textual items. So, in contrast with this model we focus on numerical aspect of posted feedback. However, the model proposed by Mukherjee et.al. is still vulnerable to some attacks. For example, if the number of attackers is much higher than honest raters on a product the model can not identify it as a potential case of collusion.

Another major difference between our work and other related work is that, we propose a graph data model and also a flexible query language [2] for better understanding, analyzing and querying collusion. This aspect is missing in almost all previous work.

8   Conclusion

In this paper we proposed a novel framework for collusion detection in online rating systems. We used two algorithms designed using frequent itemset mining technique for finding candidate collusive groups and sub-groups in our dataset. We propose several indicators showing the possibility of collusion from different aspects. We used these indicators to assign every group a rank to show their probability of collusion and also damaging impact. We evaluated our model first statically and showed the adequacy of the way we define and calculate collusion indicators. We then used precision and recall metrics to show quality of output of our method.

As future direction, we plan to identify more possible collusion indicators. We also plan to extend the implemented query language with more features and enhanced visual query builder to assist users employing our model. Moreover, we plan to generalize our model and apply it to other possible areas which are subject to collusive activities.

References

1. Agrawal, R., Srikant, R.: Fast algorithms for mining association rules in large databases. In: Proceedings of VLDB 1994, pp. 487–499 (1994)
2. Allahbakhsh, M., Ignjatovic, A., Benatallah, B., Beheshti, S.-M.-R., Foo, N., Bertino, E.: Detecting, Representing and Querying Collusion in Online Rating Systems. ArXiv e-prints (November 2012)

3. Beheshti, S.-M.-R., Benatallah, B., Motahari-Nezhad, H.R., Allahbakhsh, M.: A framework and a language for on-line analytical processing on graphs. In: Wang, X.S., Cruz, I., Delis, A., Huang, G. (eds.) WISE 2012. LNCS, vol. 7651, pp. 213–227. Springer, Heidelberg (2012)
4. Ciccarelli, G., Cigno, R.L.: Collusion in peer-to-peer systems. Computer Networks 55(15), 3517–3532 (2011)
5. Flanagin, A., Metzger, M., Pure, R., Markov, A.: User-generated ratings and the evaluation of credibility and product quality in ecommerce transactions. In: HICSS 2011, pp. 1–10. IEEE (2011)
6. Harmon, A.: Amazon glitch unmasks war of reviewers. NY Times (February 14, 2004)
7. Brown, J.M.J.: Reputation in online auctions: The market for trust. California Management Review 49(1), 61–81 (2006)
8. Kamvar, S.D., Schlosser, M.T., Garcia-Molina, H.: The eigentrust algorithm for reputation management in p2p networks. In: Proceedings of the WWW 2003, pp. 640–651 (2003)
9. Kerr, R.: Coalition detection and identification. In: The 9th International Conference on Autonomous Agents and Multiagent Systems, pp. 1657–1658 (2010)
10. Lee, H., Kim, J., Shin, K.: Simplified clique detection for collusion-resistant reputation management scheme in p2p networks. In: ISCIT 2010, pp. 273–278 (2010)
11. Leskovec, J., Adamic, L.A., Huberman, B.A.: The dynamics of viral marketing, vol. 1. ACM, New York (2007)
12. Lim, E., et al.: Detecting product review spammers using rating behaviors. In: Proceedings of the 19th ACM International Conference on Information and Knowledge Management, CIKM 2010, pp. 939–948. ACM, New York (2010)
13. Mukherjee, A., Liu, B., Glance, N.: Spotting fake reviewer groups in consumer reviews. In: Proceedings of the 21st International Conference on World Wide Web. ACM (2012)
14. Qio, L., et al.: An empirical study of collusion behavior in the maze p2p file-sharing system. In: Proceedings of the ICDCS 2007, p. 56 (2007)
15. Salton, G., Buckley, C., Fox, E.A.: Automatic query formulations in information retrieval. Journal of the American Society for Information Science 34(4), 262–280 (1983)
16. Salton, G., McGill, M.: Introduction to modern information retrieval. McGraw-Hill computer science series. McGraw-Hill (1983)
17. Sun, Y., Liu, Y.: Security of online reputation systems: The evolution of attacks and defenses. IEEE Signal Processing Magazine 29(2), 87–97 (2012)
18. Swamynathan, G., Almeroth, K., Zhao, B.: The design of a reliable reputation system. Electronic Commerce Research 10, 239–270 (2010), 10.1007/s10660-010-9064-y
19. Yang, Y., Feng, Q., Sun, Y.L., Dai, Y.: Reptrap: a novel attack on feedback-based reputation systems. In: Proceedings of SecureComm 2008, pp. 8:1–8:11 (2008)

A Recommender System Model Combining Trust with Topic Maps

Zukun Yu[1], William Wei Song[2], Xiaolin Zheng[1], and Deren Chen[1]

[1] Computer Science College, Zhejiang University, Hangzhou, China
[2] School of Technology and Business Studies, Dalarna University, Borlänge, Sweden
{zukunyu,xlzheng,drc}@zju.edu.cn,
wso@du.se

Abstract. Recommender Systems (RS) aim to suggest users with items that they might like based on users' opinion on items. In practice, information about the users' opinion on items is usually sparse compared to the vast information about users and items. Therefore it is hard to analyze and justify users' favorites, particularly those of cold start users. In this paper, we propose a trust model based on the user trust network, which is composed of the trust relationships among users. We also introduce the widely used conceptual model Topic Map, with which we try to classify items into topics for Recommender analysis. We novelly combine trust relations among users with Topic Maps to resolve the sparsity problem and cold start problem. The evaluation shows our model and method can achieve a good recommendation effect.

Keywords: Recommender Systems, Trust Model, Reputation, Trust Propagation, Topic Maps.

1   Introduction

With rapid development of Internet, more and more people use online systems to buy products and services (hereafter called items). However, with overwhelming amount of information about items available on the Internet, it is extremely difficult for users to easily find and determine what they would like to buy. Recommender systems aim to recommend the target users with the items which are considered to have high possibilities of meeting their preferences.

Given a huge number of users making commercial transactions online and an even larger amount of items available for sale online, recommender systems have to face two major challenges: **data sparsity** – the average number of ratings given by users is often very small compared to the huge number of items, and **cold start** – the "dumb" users who review few items and provide little information. It causes the problem that a recommender system cannot decide what should be recommended to the target users since it can only directly access to the users' opinions on a small proportion of items. Therefore, the data sparsity is now taken into account by many recommendation methods [9]. The cold start problem is a challenge to recommender systems due to its lack of sufficient information to justify their interests. The traditional solutions to the

Y. Ishikawa et al. (Eds.): APWeb 2013, LNCS 7808, pp. 208–219, 2013.

problems are to use a combination of content-based matching and collaborative filtering [21]. Recently, researchers have considered using trust to deal with them [10, 14].

Trust is an assumed reliance on some person or thing, namely, a confident dependence on the characteristics, ability, strength or truth of someone or something [12]. As pointed by Massa [15], recommender systems that make use of trust information are the most effective in term of accuracy while preserving a good coverage. In a trust-based recommender system, trust propagation is computed based on the trust network to derive indirect trust relationships between users, such as in the case of FOAF (friend of a friend) [7] – the framework for representing information about people and their social connections.

In general, the key concepts considered in modeling recommender systems include users, items, and the characteristics of them. Trust-based methods take only users' social links (Trust) into account but ignore the relations among items, which are helpful in predicting users' interest. In order to achieve a good recommendation method, we will combine trust relations among users with similarity relations among items derived by topic maps in an integrated model of recommender systems. This model maintains three types of relationships. The first type is the **trust** relationships among users. The second is the **rating scores** given to items by users, expressing how much users like items. The third one is called **relatedness** of an item to a topic, representing how much the item belongs to this topic, which leads to the computation of similarity relations among items. We use Topic Maps to represent the third type of relationships. Topic Maps is defined as an abstract model for semantically structured, self-describing link networks laid over a pool of addressable information objects [13]. A topic map includes three key components: topics, associations, and occurrences [22]. Using these elements, topic maps can be built in many domains.

The rest of this paper is organized as follows. We introduce the related work in recommender systems, trust and Topic Maps in Section 2. Then in Section 3 we propose our model based on trust and Topic Maps, and we novelly propose a method to propagate trust to determine users' interests and to apply Topic Maps to consider the relationship between items. Then we introduce our datasets for experiments and analyze the experiment results in Section 4. We conclude this paper and point out our future research in the last Section.

2   Related Work

Many efforts have been put in studying and developing recommender systems, aiming to support users doing business online. The major ones include content-based methods, collaborative filtering (CF) methods, hybrid methods - a combination of content-based methods, collaborative filtering methods and others. Recently, researchers are developing methods using trust to analyze users or items thereby to recommend users with items they might like.

The content-based methods analyze the items that are rated by users and use the contents of items to infer users' profiles, which is used in recommending items of interest to these users [2]. More specifically, the TF-IDF (term frequency–inverse document frequency) method, a computation method reflecting how important a word is to a document in a collection, is used to compute similarities of contents [20]. In this method, if two users collect more items with the similar content, they would be more likely to prefer the same items. These methods have been extended to consideration of using the attributes of items (for similarity computation) and the ratings given to items by users (for users profile computation) in construction of recommendations [23]. However, the content-based methods suffer from some shortcomings. Such kind of methods usually needs to collect user profiles. This is a problem of privacy. What is more, the content-based methods may cause overspecialized recommendations that only include items very similar to those of which the user already knows [3].

The CF method uses a database about users' preferences to predict more items users might like [4]. Papagelis et al. [19] propose a recommendation method based on incremental collaborative filtering of users' similarities. This method expresses the new similarity values between two users in relation to the old similarity values, so as to maintain an incremental update of their associated similarity. Zhang et al. [26] propose a recommendation algorithm based on an integrated diffusion method making use of both the user-item relations and the collaborative tagging information. They use a so called user-item-tag tripartite graph as the base of the diffusion process to generate recommendations. This method uses both the user-item relations and the collaborative tagging information to improve the algorithmic performance. The shortcoming of CF methods is that they do not explicitly incorporate feature information and face the sparsity problem and cold-start problem.

The authors of [18] propose a method using a weighted combination to fusion ratings obtained by content-based methods and CF-based methods separately. The weights are adjusted based on the strength of both the content-based method and the CF-based method. As the number of users and ratings given by them increase, the CF-based method is usually weighted more heavily, to take advantage of the wisdom of crowd via globally computing all the ratings. Melville et al. [17] propose an approach using a content-based predictor to enhance existing user data thereby to provide personalized suggestions through collaborative filtering. The content-based predictor accepts the item with a high rating score and rejects the item with a low rating score. However, these methods ignore the important social information which can reflect users' interest.

Trust among users and reputation of users are becoming important and elementary issues in social networking study. As pointed out by Guha and Kumar et al. [8], a user trust network is a fundamental buildings block in many of today's most successful e-commerce and recommendation systems. The authors propose a framework of trust propagation schemes, which appears to be the first to incorporate distrust in a computational trust propagation setting. This paper shows that a small amount of expressed trust or distrust information can be used to predict trust between any two people in the system with high accuracy. Ziegler and Lausen [27] introduce a classification scheme for trust metrics. They present some model constituents for semantic web trust infrastructure in the case FOAF (friend of a friend). However this paper has a limitation that it assumes all trust information is publicly accessible, which, in practice, is nearly impossible. Vydiswaran et al. [25] propose a trust propagation framework to compute how freetext claims of internet and their sources can be trusted by using an iterative algorithm to compute the scores of trust propagation. But this work utilizes a weak supervision at the evidence level, which makes it difficult to be used in other domains. Apart from trust, reputation - a global aggregation of the local trust scores by all the users [11] - is also important because it represents users' trustworthiness from a systemic perspective. Kamvar et al. [11] describe an algorithm to decrease the number of downloads of inauthentic files in a peer-to-peer file-sharing network. This paper assigns each peer a unique global trust value, based on the peer's history of uploads. Adler et al. [1] propose a content-driven reputation system for Wikipedia authors, which can be used to flag new contributions from low-reputation authors, or to allow only high-reputation authors to edit critical pages. Trust-based methods use trust relationships among users to build a social network to link users and use it to derive users' interest. But these methods have a shortcoming that they fail to analyze the relationships among items.

It is important for recommendation to take into account the relationships of items. We will apply Topic Maps technology to model relationships between items. Topic Maps related technologies are used in different research work. Dichev et al. [5] try to use Topic Maps to organize and retrieve online information in the context of e-learning courseware. They think Topic Maps offers a standard-based approach for expert's knowledge. This allows further reusing, sharing and interoperability of knowledge and teaching units between courseware authors and developers. Dong and Li [6] propose a new set of hyper-graph operations on XTM (XML Topic Map), called HyO-XTM, to manage the distributed knowledge resources. In the HyO-XTM, the set of vertices is the union of the vertices and the hyper-edges' sets of the hyper-graph; the set of edges is defined by the relation of incidence between vertices and hyper-edges of the hyper-graph. The hyper-graph model matches the Topic Maps with Hyper-graph vertices mapping to topic nodes and edges mapping to association nodes. Topic Maps is shown to be a new way to graphically manage the knowledge. Based on the previous work, we will first time try to use a topic map to represent relationships among items of recommender systems, so as to be aware of the relationships of items.

3   Topic Maps Based Trusted Recommender Model

We propose a graphical conceptual model for recommender systems, in which we describe three types of nodes, i.e., user nodes, item nodes, and topic nodesIn this model we also consider three types of relationships: *trust* (from one user to another), *rating* (from a user to an item), and *belonging* (from an item to a topic), see Fig. 1.

3.1   Model Description

We use a user graph $G(V,E)$ to model the user trust network, where V is a set of user nodes, representing users, and E a set of directed edges, representing *trust* relationships, see Fig. 1 (left). A trust relationship $e \in E$, is an edge in G, from a user node v_i

to another user node v_j. Each e in E is associated to a value in [-1, 1], indicating the weight of e. A negative trust value between two users means that they distrust each other and a positive trust value means that they trust each other. We use $e(v_i, v_j)$ to denote the trust value. We define a *truster* function of a user node v, yielding the set of all the users who have a trust relationship (an edge) to v, as $truster(v) = \{\forall u \in V \mid (u,v) \in E\}$. We also define a *trustee* function of v, representing the set of all the users who have a trust relationship from v, as $trustee(v) = \{\forall u \in V \mid (v,u) \in E\}$. For each user node v, we define a reputation function, denoted as $\rho(v)$, taking values from [0, 1].

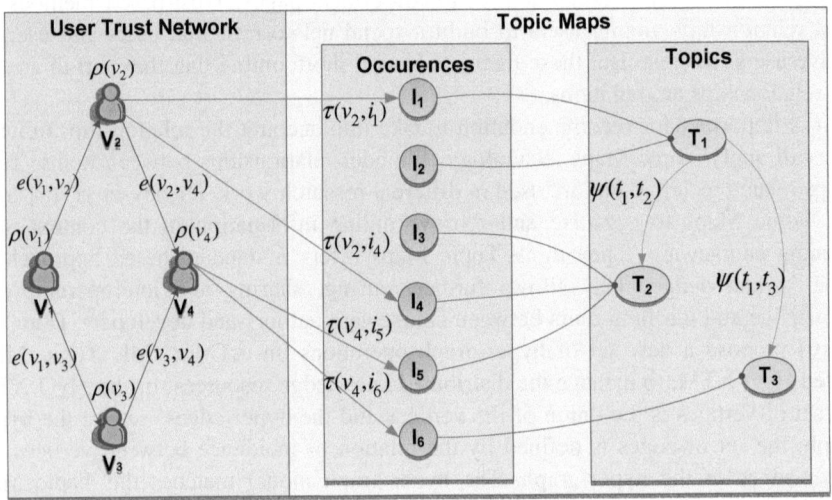

Fig. 1. Example of Topic Maps based trust recommender model – Users Trust Network (left) and Topic Maps (right)

Now we consider this type of relationships: *rating R*, from the user node set V to the item node set I. Each element $(v,i) \in R$ is an edge from a user who rates the item to the item with a rating function $\tau(v,i)$. We define the set of items rated by the user v by the function $item(v) = \{i \in I \mid (v,i) \in R\}$. Similarly, we use the function $reviewer(i) = \{v \in V \mid (v,i) \in R\}$ to represent the set of users who rate the item i.

According to the concept of Topic Maps, an item is an occurrence which belongs to a topic. We define the type of relationship, *belonging B*, from the item set I to the topic set T. Each element $(i,t) \in B$ is a direct edge from an item i to a topic t. An item might belong to a number of topics. So we use the function $topic(i) = \{t \in T \mid (i,t) \in B\}$ to define the set of all the topics the item i belongs to. We use the function $occurrence(t) = \{i \in I \mid (i,t) \in B\}$ to define the set of all the items belonging to the topic t. To model the association between topics in a topic map, we use a function $\psi(t_m, t_n)$ to represent the association degree between two topics t_m and t_n. Its value is a real number in [0, 1]. The higher the value is, the closer the two topics are.

3.2   Topic Maps Based Trusted Recommendation Methods

Using Topic Maps, we conceptually describe users, items, and topics, as well as various relationships between them. In this section we will introduce a computation method to quantitatively calculate these functions defined on the nodes and relationships.

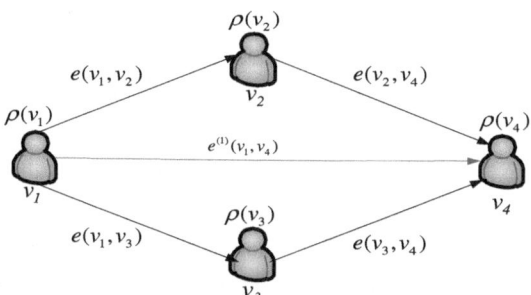

Fig. 2. Example of trust propagation

3.3   Trust Model

In the trust model, the trust relationships among users are the basic ones, from which we derive users' reputations and propagate new trust relationships among users. The reputation of a user v is computed by averaging all the trust values of the trust relationships to v from other users, i.e.

$$\rho(v) = \left(\sum_{u \in truster(v)} e(u,v) \right) \Big/ |truster(v)| \qquad (1)$$

If a user has no trust relationship from a truster, its reputation is set to 0. However, for any two users u, v if they do not have a direct edge (i.e. trust relationship) between them but have indirect edges (i.e. via one more users), we use trust propagation to determine the trust value between u, v. In other words, we aim to use existing values in the trust network to gain more trust values between the users without direct trust relationships. For simplicity, we only consider to propagate trust relationships but not distrust relationships in this paper. We use a step-by-step trust transitivity to derive indirect trust, in which a single step means to derive trust by the intermediate user nodes which have direct edge to both the start user node and the target user node in the user trust network, see Fig. 2. We use solid lines to represent existing trust relationships among users and dashed lines to represent the derived trust relationship. The trust relationships between v_1 to v_4 can be derived in the first single transitivity of trust propagation and that between v_1 and v_5 can be derived in the second single transitivity. In order to clearly explain the propagation process of trust, we use $e^{(n)}(v_i,v_j)$ to denote a *newly derived* trust value in the nth single step of trust transitivity. The original trust value $e(v_i,v_j)$ is denoted as $e^{(0)}(v_i,v_j)$, which means the existed trust edges in the original trust network.

For example, to derive a new trust value from the user node v_1 to the user node v_4, there might be a number of paths, where a path means a chain of edges from v_1 to v_4 via an intermediate user node, e.g. v_2. Here we define a function for the set of the common user nodes as $com(v_i, v_j) = trustee(v_i) \cap truster(v_j)$. We consider all the paths to derive the trust value:

$$e^{(n+1)}(v_i,v_j) = \frac{\sum\limits_{v_k \in \left(trustee(v_i) \cap truster(v_j)\right)} (e^{(\{0,1,2,...n\})}(v_i,v_k) \times \rho(v_k) + e^{(\{0,1,2,...n\})}(v_k,v_j) \times \rho(v_j))}{\sum\limits_{v_k \in \left(trustee(v_i) \cap truster(v_j)\right)} (|\rho(v_k)| + |\rho(v_j)|)} \qquad (2)$$

Here, $e^{(\{0,1,2,...n\})}(v_i,v_k)$ is a trust relationship in the set of all the trust relationships including the original trust relationships and those generated in the 1st, 2nd,..., and n^{th} step of the single step transitivity. The formula above gives a new trust value in [-1, 1] by applying once the single step transitivity. For a trust network, we can propagate trust relationships using a number of steps of the single step transitivity. We use a parameter s to control the number of steps for trust propagation.

3.4   Deriving Users' Opinion Based on the User Trust Network

In this section, we discuss how to generate a rating value between a user and an item if there was none there. We use the trust values from the user trust network and the rating values from the existing ratings given by the users to the items to derive users' new opinions on items together with new rating values. We denote the newly generated rating relationships through the user trust network as R'. We indirectly compute users' opinion on items through rating on the items from the intermediate users if they rated them directly. These intermediate users should have positive reputation values. The users with negative reputation values are thought as malicious users and not allowed to give advices to others. The newly generated rating relationships with rating values are computed by the function τ' given below.

$$\tau'(v,i) = \left(\sum\limits_{\mu \in trustee(v) \cap reviewer(i)} e^{(\{0,1,2,...n\})}(v,\mu) \times \tau(\mu,i) \times \rho(u) \right) \bigg/ \sum\limits_{trustee(v) \cap reviewer(i)} \rho(u) \qquad (3)$$

3.5   Deriving User's Opinion Based Topic Maps

In section 3.4 we generated a new set R' of rating relationships using trust propagation on the user trust network in the section 3.4. However, if the user trust network contains a lot of isolated clusters, the trust values and rating values from one cluster would not be possible to be used for other clusters. We call this the isolated trust cluster problem and will further explain it in section 4.1.

Here we consider using the Topic Maps to solve this problem. We use the topic map to derive users' opinion on topics. To do so, we first define a function $g(v, t)$ to be an user v's opinion on a topic t, deriving from users' opinion on the items and items belonging to the topic t, as follows:

$$g(v,t) = \left(\sum_{i \in item(v) \cap occurrences(t)} \tau(v,i) \right) \Big/ |item(v) \cap occurrences(t)| \quad (4)$$

Let us consider how to construct a topic map for a recommender system. We adopt association rule for building the association relationships between topics, because the association rule is frequently used to investigate sales transactions in market basket analysis and navigation paths within websites [24]. We define a function $\psi(t_m, t_n)$ to measure associations between the topics t_m and t_n, in terms of the users who have rated the topics, as follows:

$$\psi(t_m, t_n) = |user(t_m) \cap user(t_n)| / |user(t_m)| \quad (5)$$

In formula (5), the function $user(t)$ represents the set of users who rated the topic t. We set a threshold, denoted as $confidence\ \theta$, to be used in **filtering out** the associations with value less than the threshold. The associations after filtering can be used to derive users' opinion on a topic which they did not rate. The computation formula is as follows:

$$g(v,t) = \frac{\displaystyle\sum_{t_m \in \{t \in T | g(v,t) \neq 0, \psi(t_m,t) \neq 0\}} g(v,t_m) \times \psi(t_m,t)}{|\{t \in T \mid g(v,t) \neq 0, \psi(t_m,t) \neq 0\}|} \quad (6)$$

Finally, we consider how to decide a user's opinion on an item through both trust and topic map. We use a combination of users' opinion derived based on the topic map and users' opinion on items, to compute users' opinion $\tau'(v,i)$ based on the user trust network.

$$\tau'(v,i) = (1-\varphi) \times \tau(v,i) + \varphi \times \left(\left(\sum_{t \in topic(i)} g(v,t) \right) \Big/ |topic(i)| \right) \quad (7)$$

Here, φ is a weight parameter in [0, 1] to help controlling the weight of implicit opinion by themselves on an item, i.e., the weight of user's own opinion.

4   Experiments

In this section, first we describe the dataset used in our experiments, and then we discuss the experiments and their results.

4.1   Dataset

We use a data collection of the real review data from Epinions.com, provided at Massa's website [16], as the input dataset. We use two datasets, the trust relationships (and their values) between users and the users' ratings on items (and their values). In order to reduce the time and space complexities of algorithm, we use two subsets obtained by the method described below:

- Extract all the users in the first dataset and select 100 different users.
- Extract all the trust relationships among the 100 users. We obtained a subset, called subset 1. It contains 110 trust relationships, 82 of which have trust value 1 and 28 have trust value -1.
- Extract all the ratings given by the 100 users in the second dataset. We first obtain 282418 ratings on 230126 items from the 64 users who rated the items. Through sampling the ratings, we obtain the subset 2. It contains 10141 ratings, with 8858 items and 34 users. The subset 2 has 6 different levels of rating values and 27 topics. As shown in Figure 3, most items are rated 5 and 4. Only a small proportion of them are rated 1, 2, 3 and 6.

From the sample datasets we constructed for the experiments, we clearly observe the problems of "data sparsity" and "cold start users", which we discussed in Section 1. There are 10141 ratings for 100 users and 8858 items in this recommender system. So the subset of this recommender system is sparse because the ratio of the number of ratings to the total number of the matrix (number of users times number of items) is 10141/(100*8858)= 1.14%. Only one out of 100 users rated some items, so there are more than 98 cold start users. The maximum ratings for one user is 7507, on average every user 101 ratings.

The trust relationships in the subset 1 form a user trust network, as shown in Fig. 4. We use the trust propagation method discussed in section 3 to obtain more trust relationships. But this user trust network consists of many "isolated trust clusters". They do not connect to each other and contribute little to the trust propagation computation.

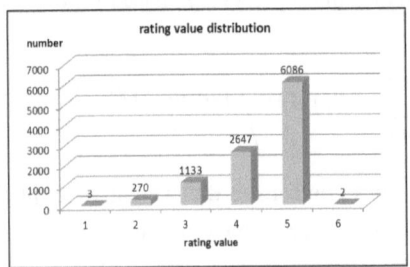   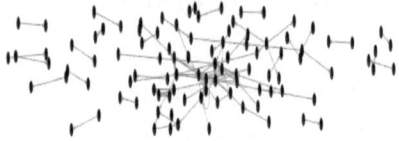

Fig. 3. The distribution of rating values   **Fig. 4.** The structure of user trust network

4.2   Results

We use MAE (Mean Absolute Error) and MAUE (Mean Absolute User Error) [15] to evaluate our recommendation method, because the MAE is the most commonly used and the easiest to be understood, and based on MAE, the MAUE provides the averages of evaluation.

Based on the observation that the most derived trust values have been obtained in the first three steps of trust propagation on the user trust network, we consider to set $s=3$. We evaluate the variation of accuracy as the weight parameter changes with a *confidence* $\theta =0.95$ in constructing a topic map. As show in Fig. 5, both the MAE and MAUE decline nearly in a straight line with the weight parameter. We can see that the MAUE is always less than the MAE due to the many cold start users in our dataset. So we find that the two metrics keep consistent in the relationship between accuracy and weight parameter.

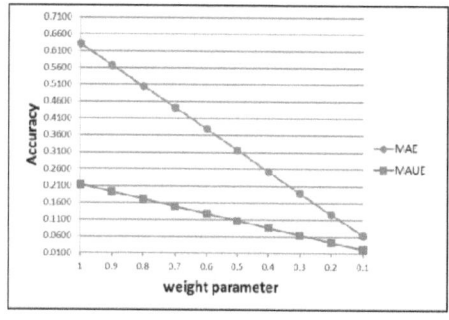 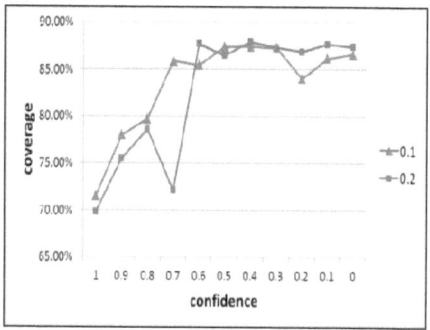

Fig. 5. The variation of accuracy with weight parameter

Fig. 6. The coverage changing as confidence of Topic Map

Furthermore, we consider using coverage to measure the performance of our method. First we set the weight parameter φ to be 0.1 and 0.2 respectively. We select the top 10 most-suited items in the recommendation lists of items. In Fig. 6, the line named 0.1 represents the coverage with the weight parameter φ to be 0.1 and the line named 0.2 represents the coverage with the weight parameter φ to be 0.2. As the confidence θ decreases from 1 to 0, the coverage rises. The coverage in the two lines rises sharply when the confidence is 0.7 and then it does not vary considerably. The reason for the changing of coverage is that when the confidence θ becomes smaller, the topic map will give more association relationships between topics to support deriving users' opinions on items. The result tells that the Topic Maps does help improve the coverage of recommender systems.

5　Conclusion and Future Work

We have three contributions in this work. First we propose a trust model based on the user trust network and a method to compute the trust propagation. We use the reputation as the weight in trust propagation, which mimics considering friends' advices before people buying products or services in the real world. That is, advices from the users with the greater reputation values are trusted by their friends with higher degree. Second we propose a recommendation method, applying Topic Maps in analyzing users' opinions on topics, which better and further supports deriving users' opinions on items. Third we evaluate our method using the two metrics accuracy and coverage.

The result tells that the recommender system based on our method provides better accuracy and coverage in coping with the problems of the sparsity and cold start users.

For the next step of study, we plan to scale up our method for recommender systems to a reasonably large number of users in the user trust network as in the reality a recommender system should be able to deal with millions of users and hundred millions of items. We also consider to include the computation of distrust relationships in the trust propagation method as we believe it will greatly contribute to the accuracy and coverage of recommender systems.

Acknowledgments. The authors would like to thank Dalarna University, Sweden for providing a visiting position and Zhejiang University, China for providing financial support of this visit.

References

1. Adler, B.T., Alfaro, L.D.: A Content-Driven Reputation System for the Wikipedia. Technical Report ucsc-crl-06-18. School of Engineering, University of California, Santa Cruz (2006)
2. Basu, C., Hirsh, H., Cohen, W.: Recommendation as Classification: Using Social and Content-Based Information in Recommendation, pp. 714–720. AAAI Press (1998)
3. Blanco-Fernández, Y., Pazos-Arias, J.J., Gil-Solla, A., Ramos-Cabrer, M., López-Nores, M., García-Duque, J., Fernández-Vilas, A., Díaz-Redondo, R.P., Bermejo-Muñoz, J.: A flexible semantic inference methodology to reason about user preferences in knowledge-based recommender systems. Knowl-Based Syst. 21, 305–320 (2008)
4. Breese, J.S., Heckerman, D., Kadie, C.: Empirical analysis of predictive algorithms for collaborative filtering. In: Proceedings of the Fourteenth Conference on Uncertainty in Artificial Intelligence, pp. 43–52. Morgan Kaufmann Publishers Inc., Madison (1998)
5. Christo, D., Darina, D., Lora, A.: Using topic maps for e-learning. In: Proceedings of International Conference on Computers and Advanced Technology in Education including the IASTED International Symposium on Web-Based Education, Rhodes, Greece, pp. 26–31 (2003)
6. Dong, Y., Li, M.: HyO-XTM: a set of hyper-graph operations on XML Topic Map toward knowledge management. Future Gener. Comp. Sy. 20, 81–100 (2004)
7. Golbeck, J., Rothstein, M.: Linking social networks on the web with FOAF: a semantic web case study. In: AAAI 2008, pp. 1138–1143. AAAI Press (2008)
8. Guha, R., Kumar, R., Raghavan, P., Tomkins, A.: Propagation of trust and distrust. In: Proceedings of the 13th International Conference on World Wide Web, pp. 403–412. ACM, New York (2004)
9. Huang, Z., Chen, H., Zeng, D.: Applying associative retrieval techniques to alleviate the sparsity problem in collaborative filtering. ACM Trans. Inf. Syst. 22, 116–142 (2004)
10. Jamali, M., Ester, M.: TrustWalker: a random walk model for combining trust-based and item-based recommendation. In: KDD 2009, pp. 397–406. ACM, New York (2009)
11. Kamvar, S.D., Schlosser, M.T., Garcia-Molina, H.: The eigentrust algorithm for reputation management in P2P networks. In: Proceedings of the 12th International Conference on World Wide Web, pp. 640–651. ACM, Budapest (2003)

12. Kini, A., Choobineh, J.: Trust in electronic commerce: definition and theoretical considerations. In: Proceedings of the Thirty-First Hawaii International Conference on System Sciences, vol. 4, pp. 51–61 (1998)
13. Knud, S.: Topic Maps-An Enabling Technology for Knowledge Management. In: International Workshop on Database and Expert Systems Applications, p. 472 (2001)
14. Massa, P., Bhattacharjee, B.: Using Trust in Recommender Systems: An Experimental Analysis, pp. 221–235 (2004)
15. Massa, P., Avesani, P.: Trust-aware recommender systems. In: RecSys 2007, pp. 17–24. ACM, New York (2007)
16. Massa, P.:
    http://www.trustlet.org/wiki/Extended_Epinions_dataset.
17. Melville, P., Mooney, R.J., Nagarajan, R.: Content-boosted collaborative filtering for improved recommendations. In: Eighteenth National Conference on Artificial Intelligence (AAAI 2002)/Fourteenth Innovative Applications of Artificial Intelligence Conference, pp. 187–192 (2002)
18. Miranda, T., Claypool, M., Gokhale, A., Murnikov, P., Netes, D., Sartin, M.: Combining Content-Based and Collaborative Filters in an Online Newspaper (1999)
19. Papagelis, M., Rousidis, I., Plexousakis, D., Theoharopoulos, E.: Incremental Collaborative Filtering for Highly-Scalable Recommendation Algorithms. In: Hacid, M.-S., Murray, N.V., Raś, Z.W., Tsumoto, S. (eds.) ISMIS 2005. LNCS (LNAI), vol. 3488, pp. 553–561. Springer, Heidelberg (2005)
20. Salton, G., McGill, M.J.: Introduction to Modern Information Retrieval. McGraw-Hill, Inc., New York (1986)
21. Schein, A.I., Popescul, A., Popescul, R., Ungar, L.H., Pennock, D.M.: Methods and Metrics for Cold-Start Recommendations, pp. 253–260. ACM Press (2002)
22. G. M. O. The Topic Maps: XML Topic Maps (XTM) 1.0. (2001)
23. Tso, K., Schmidt-Thieme, L.: Empirical Analysis of Attribute-Aware Recommendation Algorithms with Variable Synthetic Data. Data Science and Classification, 271–278 (2006)
24. Vercellis, C.: Business Intelligence: Data Mining and Optimization for Decision Making, p. 277. John Wiley & Sons, Ltd., Chichester (2009)
25. Vydiswaran, V.G.V., Zhai, C., Roth, D.: Content-driven trust propagation framework. In: Proceedings of the 17th ACM SIGKDD International Conference on Knowledge Discovery and Data Mining, pp. 974–982. ACM, San Diego (2011)
26. Zhang, Z.K., Zhou, T., Zhang, Y.C.: Personalized recommendation via integrated diffusion on user-item-tag tripartite graphs. Physica A 389, 179–186 (2010)
27. Ziegler, C., Lausen, G.: Spreading activation models for trust propagation. In: Proceedings of 2004 IEEE International Conference on e-Technology, e-Commerce and e-Service, pp. 83–97 (2004)

A Novel Approach
to Large-Scale Services Composition

Hongbing Wang* and Xiaojun Wang

School of Computer Science and Engineering, Southeast University, Nanjing, China
{hbw,seuwxj}@seu.edu.cn

Abstract. We investigate a multi-agent reinforcement learning model for the optimization of Web service composition in this paper. Based on the model, a multi-agent Q-learning algorithm was proposed, where agents in a team would benefit from one another. In contrast to single-agent reinforcement-learning, our algorithm can speed up the convergence to optimal policy. In addition, it allows composite service to dynamically adjust itself to fit a varying environment, where the properties of the component services continue changing. A set of experiments is given to prove the efficiency of the analysis. The advantages and the limitations of the proposed approach are also discussed.

Keywords: Web Service composition, multi-agent.

1 Introduction

As a common understanding, Web services are self-describing and open building blocks for rapid, low-cost composition of distributed applications [8]. In practice, a single Web service may not be sufficient at performing complex tasks. We usually need to combine multiple existing services together to meet customers' complex requests. With today's SOC technology, such composition is usually performed by human engineers. However, the Web environment is highly dynamic, and most Web services are evolving at all time. A service engineer cannot always foresee all the changes that could happen in the future. A manual service composition can also be too rigid to adapt to a dynamic environment. Therefore, dynamic service composition is regarded as a crucial functionality for the Web of the future. Different technologies of computational intelligence have been investigated for solving the problem of dynamic service composition.

AI planning is a typical type of techniques used to automate Web services composition [9] [2]. Doshi [3] and Gao [4] have studied the application of MDPs (Markov Decision Processes) in Web service composition. MDPs assume a fully observable world and require explicit reward functions and state transition functions. Such requirements are too strict to a real world scenario. Wang et al. [11] proposed to use reinforcement learning (RL) for service composition, so as to

---

* This work is partially supported by NSFC (61232007) and JSNSF of China (No.BK2010417).

Y. Ishikawa et al. (Eds.): APWeb 2013, LNCS 7808, pp. 220–227, 2013.

avoid complex modeling of the real world. Although it has been proved to be effective for small scale service compositions, it can be overly computationally expensive when working on a large number of services.

In this paper, we present a novel mechanism based on multi-agent reinforcement learning to enable adaptive service composition. The model proposed in this paper extends the reinforcement learning model that we have previously introduced in [11]. In order to reduce the time of convergence, we introduce a sharing strategy to share the policies among the agents, through which one agent can use the policies explored by the others. As the learning process continues throughout the life-cycle of a service composition, the composition can automatically adapt to the change of the environment and the evolvement of its component services. Experimental evaluation on large scale service compositions has demonstrated that the proposed model can provide good results.

2   A MAMDP Model for Service Composition

In this section we present the reinforcement learning model that we have previously introduced in [11] for solving the dynamic web service composition problem. The RL model introduced in [11] will be extended in this section towards a distributed architecture.

Reinforcement learning is the problem faced by an agent that must learn behavior through trial-and-error interactions with a dynamic environment [7]. One key aspect of reinforcement learning is a trade-off between exploitation and exploration. To accumulate a lot of reward, the learning system must prefer the best experienced actions, however, it has to try new actions in order to discover better action selections for the future. One such method is $\epsilon-$ greedy, when the agent chooses the action that it believes has the best long-term effect with probability $1 - \epsilon$, and it chooses an action uniformly at random, otherwise.

We formally define the key concepts used in the model.

Definition 1. (Web Service). *A Web service is modeled as a triple* $WS =<Pr; E; QoS >$*, where*

- *Pr represents the precondition of WS, which specifies the states of the world in which WS can be executed.*
- *E represents the effect of WS, which describes how WS changes the state of the world.*
- *QoS is a n-tuple* $< att_1; att_2; ::::; att_n >$*, where each* att_i *denotes a QoS attribute of WS.*

As mentioned earlier, we use Multi-agent Markov Decision Process (MAMDP) to model service composition. A MAMDP involves multiple actions and paths for each agent to choose. For agent m, we call our service composition model WSC-MDP, which simply replaces the actions in a MDP with Web services.

Definition 2. (Web service composition MDP (WSC-MDP). *A Web service composition MDP is a 7-tuple* $WSC\text{-}MDP =< m, s_0^m, S^m, A^m(.), P^m, R^m, s_r^m >$*, where*

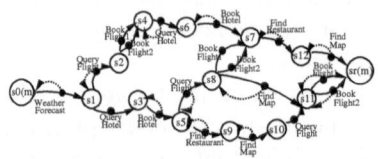

Fig. 1. The WSC-MDP of a Composite Service for Travel Plan

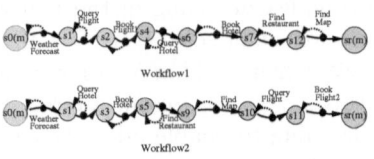

Fig. 2. Two Workflows Contained by the WSC-MDP in Fig. 1

- m: denote the agent m.
- S^m: a finite set of state spaces of agent m.
- $s_0^m \in S_m$: is the initial state of agent m and also an execution of the service compositions starts from this state.
- $s_r^m \in S_m$: is the set of terminal states for agent m. Upon arriving at one of the states, an execution of the service composition terminates.
- $A^m(.)$: represents the set of Web services that can be executed in state $s \in S^m$. A service ws belongs to A^m, only if the precondition ws P is satisfied by s.
- $P^m : [p_{iaj}^m]$ $S^m \times A^m \times S^m \rightarrow [0,1]$: The transitions function for the agent. It expresses the probability that the agent m goes to state j if it executes web service a in state i is p_{iaj}^m.
- $R^m : [r_{iaj}^m]$ $S^m \rightarrow R$: defines the rewards that agent m receives if it is in state i and goes to state j with the execution of web service a.

A WSC- MDP can be visualized as a transition graph. As illustrated by Fig. 1, the graph contains two kinds of nodes, i.e. state nodes and service nodes, which are represented by open circles and solid circles respectively. $s0(m)$ represents the initial state node. The terminal states nodes are $sr(m)$. A state node can be followed by a number of service nodes, representing the possible services that can be invoked in the state. There is at least one arrow pointing from a service node to the next state node. Each arrow is labeled with a transition probability p_{iaj}^m, and the expected reward for that transition r_{iaj}^m. (For simplicity, we omit the labels in Fig. 1.) The transition probabilities on the arrows rooted at a single action node always sum to one.

Definition 3. *Service Workflow. Let wf be a subgraph of a WSC-MDP. wf is a service workflow if and only if there is at most one service that can be invoked at each state wf. In other words, a service workflow is actually equivalent to a deterministic state machine. A tradition service composition usually builds on a single such workflow.*

Example 1. Fig. 2 shows two of multiple service workflows. Which workflow to be executed is determined by the policy of the Markov decision process.

Definition 4. *(Policy): A policy π is a mapping from state $s \in S$ to a service $ws \in A$, which tells which service $ws = (s)$ to execute when in state s.*

Each policy of a WSC-MDP can determine a single workflow. By executing a workflow, the service customer is supposed to receive a certain amount of reward, which is equivalent to the cumulative reward of all the executed services. Given a WSC-MDP, the task of our service composition system is to identify the optimal policy or workflow that offers the best cumulative reward. As the environment of a service composition keeps changing, the transition function P and the reward function R of a WSC-MDP change too. As a result, the optimal policy changes with time. If our system is able to identify the optimal policy at any given time, the service composition will be highly adaptive to the environment.

3   Algorithm for Service Composition

The proposed Distributed Approach

The Q-learning is the most popular and seems to be the effective model-free algorithm about the RL problems. It does not, however, address any of the issues involved in generalizing over large state and/ or action spaces [7]. That is why, in order to speed up the training process, we extend the proposed approach towards a distributed one, in which multiple cooperative agents learn to coordinate in order to find the optimal policy in their environment.

Experience sharing can help agents with similar tasks to learn faster and better. For instance, agents can exchange information using communication [10]. Furthermore, by design, most multiagent systems also allow the easy insertion of new agents into the system, leading to a high degree of scalability [1].

From agent-m's standpoint, its control task could be thought of as an ordinary reinforcement problem except that their action selecting strategy is dependent on other agent's optimal policies at the beginning of the learning. So at a certain state, one agent's policy may be useful to other agents which can help them to find optimal strategy quickly. But assume that each agent can simultaneously send its current policy to other agents, the communication information is huge. For the purpose of reducing the communication information we don't let the agent to communicate with each others, but introduce supervisor agent which supervises the learning process and synchronizes the computations of the individual agents. In our algorithm, each agent m use the global Q-values estimations stored in the blackboard which stores the global Q-values estimations and communicate to the supervisor agent their intention to update a Q-value estimation.

So we have two types of agents in our architecture:

- WSCA (Web Service Composition Agents). Each WSCA agent runs in a separate process or thread and is trained using the Q-learning algorithm. Each local agent performs local Q-values estimations updating from its own point of view.
  a WSCS(Web Service Composition Supervisor) agent which supervises the learning process and synchronizes the computations of the individual WSCA

agents. It keeps a blackboard [5] which stores the global Q-values estimations. The local WSCA agents use the global Q-values estimations stored
in the blackboard and communicate to the WSCS agent their intention to
update Q-value estimation. If a local agent tries to update a certain Q-value,
the WSCS agent will update the global Q-value estimation only if the new
estimation received from the local agent is greater than the Q-values estimations existing in the blackboard.

In this study, each reinforcement-learning agent uses the one-step Q-learning
algorithm. Its learning decision policy is determined by the central Q-table and
the state/action value function which estimates long-term discounted rewards for
each state/action pair. In order to get more knowledge for the agent and jump
out of the sub-superior strategy trap, searching strategy was introduced into the
Q-learning. The agent is allowed to take the action which isn't the superior at the
current view, so $\varepsilon - greedy$ strategy was proposed. Thus the agent can explore
the state-action space by choose the viable action randomly at some degree, and
avoid arriving at the local superior solution via only choosing the action with
the maximal Q-value.

For agent m, given a current state s and available actions $A(m)$, a Q-learning
agent selects action a_i with a probability given by the rule below:

$$p^m(a_i|s) = \begin{cases} (1 - \varepsilon) \text{ if } a_i = argMax_a \ Q[s, a] \\ \varepsilon \qquad\quad others \end{cases}$$

In each time step, the agent m updates $Q(s, a)$ by recursively discounting future
utilities and weighting them by a positive learning rate α :

$$Q(s, a) = (1 - \alpha) \cdot Q(s, a) + \alpha \cdot [r(s, a) + \gamma \cdot maxQ(s\prime, a\prime)]. \tag{1}$$

Here $\gamma(0 \leq \gamma < 1)$ is a discount parameter.

If $Q(s, a)$ is the biggest among the actions in state s, than the WSCS will
update $Q(s, a)$ in the blackboard.

The training process consists of three phases and will be briefly described in
the following.

Phase 1. Initial phase

The WSCS supervisor agent initializes the Q-values from the blackboard.

Phase 2. Training phase of each WSCA agent

During some training episodes, the individual WSCA agents will experiment
some paths from the initial to a final state, using the $\epsilon - greedy$ mechanism and
updating the Q-values estimations according to the algorithm described below
(algorithm 1). We denote in following by $Q(s, a)$ the Q-value estimate associated
to the state s and a, as stored by the blackboard of the WSCS agent.

Algorithm 1. Multiagent Q-learning Algorithm for Agent m

Require:
  The WSC-MDP for agent m;
  The WSCS agent;
  repeat
    evaluate the starting state s
    select action a from s using policy derived from Q ($\epsilon - Greddy$)
    repeat
      Learning:(for each step of episode)
      Take action a, observe the reward $r(s,a)$ and the next state $s\prime$.
      WSCA agent asks WSCS agent for $Q(s,a)$.
      WSCS retrieves $Q(s,a)$ from the blackboard.
      WSCS sends the retrieved $Q(s,a)$ to WSCA.
      WSCA agent updates the table entry $Q(s,a)$ ad follows

$$Q(s,a) = (1 - \alpha) \cdot Q(s,a) + \alpha \cdot [r(s,a) + \gamma \cdot maxQ(s\prime, a\prime)]. \qquad (2)$$

      WSCA sends the new $Q(s,a)$ to WSCS.
      WSCS updates the $Q(s,a)$ if it is greater than the old.
      WSCA agent go to state $s\prime$
    until s is the terminal state
  until the Q-values tiny changes

Phase 3. Executing phase

After the training of the multi-agent system has been completed, the solution learned by the WSCS supervisor agent is constructed by starting from the initial state and following the *Greedy* mechanism until a solution is reached. The system applies the solution as a service workflow to execute. At the same time, the execution is also treated as an episode of the learning process. The Q-functions are updated afterwards, based on the newly observed reward.

By combining execution and learning, our framework achieves self-adaptively automatically. As the environment changes, service composition will change its policy accordingly, based on its new observation of reward. It does not require prior knowledge about the QoS attributes of the component services, but is able to achieve the optimal execution policy through learning.

4 Experimental Evaluation

In order to evaluate the methods, we conducted simulation to evaluate the properties of our service composition mechanism based on the methods discussed in this paper. The PC configuration: Intel Xeon E7320 2.13GHZ with 8GB RAM, Windows 2003, jdk1.6.0.

We considered two QoS attributes of services. They ware service fee and execution time. We assigned each service node in a simulated WSC-MDP graph with random QoS values. The value followed normal distribution. To simulate the dynamic environment, we periodically varied the QoS values of existing services based on a certain frequency. We applied the algorithms introduced in Section 3 to execute the simulated service compositions. The reward function used by the each learner was solely based on the two QoS attributes. After an execution of a service , the learners get a reward , whose value is:

$$R(s) = \frac{fee_i^{max} - fee_i^s}{fee_i^{max} - fee_i^{min}} + \frac{time_i^{max} - time_i^s}{time_i^{max} - time_i^{min}}$$

The reward was always positive, however service consumers always prefer low execution time and service fee.

We will show that such cooperative agents can speed up learning, measured by the average cumulative values in training, even though they will eventually reach the same asymptotic performance as independent agents.

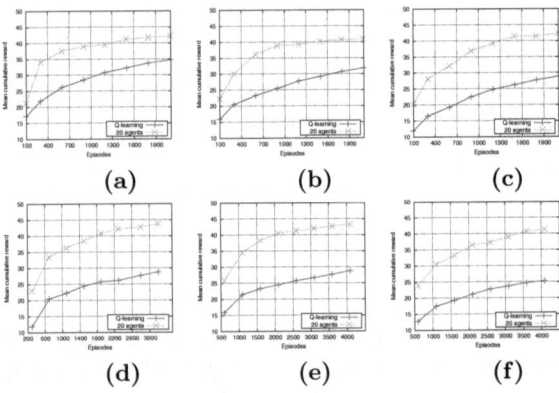

Fig. 3. (a) Results of comparison with 20 services in each state; (b) Results of comparison with 30 services in each state; (c) Results of comparison with 40 services in each state; (d) Results of comparison with 100 services in each state; (e) Results of comparison with 150 services in each state; (f) Results of comparison with 200 services in each state;

Scalability with respect to the number of services

In this stage of our evaluation, we studied the effect of distributed RL approach with varied number of services in each state. We fix the state number on 500 and vary the services from 20 to 200. As the Fig. 3 shows, the distributed RL learns more quickly than Q-learning, and when the number of services increases, the reducing of convergence time is more considerable, because they may have explored the different parts of a state space and share their knowledge. If agents perform the similar task, two agents can complement each other by exchanging their policies or use what the other agent had already learned for its own benefit. Assume that each agent can simultaneously send its current policy at some state to blackboard Q-table by WSCS agent, if some agent finds a better choice, it may update blackboard Q-table through WSCS agent, then other agents can adopt that policy with certain probability in that state.

The results in all cases clearly indicate that distributed RL approach presented in this paper learns more quickly and reduces the overall computational time compared againest the Q-learning.

5 Conclusion

This paper studied a novel framework for large scale service composition. In order to reduce the time of convergence we introduce a sharing strategy to share the policies among agents in a team. The experimental results show that the strategy of sharing state-action space improves the learning efficiency significantly. Additionally, the problem that has to be further investigated is how to reduce the communication cost between the WSCA agents and WSCS agent and explore other local search mechanisms. Next, we will concentrate on these issues and improve our algorithm further.

References

1. Busoniu, L., Babuska, R., De Schutter, B.: A comprehensive survey of multiagent reinforcement learning. IEEE Transactions on Systems, Man, and Cybernetics, Part C: Applications and Reviews 38(2), 156–172 (2008)
2. Carman, M., Serafini, L., Traverso, P.: Web service composition as planning. In: ICAPS 2003 Workshop on Planning for Web Services, pp. 1636–1642 (2003)
3. Doshi, P., Goodwin, R., Akkiraju, R., Verma, K.: Dynamic workflow composition using markov decision processes. In: IEEE International Conference on Web Services, pp. 576–582. IEEE (2004)
4. Gao, A., Yang, D., Tang, S., Zhang, M.: Web service composition using markov decision processes. In: Fan, W., Wu, Z., Yang, J. (eds.) WAIM 2005. LNCS, vol. 3739, pp. 308–319. Springer, Heidelberg (2005)
5. Gonzaga, T., Bentes, C., Farias, R., de Castro, M., Garcia, A.: Using distributed-shared memory mechanisms for agents communication in a distributed system. In: Seventh International Conference on Intelligent Systems Design and Applications, ISDA 2007, pp. 39–46. IEEE (2007)
6. Hwang, S.Y., Lim, E.P., Lee, C.H., Chen, C.H.: Dynamic web service selection for reliable web service composition. IEEE Transactions on Services Computing 1(2), 104–116 (2008)
7. Kaelbling, L., Littman, M., Moore, A.: Reinforcement learning: A survey. Arxiv preprint cs/9605103 (1996)
8. Papazoglou, M., Georgakopoulos, D.: Service-oriented computing. Communications of the ACM 46(10), 25–28 (2003)
9. Sirin, E., Parsia, B., Wu, D., Hendler, J., Nau, D.: Htn planning for web service composition using shop2. Web Semantics: Science, Services and Agents on the World Wide Web 1(4), 377–396 (2004)
10. Sutton, R., Barto, A.: Reinforcement learning. Journal of Cognitive Neuroscience 11(1), 126–134 (1999)
11. Wang, H., Zhou, X., Zhou, X., Liu, W., Li, W., Bouguettaya, A.: Adaptive service composition based on reinforcement learning. In: Maglio, P.P., Weske, M., Yang, J., Fantinato, M. (eds.) ICSOC 2010. LNCS, vol. 6470, pp. 92–107. Springer, Heidelberg (2010)

The Consistency and Absolute Consistency Problems of XML Schema Mappings between Restricted DTDs

Hayato Kuwada[1], Kenji Hashimoto[2], Yasunori Ishihara[1], and Toru Fujiwara[1]

[1] Osaka University, Japan
{h-kuwada,ishihara,fujiwara}@ist.osaka-u.ac.jp
[2] Nara Institute of Science and Technology, Japan
k-hasimt@is.naist.jp

Abstract. Consistency of XML schema mappings, which means that some document conforming to the source schema can be mapped into a document conforming to the target schema, is an essentially necessary property. It is also important for XML schema mappings to be absolutely consistent, that is, every document conforming to the source schema can be mapped into a document conforming to the target schema. As a known result, consistency of a mapping between general DTDs is EXPTIME-complete even if the class of document patterns for defining mappings is restricted to downward axes and qualifiers. In addition, the known tractability result is only on a restricted class of document patterns under restricted DTDs called nested-relational DTDs. Moreover, there are few known results on the tractability of absolute consistency. In this paper, we discuss the consistency and absolute consistency problems under restricted DTDs called disjunction-capsuled DTDs, which were proposed by Ishihara et al. We show that for many document pattern classes, both problems are solvable in polynomial time under disjunction-capsuled DTDs. Although disjunction-capsuled DTDs are an incomparable class to nested-relational DTDs, a part of our results can be extended to a proper superclass of nested-relational DTDs.

1 Introduction

XML (Extensible Markup Language) is a markup language for describing structured data and used for a variety of applications and database systems. Accordingly, many schemas are defined to specify document structures, and schema description languages such as DTD (Document Type Definition) are used for specifying schemas. A schema mapping is the basis of data transformation from one schema to another caused by data exchange or schema evolution. In particular, schema mappings for XML are called *XML schema mappings*. A mapping is a triple of a *source schema*, a *target schema*, and a set of *dependencies*, which are pairs of *document patterns* and specify correspondence of data. *Consistency* of schema mappings, which means that some document conforming to the source

Y. Ishikawa et al. (Eds.): APWeb 2013, LNCS 7808, pp. 228–239, 2013.

schema can be mapped into a document conforming to the target schema according to the given dependencies, is an essentially necessary property. It is also important for schema mappings to be *absolutely consistent*, that is, every document conforming to the source schema can be mapped into a document conforming to the target schema according to the given dependencies.

Example 1. Define the source and target schemas D_S and D_T as follows:

```
DS:                          DT:
root -> student+             root -> student+
student -> name phone        student -> name phone*
phone -> home mobile other*  phone -> home | mobile | other
```

First, consider the following dependency set Σ_1:

$$\Sigma_1 = \{\texttt{student/phone[home][mobile]} \longrightarrow \texttt{student/phone[home][mobile]}\}.$$

Σ_1 states that if there is a document of D_S such that a phone node has both home and mobile nodes as its children, then the document is mapped into a document of D_T with a phone node which has, again, both home and mobile nodes as its children. However, Σ_1 is not consistent because in any document of D_T, a phone node cannot have both home and mobile nodes as its children. Next, consider the following dependency set Σ_2:

$$\Sigma_2 = \{\texttt{student/phone[home][other]} \longrightarrow \texttt{student/phone[home][other]}\}.$$

Σ_2 is consistent because there is a document of D_S which has no other nodes. Such a document vacuously satisfies Σ_2. On the other hand, Σ_2 is not absolutely consistent because of the similar reason to the case of Σ_1. Lastly, consider the following dependency set Σ_3:

$$\Sigma_3 = \{\texttt{student/phone[home][mobile][other]} \longrightarrow$$
$$\texttt{student[phone/home][phone/mobile][phone/other]}\}.$$

Σ_3 is absolutely consistent because for every document of D_S, if it has a phone node with children home, mobile, and other, then it can be mapped into a document of D_T which has (at least three) phone nodes one of which has home, another has mobile, and the other has other as their children.

Amano et al. investigated the computational complexity of the problems of determining consistency and absolute consistency of XML schema mappings [1]. The main focus of [1] is XML schema mappings with data values, but they addressed also mappings without data values. The results of the case without data values are shown in Table 1. In this paper we use XPath expressions for specifying document patterns, although in [1] tree patterns is used. Using XPath expressions allows us to utilize the rich known results on the satisfiability problem for our problem. XPath expressions consist of $\downarrow$ (child axis), $\downarrow^*$ (descendant-or-self axis), $\uparrow$ (parent axis), $\uparrow^*$ (ancestor-or-self axis), $\rightarrow$ (next-sibling axis), $\rightarrow^*$ (following-sibling-or-self axis), $\rightarrow^+$ (following-sibling axis), $\leftarrow^+$ (preceding-sibling axis),

Table 1. Known results in [1]

$\downarrow$	$\downarrow^*$	$\uparrow$	$\uparrow^*$	$\rightarrow$	$\rightarrow^*$	$\rightarrow^+$	$\leftarrow^+$	$\cup$	$[\,]$	$*$	Consistency	Absolute Consistency
											General DTDs	
+	+							+	+		EXPTIME-complete	Π_2^p-complete
+	+			+	+			+	+		EXPTIME-complete	Π_2^p-complete

$\downarrow$	$\downarrow^*$	$\uparrow$	$\uparrow^*$	$\rightarrow$	$\rightarrow^*$	$\rightarrow^+$	$\leftarrow^+$	$\cup$	$[\,]$	$*$	Consistency	Absolute Consistency
											Nested-relational DTDs	
+								+			PTIME	PTIME
+	+							+	+		PTIME	Π_2^p
+	+			+				+			PSPACE-hard	Π_2^p

Table 2. Results of this paper

$\downarrow$	$\downarrow^*$	$\uparrow$	$\uparrow^*$	$\rightarrow$	$\rightarrow^*$	$\rightarrow^+$	$\leftarrow^+$	$\cup$	$[\,]$	$*$	Consistency	Absolute Consistency
											DC-DTDs	
+	+					+	+	+	+	+	PTIME	PTIME
+	+	+	+			+	+	+			PTIME	PTIME
+		+				+		+			PTIME	PTIME

$\cup$ (path union), $[\,]$ (qualifier), and $*$ (wild card). Table 1 describes computational complexity of each class including only axes and operators with the "+". Nested relational DTDs are non-recursive DTDs such that each content model is in the form of $\hat{l}_1 \cdots \hat{l}_m$, where all l_i's are distinct and each $\hat{l}_i$ is either l_i, l_i^*, l_i^+, or $l_i^?$.

As can be seen from Table 1, constraints under which decision problems of consistency and absolute consistency are solved in PTIME have been little known, even if data values are not concerned. Thus, the purpose of this paper is to find practically wide classes of DTDs and XPath expressions for which both problems can be solved in PTIME. As a first step for achieving this, we focus on the case without data values in this paper.

In this paper, we use a restricted class of DTDs called disjunction-capsuled DTDs (DC-DTDs), which were proposed in [2]. DC-DTDs are an important class from the theoretical point of view because many tractability results on XPath satisfiability under DC-DTDs are known. Such results can be exploited for consistency and absolute consistency problems. DC-DTDs are also a meaningful class from the practical point of view, because 7 out of 22 real-world DTDs, 853 out of 950 real-world DTD rules are actually DC [3]. Although DC-DTDs are incomparable to nested-relational DTDs, the tractability results on the satisfiability under DC-DTDs can be extended to a wider class, called DC$^{?+}$-DTDs [3], which is a proper superclass of nested-relational DTDs. 13 out of 22 DTDs, 928 out of 950 rules are $DC^{?+}$.

The results of this paper are shown in Table 2. "+" acrossing multiple cells represents "one of them". First, it is shown that consistency of mappings between DC-DTDs is solvable by deciding satisfiability and validity of document patterns *linearly many times*. Precisely speaking, we show that under DC-DTDs, satisfiability of a set of dependencies can be decomposed into satisfiability and validity of each document pattern in the dependencies. Hence, if the document patterns are in an XPath class for which satisfiability and validity are tractable, then consistency is also tractable. Next, it is shown that absolute consistency of mappings between DC-DTDs is solvable by deciding satisfiability of document patterns *linearly many times*. Again, the technically important point is that satisfiability of a set of dependencies can be done by independently checking the satisfiability of each document pattern in the dependencies. Hence, if the document patterns are in an XPath class for which satisfiability is tractable, then absolute consistency is also tractable.

The rest of this paper is organized as follows. Several definitions are given in Section 2. Sections 3 and 4 present tractability results of consistency and absolute consistency of mappings between DC-DTDs, respectively. Section 5 summarizes the paper.

2   Preliminaries

2.1   XML Documents and DTDs

In this paper, we regard elements of XML documents as labeled nodes, and entire XML documents as labeled ordered trees. In what follows, we write trees to mean labeled ordered trees. Let $\lambda(v)$ denote the label of a node v. For a sequence of nodes $v_1 v_2 \cdots v_n$, define $\lambda(v_1 v_2 \cdots v_n)$ as $\lambda(v_1)\lambda(v_2)\cdots\lambda(v_n)$.

A regular expression over Γ consists of ϵ (empty sequence), the symbols in Γ, and operators · (concatenation, usually omitted in the notation), | (disjunction), and $*$ (repetition). We exclude $\emptyset$ (empty set) because we are interested in only nonempty regular expressions. Let $L(e)$ denote the string language represented by a regular expression e.

Definition 1. *A DTD D is a triple (Γ, r, P), where*

- *Γ is a finite set of labels,*
- *$r \in \Gamma$ is the root label, and*
- *P is a mapping from Γ to the set of regular expressions over Γ. $P(l)$ is called the* content model *of label l.*

A tree T conforms to a DTD $D = (\Gamma, r, P)$ if the root label of T is r, and for each node v of T and its children sequence $v_1 v_2 \cdots v_n$, it holds that $\lambda(v_1 v_2 \cdots v_n) \in L(P(\lambda(v)))$. Let $TL(D)$ denote the set of all the trees conforming to D.

Next, we introduce DC-DTDs, a restricted class of DTDs.

Definition 2. *A regular expression e is* disjunction-capsuled *[2], or* DC *for short, if e is in the form of $e_1 e_2 \cdots e_n$ ($n \geq 1$), where each e_i is either a symbol in Γ or in the form of $(e_i')^*$ for an arbitrary regular expression e_i'.*

For example, $(a|b^*)^* b a^*$ is disjunction-capsuled, and ϵ is also disjunction-capsuled because ϵ is ϵ^*. On the other hand, $a^*|b^*$ is not disjunction-capsuled.

Definition 3. *Let $D = (\Gamma, r, P)$. D is a* disjunction-capsuled DTD, *or DC-DTD for short, if $P(l)$ is disjunction-capsuled for each $l \in \Gamma$.*

2.2   XPath Expressions

XPath is a query language for XML documents.

Definition 4. *The syntax of an XPath expression p is defined as follows:*

$$p ::= \chi :: l \mid p/p \mid p \cup p \mid p[q],$$
$$\chi ::= \cdot \mid \downarrow \mid \uparrow \mid \downarrow^* \mid \uparrow^* \mid \rightarrow^+ \mid \leftarrow^+,$$
$$q ::= p \mid q \wedge q \mid q \vee q,$$

where $l \in \Gamma$. Let $\mathcal{X}$ denote the set of all XPath expressions.

Note that $\rightarrow$ (next-sibling axes) and $\rightarrow^*$ (following-sibling-or-self axes) are excluded from $\mathcal{X}$ because they are not proper axes in W3C XPath. Moreover, $*$ (wild card) is also excluded because it can be represented by using $\cup$ (path union). On the other hand, $\mathcal{X}$ contains $\cdot$ (self axis), which is not considered in previous research. Self axis makes our discussion simpler without destroying the known tractability results of XPath satisfiability.

Definition 5. *The semantics of an XPath expression p over a tree T is defined as follows, where p and q are regarded as binary and unary predicates over nodes of T, respectively:*

- $T \models (\cdot :: l)(v, v)$ *if $\lambda(v) = l$;*
- $T \models (\downarrow :: l)(v, v')$ *if v' is a child of v and $\lambda(v') = l$;*
- $T \models (\uparrow :: l)(v, v')$ *if v' is a parent of v and $\lambda(v') = l$;*
- $T \models (\downarrow^* :: l)(v, v')$ *if v' is v or a descendant of v, and $\lambda(v') = l$;*
- $T \models (\uparrow^* :: l)(v, v')$ *if v' is v or an ancestor of v, and $\lambda(v') = l$;*
- $T \models (\rightarrow^+ :: l)(v, v')$ *if v' is a following sibling of v and $\lambda(v') = l$;*
- $T \models (\leftarrow^+ :: l)(v, v')$ *if v' is a preceding sibling of v and $\lambda(v') = l$;*
- $T \models (p/p')(v, v')$ *if there is v'' such that $T \models p(v, v'')$ and $T \models p'(v'', v')$;*
- $T \models (p \cup p')(v, v')$ *if $T \models p(v, v')$ or $T \models p'(v, v')$;*
- $T \models (p[q])(v, v')$ *if $T \models p(v, v')$ and $T \models q(v')$;*
- $T \models p(v)$ *if there is v' such that $T \models p(v, v')$;*
- $T \models (q \wedge q')(v)$ *if $T \models q(v)$ and $T \models q'(v)$;*
- $T \models (q \vee q')(v)$ *if $T \models q(v)$ or $T \models q'(v)$.*

A tree T satisfies an XPath expression p if there is a node v such that $T \models p(v_0, v)$, where v_0 is the root node of T.

Definition 6. *An XPath expression p is* satisfiable *under a DTD D if some $T \in TL(D)$ satisfies p. Moreover, an XPath expression p is* valid *under a DTD D if every $T \in TL(D)$ satisfies p.*

2.3   XML Schema Mappings

In this section, we define XML schema mappings according to [1,4,5].

Definition 7. *A* dependency *is an expression of the form* $\pi \longrightarrow \pi'$, *where* π *and* π' *are XPath expressions called* document patterns. *An* XML schema mapping $\mathcal{M}$ *is a triple* (D_S, D_T, Σ), *where* D_S *is a DTD representing a source schema,* D_T *is a DTD representing a target schema, and* Σ *is a finite set of dependencies. Hereafter, XML schema mappings are referred to as just* mappings.

Originally, in [1,4,5], document patterns are defined by means of tree patterns. However, we use XPath expressions in order to exploit many results on XPath satisfiability and validity.

Definition 8. *A pair of trees* $T_S \in TL(D_S)$ *and* $T_T \in TL(D_T)$ *satisfies a dependency* $\pi \longrightarrow \pi'$ *if the following condition is satisfied: If* T_S *satisfies* π, *then* T_T *satisfies* π'.

The set of all pairs of trees which satisfy all dependencies of $\mathcal{M}$ is denoted as $[\![\mathcal{M}]\!]$. Then, consistency and absolute consistency are defined as follows:

Definition 9. *Let* $\mathcal{M} = (D_S, D_T, \Sigma)$. $\mathcal{M}$ *is* consistent *if* $[\![\mathcal{M}]\!] \neq \emptyset$. *On the other hand,* $\mathcal{M}$ *is* absolutely consistent *if, for every tree* T_S *conforming to* D_S, *there exists a tree* T_T *conforming to* D_T *such that* $(T_S, T_T) \in [\![\mathcal{M}]\!]$.

3   Deciding Consistency

3.1   Deciding Consistency of Mappings between General DTDs

To begin with, we show that the consistency decision problem can be solved by deciding validity and satisfiability of combinations of document patterns in dependencies. The following lemma is essentially equivalent to the statement given in the proof of the EXPTIME-completeness of deciding consistency under general DTDs [5].

Lemma 1. *Let* $\mathcal{M} = (D_S, D_T, \Sigma)$ *and* $\Sigma = \{\pi_i \longrightarrow \pi_i' \mid i \in [n]\}$ *where* $[n] = \{1, 2, \ldots, n\}$. *Then,* $\mathcal{M}$ *is consistent if and only if there are two disjoint subsets* $I = \{i_1, \cdots, i_k\}$ *and* $J = \{j_1, \cdots, j_{k'}\}$ *of* $[n]$ *such that* $I \cup J = [n]$,

- $\cdot :: r_S[\pi_{i_1} \vee \cdots \vee \pi_{i_k}]$ *is not valid under* D_S, *and*
- $\cdot :: r_T[\pi_{j_1}' \wedge \cdots \wedge \pi_{j_{k'}}']$ *is satisfiable under* D_T

where r_S *and* r_T *are the root labels of* D_S *and* D_T, *respectively.*

Proof. Suppose $\mathcal{M}$ is consistent. Then, a pair of trees (T_S, T_T) belongs to $[\![\mathcal{M}]\!]$. Let $J = \{j_1, \cdots, j_{k'}\}$ be a subset of $[n]$ such that T_T satisfies π_{j_l}' for each $j_l \in J$. Then, because of the semantics of qualifier [ ], T_T satisfies $\cdot :: r_T[\pi_{j_1}' \wedge \cdots \wedge \pi_{j_{k'}}']$. On the other hand, let $I = \{i_1, \cdots, i_k\} = [n] - J$. Then, T_T does not satisfy π_i' for any $i \in I$. So, because $(T_S, T_T) \in [\![\mathcal{M}]\!]$, T_S must not satisfy π_i for any $i \in I$. Therefore, because of the semantics of $\vee$, T_S does not satisfy $\cdot :: r_S[\pi_{i_1} \vee \cdots \vee \pi_{i_k}]$.

Conversely, suppose that $\cdot :: r_S[\pi_{i_1} \vee \cdots \vee \pi_{i_k}]$ is not valid under D_S and $\cdot :: r_T[\pi'_{j_1} \wedge \cdots \wedge \pi'_{j_{k'}}]$ is satisfiable under D_T for some $I = \{i_1, \cdots, i_k\}$ and $J = \{j_1, \cdots, j_{k'}\}$ such that $I \cap J = \emptyset$ and $I \cup J = [n]$. The non-validity of $\cdot :: r_S[\pi_{i_1} \vee \cdots \vee \pi_{i_k}]$ under D_S means that there is an instance $T_S \in TL(D_S)$ which does not satisfy π_i for any $i \in I$. On the other hand, if $\cdot :: r_T[\pi'_{j_1} \wedge \cdots \wedge \pi'_{j_{k'}}]$ is satisfiable under D_T, there is $T_T \in TL(D_T)$ which satisfies π'_j for all $j \in J = [n] - I$. Therefore, the pair (T_S, T_T) belongs to $[\![\mathcal{M}]\!]$ because T_S does not satisfy π_l or T_T satisfies π'_l for each $\pi_l \longrightarrow \pi'_l \in \Sigma$ where $l \in [n]$. □

According to Lemma 1, the problem of deciding consistency of schema mappings can be solved by using the decision procedure of validity and satisfiability of XPath expressions under DTDs. However, the complexity of this decision procedure for consistency of schema mappings between general DTDs is high, because deciding validity and satisfiability of XPath expressions under general DTDs is known to be intractable [6] and combinatorial explosion occurs in the choice of disjoint subsets I and J of $[n]$. In fact, consistency of mappings between general DTDs is known to be EXPTIME-complete [5] even if the class of document patterns is restricted to downward axes and qualifiers.

3.2   Deciding Consistency of Mappings between DC-DTDs

Now we restrict source and target schemas to be DC-DTDs. We shall give tractability results of consistency of mappings between DC-DTDs by showing the following facts:

1. Satisfiability for broad XPath subclasses under DC-DTDs is decidable in polynomial time [2,3].
2. Validity for XPath expressions in $\mathcal{X}$ (i.e., the whole XPath class of this paper) under DC-DTDs is decidable in polynomial time.
3. Under DC-DTDs, we can decide consistency of a schema mapping by deciding validity or satisfiability of each individual document pattern in a mapping instead of combinations of document patterns.

First, we recall that satisfiability under DC-DTDs for several XPath subclasses can be decided in polynomial time [2,3]. Table 3 shows the known tractability results of XPath satisfiability under DC-DTDs.

Next, we show that validity of XPath expressions in $\mathcal{X}$ under DC-DTDs is tractable. For a DC regular expression $e = e_1 e_2 \cdots e_n$, let $\tilde{e}$ denote the regular expression obtained by replacing with ϵ each subexpression e_i such that $e_i = (e'_i)^*$ for some regular expression e'_i. For a DC-DTD $D = (\Gamma, r, P)$, let $\tilde{D} = (\Gamma, r, \tilde{P})$ such that $\tilde{P}(l) = \tilde{e}$ when $P(l) = e$. Note that $TL(\tilde{D})$ is a singleton because each content model $\tilde{e}$ in $\tilde{D}$ does not contain any disjunction operator and thus it represents only one sequence over Γ. The DTD $\tilde{D}$ has the following property on validity of XPath expressions under a DC-DTD D.

Lemma 2. *For an XPath expression π in $\mathcal{X}$ and a DC-DTD D, the unique tree $\tilde{t}$ in $TL(\tilde{D})$ satisfies π if and only if π is valid under D.*

Table 3. Computational complexity of the satisfiability decision under DC-DTDs

$\downarrow$	$\downarrow^*$	$\uparrow$	$\uparrow^*$	$\rightarrow^+$	$\leftarrow^+$	$\cup$	[ ]	Satisfiability
+	+			+	+	+	+	PTIME
+	+	+	+	+	+	+		PTIME
+			+		+		+	PTIME
+		+	+				+	NP-complete
+		+				+	+	NP-complete
	+		+				+	NP-complete
+			+	+	+		+	NP-complete

Proof. For the if part, assume that π is valid under D. Then, since $\tilde{t} \in TL(D)$, $\tilde{t}$ satisfies π. For the only if part, assume that $\tilde{t}$ satisfies π. From the definition of $\tilde{D}$, for each $l \in \Gamma$, the unique sequence in $L(\tilde{P}(l))$ is a subsequence of any $w \in L(P(l))$. Let t be an arbitrary tree in $TL(D)$. Then there is an injective mapping θ from the set of nodes of $\tilde{t}$ to that of t such that

- $\lambda(v) = \lambda(\theta(v))$ for each node v of $\tilde{t}$,
- for any two nodes v_1 and v_2 of $\tilde{t}$, if v_1 is a parent of v_2, then $\theta(v_1)$ is a parent of $\theta(v_2)$, and
- for any two nodes v_1 and v_2 of $\tilde{t}$, if v_1 is a following sibling of v_2, then $\theta(v_1)$ is a following sibling of $\theta(v_2)$.

For this, we can see by induction on the structure of π that $\tilde{t} \models \pi(v, v')$ implies $t \models \pi(\theta(v), \theta(v'))$. Thus, t also satisfies π. $\qquad\square$

Lemma 2 leads to tractability of XPath validity checking under DC-DTDs. That is, validity of an XPath expression π in $\mathcal{X}$ under a DC-DTD D is linear-time reducible to evaluation of π on the tree $\tilde{t} \in TL(\tilde{D})$. Since a polynomial-time algorithm for evaluation is known [7], validity of XPath expressions in $\mathcal{X}$ under DC-DTDs can be decided in polynomial time.

Lastly, we can avoid combinatorial explosion mentioned in Section 3.1 by the following lemmas.

Lemma 3. *Let π_1 and π_2 be XPath expressions in $\mathcal{X}$ and D be a DC-DTD. Either or both of π_1 and π_2 is valid under D if and only if $\cdot :: r[\pi_1 \vee \pi_2]$ is valid under D, where r is the root label of D.*

Proof. The only if part is trivial. For the if part, assume that $\cdot :: r[\pi_1 \vee \pi_2]$ is valid under D. Consider $\tilde{t} \in TL(\tilde{D})$. Because $\tilde{t} \in TL(D)$, $\tilde{t}$ satisfies $\cdot :: r[\pi_1 \vee \pi_2]$. By the semantics of $\vee$, $\tilde{t}$ satisfies at least one of π_1 and π_2. Without loss of generality, we assume that $\tilde{t}$ satisfies π_1. From Lemma 2, π_1 is valid under D. $\qquad\square$

Lemma 4. *Let π_1 and π_2 be XPath expressions in $\mathcal{X}$ and D be a DC-DTD. Both π_1 and π_2 are satisfiable under D if and only if $\cdot :: r[\pi_1 \wedge \pi_2]$ is satisfiable under D, where r is the root label of D.*

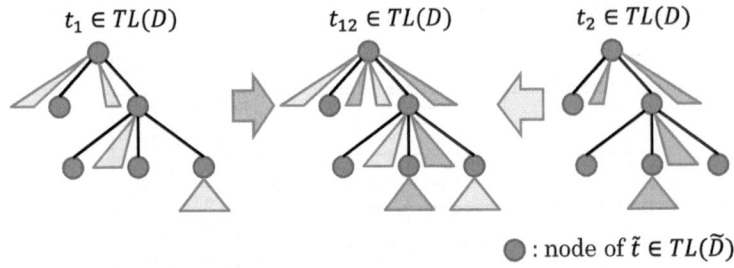

$t_1 \in TL(D)$ $t_{12} \in TL(D)$ $t_2 \in TL(D)$

●: node of $\tilde{t} \in TL(\tilde{D})$

Fig. 1. Construction of t_{12} from t_1 and t_2 by inserting subtrees

Proof. The if part is trivial. For the only if part, assume that both π_1 and π_2 are satisfiable under D. Then, there are trees t_1 and t_2 in $TL(D)$ such that t_1 and t_2 satisfy π_1 and π_2, respectively. Because $\mathcal{X}$ does not include negation operator and next-sibling axis, if a tree t satisfies XPath expression $\pi \in \mathcal{X}$, any tree obtained by inserting arbitrary subtrees to t also satisfies π [3]. Thus, if we can obtain the same tree t_{12} from t_1 and t_2 by inserting subtrees (see Fig. 1), t_{12} satisfies $\cdot :: r[\pi_1 \wedge \pi_2]$. To show that such t_{12} exists, we should prove that for any regular DC expression e, if $w_1, w_2 \in L(e)$, there is $w \in L(e)$ such that w_1 and w_2 are subsequences of w. This property holds because DC regular expressions cannot specify non-co-occurrence among labels. □

From the above, we can refine Lemma 1 for deciding consistency of mappings between DC-DTDs as follows.

Lemma 5. *Let $\mathcal{M}$ be a mapping (D_S, D_T, Σ) such that the schemas D_S and D_T are DC-DTDs. Then, $\mathcal{M}$ is consistent if and only if there is no dependency $\pi \longrightarrow \pi'$ in Σ such that π is valid under D_S and π' is not satisfiable under D_T.*

Proof. Let $\Sigma = \{\pi_i \longrightarrow \pi'_i \mid i \in [n]\}$ and r_S and r_T be the root labels of D_S and D_T, respectively. From Lemma 1, we prove this lemma by showing the next condition: there is no $\pi \longrightarrow \pi' \in \Sigma$ such that π is valid under D_S and π' is not satisfiable under D_T if and only if there are two disjoint subsets $I = \{i_1, \ldots, i_k\}$ and $J = \{j_1, \ldots, j_{k'}\}$ of $[n]$ such that $I \cup J = [n]$, $\cdot :: r_S[\pi_{i_1} \vee \cdots \vee \pi_{i_k}]$ is not valid under D_S, and $\cdot :: r_T[\pi'_{j_1} \wedge \cdots \wedge \pi'_{j_{k'}}]$ is satisfiable under D_T.

Assume that there is no $\pi \longrightarrow \pi' \in \Sigma$ such that π is valid under D_S and π' is not satisfiable under D_T. Then, for each $l \in [n]$, π_l is not valid under D_S or π'_l is satisfiable under D_T. Let $I = \{i_1, \ldots, i_k\}$ be the set of all the indexes i_l in $[n]$ such that π_{i_l} is not valid under D_S. Let $J = \{j_1, \ldots, j_{k'}\} = [n] - I$. Then π'_{j_l} is satisfiable under D_T for each $j_l \in J$. From Lemmas 3 and 4, $\cdot :: r_S[\pi_{i_1} \vee \cdots \vee \pi_{i_k}]$ is not valid under D_S and $\cdot :: r_T[\pi'_{j_1} \wedge \cdots \wedge \pi'_{j_{k'}}]$ is satisfiable under D_T.

Conversely, assume that there are two disjoint subsets $I = \{i_1, \ldots, i_k\}$ and $J = \{j_1, \ldots, j_{k'}\}$ of $[n]$ such that $I \cup J = [n]$, $\cdot :: r_S[\pi_{i_1} \vee \cdots \vee \pi_{i_k}]$ is not valid under D_S, and $\cdot :: r_T[\pi'_{j_1} \wedge \cdots \wedge \pi'_{j_{k'}}]$ is satisfiable under D_T. From Lemmas 3 and 4, π_l is not valid under D_S for any $l \in I$ and π'_l is satisfiable under D_T for any $l \in J$. Since $I \cup J = [n]$, there is no $\pi \longrightarrow \pi' \in \Sigma$ such that π is valid under D_S and π' is not satisfiable under D_T. □

Here, in order to decide consistency of mappings between DC-DTDs, we just have to check validity of π under D_S and satisfiability of π' under D_T for each $\pi \longrightarrow \pi'$ in Σ. Therefore, if we choose as the class of document patterns the XPath subclasses in which validity and satisfiability under DC-DTDs are tractable, deciding consistency of schema mappings is tractable.

Theorem 1. *Let $\mathcal{M}$ be a mapping (D_S, D_T, Σ), where D_S, D_T are DC-DTDs, such that for each $\pi \longrightarrow \pi' \in \Sigma$, $\pi \in \mathcal{X}$ and $\pi' \in \mathcal{X}_T$, where $\mathcal{X}_T$ is an XPath subclass for which satisfiability is decidable in polynomial time under DC-DTD (see Table 3). Then, consistency of $\mathcal{M}$ can be decided in polynomial time.*

4   Deciding Absolute Consistency

4.1   Deciding Absolute Consistency of Mappings between General DTDs

We begin with a lemma stating that absolute consistency of a mapping $\mathcal{M} = (D_S, D_T, \Sigma)$ can be determined by checking satisfiability of all XPath expressions induced by all subset of Σ. Again, this lemma is essentially equivalent to the statement in the explanation of Π_2^p-completeness of absolute consistency of mappings between general DTDs [1].

Lemma 6. *Mapping $\mathcal{M} = (D_S, D_T, \Sigma)$ is absolutely consistent if and only if the following condition holds: For every subset $\Sigma' = \{\pi_1 \longrightarrow \pi'_1, \ldots, \pi_k \longrightarrow \pi'_k\}$ of Σ, if $\cdot :: r_S[\pi_1 \wedge \cdots \wedge \pi_k]$ is satisfiable under D_S, then $\cdot :: r_T[\pi'_1 \wedge \cdots \wedge \pi'_k]$ is satisfiable under D_T, where r_S and r_T are the root labels of D_S and D_T, respectively.*

Proof. Suppose that mapping $\mathcal{M} = (D_S, D_T, \Sigma)$ is absolutely consistent, and consider an arbitrary subset $\Sigma' = \{\pi_1 \longrightarrow \pi'_1, \ldots, \pi_k \longrightarrow \pi'_k\}$ of Σ. Also suppose that $T_S \in TL(D_S)$ satisfies $\cdot :: r_S[\pi_1 \wedge \cdots \wedge \pi_k]$. Since $\mathcal{M}$ is absolutely consistent, there is $T_T \in TL(D_T)$ such that $(T_S, T_T) \in [\![\mathcal{M}]\!]$. Hence, T_T satisfies $\cdot :: r_T[\pi'_1 \wedge \cdots \wedge \pi'_k]$ and the only if part holds.

Conversely, suppose that for every subset $\Sigma' = \{\pi_1 \longrightarrow \pi'_1, \ldots, \pi_k \longrightarrow \pi'_k\}$ of Σ, if $\cdot :: r_S[\pi_1 \wedge \cdots \wedge \pi_k]$ is satisfiable under D_S, then $\cdot :: r_T[\pi'_1 \wedge \cdots \wedge \pi'_k]$ is satisfiable under D_T. Consider an arbitrary tree $T_S \in TL(D_S)$. Let $\Sigma' = \{\pi_1 \longrightarrow \pi'_1, \ldots, \pi_k \longrightarrow \pi'_k\}$ be a subset of Σ such that T_S satisfies π_i for each i ($1 \leq i \leq k$). Then, T_S satisfies $\cdot :: r_S[\pi_1 \wedge \cdots \wedge \pi_k]$, and therefore, $\cdot :: r_S[\pi_1 \wedge \cdots \wedge \pi_k]$ is satisfiable under D_S. Then, by assumption, there must be $T_T \in TL(D_T)$ such that T_T satisfies $\cdot :: r_T[\pi'_1 \wedge \cdots \wedge \pi'_k]$, and hence, $(T_S, T_T) \in [\![\mathcal{M}]\!]$. Consequently, $\mathcal{M}$ is absolutely consistent.   □

Similarly to the case of consistency, the above lemma tells us that the hardness of deciding absolute consistency stems from the combinatorial explosion caused by checking all the subset of dependencies as well as the hardness of deciding XPath satisfiability.

4.2   Deciding Absolute Consistency of Mappings between DC-DTDs

To obtain tractability results on absolute consistency, we focus on DC-DTDs again. Similarly to the case of consistency, the combinatorial explosion of subsets of dependencies can be avoided, that is, satisfiability can be independently checked for each dependency.

Lemma 7. *Let D_S and D_T be DC-DTDs. Mapping $\mathcal{M} = (D_S, D_T, \Sigma)$ is absolutely consistent if and only if for every dependency $\pi \longrightarrow \pi' \in \Sigma$, if π is satisfiable under D_S, then π' is satisfiable under D_T.*

Proof. Suppose that mapping $\mathcal{M} = (D_S, D_T, \Sigma)$ is absolutely consistent. Then, by Lemma 6, for every subset $\Sigma' = \{\pi_1 \longrightarrow \pi'_1, \ldots, \pi_k \longrightarrow \pi'_k\}$ of Σ, if $\cdot :: r_S[\pi_1 \wedge \cdots \wedge \pi_k]$ is satisfiable under D_S, then $\cdot :: r_T[\pi'_1 \wedge \cdots \wedge \pi'_k]$ is satisfiable under D_T. This immediately implies the only if part of the lemma.

Conversely, suppose that for every dependency $\pi \longrightarrow \pi' \in \Sigma$, if π is satisfiable under D_S, then π' is satisfiable under D_T. By the property of DC-DTDs (Lemma 4), two XPath expressions π_1 and π_2 are satisfiable under DC-DTD D if and only if $\cdot :: r[\pi_1 \wedge \pi_2]$ is satisfiable under D, where r is the root label of D. Hence, for any subset $\Sigma' = \{\pi_1 \longrightarrow \pi'_1, \ldots, \pi_k \longrightarrow \pi'_k\}$ of Σ, $\pi_1, \ldots, \pi_k$ are satisfiable under D_S if and only if $\cdot :: r_S[\pi_1 \wedge \cdots \wedge \pi_k]$ is satisfiable under D_S. On the other hand, by the assumption, if $\pi_1, \ldots, \pi_k$ are satisfiable under D_S, then $\pi'_1, \ldots, \pi'_k$ are satisfiable under D_T, and hence, $\cdot :: r_T[\pi'_1 \wedge \cdots \wedge \pi'_k]$ is satisfiable under D_T. Therefore, the if part holds.   □

Now, the next theorem immediately follows from the above lemma:

Theorem 2. *Let $\mathcal{M}$ be a mapping (D_S, D_T, Σ), where D_S, D_T are DC-DTDs, such that for each $\pi \longrightarrow \pi' \in \Sigma$, $\pi, \pi' \in \mathcal{X}_T$, where $\mathcal{X}_T$ is an XPath subclass for which satisfiability is decidable in polynomial time under DC-DTD (see Table 3). Then, absolute consistency of $\mathcal{M}$ can be decided in polynomial time.*

5   Conclusion

This paper has discussed the consistency and absolute consistency problems of XML schema mappings between DC-DTDs. First, we have shown that consistency of mappings between DC-DTDs can be solved by deciding validity and satisfiability of XPath expressions linearly many times. Moreover, we have proved that the XPath validity problem in the presence of DC-DTDs can be solved in polynomial time. From these facts, we have proved that the consistency of mappings between DC-DTDs can be solved in polynomial time if the document patterns are in an XPath class for which satisfiability and validity are tractable under DC-DTDs. Next, we have shown that absolute consistency of the mappings between DC-DTDs can also be solved by deciding satisfiability of XPath expressions in the presence of DC-DTD linearly many times. So the absolute consistency of mappings between DC-DTDs can also be solved in polynomial time if the document patterns are in an XPath class for which satisfiability is tractable under DC-DTDs.

As stated in Section 1, there is a superclass of DC-DTDs, called $DC^{?+}$-DTDs [3]. They allow two operators ? (zero or one occurrence) and + (one or more occurrences) of regular expressions in a restricted manner. Actually, $DC^{?+}$-DTDs are also a proper superclass of nested-relational DTDs. The tractability results of absolute consistency of mappings between DC-DTDs can be extended to mappings between $DC^{?+}$-DTDs because $DC^{?+}$-DTDs inherit all the tractability of XPath satisfiability of DC-DTDs. However, as for consistency, the tractability results do not seem to be extended to mappings between $DC^{?+}$-DTDs because we are conjecturing that XPath validity under $DC^{?+}$-DTDs is coNP-hard. Now we are trying to prove the coNP-hardness and then propose a DTD class which is a superclass of both nested-relational DTDs and DC-DTDs but under which XPath validity is tractable.

We are also planning to incorporate data values into mappings. As stated in [1], consistency of mappings with data values can be reduced to consistency of mappings without data values. However, such reduction is not possible for absolute consistency. It is interesting to find XPath classes such that absolute consistency of mappings between DC-DTDs with data values becomes tractable.

Acknowledgment. The authors thank Mr. Yohei Kusunoki of Osaka University for his insightful comments and suggestions on deciding XPath validity under DC-DTDs. The authors are also thankful to the anonymous reviewers for their helpful comments. This research is supported in part by Grant-in-Aid for Scientific Research (C) 23500120 from Japan Society for the Promotion of Science.

References

1. Amano, S., Libkin, L., Murlak, F.: XML schema mappings. In: Proceedings of the Twenty-Eigth ACM SIGMOD-SIGACT-SIGART Symposium on Principles of Database Systems, pp. 33–42 (2009)
2. Ishihara, Y., Morimoto, T., Shimizu, S., Hashimoto, K., Fujiwara, T.: A tractable subclass of DTDs for XPath satisfiability with sibling axes. In: Gardner, P., Geerts, F. (eds.) DBPL 2009. LNCS, vol. 5708, pp. 68–83. Springer, Heidelberg (2009)
3. Ishihara, Y., Shimizu, S., Fujiwara, T.: Extending the tractability results on XPath satisfiability with sibling axes. In: Lee, M.L., Yu, J.X., Bellahsène, Z., Unland, R. (eds.) XSym 2010. LNCS, vol. 6309, pp. 33–47. Springer, Heidelberg (2010)
4. Arenas, M., Libkin, L.: XML data exchange: Consistency and query answering. Journal of the ACM 55(2) (2008)
5. Arenas, M., Barcelo, P., Libkin, L., Murlak, F.: Relational and XML Data Exchange. Morgan & Claypool (2010)
6. Benedikt, M., Fan, W., Geerts, F.: XPath satisfiability in the presence of DTDs. Journal of the ACM 55(2) (2008)
7. Bojańczyk, M., Parys, P.: XPath evaluation in linear time. Journal of the ACM 58(4), 17 (2011)

Linking Entities in Unstructured Texts
with RDF Knowledge Bases

Fang Du[1,2], Yueguo Chen[1], and Xiaoyong Du[1]

[1] School of Information, Renmin University of China, Beijing, China
[2] School of Mathematics and Computer Science, Ningxia University, China
{dfang,chenyueguo,duyong}@ruc.edu.cn

Abstract. Entity linking (entity annotation) is the task of linking named entity mentions on Web pages with the entities of a knowledge base (KB). With the continued progress of information extraction and semantic search techniques, entity linking has received much attention in both research and industrial communities. The challenge of the task is mainly on entity disambiguation. To our best knowledge, the huge existing RDF KBs have not been fully exploited for entity linking. In this paper, we study the entity linking problem via the usage of RDF KBs. Besides the accuracy of entity linking, the scalability of handling huge Web corpus and large RDF KBs are also studied. The experimental results show that our solution on entity linking achieves not only very good accuracy but also good scalability.

Keywords: RDF knowledge base, entity linking, named entity disambiguation.

1 Introduction

Nowadays, search engines have been widely used as the most convenient way of accessing information within the huge amount of unstructured texts on the Web. However, techniques adopted by most existing search engines are based on straightforward matches of terms within queries and those within the unstructured texts. This often results in that, users are frustrated in frequently adjusting their query terms to retrieve the desired results. To address this problem, semantic search has been proposed and widely studied [1]. In semantic search, the contextual meaning of terms in the unstructured texts is very important to enrich the semantics of terms. In this paper, we study the problem of enriching contexts of named entities within unstructured texts, by linking the detected named entities with existing RDF knowledge bases. By doing this, we are able to extract important concepts and topics out of the plain texts, and effectively support the semantic search over the unstructured texts.

With the continued progress of Semantic Web and information extraction techniques, more and more RDF data emerge on the web. They form many huge RDF knowledge bases (KBs) such as Yago[2], Freebase[3], DBPedia[4] et al. Such RDF KBs contain billions of RDF triple facts either extracted from Web pages or contributed manually by users, describing information about hundreds of millions of entities. The entity information in these KBs is so abundant and diverse that they are very suitable for enriching the contexts of named entity mentions within the unstructured texts.

Y. Ishikawa et al. (Eds.): APWeb 2013, LNCS 7808, pp. 240–251, 2013.

The entity linking (also called entity annotation in the paper) task is defined as to map a named entity mention m in the free texts of a Web page to a corresponding entity e in a KB. In our work, a KB refers to an RDF KB without further specification. We list some applications that such a way of entity linking can be applied as follows:

- Enhancing search results of semantic search. Through linking entities with RDF KBs, the search results of unstructured texts can be expanded by utilizing relevant semantic information derived from RDF KBs.
- Cleaning the extracted named entities. The named entity mentions extracted by automatic information extraction tools often have quality issues such as errors or duplicates. Linking entities with proper entities in RDF KBs can help to clean the extracted named entities.
- Finding relations between entities. Given a pair of entities, their relations can be learned from the contexts. The accuracy of entity relation discovery will highly depend on the accuracy of entity resolution during the annotation process.

One major challenge of entity linking is the disambiguation problem. A named entity mention has the disambiguation problem because a number of entities (of different types) in the KB can have same name mention. When this happens, we need to accurately link the entity mention to a proper KB entity. This is called entity resolution (or entity disambiguation) problem. For example, a mention "Michael Jordan" may map to at least two entities in a KB. One is a famous retired NBA basketball player, and the other is a famous computer scientist in U.C. Berkeley. Many existing approaches use a lot of features from contexts to address the entity disambiguation problem. However, an RDF KB can be treated as a huge labeled graph, by considering the properties of the graph, we observe that the relevant entities of mentions in the same context often are close pairs in the KB graph. Based on the observation, we propose an effective approach that accomplish the entity disambiguation problem with only two proposed features.

An open domain RDF KB has a large number of entities. However, many existing approaches[8,9] are designed on small-scale data set, or can only process one document a time[21], which are not scalable. For a huge RDF KB, entity linking will have to be done in a parallel and distributed fashion for guaranteeing the efficiency and scalability. Fortunately, the MapReduce[6] framework on cloud computing provides an easy-to-use way of dispatching the expensive annotating and disambiguating tasks over clusters. We therefore propose an efficient entity disambiguation algorithm that is able to conduct the entity linking task in parallel based on the MapReduce framework.

The main contributions of the paper are as follows:

- We propose a simple but effective algorithm to accurately link named entity mentions in the unstructured texts with proper entities in an RDF KB.
- We propose an MapReduce based approach to linking mentions of unstructured texts with entities in the RDF KB. The annotation of multiple documents can be conducted parallelly and in batch.
- We test the performance of our approaches over two real-life RDF KBs. The results show that our solution can achieve high accuracy on both KBs. Meanwhile, by designing algorithms running over MapReduce framework, the solution is very efficient and scalable.

The rest of the paper is organized as follows. Section 2 reviews the related work. Section 3 describes the solution of entity disambiguation using RDF KBs. In Section 4, we propose an efficient framework of conducting the entity linking task on MapReduce framework. Section 5 presents the results of experimental study. Finally, the paper ends with conclusion and future work in Section 6.

2   Related Work

Entity linking has received much attention recently[7,15,14], especially after the emerging of large knowledge bases such as Yago[2], Freebase[3], DBPedia[4]. Most of these works[8,9] link free texts to real world knowledge bases. The most important and challenging issue in entity linking task is the named entity disambiguation problem.

Named entity disambiguation is also called co-reference resolution or word sense disambiguation in some other studies. Most studies in this area tackle the challenges based on rich contexts where named entity mentions occur. Bagga and Baldwin[11] use vector space model to resolve the ambiguities in persons' names. They represent the context of the entity mentions with the bag of words model and compute similarity between vectors. An early work [7] identifies the most proper meaning of ambiguous words by measuring their context overlaps. In [10], the authors derive a set of features from Wikipedia to compute similarity between context of mentions and the texts of Wikipedia.

In word sense disambiguation, there are two main streams of solutions[17,18]. Some researchers use knowledge extracted from dictionaries to identify the correct word in a given context[8,16]. The others collect probabilities from large amounts of sense-annotated data, and then use machine learning approaches to solve the problem[17,18]. The authors of [12] implement and combine these two approaches. They choose the right and left word of the ambiguous word as its context. For each word, they extract a training feature vector from Wikipedia links, and then integrate the features in Naive Bayes classifier. These works in word sense disambiguation are similar to the entity linking task. However, they are more likely to assign dictionary meanings to the polysemous words, which is not as difficult as the entity linking task.

Weikum and Theobald[19] propose their named entity disambiguation approach in facts harvesting research works. In [20], they interconnect RDF data and Web contents via LOD (Linked Open Data)[5]. They construct an entity mention graph and use a coherence graph algorithm to solve the entity disambiguation problem. Through computing the coherence between ambiguous entities, the mentions are connected to at most one entity in a KB.

The most similar work to ours are Linden[21]. Linden first builds a dictionary based on four kinds of Wikipedia pages: entity page, redirect page, disambiguation page and hyper links. By using the dictionary, Linden generates a candidate linking list for each mention. In order to conduct entity disambiguation, it creates feature vectors of four dimensions to rank the entities in candidate list. However, Linden can process only one document at the same time. For real world applications, there will be much more mentions from more than one document to be processed simultaneously. Moreover, it generates many features to rank the candidate linking entities. Some of them (e.g., the global coherence feature) are relatively hard to be accurately computed.

RDF (Resource Description Framework) is recommended by W3C Consortium to describe information on the Web. The basic information unit of RDF data is a triple $t = <s, p, o>$ which stands for subject, predicate and object respectively. An RDF KB contains a finite set of RDF triples. Assuming subjects and objects are entities, predicates are relations between entities, an RDF KB then forms labeled entity relation graphs. An example is shown in Fig. 1.

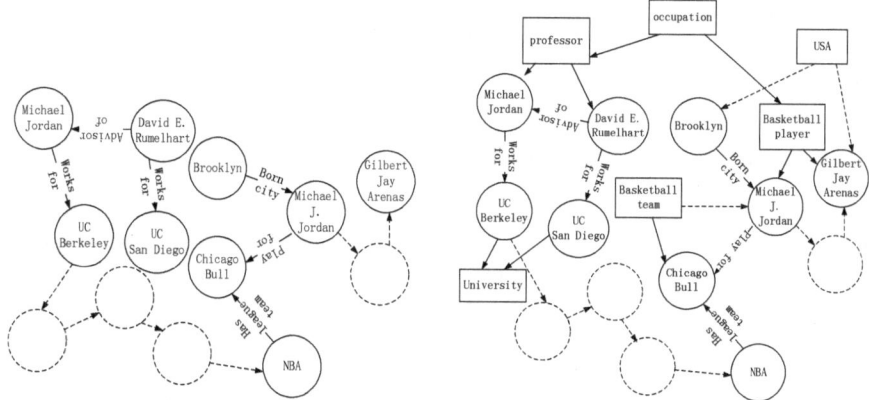

Fig. 1. An example of RDF KB graph. Circles refer to entities in RDF KBs.

Fig. 2. An example of semantic KB graph. Square nodes refer to the classes of entities.

To the best of our knowledge, there is no entity linking work based on RDF knowledge base over MapReduce framework. Because RDF data can be represented as graphs, we can use the properties of the graph to design a simple but effective approach.

The frequently used notations in this paper are summarized in Table 1.

Table 1. Frequently Used Notations

$\mathbb{D}$	all documents need to be processed
D	one document in $\mathbb{D}$
$\mathbb{M}$	all mentions in $\mathbb{D}$
M	the set of mentions in D
$\mathcal{C}$	the set of contexts of mention $m \in M$
$\mathcal{L}$	the candidate linking entity list of $m \in M$
$\mathcal{E}$	the set of linking entities of each $c \in \mathcal{C}$

3 Entity Disambiguation via RDF Knowledge Bases

The entity linking task basically links named entity mentions with entities in a KB. If it is naïvely conducted based on string match, quite often, a mention will be mapped to more than one entities in the KB because of the homonymy problem. In our work, we use Jaccard to compute the similarity between two strings (entity mention and entity in

KB), an entity mention will be mapped to an entity in KB if the similarity between them is higher than some threshold. For example, a mention "Michael Jordan" in unstructured texts will be mapped to more than one entity in KB, for instance, "Michael I. Jordan" and "Michael J. Jordan". This requires a solution of entity disambiguation.

In this study, we take two measures into account for addressing the entity disambiguation problem. One is the relevance, which measures the relation between a mention and an entity. The other one is the homogeneity, that is to find the proper linking entity which has closer hyponyms with the mention. By studying the properties of existing RDF KBs, we introduce two features to measure the relevance and the homogeneity of an entity in a KB to a given mention of certain context.

3.1   Construct Semantic RDF KB Graph

An RDF KB can be represented as a graph, which is called RDF KB graph. Based on the RDF KB graph G, we construct a semantic KB graph SG. Through SG we propose approaches to linking mentions with proper entities in an RDF KB.

Definition 1 (RDF KB Graph). *Given an RDF KB, let κ be the set of all entities in the KB, R be the set of relations between entities in the KB. An RDF KB Graph is $G = (V, E, L)$, where V is the set of vertexes, E is the set of edges and L is the set of literals. $V \in \kappa$, $E \in R$, L_v is the set of labels on vertex, while L_e is the set of labels on edges. $L = L_v \cup L_e$.*

Definition 2 (Semantic KB Graph). *Given an RDF KB graph, a semantic KB graph (SG) is $SG = (V, E, L, C)$, where V is the set of vertexes, E is the set of edges, L is the set of literals and C is the set of classes. $\forall v \in V, C_v \in C$.*

To construct the semantic KB Graph, we firstly recognize entities and relations in a given RDF KB, where an entity is the triples with same subject(s), and a relation between entities is the predicate (p) which connects two entities. Based on entities and relations, we can easily construct an RDF KB graph. Then, we pick predicates such as "rdf:type" in the RDF KB graph as initial class nodes set. Duplicates and noises in the set are cleaned. Some new class nodes are added based on WordNet. We finally built the taxonomy of the RDF KB. For pages limitation, we will not discuss the details of taxonomy construction. An RDF KB graph expanded by the taxonomy is called semantic KB graph. Fig. 1 is an example of an RDF KB graph, while Fig. 2 is an example of the corresponding semantic KB graph. For instance, after expanded as Semantic Graph, node "Michael Jordan" and node "David E. Rumelhart" in Fig. 1 have class node as "Professor" in Fig. 2.

3.2   Context Dependency

The mentions appearing in the same context often have a high probability to be under the same topic. Meanwhile, in an RDF KB graph, the relevant entities (e.g., under the same topic) are likely to be close with each other in terms of the graph distances. For example, if "Michael Jordan" refers to the famous basketball player, mentions such as "NBA" and "Chicago Bulls" are more likely to be found than mentions such as "machine learning" in the context. Moreover, in the KB graph, the shortest path distance between the entities "Michael Jordan" (basketball player) and "NBA" is likely to be shorter than the distance between "Michael Jordan" (computer scientist) and "NBA". Accordingly, we are able to measure the relevance of entities to a given mention.

We denote the mentions in same document with m as the contexts of m, that is C. An entity $e \in \mathcal{E}$ is the linking entity of a named entity mention $m \in \mathbb{M}$. Suppose $\mathfrak{c}_k \in C$ is the kth context mention, $\mathfrak{e}_k \in \mathcal{E}$ is the linking entity of $\mathfrak{c}_k$ (where we choose the cs from C which have only one linking entity. Mentions have no linking entity are useless. Mentions with more than one linking entities will lead to a recursive processing.), we formally define the context dependency of an entity e, denoted as $CD(e)$ as:

$$CD(e) = \frac{1}{N} \sum_{k=1,k\neq i}^{n} {}_{\mathfrak{c}_k \in \mathcal{D}} Dist(\mathfrak{e}_k, e), \tag{1}$$

where $Dist(\mathfrak{e}_k, e)$ is a function measuring the distance between $\mathfrak{e}_k$ and e as

$$Dist(\mathfrak{e}_k, e) = \frac{1}{D_k} \tag{2}$$

where D_k is the shortest path distance from e to $\mathfrak{c}_k$. If the average distance between $e_i \in \mathcal{L}$ and all $\mathfrak{e}_k \in \mathcal{E}$ is shorter than some other $e \in \mathcal{L}$, the value of $CD(e)$ for e_i will be the highest.

3.3   Semantic Similarity

Most entity nodes in an RDF KB have at least one corresponding class node (see square nodes in Fig.2). A class node may have a super class node as its ancestors. Therefore, when a mention $m \in \mathbb{M}$ is linked to an entity e, m will have the same class node and super class nodes as e. For these mentions without direct connected class node, we choose the class with the shortest path to m as its class node. As shown in Fig.2, if a mention links to an entity "Michael Jordan" (the basketball player), it will have a "basketball player" as its class node, and "sportsman" as a super class node.

Assuming that the taxonomy of the KB is a tree, for two entities e_i and e_j in KB, the classes in taxonomy for them are C_i and C_j respectively, where $C_0(i,j)$ is the lowest common ancestor (LCA) class (super class) of C_i and C_j. Suppose e_i is the linking entity of m_i, while $\mathfrak{c}_{ik}$ is one of the context mention of m_i, e_j is the linking entity of $\mathfrak{c}_{ik}$. The semantic similarity between e_i and e_j can be computed by equation 3.

$$Sim(C_i, C_j) = \frac{r(i,j)}{2} \tag{3}$$

where $r(i,j)$ is the reverse weight of class C. $r(i,j) = L_{C_0}^{-P_{ij}}$, Suppose $C_0(i,j)$ is at the level L in the taxonomy tree, P is the shortest path distance between $C_0(i,j)$ and C_i in the tree. Larger P leads to less $r(i,j)$, which means less similarity of C_j and C_i. Therefore, the semantic similarity for entity e can be defined as:

$$SS(e) = \frac{1}{N} \sum_{j=1,j\neq i}^{n} Sim(C, C_j), \tag{4}$$

$SS(e)$ shows the type similarity between the context mention c and the mention m, it measures the homogeneity of e. For example, e_1 and e_2 are two candidate linking entities of m, if the class nodes of linking entities (e) for c are more closer to e_1 than e_2, e_1 is more likely to be the proper linking entity of m. Higher value of $SS(e)$ means closer class structure to the candidate linking entity.

3.4 Ranking Candidate Entities

The score of each candidate entity in the candidate list of a mention m can be computed by combining $CD(e)$ and $SS(e)$:

$$Score(e) = \alpha \cdot CD(e) + (1 - \alpha) \cdot SS(e) \tag{5}$$

where α are weights of $CD(e)$ and $SS(e)$. A max-margin technique introduced in [23,21] is applied to learn the weights.

When there are more than one entities in the candidate linking list (CL) for a mention m, the one with the highest score $e_{top} = \max_{e \in CL} Score(e)$ is chosen as the proper linking entity of m. Algorithm 1 provides the details of linking entity disambiguation.

Algorithm 1. Linking Entity Disambiguation (LED)

Input: $CL_1, \ldots, CL_t$: A set of candidate entity lists, each CL has a corresponding mention m
Input: SG: A Semantic KB graph
Input: $\bar{\alpha}, \bar{\beta}$: weight parameters
Output: $ME_1, \ldots, ME_n$: a list of $< m, e >$ pairs, where m is the mention and e is the proper corresponding linking entity
1. $\mathcal{ME} = \emptyset$;
2. **for** each mention m and corresponding CL **do**
3.   map CL of m and CL of context mention $\mathcal{C}$ of m to same node;
4.   **for** each key value pair (e, e_k) **do**
5.   compute $Dist(e_k, e)$;
6.   reduce e as key, $\sum Dist(mathfrake_k, e)$ as value;
7.   compute $CD(e)$;
8. **for** each class of e_i and class of e_j **do**
9.   compute $Sim(C_i, C_j)$
10.   **for** each key value pair $(e, \sum Sim(C_i, C_j))$ **do**
11.   compute $SS(e)$;
12. **for** each key value pair $(e, (\text{list of } CD(e), SS(e)))$ **do**
13. $Score(e) = \alpha \cdot CD(e) + \beta \cdot SS(e)$;
14. $e = \max Score(e)$;
15.   add e into ME;
16. **return** a set of ME;

---

4 Fast Entity Linking

Nowadays, as numerous data emerge both on Web pages and RDF KBs, the entity linking task faces with the big data problem. Therefore, efficiency and scalability become important issues of the entity linking solution. In this section, we propose a framework for efficient entity linking based on the entity disambiguation algorithm proposed above. We assume that the mentions need to be linked to entities have been extracted and recognized from web pages. We also assume the mentions have potential linking entities in the KB. In order to linking a mention to an RDF KB, we propose a framework including the following modules:

- **Generate Candidate Entity Linking List**. For each mention $m \in M$, obtain the list of entities having the name m (in the KB) as the candidate linking list. For the mentions that do not have any linking entity in the KB, NIL will be returned as the result.
- **Linking Entities Disambiguation**. If the candidate linking list has more than one candidate, we need to find the proper linking entity of m. For each mention m, we use a defined score function to compute the score of each entities e in the candidate linking list. Firstly, we compare the context dependency to see which candidate linking entity is more relevant to the context mentions. Secondly, we compute semantic similarity to check the similarity of candidate entities in terms of their semantic types. Finally, a score is calculated for each entity in candidate linking list, the one with the maximal score will be the proper entity to mention m.

Each mention m in Web pages will be compared with the entities e in KB, if they match, e will be add into the candidate linking list of m. The length of linking list, $|L|$, must be in one of the following three cases. (1) $|L| = 0$, which means no matching entity exists in the knowledge base. (2) $|L| = 1$, which means there is only one match in KB for m. (3) $|L| \geq 1$, which means two or more e in knowledge base are the possible entity links of m. For case 1, NIL will be returned to m; for case 2, the only e is returned to show there is a link between m and e; only for case 3 we need to use the disambiguation algorithm described in Section3 to find the proper linking entity of m. The algorithm is executed on MapReduce framework. In the map phrase, mentions and entities with the same key are dispatched to the same reduce node; In the reduce phrase, mentions(entities) and their neighbors in the semantic KB graph are gathered to compute CD and SS. After several iterative map and reduce phases, the algorithm can output the proper linking entity of m. With MapReduce framework, the algorithm is conducted in a parallel fashion, which makes the approach more scalable.

5   Experimental Study

5.1   Experimental Setup

All the experiments are run on the Renda-Cloud platform, which is a *Hadoop* cluster. Each node in the cluster has 30GB memory, 2TB disk space and 2.40 GHz core 24 processor, running *Hadoop* 0.20.2 under *Ubuntu* 10.04 Linux 2.6.32-24.

In order to evaluate our approach, we choose two real world RDF KBs, YAGO and FreeBase, to conduct the experimental study. YAGO is a large knowledge base developed at *Max-Planck-Institute Sarrbrücken*. It contains information from *Wikipedia* and linked to *Wordnet*. There are more than 10 million entities and 120 million facts about the entities. YAGO contains not only entities and relations between entities but also semantics of entities. The semantics relations include TYPE, subClassOf, domain et al. Based on the structure, we can build semantic KB graph over YAGO. FreeBase is a large collaborative knowledge base developed by the Metaweb company. It contains data from Wikipedia, NNDB et al., with 125M tuples and more than 4000 types of entities. The facts in FreeBase are on different fields, such as sports, arts and entertainment.

In the experiments, we use Accuracy(Accu.) to evaluate the performance of our solution. It is calculated as in Equation 6:

$$Accu. = \frac{N_l}{N_m} \tag{6}$$

where N_l is the number of mentions which are correctly linked to entities, N_m is the total number of mentions.

Sections 5.2 and 5.3 give the experimental results on accuracy and scalability over the two RDF KBs respectively.

5.2 Experiments on YAGO

Accuracy. In order to evaluate the performance, we firstly extract 1000 mentions from 50 Wiki pages as 50 sets of Mentions. mentions in each set are contexts for each other. We test *Accu.* by using different features in score function. Table 2 shows the highest, lowest and average *Accu.* with different features for the 50 groups of mentions. Then we compare our solution with Linden, which is the most similar work with ours in the existing solutions. we use the 1000 mentions in one group and test our solution on 2 Hadoop nodes, while test Linden on a PC with a 1.8Ghz Core 4 Duo processor, 10 GB memory and running 64-bit Linux Redhat kernel. Table 3 shows the results.

Table 2. Experimental results on YAGO: Accu.

feature set	Accu.		
	Max	Min	Avg.
CD	0.9406	0.9198	0.9325
SS	0.9264	0.8813	0.9054
$CD + SS$	0.9487	0.9232	0.9411

Table 3. Experimental results on YAGO: compared with Linden

our solution		Linden	
feature set	Accu.	feature set	Accu.
CD	**0.9395**	LP	0.8665
SS	**0.9240**	LP+SA	0.9189
$CD + SS$	**0.9436**	LP+SA+SS	0.9391

From Table 2 we can see that the accuracy of CD is higher than SS. By combining the two features, our solution achieves the best accuracy. It indicates that in most cases, relatedness is more useful in dealing with entity linking problem. Although Linden has good performance, the comparison results in Table 3 shows that the accuracy of our solution is better than that of Linden.

Scalability. In order to test the scalability of the algorithm, we varying the number of mentions from 1000 to 100,000 and execute the algorithms on 2 Hadoop nodes, 8 Hadoop nodes, 20 Hadoop nodes respectively. The results are shown in Table 4. The time consumption includes generating candidate linking list and scoring candidate linking entities, for these are the main jobs in linking entity task.

Table 4. Time Consumption in Seconds over different number of mentions

# of cluster nodes	# of mentions			
	1000	5000	10000	100000
2	23.39	26.27	38.61	108.05
8	20.45	23.14	29.38	42.06
20	16.07	18.27	22.75	35.11

Results in Table 4 show the scalability of our solution. With more nodes in Hadoop cluster, the computing capability are efficiently enhanced. The minimum execution time in our solution is above 10 seconds. Note that the initial time for MapReduce job is at least 10 seconds.

5.3   Experiments on Freebase

In this part of experiment, we choose 30GB Freebase KB data and more than 250,000 mentions to evaluate our solution in processing a large amount of data. The mentions are extracted from IMDB, which is a database about movies, directors, actors and rating from users et al. We store the data from IMDB into a flat table, each row in table are extracted from one page, therefore the cell contents in same row are treated as context to each other. We also test the accuracy and the scalability to see the performance of our solution.

Accuracy. In the experiment, we firstly process 250,000 mentions in 50 group, the results in Table 5 show the maximum, the minimum and the average accuracy with different features used in the score function. Then we compare our solution of feature SS with baseline: LCA(Lowest Common Ancestor) with one group of 5,000 mentions (results are shown in Table 6), for both SS in our approach and LCA are annotated entities by using the information of their classes (types). Where LCA is simply assign all cells in one column with their lowest common ancestor.

Table 5. Experimental results on Freebase: Accu.

feature set	Accu.		
	Max	Min	Avg.
CD	0.9377	0.9104	0.9265
SS	0.9212	0.8795	0.9033
$CD + SS$	0.9407	0.9185	0.9366

Table 6. Experimental results on Freebase: Compare with Baseline

Our solution	Baseline
Accu.	Accu.
SS	LCA
0.9033	0.6680

From Table 5 we can see that the accuracy results are worse than that in Table 2. We just store data extracted from IMDB without any preprocessing, because they have little noisy than data form Wikipedia. However, the accuracy is still above 0.93. The results in Table 6 show that with only the feature SS our solution still outperforms the baseline.

Table 7. RDF KB preprocessing run-time in seconds over different nodes of Hadoop cluster.

Size of KB	# of nodes in Hadoop cluster.		
	2 nodes	8 nodes	20 nodes
5M	42	17	13
600M	286	134	58
30G	$> 30min$	1305	392

Scalability. In this part of experiments, we test the time consumption in KB preprocessing to see the scalability of our solution. The main work of KB preprocessing includes construction of RDF KB graph and semantic KB graph. We treat it as a preprocessing step, because once the RDF KB graph and semantic KB graph has been constructed, we can do entity linking on top of them when needed without of reconstruction.

We choose part of connected data from Freebase, so that we can vary the KB size from 5M to 30G. We execute KB preprocessing steps on 2 Hadoop nodes, 8 Hadoop nodes, 20 Hadoop nodes respectively. The time consumption results are shown in Table 7. The results show that with more nodes in Hadoop cluster ,we can handle the linking entity task over KB size above tens of Gigabytes. It also shows that the algorithm we proposed in the paper has high scalability. Our solution gives an efficient way to handle the linking entity problem on big data.

6   Conclusions

In this paper, we propose a simple but effective algorithm to accurately link entity mentions with proper entities in an RDF KB. The algorithm is designed on the MapReduce framework, which can annotate multiple documents in parallel. The experimental study over huge real-life RDF knowledge bases demonstrates the accuracy and scalability of the approach.

Based on our solution we can study semantic search techniques using the annotated unstructured texts. For example, by linking unstructured texts to RDF KB, we are able to build a huge extended labeled graph, through which we are able to apply and extend semantic keyword search over text graphs.

Acknowledgements. This work is supported by the National Science Foundation of China under grant No. 61170010 and No. 61003085, and HGJ PROJECT 2010ZX01042-002-002-03.

References

1. Guha, R.V., McCool, R., Miller, E.: Semantic search. In: WWW, pp. 700–709 (2003)
2. Suchanek, F.M., Kasneci, G., Weikum, G.: Yago: A core of semantic knowledge unifying wordnet and wikipedia. In: WWW, pp. 697–706 (2007)
3. Bollacker, K.D., Evans, C., Paritosh, P., Sturge, T., Taylor, J.: Freebase: a collaboratively created graph database for structuring human knowledge. In: SIGMOD, pp. 1247–1250 (2008)

4. Auer, S., Bizer, C., Kobilarov, G., Lehmann, J., Cyganiak, R., Ives, Z.G.: DBpedia: A Nucleus for a Web of Open Data. In: Aberer, K., Choi, K.-S., Noy, N., Allemang, D., Lee, K.-I., Nixon, L.J.B., Golbeck, J., Mika, P., Maynard, D., Mizoguchi, R., Schreiber, G., Cudré-Mauroux, P. (eds.) ISWC/ASWC 2007. LNCS, vol. 4825, pp. 722–735. Springer, Heidelberg (2007)
5. Linking Open Data, http://www.w3.org/wiki/SweoIG/TaskForces/CommunityProjects/LinkingOpenData
6. Dean, J., Ghemawat, S.: MapReduce: Simplified Data Processing on Large Clusters. In: OSDI 2004 (2004)
7. Lesk, M.: Automatic sense disambiguation using machine readable dictionaries: How to tell a pine cone from an ice cream cone. In: SIGDOC, Toronto (June 1986)
8. Mihalcea, R.: Large vocabulary unsupervised word sense disambiguation with graph-based algorithms for sequence data labeling. In: Proceedings of the Human Language Technology/Empirical Methods in Natural Language Processing Conference, Vancouver (2005)
9. Navigli, R., Velardi, P.: Structural semantic interconnections: a knowledge-based approach to word sense disambiguation. IEEE Transactions on Pattern Analysis and Machine Intelligence (PAMI) 27 (2005)
10. Bunescu, R., Pasca, M.: Using Encyclopedic Knowledge for Named Entity Disambiguation. In: EACL, pp. 9–16 (2006)
11. Bagga, A., Baldwin, B.: Entity-based cross-document coreferencing using the vector space model. In: COLING, pp. 79–85 (1998)
12. Dredze, M., McNamee, P., Rao, D., Gerber, A., Finin, T.: Entity disambiguation for knowledge base population. In: COLING, pp. 277–285 (2010)
13. Hasegawa, T., Sekine, S., Grishman, R.: Discovering relations among named entities from large corpora. In: ACL, pp. 415–422 (2004)
14. Demartini, G., Difallah, D.E., Cudré-Mauroux, P.: ZenCrowd: leveraging probabilistic reasoning and crowdsourcing techniques for large-scale entity linking. In: WWW, pp. 469–478 (2012)
15. Stoyanov, V., Mayfield, J., Xu, T., Oard, D.W., Lawrie, D., Oates, T., Finnin, T.: A context aware approach to entity linking. In: The Joint Workshop on Automatic Knowledge Base Construction and Web-scale Knowledge Extraction, NAACL-HLT 2012 (2012)
16. Navigli, R., Velardi, P.: Structural semantic interconnections: a knowledge-based approach to word sense disambiguation. IEEE Transactions on Pattern Analysis and Machine Intelligence 27, 1075–1086 (2005)
17. Gliozzo, A., Giuliano, C., Strapparava, C.: Domain kernels for word sense disambiguation. In: ACL (2005)
18. Ng, H., Lee, H.: Integrating multiple knowledge sources to disambiguate word sense: An examplar-based approach. CoRR, vol. 9606032 (1996)
19. Weikum, G., Theobald, M.: From information to knowledge: harvesting entities and relationships from web sources. In: PODS, pp. 65–76 (2010)
20. Hoffart, J., Yosef, M.A., Bordino, I., Fürstenau, H., Pinkal, M., Spaniol, M., Taneva, B., Thater, S., Weikum, G.: Robust Disambiguation of Named Entities in Text. In: Proceedings of EMNLP, pp. 782–792 (2011)
21. Shen, W., Wang, J.Y., Luo, P., Wang, M.: LINDEN: linking named entities with knowledge base via semantic knowledge. In: WWW 2012, pp. 449–458 (2012)
22. Shen, W., Wang, J.Y., Luo, P., Wang, M.: LIEGE: Link Entities in Web Lists with Knowledge Base. In: Proceedins of KDD 2012 (2012)
23. Carlos, B.T., Guestrin, C., Koller, D.: Max-margin markov networks. In: NIPS (2003)
24. Nadeau, D., Turney, P.D., Matwin, S.: Unsupervised Named-Entity Recognition: Generating Gazetteers and Resolving Ambiguity. In: Lamontagne, L., Marchand, M. (eds.) Canadian AI 2006. LNCS (LNAI), vol. 4013, pp. 266–277. Springer, Heidelberg (2006)
25. Stanford NER, http://nlp.stanford.edu/ner/index.shtml
26. Hadoop, http://hadoop.apache.org/

An Approach to Retrieving Similar Source Codes
by Control Structure and Method Identifiers

Yoshihisa Udagawa

Faculty of Engineering, Tokyo Polytechnic University
1583 Iiyama, Atsugi-city, Kanagawa, 243-0297 Japan
udagawa@cs.t-kougei.ac.jp

Abstract. In this paper, we present an approach to improve source code retrieval using the structure of control statements. We develop a lexical parser that extracts structural information for each method including control statements and method identifiers. We present a source code retrieval model, named "derived structure retrieval model," in which a retrieval conditions are defined as a sequence of control statements and/or method identifiers, and meaningful search conditions are derived from the given sequence. Experiments on the source code of Struts 2 Core show that the derived structure retrieval model outperforms the vector space retrieval model.

Keywords: Derived structure retrieval model, Control statement, Method identifier, Vector space retrieval model.

1   Introduction

Source code retrieval is an important task in order to maintain the quality of software in the development cycle. Duplicated code complicates this activity and leads to higher maintenance cost because the same bugs will need to be fixed and consequently more code will need to be tested. Many developers put significant effort into finding software defects. However, due to the vast amounts of source codes, it is difficult to find efficiently the code segments that we want.

Various techniques have been proposed to collect similar source codes. These techniques can be classified into four categories:

- Text-based comparison

This approach compares source codes in the same partition. The key idea of this approach is to identify source-code fragments using similar identifiers [3].

- Token comparison

In this approach, before comparison, tokens of identifiers (data type names, variable names, etc.) are replaced by special tokens, and then similar subsequences of tokens

Y. Ishikawa et al. (Eds.): APWeb 2013, LNCS 7808, pp. 252–259, 2013.

are identified [2]. Because the encoding of tokens abstracts from their concrete values, code fragments that are different only in parameter naming can be detected.

- Metrics comparison

This approach characterizes code fragments using some metrics, and compares these metric vectors instead of directly comparing the code [4].

- Structure-based comparison

This approach applies pattern matching and complex algorithms on abstract syntax trees or dependency graphs. Baxter et al. [1] propose a method using abstract syntax trees for detecting exact and near-miss program source fragments. Ouddan et al. [5] propose a multi-language source code retrieval system using the structural content of the source code and the semantic one. Their approach is focused on detecting plagiarism in source code between programs written in different programming languages, such as C, C++, Java, and C#.

Our approach is the structure-based comparison that takes a sequence of statements as a retrieval condition. We developed a lexical parser and extracted structures of source codes for control statements and method identifiers. The extracted structural information is an input to the vector space model and a proposed source code retrieval model, named "derived structure retrieval model." Our retrieval model takes a sequence of statements as a retrieval condition and derives meaningful search conditions from the given sequence. Because a program is composed of a sequence of statements, our retrieval model practically improves the performance of source code retrieval.

The rest of this paper is organized as follows. In Section 2, we present architecture of the developed tools. In Section 3, we present overview of the Struts 2 framework [7] and extracted structures. In Section 4, we discuss a similarity retrieval approach based on the vector space model. In Section 5, we discuss the derived structure retrieval model and results of code retrieval. Section 6 concludes the paper.

2   Overview --- Developed Tools

Figure 1 illustrates a high-level architecture of the tools we developed. The structure extraction tool is implemented in C-language. It extracts control statements and method identifiers from every method of a class in Java. Then, these extracted structures are inputted to a statistic tool, a vector space retrieval tool, and a derived structure retrieval tool, which are written in Visual Basic language. Finally, the outputs of those modules are fed into a source code viewer.

The vector space retrieval tool retrieves a set of similar methods by assigning control statements and method identifiers to a document vector, and then computing the cosine of the angle between a document vector and a query vector as similarity. Details are discussed in Section 4.

The derived structure retrieval tool takes a sequence of control statements and/or method identifiers as a retrieval condition. Then, it generates derived sequences by

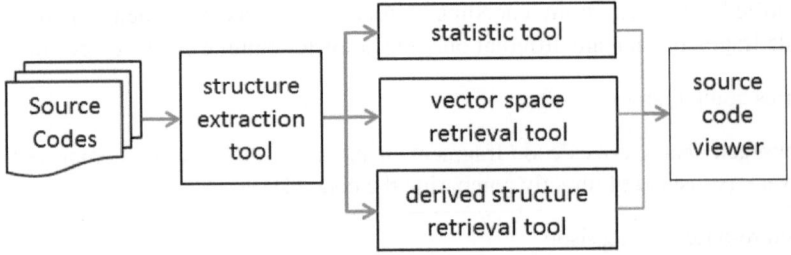

Fig. 1. Architecture of developed tools

replacing each element of the given sequence with a wildcard.   In case, the sequence consists of n elements, 2^n - 2 sequences are derived to retrieve a set of similar methods.   Details are discussed in Section 5.

3   Struts 2 Core and Extracted Structures

3.1   Framework and Struts 2

A framework automates common tasks, and thereby providing a user platform that simplifies web development. The Struts 2 framework [7] implements the model-view-controller (MVC) design pattern.

We estimated the volume of the source codes using file metrics. Struts 2.3.1.1 Core consists of 46,100 lines in source code including comment lines and blank lines. As for the number of lines, Struts 2.3.1.1 Core is a middle scale application in industry.

We can estimate the volume of the source codes using file metrics. Typical file metrics used are as follows:

Number of Java Files	----	368
Number of Classes	----	414
Number of Methods	----	2,667
Number of Code Lines	----	21,543
Number of Comment Lines	----	17,954
Number of Total Lines	----	46,100

The number of Java files are 368, which differs from the number of declared classes 414 because some java files include definitions of inner classes and anonymous classes.

3.2   Extracted Control Statements and Method Identifiers

Fig. 2 shows the extracted structures of the *evaluateClientSideJsEnablement()* method in *Form.java* file in *org.apache.struts2.components* package. Three numbers preceded by # symbol are the number of comment lines, the number of blank lines and the number of code lines, respectively.   Because the extracted structures include the depth of nesting of the control statements, they supply enough information for retrieving methods using a substructure of source codes.

```
Form::void evaluateClientSideJsEnablement(String, String, String)
#       7   1   22
{
        if{
        addParameter
        Configuration.getRuntimeConfiguration
        RuntimeConfiguration.getActionConfig
            if{
            ActionConfig.getInterceptors
                for{
                    if{
                    interceptorMapping.getInterceptor
                    ValidationInterceptor.getExcludeMethodsSet
                    ValidationInterceptor.getIncludeMethodsSet
                        if{
                        addParameter
                        }
                    }
                }
            }
        }
}
```

Fig. 2. An example of extracted control statements and method identifiers

4   Code Retrieval Using the Vector Space Model

4.1   Structural Metrics as Vector Components

The vector space model [6] is an algebraic model for representing text documents as vectors of identifiers or terms. Given a set of documents D, a document d_j in D is represented as a vector of term weights:

$$d_j = (w_{1,j}, w_{2,j}, \ldots, w_{N,j}) \tag{1}$$

where N is the total number of terms in document d_j and $w_{i,j}$ is the weight of the i-th term. A user query can be similarly converted into a vector q:

$$q = (w_{1,q}, w_{2,q}, \ldots, w_{N,q}) \tag{2}$$

The similarity between document d_j and query q can be computed as the cosine of the angle between the two vectors d_j and q in the N-dimensional space:

$$\text{Similarity}(d_j, q) = \cos(d_j, q) = \frac{\sum_{i=1}^{N} w_{i,j} * w_{i,q}}{\sqrt{\sum_{i=1}^{N} w_{i,j}^2} * \sqrt{\sum_{i=1}^{N} w_{i,q}^2}} \tag{3}$$

4.2   Results of Code Retrieval by the Vector Space Model

The left part of Table 1 shows the top 27 methods obtained by a query vector with the components consisted by one *if*-statement and one *addParameter* method identifier. The *addParameter* method maintains a pair of a parameter identifier and its value in an internal list of Struts 2 Core. It is typically used after checking the existence of a parameter identifier.   Thus, one *if*-statement and one *addParameter* method identifi-er are a reasonable retrieval condition in the vector space retrieval model.

392 methods are retrieved with the similarity between 0.997 and 0.009 because the 392 methods include at least one *if*-statement. However, only 21 methods whose No columns are shaded in Table 1 are valid candidate methods in the sense that an *add-Parameter* method is called after an *if*-statement. Details are discussed in Section 5.

Table 1. The top 27 retrieved methods

No	Method Name	Vector Space Model Similarity	Derived Structure Retrieval Model		
			Similarity	Exact Match	Partial Match	Code Lines
1	Form::void evaluateExtraParams()	0.997	0.952	20	1	36
2	DoubleListUIBean::void evaluateExtraParams()	0.990	0.880	66	9	128
3	TextArea::void evaluateExtraParams()	0.985	1.000	8	0	15
4	Select::void evaluateExtraParams()	0.982	0.889	8	1	16
5	ComboBox::void evaluateExtraParams()	0.981	0.762	16	5	39
6	TextField::void evaluateExtraParams()	0.973	1.000	6	0	12
7	InputTransferSelect::void evaluateExtraParams()	0.967	0.737	14	5	50
8	UpDownSelect::void evaluateParams()	0.967	0.842	16	3	41
9	OptionTransferSelect::void evaluateExtraParams()	0.938	0.667	16	8	106
10	FormButton::void evaluateExtraParams()	0.913	0.200	2	8	31
11	ActionError::void evaluateExtraParams()	0.894	0.000	0	4	14
12	ActionMessage::void evaluateExtraParams()	0.894	0.000	0	4	14
13	ListUIBean::void evaluateExtraParams()	0.874	0.235	4	13	46
14	Checkbox::void evaluateExtraParams()	0.866	0.667	2	1	7
15	FieldError::void evaluateExtraParams()	0.866	0.667	2	1	6
16	Label::void evaluateExtraParams()	0.862	0.571	4	3	18
17	Form::void populateComponentHtmlId()	0.816	1.000	2	0	6
18	Password::void evaluateExtraParams()	0.816	1.000	2	0	6
19	Reset::void evaluateExtraParams()	0.816	1.000	2	0	5
20	Submit::void evaluateExtraParams()	0.816	1.000	2	0	5
21	StrutsVelocityContext::Object internalGet()	0.816	0.000	0	6	24
22	AdapterFactory::Node proxyNode()	0.816	0.000	0	5	13
23	StrutsVelocityContext::Object internalGet()	0.802	0.000	0	6	23
24	Form::void evaluateClientSideJsEnablement()	0.756	0.667	4	2	22
25	Token::void evaluateExtraParams()	0.707	0.000	0	6	21
26	File::void evaluateParams()	0.707	0.444	4	5	24
27	Anchor::void evaluateExtraParams()	0.707	1.000	4	0	14

5   Code Retrieval Using Derived Structure Retrieval Model

5.1   Derived Structure Retrieval Model

In the vector space retrieval model, a document is represented as a vector of terms that comprise the document. Similarity of a document and a query is calculated as the cosine of the angle between the document vector and the query vector. This means that the order in which the terms appear in the document is lost in the vector space model. On the other hand, a program is essentially a sequence of statements. So, the vector space model has limitations in performance when it is applied to retrieving program source codes.

The derived structure retrieval model, a source code retrieval model proposed in the paper, takes a sequence of statements as a retrieval condition, and derives meaningful search conditions from the given sequence. Let S_1 and S_2 are statements extracted by the structure extraction tool. $[S_1 \rightarrow S_2]$ denotes a sequence of S_1 followed by S_2. In general, for a positive integer n, let Si (i ranges between 1 and n) be a statement. $[S_1 \rightarrow S_2 \rightarrow ... \rightarrow S_n]$ denotes a sequence of n statements, i.e., S_1, S_2, and S_n.

The essential idea of the derived structure retrieval model is matching on wildcard characters. Given a sequence $[S_1 \rightarrow S_2 \rightarrow ... \rightarrow S_n]$ and a method structure m, our model first try to search structure of m that matches the sequence $[S_1 \rightarrow S_2 \rightarrow ... \rightarrow S_n]$. Next, derive a sequence, such as $[* \rightarrow S_2 \rightarrow ... \rightarrow S_n]$, by replacing one of elements in $[S_1 \rightarrow S_2 \rightarrow ... \rightarrow S_n]$, and try to search structure of m that matches the derived sequence. The algorithm continue to evaluate structure of m until all of derivable sequences, except for the sequence $[* \rightarrow * \rightarrow ... \rightarrow *]$, are evaluated.

The sequence $[S_1 \rightarrow S_2 \rightarrow ... \rightarrow S_n]$ is more "restrictive" than $[* \rightarrow S_2 \rightarrow ... \rightarrow S_n]$ and $[S_1 \rightarrow * \rightarrow ... \rightarrow S_n]$, etc. In general, $2^n - 1$ sequences are derived from an n-element sequence including the sequence whose all elements are replaced by wildcards. Because the sequence $[* \rightarrow * \rightarrow ... \rightarrow *]$ matches all substructures, it does not work as a retrieval condition. A given sequence $[S_1 \rightarrow S_2 \rightarrow ... \rightarrow S_n]$ and $2^n - 2$ derived sequences define a domain of discourse for retrieval conditions.

Let Count (m, $[S_1 \rightarrow S_2 \rightarrow ... \rightarrow S_n]$, r) is a function that counts the number of substructures of m that match derived sequences of $[S_1 \rightarrow S_2 \rightarrow ... \rightarrow S_n]$ whose r elements are replaced by the wildcard. Fig.3 shows an algorithm to compute the similarity of methods that near match a retrieval condition $[S_1 \rightarrow S_2 \rightarrow ... \rightarrow S_n]$. In Fig.3, it is assumed that *getMethodStructure(j)* returns a structure of the j-th method extracted by the structure extraction tool. Note that the similarity is 1.0 when a method includes the sequence $[S_1 \rightarrow S_2 \rightarrow ... \rightarrow S_n]$ and it does not include any of derived sequences from $[S_1 \rightarrow S_2 \rightarrow ... \rightarrow S_n]$.

5.2   Results of Code Retrieval by Derived Structure Retrieval Model

A sequence [*if* → *addParameter*] is used as a retrieval condition in order to compare the result obtained in Subsection 4.2. The right part of Table 1 shows retrieved methods that match sequences [*if* → *addParameter*], [* → *addParameter*], and [*if* → *]].

FormButton::void evaluateExtraParams() method is ranked 21th with similarity 0.200 in the derived structure retrieval model, whereas it is ranked 10th with similarity 0.913 in the vector space model. The method includes five *if*-statements, three *addParameter* identifiers, and one [*if* → *addParameter*] sequence.   Because the similarity in the vector space model is computed by the number of statements and/or method identifiers, *FormButton::void evaluateExtraParams()* method is ranked higher than it is ranked in the derived structure retrieval model.

```
Input:  set_of_structure M;
Input:  sequence [S₁→S₂→...→Sₙ];
Output: Sim[M.length];
  // Declare array Sim[].
double Sim[M.length];
  // Compute Similarity for each method in M.
int Nume;  int Deno;
for ( int j=0; j < M.length; j++ ) {
  Nume= Count(getMethodStructure(j),[S₁→S₂→...→Sₙ],0);
  Deno= 0;
  for ( int r=1; r < [S₁→S₂→...→Sₙ].length; r++ ){
    Deno= Deno+ Count(getMethodStructure(j),[S₁→S₂→...→Sₙ],r);
  }
  Sim[j]= (double) Nume / (double)( Nume + Deno );
}
```

Fig. 3. Algorithm to compute the similarity of methods for a sequence $[S_1 \to S_2 \to ... \to S_n]$

The similarity of *ActionError::void evaluateExtraParams()* method is 0.000 in the derived structure retrieval model, whereas it is 0.894 and is ranked 11th in the vector space model. The method includes two *if*-statements, two *addParameter* identifiers and zero [*if→addParameter*] sequences.

A program is essentially represented by a sequence of statements. Because the vector space model computes similarity of two document vectors based on a set of terms that comprise the document vectors, it theoretically has a limitation on performance in source code retrieval. On the other hand, because the derived structure model computes similarity based on a sequence of statements, it achieves higher performance than the vector space model.

As shown in Table 1, 6 methods do not satisfy the [*if→addParameter*] sequence condition, the derived structure retrieval model achieves (27-21) / 27= 22.2% improvement.   Table 2 shows a summary of performance improvement of other experiments on source code retrieval.

Table 2. Summary of performance improvement on source code retrieval

No	Retrieval Sequence	Number of Methods by Vector Space Model	Number of Methods by Derived Structure Model	Degree of improvement
1	for{ → List<String>.add	5	4	20.0%
2	if{ → addParameter	27	21	22.2%
3	catch{ → Logger.error	41	18	56.1%
4	if{ → Logger.debug	72	31	56.9%
5	for{ → iterator.next	21	9	57.1%
6	while{ → Enumeration.nextElement	13	5	61.5%
7	while{ → StringTokenizer.nextToken	16	6	62.5%
8	for{ → StringBuilder.append	10	3	70.0%
9	if{ → Logger.warn	45	13	71.1%
10	while{ → Iterator.next	28	4	85.7%

6   Conclusions

In this paper, we presented an approach that improves source code retrieval using the structural information of control statements. Our key contribution in this research is the development of an algorithm that derives meaningful search conditions from a given sequence, and then performs retrieval using all of the derived conditions.   Thus our source code retrieval model retrieves all of source codes that partially match the given sequence.

The results are promising enough to warrant future research. In future work, we will work on improving our methods by combining information, such as an inheritance of a class and implementation of an abstract class, etc. into structural information. We will also conduct experiments on various types of open source programs available on the Internet.

References

1. Baxter, I.D., Yahin, A., Moura, L., Sant'Anna, M., Bier, L.: Clone detection using abstract syntax trees. In: Proc. of the 14th International Conference on Software Maintenance, pp. 368–377 (November 1998)
2. Kamiya, T., Kusumoto, S., Inoue, K.: CCFinder: A multi-linguistic tokenbased code clone detection system for large scale source code. IEEE Transactions on Software Engineering 28(7), 654–670 (2002)
3. Marcus, A., Maletic, J.: Identification of high-level concept clones in source code. In: Proc. of the 16th International Conference on Automated Software Engineering, pp. 107–114 (November 2001)
4. Mayland, J., Leblanc, C., Merlo, E.M.: Experiment on the automatic detection of function clones in a software system using metrics. In: Proc. of the 12th International Conference on Software Maintenance, pp. 244–253 (November 1996)
5. Ouddan, A.M., Essafi, H.: A Multilanguage Source Code Retrieval System Using Structural-Semantic Fingerprints. International Journal of Computer and Information Science and Engineering, 96–101 (Spring 2007)
6. Salton, G., Buckley, C.: Term-Weighting approaches in automatic text retrieval. Information Processing and Management 24(5), 513–523 (1988)
7. Struts - The Apache Software Foundation (2012), http://struts.apache.org/

Complementary Information for Wikipedia by Comparing Multilingual Articles

Yuya Fujiwara[1], Yu Suzuki[2], Yukio Konishi[1], and Akiyo Nadamoto[1]

[1] Konan University, 8-9-1 Okamoto, Higashi-Nada, Kobe, Hyogo 6588501, Japan
[2] Nagoya University, Furo, Chikusa, Nagoya, Aichi 4648601, Japan

Abstract. Information of many articles is lacking in Wikipedia because users can create and edit the information freely. We specifically examined the multilinguality of Wikipedia and proposed a method to complement information of articles which lack information based on comparing different language articles that have similar contents. However, much non-complementary information is unrelated to a user's browsing article in the results. Herein, we propose improvement of the comparison area based on the classified complementary target.

1 Introduction

Wikipedia has become an extremely popular website, with over 284 language versions in 2012. Nevertheless, information of many articles is lacking because users can create and edit the information freely. Furthermore, Wikipedia has different levels of value of its information depending on the language version of the site because different users create and edit articles of the respective language versions. We specifically examine the multilinguality of Wikipedia and propose a method that complements information of articles that lacks information based on comparing different language articles that have the similar contents[3].

In our proposed method, a user first browses an article in Wikipedia of the user's language. Then the system extracts complementary information from user-specified articles that have been composed in other languages. As described in our research, we designate the lacking information as "complementary information", an article that a user is browsing and which lacks information as a "browsing article", its Wikipedia as a "browsing Wikipedia", and its language as a "browsing language". We also designate a different language article as a "comparison article", its Wikipedia as a "comparison Wikipedia", and its language as a "comparison language".

When we compare a browsing article with a comparison article, the information granularity is shown to differ. The browsing article is only one page, but the comparison article might have multiple pages because these articles are written by intercultural authors. Our proposed method has been used to extract lacking information from all content in target articles in comparison Wikipedia.

Y. Ishikawa et al. (Eds.): APWeb 2013, LNCS 7808, pp. 260–267, 2013.

However, much un-complementary information is unrelated to a user's browsing article in the results.

In this paper, we propose improvement of the method, which is to extract complementary information from different language articles in Wikipedia. The improvement points are the following.

- Classifying complementary target articles.
- Improving the comparison area based on a classified complementary target.
- Experimenting to ascertain the usefulness of our proposed method.

The process used for our proposed system is (1)A user who browses the Wikipedia article clicks the complement button on the web page, and inputs the comparison language. (2)The system extracts comparison articles from the comparison Wikipedia using a link graph of Wikipedia. (3)It classifies the comparison articles and extracts a comparison area from each comparison article. (4)It compares the browsing article with the comparison articles extracted in step (3), and extracts complementary information. (5)It browses the browsing article and the complement information.

In our prototype system, we use the English version of Wikipedia as the browsing Wikipedia, and the Japanese version of Wikipedia as the comparison Wikipedia. Nevertheless, our system is language-independent. When we use other versions of Wikipedia, we can develop a version for any language. In addition, our target users are people who browse Wikipedia articles.

2   Related Work

Our proposed approach uses link structure analysis techniques. Many studies use the Wikipedia link structure. Milne,[4] and Strub et al.[7] extract measures of semantic relatedness of terms using the Wikipedia link structure.

Some studies have been conducted to examine complementary information. Ching et al. [1] propose a framework to assist Wikipedia editors to enrich Wikipedia articles. They extract complementary information by comparing multilingual Wikipedia. However, they are not interested in differences of information granularity between the languages. In contrast, we consider the coverage of comparison articles and create a Wikipedia article to extract complementary information from comparison articles. Eklou et al.[2] proposed a method of information complementation of Wikipedia by the web. They extract complementary information from web pages using LDA. Their goal is similar to ours, but the target complement information is different. Ma et al.[6] proposed integrating cross-media news contents such as TV programs and Web pages to provide users with complementary information. They extract complementary information from web pages using their proposed topic structure. Their research issue, which is complement information, is similar to our issue. However, the target of the content is different.

3   Extraction of Complementary Information from Articles

3.1   Extraction of Comparison Articles from Comparison Wikipedia

We extract comparison articles based on the Wikipedia link graph and our proposed relevance degree.

Link Structure Analysis
We create a link graph for a comparison Wikipedia based on the user's browsing title. First, the system extracts an article having the same title as the user's input from the comparison Wikipedia. We designate the article as "the basic article". We regard the basic article as the root node of the link graph. Next, we regard the interactive linked articles which are the subjects of link-out and link-in connections with the basic article as important and relevant articles to the basic article. The system extracts all interactive linked articles from the comparison Wikipedia. It includes the articles as nodes and thereby creates the link graph.

Calculating Relevance Degree
The article of an interactive link does not always have a high relation to a basic node. We calculate the relevance degree between a root node (basic article) and other nodes (interactive linked articles) in the link graph. An important anchor text appears many times in a basic article, and appears also in the summary area in Wikipedia. We use that structural feature and calculate the relevance degree between the root node and other node in the link graph as described below.

1. The system divides the basic article according to the section which is the structure of the table of contents of the basic article. We designate the divided parts as segments.
2. From the link graph, the system extracts a title from a interactive linked article. The title becomes the keyword when we extract the anchor text from the basic article. We designate the title as the "target title".
3. The system counts the anchor text of the target title in the summary area of the basic article. The number of the anchor text is $TFsum_i$. Therein, i is the identification number of the interactive link article.
4. It also counts of the anchor text in each segment of the basic article. The number of the anchor text is TF_{ik}. Therein, k demotes the segment number.
5. It calculates the relevance degree between each segment and the interactive linked article for which the title is the target title. We use the following expression to calculate the degree of relevance.

$$R_i = (\alpha \cdot TFsum_i \cdot Ssum_i + \sum_{k=1}^{n}(TF_{ik} \cdot S_{ik}))/\max(R_{im}) \qquad (1)$$

In the equation presented above, R_i is the relevance degree of each interactive link article i. $Ssum_i$ represents the cosine similarity between the summary area of the basic article and an interactive link article i. n is the number of segment in the basic article. S_{ik} represents the cosine similarity between segment k of the basic article and an interactive link article i. $\max(R_{im})$ signifies the maximum value in all R_i.

The system calculates from 2. to 5. for all interactive linked articles in the link graph. When relevance degree R_i is less than the threshold β value, we delete the article from the link graph. When the system extracts complementary information from the comparison articles, nodes that consist of a root node and other nodes become comparison articles.

3.2   Comparison Area in Comparison Articles

Then some comparison articles have information that is unrelated to the title of browsing article. For example, we extract "My Neighbor Totoro", which is famous Japanese animated movie as a basic article, then we extract "Sayama Hill" as a comparison article. (Sayama Hill is the setting of the movie. The hill actually exists in Japan.) However, the Sayama Hill article invariably mentions the geographic information and the animals that live on the hill. The information about My Neighbor Totoro is only a small amount of the information included in the Sayama Hill article. In this case, our past proposed method extracted information such the geographically information and the animals in Sayama Hill as complementary information for My Neighbor Totoro. However this information is not related to My Neighbor Totoro. In this paper, we propose a new method that incorporates consideration of the comparison area of comparison articles. In our new method, we first cluster three kinds of articles. Next we extract comparison areas from articles of each kind and extract complementary information from them.

3.2.1 Clustering of Articles

We divide the comparison articles into three types.

- Basic article
  A basic article has the same title as that of a browsing article. It is a root node in the link graph.
- Inclusive relation article
  When a comparison article has an inclusive relation to a basic article, it becomes an inclusive relation article. That is, an inclusive relation article is a kind of "is-a" relation to a basic article. For example, if Sushi is a basic article, then Inari-sushi and Maki-sushi are inclusive relation articles because they are kinds of sushi. When we extract the inclusive relation article, we use Lead Sentence Parsing (LSP) method[5]. LSP method is that the lead part is often defined "is-a" relation to other articles in Wikipedia. That is, when there is the anchor text that links to the basic article in a lead part of comparison article, the article becomes an inclusive relation article.
- Partial match article
  When a partial content in a comparison article is related to the basic article, it becomes a partial match article. We define a partial match article as a

comparison article that has no anchor text which links to the basic article in the lead part, but it has anchor text in other parts of comparison articles. That is, the partial match article excludes inclusively related articles from comparison articles.

3.2.2 Comparison Area

We extract complementary information from the comparison area, which is based on articles of three types described above.

Basic Article

A comparison area of the basic article is all contents of a page (article) because a basic article has the same title as the browsing article. It is strongly related to the browsing article. (Fig. 1(a)).

Inclusive Relation Article

An inclusive relation article is a subset of a basic article. We regard a comparison area of the inclusive relation article as all contents of a article. That is, the comparison area is the same as its basic article. (Fig. 1(a)).

Partial Match Article

We specifically examine the position of the anchor text which links to the basic article, in a partial match article. We extract a comparison area as follows:

– An anchor text in a title area of section or sub-section.
  - Anchor text in a title of section
    We consider that the section and its sub-section are related to a basic article. A comparison area is both section and sub-section (Fig. 1(b)).
  - Anchor text in a title of sub-section
    We consider that the sub-section and its parent section are related to the basic article. A comparison area is both its parent section and the sub-section of itself. In this case, we regard sibling sub-sections as not related to the sub-section. They are not a comparison area (Fig. 1(c)).
– Anchor text only in the context area of a partial match article.
  We consider a paragraph that includes an anchor text is related to a basic article. We regard only the paragraph as a comparison area (Fig. 1(d)).

3.3 Extracting Complementary Information from Comparison Articles

After the system extracts comparison articles, it calculates and extracts complementary information by comparing the browsing article with each comparison article. Almost all Wikipedia articles are divisible into segments based on the table of contents, which means that the segments are divided semantically. When comparing the similarity of multilingual Wikipedia, we specifically examine the segments of the table of contents of Wikipedia specifically. Particularly, we compare each article based on each segment.

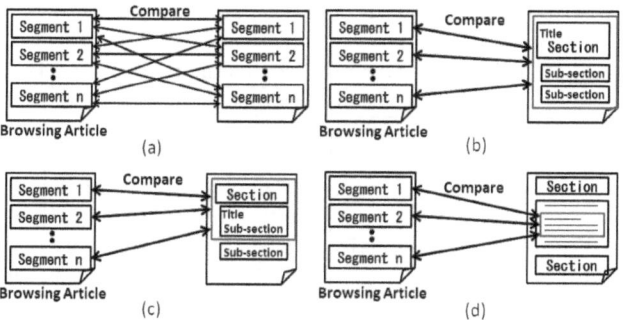

Fig. 1. Comparing browsing article with comparison article

1. We extract nouns from respective articles.
2. We translate comparison articles to browsing language using dictionaries. At this time, the translation is insufficient because it is done solely using the dictionary. We use a translation API such as Google API. Furthermore, we translate proper nouns using Wikipedia's language link. As described in this paper, we are unconcerned about word sense disambiguation. That is left as a subject for future work.
3. We extract complementary information from each type of comparison article.

Basic Article and Inclusive Relation Article

The comparison area of these articles includes all contents of the article. However, it is difficult to extract complementary information based on comparing whole contents of the browsing article with whole contents of the comparison area. We specifically examine the structure of an article. Almost all Wikipedia articles are divided into segments based on the table of contents, meaning that the segments are divided semantically. We calculate the similarity between the browsing article and each comparison article using cosine similarity. When comparing the similarity of multilingual Wikipedia, we examine the segment of the table of contents of Wikipedia specifically(show Fig. 1(a)). Particularly, we compare each article based on each segment.

$$\cos(x_k, y_l) = \frac{\sum x_i \cdot y_i}{\sqrt{\sum x_i^2 \cdot \sum y_i^2}} \qquad (2)$$

In that equation, x_k signifies a segment of the browsing article k; y_l denotes a segment of comparison article l. In addition, x_i represents the term frequency of noun i in a browsing article's segment and y_i stands for the term frequency of noun i in a comparison target article's segment.

If the cosine similarity is less than the threshold γ value, then it is regarded as different information.

Partial Match Article

We extract complementary information by comparing the browsing article with the comparison area in section 3.2.2. When we compare articles, we

Table 1. Dataset and results of experiment

Query	Dataset			Our proposed method			Baseline		
	A[1]	B[2]	C[3]	Precision	Recall	F-measure	Precision	Recall	F-measure
My Neighbor Totoro	9	2	25	0.82	0.86	0.84	0.75	0.85	0.8
Doraemon	9	5	38	0.65	0.77	0.71	0.76	0.7	0.74
Iaido	10	4	24	1	0.82	0.9	0.45	0.82	0.58
Manzai	3	4	22	1	0.48	0.65	1	0.85	0.92
Yukata	1	3	7	0.75	0.5	0.6	0.8	0.67	0.72
Urashima Taro	5	5	39	1	0.48	0.64	0.48	0.57	0.52
Pikachu	6	2	25	1	0.55	0.71	0.46	0.55	0.5
Kinkaku-ji	24	6	28	0.78	0.88	0.82	0.33	0.88	0.48
Hello Kitty	15	2	60	0.71	0.48	0.57	0.6	0.48	0.53
Kyudo	16	3	47	0.92	0.45	0.69	0.37	0.45	0.41
Average	-	-	-	0.86	0.62	0.71	0.6	0.68	0.62

use cosine similarity, as shown in equation (2). At this time, x_k signifies a segment of the browsing article k; y_l denotes a paragraph of comparison article l which is comparison area in section 3.2.2. In addition, x_i represents the term frequency of noun i in a browsing article's segment and y_i stands for the term frequency of noun i in a comparison area in the partial match article.

4   Experimental Evaluation

We assessed the availability of extracting complementary information based on comparing our proposed method which is a calculated comparison area with a baseline which is not a calculated comparison area. In both of our experiments, we set an appropriate weight $\alpha=3.0$ and threshold $\beta=0.2$ of the relevance degree described in equation (1) in section 3, and threshold $\gamma=0.2$ of cosine similarity in equation (2) in our earlier experiment[3]. Table 1 presents results of the experiment. The precision is 0.26 higher than the baseline, which means that our proposed method is useful for extracting complementary information. An example of a good result is the browsing article is written for the plot and cast of My Neighbor Totoro. One comparison article is written about Sayama Hill, the geographic information related to Sayama Hill differs, but it is unrelated to the movie. This time, in our proposed method, the information is not a comparison area. We do not extract the information as complementary information. However, the exemplary bad results are those for Manzai, which is a type of Japanese comedy. We can not extract the information written about the association of Manzai, Roukyoku Manzai which is a kind of manzai, and so on because the anchor text in each case is in an itemization area. Our proposed method

---

[1] The total number of section in browsing article.
[2] The total number of comparison articles.
[3] The total number of sections in comparison articles.

processes an itemization area is a segment. Few contents are included in segments. Therefore, we can not extract such information. This is left as a problem for future study.

5 Conclusion

As described in this paper, we proposed a method for extracting complementary information from Wikipedia by comparing multilingual Wikipedia. Specifically, we improve our proposed method to examine the comparison area based on a classified complementary target. This paper presented three important technical points, (1)Classifying complementary target articles, (2)Improving the comparison area based on a classified complementary target, and (3)Experiment to measure the usefulness of our new proposed method.

Future work is that we should consider word sense disambiguation, because many instances of word sense disambiguation exist, but we have ignored such cases in the analyses presented in this paper.

Acknowledgments. This work was partially supported by Research Institute of Konan University.

References

1. Ching-Man, A.Y., Duh, K., Nagata, M.: Assisting wikipedia users in cross-lingual editing. In: WebDB Forum 2010, pp. 11–12 (2010)
2. Eklou, D., Asano, Y., Yoshikawa, M.: How the web can help wikipedia: a study on information complementation of wikipedia by the web. In: Proceedings of the 6th International Conference on Ubiquitous Information Management and Communication, ICUIMC 2012 (2012)
3. Fujiwara, Y., Suzuki, Y., Konishi, Y., Nadamoto, A.: Extracting Difference Information from Multilingual Wikipedia. In: Sheng, Q.Z., Wang, G., Jensen, C.S., Xu, G. (eds.) APWeb 2012. LNCS, vol. 7235, pp. 496–503. Springer, Heidelberg (2012)
4. Milne, D.: Computing semantic relatedness using wikipedia link structure. In: Proc. of New Zealand Computer Science Research Student Conference, NZCSRSC 2007. CDROM (2007)
5. Nakayama, K.: Wikipedia mining for triple extraction enhanced by co-reference resolution. In: Proceedings of the 1st International Workshop on Social Data on the Web, SDoW 2008 (2008)
6. Qiang, M., Nadamoto, A., Tanaka, K.: Complementary information retrieval for cross-media news content. Elsevier ARTICLE Information Systems 31(7), 659–678 (2006)
7. Strube, M., Ponzetto, S.P.: WikiRelate! computing semantic relatedness using wikipedia. In: Proceedings of the 21st International Conference on Artificial Intelligence, AAAI 2006, pp. 1419–1424 (2006)

Identification of Sybil Communities Generating Context-Aware Spam on Online Social Networks

Faraz Ahmed and Muhammad Abulaish

Center of Excellence in Information Assurance
King Saud University, Riyadh, Saudi Arabia
{fahmed.c,mabulaish}@ksu.edu.sa

Abstract. This paper presents a hybrid approach to identify coordinated spam or malware attacks carried out using sybil accounts on online social networks. It also presents an online social network data collection methodology, with a special focus on Facebook social network. The pages crawled from Facebook network are grouped according to users' interests and analyzed to retrieve users' profiles from each of them. As a result, based on the users' page-likes behavior, a total number of six groups has been identified. Each group is treated separately and modeled using a graph structure, termed as *profile graph*, in which a node represents a profile and a weighted edge connecting a pair of profiles represents the degree of their behavior similarity. Behavior similarity is calculated as a function of *common shared links*, *common page-likes*, and *cosine similarity of the posts*, and used to determine weights of the edges of the profile graph. Louvain's community detection algorithm is applied on the profile graphs to identify various communities. Finally, a set of statistical features identified in one of our previous works is used classify the obtained communities either as malicious or benign. The experimental results on a real dataset show that profiles belonging to a malicious community have high closeness-centrality representing high behavioral similarity, whereas those of a benign community have low closeness-centrality.

Keywords: Social network analysis, social network security, sybil community detection.

1 Introduction

Online social networking sites have attracted a large number of internet users. Among many existing Online Social Networks (OSNs), Facebook and Twitter are the most popular social networking sites with over 800 million and 100 million active users, respectively. However, due to this popularity and existence of a rich set of potential users, malicious third parties have also diverted their attention towards exploiting various features of these social networking platforms. Though, the exploitation methodologies vary according to the features provided by the social networking platforms, malware infections, spam, and phishing are the most common security concerns for all of these platforms. In addition, a number of social botnets have emerged that utilize social networking features to

Y. Ishikawa et al. (Eds.): APWeb 2013, LNCS 7808, pp. 268–279, 2013.
© Springer-Verlag Berlin Heidelberg 2013

spreading infections as command and control channels [1], [2], [3]. The root cause of all these security concerns is the social network sybils or fake accounts created by malicious users to increase the efficacy of their attacks that are commonly known as *sybil attacks* [4]. Generally, an attacker uses multiple fake identities to unfairly increase its ranking or influence in a target community. Moreover, several underground communities exist, which trade sybil accounts with users and organizations looking for online publicity [5], [6]. Recent studies have shown that with the increase in the popularity of social media, sybil attacks are becoming more widespread [7]. Several sybil communities have been reported so far that forward spam and malwares on Facebook [8] network. In online social networks, third-party nodes are most vulnerable to sybil attacks, where the third-party nodes are communities and groups on OSN platforms which bring together users from different real-world communities on the basis of their interests. In case of Facebook, a third-party node can be defined as a Facebook community page which is used to connect two users from entirely different regions. Sybil accounts hired for carrying out spam campaigns target such vulnerable nodes. Recently, the rapid increase in the number of spam on popular online social networking sites has attracted the attention of researchers from security and related fields.

Though a significant amount of research works has been reported for the detection and characterization of spam on Facebook and Twitter networks [9], [10], [11], [12], [13], the existing techniques do not focus on the detection of coordinated spam campaigns carried out by the communities of sybil accounts. Similarly, several techniques have been presented for the identification of sybil communities [4], [14], [15],[16], [17], but all of them focus on the decentralized detection of sybil accounts. Moreover, the existing techniques are based on two common assumptions about the behavior of sybil nodes. Firstly, sybil nodes can form edges between them in a social graph and secondly, the number of edges connecting sybil and normal nodes is less as compared to the number of edges connecting either only normal nodes or only sybil nodes. These assumptions were based on the intuition that normal users do not readily accept friendship requests from seemingly unknown users. Although empirical studies from [17] showed existence of such sybil communities in the Tuenti social network, another study of Renren social network [7] showed that sybil nodes rarely created edges between themselves. This implies that the community behavior of sybil nodes in a social graph is mercurial and the assumption that sybil nodes form communities cannot be generalized [18].

In this work, the authors utilize the rich corpus of prior research works on spam detection and sybil community identification as a basis and present a hybrid approach to identify coordinated spam or malware attacks carried out using sybil accounts. The proposed approach is independent of the assumptions discussed above by the previous researchers. Although the proposed approach is generic in nature, this paper focuses on the sybil accounts present on Facebook social network for experiment and evaluation purposes. The contributions of this paper can be summarized as follows:

- An online social network data collection methodology is introduced which is based on the intuition that sybil accounts under the control of a single user tend to attack different nodes of the same community; they may not be connected to each other, but may have a common target.
- A new social graph generation mechanism is presented, in which a node represents a profile and an edge represents an association between a pair of connecting profiles. The weight of an edge is determined as a function of the features extracted from the content of linked profiles. In this way, the weight of an edges is independent of the actual friendship link between the profiles, and consequently profiles with similar behavior are interlinked together to form a single group.
- Each group of related profiles is modeled as a social graph and analyzed independently using a community detection algorithm.
- A statistical approach is applied on the obtained communities from each profile group to identify sybil communities.

The rest of the paper is organized as follows. After a brief review of the existing state-of-the-art techniques for spam identification in online social networks in Section 2, Section 3 presents a data collection methodology from Facebook social network. Section 4 presents the profile grouping methodology to generate various groups of similar profiles in the original social network dataset. This Section also presents the experimental results obtained from a real dataset and their analyses. Finally, Section 5 presents conclusions.

2   Related Work

A significant number of research works has been reported in last few years for spam detection on online social networks. In [19], the authors proposed a real time URL-spam detection scheme for Twitter. They proposed a browser monitoring approach, which takes into account a number of details including HTTP redirects, web domains contacted while constructing a page, HTML content being loaded, HTTP headers, and invocation of JavaScript plug-ins. In [11], the authors created honey-profiles representing different age, nationality, and so on. Their study is based on a dataset collected from the profiles of several regions, including USA, Middle-East, and Europe. They logged all types of requests, wall posts, status updates, and private messages on Facebook. Based on the users' activities over social networking sites, they distinguished spam and normal profiles. The authors in [12] utilized the concept of *social honeypot* to lure content polluters on Twitter. The harvested users are analyzed to identify a set of features for classification purpose. The technique is evaluated on a dataset of Twitter spammers collected using the *@spam* mention to flag spammers. In [8], the authors analyzed a large dataset of wall posts on Facebook user profiles to detect spam accounts. They built wall posts similarity graph for the detection of malicious wall posts. Similarly, in [13] the authors presented a thorough analysis of profile-based and content-based evasion tactics employed by Twitter spammers. The authors proposed a set of 24 features consisting of graph-, neighbor-, automation-, and timing-based features that are evaluated using different machine learning techniques. In [20] and [10], the authors proposed combination of content-based and user-based features for the detection of spam profiles on Twitter. In order to evaluate the importance of these features, the collected dataset is fed into traditional classifiers.

Similar to spam detection on online social networks that has received a lot of attention from researchers, a significant effort have been diverted towards the detection of sybil accounts. Initial studies [4], [14], [15], [16], [7] focus on detecting sybil users. However, individual users do not pose a great threat to normal users of OSNs. The situation becomes alarming when a large number of sybil accounts generate a coordinated attack. Several techniques have been presented to detect groups of accounts coordinating with each other [4], [14], [15],[16], [17]. All these techniques focus on the decentralized detection of sybil accounts. Moreover, they are based on two common assumptions: i) sybil nodes can form edges between them in a social graph, and ii) the number of edges connecting sybil and normal nodes is less as compared to the number of edges connecting either only normal nodes or only sybil nodes. However, later studies have shown that these assumptions cannot be applied in general [7]. Despite the presence of rich amount of works for spam and sybil detection, there has been little attention towards the identification of sybil accounts that are particularly responsible for spam proliferation. Therefore, this paper focuses on the detection of coordinated spam campaigns that are carried out by sybil accounts under the control of a single user.

3 Dataset

Based on the analyses reported in [21], it is found that a significant amount of spam posts on Facebook are directed towards those Facebook pages that are publicly accessible and any user in the network can post on them. Spammers generally utilize such openly accessible public pages to spread spam in the network. This type of spam spreading mechanism not only relieves the spammers from their dependence on friendship requests, but also increases the number of target users. Once a spam post is visible on a page's wall, it can be visible to every user who *likes* that page. In addition, users' page-like information can help spammers to spread context-aware spam through Facebook pages in normal user communities. Recently, there has been an increasing number of evidence about the existence of underground communities trading groups of accounts that carry spam campaigns [18]. Therefore, a group of accounts bought by a party could be used for a single purpose, resulting in a high correlation in their behavior.

This work exploits the intuition that a spam targeting a community is most likely the spam generated by a community. A dataset [21] containing Facebook spam profiles is analyzed to identify Facebook pages that have been mostly

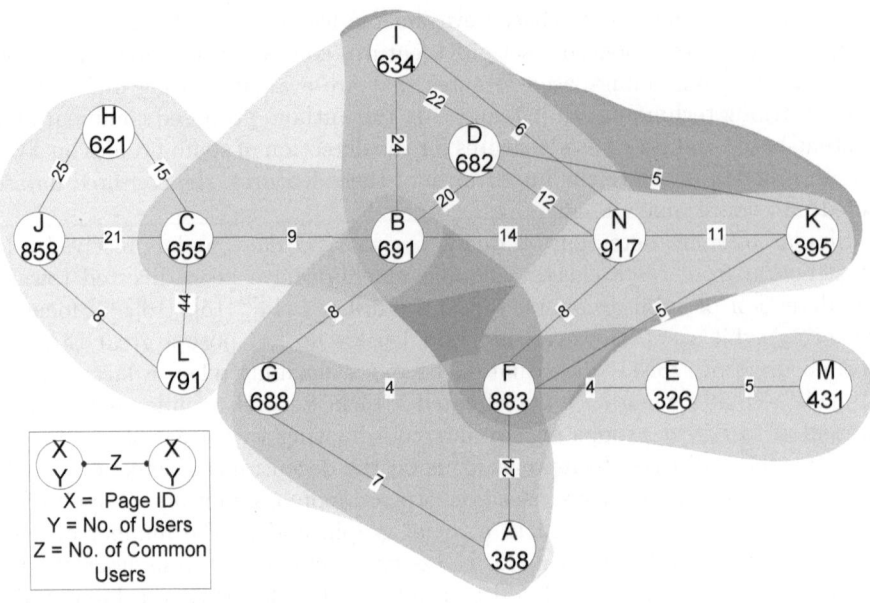

Fig. 1. Graphical illustration of Facebook pages and users

targeted by spammers. As a result, a total number of 14 Facebook pages is found that are heavily spammed by the spam profiles identified in [21]. All these pages are accessed to identify active users and to group profiles based on the number of their common pages-likes. Figure 1 shows a graph in which each node represents a Facebook page and the weight associated to an edge represents the number of users commonly shared by the connected pair of nodes. The weights of the edges can be used to divide the users into various groups based on their interests in the network. In Figure 1, there are six groups of pages that are close to each other in terms of their common interests, and each group is treated separately for detection of profiles that are under the control of a single spammer and generate context-aware spam towards a community of normal users. Table 1 shows the various groups and the number of users belonging to each of them. The names of the groups in Table 1 has been derived from the node levels used in Figure 1. The next Section explains the methodology used to identify the groups of sybil accounts.

Table 1. Various profile groups along with the number of users

Groups	FEM	BGFA	BNDIC	AFGNK	JHLCB	DBINKF
n	1	2	3	4	5	6
No. of users	1631	2575	3465	3166	3482	4059

4   Methodology

To detect communities of sybil accounts generating context-aware spam, the rich amount of textual information contained in each profile is used to generate an undirected-weighted social graph, in which a node represents a profile and an edge connecting a pair of nodes represents a link between them. The connections initiated through a friendship request are independent of the links created in the actual social graph. A total number of three important features has been used to determine the weight of an edge in the social graph. For each group of profiles identified in Section 3, a social graph is generated as $G = (V_n, E, W)$, where n represent the group id, V_n is the set of profiles in group n, $E \subseteq V \times V$ is the set of edges, and $W \subseteq \Re$ is the set of weights assigned to edges. For each node $v \in V_n$, a 3-dimensional feature vector comprising *profile similarity*, *page likes*, and *URLs shared* is generated, which is then used to calculate the weight of an edge $e_{ij} = (v_i, v_j)$. Further details about the features and weight calculation process are presented in the following Subsections.

4.1   Social Graph Generation

To generate social graph, a set of features has been identified to determine the weight of an edge highlighting the degree of similarity of the connected profiles. The following paragraphs present a detailed discussion on the identified features and edge's weight calculation mechanism.

Profile Similarity: The profile similarity of a pair of connected users represents the degree of match in their posts. This is calculated as a similarity index, I_s, for each edge $e_{ij} = (v_i, v_j)$ that connects a pair of nodes. The similarity index uses vector-space model to represent users' posts and applies *cosine* function to measure their degree of similarity. The first criteria for two profiles to be similar is the number of times a profile has posted on its own wall. For example, a profile v_i with a large number of posts as compared to a profile v_j with a small number of posts on their own walls is clearly dissimilar. In this elimination process, the posts from other profiles on the subject profile's wall are not considered. For two profiles v_i and v_j containing x and y number of posts, respectively, a *squareness measure*, as shown in Equation 1, is used to determine the eligibility of the two profiles to be considered for further comparison. In Equation 1, S_{ij} is the squareness measure of nodes v_i and v_j, which must be greater than or equal to 4 before considering them for similarity index calculation.

$$S_{ij} - x/y | x > y \qquad (1)$$

For the nodes v_i and v_j that fulfil the squareness measure criterion, the similarity index is calculated as follows. Considering x and y as the number of posts of v_i and v_j on their own walls, respectively, a cosine similarity matrix C of dimensions $x \times y$ is created in such a way that each post of v_i is compared with all the posts of v_j. For cosine similarity, each post is converted into a *tf-idf* feature vector, where *tf-idf* of a term t is calculated using Equations 2 and 3. In these equations,

d is the post under consideration, D is the set of posts present in nodes v_i and v_j, and $tf(t, d)$ is calculated as the number of times t appears in d.

$$tf\text{-}idf(t, d, D) = tf(t, d) \times idf(t, D) \tag{2}$$

$$idf(t, D) = log\frac{|D|}{|\{d \in D : t \in d\}|} \tag{3}$$

For any two posts a and b with their corresponding tf-idf feature vectors A and B, the value of an element c_{ij} of the matrix C is calculated as a cosine similarity using Equation 4, where l is the length of feature vectors.

$$c_{ij} = \frac{\sum_{k=1}^{l} A_k \times B_k}{\sqrt{\sum_{k=1}^{l}(A_k)^2} \times \sqrt{\sum_{k=1}^{l}(B_k)^2}} \tag{4}$$

Finally, after smoothing the values of the matrix C using Equation 5, the similarity index for the edge $e_{ij} = (v_i, v_j)$ is calculated as the normalized cardinality of the set of non-zero elements in C, as shown in Equation 6.

$$c_{ij} = \begin{cases} 1 & \text{if } c_{ij} > 0.1 \\ 0 & \text{if } c_{ij} < 0.1 \end{cases} \tag{5}$$

$$I_s = \frac{|\{c_{ij} \in C | c_{ij} = 1\}|}{x \times y} \tag{6}$$

Page-Likes: This feature is similar to the feature considered in [21]. However, in this work, the value of this feature is normalized along the lines of the similarity index normalization process. This feature captures the *page-likes* behavior of the users in a social network. For an edge $e_{ij} = (v_i, v_j)$, the common *page-likes* of v_i and v_j, P_{ij}, is calculated as a fraction of the *page-likes* commonly shared by them, as given in Equation 7. In this equation, P_i and P_j represent the set of page-likes by nodes v_i and v_j, respectively.

$$P_{ij} = \frac{|P_i \cap P_j|}{|P_i \cup P_j|} \tag{7}$$

URL Sharing: Like page-likes feature, the value of URL sharing feature of nodes v_i and v_j is calculated as the fraction of the URLs commonly shared by them, as shown in Equation 8. In this equation, U_i and U_j represent the set of URLs shared by nodes v_i and v_j, respectively.

$$U_{ij} = \frac{|U_i \cap U_j|}{|U_i \cup U_j|} \tag{8}$$

Based on the values of the features discussed above, the final weight of edge $e_{ij} = (v_i, v_j)$, $\omega(e_{ij})$, is calculated using Equation 9, where α_1, α_2, and α_3 are constants such that each $\alpha_i > 0$ and $\alpha_1 + \alpha_2 + \alpha_3 = 1$.

$$\omega(e_{ij}) = \alpha_1 \times I_s + \alpha_2 \times P_{ij} + \alpha_3 \times U_{ij} \tag{9}$$

4.2   Community Detection

To identify the communities in a social graph, the proposed approach uses the *Louvain* algorithm, which has been implemented as a part of an open source social network analysis tool `Gephi 0.8.1` [22]. It has been widely used for social network analysis [23]. The algorithm supports community detection in various types of graphs and provides the flexibility to identify communities at different levels of granularity. It implements a greedy approach for optimizing *modularity* of a network divisions. The modularity measures the strength/ability of a network to be divided into groups or communities. Initially, the algorithm optimizes the modularity of smaller individual communities, then nodes from the same communities are added to form a new network in which each node represents a community. This process is repeated until maximum possible modularity is obtained. The result is a hierarchy of communities present in the underlying social graph.

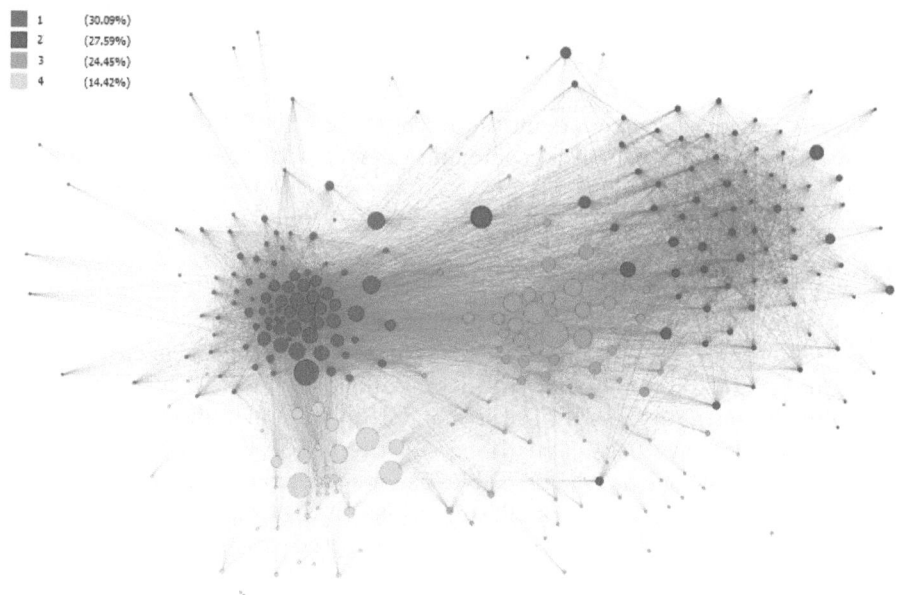

Fig. 2. Community structures in FEM group of profiles

Figure 2 shows a subgraph of the *FEM* group present in the dataset. The graph shows 4 major communities out of total 14 communities obtained through *Louvain* algorithm. In the experiment, the default resolution values of Louvain's implementation in Gephi has been used. In Figure 2, the legend describes the percentage of nodes in each community. It can be observed in Figure 2 that nodes with modularity class 2 are dispersed, whereas nodes of classes 1, 3 and 4 are more closely related. In Section 4.3, the analysis has been further extended

Table 2. Modularity percentages of communities identified for each group of the dataset

Groups	FEM	BGFA	BNDIC	AFGNK	JHLCB	DBINKF
Class-1	30.09	18.25	17	16.33	10.60	20.92
Class-2	27.59	10.72	12.35	11.37	10.43	11.06
Class-3	24.45	9.28	7.22	11.24	7.96	6.48
Class-4	14.42	6.8	4.7	9.79	6, 15	3.03
Class-5	0.18	0.08	3.17	0.25	5.14	2.81

Table 3. Communities with closeness-centrality values

Groups	FEM	BGFA	BNDIC	AFGNK	JHLCB	DBINKF
Class-1	0.651	0.573	0.421	0.546	0.592	0.589
Class-2	0.549	0.600	0.596	0.548	0.609	0.611
Class-3	0.562	0.620	0.582	0.545	0.644	0.590
Class-4	0.628	0.568	0.592	0.548	0.610	0.570
Class-5	0.621	0.448	0.574	0.451	0.601	0.575

to classify the identified communities as sybils or normal. Table 2 shows details about the percentage of nodes in communities along with the highest modularity in each group of the dataset.

4.3   Sybil Community Identification

Once the communities are identified, profiles of each community with the highest closeness-centrality have been processed separately to classify them either as malicious or benign. Table 3 provides the details about the nodes with highest values of closeness-centrality. A set of features and JRip rules identified from a locally crawled dataset have been used to classify the nodes with highest closeness-centrality as malicious or benign. Table 4 shows the final results obtained after identifying communities as normal or malicious on the basis of the nodes' closeness centrality values. After having a close look at the closeness centrality values and the final results, it can be found that, in most of the cases, the nodes of a normal user communities have low closeness centrality values. This mainly happens because the weights assigned to the edges are according to the degree of similarity among the nodes. A higher similarity between a pair of nodes produces a higher weight for the edge connecting them in the social graph. Therefore, in the generated social graph, nodes with high closeness centrality values are similar to the majority of the nodes in the set, and as a result, a higher weight is assigned to all the edges connecting the similar nodes. Moreover, because the sybil accounts are controlled by a single spammer, they have high similarity among them as compared to normal users. Hence, nodes belonging to sybil communities have higher closeness centrality values in comparison to normal users.

Table 4. Communities identified as malicious (M) or benign (B)

Groups	FEM	BGFA	BNDIC	AFGNK	JHLCB	DBINKF
Class-1	M	B	B	B	B	B
Class-2	B	M	M	M	M	M
Class-3	B	M	B	B	M	M
Class-4	M	M	M	M	M	B
Class-5	M	B	M	B	B	B

5   Conclusions

Along the lines of the previous research works, this paper has presented a hybrid approach to detect communities of sybil accounts that are under the control of spammers and generate context-aware spam towards normal user communities. The proposed approach is independent of the assumptions made by the previous efforts and identifies six different profiles groups in the dataset based on the users' interests on Facebook network. The users with most common page-likes have been grouped together for further analysis. Three different types of features have been identified and used to model each group as a social graph in which profiles are represented as nodes and their links as edges. The weight of a link is calculated as a function of the degree of similarity of the nodes. Louvain community detection algorithm is applied on the social graphs to identify communities embedded within them. Thereafter, based on the class (malicious or benign) of the nodes with high closeness-centrality values, the underlying community is marked either as malicious or benign. The obtained results highlight that generally nodes with high closeness-centrality values are malicious and belong to sybil communities, whereas nodes with low closeness-centrality values are benign and constitute normal user communities.

Acknowledgment. The authors would like to thank King Abdulaziz City for Science and Technology (KACST) and King Saud University for their support. This work has been funded by KACST under the NPST project number 11-INF1594-02.

References

1. Boshmaf, Y., Muslukhov, I., Beznosov, K., Ripeanu, M.: Key challenges in defending against malicious socialbots. In: Proceedings of the 5th USENIX Conference on Large-scale Exploits and Emergent Threats, LEET, vol. 12 (2012)
2. Nagaraja, S., Houmansadr, A., Piyawongwisal, P., Singh, V., Agarwal, P., Borisov, N.: Stegobot: A covert social network botnet. In: Filler, T., Pevný, T., Craver, S., Ker, A. (eds.) IH 2011. LNCS, vol. 6958, pp. 299–313. Springer, Heidelberg (2011)
3. Thomas, K., Nicol, D.: The koobface botnet and the rise of social malware. In: IEEE 2010 5th International Conference on Malicious and Unwanted Software, MALWARE, pp. 63–70 (2010)

4. Yu, H., Kaminsky, M., Gibbons, P., Flaxman, A.: Sybilguard: defending against sybil attacks via social networks. ACM SIGCOMM Computer Communication Review 36, 267–278 (2006)
5. Boyd, S., Ghosh, A., Prabhakar, B., Shah, D.: Gossip algorithms: Design, analysis and applications. In: Proceedings IEEE 24th Annual Joint Conference of the IEEE Computer and Communications Societies, INFOCOM 2005, vol. 3, pp. 1653–1664. IEEE (2005)
6. Danezis, G., Lesniewski-Laas, C., Frans Kaashoek, M., Anderson, R.: Sybil-resistant DHT routing. In: De Capitani di Vimercati, S., Syverson, P.F., Gollmann, D. (eds.) ESORICS 2005. LNCS, vol. 3679, pp. 305–318. Springer, Heidelberg (2005)
7. Yang, Z., Wilson, C., Wang, X., Gao, T., Zhao, B., Dai, Y.: Uncovering social network sybils in the wild. In: Conference on Internet Measurement (2011)
8. Gao, H., Hu, J., Wilson, C., Li, Z., Chen, Y., Zhao, B.: Detecting and characterizing social spam campaigns. In: Proceedings of the 10th Annual Conference on Internet Measurement, pp. 35–47. ACM (2010)
9. Lee, K., Caverlee, J., Cheng, Z., Sui, D.: Content-driven detection of campaigns in social media (2011)
10. Jin, X., Lin, C., Luo, J., Han, J.: A data mining-based spam detection system for social media networks. Proceedings of the VLDB Endowment 4(12) (2011)
11. Stringhini, G., Kruegel, C., Vigna, G.: Detecting spammers on social networks. In: Proceedings of the 26th Annual Computer Security Applications Conference, pp. 1–9. ACM (2010)
12. Lee, K., Eoff, B., Caverlee, J.: Seven months with the devils: A long-term study of content polluters on twitter. In: Int'l AAAI Conference on Weblogs and Social Media, ICWSM (2011)
13. Yang, C., Harkreader, R.C., Gu, G.: Die free or live hard? Empirical evaluation and new design for fighting evolving twitter spammers. In: Sommer, R., Balzarotti, D., Maier, G. (eds.) RAID 2011. LNCS, vol. 6961, pp. 318–337. Springer, Heidelberg (2011)
14. Yu, H., Gibbons, P., Kaminsky, M., Xiao, F.: Sybillimit: A near-optimal social network defense against sybil attacks. In: IEEE Symposium on Security and Privacy, SP 2008, pp. 3–17. IEEE (2008)
15. Danezis, G., Mittal, P.: Sybilinfer: Detecting sybil nodes using social networks. NDSS (2009)
16. Tran, N., Min, B., Li, J., Subramanian, L.: Sybil-resilient online content voting. In: Proceedings of the 6th USENIX Symposium on Networked Systems Design and Implementation, pp. 15–28. USENIX Association (2009)
17. Cao, Q., Sirivianos, M., Yang, X., Pregueiro, T.: Aiding the detection of fake accounts in large scale social online services. Technical Report (2011), http://www.cs.duke.edu/~qiangcao/publications/sybilrank_tr.pdf
18. Wang, G., Mohanlal, M., Wilson, C., Wang, X., Metzger, M., Zheng, H., Zhao, B.: Social turing tests: Crowdsourcing sybil detection. Arxiv preprint arXiv:1205.3856 (2012)
19. Thomas, K., Grier, C., Ma, J., Paxson, V., Song, D.: Design and evaluation of a real-time url spam filtering service. In: IEEE Symposium on Security and Privacy (2011)

20. McCord, M., Chuah, M.: Spam detection on twitter using traditional classifiers. In: Calero, J.M.A., Yang, L.T., Mármol, F.G., García Villalba, L.J., Li, A.X., Wang, Y. (eds.) ATC 2011. LNCS, vol. 6906, pp. 175–186. Springer, Heidelberg (2011)
21. Ahmed, F., Abulaish, M.: An mcl-based approach for spam profile detection in on-line social networks. In: The 11th IEEE International Conference on Trust, Security and Privacy in Computing and Communications, TrustCom 2012, IEEE (2012)
22. Blondel, V., Guillaume, J., Lambiotte, R., Lefebvre, E.: Fast unfolding of communities in large networks. Journal of Statistical Mechanics: Theory and Experiment 2008, 10008 (2008)
23. Blondel, V.: The Louvain method for community detection in large networks (2011), http://perso.uclouvain.be/vincent.blondel/research/louvain.html (accessed July 11, 2012)

Location-Based Emerging Event Detection in Social Networks

Sayan Unankard, Xue Li, and Mohamed A. Sharaf

School of Information Technology and Electrical Engineering,
The University of Queensland, Australia
{uqsunank,m.sharaf}@uq.edu.au, xueli@itee.uq.edu.au

Abstract. This paper proposes a system for the early detection of emerging events by grouping micro-blog messages into events and using the message-mentioned locations to identify the locations of events. In our research we correlate user locations with event locations in order to identify the strong correlations between locations and events that are emerging. We have evaluated our approach on a real-world Twitter dataset with different granularity of location levels. Our experiments show that the proposed approach can effectively detect the top-k ranked emerging events with respect to the locations of the users in the different granularity of location scales.

Keywords: Emerging event detection, Location-based social networks, Micro-blogs.

1 Introduction

The detection of emerging events with respect to the locations and participants of events provides a better understanding of ongoing events. Emerging events like natural disasters may need to be reported in real time when they are observed by many people. Emerging events such as infectious diseases and cyberspace-initiated/plotted attacks/unrest need to be detected at their early stages.

Currently, a User Generated Content (UGC) system, such as micro-blog services, provides a wealth of online discussions about real-world events. Micro-blog, e.g. Twitter is being considered as a powerful means of communication for people who are looking for sharing and exchanging information over the social media. The social events range from popular, widely known ones such as events concerning well-known people or political affairs to local events such as accidents, protests or natural disasters. Moreover social events could be malicious, dangerous, or disastrous. For instance, in August 2011, rioters used instant messaging and social network services to have arranged meetings and agitated people across England. Our research interest is to understand where, when, and what is happening (emerging), so to predict its occurrence via the real-time monitoring of the social networks. More specifically, our research interest is focused on the emerging "hotspot" events. We define that a hotspot event is a tuple (location, time, topic) that a social network user is associated with through the posting of a micro-blog.

Y. Ishikawa et al. (Eds.): APWeb 2013, LNCS 7808, pp. 280–291, 2013.

Table 1. An example of Twitter data

tweet id	created time	user id	user location	geo-tagged	content
1	2012-08-21 08:26:45	22855027	Brisbane, Australia	-27.480835, 153.030392	STOP THE CUTS: march on Qld Parliament. Meet 5pm at **King George Square, Brisbane** this Thursday 23rd
2	2012-08-21 08:32:55	19573441	Queensland	-	Stop the Cuts Rally this afternoon 5pm **King George Square**. I'll be there. http://t.co/UqH3EMOl

In this paper, we developed a Location-based Emerging Event Detection (LEED) system in social networks. We used this system to detect hotspot events from micro-blog messages that can help government or organizations prepare for and response to unexpected events. With the large range of events discussed on social networks, we do not know how many emerging hotspot events are likely to happen. Moreover, dealing with a very large number of messages and noisy data is difficult. The research challenges of our study are: (1) how to effectively detect events in terms of keywords in micro-blogs? (2) how to detect hotspot events (i.e. associate the message-mentioned location(s) to an event)? (3) how do we know a hotspot event is emerging?

Our approach has five stages. Firstly, the pre-processing is performed to improve quality of the dataset. Secondly, we propose a clustering approach to automatically group the messages into events. Thirdly, we propose a hotspot event detection method. Fourthly, emerging hotspot event detection is performed. Finally, we develop a visualization model for representing emerging events. The experiments are conducted to demonstrate the effectiveness of our approach.

This paper is organised as follows, Section 2 is about micro-blog data and events. The architecture of proposed system is presented in Section 3. In Section 4, we present the experimental setup and results, Section 5 describes the related work. The conclusions are given in Section 6.

2   Micro-blog Data and Events

A micro-blog message is a short text message such as a tweet that is restricted to 140 characters and therefore is much concise than a blog post. The twitter data consists of tweet ID, creating time, user ID, user location, tweet location, text content and the message-mentioned locations. A tweet location is known when a user posts the message using a smart phone. An example of Twitter data is shown in Table 1.

An event is something that occurs in a certain place during a particular interval of time[1]. An event location is a place where the event will happen or is happening while a topic location is a place that is included in the topic content. In real-world, an event location can also be a topic location, or they can be different to each other. In this research we only consider the event locations. The topic locations are not considered exclusively, e.g., an "earthquake" location is

---

[1] http://dictionary.reference.com/browse/event?s=t

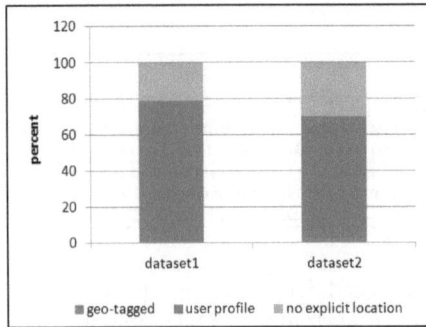

Fig. 1. A bar chart of micro-blog message's location

Table 2. Statistics of message locations

List	Dataset1		Dataset2	
	#msgs	%	#msgs	%
Messages with geo-tagged location	3,121	1.42	5,055	2.57
Messages with user profile location	170,334	77.45	132,661	67.40
Messages with IP address based location	46,478	21.13	59,118	30.03
Messages that contain geographical location in contents	16,875	7.67	19,529	9.92
Messages that contain geographical location in contents more than 1 location	744	4.41	1,337	6.85

both the topic as well as the event location. An emerging event is an event that has significantly increased in the number of messages but rarely posted in the past. A hotspot event is an event where there is a strong association between event location and user location. User location is the location where the message is sent from. Message-referred location is the location mentioned in the message. It could be event location which is the location where the event occurs or other locations referred to within the messages.

In order to understand where the locations of messages are found in social media, we conducted statistical studies. Firstly, we try to understand the availability of the locations within the micro-blogs. We used Twitter API[2] to have crawled the micro-blog datasets. Dataset1 is crawled from the messages sent by users around Brisbane Australia, from the dates 17 March 2012 to 25 March 2012 with 219,933 messages while Dataset2 is crawled from the messages sent by users around the USA, from the dates 21 June 2012 and 27 June 2012 with 196,834 messages. The statistical information of locations from the two datasets is shown in Figure 1 and Table 2.

As we can see from Datasets 1 and 2 that, the user locations are divided into three types i.e. the geo-tagged locations, from the user profiles, and implied by IP addresses (first three rows in the Table 2). The majority of user locations come from user profiles, approximately 77% of Dataset1 and 67% of Dataset2. Also from Datasets 1 and 2 we can see that the message-mentioned locations are only available within a small proportion of the micro-blogs and appeared in one or two times in average in the micro-blog messages. Our most important observation on the locations of micro-blogs is that most of the micro-blog messages contain only one geographical location while messages that contain geographical locations of more than one constituted approximately 4% of Dataset1 and 7% of Dataset2. When a location is mentioned in the micro-blog message, it can be either an event location or a topic location.

In Table 3, we also try to understand what the locations used for in the messages contents. Dataset3 is downloaded from G. R. Boynton, University of

---

[2] https://dev.twitter.com/docs/api/1.1

Table 3. Message-mentioned locations

Dataset	Event	Total No. of msgs	Messages contain mentioned location		#of times that event location occur	#of times that other locations occur	Confidence of location mentioned in the content as event location
			#of msgs	%			
3	H1N1	958	79	8.25	73	6	92.41%
	Earthquake	801	798	99.66	846	37	100%
4	QLD Election 2012	629	87	13.83	86	1	98.85%

Iowa which consists of two events: USA H1N1 and the Indonesia Earthquake[3]. Dataset4 is crawled by using hashtag "#qldvote" for Queensland Election 2012. As we can see from Datasets 3 and 4 that the message-mentioned locations are mostly the event locations. The message-mentioned location can appear more than one time in the message. Our most important observation on the message-mentioned locations is that the more frequently a message-mentioned location, the more likely it is the event location. Therefore, the use of message-mentioned locations to identify event location is the key goal behind this study.

3 Proposed System Architecture

In order to provide a complete coverage of emerging hotspot event detection we proposed our system which has five stages presented in Figure 2. The following information provides detail of each stage.

3.1 Data Pre-processing for Events Detection

In particular dealing with micro-blog messages, the message is short and often noisy. In order to improve the quality of our dataset and the performance of the subsequent steps, the pre-processing was designed to ignore common words that carry less important meaning than keywords and remove irrelevant data e.g re-tweet keyword, web address and message-mentioned username. A micro-blog loader is developed to collect the Twitter data from public users via the Twitter API service. The messages are converted into lower case and are removed by web addresses and a single character word. We also remove the keyword RT(ReTweet) and the message-mentioned username such as @Sayan. The stop words are removed and all words are converted into a seed word (stemming word) by using Lucene 3.1.0 Java API[4]. All messages after pre-processing are stored in the database.

3.2 Text Stream Clustering

The problem that we address in this paper is how to identify events from a given set of micro-blog messages. We consider a set of messages where each message is associate with an event. With the large range of events discussed on social

---

[3] http://ir.uiowa.edu/polisci_nmp/2/ and http://ir.uiowa.edu/polisci_nmp/5/
[4] http://lucene.apache.org

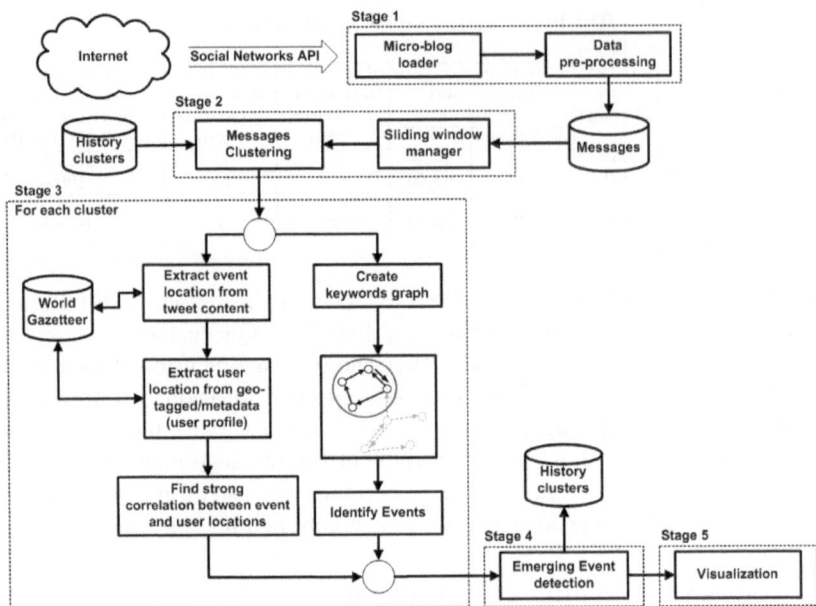

Fig. 2. Architecture of LEED system

networks, we do not know the number of events in advance. On the other hand, our approach requires no prior knowledge of the number of events. Therefore, the hierarchical clustering is used. In this stage, we aim to automatically group messages into the same event. We also need a fast and efficient message clustering system to overcome the problem of the high arrival rate of messages. In order to dealing with the high incoming rate of messages, we use a sliding window manager to keep track of messages arriving in the system. The size of the sliding window can be defined as the number of messages or time interval. In our case, we use time intervals such as one hour, two hours or one day depending on the user given preference. Additionally, the number of previous time slots needs to be specific because it is not necessary to consider the complete usage history of data to compute an emerging of the event.

In order to find the best representation of term weight for tweet messages, we compare four different term weight formulas (i.e. Term frequency, Augmented Normalized Term Frequency[13], TFIDF[13], and Smooth-TFIDF[10]). We also compare four different similarity functions (i.e. Jaccard index, Euclidean distance, Manhattan distance and Cosine similarity) for finding the best similarity function. To evaluate an effective clustering method, we manually label 16,144 tweets into 17 topics. We evaluate our algorithm by using pair-wise precision, recall and F1-score. Due to space limit, we do not show our preliminary experiments. The clustering method performs well when using the augmented normalized term frequency and cosine similarity function. Therefore, we calculate the weight $w_{i,t}$ of term t in message i by using augmented normalized term frequency:

$$w_{i,t} = 0.5 + 0.5 \times \frac{tf_{i,t}}{tf_i^{max}} \tag{1}$$

where $tf_{i,t}$ is the term frequency value of the term t in message i, tf_i^{max} is the highest term frequency value of the message i. The cosine similarity function is used to calculate the similarity between the existing cluster and the new message:

$$cos(m,c) = \frac{\sum_i (w_{m,t_i} \times w_{c,t_i})}{\sqrt{\sum_j w_{m,t_j}^2} \times \sqrt{\sum_j w_{c,t_j}^2}} \tag{2}$$

where m is a message, c is a cluster centroid, and w_{m,t_i} is the weight of term t_i in message m.

We use a clustering method called leader-follower clustering [5]. Message clustering is executed when the timestamp of the coming message is greater than the sliding window size. The system keeps history clusters within previous time slots to decrease time of computation. Every new message is compared with previous clusters. The algorithm creates a new cluster for the message if there is no cluster whose similarity to the message is greater than threshold. We use a centroid representation of the cluster because centroid is agglomerative, using the mean, which trades memory use for speed of clustering. In the final step, we find the most similar pair of clusters and merge them into the same cluster. In order to find the most similar cluster, we calculate cosine similarity of two clusters. If the cosine similarity value exceeds the margin threshold, we merge them. The detailed algorithm is not presented here due to the limitation of the space.

3.3   Hotspot Event Identification

From the previous stage results, all clusters can not be assigned as event clusters because it can be private conversations, advertisements or others. In this step, we focus on how to identify the hotspot event clusters. A cluster will be assigned as the hotspot event if there is strong correlation between the event location and user location. The tasks are divided into two parts; to find a correlation between the event location and user location in the cluster and to extract the event topic.

Find a Correlation between the Event Location and User Location: we calculate the correlation scores between user locations and event locations for identifying the hotspot event. In order to calculate a correlation score, we need to extract the user locations and event location first.

Firstly, for user location, we extract from the geo-tag and from the user profile. The geo-tagged information is generated from smart phone applications while the location from user profile is a free format text where the user fills in the user profile. For those users who can post messages from different locations, for a given message we use geo-tagged information to locate the message location firstly, because it can provide precise location of the user. In order to convert a latitude/longitude pair into an address, we use Google Maps API[5]. If geo-tagged information is not available we use user profile to query the gazetteer

---

[5] https://developers.google.com/maps/documentation/geocoding/

database, the database of geographic locations, for getting the location. Finally, if neither of them is not available we set message location equal to "World". For the gazetteer database, we downloaded a list of geographic locations from GeoNames[6] and stored a local copy of the gazetteer in a database.

Secondly, for event location, we find all terms or phrases reference to geographic location (e.g. country, state and city) from tweet contents. Since the location extraction from text is one of the challenging problems of this research area, this task will be more focused in future work. In this research, we simply extract the message-mentioned locations via Named Entity Recognition (NER). We use the Stanford Name Entity Recognizer [7] to identify locations within the messages. We also use the Part-of-Speech Tagging for Twitter which is introduced in [6] to extract proper nouns. We use extracted terms query into the gazetteer database to obtain candidate locations of the event. We find the most possible location of the event using the frequency of each location in the cluster. The location which has the highest frequency is assigned as event location.

Finally, for finding the correlation between event location and user location, a correlation score is computed by comparing the level of location granularity. The granularity level is defined as Country>State>City>PlaceName. We assign scores for each level if both of them have the same value. The equation is shown below:

$$CorrelateScore = \alpha 1(F(uCountry, eCountry)) + \alpha 2(F(uState, eState))$$
$$+ \alpha 3(F(uCity, eCity)) + \alpha 4(F(uPlace, ePlace)) \quad (3)$$

where $\alpha 1$ - $\alpha 4$ are the weight of granularity levels, $\alpha 1 = \alpha 2 = \alpha 3 = \alpha 4 = 0.25$, $uCountry, uState, uCity$ and $uPlace$ are user location, $eCountry, eState, eCity$ and $ePlace$ are event location, and $F(x, y) = 1$; if x = y and the higher granularity level has the same value otherwise $F(x, y) = 0$.

To identify which cluster is a hotspot event, the $LocScore$ is computed. The range of $LocScore$ is 0 to 1. It will be used to compute emerging score in the next section and the top k ranking emerging hotspot events will be selected.

The $LocScore$ of cluster c is defined as:

$$LocScore_c = \frac{\sum_{u \in U} CorrelateScore_u}{Np} \quad (4)$$

where Np is the number of users who post messages in cluster c and U represents users who post messages in cluster c.

Event Topic Extraction: In order to understand what the event cluster is about, we need to find the set of keywords to represent the event topic. Our intuition is that keywords co-occur when there is a meaningful topical relationship between them. To extract the set of co-occurring keywords, firstly, we create a directed, edge-weighted graph. The edge is created if the correlation weight between the two terms exceeds the threshold. The threshold is defined as the average of correlation weights in the cluster. We adopt the smoothed correlation

---

[6] http://www.geonames.org

weight function which is introduced in [12], to calculate the semantic correlation weight between terms. The function is shown below:

$$c_{k,z} = log(\frac{(n_{k,z} + n_k/N)/(n_z - n_{k,z} + 1)}{(n_k - n_{k,z} + n_k/N)/(N - n_k - n_z + n_{k,z} + 1)}) \qquad (5)$$

where n_k is number of posts containing term k, n_z is number of posts containing term z, $n_{k,z}$ is number of posts containing the terms k and z, and N is the total number of posts. We identify the event topic by extracting the Strongly Connected Components (SCCs) from the graph. In the case of SCCs' extraction, if the number of SCCs is more than one sub-graph we calculate the sum of edge weight on each SCC sub-graph. The SCC which has the highest score is defined as an event topic.

3.4 Emerging Hotspot Event Detection

As our research interest is to detect emerging hotspot events, all events from previous stage are not emerging hotspot events. We need to extract only emerging hotspot events. We find the event that has significantly increased in the amount of messages but rarely posted in the past. A burst in data is one technique for detecting emerging events. It is calculated by comparing the number of messages in the current time slot with the mean and the standard deviation of the number of messages in the previous time slots. Any data point which is higher than the sum of mean and two standard deviations can be considered as an emerging point.

In order to compute an emerging score for a given event's cluster we calculate the mean and the standard deviation of the number of messages in the previous time slots. If the number of messages in the current time slot is greater than the sum of mean and two standard deviations, we calculate the emerging score of this event. The emerging score of the event e in the current slot is computed by the following equation:

$$EmergScore_e = (1 + LocScore_e) \times \frac{N_e}{(Mean_{prev} + 2SD_{prev})} \qquad (6)$$

where $LocScore_e$ is the location score of event e, N_e is the number of messages of event e in the current time slot, and $Mean_{prev}$ and SD_{prev} are the mean and standard deviation of the number of messages in the previous time slots of the given event, respectively.

To detect emerging hotspot event in different location granularity e.g. state and city. We firstly segment event clusters into user's location groups according to location granularity and follow all of the steps above.

3.5 Visualization

For usability and understanding issues of visualizing model, we use a motion chart[7] to represent emerging hotspot events. Google Map[8] is used to represent

---

[7] http://code.google.com/apis/chart/interactive/docs/
gallery/motionchart.html

[8] https://developers.google.com/maps/documentation/javascript/examples

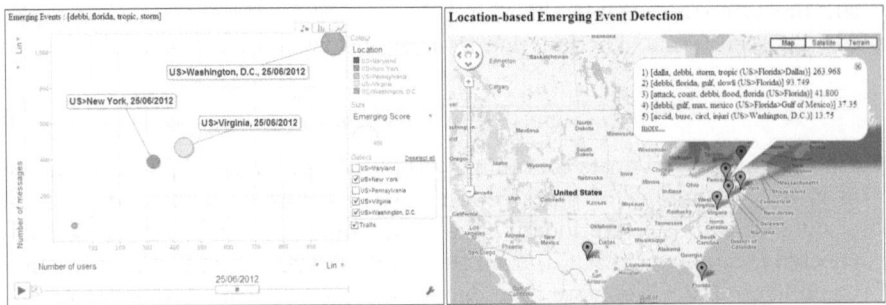

Fig. 3. Example of a motion chart (left) and a geo-map (right)

Table 4. Detection results of KeyGraph and LEED in Country level

Method	No. of detected events	No. of detected real-world events	Precision	Recall
KeyGraph	54	23	0.426	15
Hashtags	949	23	0.024	17
LEED	151	121	0.801	95

the top k emerging hotspot events for a specific location. Examples of a motion chart and a geo-map are shown in Figure 3. Figure 3 (left) shows a given emerging hotspot event within a time period (x-axis is the number of users who post massages; y-axis is the number of messages; colour represents location and bubble size is emerging score). Figure 3 (right) shows the top five events in a selected State.

4   Experiments and Evaluation

In order to evaluate our approach, we use search API from Twitter to collect the messages sent by users around the USA, from the dates 21 June 2012 and 27 June 2012 with 196,834 messages. Since no ground-truth labels are available for us on realistic events within the data collection period, we manually search local news from Google to check the events detected by our system. It is impractical to manually label the overly large number of tweets in the dataset. We follow the definition of Precision used in [17], which is defined as follows:

$$Precision = \frac{detected\_realworld\_events}{total\_detected\_events} \qquad (7)$$

However, Recall was not defined in [17] because it is not feasible to enumerate all the real-life events which happened in the dataset. Therefore, we follow the definition of Recall used in [8], which is defined as the number of distinct realistic events detected from the dataset. Note that, more than one detected event can relate to the same real-world event, then they are considered correct in terms of precision but only one event is considered in counting recall.

$$Recall = no\_distinct\_realworld\_events \qquad (8)$$

Table 5. Sample of 5 events detected by the KeyGraph approach on 25/06/2012

Detected events	Description
1) thunder,spawns,stalls,tropical,storm, tornadoes,threatened	Tropical Storm Debby stalls in Gulf of Mexico
2) park,police,office	Police officer shot dead at jazz concert in Denver park
3) truck,sell,right	City's red fire trucks to be transformed into billboards
4) township,vehicle,south,unknown	Vehicle accident, Hempfield and Manheim Township
5) alarm,calories,brush,burn,acres	Brush fire burns 50 acres on Kent Island

Table 6. Sample of 5 events detected by the Hashtags approach on 25/06/2012

Detected events	Description
1) #Debby,#debbie,#dc	Tropical Storm Debby
2) #nascar,#teamlaurenscartomney,#winning	Car racing
3) #phillies,#debby,#philly	Tropical Storm Debby
4) #ripmj,#legend,#timeflys	Michael Jackson has been dead 3 years today
5) #sussexde,#pasco,#florida	Strong thunderstorm heading into SW, Sussex

Table 7. Sample of top 5 events detected by the LEED approach on 25/06/2012. Data is represented as event topic (event location).

Detected events	Description
1) behold, dalla, debbi, editor, storm, tropic (US>Texas>Dallas)	A tropical storm named Debbie, headed for Dallas
2) debbi, florida, gulf, move,slowli, soak, storm, tropic (US>Florida)	Moving Slowly in Gulf,Tropical Storm Debby Soaks Florida (Related to event 1)
3) attack, coast, debbi, flood, florida, lead, neighborhood, storm, tornado, touch, tropic, weather, latest, listen, newscast (US>Florida)	Tropical Storm Debby attacks Florida's coast, flooding neighborhoods (Related to event 1)
4) concert, denver, offic, park, polic, shot (US>Colorado>Denver)	Police officer shot dead at jazz concert in Denver
5) accid, buse, circl, injuri, involv, metro, report, washington (US>Washington, D.C.)	Accident Washington Circle 2 Metro buses involved.

We compare the performance of our approach with the KeyGraph approach[9] described in [15] and the Semantic Expansion of Hashtags approach described in [11]. Our proposed approach can effectively detect emerging hotspot events with a precision of 0.801 which is significantly larger than the baselines. On the other hand, the correlation between user location and event location can filter non real-world event clusters out from our system. Our approach can also detect the larger number of real-world events than the baselines. However, problems with using a gazetteer are the granularity of locations required and speed of processing required. For illustration purposes, we present the precision and recall in Table 4. Examples of detected events are shown in Table 5, 6 and 7.

5 Related Work

5.1 Event Detection in Social Streams

Event detection on micro-blogs as a challenging research topic has been increasingly reported recently [1,3,8,11,15,17]. Sayyadi et al. in [15] developed a new event detection algorithm by using a keyword graph and community detection algorithm to discover events from blog posts. Keywords with low document

---

[9] http://keygraph.codeplex.com

frequency are filtered. An edge is removed if it does not satisfy the conditions. Communities of keywords are detected by removing the edges with a high betweenness centrality score. However, the number of detected events depends on the threshold parameters and there was no evaluation conducted. Ozdikis et al. in [11] proposed an event detection method in Twitter base on clustering of hashtags, the # symbol is used to mark keywords or topics in Twitter, and applied a semantic expansion to message vectors. For each hashtag the most similar three hashtags are extracted by using cosine similarity. A tweet vector with a single hashtag is expanded with three similar hashtags and then used in the clustering process. However, using only messages with a single hashtag can lead to ignore some important events. Also they did not implement any credibility filter in order to decide whether a tweet is about an event or not.

5.2   Emerging Topic Detection in Social Networks

A significant amount of research has previously been conducted on emerging topic detection [2,4,9]. Cataldi et al. presented the emerging topic detection on Twitter [4]. Their work use aging theory to model the life cycle of each term. The importance of Twitter users is studied, which represents an important weighting of contents. Emerging terms are identified within a given time interval. Terms that frequently correlated with emerging terms are extracted and are reported together as emerging topics. However, the system need user-authority for weighting term which is difficult to collect complete user-network. More recent work is presented by Alvanaki et al. [2]. Correlation of tag pairs which contains at least one seed tag are tracked. The seed tags are selected from current sliding window based on term frequency. Emerging events are identified as top k highest shift scores of tag pairs. However, the performance is highly sensitive to the selected seed tags and it is focused on detecting emerging topics rather than emerging events.

5.3   Location-Based Social Network

The geographical scope of social networks content has been studied in the last decade. TwitterStand has been proposed by Sankaranarayanan et al. in [14]. The 2,000 handpicked users of Twitter are used as seeders who are known to publish news. The online clustering method is used to group the messages into the news topic. User location and content location are used to locate geographic content from each news topic. However, the results relied on handpicked users and there was no evaluation conducted. Watanabe et al. proposed a real-time local-event detection system [16]. Local events are detected using geo-tagged from Twitter data. The place name are extracted from check-in messages such as "I'm at Time Square". Watanabe et al. searched for non geo-tagged messages which contained a distinct place name and allocated a geo-tagged to the location. However, the results relied on the number of geo-tagged data they had.

6   Conclusions

In this paper, we develop a system namely LEED, to automatically detect emerging hotspot events over micro-blogs. Our contributions can be summarized as: (1) We proposed a solution to detect the emerging hotspot events which can help governments or organizations prepare for and respond to unexpected events such as natural disasters and protests. (2) We correlate user location with event location to establish a strong correlation between them. (3) We provide an evaluation for the effective event detection on a real-world Twitter dataset with different granularity of location levels. Our experiments show that the proposed approach can effectively detect emerging events over the baselines. However, some clusters that we found are not belong to the real-world events. It can be chat topic. In future work, the differentiation between event location and topic location which are mentioned in the text will be studied. Moreover, each component may affect the final detection results so a break-down staged evaluation will be assessed.

References

1. Aggarwal, C., Subbian, K.: Event detection in social streams. In: SDM (2012)
2. Alvanaki, F., Michel, S., Ramamritham, K., Weikum, G.: See what's enblogue: real-time emergent topic identification in social media. In: EDBT (2012)
3. Becker, H., Naaman, M., Gravano, L.: Beyond trending topics: Real-world event identification on twitter. In: ICWSM (2011)
4. Cataldi, M., Caro, L.D., Schifanella, C.: Emerging topic detection on twitter based on temporal and social terms evaluation. In: MDMKDD (2010)
5. Duda, R.O., Hart, P.E.: Pattern Classification and Scene Analysis, 1st edn. John Wiley & Sons Inc. (February 1973)
6. Gimpel, K., Schneider, N., O'Connor, B., Das, D., Mills, D., Eisenstein, J., Heilman, M., Yogatama, D., Flanigan, J., Smith, N.A.: Part-of-speech tagging for twitter: Annotation, features, and experiments. In: ACL (2011)
7. Jurafsky, D., Martin, J.H.: Speech and Language Processing: An Introduction to Natural Language Processing, Computational Linguistics, and Speech Recognition, vol. 163. Prentice Hall (2000)
8. Li, C., Sun, A., Datta, A.: Twevent: segment-based event detection from tweets. In: CIKM (2012)
9. Mathioudakis, M., Koudas, N.: Twittermonitor: trend detection over the twitter stream. In: SIGMOD (2010)
10. Ni, X., Quan, X., Lu, Z., Wenyin, L., Hua, B.: Short text clustering by finding core terms. Knowl. Inf. Syst. 27(3), 345–365 (2011)
11. Ozdikis, O., Senkul, P., Oguztuzun, H.: Semantic expansion of hashtags for enhanced event detection in twitter. In: VLDB-WOSS (2012)
12. Ruthven, I., Lalmas, M.: A survey on the use of relevance feedback for information access systems. Knowl. Eng. Rev. 18(2), 95–145 (2003)
13. Salton, G., Buckley, C.: Term-weighting approaches in automatic text retrieval. Inf. Process. Manage. 24(5), 513–523 (1988)
14. Sankaranarayanan, J., Samet, H., Teitler, B.E., Lieberman, M.D., Sperling, J.: Twitterstand: news in tweets. In: GIS (2009)
15. Sayyadi, H., Hurst, M., Maykov, A.: Event detection and tracking in social streams. In: ICWSM (2009)
16. Watanabe, K., Ochi, M., Okabe, M., Onai, R.: Jasmine: a real-time local-event detection system based on geolocation information propagated to microblogs. In: CIKM (2011)
17. Weng, J., Lee, B.-S.: Event detection in twitter. In: ICWSM (2011)

Measuring Strength of Ties in Social Network[*]

Dakui Sheng, Tao Sun, Sheng Wang, Ziqi Wang, and Ming Zhang[**]

School of Electronic Engineering and Computer Science
Peking University
Beijing 100871, China
{shengdakui,suntao,wangshengpkucn,wangziqi,mzhang_cs}@pku.edu.cn

Abstract. Measuring the strength of ties has been a fundamental task for a long time. However, most of previous work treated it as classifying a tie as strong or weak and were not able to quantitatively estimate the strength, which limits their scope of contribution. To tackle the problem, through leveraging user similarities and social interactions, we propose a latent variable model to calculate a continuous value that measures the strength. By bringing real users as judge, we demonstrate that the proposed method can outperform previous methods. Further, we utilize it to measure the strength of ties among a large set of microblogging users, and conduct statistical analysis on triads. We find that comparing to other types of triads, the one with three significantly strong ties are more likely to be created, which verifies the theory of sociology.

1   Introduction

Granovetter [4] introduced the concept of strength of ties in his landmark paper "The Strength of Weak Ties". Strong ties connect people you really trust, and those whose social circles tightly overlap with yours. Often, they are also the people who are most like you while weak ties are merely acquaintances.

Weak ties often provide people with access to information and resources out of their own circles [3] while strong ties have greater motivation to be of assistance and are typically more easily available [3]. So the study of social ties can bring lots of benefits to daily life [10, 5] which is well worth working on. However, in most studies of social network, social connections are often presented as being friends or not, neglecting the difference between strong and weak ties. And the problem hasn't been well explored yet.

Previous studies have investigated the strength of ties, e.g., by exploiting human annotation as training data [2, 6]. The problem is that performance largely depends on the quality of annotation. At the same time, they treat the problem as a binary classification task, i.e., to classify a relationship as strong or

---

[*] This work is partially supported by the the National Natural Science Foundation of China (NSFC Grant No. 61272343), as well as The Specialized Research Fund for the Doctoral Program of Higher Education of China under Grant ("FSSP" Grant No. 20120001110112).

[**] Corresponding author.

Y. Ishikawa et al. (Eds.): APWeb 2013, LNCS 7808, pp. 292–300, 2013.

weak and were not able to quantitatively estimate the strength, which limits the scope of their contribution in real world. In this paper, we harness the homophily principles and empirical analysis to develop a latent variable model for estimating tie strength based on user similarities and social interactions. Comparing to the existing methods, the proposed method requires no human annotation, and the outputs are real numbers (instead of 0 or 1) representing the strength of ties.

To evaluate the proposed method, we conduct experiments on data collected from Sina Weibo[1], and experimental results demonstrate that our model outperform previous methods. Further, with the method to calculate tie strength, we investigate the phenomenon of social triads. In sociology a triad is a group of three people which is mostly investigated by microsociology. We examine social triads and our findings are consistent with Granovetter's conclusion [3].

The rest part of this paper is organized as follows. We review the related works in Section 2. In Section 3, we introduce the framework of measuring the strength of ties in social network. In Section 4, we show the experimental results of our work. Finally, in Section 5, we conclude and talk about future work.

2   Related Work

Some recent efforts have been made on the task of measuring the strength of ties. Most of them focus on supervised model which involves efforts on human annotation [2, 6]. However, measuring the strength of ties is only treated as a binary classification task by these works, i.e., to classify a relationship as strong or weak and they were unable to quantitatively estimate the strength between different users. Zhao et al. [13] proposed a novel framework for measuring the relationship strength between different users on various activity fields in online social networks. However, it only represented the strength of ties on various activity fields rather than the overall relationship strength.

The most relevant work was proposed by Xiang et al. [12] which is a latent variable model that measures the strength of ties by outputting continuous values without human annotation. User similarity and social interactions are exploited while constructing the model. Our model is the same with it in the aspects mentioned above. However, there are several differences: 1)we consider both user similarities and social interactions as decisive factor of tie strength while they only considered user similarities. 2)The way in which we utilize social interactions is rather different. In our work, we believe that social interactions directly impact on the strength of ties and it in turn impacts on the interaction probability of the user pairs in the future. In the previous work, researchers believe that the strength of ties impacts on the interaction probability of user pairs all the time. Experimental results confirm that our model outperforms the previous methods.

At present, there has been work concerning about triads. However these works either investigated the network structure using social triad theory [11, 1] or

---

[1] http://weibo.com. It is one of the most popular microblog systems in China, which is similar to twitter.

studied the triads composed of positive and negative edges(friends and ene-
mies) [8] in social media . To the best of our knowledge, our work is the first
that investigates the triads whose edges are measured by strength of ties.

3   Model

One key assumption is the theory of homophily [9]. It postulates that people
tend to form ties with other people who have similar characteristics. Moreover,
it is likely that the stronger the tie, the higher the similarity. Therefore, we can
model the strength of ties as a hidden effect of user similarities.

Another assumption is that the frequency of social interactions between user
pairs directly impact on their strength of ties. Since each user has a finite amount
of resources (e.g., time) to form and maintain relationships, it is likely that they
direct the sources(interacts) towards the ties that they deem more important [12].
As a result, user pairs with strong ties are more likely to interact in the future. On
the other hand, it can be inferred that user pairs with more social interactions in
the past are more likely to form strong ties between them. In this way, we model
the strength of ties as the hidden cause of users' future interaction probability
and the hidden effect of the users' historical social interactions. Graphical model
representation of the tie strength model is shown in Fig. 1.

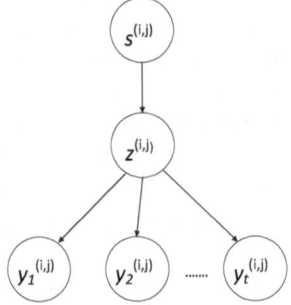

Fig. 1. Graphical model representation of the tie strength model

3.1   Model Specification

Formally, let $s^{(i,j)}$ be the attribute vectors of two individuals i and j which represents the user similarity and social interactions between them in the past. Let
$z^{(i,j)}$ be the hidden strength of ties between i and j. Interactions between them
are divided into t categories(e.g., in weibo system, interactions can be divided
into mention, retweet, comment, reply and so on), and let $y_m^{(i,j)}(1 \leq m \leq t)$
to be the probability of whether the m th kind of interaction would happen
in the last month. The model can be viewed as a hybrid of discriminative
and generative models: the upper part is discriminative ($p(Z|S)$), while $p(Y|Z)$

is generative. Our model represents the possible casual relationships among these variables by modeling the conditional dependencies, so it decomposes as follows:

$$P(z^{(i,j)}, y^{(i,j)}|s^{(i,j)}) = P(z^{(i,j)}|s^{(i,j)}) \prod_{m=1}^{t} P(y_m^{(i,j)}|s^{(i,j)}) \tag{1}$$

The general latent variable model of tie strength can be instantiated in different ways and we adopt the Gaussian distribution to model the conditional probability of the tie strength given the vectors of the similarity and historical interactions information. Then the dependency between $z^{(i,j)}$ and $s^{(i,j)}$ is :

$$P(z^{(i,j)}|s^{(i,j)}) = N(w^T s^{(i,j)}, v) \tag{2}$$

where $s^{(i,j)}$ is n-dimensional, w is a n-dimensional weight vector to be estimated, and v is the variance in Gaussian model which is configured to be 0.5.

The probability distribution of each $y_m^{(i,j)}$ given $z^{(i,j)}$ is considered to be conditionally independent. $y_m^{(i,j)}$ ($1 \leq m \leq t$) is the probability of whether the m th kind of interaction would happen in the last month. For example, $y_m^{(i,j)}$ may denote whether i has commented on j's statuses in the last month. We use logistic function to model the conditional probability of $y_m^{(i,j)}$ given $z^{(i,j)}$:

$$P(y_m^{(i,j)} = 1|z^{(i,j)}) = \frac{1}{1 + e^{-(\theta_m z^{(i,j)} + b)}} \tag{3}$$

where $\theta_m (1 \leq m \leq t)$ and b are the parameters estimated to adjust the probability of $y_m^{(i,j)}$ given $z^{(i,j)}$. b is defined to be zero for simplicity.

The data are represented as N user pair samples, denoted by $D = \{(i_1, j_1), (i_2, j_2)....(i_N, j_N)\}$. During training, $y_m^{(i,j)}$ and $s^{(i,j)}$ can be observed. Given all the observed variables, based on Eq.(1), the joint probability is as follows:

$$P(D|w,\theta)P(w,\theta) = P(y,z,w,\theta|s)$$

$$= \left(\prod_{(i,j)\in D} (P(z^{(i,j)}|s^{(i,j)}, w) \prod_{m=1}^{t} P(y_m^{(i,j)}|z^{(i,j)}, \theta_m)) \right) P(w)P(\theta)$$

$$\propto \left(\prod_{(i,j)\in D} (e^{-\frac{1}{2v}(w^T s^{(i,j)} - z^{(i,j)})^2} \sum_{m=1}^{t} \frac{e^{-(\theta_m^T z^{(i,j)} + b)}(1 - y_m^{(i,j)})}{1 + e^{-(\theta_m z^{(i,j)} + b)}} \cdot e^{-\frac{\lambda_w}{2} w^T w} \tag{4} \right)$$

$$\prod_{m=1}^{t} e^{-\frac{\lambda_\theta}{2}\theta_m^T \theta_m} \right)$$

Finally, to avoid over-fitting, we put L2 regularizers(i.e., λ_θ and λ_w) on the parameters w and θ, which can be regarded as Gaussian priors.

$z^{(i,j)}$, w and θ are unvisible, so the task can be modified as to find point estimates $\widehat{w}, \widehat{z}, \widehat{\theta}$ that maximize the likelihood $P(y, \widehat{z}, \widehat{w}, \widehat{\theta}|s)$

3.2   Inference

Taking the logarithm of Eq.(4), we get the log-likehood:

$$L = \log(P(y, z, w, \theta | s))$$

$$= \sum_{(i,j)\in D} \left(-\frac{1}{2v}(w^T s^{(i,j)} - z^{(i,j)})^2 \right) - \frac{\lambda_w}{2} w^T w - \sum_{m=1}^{t} \frac{\lambda_\theta}{2} \theta_m^T \theta_m + \tag{5}$$

$$\sum_{m=1}^{t} \left(-(1 - y_m^{(i,j)})(\theta_m z^{(i,j)} + b) - \log(1 + e^{-(\theta_m z^{(i,j)}+b)}) \right) + C$$

Note that both the quadratic terms and the logarithm of logistic function are concave. Since the sum of concave functions is concave, the function L is concave. Therefore, a gradient based method will allow us to optimize over the parameters $w, \theta_m (m = 1, 2, \ldots, t)$ and the latent variables $z^{(i,j)}, (i, j) \in D$, to find the maximum of L. We derive a coordinate ascent method for the optimization.

The coordinate-wise gradients are:

$$\frac{\partial L}{\partial z^{(i,j)}} = \frac{1}{v}(w^T s^{(i,j)} - z^{(i,j)}) + \sum_{m=1}^{t} \left(y_m^{(i,j)} - \frac{1}{1 + e^{-(\theta_m z^{(i,j)}+b)}} \right)\theta_m \tag{6}$$

$$\frac{\partial L}{\partial \theta_m} = \sum_{(i,j)\in D} \left(y_m^{(i,j)} - \frac{1}{1 + e^{-(\theta_m z^{(i,j)}+b)}} \right)z^{(i,j)} - \lambda_\theta \theta_t \tag{7}$$

$$\frac{\partial L}{\partial \theta_m} = \frac{1}{v} \sum_{(i,j)\in D} (z^{(i,j)} - w^T s^{(i,j)})s^{(i,j)} - \lambda_w w \tag{8}$$

A coordinate ascent optimization scheme will update w, $z^{(i,j)}$, and θ_m iteratively until convergence. For $z^{(i,j)}$ and θ_m, we use Newton-Raphson update in each iteration, where the 2nd order derivatives are given by:

$$\frac{\partial^2 L}{\partial (z^{(i,j)})^2} = -\frac{1}{v} - \prod_{m=1}^{t} \frac{\theta_m^2 e^{-(\theta_m^T z^{(i,j)}+b)}}{(1 + e^{-(\theta_m z^{(i,j)}+b)})^2} \tag{9}$$

$$\frac{\partial^2 L}{\partial \theta_t \partial \theta_t^T} = -\prod_{(i,j)\in D} \frac{(z^{(i,j)})^2 e^{-(\theta_m^T z^{(i,j)}+b)}}{(1 + e^{-(\theta_m z^{(i,j)}+b)})^2} - \lambda_\theta I \tag{10}$$

For w, the root of Eq.(8) can be found analytically as in usual ridge regression:

$$W = (\lambda_w I + S^T S)^{-1} S^T Z \tag{11}$$

where $S = (s^{(i_1,j_1)}, s^{(i_2,j_2)} \ldots s^{(i_N,j_N)})^T$, and $Z = (z^{(i_1,j_1)}, z^{(i_2,j_2)} \ldots z^{(i_N,j_N)})^T$.

Implementing the model , given $y^{(i,j)}$ and $s^{(i,j)}$, we can gradually obtain the parameter w, θ and $z^{(i,j)}$ (continuous). So the problem can be solved.

Algorithm 1. The process of the Newton-Raphson update

1: **while** not converged: **do**
2:    **for** $m = 1, 2...t$: **do**
3:       update θ_m based on Eq.(7) and Eq.(10).
4:    **end for**
5:    **for** $(i, j) \in D$: **do**
6:       update $z^{(i,j)}$ based on Eq.(6) and Eq.(9).
7:    **end for**
8:    update W based on Eq.(11).
9: **end while**
10: return Z, W, θ.

4 Experiment

4.1 Model Evaluation

The experiments are conducted on a real-world dataset crawled from Sina Weibo using API[2]. Since two-way communication are mostly be observed between bi-followers[3], we focus on ties between bi-followers in the experiments. We recruits several people to label the category of ties(strong or weak) between them and their bi-followers. The criterion of human annotation is based on the definition of strength of ties proposed by Granovetter [4] talked about in Section 1. Then we crawl data, extract features, and apply our model to the dataset. Meanwhile, the human annotation results are not used while calculating the strength of ties.

To evaluate the model, we use the continuous value obtained to identify whether a tie is strong or weak. The accuracy and AUC of the classification task are used to evaluate the model since there are no other criterion suitable for it. Several baselines are compared with our model. One is the model proposed by Xiang et al. [12] referred as UBL model here. One uses social interaction features to classify while the other exploits both social interaction and user similarity features. They are referred as UC model and UIC model respectively while our model is referred as UIBL model. The experiment results are shown in Fig. 2(a).

Models	Accuracy	AUC
UIBL	**0.754**	**0.638**
UBL	0.695	0.545
UIC	0.714	0.609
UC	0.697	0.571

(a) Evaluation results

	Sports	Technology	Entertainment
Nodes	9,798	23,358	27,728
Edges	176,276	811,410	991,759
Triads	688,224	4,906,228	4,759,689

(b) Dataset statistics

Fig. 2. Experiment information

From the result, it can be seen that our model outperforms the others on both criterions. Despite of the fact that our model and UBL model are quite similar[4]

---

[2] http://open.weibo.com/.
[3] Your bi-followers follow you and you follow them as well.
[4] We have discussed about the differences and common point in Section 2.

to some extent, our model increases the accuracy of it by 6 percent and achieves much better in AUC mainly for the effect of historical social interactions on tie strength that has been into account. Although UIC model has exploited all the features of our model as well as the results of human annotation, it is still beat by our model on both criterions. We explain it for that theory of homophily and essence of tie strength are taken into account while proposing our model. In other words, we exploited the inner relation between features and tie strength rather than just training based on the features and training set.

4.2   Statistically Analysis on Social Triads

Relations between social media sites often reflect a mixture of friendly and antagonistic interactions [7] while in social network, it often reflects a mixture of strong and weak ties. Different from triads signed in social media [8], we classify edges of triads based on the strength of ties. The edge(tie) in social triads can be classified to three categories, strong,weak and atypical[5]. Triads with atypical ties are not considered in experiments. In this way, social triads composed of strong and weak ties can be classified into four categories shown in Fig. 3.

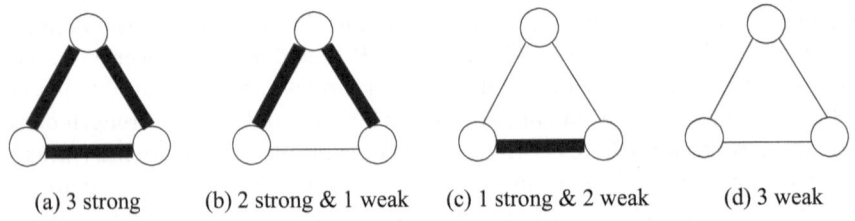

(a) 3 strong   (b) 2 strong & 1 weak   (c) 1 strong & 2 weak   (d) 3 weak

Fig. 3. Four types of social triads

We use snowball sampling to crawl the data from Sina Weibo. We randomly pick up several celebrities as seed , utilize snowball sampling to crawl all their bi-followers and repeat it several times. Since users are more likely to be friends with people of common interests, the datasets can be seen as communities composed of celebrities in one field to some extent. By choosing celebrities from Sports, Technology and Entertainment field as seed respectively, we obtain three datasets of celebrities in different fields which are referred to as dataset Sports, Technology and Entertainment. Statistics of them are shown in Fig. 2(b).

We conduct several experiments to reveal the phenomena behind social triads. Significance level(α) are defined to classify between strong , weak ties and atypical ties. It is denoted by k%($0 \leq k \leq 100$). We sort ties by their strength in ascending order, and considere that the first $k\%$ ties are strong ties while the last $k\%$ ones are weak ties. The others are atypical ties. In this way, the strong

---

[5] Considering that it is rather hard to define the boundary between strong ties and weak ties, we add another type of tie which is not significant enough to be a strong tie or weak tie and refer it as atypical tie. They are not considered to be statistically meaningful here.

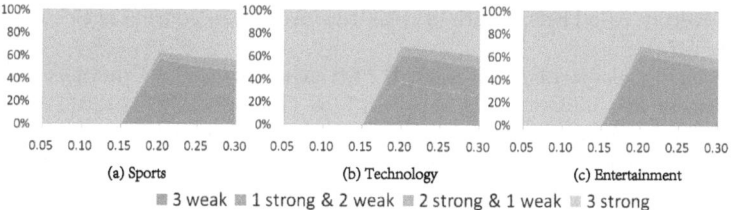

■ 3 weak ■ 1 strong & 2 weak ■ 2 strong & 1 weak ■ 3 strong

Fig. 4. The proportions of four triads change with the level of significance, in data set Sports, Technology and Entertainment, respectively. The horizontal ordinate here is the significance level.

ties and weak ties we defined here are significantly strong or weak ties when α is small. In the experiments, we gradually take k as 5, 10, 15, 20, 25, 30 for study.

Fig. 4 shows changes of triads in proportion when α varies. They appear to be similar for different datasets. When α is less than 0.15, there are only triads with three strong ties since few weak ties make up triads then. When α increases from 0.15 to 0.30, triads with three strong ties remains the most. So we can infer that triads with three strong ties are more likely to form. Meanwhile, the number of triads with two strong and one weak tie is much smaller which matches with the theory of sociology that your friend's friend are more likely to be your friend.

5   Conclusions and Future Work

In this paper, through leveraging user similarities and social interactions, we propose a latent variable model to calculate a continuous value that measures the strength of ties. Our experiments show that using this model to identify whether a tie is strong or weak gives rise to better performance than baselines. Further, we utilize our method to measure the strength of ties among a large set of microblogging users, and conduct statistical analysis on triads. We find that comparing to other types of triads, the one with three significantly strong ties are more likely to be created. It is consistent with Granovetter's conclusion, which confirms the effectiveness of our model from an alternative point of view.

In the future, we will consider about applying it to different social networks, constructing bigger data set and proposing a more effective model.

References

[1] Cartwright, D., Harary, F.: Structural balance: a generalization of heider's theory. Psychological Review 63(5), 277 (1956)
[2] Gilbert, E., Karahalios, K.: Predicting tie strength with social media. In: Proceedings of the 27th International Conference on Human Factors in Computing Systems, pp. 211–220. ACM (2009)
[3] Granovetter, M.: The strength of weak ties: A network theory revisited. Sociological Theory 1(1), 201–233 (1983)

 [4] Granovetter, M.: The strength of weak ties. American Journal of Sociology, 1360–1380 (1973)
 [5] Harrison, F., Sciberras, J., James, R.: Strength of social tie predicts cooperative investment in a human social network. PloS One 6(3), e18338 (2011)
 [6] Kahanda, I., Neville, J.: Using transactional information to predict link strength in online social networks. In: Proceedings of the Third International Conference on Weblogs and Social Media, ICWSM (2009)
 [7] Leskovec, J., Huttenlocher, D., Kleinberg, J.: Predicting positive and negative links in online social networks. In: Proceedings of the 19th International Conference on World Wide Web, pp. 641–650. ACM (2010a)
 [8] Leskovec, J., Huttenlocher, D., Kleinberg, J.: Signed networks in social media. In: Proceedings of the 28th International Conference on Human Factors in Computing Systems, pp. 1361–1370. ACM (2010b)
 [9] McPherson, M., Smith-Lovin, L., Cook, J.: Birds of a feather: Homophily in social networks. Annual Review of Sociology, 415–444 (2001)
[10] Montgomery, J.: Job search and network composition: Implications of the strength-of-weak-ties hypothesis. American Sociological Review, 586–596 (1992)
[11] Weber, C., Weber, B.: Exploring the antecedents of social liabilities in cvc triads–a dynamic social network perspective. Journal of Business Venturing 26(2), 255–272 (2011)
[12] Xiang, R., Neville, J., Rogati, M.: Modeling relationship strength in online social networks. In: Proceedings of the 19th International Conference on World Wide Web, pp. 981–990. ACM (2010)
[13] Zhao, X., Li, G., Yuan, J., et al.: Relationship strength estimation for online social networks with the study on facebook. Neurocomputing (2012)

Finding Diverse Friends in Social Networks

Syed Khairuzzaman Tanbeer and Carson Kai-Sang Leung*

Department of Computer Science, University of Manitoba, Canada
kleung@cs.umanitoba.ca

Abstract. Social networks are usually made of users linked by friendship, which can be dependent on (or influenced by) user characteristics (e.g., connectivity, centrality, weight, importance, activity in the networks). Among many friends of these social network users, some friends are more diverse (e.g., more influential, prominent, and/or active in a wide range of domains) than other friends in the networks. Recognizing these diverse friends can provide valuable information for various real-life applications when analyzing and mining huge volumes of social network data. In this paper, we propose a tree-based mining algorithm that finds diverse friends, who are highly influential across multiple domains, in social networks.

1 Introduction and Related Works

Social networks [9,11] are made of social entities (e.g., users) who are linked by some specific types of relationships (e.g., friendship, common interest, kinship). Facebook, Google+, LinkedIn, Twitter and Weibo [13,16] are some examples of social networks. Within these networks, a user f_i usually can create a personal profile, add other users as friends, endorse their skills/expertise, exchange messages among friends. These social networks may consist of thousands or millions of users, and each user f_i can have different number of friends. Among them, some are more important or influential than others [2,6,7,15,17].

Over the past few years, several data mining techniques [5,8,12,14] have been developed to help users extract implicit, previously unknown, and potentially useful information about the important friends. Recent works on social network mining include the discovery of strong friends [3] and significant friends [10] based on the degree of one-to-one interactions (e.g., based on the number of postings to a friend's wall).

However, in some situations, it is also important to discover users who (i) are influential in the social networks, (ii) have high level of expertise in some domains, and/or (iii) have diverse interest in multiple domains. In other words, users may want to find important friends based on their influence, prominence, and/or diversity. For instance, some users may be narrowly interested in one specific domain (e.g., computers). Other users may be interested in a wide range of domains (e.g., computers, music, sports), but their expertise level may vary from one domain to another (e.g., a user f_i may be a computer expert but only a beginner in music).

---

* Corresponding author.

Y. Ishikawa et al. (Eds.): APWeb 2013, LNCS 7808, pp. 301–309, 2013.
© Springer-Verlag Berlin Heidelberg 2013

Table 1. Prominence values & lists of interest groups

(a) Prominence of friends

Friend (f_i)	Prominence $Prom(f_i)$		
	Domain D_1	Domain D_2	Domain D_3
Amy	0.45	0.60	0.50
Bob	0.90	0.70	0.30
Cathy	0.20	0.60	0.70
Don	0.30	0.50	0.40
Ed	0.50	0.40	0.45
Fred	0.42	0.24	0.70
Greg	0.57	0.10	0.20

(b) Lists of interest groups in F_{SN}

Domain	Interest-group list L_j
D_1	$L_1 = \{Amy, Bob, Don\}$
	$L_2 = \{Cathy, Don\}$
	$L_3 = \{Amy, Bob\}$
D_2	$L_4 = \{Bob, Greg\}$
	$L_5 = \{Bob, Cathy, Don\}$
	$L_6 = \{Cathy, Ed\}$
	$L_7 = \{Bob, Cathy, Ed\}$
D_3	$L_8 = \{Amy, Cathy, Ed\}$
	$L_9 = \{Amy, Fred\}$
	$L_{10} = \{Amy, Cathy\}$

In this paper, one of our *key contributions* is an efficient tree-based algorithm called **Div-growth** for mining diverse friends from social networks. Div-growth takes into account multiple properties (e.g., influence, prominence, and/or diversity) of friends in the networks. Another *key contribution* is a prefix-tree based structure called **Div-tree** for capturing the social network data in a memory-efficient manner. Once the Div-tree is constructed, Div-growth computes the diversity of users based on both their influence and prominence to mine diverse groups of friends.

The remainder of this paper is organized as follows. We introduce the notion of diverse friends in the next section. Section 3 introduces our Div-growth algorithm, which mines diverse friends from our Div-tree. Experimental results are reported in Section 4; conclusions are presented in Section 5.

2   Notion of Groups of Diverse Friends

Consider a social network on three different domains (domains D_1, D_2, D_3) and seven individuals (*Amy, Bob, Cathy, Don, Ed, Fred & Greg*) with prominence values in each domain, as shown in Table 1(a). Each *domain* represents a sub-category (e.g., sports, arts, education) of interest. The **prominence value** of an individual reveals his level of expertise (e.g., importance, weight, value, reputation, belief, position, status, or significance) in a domain. In other words, the prominence value indicates how important, valued, significant, or well-positioned the individual is in each domain. The prominence value can be measured by using a common scale, which could be (i) specified by users or (ii) automatically calculated based on some user-centric parameters (e.g., connectivity, centrality, expertise in the domain, years of membership in the domain, degree of involvement in activities in the domain, numbers of involved activities in the domain). In this paper, the prominence value is normalized into the range $(0, 1]$. As the same individual may have different levels of expertise in different domains, his corresponding prominence value may vary from one domain to another. For example, prominence value $Prom_{D_1}(Amy)$ of *Amy* in domain D_1 is 0.45, which (i) is different from $Prom_{D_2}(Amy)=0.60$ and (ii) is higher than $Prom_{D_1}(Cathy)=0.20$ (implying that *Amy* is more influential than *Cathy* in D_1).

Consistent with existing settings of a social network [3,5,10], let $F = \{f_1, f_2, \ldots, f_m\}$ be a set of individuals/friends in a social network. An *interest-group list* $L \subseteq F$ is a list of individuals who are connected as friends due to some common interests. Let $G = \{f_1, f_2, \ldots, f_k\} \subseteq F$ be a **group of friends** (i.e., friend group) with k friends. Then, $Size(G) = k$, which represents the number of individuals in G. A *friend network* $F_{SN} = \{L_1, L_2, \ldots, L_n\}$ is the set of all n interest-group lists in the entire social network. These lists belong to some domains, and each domain contains at least one list. The set of lists in a particular domain D is called a *domain database* (denoted as F_D). Here, we assume that there exists an interest-group list in every domain. The *projected list* F_D^G of G in F_D is the set of lists in F_D that contains group G. The frequency $Freq_D(G)$ of G in F_D indicates the number of lists L_j's in F_D^G, and the frequencies of G in multiple domains are represented as $Freq_{D_{1,2,\ldots,d}}(G) = \langle Freq_{D_1}(G), Freq_{D_2}(G), \ldots, Freq_{D_d}(G)\rangle$.

Example 1. Consider F_{SN} shown in Table 1(b), which consists of $n=10$ interest-group lists $L_1, \ldots, L_{10}$ for $m=7$ friends in Table 1(a). Each row in the table represents the list of an interest group. These 10 interest groups are distributed into $d=3$ domains D_1, D_2 and D_3. For instance, $F_{D_1} = \{L_1, L_2, L_3\}$. For group $G = \{Cathy, Ed\}$, its $Size(G)=2$. As its projected lists on the 3 domains are $F_{D_1}^G=\emptyset$, $F_{D_2}^G=\{L_6, L_7\}$ and $F_{D_3}^G=\{L_8\}$, its frequencies $Freq_{D_{1,2,3}}(G) = \langle 0, 2, 1\rangle$. □

Definition 1. The **prominence value** $Prom_D(G)$ of a friend group G in a single domain D is defined as the average of all prominence values for all the friends in G: $Prom_D(G) = \frac{\sum_{i=1}^{Size(G)} Prom_D(f_i)}{Size(G)}$. Prominence values of G in multiple domains are represented as $Prom_{D_{1,2,\ldots,d}}(G) = \langle Prom_{D_1}(G), Prom_{D_2}(G), \ldots, Prom_{D_d}(G)\rangle$. □

Definition 2. The **influence** $Inf_D(G)$ of a friend group G in a domain D in F_D is defined as the product of the prominence value of G in the domain D and its frequency in the domain database F_D, i.e., $Inf_D(G) = Prom_D(G) \times Freq_D(G)$. For multiple domains, $Inf_{D_{1,2,\ldots,d}}(G) = \langle Inf_{D_1}(G), Inf_{D_2}(G), \ldots, Inf_{D_d}(G)\rangle$. □

Definition 3. The **diversity** $Div(G)$ of a friend group G among all d domains in F_{SN} is defined as the average of all the influence values of G in all domains in the social network: $Div(G) = \frac{\sum_{j=1}^{d} Inf_{D_j}(G)}{d}$. □

Example 2. Revisit F_{SN} in Table 1(b). The prominence value of friend group $G = \{Cathy, Ed\}$ in $D_1 = \frac{Prom_{D_1}(Cathy)+Prom_{D_1}(Ed)}{Size(G)} = \frac{0.20+0.50}{2} = 0.35$. We apply similar computation and get $Prom_{D_{1,2,3}}(G) = \langle 0.35, \frac{0.60+0.40}{2}, \frac{0.70+0.45}{2}\rangle = \langle 0.35, 0.5, 0.575\rangle$. Recall from Example 1 that $Freq_{D_{1,2,3}}(G) = \langle 0, 2, 1\rangle$. So, the overall influence of G in all 3 domains can be calculated as $Inf_{D_{1,2,3}}(G) = \langle 0.35\times 0, 0.5\times 2, 0.575\times 1\rangle = \langle 0, 1, 0.575\rangle$. Thus, the diversity of G in these $d=3$ domains in F_{SN} is $Div(G)=\frac{0+1+0.575}{3}=0.525$. □

A group G of friends in a social network F_{SN} is considered **diverse** if its diversity value $Div(G) \geq$ the user-specified minimum threshold $minDiv$, which can be expressed as an absolute (non-negative real) number or a relative percentage (with respect to the size of F_{SN}). Given F_{SN} and $minDiv$, the research problem of **mining diverse friends from social networks** is to find every group G of friends having $Div(G) \geq minDiv$.

Example 3. Recall from Example 2 that $Div(\{Cathy, Ed\})$=0.525. Given (i) F_{SN} in Table 1(b) and (ii) the user-specified $minDiv$=0.5, group $G=\{Cathy, Ed\}$ is *diverse* because $Div(G)$=0.525 $\geq$ 0.5=$minDiv$. But, group $G'=\{Ed\}$ is *not* diverse because $Div(G') = \frac{(0.5 \times 0)+(0.4 \times 2)+(0.45 \times 1)}{3} = \frac{0+0.8+0.45}{3} = 0.417 < minDiv$. $\square$

3   Our Div-growth Algorithm for Mining Diverse Friends

When mining frequent patterns, the frequency/support measure [1,4] satisfies the downward closure property (i.e., all supersets of an infrequent patterns are infrequent). This helps reduce the search/solution space by pruning infrequent patterns, which in turn speeds up the mining process. However, when mining diverse friends, diversity does *not* satisfy the downward closure property. Recall from Example 3, group $G'=\{Ed\}$) is not diverse but its super-group $G=\{Cathy, Ed\}$) is diverse. As we cannot prune those groups that are not diverse, the mining of diverse friends can be challenging.

To handle this challenge, for each domain D, we identify the **(global) maximum prominence value** $GMProm_D$ among all friends. Then, for each friend f_i, we calculate an upper bound of the influence value $Inf_D^U(f_i)$ by multiplying $GMProm_D$ (instead of the actual $Prom_D(f_i)$) with the corresponding frequency $Freq_D(f_i)$. The upper bound of diversity value $Div^U(f_i)$ can then be computed by using $Inf_D^U(f_i)$.

Lemma 1. Let G be a group of friends in F_{SN} such that a friend $f_i \in G$. If $Div^U(f_i) < minDiv$, then $Div(G)$ must also be less than $minDiv$. $\square$

Example 4. Revisit F_{SN} in Table 1(b). Note that $GMProm_{D_1}$=0.90, $GMProm_{D_2}$= 0.70, and $GMProm_{D_3}$=0.70. Recall from Example 3 that $Freq_{D_{1,2,3}}(\{Ed\})=\langle 0, 2, 1\rangle$. Then, we can compute $Div^U(\{Ed\})=\frac{(0.90 \times 0)+(0.70 \times 2)+(0.70 \times 1)}{3}$=0.7 $\geq minDiv$. So, we do not prune $\{Ed\}$ to avoid missing its super-group $\{Cathy, Ed\}$, which is diverse. Similarly, we can compute $Div^U(\{Fred\})=\frac{(0.90 \times 0)+(0.70 \times 0)+(0.70 \times 1)}{3}$=0.23 $< minDiv$. Due to Lemma 1, we prune *Fred* as none of its super-groups can be diverse. $\square$

3.1   Phase 1: Constructing a Div-tree Structure

Given F_{SN} and $minDiv$, our proposed Div-growth algorithm constructs a Div-tree as follows. It first scans F_{SN} to calculate $Freq_{D_j}(f_i)$ for each friend f_i in each domain D_j. For each f_i, Div-growth then uses $GMProm_D$ to compute the upper bound of the diversity value $Div^U(f_i)$, which is used to prune groups of friends who are not potentially diverse. Every potentially diverse friend f_i, along with its $Freq_{D_{1,...,d}}(f_i)$, is stored in the header table.

Afterwards, Div-growth scans F_{SN} the second time to capture the important information about potentially diverse friends in a user-defined order in the Div-tree. Each tree node consists of (i) a friend name and (ii) its frequency counters for all d domains in the respective path. The basic construction process of a Div-tree is similar to that of the FP-tree [4]. A key difference is that, rather than using only a single frequency counter capturing either the maximum or average frequency for all domains (which may lead to loss of information), we use d frequency counters capturing the frequency for all d domains. See Example 5.

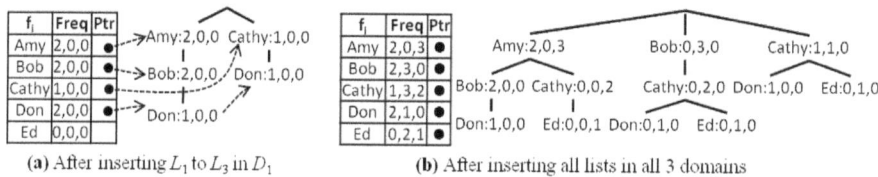

(a) After inserting L_1 to L_3 in D_1 (b) After inserting all lists in all 3 domains

Fig. 1. Construction of a Div-tree

Example 5. To construct a Div-tree for F_{SN} shown in Table 1(b) when $minDiv=0.5$, Div-growth scans F_{SN} to compute (i) $GMProm_{D_{1,2,3}} = \langle 0.9, 0.7, 0.7 \rangle$ for all $d=3$ domains, (ii) frequencies of all 7 friends in $d=3$ domains (e.g., $Freq_{D_{1,2,3}}(\{Amy\}) = \langle 2,0,3 \rangle$), (iii) upper bound of diversity values of all 7 friends (e.g., $Div^U(\{Amy\}) = \frac{(0.9 \times 2)+(0.7 \times 0)+(0.7 \times 3)}{3} = 1.3$ using $Inf^U_{D_{1,2,3}}(\{Amy\})$). Based on Lemma 1, we safely remove $Fred$ and $Greg$ having $Div^U(\{Fred\})=0.23$ and $Div^U(\{Greg\})=0.23$ both below $minDiv$ as their super-groups cannot be diverse. So, the header table includes only the remaining 5 friends—sorted in some order (e.g., lexicographical order of friend names)—with their $Freq_{D_{1,2,3}}(\{f_i\})$. To facilitate a fast tree traversal, like the FP-tree, the Div-tree also maintains horizontal node traversal pointers from the header table to nodes of the same f_i.

Div-growth then scans each $L_j \in F_{SN}$, removes any friend $f_i \in L_j$ having $Div^U(f_i) < minDiv$, sorts the remaining friends according to the order in the header table, and inserts the sorted list into the Div-tree. Each tree node captures (i) f_i representing the group G consisting of all friends from the root to f_i and (ii) its frequencies in each domain $Freq_{D_{1,2,3}}(G)$. For example, the rightmost node Ed:0,1,0 of the Div-tree in Fig. 1(b) captures $G=\{Cathy, Ed\}$ and $Freq_{D_{1,2,3}}(G)=\langle 0,1,0 \rangle$. Tree paths of common prefix (i.e., same friends) are shared, and their corresponding frequencies are added. See Figs. 1(a) and 1(b) for Div-trees after reading all interest-group lists in domain D_1 and the entire F_{SN}, respectively.   □

With this tree construction process, the size of the Div-tree for F_{SN} with a given $minDiv$ is observed to be bounded above by $\sum_{L_j \in F_{SN}} |L_j|$.

3.2 Phase 2: Mining Diverse Friend Groups

Once the Div-tree is constructed, Div-growth recursively mines diverse friend groups by building projected and conditional trees in a fashion similar to that of FP-growth [4].

Recall that $Div(G)$ computed based on $Prom_D(G)$ does *not* satisfy the downward closure property. To facilitate pruning, we use $GMProm_D(f_i)$ to compute $Div^U(f_i)$, which then satisfies the downward closure property. However, if $Div^U(G)$ was computed as an upper bound to super-group G of f_i, then it may overestimate diversity of G and may lead to false positives. To reduce the number of false positives, Div-growth uses the **local maximum prominence value** $LMProm_D(G) = \max_{f_i \in F_D^G}\{Prom_D(G)\}$ for the projected and conditional trees for G. See Lemma 2 and Example 6.

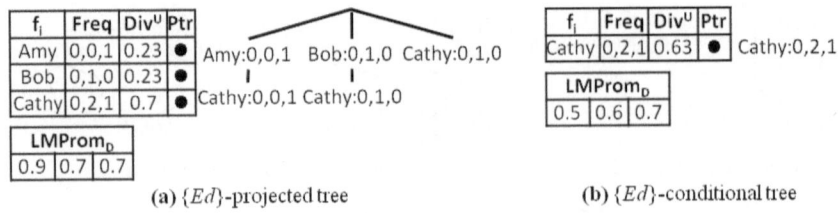

(a) $\{Ed\}$-projected tree

(b) $\{Ed\}$-conditional tree

Fig. 2. Tree-based mining of diverse friend groups

Lemma 2. *The diversity value of a friend group G computed based on $LMProm_D(G)$ is a tighter upper bound than $Div^U(G)$ computed based on $GMProm_D$.* □

Example 6. To mine potentially diverse friend groups from the Div-tree in Fig. 1(b) using $minDiv$ =0.5, Div-growth first builds the $\{Ed\}$-projected tree—as shown in Fig. 2(a)—by extracting the paths $\langle Amy, Cathy, Ed\rangle$:0,0,1, $\langle Bob, Cathy, Ed\rangle$:0,1,0 and $\langle Cathy, Ed\rangle$:0,1,0 from the Div-tree in Fig. 1(b). For $F_{D_{1,2,3}}^{Ed}=\{Amy, Bob, Cathy, Ed\}$, Div-growth also uses $LMProm_{D_{1,2,3}}(F_{D_{1,2,3}}^{Ed}) = \langle 0.9, 0.7, 0.7\rangle$ to compute the tightened $Div^U(G)$ such as tightened $Div^U(\{Amy, Ed\})=\frac{(0.9\times0)+(0.7\times0)+(0.7\times1)}{3}=0.23 < minsup$.

As $Div^U(\{Amy, Ed\})$ and $Div^U(\{Bob, Ed\})$ are both below $minsup$, Div-growth prunes Amy and Bob from the $\{Ed\}$-projected tree to get the $\{Ed\}$-conditional tree as shown in Fig. 2(b). Due to pruning, Div-growth recomputes $LMProm_{D_{1,2,3}}(F_{D_{1,2,3}}^{Ed})$ $=\langle 0.5, 0.6, 0.7\rangle$ and the tightened $Div^U(\{Cathy, Ed\})=\frac{(0.5\times0)+(0.6\times2)+(0.7\times1)}{3}=0.63$ for the updated $F_{D_{1,2,3}}^{Ed}=\{Cathy, Ed\}$. This completes the mining for $\{Ed\}$.

Next, Div-growth builds $\{Don\}$-, $\{Cathy\}$- & $\{Bob\}$-projected and conditional trees, from which potentially diverse friend groups can be mined. Finally, Div-growth computes the true diversity value $Div(G)$ for each of these mined groups to check if it is truly diverse (i.e., to remove all false positives). □

4   Experimental Results

To evaluate the effectiveness of our proposed Div-growth algorithm and its associated Div-tree structure, we compared them with a closely related *weighted* frequent pattern mining algorithm called Weight [18] (but it does not use different weights for individual items). As Weight was designed for frequent pattern mining (instead of social network mining), we apply those datasets commonly used in frequent pattern mining for a fair comparison: (i) IBM synthetic datasets (e.g., T10I4D100K) and (ii) real datasets (e.g., mushroom, kosarak) from the Frequent Itemset Mining Dataset Repository fimi.cs.helsinki.fi/data. See Table 2 for more detail. Items in transactions in these datasets are mapped into friends in interest-group lists. To reflect the concept of *domains*, we subdivided the datasets into several batches. Moreover, a random number in the range (0, 1] is generated as a prominence value for each friend in every domain.

All programs were written in C++ and run on the Windows XP operating system with a 2.13 GHz CPU and 1 GB main memory. The runtime specified indicates the total execution time (i.e., CPU and I/Os). The reported results

Table 2. Dataset characteristics

Dataset	#transactions	#items	maxL	avgTL	Density
mushroom	8,124	119	23	23.0	Dense
T10I4D100K	100,000	870	29	10.1	Sparse
kosarak	990,002	41,270	2498	8.1	Sparse

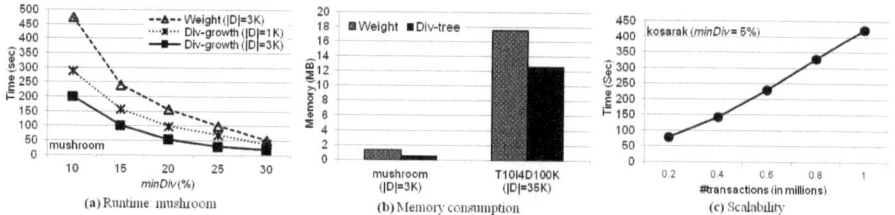

(a) Runtime: mushroom   (b) Memory consumption   (c) Scalability

Fig. 3. Experimental results

are based on the average of multiple runs for each case. We obtained consistent results for all of these datasets.

Runtime. First, we compared the runtime of Div-growth (which includes the construction of the Div-tree, the mining of potentially diverse friend groups from the Div-tree, and the removal of false positives) with that of Weight. Fig. 3(a) shows the results for a dense dataset (mushroom), which were consistent with those for sparse datasets (e.g., T10I4D100K). Due to page limitation, we omit the results for sparse datasets. Runtimes of both algorithms increased when mining larger datasets (social networks), more batches (domains), and/or with lower $minDiv$ thresholds. Between the two algorithms, our tree-based Div-growth algorithm outperformed the Apriori-based Weight algorithm. Note that, although FP-growth [4] is also a tree-based algorithm, it was *not* design to capture weights. To avoid distraction, we omit experimental results on FP-growth and only show those on Weight (which captures weights).

Compactness of the Div-tree. Next, we evaluated the memory consumption. Fig. 3(b) shows the amount of memory required by our Div-tree for capturing the content of social networks with the lowest $minDiv$ threshold (i.e., without removing any friends who were not diverse). Although this simulated the worst-case scenario for our Div-tree, Div-tree was observed (i) to consume a reasonable amount of memory and (ii) to require less memory than Weight (because our Div-tree is compact due to the prefix sharing).

Scalability. Then, we tested the scalability of our Div-growth algorithm by varying the number of transactions (interest-group lists). We used the kosarak dataset as it is a huge sparse dataset with a large number of distinct items (individual users). We divided this dataset into five portions, and each portion is subdivided into multiple batches (domains). We set $minDiv=5\%$ of each portion. Fig. 3(c) shows that, when the size of the dataset increased, the runtime also increased proportionally implying that Div-growth is scalable.

Additional Evaluation. So far, we have evaluated the *efficiency* (e.g., runtime, compactness or memory consumption, as well as scalability) of our Div-growth algorithm. Experimental results show that Div-growth is time- and space-efficient as well as scalable. As ongoing work, we plan to evaluate the *quality* (e.g., precision) of Div-growth in finding diverse friend groups. Moreover, for a fair comparison with Weight, we have used those datasets that are commonly used in frequent pattern mining. As ongoing work, we plan to evaluate Div-growth using real-life social network datasets.

5   Conclusions

In this paper, we (i) introduced a new notion of *diverse friends* for social networks, (ii) proposed a compact tree structure called *Div-tree* to capture important information from social networks, and (iii) designed a tree-based mining algorithm called *Div-growth* to find diverse (groups of) friends from social networks. Diversity of friends is measured based on their prominence, frequency and influence in different domains on the networks. Although diversity does not satisfy the downward closure property, we managed to address this issue by using the global and local maximum prominence values of users as upper bounds. Experimental results showed that (i) our Div-tree is compact and space-effective and (ii) our Div-growth algorithm is fast and scalable for both sparse and dense datasets. As ongoing work, we conduct more extensive experimental evaluation to measure other aspects (e.g., precision) of our Div-growth algorithm in finding diverse friends. We also plan to (i) design a more sophisticated way to measure influence and (ii) incorporate other computational metrics (e.g., popularity, significance, strength) with prominence into our discovery of useful information from social networks.

Acknowledgements. This project is partially supported by NSERC (Canada) and University of Manitoba.

References

1. Agrawal, R., Srikant, R.: Fast algorithms for mining association rules in large databases. In: VLDB 1994, pp. 487–499 (1994)
2. Anagnostopoulos, A., Kumar, R., Mahdian, M.: Influence and correlation in social networks. In: ACM KDD 2008, pp. 7–15 (2008)
3. Cameron, J.J., Leung, C.K.-S., Tanbeer, S.K.: Finding strong groups of friends among friends in social networks. In: IEEE DASC/SCA 2011, pp. 824–831 (2011)
4. Han, J., Pei, J., Yin, Y.: Mining frequent patterns without candidate generation. In: ACM SIGMOD 2000, pp. 1–12 (2000)
5. Jiang, F., Leung, C.K.-S., Tanbeer, S.K.: Finding popular friends in social networks. In: CGC/SCA 2012, pp. 501–508. IEEE (2012)
6. Kamath, K.Y., Caverlee, J., Cheng, Z., Sui, D.Z.: Spatial influence vs. community influence: modeling the global spread of social media. In: ACM CIKM 2012, pp. 962–971 (2012)

7. Lee, W., Leung, C.K.-S., Song, J.J., Eom, C.S.-H.: A network-flow based influence propagation model for social networks. In: CGC/SCA 2012, pp. 601–608. IEEE (2012)
8. Lee, W., Song, J.J., Leung, C.K.-S.: Categorical data skyline using classification tree. In: Du, X., Fan, W., Wang, J., Peng, Z., Sharaf, M.A. (eds.) APWeb 2011. LNCS, vol. 6612, pp. 181–187. Springer, Heidelberg (2011)
9. Leung, C.K.-S., Carmichael, C.L.: Exploring social networks: a frequent pattern visualization approach. In: IEEE SocialCom 2010, pp. 419–424 (2010)
10. Leung, C.K.-S., Tanbeer, S.K.: Mining social networks for significant friend groups. In: Yu, H., Yu, G., Hsu, W., Moon, Y.-S., Unland, R., Yoo, J. (eds.) DASFAA Workshops 2012. LNCS, vol. 7240, pp. 180–192. Springer, Heidelberg (2012)
11. Peng, Z., Wang, C., Han, L., Hao, J., Ou, X.: Discovering the most potential stars in social networks with infra-skyline queries. In: Sheng, Q.Z., Wang, G., Jensen, C.S., Xu, G. (eds.) APWeb 2012. LNCS, vol. 7235, pp. 134–145. Springer, Heidelberg (2012)
12. Sachan, M., Contractor, D., Faruquie, T.A., Subramaniam, L.V.: Using content and interactions for discovering communities in social networks. In: ACM WWW 2012, pp. 331–340 (2012)
13. Schaal, M., O'Donovan, J., Smyth, B.: An analysis of topical proximity in the twitter social graph. In: Aberer, K., Flache, A., Jager, W., Liu, L., Tang, J., Guéret, C. (eds.) SocInfo 2012. LNCS, vol. 7710, pp. 232–245. Springer, Heidelberg (2012)
14. Sun, Y., Barber, R., Gupta, M., Aggarwal, C.C., Han, J.: Co-author relationship prediction in heterogeneous bibliographic networks. In: ASONAM 2011, pp. 121–128. IEEE (2011)
15. Tanbeer, S.K., Leung, C.K.-S., Cameron, J.J.: DIFSoN: discovering influential friends from social networks. In: CASoN 2012, pp. 120–125. IEEE (2012)
16. Yang, X., Ghoting, A., Ruan, Y., Parthasarathy, S.: A framework for summarizing and analyzing twitter feeds. In: ACM KDD 2012, pp. 370–378 (2012)
17. Zhang, C., Shou, L., Chen, K., Chen, G., Bei, Y.: Evaluating geo-social influence in location-based social networks. In: ACM CIKM 2012, pp. 1442–1451 (2012)
18. Zhang, S., Zhang, C., Yan, X.: Post-mining: maintenance of association rules by weighting. Information Systems 28(7), 691–707 (2003)

Social Network User Influence Dynamics Prediction

Jingxuan Li[1], Wei Peng[2], Tao Li[1], and Tong Sun[2]

[1] School of Computer Science, Florida International University
{jli003,taoli}@cs.fiu.edu
[2] Xerox Innovation Group, Xerox Corporation
{Wei.Peng,Tong.Sun}@xerox.com

Abstract. Identifying influential users and predicting their "network impact" on social networks have attracted tremendous interest from both academia and industry. Most of the developed algorithms and tools are mainly dependent on the static network structure instead of the dynamic diffusion process over the network, and are thus essentially based on descriptive models instead of predictive models. In this paper, we propose a dynamic information propagation model based on *Continuous-Time Markov Process* to predict the influence dynamics of social network users, where the nodes in the propagation sequences are the users, and the edges connect users who refer to the same *topic* contiguously on time. Our proposed model is compared with two baselines, including a well-known time-series forecasting model – Autoregressive Integrated Moving Average and a widely accepted information diffusion model – Independent cascade. Experimental results validate our ideas and demonstrate the prediction performance of our proposed algorithms.

Keywords: Social media, Influential users, Dynamic information diffusion, Continuous-Time Markov Process.

1 Introduction

1.1 Identifying Influential Users

One of most popular topics of the social network analysis is identifying influential users and their "network impact". More interestingly, knowing the influence of users and being able to predict it can be leveraged for many applications. The most famous application to researchers and marketers is the viral marketing [5], which aims at targeting a group of influential users to maximize the marketing campaign ROI (Return of Investment). Other interesting applications include search [1], expertise/tweets recommendation [22], trust/information propagation [9], and customer handling prioritization in social customer relationship management.

1.2 Limitations of Current Research Efforts

There is one main limitation of current research efforts on identifying influential users in social network analysis: lacking of an effective approach for modeling, predicting, and measuring the influence.

Y. Ishikawa et al. (Eds.): APWeb 2013, LNCS 7808, pp. 310–322, 2013.

Influence Models and Measures. Currently most applications and tools compute user influence based on their static network properties, such as, the number of friends/followers in the social graph, the number of posted tweets/received retweets/mentions/replies in the activity graph, or users' centrality (e.g. PageRank, Betweeness-centrality, etc.). All of them make implicit assumptions about the underlying dynamic process that social network users can only influence their followers/friends, with equal impact strength, equal "infected" probabilities, and at the same propagation rate; or the influence is assumed to take a "random walk" on the corresponding static network.

A few works investigate adoption behaviors of social network users as the dynamic influence propagation [1] or diffusion process [18]. The adoption behaviors refer to some activities or topics (tweets, products, Hashtags, URLs, etc.) shared among users implicitly and explicitly such as users forwarding a message to their friends, recommending a product to others, joining some groups with the similar musical favor, and posting messages about the same topics, etc. According to the diffusion theory, the information cascades from social leaders to followers. In most diffusion models, propagators have certain probabilities to influence their receivers, and the receivers also have certain thresholds to be influenced. Finding the social leaders or the users who can maximize the influence coverage in the network is the major goal of most diffusion models. Targeting these influential users is the idea behind viral marketing, aiming to activate a chain-reaction of influence, with a small marketing cost.

Some drawbacks of existing social network influence models based on either static networks or the "influence maximization" diffusion process are: (1) Most existing models are descriptive models rather than predictive models. For example, the number of friends or the centrality score of a given user describes his/her underlying network connectivity. The number of tweets that a user posted or get retweeted indicates the trust/insterest that his/her followers have on his/her tweets. All these measures/models are descriptive and very few models are able to predict users' future influence. (2) Existing "influence maximization" diffusion process is often modeled by discrete-time models such as Independent Cascade Model or Linear Threshold Model. Using discrete-time models to model diffusion process in continuous time is very computationally expensive. The time step t needs to be defined as a very small number. Thus parameter estimation (e.g. activation probability) tends to come out very small numbers as well, and diffusion process simulation could take forever to run.

1.3   Content of the Paper

The aforesaid limitation motivates our study on social network user influence and dynamics prediction in this paper. In particular,

To address the limitation, we propose a dynamic information diffusion model based on *Continuous-Time Markov Process* (CTMP) to predict the influence dynamics of social network users. CTMP assumes that the number of activations

---

[1] We use "information/influence propagation", "information/influence diffusion", and "information cascade", interchangeably to represent the same concept.

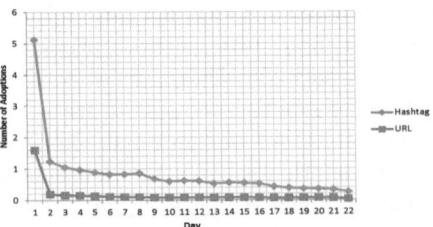

Fig. 1. The average number of topic adoptions over the time on our Twitter dataset. The x axis indicates the time, while the y axis denotes the average number of the summation of the adoption frequency of all topics at the specific time point.

from a given node is following an exponential distribution over the time, which can be observed in the real-world data [12]. Figure 1 shows that the average number of topic adoptions decreases exponentially over the time. Hashtags receive more adoptions compared with URLs, and the number of Hashtag adoptions decreases more slowly. Furthermore, transition rates q are calculated and treated as the transition probabilities (or activation probability) of the embedded Markov chain in CTMP. Then the transition probability $P(t)$ can be computed from q, given any time t. In this paper, the nodes in the propagation sequences are the users, and the edges connect users who refer to the same topic contiguously on time. Topics here particularly refer to Hashtags (expressed as # followed by a word) and short URLs (e.g. bit.ly) on twitter, which is one of the most popular microblog services, was launched since July 13, 2006. Hashtags and URLs are both unique identifiers tagging distinct tweets with certain "topic" labels. We regard the temporal sequences of Hashtags and URLs as the diffusion paths, where the topics are reposted subsequently. Although retweeting is not included in our paper as a diffusion approach, it is implicitly considered because the retweets would usually contain the same Hashtags and URLs as in the original tweets. Essentially marketers hope by targeting a small set of influential users not only can their marketing campaign messages receive many retweets, but also can the influential users create topics about their products and stimulate a lot of tweets about them continuously. Our experimental results on a large-scale twitter dataset demonstrates a promising prediction performance on estimating the number of influenced users within a given time.

The contributions of this paper can be summarized as: (1) A dynamic information diffusion model based on the *Continuous-Time Markov Process* is proposed to model and predict the influence coverage of social network users (See Section 4). (2) The proposed model is compared with two other baseline models. In addition, the experiment is conducted over a large-scale twitter dataset, which validates the capabilities of our model in the real world.

2   Related Work

Some recent research aims at studying the content in the social network, e.g., Petrovic et al [16] studied the problem of deciding whether a tweet can be retweeted, Tsur and Rappoport [23] did content based prediction using hash-tags provided by microblogging websites. Meanwhile, a number of works have addressed the matter of user influence on social network. Many of them regard user influence as their network metrics. Kwak et al. [12] found the difference between three influence measures: number of followers, page-rank, and number of retweets. Cha et al. [4] also compared these three measures, and discovered that the number of retweets and the number of mentions are correlated well with each other while the number of friends does not correlated well with the other two measures. Their hypothesis is that the number of followers of user may not be a good influence measure. Weng et al. [24] regarded the central users of each topic-sensitive subnetwork of the follower-and-followee graph as influential users. Other work such as [7], mined users influence from their static network properties derived from either their social graphs or activity graphs.

Various dynamic diffusion models have also been proposed to discover the influential users. They are shown to outperform influence models based on static network metrics [17,7]. A lot of work in this direction are devoted to *viral marketing*. Domingos and Richardson [5,17] were the first to mine customer network values for 'influence maximization' for viral marketing in data mining domain. The proposed approach is a probability optimization method with the hill-climbing heuristics. Kemper et al. [11] further showed that a natural greedy strategy can achieve 63% of optimal for two fundamental discrete-time propagation models - Independent Cascade Model (IC) and Linear Threshold Model (LT) . Many diffusion models assume the influence probabilities on the edges or the probability of acceptance on the nodes are given or randomly simulated. Goyal et al. [8] proposed to mine these probabilities by analyzing the past behavior of users. Saito et al. [19,20] extend IC model and LT model to incorporate asynchronous time delay. Model parameters including activation probabilities and continuous time delay are estimated by Maximum Likelihood. Our proposed diffusion model is different from the above discussed models: (1) We model the dynamic probabilities of edge diffusion and node threshold changing over the time, rather than computing the static probabilities. (2) Our model is a Continuous-Time diffusion model instead of a discrete-time one. Although Saito et al. also proposed Continuous-Time models, the fundamental diffusion process of their models are following LT and IC models. For example, in asynchronous IC, an active node can only infect one of its neighbors in one iteration, while our proposed models does not assume iterations so that nodes can be activated at any time without resetting the clock in the new iteration. Moreover, the models proposed by Saito et al. supposed only one initial active user and focused on model parameter estimation, not much on prediction. The experiments are evaluated on simulated data from some real network topology. Our proposed model estimates the model parameters from the real large-scale social network data, allows many initial active users asynchronously or simultaneously to influence other users, and predicts the real diffusion sizes in the future.

In addition, most of influence models are basically descriptive models instead of predictive models. Bakshy et al. [3] studied the diffusion tree of URLs on twitter, and train a regression tree model on a set of user network attributes, user past influence, and URL content to predict users' future influence. Our work is quite different from the work of Bakshy et al. in the following aspects : (1) They predict users average spreading size in the next month based on the data from the previous month. However, the dynamic nature of word-of-mouth marketing determines that the influence coverage vary over the time. Thus our work aims at predicting the spreading size of each individual user within a specific given date, so we can answer "what is the spreading size of user A within 2 hours, 1 day, or 1 month, etc.". (2) Their work is built on top of a regression model, while our work proposes a real-time stochastic model. The input and output from these two models are quite different. (3) Besides URLs diffusion, we also study the diffusion of Hashtags on twitter, which usually have longer lifetime.

Continuous-Time Markov Process (CTMP) has showed its applicability for various research topics, including simulation, economics, optimal control, genetics, queues, and etc.. The detailed description of CTMP can be found in the book by Norris [15]. Huang et al [10] adopted it to model the web user visiting patterns. Liu et al. [13] also utilized CTMP to model user web browsing patterns for ranking web pages. Song et al. [21] employed CTMP to mine document and movie browsing patterns for recommendation. To the best of our knowledge, our work is the first to construct influence diffusion model based on CTMP for spreading coverage prediction and user influence on social networks.

3   Influence Network and Influence Models

3.1   Influence Network

A social graph can be denoted as $G(V, E)$, where V represents social network users, and E is the set of edges/relations between users. The follower-followee graph is one type of social graphs, where the edges indicate following relations. *Activity graphs* are another type of social graphs, which are extracted from users tweeting behaviors, e.g., the tweet-retweet graph.

The above networks can be viewed as static networks, which do not demonstrate the dynamic propagation process over the time. In order to analyze how topics are passing on social networks progressively, we construct a temporal influence network by considering the continuous time. Given a Hashtag/URL (*topic*), a group of users can be ordered based on the time when they post this topic. As shown in Figure 2, user i is linked to user j if they post the same topic contiguously and user j follows/friend with user i. The number on the top of each arrow is the time taken to transfer a topic from a user to another user.

Definition 1 (temporal influence network). *The temporal influence network is $G(V, E, T(E))$, where $V = \{V_0, V_1, \cdots, V_n\}$ contains all users who posted at least one Hashtag or URL, $E = \{V_i \leftarrow V_j | V_i$ posted a topic earlier than $V_j\}$, where edges can be constrained to only exist between followers and followees or between friends. So the propagation is along the paths from followees to followers over continuous time. The function $T(V_i \leftarrow V_j) = \{t_{ij}^0, t_{ij}^1, \cdots, t_{ij}^l\}$. $t_{ij}^m \in \{t_{ij}^0, t_{ij}^1, \cdots, t_{ij}^l\}$ is the time difference between user i posting a topic and user j posting the same topic.*

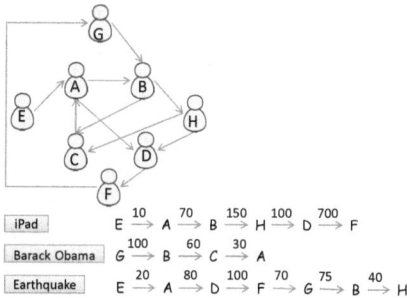

Fig. 2. The example of Temporal Influence Network construction

There can be multiple entries in $T(V_i \leftarrow V_j)$ since user i and user j can post the same set of topics or one topic at multiple times. Note that we aggregate all topics together to form this temporal influencer network in this paper.

4   Information Diffusion Model Based on Continuous-Time Markov Process

The descriptive influence models answer questions such as "How many followers that user A has?" and "How many followers post the topic 'ipad' after user A?", etc. In this section, we introduce our proposed predictive Information Diffusion Model based on Continuous-Time Markov Process, abbreviated as IDM-CTMP for convenience. IDM-CTMP is able to answer the following question, "In the next month, how many users would post the topic 'ipad' estimably if user A posts it now.", or even a harder question "In order to make a maximal number of people to talk about our product in the next week, who are the seed users we should target?". Note the influential users discovered by IDM-CTMP maximize not only the information *coverage*, but also the *rate* of information cascade given a certain period of time.

4.1   Model Formulation

A trending topic (a Hashtag/URL) is propagated by social network users within the temporal influence network defined in Definition 1. Suppose $X(t)$ denotes the user who posts a specific topic at time point t, $X = X(t), t \geq 0$ forms a Continuous-Time Markov Process (CTMP) [2], in which the user who will discuss this topic next only depends on the current user given the whole history of the topic propagation. Formally, this markov property can be defined by:

$$P_{ij}(t) = P\{X(t+\gamma) = j | X(\gamma) = i, X(\mu) = x(\mu), 0 \leq \mu < \gamma\} = P\{X(t+\gamma) = j | X(\gamma) = i\}, \quad (1)$$

where $P_{ij}(t)$ is the transition probability from i to j within time t, i is the current user who discusses the trending topic, j is the next user who posts the topic following i, and $x(\mu)$ denotes the history of the topic propagation before the time point γ. We assume that the transition probability $P_{ij}(t)$ does not depend on the actual starting time of the propagation process, thus the CTMP is time-homogeneous:

$$P_{ij}(t) = P\{X(t+\gamma) = j | X(\gamma) = i\} = P\{X(t) = j | X(0) = i\}. \qquad (2)$$

In order to estimate the diffusion size of user i given a pre-defined time window t, we need to compute the transition probability from user i to all the others, then determine the number of users being affected by i at the end of t. The diffusion size of user i over t based on CTMP can be defined as

$$DS_{i,t} = \sum_j P_{ij}(t) \cdot n_i, \qquad (3)$$

where n_i is the number of times that user i occurs at time t. It can be estimated by supposing that it linearly increments on t. However, it is impractical to estimate the transition probability matrix $P(t)$ with infinite possible t. Thus instead of estimating $P(t)$ directly, we calculate the transition rate matrix Q, and then $P(t)$ can be estimated from Q.

Before jumping into the details of Q matrix and $P(t)$ matrix, a simple example is presented in Figure 3 to illustrate the procedure of predicting the diffusion size of a particular user.

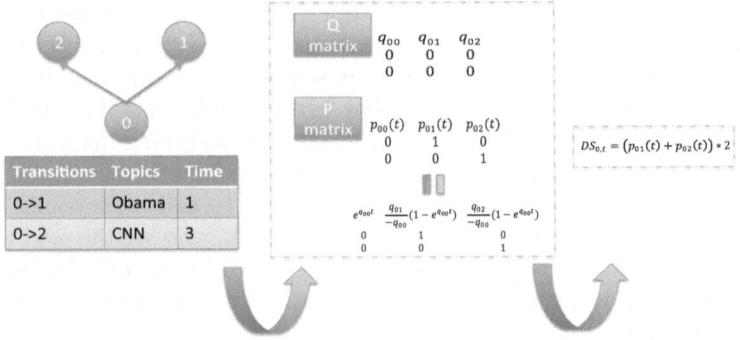

Fig. 3. An example to show the application of our model over a simple dataset which composed of only three users and two topics. Based on the two topic transitions in the dataset, we can calculate the Q matrix. Then, the P(t) matrix can be obtained via the assistance of Q matrix, assuming the predicting time is t. Once P(t) matrix is ready, the diffusion size can be predicted.

4.2   Estimation of Transition Rate Matrix

The transition rate matrix Q is also called the infinitesimal generator of the Continuous-Time Markov Process. It is defined as the derivative of $P(t)$ when t goes to 0. The entry q_{ij} is the transition rate to propagate a topic from user i to user j. The sum of the rows in Q is zero, with $\sum_{j,j\neq i} q_{ij} = -q_{ii}$.

$$q_{ij} = \lim_{t\to 0} \frac{P_{ij}(t)}{t} = P'_{ij}(0) \quad (i \neq j). \qquad (4)$$

Note that q_{ij} reflects a change in the transition probability from user i to user j. q_i, namely out-user transition rate in this paper, is equal to $-q_{ii}$. It indicates the rate of user i propagating topics to any other users. As shown in Figure 1, the average number of topic adoptions decreases exponentially over the time. Thus, in order to compute q_i, we assume that the time for user i to propagate a topic to all the other users is following an exponential distribution as observed for many users in our data, where the rate parameter is q_i. It is known that the expected value of an exponentially distributed random variable T_i (in this case, the topic propagation time for user i) with rate parameter q_i is given by:

$$E[T_i] = \frac{1}{q_i}. \tag{5}$$

Thus q_i is one divided by the mean of $\cup_j(T(V_i \leftarrow V_j))$, which is defined in the temporal influence network.

According to the theory of Continuous-Time Markov Process, if a propagation occurs on user i, the probability that the other user j would post the topic forms an embedded Markov chain. The transition probability is S_{ij}, and $\sum_j S_{ij} = 1$ ($i \neq j$) and $S_{ii} = 0$. One important property is that $q_{ij} = q_i S_{ij}$. Then, the transition rate from user i to j can be estimated by:

$$q_{ij} = \sum_m q_i^2 \cdot exp(-q_i \cdot t_{ij}^m), \tag{6}$$

where m is the number of topics diffused from use i to j, and t_{ij}^m denotes the transition time from user i to j on the m-th topic.

4.3 Estimation of Transition Probability Matrix

Now we obtain all the entries of the transition rate matrix Q. Next, we will specify how to derive the transition probability matrix $P(t)$. The well accepted Kolmogorov's Backward Equations [6] in the Continuous-Time Markov Process can be utilized:

$$P'_{ij}(t) = q_i \times \sum_{i \neq k} P_{ik}(t) \times P_{kj}(t) - q_i \times P_{ij}(t). \tag{7}$$

By performing some algebra operations, the above equation can be written as the following matrix form:

$$P'(t) = QP(t). \tag{8}$$

The general solution for this equation is given by:

$$P(t) = e^{Qt}. \tag{9}$$

$P(t)$ is a stochastic, irreducible matrix for any time t. We approximate it using Taylor expansion, so that $P(t)$ can be estimated by [10]:

$$P(t) = e^{Qt} = \lim_{n \to \infty} (I + Qt/n)^n. \tag{10}$$

We raise the power of $(I + Qt/n)$ to a sufficiently large n.

5 Experiment

5.1 The Dataset Description

We continuously collected 22-day twitter data, ranging from March 2 to March 24, 2011, using Twitter Gardenhose streaming API. The first 12-day data is used for our training purpose, and the remaining 10-day data is for testing and validation. We removed all non-English tweets and those tweets posted by users with less than 20 followers (authors of these tweets are less likely to be influential users.) Notice that the data is represented by tweet records, one of which may include the tweet and the corresponding retweets, thus more than one tweet. Then there are 48,113,490 tweet records in the 12-day training data and 27,237,631 tweet records in the 10-day testing data (tweets without Hashtags or URLs are also removed from the testing data). After, those tweets with more than three Hashtags or URLs are filtered, since they tend to be spam ones as introduced in [12]. Finally, we obtained totally 78,858,046 tweets, in which there are 9,431,404 unique users, 3,209,330 Hashtags, and 21,107,164 URLs.

5.2 Predicting Spreading Size Using IDM-CTMP

This paper aims at modeling, predicting, and measuring users' influence in the social network. Hence, to validate the dynamic metric and the predictive model – IDM-CTMP, instead of focusing on identifying influential users like the previous work, we mainly evaluate its capability of predicting how many users would adopt one topic given a certain period of time after a user posts it. Specifically, in our experiment, we first train IDM-CTMP model on the first 12-day training data, then calculate the spreading coverage of each user for each day from day 1 to day 22, including both training and testing periods.

The Ground Truth. In order to evaluate the prediction performance of IDM-CTMP and present its feasibility in real world applications, we need to provide the ground truth. Suppose user u posts a topic τ at time t_1, and subsequently n users post the same topic τ till time t_2, then the ground-truth spreading size of u from t_1 to t_2 with regards to topic τ, denoted as $DS_{u,\tau}^{t_1 \sim t_2}$, is n. For example, we know that user B first posts a topic #ipad in day 2, afterwards there are 10 users posting #ipad in day 2, and 20 users posting #ipad in day 3. Then the ground-truth spreading size of B is 0 for day 1, 10 for day 2, and 20 for day 3.

After a user's spreading sizes over different topics in a particular period of time are computed, we can obtain the average spreading size over all topics in that time period by dividing the number of involved topics:

$$DS_u^{t_1 \sim t_2} = \frac{\sum_\tau DS_{u,\tau}^{t_1 \sim t_2}}{\#(\tau)}, \tag{11}$$

where DS is the spreading size, u denotes a user, $t_1 \sim t_2$ indicates a time window, and τ is a topic.

Baselines. To our best knowledge, this is the first attempt to predict the continuous-time spreading coverage of social network users. Therefore, we employ the *Autoregressive Integrated Moving Average* (ARIMA) model [14], which is widely used for fitting and forecasting the time series data in the area of statistics and econometrics, as one baseline. This model can first fit to time series data (in our case, a user's spreading sizes of different days in the history), then predict this user's spreading size in the future. Thus, the spreading sizes of first 12 days are used to build the ARIMA model. Then, it predicts the entire 22 days. Note that the optimal ARIMA is always selected based on Akaike information criterion (AIC) and Bayesian information criterion (BIC) for comparison.

In addition to ARIMA, one of the basic information diffusion models – *Independent Cascade* (IC) [11] is used as the second baseline. In the IC model, a user u who mentions a topic in his/her tweets at the current time step t is treated as a new activated user. An activated user has one chance to "activate" each of his/her neighbors (i.e., make them adopt this topic) with a certain probability. If a neighbor v posts the same topic after time step t, it is said that v becomes active at time step $t+1$. If v becomes active at time step $t+1$, u cannot activate v in the subsequent rounds.

In order to apply the IC model to calculate users' spreading sizes, the activation probability for every pair of users needs to be estimated. Specifically, for a user u, we first obtain the spreading size of each of his/her topics during the first 12 days, thus we can get average spreading size over all of his/her topics. Then, the daily average spreading size (DDS) is computed from dividing the average spreading size by 12 days. Finally, $1/DDS$ is taken as the activation probability of u and each of his/her neighbors.

Prediction. To compare the performance of IDM-CTMP, we choose $10,000$ top users computed by IDM-CTMP given the entire 22 days and $10,000$ random users. Three well-known metrics for measuring prediction accuracy are utilized in our experiment for evaluation: MAE (Mean Absolute Error), RMSE (Root Mean Square Error), and MASE (Mean Absolute Scaled Error). The average values of three metrics for IDM-CTMP, ARIMA, and IC are listed in Table 1 and Table 2. It can be seen that our proposed method IDM-CTMP performs better than baseline methods ARIMA and IC, because ARIMA fits the overall trend of the time series data and does not consider the underlying network cascading causing the change of the spreading sizes. The basic IC model needs predefined time step, which is set to be 1 day. It might be too large to capture the real-time topic propagation. However, if setting it to be small, it would take long time to run. The parameter estimation of the basic IC model assumes the constant activation probability for all neighbors, which could be another reason of poor performance.

In Figure 4, we plot the ground truth spreading sizes and the predicted spreading sizes, using IDM-CTMP, ARIMA, and IC for both top 5 users and 5 random users. We can observe that even though the predicted results by IDM-CTMP are

Table 1. The comparison over the 10,000 top users

Methods	MAE	RMSE	MASE
IDM-CTMP	3.290	4.231	0.714
ARIMA	4.369	5.470	1.294
IC	5.858	7.209	2.355

Table 2. The comparison over the 10,000 random users

Methods	MAE	RMSE	MASE
IDM-CTMP	1.686	2.055	0.702
ARIMA	2.026	2.855	0.764
IC	3.928	4.834	2.091

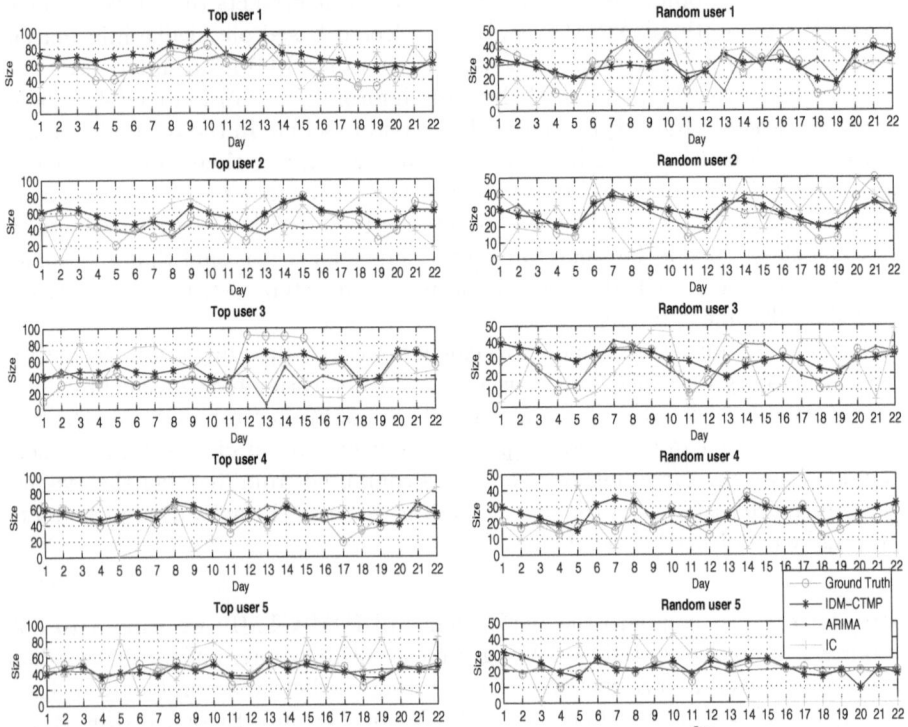

Fig. 4. The comparison between the predicted spreading size of top ranked 5 users (left side) and randomly picked 5 users (right side) by IDM-CTMP and baseline against the ground truth

not exactly as same as the ground truth, most of the predicted curves fit very close to the true curves. In particular, most of the "peaks" and "valleys" can be well captured by our proposed method. However, ARIMA and IC does not perform well, missing many "peaks" and "valleys" and having wrong predictions.

In addition to the prediction capability, we also conduct experiments to compare the running time of three models. It turns out that IDM-CTMP spends similar running time as IC model for prediction tasks, while ARIMA requires the least running time. However, to achieve the best performance of ARIMA, a lot of parameter tuning work needs to be done beforehand. Considering this, ARIMA cannot be the best choice in terms of the running time.

6   Conclusion

In this paper, we propose IDM-CTMP, an information diffusion model based on Continuous-Time Markov Process. IDM-CTMP is able to predict the influence dynamics of social network users, i.e., it can predict the spreading coverage of a user within a given period of time. Our experiment results show that IDM-CTMP achieves very promising performance as its predicted spreading size demonstrated can fit closely to the ground truth.

Acknowledgements. The work is partially supported by a Xerox Universit Affair Committee (UAC) Award and by US National Science Foundation under grants IIS-0546280, CCF-0830659, HRD-0833093, and IIS-1213026.

References

1. Adamic, L.A., Adar, E.: How to search a social network. Social Networks 27, 2005 (2005)
2. Anderson, W.J., James, W.: Continuous-time Markov chains: An applications-oriented approach, vol. 7. Springer, New York (1991)
3. Bakshy, E., Hofman, J.M., Mason, W.A., Watts, D.J.: Everyone's an influencer: quantifying influence on twitter. In: WSDM, pp. 65–74 (2011)
4. Cha, M., Haddadi, H., Benevenuto, F., Gummadi, K.P.: Measuring user influence in twitter: The million follower fallacy. In: AAAI, ICWSM (2010)
5. Domingos, P., Richardson, M.: Mining the network value of customers. In: SIGKDD, pp. 57–66. ACM (2001)
6. Gardiner, C.W.: Handbook of stochastic methods. Springer, Berlin (1985)
7. Ghosh, R., Lerman, K.: Predicting influential users in online social networks. CoRR, abs/1005.4882 (2010)
8. Goyal, A., Bonchi, F., Lakshmanan, L.V.S.: Learning influence probabilities in social networks. In: WSDM, pp. 241–250 (2010)
9. Gruhl, D., Guha, R., Liben-Nowell, D., Tomkins, A.: Information diffusion through blogspace. In: WWW, pp. 491–501. ACM (2004)
10. Huang, Q., Yang, Q., Huang, J.Z., Ng, M.K.: Mining of web-page visiting patterns with continuous-time markov models. In: Dai, H., Srikant, R., Zhang, C. (eds.) PAKDD 2004. LNCS (LNAI), vol. 3056, pp. 549–558. Springer, Heidelberg (2004)
11. Kempe, D., Kleinberg, J., Tardos, É.: Maximizing the spread of influence through a social network. In: SIGKDD, pp. 137–146 (2003)
12. Kwak, H., Lee, C., Park, H., Moon, S.: What is twitter, a social network or a news media? In: WWW, pp. 591–600 (2010)
13. Liu, Y., Gao, B., Liu, T.Y., Zhang, Y., Ma, Z., He, S., Li, H.: Browserank: letting web users vote for page importance. In: SIGIR, pp. 451–458 (2008)
14. Mills, T.C.: Time series techniques for economists. Cambridge Univ. Pr. (1991)
15. Norris, J.R.: Markov chains. Number 2008. Cambridge University Press (1998)
16. Petrovic, S., Osborne, M., Lavrenko, V.: Rt to win! predicting message propagation in twitter. In: 5th ICWSM (2011)
17. Richardson, M., Domingos, P.: Mining knowledge-sharing sites for viral marketing. In: SIGKDD, pp. 61–70 (2002)
18. Rogers, E.M.: Diffusion of Innovations, vol. 27. Free Press (2003)

19. Saito, K., Kimura, M., Ohara, K., Motoda, H.: Efficient estimation of cumulative influence for multiple activation information diffusion model with continuous time delay. In: Zhang, B.-T., Orgun, M.A. (eds.) PRICAI 2010. LNCS, vol. 6230, pp. 244–255. Springer, Heidelberg (2010)
20. Saito, K., Kimura, M., Ohara, K., Motoda, H.: Generative models of information diffusion with asynchronous timedelay. JMLR - Proceedings Track 13, 193–208 (2010)
21. Song, X., Chi, Y., Hino, K., Tseng, B.L.: Information flow modeling based on diffusion rate for prediction and ranking. In: WWW, pp. 191–200 (2007)
22. Song, X., Tseng, B.L., Lin, C.-Y., Sun, M.-T.: Personalized recommendation driven by information flow. In: SIGIR, pp. 509–516 (2006)
23. Tsur, O., Rappoport, A.: What's in a hashtag?: content based prediction of the spread of ideas in microblogging communities. In: Proceedings of the Fifth ACM International Conference on Web Search and Data Mining. ACM (2012)
24. Weng, J., Lim, E.P., Jiang, J., He, Q.: Twitterrank: finding topic-sensitive influential twitterers. In: WSDM, pp. 261–270 (2010)

Credibility-Based Twitter Social Network Analysis

Jebrin Al-Sharawneh[1], Suku Sinnappan[2], and Mary-Anne Williams[1]

[1] Centre for Quantum Computation & Intelligent Systems (QCIS),
University of Technology Sydney, Sydney, Australia
Jebrin@it.uts.edu.au, mary-anne.williams@uts.edu.au
[2] Swinburne University, Australia
ssinnappan@swin.edu.au

Abstract. Social Network (SN) members in Twitter communicate in varied contexts such as crisis. Formations within social networks are unique as some members have more influence over other members; members with more influence are known as leaders or pioneers. Finding leaders in a social network is the central factor in information dissemination; therefore, the main objective of this paper is to utilize social network information to identify leaders in Twitter during crisis situations. We propose a new approach to identify social network leaders based on their credibility which is drawn from their trustworthiness and expertise utilizing the "Follow the Leader" model based on the content quality generated by SN members. Our approach is highly innovative and the empirical results demonstrate that our approach can identify SN community leaders in a specific context (crisis) and has significant practical benefits.

Keywords: social network, Twitter, crisis, leaders, credibility, follow the leader, content quality.

1 Introduction

The immediacy of social networks play an excellent host to various social networks formed at times of crisis. Examining complex Twitter communication can be challenging; however, it is known that in Social Networks (SNs), some members have more influence over other members; they are known as leaders or pioneers [6, 13]. Finding a credible member within an SN is a challenging issue, therefore, the main objective of this paper is to utilize social network information to identify members' roles as Leader or Follower in Twitter Social Network Service (SNS), based on their credibility and to assess the impact of Leaders in crisis situations.

Credibility refers to the objective and subjective components of the believability of an agent. A credible SN member is defined as one who has performed consistently, accurately, and its generated content is useful over a period of time in a specific context. The credibility of an agent refers to its quality of being believable or trustworthy and can be measured by its trustworthiness, expertise, and dynamism [8].

In SNSs members interact in varied contexts. We consider Victorian bushfires [11] as the context of the interaction between Twitter members, where the worst day known as Black Saturday when 173 people died.

Y. Ishikawa et al. (Eds.): APWeb 2013, LNCS 7808, pp. 323–331, 2013.

1.1   Motivations and Contributions

Identifying natural leaders in large SNSs such as Twitter is a challenging task. Making assumptions from the number of followers and the number of tweets on Twitter profiles can be deceiving and may not be accurate as both are cumulative figures. Communication in Twitter is best analysed and understood in a context [11]. The Goldbaum's "Follow the Leader" model [6] provides us with insights to identify members' roles in the SNS whether they are leaders or followers. Enriching the "Follow the Leader" model with trust provides us with the potential to analyse SNSs based on members' credibility. Member credibility level is a measure that reflects their trustworthiness and expertise, and provides the means to classify members in a specific context; some members can be classified either leaders or followers.

Our key contributions in this paper as follows:
- A credibility based approach that identifies SNS members' roles.
- An answer to what extent is leaders' prominent and influential members in SNS.
- An answer to what extent new tweets spread during crisis based on their quality.

The rest of this paper is organized as follows: Section 2 provides a brief overview of some related research work followed by the proposed SN members' credibility based model in Section 3. Sections 4 and 5 report, analyse and evaluate the empirical study of the proposed "Follow the Leader" model.

2   Related Works

2.1   Social Networks and Trust

In SNSs, members can express how much they trust other members through the adoption of their postings [14]. Since most SN participants do not know each other in real life, the trust inference mechanism is used when participants want to establish a new trust relationship or measure trust values between connected members. The connection between member similarity and trust was established in [14]. They demonstrated that there is a significant correlation between the trust expressed by the members and their similarity. The more similar two people are, the greater the trust between them. Similarity can be interpreted as similarity in interests or opinions.

2.2   Follow the Leader

Social psychology theory [10] points out that a person's role in a specific context has a significant influence on trust evaluation if that person refers a person or object. The Follow the Leader phenomenon in dynamic social networks [6] can be used as a model of opinion formation with dynamic confidence in social networks; in which the profiling of agents as leaders or followers is possible. Goldbaum [6] defines the leader as somebody who possesses better knowledge and information than the general public in a specific context.

3   SNS Members' Credibility Model

In SNS, the number and types of links that members have are keys to determining their importance in the network, which constrains the range of opportunities, influence and power that they have [7]. A Twitter member credibility reflects their quality of being believable or trustworthy and can be measured by their trustworthiness, expertise, and dynamism [8]. We adopt the credibility computation from our previous work [2] with variation in the expertise computation.

3.1   SNS Members Trustworthiness

Trustworthiness is an attribute of individual exchange partners; a trustworthy member is a member in which others place their trust. SNS members gain trustworthiness from their direct followers and indirect followers using the trust transitivity feature of trust; friends of friends who trust their nearest friend and also trust their friends' next friend, yield a trust score which is the product of the two scores.

Direct and indirect followers (trusters) of a member provide implicit trust indicating their trust in that member. Members' similarity in their opinions implies their implicit trust. Trust in Twitter is considered as binary construct (i.e., Trust score = 1, if there is a trust statement including @ tag referring a member, and 0 otherwise). The aggregation of trust scores is a measure of the member's trustworthiness; consequently it is a measure of member credibility in the SNS. Formally, member m credibility from followers' trust type x *(direct or indirect)* is denoted by $Cr_t(T_x^m)$, and defined as follows:

$$Cr_t(T_x^m) = \frac{N_x^m}{Max(N_x^m)} \tag{1}$$

where $Cr_t(T_x^m)$ value is a real number in [0, 1] representing member m credibility gained from direct\indirect followers N_x^m at time t, $Max(N_x^m)$ refers to number of direct\indirect followers of the member with maximum direct\indirect followers in the context, considered as a reference point. Notably, credibility from followers is normalized, hence if the most trustworthy member received the highest trust score from all friends, then $Cr_t(T_x^m) = 1$. Notably, the credibility gained from lower level followers in the hierarchy is ignored, since their contribution to the members' credibility is small compared to first and second level followers.

3.2   Expertise Component

Expertise, a key dimension of credibility, is defined as the degree of a member's competency to provide valuable quality tweets and exhibit high activity [9]. The expertise dimension of the member credibility captures their perceived knowledge and skills in a given context. SNS members vary in their knowledge and consequently their level of expertise; the more knowledge a member has, the more power they possess, to the extent that "knowledge is power" [7]. The credibility of a member coincides with their generated tweets quality which consequently coincides with their level of expertise in a particular context. In Twitter, members who are seen to be disseminating information laden have more followers and are more trusted.

Member m expertise in SNS Twitter is drawn from the following three types of messages a member can post:

- A new message originated by member m.
- A retweet message adopting other member opinion and contains (RT @, or Retweet, via, follow or similar expressions).
- A mention message [3] or a reply message to another member with (@ sign).

Each type of message has its influence based on its content quality. We believe that new messages are the most important messages, followed by retweets and replies. Thus, member m expertise $Cr_t(E_y^m)$ in posting type y messages, (new, retweet, mention), with accumulated quality Q_y^m is given by the following relationship:

$$Cr_t(E_y^m) = \frac{Q_y^m}{Max(Q_y^m)} \qquad (2)$$

Each type y tweets accumulated quality posted by member m defined as $Q_y^m = \sum_{i=1}^{ny} Q_i$ where Q_i refers to type y tweet i quality and $Max(Q_y^m)$ represents the member with the maximum type y tweets accumulated quality, i.e., the member who has the highest quality of type y tweets issued in the context, considered as a reference point.

Quality of a tweet message Q_i in the range [0, 1] is drawn from its content that reflects its importance and influence in the context based on the following assessment components: (1) number of keywords KN, (2) number of hash tags #N, (3) number of other Twitter references indicated with @ signs @N, and (4) number of URLs included in the tweet UN; each with a corresponding weight factor W in that component. For tweet i, Quality Q_i, is given by:

$$Q_i = 0.4 \frac{\sum_{j=1}^{KN} K_j W_j}{Max_i \{\sum_{j=1}^{KN} K_j W_j\}} + 0.1 \frac{\sum_{j=1}^{\#N} \#_j W_j}{Max_i \{\sum_{j=1}^{\#N} H_j W_j\}} + 0.1 \frac{\sum_{j=1}^{@N} @_j W_j}{Max_i \{\sum_{j=1}^{@N} @W_j\}} + 0.1 \frac{\sum_{j=1}^{UN} U_j W_j}{Max_i \{\sum_{j=1}^{UN} U_j W_j\}} \qquad (3)$$

where $W = (0.4, 0.1, 0.1, 0.1)$ represent each component weight impact in tweet quality respectively, and finally Q_i is normalized to the max Q_i in the context. Notably, the @ parameter does not appear in new messages, so we need to consider this impact when computing new tweet quality by scaling new tweets quality by 7/6.

3.3 Tweet Message Quality Computation

Quality of a message in Twitter should reflect its influence and importance in the context. Most of SNS contributors in crisis are from general public who want to provide information that help to minimize the impact and consequences of the crisis. Thus, our concern is to assess message quality based on the content of the message which is reflected in the above four components. Although other researchers such as [1] use textual features and semantic features as basis to assess content quality; we believe that semantic metrics can be considered as added-value to the content quality.

Our approach as follows: (1) For each assessment component, we build a list of keywords and items from tweets after removing stop words from each tweet message. (2) Ignore each keyword with a frequency less than four. (3) Assign a weight for each item based on its frequency in the context as the standard deviation of the keywords set. (4) Compute quality of a tweet message as shown in equation 3.

3.4   Computing Member Credibility

By aggregating the credibility components, *i.e.* trustworthiness component credibility from (direct followers $Cr_t(T_D^m)$ and indirect followers $Cr_t(T_I^m)$), and expertise component (credibility gained from new tweets, retweets and replied tweets), thus member m credibility $Cr_t(m)$ at time t is given by:

$$Cr_t(m) = \alpha Cr_t(T_D^m) + \beta Cr_t(T_I^m) + \gamma Cr_t(E_T^m) + \delta Cr_t(E_R^m) + \varepsilon Cr_t(E_Y^m) \qquad (4)$$

where α, β represent trustworthiness credibility weighted importance factors gained from direct followers and indirect followers respectively, and $\gamma, \delta, \varepsilon$ represent expertise credibility weighted importance factors gained from new tweets, retweets and mention/replied tweets respectively. Where $\alpha, \beta, \gamma, \delta, \varepsilon$ are model tuning parameters, and $\alpha + \beta + \gamma + \delta + \varepsilon = 1$. These parameters can be normally set or using automated machine learning techniques. In our experiment, we set the values to (0.3, 0.2, 0.25, 0.15, 0.10) for model tuning parameters respectively, where trustworthiness and expertise components each assigned equal weights 0.5.

3.5   Classifying Members Based on Credibility

To find the most trustworthy and expert SN members who can act as leaders, we define a *credibility threshold* from which leaders and followers are identified as:

$$if\ Cr_t(u) \geq Cr_{t_{Threshold}},\ \text{then member is a leader, otherwise member is a follower} \qquad (6)$$

Credibility Threshold is a system parameter and is used to promote an adequate number of leaders from the crisis dataset based on the power-law degree distribution.

4   Experiments and Results

To demonstrate the feasibility and effectiveness of the proposed approach to identify SNS leaders, in this section, we discuss the dataset and the results which demonstrate that the notion of "Follow the Leader" in assessing credibility can be used as a means to identify leaders and followers in the context.

4.1   SNS Crisis Dataset

Our Victorian bushfire crisis dataset was borrowed from [11]. The dataset contains 1684 tweets representing members' communication in one week; as the event unfolded from February 6-14, 2009. 697 unique users are identified in the dataset.

In Victorian bushfire crisis, 71% of the tweets are new tweets to disseminate new information, followed by mentions/replies (21%) and retweets (8%). Although the retweets ratio is less than the number of mentions/replies, retweets have more impact on information diffusion than replies to spread critical crisis information among SNS members. This is because retweets reinforce a message and as such are considered important, and hence they are passed on down the network.

4.2   Leaders and Followers Analysis

In the crisis dataset, about 78% of the population are inactive members who listen to active members. They passively contribute to the conversation by posting limited number of new messages, and they do not retweet messages of active members. Similar findings were reported by Starbird et al. [12] in their study covering the Red River flooding in America. Table 1 below provides a comparison between leaders' community and followers' community for the active subset, i.e., after removal of inactive members who do not have from/to links.

Table 1. Active Members: Leaders vs. Followers community
(NW : New, RP: Mention/Replies, RT: Retweets)

Active	Members	Trustworthiness					Expertise				
Members	Number	Followers	IndFlwrs	RT 2 Me	RP 2 Me	TweetsNo	NW	RP	RT	CRED
Leaders	31	120	83	80	40	463	296	92	75	4.82
Community	20.0%	67.0%	72.8%	84.2%	47.6%	59.3%	63.8%	46.9%	62.0%	63.7%
Followers	124	59	31	15	44	318	168	104	46	2.75
Community	80.0%	33.0%	27.2%	15.8%	52.4%	40.7%	36.2%	53.1%	38.0%	36.3%
Total	155	179	114	95	84	781	464	196	121	
							100.0%	59.4%	25.1%	15.5%	

From Table 1, for active members; we note the following observations: (1) 59.4% of the tweets are new, of which 15.5% of the new tweets are retweeted again; furthermore, 25.1% of the tweets are mentions or replies. This indicates new tweets are vital source of information distribution. (2) Leaders who represent 20% of the population in the active set generate 59.3% of all tweets, they retweet 62% of all retweet posts, and this for certain extent reflects power law distribution. (3) As shown in (RT 2 Me) column, 84.2% of the Retweets in the communication are generated by members addressing leaders' content, i.e., originally leaders' generated content as new messages. On the other hand 15.8% of the Retweets are generated by members addressing followers' content. (4) 67% of followers follow leaders while 33% of followers follow other members in the SNS. (5) Leaders' community possess 72.8% of indirect followers, on other hand, 27.2% of indirect followers follow other followers' members, and those appear in the fourth level of the "Follow the Leader" hierarchy. (6) Followers are more tweet repliers than leaders; they generate 53.1% of reply messages. Leaders usually create new messages and pass (retweet) messages more than followers. (7) Leaders' community possesses 63.7% of SNS credibility, while followers' community possess 37.3% of SNS credibility; this tends to align with the standard power law distribution.

4.3   SNS Leaders Influence Evaluation

To determine the efficiency of the proposed Credibility approach in identifying leaders properly: first, we adopted centrality measure [4] which have been used as a proxy for power and influence; second, we conducted quality and spread analysis on the new tweets and compare between leaders and followers influence.

4.3.1 Centrality Measure

Centrality reveals how influential and powerful a node is, which reflects the roles of individuals in the network. Degree centrality is measured as the number of links of a given node [5]. According to [7] if an agent receives many links, they are said to be *prominent*; this justifies why many other agents seek to direct links to them and this indicate their importance. Agents who have a significant high out-degree are able to exchange opinions with many other members, or make many others aware of their views and opinions. Agents who show high out-degree centrality are often said to be *influential* agents [7]. So, to what extent leaders are prominent and *influential* agents?

Table 2. Centrality Measure and New Content Analysis

a. Centrality degree

Category	Avg In-Degree	Avg Out-Degree
Leaders	3.87	5.39
Followers	0.48	1.21
All Members	1.15	2.05

b. New content per quality category

Quality Category	Cat. Ratio%	LdrsContent%	Retweeted%
Low Quality	17.77%	28.71%	1.44%
BelowAvg	36.82%	28.64%	3.46%
AboveAvg	28.66%	37.98%	5.34%
High Quality	16.75%	41.62%	4.06%

c. New content spread time

Quality Category	1st Retweet Avg Time (Min)	Last Retweet Avg Time (Min)
Low Quality	58.7	59.0
BelowAvg	94.0	138.8
AboveAvg	9.4	34.3
High Quality	8.5	195.0

Table 2.a presents the centrality measures (i.e., in-degree and out-degree) for leaders, followers and all members in the dataset. The in-degree centrality of a particular node is defined as the number of in-links (trust statements this node received); it represents the number of retweets and replied tweets addressing that member. The out-degree centrality of a node is defined as the number of out-links (trust statements issued from this node), it represents the number of retweets and replied tweets issued by that member to other members in SNS. Table 2.a shows that leaders' in-degree 3.87 which is about 8 times followers' in-degree (0.48). Furthermore, leaders' out-degree 5.39 is about 4 times followers' out-degree (1.21).

Since leaders possess high average in-degree and high average out-degree centrality, which in both cases is more than the followers and more than the average member in the SNS, leaders are the most prominent and *influential* members in the SNS. This outcome is drawn from their trustworthiness and expertise; leaders possess the highest trustworthiness and expertise level among all members in the context.

4.3.2 New Content Quality and Spread Analysis

We conducted quality analysis on the new tweets for all population taking 10% as leaders using new content quality computation described in section 3.2.1. We define new content quality based on the quality distribution considering Quality Mean (QMean) and Quality Standard Deviation (QSdev) from the mean of the new content categorised as follows: (1) Low Quality: below QMean - QSdev (2) BelowAvg: equal or greater than QMean - QSdev and less than QMean, (3) AboveAvg: equal or greater than QMean and less than QMean + QSdev, and finally (4) High Quality: equal or greater than QMean + QSdev.

1. Quality Analysis Results: from Table 2.b, we note that: (1) High quality content represents 16.75% of all new tweets content, of which 41.6% of this content is generated by leaders. (2) As shown in column 3 (LdrsContent%), the percentage of new postings content generated by leaders for each category is increasing by the increase of the new content quality. This indicates that leaders are capable of generating the most credible content. (3) As shown in column 4 (Retweeted%), retweeting is increasing with the increase of the content quality; moreover, multiple retweets correspond to quality above the average.

2. Spread / Diffusion Time Analysis Based on Quality Category Results: from Table 2.c, we note that the elapsed time between new content posting time and first time retweeted as shown in (1st Retweet Avg Time (Min)) column is decreasing with the increase of new content quality. As shown for the High quality category, the time elapse from new content posting to the first retweet time is 8.5 minutes, with overall elapse time for all retweeting of 195.0 minutes. One notable phenomenon is that tweets which carry bad news spread faster than other tweets.

5   Results Summary

In a Twitter SNS, members' behaviour is the determinant of their credibility; different members undertook varied activities and possess different interests. In a SN, the credibility of a member acts as the predictor of their role. Leaders are the most prominent and influential members in Twitter SNS. They generate the highest quality content to be adopted and retweeted by other members.

SN leaders are aware and become involved in the crisis early; they produce information-laden content and maintain high activity all times. Thus, their postings are an important resource for information management and decision making.

In Twitter SNS, context leaders differ from Twitter leaders. In each context, there are a set of leaders who are considered experts in the context, thus, it is difficult to find same leaders across all contexts, simply because each Twitter member has varied interests and expertise in varied contexts.

6   Conclusions and Future Work

Finding leaders who have more influence in a SNS context is a challenging issue. In this paper, we proposed a new credibility based approach to identify leaders in SNSs specific context utilizing the "Follow the leader" strategy. In SNS members' behaviour is the key determinant of their credibility; members' credibility is drawn from their trustworthiness and expertise. Members' trustworthiness is assessed based on other members' trust in target member postings and their corresponding followers. Members' expertise is assessed using their activities and the type and quality of the content they generate.

The experimental results demonstrated that the proposed credibility based approach is a significantly innovative approach to identifying Twitter SNS community leaders in a specific context such as crisis or terrorism. New content' quality plays an important role in information dissemination and diffusion.

References

1. Agichtein, E., Castillo, C., Donato, D., Gionis, A., Mishne, G.: Finding high-quality content in social media. In: WDSM 2008, pp. 183–194. ACM (2008)
2. Al-Sharawneh, J., Williams, M.-A.: Credibility-based Social Network Recommendation: Follow the Leader. In: ACIS 2010 Proceedings, Paper 24, pp. 1–10 (2010),
   http://aisel.aisnet.org/acis2010/24
3. Bruns, A., Burgess, J.E., Crawford, K., Shaw, F.: # qldfloods and@ QPSMedia: Crisis Communication on Twitter in the 2011 South East Queensland Floods (2012)
4. Burt, R.S.: Brokerage and closure: An introduction to social capital. Oxford University Press, USA (2005)
5. Freeman, L.: Centrality in social networks conceptual clarification. Social Networks 1(3), 215–239 (1979)
6. Goldbaum, D.: Follow the Leader: Simulations on a Dynamic Social Network, UTS Finance and Economics Working Paper No 155 (2008),
   http://www.business.uts.edu.au/finance/
   research/wpapers/wp155.pdf
7. Hanneman, R., Riddle, M.: Introduction to social network methods. University of California, Riverside (2005)
8. Kouzes, J.M., Posner, B.Z.: Credibility: How Leaders Gain and Lose It, Why People Demand It, Revised Edition. Jossey-Bass, San Francisco (2003)
9. Kwon, K., Cho, J., Park, Y.: Multidimensional credibility model for neighbor selection in collaborative recommendation. Expert Systems with Applications 36(3), 7114–7122 (2009)
10. Liu, G., Wang, Y., Orgun, M.: Trust Inference in Complex Trust-Oriented Social Networks. In: Proceedings of the International Conference on Computational Science and Engineering, Vancouver, Canada, vol. 4, pp. 996–1001. IEEE (2009)
11. Sinnappan, S., Farrell, C., Stewart, E.: Priceless tweets! A study on Twitter messages posted during crisis: Black Saturday. In: ACIS 2010 Proceedings, Paper 39 (2010)
12. Starbird, K., Palen, L., Hughes, A.L., Vieweg, S.: Chatter on the red: what hazards threat reveals about the social life of microblogged information. In: Proceedings of the ACM 2010 Conference on Computer Supported Cooperative Work (CSCW 2010), pp. 241–250 (2010)
13. Xu, Y., Ma, J., Sun, Y., Hao, J., Zhao, Y.: Using Social Network Analysis as a Strategy for e-Commerce Recommendation. In: PACIS 2009 Proceedings, p. 106 (2009)
14. Ziegler, C.N., Golbeck, J.: Investigating interactions of trust and interest similarity. Decision Support Systems 43(2), 460–475 (2007)

Design and Evaluation of Access Control Model Based on Classification of Users' Network Behaviors[*]

Peipeng Liu[1,2,3,4], Jinqiao Shi[3,4], Fei Xu[3,4], Lihong Wang, and Li Guo[3,4]

[1] Institute of Computing Technology, CAS, China
[2] Graduate University, CAS, China
[3] Institute of Information Engineering, CAS, China
[4] National Engineering Laboratory for Information Security Technologies, IIE, China
liupeipeng@gmail.com

Abstract. Nowadays, the rapid development of the Internet brings great convenience to people's daily life and work, but the existence of pornography information seriously endangers the health of youngsters. So researches on network access control model, aiming at preventing youngsters from accessing pornography information, have become a hotspot. However, existing models usually don't distinguish between adults and youngsters, and they block all the accesses to pornography information, which does not meet the actual demand. So preventing youngsters from accessing pornography information, while not affecting the freedom of expressions in adults, has become a reasonable yet challenging alternative. In this paper, an access control model based on classification of users' network behaviors is proposed, and this model can effectively make a decision whether to allow one's access to pornography information according to its age. Besides, an evaluation framework is also proposed, through which the model's controllability, cost-effectiveness and users' accessibility are analyzed respectively.

Keywords: Access Control, Evaluation, Controllability, Cost-effectiveness.

1 Introduction

With the rapid development of the Internet, more and more network services are provided for people's daily life and work, but the Internet is also becoming flooded with services involving pornography, which seriously endangers youngsters' physiological and mental health [8]. Therefore, researches on access control model aiming at preventing youngsters from accessing pornography information have attracted more and more attentions.

[*] This work is supported by National Natural Science Foundation of China (Grant No. 61100174 and No.61272500), National High Technology Research and Development Program of China, 863 Program (Grant No.2011AA010701 and No.2011AA01A103), National Key Technology R&D Program (Grant No.2012BAH37B04).

Y. Ishikawa et al. (Eds.): APWeb 2013, LNCS 7808, pp. 332–339, 2013.

There have been raised many technical models for network access control [1, 7]. However, it has been shown that most of these control models work in an assertive manner by which once the content or resource is identified as pornography, the model refuses all accesses to it. Obviously, this does not meet the actual demand. So it is necessary to develop a systematic model that can effectively prevent youngsters' pornography access, while not affect the freedom of expression in adults.

In this paper, an access control model based on classification of users' network behaviors is proposed, and this model first judges whether the user is a youngster or not, and then based on the judgment, determines whether to allow the access. This paper also proposed an evaluation framework, through which the model's controllability, users' accessibility and the cost-effectiveness are analyzed respectively.

The rest of the paper is organized as follows. Section 2 discusses the existing works related to access control model and its evaluation. In section 3, the proposed access control model is described in detail. And section 4 is about the evaluation framework. An example is given in section 5. At last, we summarize the paper and give a conclusion in section 6.

2 Related Works

In [1], the authors proposed the control model (ICCON) for the ICS (Internet content security) on the Internet. Although it involves the communicators' identities, it mainly takes the coarse-grained properties, such as IP address and URL, into consideration, excluding the user's fine-grained properties, such as the ages. As a result, it cannot distinguish between adults and youngsters.

Another kind of popular access control model is based on the resource, that is once a resource on the Internet is identified as pornography, and then all access to it is refused. Otherwise, the resource is open to everyone.

As for evaluation, previous researches, such as [2, 6, 9, 10], usually focus on the controllability from a technical point, which are lack of assessment of the users' experience and the model's cost-effectiveness.

All these works have played important roles in preventing the spread of the pornography information on the Internet and evaluating the existing models, but they are sometimes not comprehensive enough. So it's necessary to develop a more comprehensive access control model.

3 Access Control Model Based on Classification of Users' Network Behaviors

3.1 General Framework for Information Access Control

As described in [5], Internet information access control mechanism controls access actions with static or dynamic control rules by RM (Reference Monitor). RM is a core component of access control mechanism, which decides whether to allow the continuance of an access. In our model, RM is deployed on the transmission channel (e.g., on the router connected to the Internet in primary/middle schools), and we call RM of this deployment NRM (network-side reference monitor). To describe the general framework for information access control, two definitions are given below.

Definition 1: Network Reference Monitor.
A network reference monitor is a 4-tuple:
$$NRM = <M,P,S,O>$$
Where M is set of hardware and software, P is set of access control strategies, S is set of access subjects, and O is set of network resource objects. NRM controls actions of subjects S accessing objects O based on the access control strategies P.

Definition 2: Internet Information Access Control.
Whether an access action a is permitted or not is determined by satisfaction between control rules $P(x)$ and network resource O which subject S wants to access. If resource O matches $P(x)$, then the action a cannot be permitted, else action a is permitted.

3.2 Access Control Model Based on Classification of Users' Network Behaviors

According to the general framework, an access control model for pornography information, based on classification of users' network behaviors, is proposed. This model consists of three parts and each of them will be described in detail below.

Framework
At first, we give the framework for the access control model based on classification of users' network behaviors as in Fig. 1.

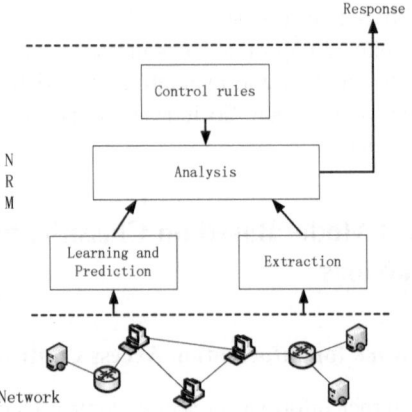

Fig. 1. Framework for access control model based on classification of users' network behaviors

The framework works as follows. At first, the model learns users' surfing habits and predicts the users' ages depending on the learned knowledge, and secondly, when an access action is captured, the model analyzes it together with the prediction results and the control strategies. At last, based on the analysis, the model makes a decision whether to allow the access action.

Learning and Prediction Module

There have been a lot of research on demographic prediction, and many of them have achieved satisfactory results [3, 4]. In our model, a simple age-prediction method is adopted. Firstly, based on users' profiles and their browsing history, the user's age information is propagated to the browsed pages, and then, a supervised regression model is trained to predict the probability distribution of the ages of a given Web-page's readers. Secondly, within Bayesian framework, an internet user's age is predicted based on the age tendency of the WebPages that he/she has browsed.

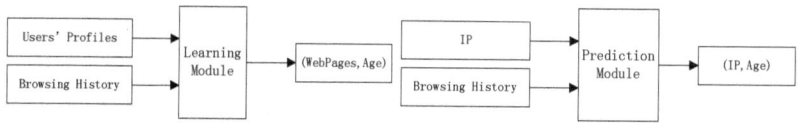

Fig. 2. Learning Module and Prediction Module

In Fig. 2, the learning module is in a training process, through which we can get the probability distribution of ages in WebPages. And based on the results of learning module, the prediction module takes the user's ip and historical browsing behaviors as inputs, and gives the prediction results in the form of (IP, Age).

Action Discovering and Analysis

The action discovering and analysis module of our model is shown in Fig.3.

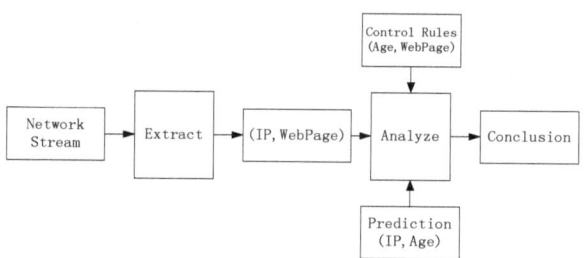

Fig. 3. Action Discovering and Analysis

In Fig. 3, extraction is applied to the network stream to get the $(IP, Webpage)$ pairs, and then together with prediction results and control rules, namely, (IP, Age) and $(Age, Webpage)$ are supplied to the analyze module as inputs, then the analyze module analyzes all the inputs comprehensively, and at last gives the control rules' satisfaction, i.e., whether the access action should be allowed or not.

The analysis process can be denoted by a tuple, that is $< L_u, L_p, P, T, \beta, D, Auth >$, where L_u is the label of a requester; L_p is the label of the webpage requested; P is the prediction results according to L_u, existing in the form likes (L_i, Age_i), where Age_i is the predicted age for L_i; T is a blacklisted control strategies, existing in the form likes $\{(Age_1, Webpage_1), (Age_2, Webpage_2)...(Age_n, Webpage_n)\}$, where each pair $(Age_i, Webpage_i)$ means users at Age_i should not be allowed to access $Webpage_i$; β is a set of non-negative real numbers; $D: (L_u, L_p) \times P \times T \rightarrow \beta$, is a distance function, and the detailed calculation method for distance varies with different applications. The smaller the distance is, the more likely that the user L_u should not be allowed to access the service L_p; $Auth: \beta \rightarrow [0,1]$, is an authentication function, which is employed to compute the probability of the response. The input of the function is the distance calculated above. The greater $Auth$ is, the greater the probability with which a response should be made is.

Response
At last, in the stage of response, the active response is chosen for the proposed model. Active response refers to taking some approaches such as blocking and cutting to refuse the access request.

4   Evaluation

An evaluation framework is also proposed to quantitatively evaluate the proposed control model from several aspects, including controllability, accessibility and cost-effectiveness.

4.1   Controllability

We firstly give the formal definition of controllability. According to the Internet information access control model, we suppose extract, analyze, match and control as atomic operation, which denote the process of extracting information from Internet, analyzing information, matching information with control rules, and controlling specific information or access action correspondingly.

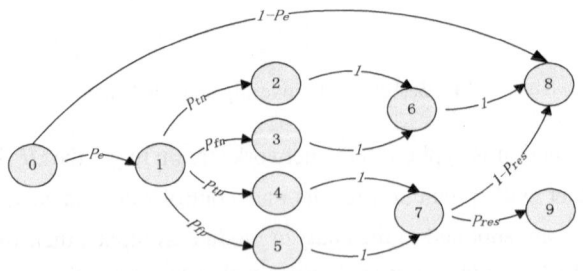

Fig. 4. Evaluation Model for Controllability

To quantitatively analyze the controllability of network access actions, we model the network information access control by finite state machine as shown in Fig.4.

In Fig. 4, P_e is the probability that information is extracted correctly from network streams; P_{tn} is the probability that information which actually mismatches the rules is judged to mismatch; P_{fn} is the probability that information which actually matches the rules is judged to mismatch; P_{tp} is the probability that information which actually matches the rules is judged to match; P_{fp} is the probability that that is, information which actually mismatches the rules is judged to match; P_{res} is the probability that response is applied to the access actions correctly.

The evaluation model works as follows. State 0 is an original state, and state 6 denotes actions not matching control rules and state 7 are actions which match control rules. State 8 is without response that actions are not extracted correctly or mismatching with control rules or being unsuccessful controlled. State 9 denotes the situation that actions are processed with a response. In this model, the transitions failures from state 1 to state 3 and state 5 correspond to two possibilities, one is NRM determines actions satisfy rules $P(x)$ but they do not in fact, and the other is NRM determines actions do not satisfy rules $P(x)$ but they actually do.

Then controllability can be defined as follow.

$$P_{con} = (((extract \wedge \neg match) \vee ((extract \wedge match) -> control)) \tag{1}$$

That is, controllability is defined to contain two cases, that is, information extracted from Internet is mismatching control rules or information must be controlled if it matches control rules.

So controllability can be expressed by P_i in Fig. 4 as $P_{con} = P_e \bullet P_{tn} + P_e \bullet P_{tp} \bullet P_{res}$.

4.2   Accessibility

Accessibility is proposed from a user's point, namely, the impact brought to users by the control model. The accessibility is divided into two types according to users' different access requests. For the pornography request, the accessibility is expressed as follow.

$$P_{pornography} = (1 - P_e) + P_e \bullet [P_{fn} + P_{tp} \bullet (1 - P_{res})] \tag{2}$$

Formula (2) expresses the probability that a user's request which actually is pornography is allowed.

While for the non-pornography requests, the accessibility is expressed as follow.

$$P_{non-pornography} = (1 - P_e) + P_e \bullet [P_{tn} + P_{fp} \bullet (1 - P_{res})] \tag{3}$$

Formula (3) expresses the probability that a user's request which actually is not pornography is allowed.

For accessibility, absolutely, the aim of the control model is to increase $P_{non\text{-}pornography}$ while decrease $P_{pornography}$.

4.3   Cost-Effectiveness

In order to achieve the best performance while decrease the costs as much as possible, an evaluation method based on linear programming is proposed. Here, we suppose it will cost W_i to increase one percentage point of probability P_i in Fig. 4. And for each component in the system, its ceiling of cost is limited, and we suppose for each P_i , its costs can take up to C_i . Then the problem can be described as follows.

$$Minimize : P_{pornography} = (1 - P_e) + P_e \cdot [P_{fn} + P_{tp} \cdot (1 - P_{res})]$$
$$Subject\ to:$$
$$100 P_i \cdot W_i \leq C_i\ ,\ P_i > 0$$

Here, according to different requirements, the aim can be set to maximize $P_{non\text{-}pornography}$ or minimize $P_{pornography}$. With the solution to this linear problem, the control model not only achieves the optimization targets, but also meets the requirements in cost.

5   Example

Considering a network management system for a community or a school, and here we mainly discuss the actions' analysis process. Suppose blacklist is used by the control model. Now we give the corresponding model.

T is a blacklist in the form of $(Age, Webpage)$ pairs, and each $(Age, Webpage)$ pair, means that users at such ages should not be allowed to access such WebPages. P , in the form of (IP, Age) , is the prediction result when an access action is captured, and A , in the form of $(IP, Webpage)$, is the access action captured.

D is a distance function defined as follows:

$$\beta = D((P, A), T) = \begin{cases} 0, if\ (P, A)\ is\ in\ T, \\ 1, otherwise. \end{cases} \tag{4}$$

$Auth$ is an authentication function:

$$Auth(\beta) = \begin{cases} 1, if\ \beta = 0, \\ undefined, \beta = 1. \end{cases} \tag{5}$$

The authentication function means that if the distance between the access actions and the blacklist is 0, then the access should be responded with probability 1, otherwise, the authentication is undefined.

As for evaluation, in order to quantitatively evaluate the network management system, we need to discuss controllability, accessibility and cost-effectiveness. Based on practical statistic or experiments, the values of these indicators can be analyzed by the way shown in section 4.

6   Conclusions and Future Work

In this paper, a fine-grained network control model based on classification of users' network behaviors is proposed. And this control model decides whether an access action is allowed or not by analyzing the ages of users. Besides, a framework to evaluate the model is also proposed, through which the model's controllability, users' accessibility and the cost-effectiveness are analyzed respectively.

However, several problems still need to be focused on in the future work. Firstly, our learning and prediction module needs a period of time to achieve certain accuracy, so for situations where users' ip changes frequently, the model is not very suitable. Besides, the widely used encryption and anonymity also make the prediction process difficult, because it becomes harder to realize which webpage the user want to access and what the contents of the webpage are.

References

1. Fang, B., Guo, Y., Zhou, Y.: Information content security on the Internet: the control model and its evaluation. Science China Information Sciences (2010)
2. Hongli, Z.: A Survey on Internet Measurement and Analysis. Journal of Software (2003)
3. Fortuna, B., Mladenic, D., Grobelnik, M.: User Modeling Combining Access Logs, Page Content and Semantics. In: 1st International Workshop on Usage Analysis and the Web of Data (USEWOD 2011) in the 20th International World Wide Web Conference (WWW 2011), Hyderabad, India, March 28 (2011)
4. Hu, J., Zeng, H.-J., Li, H., Niu, C., Chen, Z.: Demographic prediction based on user's browsing behavior. In: Proceedings of the 16th International Conference on World Wide Web, Banff, Alberta, Canada, pp. 151–160. ACM (2007)
5. Lihua, Y., Binxing, F.: Research on controllability of internet information content security. Networked Digital Technologies (2009)
6. Liu, Y.-Y.: Controllability of complex networks. Nature 473, 167–173 (2011)
7. Park, J., Sandhu, R.: The UCONabc usage control model. ACM Transactions on Information and System Security 7(1), 128–174 (2004)
8. Peter, R.W.: There is a need to regular indecency on the internet. 6 CORNELL J.L. & PUB.POL'y 363 (1997)
9. Wiedenbeck, S., Waters, J., et al.: PassPoints: Design and longitudinal evaluation of a graphical password system. International Journal of Human-Computer Studies 63(1-2), 102–127 (2005)
10. Wei, J.: Evaluating Network Security and Optimal Active Defense Based on Attack-Defense Game Model. Chinese Journal of Computers (2008)

Two Phase Extraction Method
for Extracting Real Life Tweets Using LDA

Shuhei Yamamoto and Tetsuji Satoh

Graduate School of Library, Information and Media Studies,
University of Tsukuba
{yamahei,satoh}@ce.slis.tsukuba.ac.jp
http://ce.slis.tsukuba.ac.jp

Abstract. Nowadays, many twitter users tweet their personal affairs. Some of these posts can be quite beneficial for real life, for example, Eating, Appearance, Living, Disasters, and so on. In this paper, we propose a two phase extracting method for selecting beneficial tweets. In the first phase, many topics are extracted from a sea of tweets using Latent Dirichlet Allocation (LDA). In the second phase, associations between many topics and fewer aspects is built using a small set of labeled tweets. To enhance accuracy, the weight of feature words is calculated by information gain. Our prototype system demonstrates that the proposed method can extract the aspects of each unknown tweet.

Keywords: Twitter, Real Life, LDA, Two Phase Extraction.

1 Introduction

Many information sharing services currently exist, such as community knowledge sharing sites, blogs, and microblogs. Twitter is one of the most widely spread microblogs. Articles in Twitter are posted very easily within 140 characters. Users post their experiences, opinions, and other events that arise in their daily life.

Most articles in Twitter are both useful and timely because they are written based on current events. For example, commuter train delay information posted by a passenger is timely and useful for waiting passengers. Supermarket sales and bargain information is also useful for neighborhood consumers. Such information has highly regionality and freshness. Thus, we call such posts "Real Life Tweets". Extracting real life tweets from a sea of tweets is quite an important research issue for supporting life activities.

The Great East Japan Earthquake Disaster, which occurred in March of 2011, is a perfect example of the benefits of real life tweets. There was a great amount of confusion in the stricken area immediately following the earthquake. There was a lack of food, suspension of water supply, and train service cancellations. At that time, useful tweets reported the location of water supplies and food distribution, as well as the service status of trains, demonstrating that such real life tweets helped the users in the devastated region[1].

Y. Ishikawa et al. (Eds.): APWeb 2013, LNCS 7808, pp. 340–347, 2013.

Table 1. Aspects of Real Life

Aspect	typical terms
Appearance	clothes, dress, wearing, decoration, makeup, barber ...
Contact	appointment, set a meeting, invitation, family, friend ...
Disaster	flood, tornado, earthquake, hazard, secondary disaster ...
Eating	cooking food, dining out, eating, restaurant, junk food ...
Event	festival, ceremonial functions, project, event schedules ...
Expense	shopping, order, advertisement, discount, bargain sale ...
Health	cold, physical condition, hurt, hospital, prophylaxis ...
Hobby	leisure-time, pastime, entertainment, hobby, interest ...
Living	home, lodgings, furniture, cleaning, washing, living ...
Locality	sightseeing, regionally specific, local information ...
School	study, class, examination, education, research, homework ...
Traffic	train, bus, airplane, timetables, traffic information ...
Weather	weather forecast, temperature, humidity, wind, pollen ...
Working	job hunting, part-timer, coursework, open a store, sales ...

As mentioned above, highly useful real life tweets are increasingly posted on Twitter. However, users posts various types of tweets. "Nod" and sympathetic phrases frequently appear on Twitter. For example, "Thank you" and "I see" often appear in posts. These posts do not directly support the real life situations of other users. We believe that users want a method of locating beneficial tweets on Twitter. These types of nods and sympathies simply impede the discovery of substantive tweets.

Information is used in various aspects of life. Real life tweets can accommodate such aspects. For example, tweets such as "The train is not coming!" are categorized in the "Traffic" aspect and will support users who want to ride the train. Posts such as "Today, bargain sale items are 50% off!" are categorized as "Expense" aspects and will support users who are going shopping. In our previous research, we detected 14 aspects of real life[2]. These 14 aspects, listed in Table 1, are obtained from "Local Community"[1] and "Life"[2] in the Japanese version of Wikipedia.

In this paper, we propose a two phase extraction method for extracting real life tweets. In the first phase, many topics are extracted from a sea of tweets using Latent Dirichlet Allocation (LDA). In the second phase, associations between many topics and fewer aspects, shown in Table 1, are constructed using the weight of feature terms calculated with information gain.

2 Related Works

Real life tweets consist of both the experiences and knowledge of users and regional information. Several studies with experience mining have been conducted

---

[1] http://ja.wikipedia.org/wiki/nR~jeß
[2] http://ja.wikipedia.org/wiki/

to extract experiences from documents. Kurashima et al. [3] reported that human experience can be divided into five areas: Time, space, action, object, and feeling. Inui et al. [4] described a method of indexing personal experience information from the viewpoint of time, polarity, and speaker modality. This information is indexed as Topic object, Experiencer, Event expression, Event type, and Factuality. These mining methods are effective for relatively long documents such as blogs. Hence, these methods are not appropriate for Twitter posts, which consist of many short sentences. In addition, experience mining would be much more difficult because subjects and objects in sentences are often omitted in Twitter.

The study of Twitter is flourishing. Ramage et al. [5] used large scale topic models to represent Twitter feeds and users, showing improved performance on tasks, such as post and user recommendations. Bollen et al. [6] analysed sentiment on Twitter according to a six-dimensional mood (tension, depression, anger, vigor, fatigue, and confusion) representation, determining that sentiment on Twitter correlates with real-world values such as stock prices and coincides with cultural events. Diakopoulous and Shamma [7] conducted inspirational work in this vein, demonstrating the use of timeline analytics to explore the 2008 Presidential debates through Twitter sentiment. Sakaki et al. [8] assumed that Twitter users act as sensors, discovering an event occurring in real time in the real world. Zhao et al. [9] suggested a model, called Twitter-LDA, based on the hypothesis that one tweet expresses one content of a topic. They classified tweets into appropriate topics and extracted keywords to express the contents of the topic. Mathioudakis et al. [10] extracted burst keywords in tweets collected automatically. They found a trend fluctuating in real time by creating groups using the co-occurrence of keywords.

Our paper contends that the information is not only based on user experience, but also user knowledge, which we believe to be useful in real life.

3   Two Phase Extraction Method

Real life tweets contain the various aspects mentioned in Section 1. Therefore, it is difficult to enumerate keywords related into whole aspects. Moreover, the rule-based parsing approach using experience mining does not work well because twitter posts are comprised of very short sentences.

We propose a two phase extraction method, shown in Figure 1. In the first phase, a large number of topics are extracted from a sea of tweets using LDA. LDA is an unsupervised learning model for clustering large amounts of documents [11]. In the second phase, we construct an association between the topics and aspects.

3.1   Association Building Using Labeled Tweets

For building associations, we prepare a small set of labeled tweets. A set of extraction terms from tweets labeled aspect ap is W_{ap}. To calculate the weight of terms that represent each aspect, we use information gain. Information gain

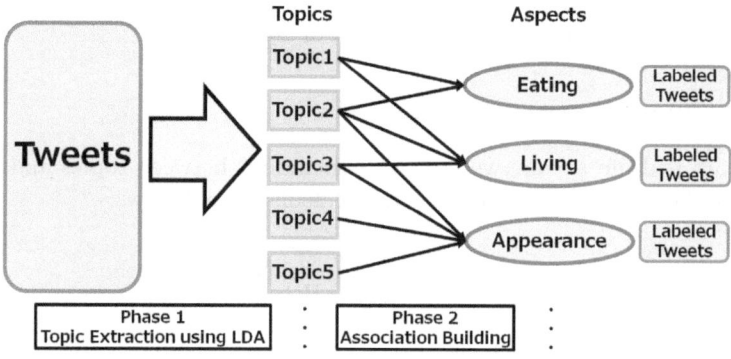

Fig. 1. Two Phase Extraction Method

provides feature choice and represents a good feature. The information gain of term w is calculated as follows:

$$IG(w) = H(AP) - (P(w)H(AP|w) + P(\bar{w})H(AP|\bar{w})), \qquad (1)$$

where AP denotes all aspects. $P(w)$ is the probability of term w's appearance in all tweets. $P(\bar{w})$ is the probability of term w not appearing in all tweets. $H(AP|w)$ is the conditional entropy of all aspects AP that appear in term w. $H(AP|\bar{w})$ is the conditional entropy of all aspects AP that do not appear in term w. When $IG(w)$ has a high value, w is a good feature.

Relevance $R(ap, k)$ between topics k and aspects ap is then calculated as follows:

$$R(ap, k) = \frac{1}{length(W_{ap})} \sum_{w \in W_{ap}} IG(w) * p_{w,k}, \qquad (2)$$

where $p_{w,k}$ denotes term w of occurrence probability in topic k. $length(W_{ap})$ is the size of the terms set. Note that this equation calculates the relevance using the occurrence probability and information gain of terms.

To fold in value 0 to 1, we normalize $R(ap, k)$ in each aspect. Normalized relevance $\hat{R}(ap, k)$ is shown as follows:

$$\hat{R}(ap, k) = \frac{R(ap, k)}{\sum_{k \in T} R(ap, k)}, \qquad (3)$$

where T denotes all topics extracted using LDA.

We make an association between topics and aspects when $\hat{R}(ap, k)$ exceeds a threshold $\gamma(ap)$. The threshold value in each aspect is calculated as follows:

$$\gamma(ap) = \max_{k \in T} \hat{R}(ap, k) - std(ap) * d, \qquad (4)$$

where $std(ap)$ denotes the standard variation in $\hat{R}(ap, k)$ for all topics. Parameter d varies for maximum extraction precision in Section 4.

3.2   Real Life Tweet Extraction

To extract real life tweets, we use the associations between topics and aspects. The score between tweets and aspects is calculated as follows:

$$Score(ap) = \frac{\hat{R}(ap, k)}{length(W)} \sum_{w \in W} p_{w,k}, \tag{5}$$

where W denotes a set of terms extracted from an unknown tweet and $p_{w,k}$, denotes the term w of occurrence probability in topic k.

The selected aspect in each unknown tweet is as follows:

$$Aspect = \arg \max_{ap \in AP} Score(ap) \tag{6}$$

4   Experimental Evaluations

We evaluate the precision of the aspect selected by equation (6). After that, we should clarify why we drew a difference between high precision and lower precision. Because of this, we carefully observe the association between topics and aspects of varying thresholds d.

4.1   Datasets and Parameters Setting

Datasets: We used datasets of tweets posted from April 15, 2012 to August 14, 2012 in the Japanese location information of "Tsukuba" or "". The number of tweets is 2,900,819. This large dataset is classified into topics using LDA.

We also prepared a small set of tweets labeled with each aspect. The collection term and condition are the same as above. The number of tweets in each aspect is 200. We also prepared 200 un-real life tweets to confirm classification. These sets of tweets are used for building the association.

Parameters Setting: LDA requires some hyper parameters. According to related works[12], we set α at $50/T$ and β at 0.1. The iterative calculation count in LDA is 100 times in every case. We set the number of topics at 500.

Extraction Precisions: We evaluate the precisions by using 10-fold cross-validation. Thus, 180 tweets in each aspect are used for building association and the remaining 20 tweets are used for testing.

4.2 Experimental Results

Extraction Precisions: The extraction precisions of every aspect are shown in Figure 2. The horizontal axis is threshold d that decides the association between topics and aspects. In some aspects, i.e., Appearance, Eating, and Traffic, maximum extraction precision is achieved at the minimum value of $d(d = 1)$. In some aspects, i.e., Contact, Event, and Health, maximum extraction precision is achieved at the maximum value of $d(d = 20)$.

To analyze the association between topics and aspects, we evaluate the number of connections from topics to each aspect. The number of topics against threshold d is shown in Figure 3. In every topic, the number of topics is increased according to d. The five aspects of Eating, Living, Expense, Disaster, and Working mostly connect one topic until $d \leq 12$. On the other hand, the Contact and Event aspects are rapidly reached to connect whole topics.

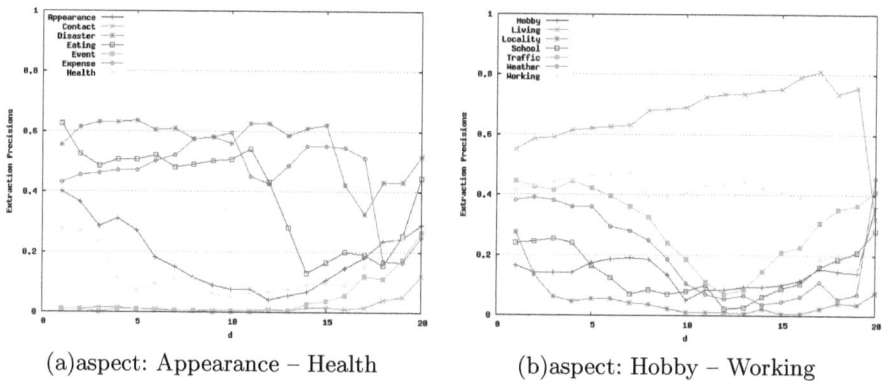

(a)aspect: Appearance – Health   (b)aspect: Hobby – Working

Fig. 2. Extraction Precisions

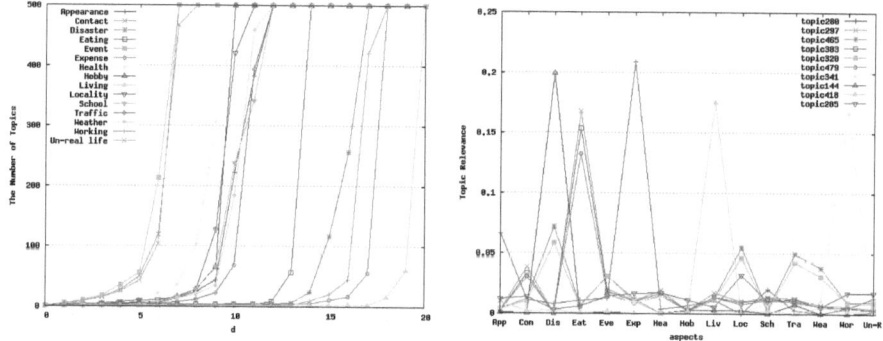

Fig. 3. Number of Topics in each Aspect   **Fig. 4.** Top 10 Topics

5   Discussion

According to Figure 2, Living, Eating, Disaster, and Expense achieved the highest precisions. These four aspects are associated with fewer topics, as shown in Figure 3, even if parameter d was increased. Association details between each aspect and topics are shown in Figure 4. These ten topics in the figure have a larger sum of relevances $\hat{R}$ from topics to aspects. In this figure, we confirm that these four aspects have strong relevance to the specific topics, Living to Topic418, Disaster to Topic144, and so on, respectively.

The worst cases, Event and Contact, have a lower precision value as shown in Figure 2. These aspects were associated with many topics even if d was small. In particular, Contact is associated with the same topics, Topic297, Topic383, and Topic479, as Eating. Moreover, the values of $\hat{R}$ are smaller than Eating. Event is associated with the same topics, Topi465 and Topic320, as Locality.

6   Conclusion

In this paper, we propose a two phase extraction method for extracting real life tweets. In the first phase, many topics are extracted from a sea of tweets using LDA. In the second phase, associations between many topics and fewer aspects are constructed using a small set of labeled tweets. To enhance accuracy, the weight of the feature terms is calculated with information gain.

Based on the experimental evaluation results, our prototype system demonstrates that our proposed method can extract aspects of each unknown tweet. We confirm that high precision aspects are associated with fewer topics that are similar to the aspects. However, low precision aspects are associated with many topics. In this case, many topics are associated with many aspects. In the future, we should consider evaluating the KL Divergence from aspect to aspect in order to enhance separation among the aspects.

Acknowledgement. This work was partly supported by Research Projects of Faculty of Library and Information and Media Science, University of Tsukuba.

References

1. Yamamoto, M., Ogasawara, H., Suzuki, I., Furukawa, M.: Tourism informatics:9. information propagation network for 2012 tohoku earthquake and tsunami on twitter. IPSJ Magazine 53(11), 1184–1191 (2012) (in Japanese)
2. Yamamoto, S., Satoh, T.: Real life information extraction method from twitter. In: The 4th Forum on Data Engineering and Information Management (DEIM 2012) F3-4 (2012) (in Japanese)
3. Kurashima, T., Tezuka, T., Tanaka, K.: Blog map of experiences: Extracting and geographically mapping visitor experiences from urban blogs. In: Ngu, A.H.H., Kitsuregawa, M., Neuhold, E.J., Chung, J.-Y., Sheng, Q.Z. (eds.) WISE 2005. LNCS, vol. 3806, pp. 496–503. Springer, Heidelberg (2005)

4. Inui, K., Abe, S., Morita, H., Eguchi, M., Sumida, A., Sao, C., Hara, K., Murakami, K., Matsuyoshi, S.: Experience mining: Building a large-scale database of personal experiences and opinions from web documents. In: Proceedings of the 2008 IEEE/WIC/ACM International Conference on Web Intelligence, pp. 314–321 (2008)
5. Ramage, D., Dumais, S., Liebling, D.: Characterizing microblogs with topic models. In: Proceedings of ICWSM 2010, pp. 130–137 (2010)
6. Bollen, J., Pepe, A., Mao, H.: Modeling public mood and emotion: Twitter sentiment and socio-economic phenomena. In: Proceedings of WWW 2010, pp. 450–453 (2010)
7. Diakopoulous, N.A., Shamma, D.A.: Characterizing debate performance via aggregated twitter sentiment. In: Proceedings of CHI 2010, pp. 1195–1198 (2010)
8. Sakaki, T., Okazaki, M., Matsuo, Y.: Earthquake shakes twitter users: Real-time event detection by social sensors. In: Proceedings of 18th International World Wide Web Conference, WWW 2010, pp. 851–860 (2010)
9. Zhao, X., Jiang, J., He, J., Song, Y., Achananuparp, P., Lim, E.P., Li, X.: Topical key phrase extraction from twitter. In: The 49th Annual Meeting of the Association for Computational Linguistics, pp. 379–388 (2011)
10. Mathioudakis, M., Koudas, N.: Twittermonitor: trend detection over the twitter stream. In: Proceedings of the 2010 International Conference on Management of Data, pp. 1155–1158 (2010)
11. Blei, D.M., Ng, A.Y., Jordan, M.I.: Latent dirichlet allocation. The Journal of Machine Learning Research 3, 993–1022 (2003)
12. Griffiths, T.L., Steyvers, M.: Finding scientific topics. Proceedings of the National Academy of Science 101, 5228–5235 (2004)

A Probabilistic Model
for Diversifying Recommendation Lists

Yutaka Kabutoya[1], Tomoharu Iwata[2],
Hiroyuki Toda[1], and Hiroyuki Kitagawa[3]

[1] NTT Service Evolution Laboratories, NTT Corporation, Japan
[2] NTT Communication Science Laboratories, NTT Corporation, Japan
[3] Faculty of Engineering, Information and Systems, University of Tsukuba, Japan
{kabutoya.yutaka,iwata.tomoharu,toda.hiroyuki}@lab.ntt.co.jp,
kitagawa@cs.tsukuba.ac.jp

Abstract. We propose a probabilistic method to diversify the results of collaborative filtering. Recommendation diversity is being studied by many researchers as a critical factor that significantly influences user satisfaction. Unlike conventional approaches to recommendation diversification, we theoretically derived a diversification method. Specifically, our method naturally diversifies a recommendation list by maximizing the probability that a user selects at most one item from the list. For enhanced practicality, we formulate a model for the proposed method on three policies — robust estimation, the use of only purchase history, and the elimination of any hyperparameters controlling the diversity. In this paper, we formally demonstrate that our method is practically superior to conventional diversification methods, and experimentally show that our method is competitive with conventional methods in terms of accuracy and diversity.

1 Introduction

Many online services, such as Amazon[1], Netflix[2], and YouTube[3], employ recommender systems because they are expected to improve user convenience and service provider profit [6, 10, 12, 18]. This is promoting a lot of research on recommendation methods [1]. Many conventional recommendation methods are *accuracy*-oriented, i.e., their goal is to accurately predict items that the user will purchase in the future.

These accuracy-oriented algorithms, however, sometimes degrade user satisfaction. The recommendation results are often presented to users in the form of a multiple item list, not one single item. The recommended items in a list generated by an accuracy-oriented method often have similar content. According to [21], since such a list reflect only partial aspects of the user's interests, it cannot necessarily satisfy the user.

---

[1] Amazon - Online shopping, http://www.amazon.com
[2] Netflix - Online movie rentals, http://www.netflix.com
[3] YouTube - Online video service, http://www.youtube.com

Y. Ishikawa et al. (Eds.): APWeb 2013, LNCS 7808, pp. 348–359, 2013.

Researchers have recently addressed this problem by focusing on the *diversity* of recommendation lists [9, 11, 17, 19, 21]. This novel evaluation metric indicates balance of recommendation list entries, i.e. how well they satisfy the active user's full range of interests.

This paper proposes a theoretical method to diversify recommendation lists. Our method make a recommendation list so as to maximize the probability that a user purchases at most *one* item from the list. While exactly optimizing our objective function is difficult, we derive a greedy heuristic for approximately optimizing it. Consequently, the diversity of the recommendation lists offered by our method is naturally enhanced. Specifically, estimating the probability that a user will purchase the lth item from the list requires us to consider, in addition to the user's preferences, the additional condition that the first $l-1$th items are not purchased. This additional condition suggests that the lth item should be dissimilar to the first $l-1$ items.

We formulate a model and its parameter estimation for our method by setting the following three policies as: First, we reduce the number of parameters for the robust estimation. Second, parameters are estimated only from purchase history, which can be automatically collected. Collecting the other kinds of information used by conventional methods, such as item taxonomy [21] and ratings [11, 17, 19] requires some manual tasks. Finally, we avoid the use of hyperparameters in controlling the trade-off between accuracy and diversity, unlike [9, 17, 21]. Determining such hyperparameters is difficult, since we do not know the optimal balance between them that will maximize end-user satisfaction in advance.

The rest of this paper is organized as follows. In § 2, we explain the detail of our method. In § 3, we formulate a model and its parameter estimation for our method. In § 4, we briefly review related work. In § 5, we formally and experimentally evaluate our method, by comparing it with conventional methods. Finally, we offer concluding remarks and a discussion of future work in § 6.

2   Proposed Method

Our method offers a recommendation list consisting of L items maximizing the probability that a user u purchases an item from the list. Our objective function for determining the recommendation list is as follows:

$$P\left(y_1 = 1 \vee \ldots \vee y_L = 1 | u, i_1, \ldots, i_L\right), \tag{1}$$

where i_l represents the lth item in the list, y_l is a binary variable indicating that the lth item is purchased or not ($y_l = 1$ if purchased, $y_l = 0$ if not). On the other hand, the objective function of a conventional accuracy-oriented method is individually set for each item in the list without considering a purchase from the whole list [2].

Here, we assume that the target user purchases at most *one* item from the list. This assumption is based on the behavior of many actual recommendation services, such as on Amazon.com. On these services, a user cannot purchase multiple items from a list, since the screen is transited when the user selects (click a link) an item from a recommendation list.

We use the greedy algorithm because exactly maximization of the objective function expressed by Eq. (1) falls into a combinatorial optimization problem, which is difficult in general. Since many users decide whether to purchase each item in the list in order of rank, applying the greedy approach seems reasonable.

The first item can be easily obtained by maximizing $P(y_1 = 1|u, i_1)$ as is done in conventional methods. Having chosen the first item i_1, we determine the second item i_2 so as to maximize $P(y_1 = 1 \vee y_2 = 1|u, i_1, i_2)$. For simplicity, we expand $P(y_1 = 1 \vee y_2 = 1|u, i_1, i_2)$ as:

$$
\begin{aligned}
&P\left(y_1 = 1 \vee y_2 = 1|u, i_1, i_2\right) \\
&= P\left(y_1 = 1|u, i_1, i_2\right) + P\left(y_1 = 0 \wedge y_2 = 1|u, i_1, i_2\right) \\
&\simeq P\left(y_1 = 1|u, i_1\right) + P\left(y_2 = 1|u, i_1, i_2, y_1 = 0\right) P\left(y_1 = 0|u, i_1\right),
\end{aligned}
$$

where the partitioning of the event of $y_1 = 1 \vee y_2 = 1$ into the disjoint events $y_1 = 1$ and $y_1 = 0 \wedge y_2 = 1$ in the second line, and the simplification in the last line is based on the independence of y_1 and i_2. Since only the second factor in this equation depends on i_2, we can formulate the probability that the user u selects the second item i_2 from the list as follows:

$$
P\left(y_2 = 1|u, i_2, y_1 = 0, i_1\right).
$$

Recursively, we can express the probability that the user u selects the lth item i_l from the list as follows:

$$
P\left(y_l = 1|u, i_l, y_1 = 0, \ldots, y_{l-1} = 0, i_1, \ldots, i_{l-1}\right). \tag{2}
$$

This probability means that for choosing i_l according to our two assumptions, we should take into account not only the target user u but also the condition that items $i_1, \ldots, i_{l-1}$ are not purchased.

3   Model and Parameter Estimation

We formulate a model and its parameter estimation for our method based on the following three policies:

1. We reduce the number of parameters for the robust estimation. The direct robust estimation of this model is difficult because the number of parameters is of the order of $O\left(UN^l\right)$, where U is the number of unique users and N is the number of unique items, and so becomes excessive compared with the size of purchase history for training.
2. We use only purchase history, which can be automatically collected. This avoids the expensive manual tasks needed to acquire the other kinds of information needed for recommendations, such as item taxonomy and ratings.

3. We avoid the use of hyperparameters in controlling the balance between accuracy and diversity. According to [21], there is a trade-off between the two, and end-user satisfaction demands that they be well balanced. Parameter tuning is, however, difficult.

We expand Eq. (2) in accordance with Bayes' rules as follows:

$$
\begin{aligned}
&P\left(y_l = 1 | u, i_l, y_1 = 0, \ldots, y_{l-1} = 0, i_1, \ldots, i_{l-1}\right)\\
&= \frac{P\left(y_l = 1 \wedge y_1 = 0 \wedge \ldots \wedge y_{l-1} = 0 | u, y_l = 1, i_1, \ldots, i_l\right)}{\sum_{y_l'=0}^{1} P\left(y_l' \wedge y_1 = 0 \wedge \ldots \wedge y_{l-1} = 0 | u, y_l', i_1, \ldots, i_l\right)}\\
&= \frac{P\left(y_l = 1 | u, i_1, \ldots, i_l\right) P\left(y_1 = 0 \wedge \ldots \wedge y_{l-1} = 0 | u, y_l = 1, i_1, \ldots, i_l\right)}{\sum_{y_l'=0}^{1} P\left(y_l' | u, i_1, \ldots, i_l\right) P\left(y_1 = 0 \wedge \ldots \wedge y_{l-1} = 0 | u, y_l', i_1, \ldots, i_l\right)}. \quad (3)
\end{aligned}
$$

We approximate the numerator of Eq. (3) so as to satisfy the policies. First, to satisfy the first policy, we assume that y_l (the event that i_l is purchased) is independent of the items shown before, and the first factor becomes as follows:

$$
P\left(y_l = 1 | u, i_1, \ldots, i_l\right) \simeq P\left(y_l = 1 | u, i_l\right).
$$

The probability expressed as the right hand side can be equivalent to the recommendation score of conventional accuracy-oriented methods, and thus can be regarded as a personalization factor focused on accuracy. To satisfy the first and the third policies, we approximate the second factor by assuming the independence of $y_1 = 0, \ldots, y_{l-1} = 0$ (the event that the first $l-1$ items are not purchased) and u (the target user) as:

$$
\begin{aligned}
&P\left(y_1 = 0 \wedge \ldots \wedge y_{l-1} = 0 | u, y_l = 1, i_1, \ldots, i_l\right)\\
&\simeq P\left(y_1 = 0 \wedge \ldots \wedge y_{l-1} = 0 | y_l = 1, i_1, \ldots, i_l\right).
\end{aligned}
$$

The probability expressed as the second line is high when i_l is dissimilar to the past items, $i_1, \ldots, i_{l-1}$, and thus can be regarded as a diversification factor that ensures diversity. This probability does not depend on any hyperparameters controlling diversity. Additionally, to satisfy the first policy, we assume the conditional independence of the event that the l'th item is not purchased and the event that the l''th item is not purchased given the event that lth item is purchased as:

$$
P\left(y_1 = 0 \wedge \ldots \wedge y_{l-1} = 0 | y_l = 1, i_1, \ldots, i_l\right) \simeq \prod_{l'=1}^{l-1} P\left(y_{l'} = 0 | y_l = 1, i_{l'}, i_l\right).
$$

We also approximate the denominator of Eq. (3) in the same manner as the numerator.

In summary, the probability that u will purchase the lth item from the list can be expressed as follows:

$$
\begin{aligned}
&P\left(y_l = 1 | u, i_l, y_1 = 0, \ldots, y_{l-1} = 0, i_1, \ldots, i_{l-1}\right)\\
&\simeq \frac{P\left(y_l = 1 | u, i_l\right) \prod_{l'=1}^{l-1} P\left(y_{l'} = 0 | y_l = 1, i_{l'}, i_l\right)}{\sum_{y_l'=0}^{1} P\left(y_l' | u, i_l\right) \prod_{l'=1}^{l-1} P\left(y_{l'} = 0 | y_l', i_{l'}, i_l\right)}. \quad (4)
\end{aligned}
$$

The number of parameters in this model is $O\left(UN + N^2\right)$, which is far smaller than $O\left(UN^l\right)$, the number in the model expressed by Eq. (2).

We approximate the personalization factor as follows without considering position l:

$$P\left(y_l = 1 | u, i_l = i\right) \simeq P\left(y = 1 | u, i\right), \tag{5}$$

where y is a binary variable representing the event that item i is purchased or not ($y = 1$ if purchased, $y = 0$ if not). For the second policy, we estimate the personalization factor by analyzing purchase history via *Latent Dirichlet Allocation* (*LDA*) [2], since it can offer high recommendation accuracy with only purchase history. With LDA, $P(i|u)$, which is the probability that the user will choose i from items $1, \ldots, N$, is estimated instead of $P(y = 1 | u, i)$ for recommendations. For estimating the probability expressed as Eq. (4), we formulate the estimation of $P(y = 1 | u, i)$ using $P(i|u)$. Since the order of the recommendation list for each user must remain after the conversion, we assume that the ratio between $P(y = 1 | u, i)$ and $P(i|u)$ depends on only the user and is independent of each item as follows:

$$P\left(y = 1 | u, i\right) = \lambda_u \cdot P\left(i | u\right), \tag{6}$$

In accordance with Bayes' rule, λ_u becomes as follows:

$$\lambda_u = \frac{P\left(y = 1 | u\right)}{\sum_{i=1}^{I} P\left(i | u\right)^2}, \tag{7}$$

where, for the readability, we skip the detailed derivation of λ_u. Once $P(y = 1 | u, i)$ is estimated, we can easily estimate $P(y = 0 | u, i)$ included in the denominator of the Eq. (4) as $1 - P(y = 1 | u, i)$. With LDA, first we assign latent topics to the past purchases using collapsed Gibbs sampling [7], and then $P(i|u)$ is calculated as:

$$P\left(i | u\right) = \sum_{z=1}^{Z} \frac{\sum_{m=1}^{M_u} \mathbb{I}\left[z_{um} = z\right] + \beta}{M_u + Z\beta} \cdot \frac{\sum_{u'=1}^{U} \sum_{m=1}^{M_{u'}} \mathbb{I}\left[z_{u'm} = z \wedge i_{u'm} = i\right] + \eta}{\sum_{u'=1}^{U} \sum_{m=1}^{M_{u'}} \mathbb{I}\left[z_{u'm} = z\right] + N\eta}, \tag{8}$$

where Z is the given number of topics, M_u is the number of u's purchases, z_{um} is the assigned latent topic for the mth purchase of user u, i_{um} is the item in the mth purchase of user u, $\mathbb{I}[\cdot]$ is the indicator function, and β and η are Dirichlet parameters. Z can be easily determined by cross validation as minimizing the prediction error metric, which is called *perplexity*, and Dirichlet parameters β and η can be inferred via Minka's fixed point iterations [13]. An expression for the MAP estimation of $P(y = 1 | u)$ using purchase history can be written as:

$$P\left(y = 1 | u\right) = \frac{\sum_{i=1}^{N} \mathbb{I}\left[i \in \boldsymbol{i}_u\right] + \gamma}{N + 2\gamma}, \tag{9}$$

where γ is a hyperparameter for smoothing. The numerator of this expression is the number of items which u purchased.

We approximate the diversification factor without considering positions l and l' as follows:

$$P\left(y_{l'} = 0 | y_l = 1, i_{l'} = i', i_l = i\right) \simeq P\left(y' = 0 | y = 1, i', i\right), \tag{10}$$

where y' represents the event that item i' is purchased or not ($y' = 1$ if purchased, $y' = 0$ if not).

For the second policy, we formulate the MAP estimation of this approximated diversification factor from the purchase history as follows:

$$P\left(y' = 0 | y = 1, i', i\right) = \frac{\sum_{u=1}^{U} \mathbb{I}\left[i \in \boldsymbol{i}_u \wedge i' \notin \boldsymbol{i}_u\right] + \delta}{\sum_{u=1}^{U} \mathbb{I}\left[i \in \boldsymbol{i}_u\right] + 2\delta}, \tag{11}$$

where δ is a hyperparameter for smoothing. The denominator of this expression is the number of users who have purchased item i, and the numerator is the number of users who have purchased item i but not item i'.

4   Related Work

Recently, diversification of recommendation lists has targeted by many researchers [9, 11, 17, 19, 21] since it is known to be an important factor in improving user satisfaction through recommendations. [21] proposed a re-ranking algorithm that uses item taxonomy to realize topic diversification. Their method can control the trade-off between accuracy and diversity by using a hyperparameter determined in an actual user experiment. [9] formalized the intra-list topic diversification problem by addressing a multi-objective optimization problem on diversity and preference similarity. [11] proposed a method to make diversified recommendations over time, and discovered the negative correlation between user profile length and the degree of recommendation diversity. [19] raised the issue of evaluating the novelty and diversity of recommendations. [17] studied the adaptive diversification problem for recommendations by connecting latent factor models with portfolio retrieval.

Additionally, in the area of information retrieval and Web search, many researchers have studied the diversity of search results [3–5, 8, 14–16]. In particular, [4] is strongly related to our work since they are pioneers in the use of the probabilistic approach to list diversification. Our method can be positioned as an extension of [4] to recommendations. We expand Eq. (1) to Eq. (2) in the same manner as the greedy approach to maximize the probability that least one document in the search results is relevant.

5   Evaluation

5.1   Formal Evaluation

First, we evaluated our method formally in terms of practicality, by comparing it to conventional methods for diversifying recommendation lists. We pick up three conventional approaches, which we label *Ziegler* [21], *Hurley* [9], and *Shi* [17].

First, our method and *Hurley* do not require any manual tasks for generating information sources, while *Ziegler* and *Shi* do. *Ziegler* uses item taxonomy, which is manually provided by service developer. *Shi* uses ratings, which are manually given by users. On the other hand, our method and *Hurley* work even if only the purchase history is available, and thus allow us to skip such manual tasks.

Second, our method and *Shi* do not use any hyperparameters for controlling the balance between accuracy and diversity, while *Ziegler* and *Hurley* do. According to [21], there is a trade-off between accuracy and diversity, and balancing the two maximizes end-user satisfaction. This balancing operation varies with the service and/or the dataset, so a parameter tuning is needed to determine the optimal balance. In general, such tuning is difficult since it requires end-user experiments.

To summarize, we formally demonstrated that our method is superior to conventional methods in terms of practicality.

5.2 Experimental Evaluation

Dataset. We experimentally evaluated our method using the MovieLens-1M dataset[4]. This publicly available dataset contains 1M ratings from about 6K users on about 3.7K items. The data sparseness is 95.5%. Each user in this dataset has at least 20 ratings. Additionally, genre labels of the items are provided. There are 18 genres, and each item can be associated with multiple genres. The average number of genres per items is 1.62. Note that we interpreted the MovieLens data as purchase data and discarded the ratings.

We evaluated our method by 5-fold cross validation. That is, we randomly partitioned the data into 5 subsets. Of the 5 subsets, we held-out a single subset as the test data (for evaluation), and we used the remaining 4 subsets as training data (for estimating parameters), where we removed movies not appearing in the training data from the test data and user-item pairs contained in the training data from the test data. Then we repeated this cross-validation process 5 times, with each of the 5 subsets used exactly once as the test data. Finally, we averaged the 5 results to produce a single estimation.

Compared Methods. We compared the following recommendation methods in terms of recommendation accuracy and diversity:

1. *Proposed*: This is the method we propose. In this paper, we used hyperparameters $Z = 50$, $\gamma = 0.1$, and $\delta = 0.1$.
2. *Non-diversified*: Straight collaborative filtering using LDA [2]. The list offered by this method is the same as the list offered by the personalization factor of our method. In this method, we used the same number of topics $Z = 50$.
3. *Ziegler*: The diversification method proposed by Ziegler et al. [21]. We used the lists generated by LDA [2] where $Z = 50$ as the original lists, the metric

---

[4] MovieLens dataset, http://www.grouplens.org/node/73

proposed in [22] to measure similarity between two items, and genre information for item taxonomy in calculating the similarity. In this paper, we used
the best settings yielded by the empirical analysis in [21, 22]. Specifically,
propagation factor $\kappa = 0.75$, and diversification factor $\Theta_F = 0.3$.

Result. In our experiments, we evaluated the accuracy and the diversity of the
recommendation lists produced by the three methods.

We used *precision, micro-averaged recall*, and *macro-averaged recall curves* to
measure accuracy. It is important to use both micro- and macro-averaged recall
[20], as micro-averaged metrics stress performance on prolific users, while macro-
averaged ones target those with few transactions. Both are used in this paper for
fairness and for detail. The formulae of precision@K, micro-averaged recall@K,
and macro-averaged recall@K are given below:

$$\text{precision@}K = 100 \cdot \frac{\sum_{u=1}^{U} \sum_{i=1}^{N} \mathbb{I}[i \in i_u^{\text{test}} \wedge i \in \hat{i}_u^{\langle K \rangle}]}{UK},$$

$$\text{micro-averaged recall@}K = 100 \cdot \frac{\sum_{u=1}^{U} \sum_{i=1}^{N} \mathbb{I}[i \in i_u^{\text{test}} \wedge i \in \hat{i}_u^{\langle K \rangle}]}{\sum_{u=1}^{U} \sum_{i=1}^{N} \mathbb{I}[i \in i_u^{\text{test}}]},$$

$$\text{macro-averaged recall@}K = 100 \cdot \frac{1}{U} \sum_{u=1}^{U} \frac{\sum_{i=1}^{N} \mathbb{I}[i \in i_u^{\text{test}} \wedge i \in \hat{i}_u^{\langle K \rangle}]}{\sum_{i=1}^{N} \mathbb{I}[i \in i_u^{\text{test}}]},$$

where i_u^{test} is the data for u appearing in the test set, and $\hat{i}_u^{\langle K \rangle}$ is top-K recommended items to u. Figure 1 show the recommendation accuracy of compared
methods via precision, micro-averaged recall, and macro-averaged recall curves.
These results reveal that our diversification also degrades the accuracy as does
Ziegler. The precision curves show that our method is of practical use because
its loss in accuracy at high rank is less pronounced than is true for *Ziegler*.
Specifically, our method degrades the accuracy less than *Ziegler* at the high
rank ($K \leq 5$), but the accuracy curves for our method drastically drop, and
our method is then outperformed by *Ziegler* at the low rank ($K > 8$). Comparing micro-averaged and macro-averaged recall, the drop of our method is less
for macro-averaged recall than for micro-averaged recall. This result means that
our method can offer more accurate recommendations to the users who make
few purchases.

To measure diversity, we employed *intra-list similarity* and *original list overlap*
as used in [21]. Intra-list similarity is calculated as follows:

$$\text{ILS@}K = \frac{1}{U} \sum_{u=1}^{U} \sum_{k=1}^{K} \sum_{k'=k+1}^{K} \text{sim}(\hat{i}_{uk}, \hat{i}_{uk'}),$$

where $\hat{i}_{uk}$ is the kth recommendation to u, and $\text{sim}(\cdot)$ is the item similarity
measured by the metric proposed in [22]. The original list overlap can be calculated by counting the number of recommended items that stay the same after

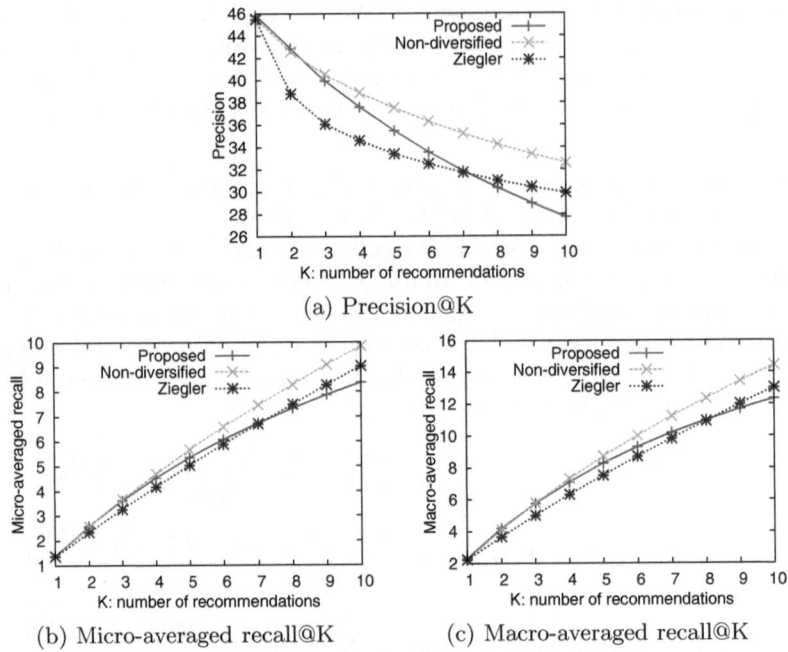

Fig. 1. Recommendation accuracy of compared methods

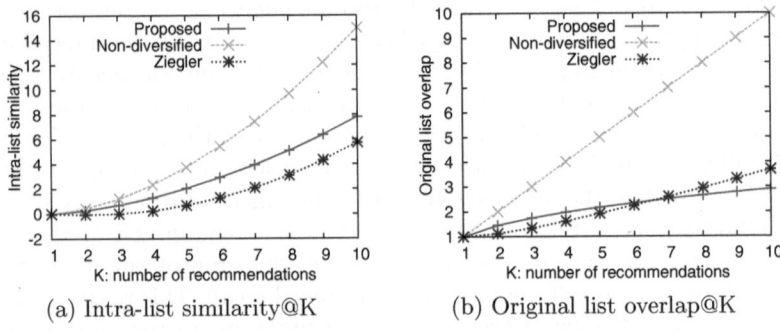

Fig. 2. Recommendation diversity of compared methods

the diversification. Higher values of these metrics denote lower diversity. Figure 2 shows recommendation diversity of the compared methods based on intra-list similarity and original list overlap. These figures reveal that our method can increase diversity even though it uses only purchase history. From Figure 2 (a), we found that our method can significantly lower the taxonomy-driven pairwise similarity in lists without recourse to item taxonomy. Specifically, in Figure 2 (a), the curve of our method is closer to the *Ziegler* curve than to the

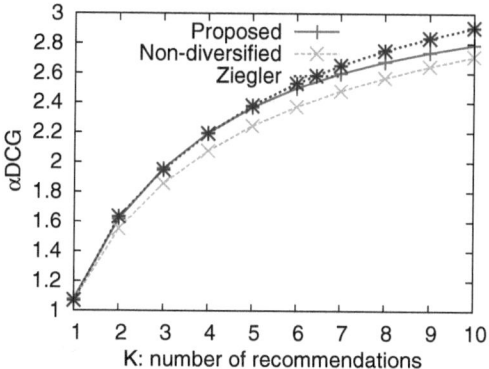

Fig. 3. Combination of accuracy and diversity based on αDCG

Non-diversified curve. Figure 2 (b) demonstrates that both diversification methods change recommended items drastically. For the top-10 recommended items, only 3 items remain after diversification.

Finally, we investigate the combination of accuracy and diversity via αDCG [5] using the genres in MovieLens dataset:

$$\alpha\text{DCG@}K = \frac{1}{U}\sum_{u=1}^{U}\sum_{k=1}^{K}\frac{\sum_{g=1}^{G}\mathbb{I}[\hat{i}_{uk} \in i_u^{\text{test}} \wedge g \in g_{\hat{i}_{uk}}](1-\alpha)^{q_{g,k-1}^u}}{\log_2(1+k)},$$

where G is the number of unique genres, $\hat{i}_{uk}$ is the kth recommendation to u, g_i are the genres assigned to item i, $q_{g,k-1}^u$ denotes the number of items ranked up to position $k-1$ that contain genre g in the recommended items to u. α is a constant controlling the magnitude of the penalty placed on recommendation redundancy. Higher values of α indicate larger penalties. In our experiments, we used the setting discussed based on the official TREC judgments for QA tasks in [5]: $\alpha = 0.36$. Although αDCG is originally normalized via αIDCG as αNDCG, we skipped normalization because our focus in this experiment was to compare diversification methods. Figure 3 shows the αDCG scores of the compared methods. This figure shows that our method is competitive with *Ziegler* at the high rank ($K \leq 5$).

Table 1 illustrates three sampled items from this user profile and an example of the recommendation lists made by the three compared methods. From the profile, we can see that this user watches various kinds of action movies. Intuitively, this example shows that our method produced a moderately diversified recommendation list: The list includes three action movies, two comedy movies. *Ziegler* yielded a similar recommendation list. On the other hand, the list offered by *Non-diversified* is overly focused. Specifically, it provides only similar action movies.

Table 1. Example recommendation results for five items using a user profile. We refer "Ac" to *Action*, "Ad" to *Adventure*, "An" to *Animation*, "Ch" to *Children's*, "Co" to *Comedy*, "Cr" to *Crime*, "H" to *Horror*, "D" to *Drama*, "R" to *Romance*, "SF" to *Sci-Fi*, "T" to *Thriller*, and "W" to *War*. ILS means intra-list similarity.

	Three samples from this user profile		
	Aliens (Ac,SF,T,W)		
	Speed (Ac,R,T)		
	Jurassic Park (Ac,Ad,SF)		

Rank	*Proposed*	*Non-diversified*	*Ziegler*
1	Raiders (Ac,Ad)	Raiders (Ac,Ad)	Raiders (Ac,Ad)
2	Star Wars V (Ac,Ad,D,SF,W)	Star Wars V (Ac,Ad,D,SF,W)	Toy Story (An,Ch,Co)
3	Braveheart (Ac,D,W)	Indiana Jones (Ac,Ad)	Ghostbusters (Co,H)
4	American Beauty (Co,D)	Lethal Weapon (Ac,Co,Cr,D)	Star Wars V (Ac,Ad,D,SF,W)
5	Ghostbusters (Co,H)	Hunt for Red Oct. (Ac,T)	Hunt for Red Oct. (Ac,T)
ILS@5	1.998	4.239	0.549

To summarize, we experimentally demonstrated that end-user satisfaction for the recommendations via our method may be competitive with *Ziegler* in terms of accuracy and diversity.

6   Conclusion and Future Work

We developed a new method to diversify recommendation lists. Our key advances were to naturally derive a diversification method by determining a recommendation list so as to maximize the probability that a user purchases at most one item from a list, and and formulate a model and its parameter estimation based on several policies. In this paper, we formally showed that our method can, unlike conventional methods, avoid manual tasks such as the collection of information and the control of the balance between accuracy and diversity. Additionally, experiments on a real-world data set demonstrated that our method is competitive with the conventional methods in terms of accuracy and diversity. These results suggest that our method is practically promising.

Future work involves three issues. First, we would like to extend our method to use recommendation information other than purchase history such as item metadata automatically drawn from external knowledge bases. Second, we will examine the effectiveness of diversification parameters, which are not used in the current model. We would like to formulate a new model that uses a parameter to control the trade-off between accuracy and diversity, and automatically estimate this parameter from user behavior logs. Finally, we would like to apply our method to an online store and examine how well it can enhance end-user satisfaction.

Acknowledgements. We would like to thank the GroupLens Research Group for the use of the MovieLens 1M dataset.

References

1. Adomavicious, G., Tuzhilin, A.: Toward the next generation of recommender systems: a survey of the state-of-the-art and possible extensions. IEEE TKDE 17(6), 734–749 (2005)
2. Blei, D., Ng, A., Jordan, M.: Latent Dirichlet allocation. J. Machine Learning Research 3, 993–1022 (2003)
3. Carbonell, J., Goldstein, J.: The use of MMR, diversity-based reranking for reordering documents and producing summaries. In: SIGIR 1998, pp. 335–336 (1998)
4. Chen, H., Karger, D.: Less is more: probabilistic models for retrieving fewer relevant documents. In: SIGIR 2006, pp. 429–436 (2006)
5. Clarke, C., Kolla, M., Cormack, G., Vechtomova, O., Ashkan, A., Büttcher, S., MacKinnon, I.: Novelty and diversity in information retrieval evaluation. In: SIGIR 2008, pp. 659–666 (2008)
6. Davidson, J., Liebald, B., Liu, J., Nandy, P., Vleet, T.V.: The YouTube video recommendation system. In: RecSys 2010, pp. 293–296 (2010)
7. Griffiths, T., Steyvers, M.: Finding scientific topics. PNAS 101, 5228–5235 (2004)
8. Guo, S., Sanner, S.: Probabilistic latent maximal marginal relevance. In: SIGIR 2010, pp. 833–834 (2010)
9. Hurley, N., Zhang, M.: Novelty and diversity in top-N recommendation – analysis and evaluation. ACM Trans. Internet Technol. 10(4), 14:1–14:30 (2011)
10. Koren, Y.: Collaborative filtering with temporal dynamics. In: KDD 2009, pp. 447–456 (2009)
11. Lathia, N., Hailes, S., Capra, L., Amatriain, X.: Temporal diversity in recommender systems. In: SIGIR 2010, pp. 210–217 (2010)
12. Linden, G., Smith, B., York, J.: Amazon.com recommendations: item-to-item collaborative filtering. IEEE Internet Computing 7(1), 76–80 (2003)
13. Minka, T.: Estimating a Dirichlet distribution. Tech. rep., M. I. T. (2000)
14. Radlinski, F., Dumais, S.: Improving personalized web search using result diversification. In: SIGIR 2006, pp. 691–692 (2006)
15. Santos, R., Macdonald, C., Ounis, I.: Selectively diversifying web search results. In: CIKM 2010, pp. 1179–1188 (2010)
16. Santos, R., Macdonald, C., Ounis, I.: Intent-aware search result diversification. In: SIGIR 2011, pp. 595–604 (2011)
17. Shi, Y., Zhao, X., Wang, J., Larson, M., Hanjalic, A.: Adaptive diversification of recommendation results via latent factor portfolio. In: SIGIR 2012, pp. 175–184 (2012)
18. Tösher, A., Jahrer, M., Bell, R.: The BigChaos solution to the Netflix grand prize (2009), http://www.netflixprize.com/assets/GrandPrize2009_BPC_BigChaos.pdf
19. Vargas, S., Castells, P.: Rank and relevance in novelty and diversity metrics for recommender systems. In: RecSys 2011, pp. 109–116 (2011)
20. Yang, Y., Liu, X.: A re-examination of text categorization methods. In: SIGIR 1999, pp. 42–49 (1999)
21. Ziegler, C.N., McNee, S.M., Konstan, J.A., Lausen, G.: Improving recommendation lists through topic diversification. In: WWW 2005, pp. 22–32 (2005)
22. Ziegler, C., Lausen, G., Schmidt-Thieme, L.: Taxonomy-driven computation of product recommendations. In: CIKM 2004, pp. 406–415 (2004)

A Probabilistic Data Replacement Strategy for Flash-Based Hybrid Storage System

Yanfei Lv, Xuexuan Chen, Guangyu Sun, and Bin Cui

School of Electronics Engineering and Computer Science, Peking University
Key Lab of High Confidence Software Technologies (Ministry of Education),
Peking University
{lvyf,bin.cui,gsun,xuexuan}@pku.edu.cn

Abstract. Currently, the popularization of flash memory is still limited by its high price and low capacity. Thus, the magnetic disk and flash memory will coexist over a long period of time. How to design an effective flash-hard disk hybrid storage system emerges as a critical issue. Most of the existing works are designed based on traditional cache management approaches by taking the characteristics of flash into consideration. In this paper, we revisit the existing hybrid storage approaches and propose a novel probabilistic data replacement strategy for flash-based hybrid storage system, named HyPro. Different from traditional deterministic approaches, our approach moves the data probabilistically based on the data access pattern. Such a method can statistically achieve a good performance over massive memory operations of modern workloads. We also present the detailed data replacement algorithm and discuss how to determine the probability of data migration in the storage hierarchy consisting of main memory, flash, and hard disk. Extensive experimental results on various hybrid storage systems show that our method can yield better performance and achieve up to 50% improvements against the competitors.

1 Introduction

Although most of the people believe that the magnetic disk will be replaced by flash-based solid state drives (SSD) in the future, currently the popularization of flash memory is still limited by its high price and low capacity. Hence magnetic disk and flash memory will coexist over quite a long period of time. From the comparison in Table 1, flash displays a moderate I/O performance and price per GB between DRAM and hard disk. Consequently, it is straightforward to adopt flash memory as a level of memory between the HDD and main memory because of its advantages in performance [18,19]. The flash-hard disk hybrid storage is more and more adopted. Seagate provides a mixed storage hard disk with 4GB flash chip to improve the overall performance [2]. Windows operating system support Quick Boost from vista to accelerate the booting [1]. In addition, some companies start to replace some of the hard disk to SSD to build a hybrid storage system. In this case, how to design an effective flash-hard disk hybrid storage system emerges as a critical issue.

Y. Ishikawa et al. (Eds.): APWeb 2013, LNCS 7808, pp. 360–371, 2013.

Table 1. Comparison on different storage media

	Price($)	Capacity(GB)	Price($)/GB	Read(μs)	Write(μs)
DRAM(DDR3)	23.99	4GB	6.00	1	1
SSD	114.00	64GB	1.78	271	2012
Disk(7.2K)	119.99	1TB	0.12	12700	13700

The data migration among main memory, flash and disk is the most important issue in hybrid storage design. Many works has been done on this problem. TAC [7](Temperature-Aware Caching) adopts temperature to determine page placement. In TAC, a global temperature table is maintained for each page. The temperature of a page is decided by its access numbers and patterns in a period. The pages with higher temperature are placed in main memory and flash memory while a cold page evicted from main memory will be replaced to disk.

In contrast, the LC [11](Lazy Cleaning) and FaCE [13](Flash as Cache Extension) always cache a page exit from main memory to flash memory. LC handles flash memory as a write-back cache. The dirty pages are kept on flash first and if the percentage of dirty pages exceeds a threshold, these pages will be flushed to disk. Whereas FaCE proposes FIFO replacement for flash memory management which is an ideal pattern for flash write. In this way, FaCE can improve the throughput and shortens the recovery time of database.

All the above methods are all in deterministic way. In some cases, such deterministic migration policy is really inefficient. For example, in some cases, pages are only accessed once, and thus it is suboptimal to keep these cold pages on the flash memory as designed in LC and FaCE. The temperature method in TAC works well on hot page detection with stable pattern. However, on a workload changing, TAC takes a rather long time to forget the history and learn the new pattern. Another disadvantage of TAC is its high time and space consumption.

In order to overcome the problems in prior approaches, we propose a probability-based policy named HyPro to manage data storage and migration in the storage hierarchy, which is composed of main memory, SSD and HDD. In our approach, the priority of data in each level of the hierarchy is maintained separately. The key difference from prior work is that the data migration among different levels are no longer deterministic but based on probabilities. Compared to prior deterministic approaches, HyPro has several advances:

- Better management efficiency. The probabilistic data migration can be considered as a statistical frequency-based implementation. Since existing cache policies can also be applied to SSD in our approach, HyPro is a seamless combination of the cache management and frequency-based migration policies, which can achieve better management efficiency so that the total I/O performance can be improved.
- Less unnecessary data replacement. For deterministic approaches, there may be some unnecessary data movements for cold pages. In our stochastic based

scheme, such unnecessary data replacement can be effectively eliminated with the control of probabilities.

– Lower overhead. To achieve the hot page detection, prior work needs to maintain a global list to identify the hottest and coldest pages. The space complexity is $O(n)$ and the time complexity of the maintainable no less than $O(log(n))$, where n is the number of all pages accessed including those on the disk. In our approach, the space complexity is negligible and the time complexity is $O(1)$.

The remainder of the paper is organized as follows. Related work is described in Section 2. Section 3 describes our framework. Parameter tuning method is presented in Section 4. Experimental results are shown in Section 5, and we make a conclusion in Section 6.

2   Related Work

Nowadays, flash-based hybrid storage has been gradually recognized as an economical way for a practical system by more and more people. Some hard disks leverage a small flash memory to improve I/O performance [5]. With the increment on the capacity, SSD is more and more widely deployed in storage systems. The early SSD is skilled in reading but uncompetitive in writing. Thus migration methods [19,14] are proposed to dynamically transfer read intensive pages to flash and write intensive ones to disk. Recently, SSD thoroughly surpasses disk on both read and write speed, and hence the popular method is to adopt flash as a middle-level cache between disk and main memory. Existing work includes static deployment and dynamic loading. An object placement method [6] is developed to give a proper deployment for the components of DBMS. By comparing the object performance on SSD and disk beforehand, those with higher benefit per size are chosen to place on SSD. Other methods suggest putting certain part of the system to flash. FlashLogging [8] illustrates that storing the log of DMBS to flash can largely improve the overall performance. Debnath et al. propose [9,10] FlashStore and SkimpyStash to discuss the proper way to put the key-value pair to SSD. The static methods need to know the specific information about the application and is not self-adaptive to various environments. Some methods based on dynamically page transferring are proposed. TAC(Temperature-Aware Caching) [7] is the dynamic version of object placement strategy. It allocates temperature to the extents according to access pattern and I/O cost and keeps the data with higher temperature to higher level of the storage structure. Researchers from Microsoft [11] discuss several possible designs for hybrid storage methods. According to testing the LC(Lazy-Cleaning) is the best design. LC method shows better performance than TAC on write intensive traces and similar on read-intensive traces. hStorageDB [16] adopts semantic information to exploit the capability of hybrid storage system, which is from another aspect to solve the hybrid storage problem. FaCE [13] proposes to use the flash in FIFO manner to improve throughput and provide faster recovery.

3   Probabilistic Framework

3.1   The HyPro Approach

The typical structure of a hybrid storage system is illustrated in Figure 1. All the data is stored on hard disk and organized as data pages. A page need to be loaded into main memory before being accessed. Since flash has better performance compared to disk, it works as the level between main memory and disk. When a page miss happens in main memory, the flash will be checked first. The disk is only accessed when the page is not found in the flash.

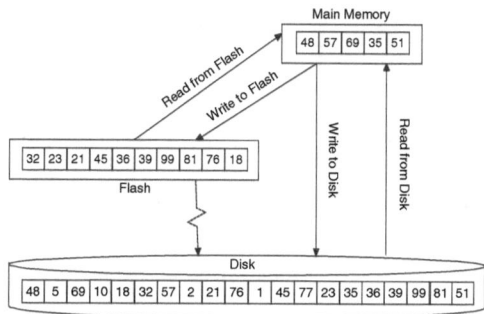

Fig. 1. Illustration of hybrid storage system

In this system, the data placement and migration is a critical issue to achieve better performance. In this part, we introduce our probabilistic approach for hybrid storage management, named HyPro. The overall structure of HyPro is shown in Fig 2. In our framework, the pages in main memory and flash memory are exclusive from each other. In other words, we do not keep a page in flash memory if it is already in main memory.

In the HyPro, we adopt two probabilities to control the data migration. If some of the pages on flash are frequently accessed, it's better to be elevated to main memory. We call this process *elevation*. Once the elevation happens we have to evict a page from main memory. Obviously, the elevation should be managed carefully so that the benefits of accessing hot pages can offset the I/O cost overhead caused by data movement. In our probabilistic data management, we use a probability named $p_{elevate}$ to control the elevation frequency. As shown in Fig 3, when a page is accessed, it has the chance of $p_{elevate}$ to be kept in main memory; otherwise, this page will be evicted on the next data access. It is obvious that the page has more chance to be elevated if it is more frequently used. Hence, real hot pages are detected and promoted into main memory statistically during a long runtime. In each elevation, a cold page in main memory need to be evicted to flash and placed in the original space of the elevated page.

In the HyPro, some page may be evicted from the main memory, which we name *sinking* operation in this paper. At first glance, this page is likely to be

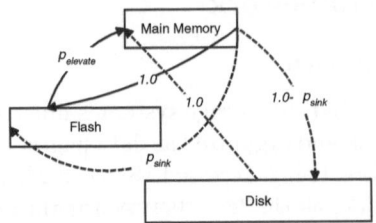

Fig. 2. Overall of Probabilistic Framework

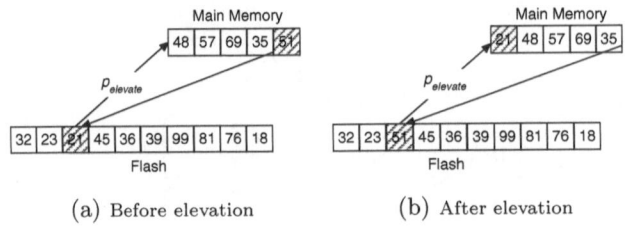

(a) Before elevation   (b) After elevation

Fig. 3. An elevation example

hotter than pages on flash memory, and should replace one flash page. However, sinking to flash incurs a flash write. Whether the future benefit is worth this write depends on the cost ratio between flash read and write as well as the hotness of the page being sunk. An example of sinking is illustrated in Fig 4. We take a probability p_{sink} to control the ratio of sinking to flash. A main memory evicted page has chance of p_{sink} to replace a flash page, otherwise, it will be discarded directly or written back to disk if it is dirty. Let's see why this works. We assume the main memory evicted page is M (page 51 in Fig 4 (a)), and the replaced page from flash is F(page 18 in Fig 4 (a)), respectively. The larger p_{sink} is, the more evicted pages are sunk to flash, and the closer the hotness of M and F are. Consequently, the benefit of sinking to flash will be small for close hotness of M and F. By setting proper value of $p_{elevate}$ and p_{sink}, we can achieve a better trade-off between main memory and flash accesses and the overall I/O cost diminishes. We will talk about the parameter tuning in the Section 4.

The pseudocode of HyPro are listed in Algorithm 1. A structure named frame is used to store the position of each page, and the frames are organized in a hash table to facilitate searching. The Algorithm 1 illustrates the routine of page access. First, the position of the page is determined, and then different operations are conducted according to the page position. The Algorithm 1(line 13) may invoke Algorithm 2. Algorithm 2 loads a page from disk and puts this page to the right position according to the p_{sink}. Although LRU is adopted in our experiments for main memory and flash memory management, HyPro can support other strategies such as LIRS [12] and ARC [17]. The HyPro is easy to

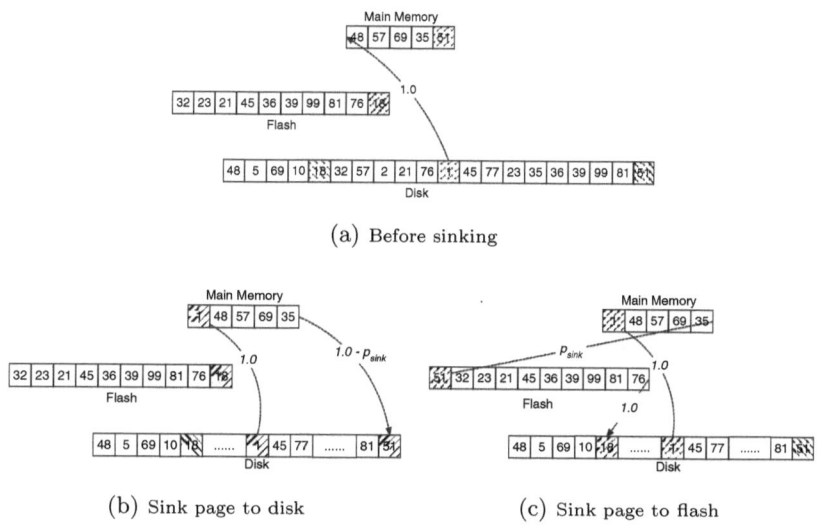

Fig. 4. A sink example

implement and quick enough for online processing. The time complexity is $O(1)$ for each page access.

In this paper, we focus on the data migration design. Nevertheless, some optimizations can be supported in HyPro applied to further improve the performance e.g. by considering the asymmetric I/O and the access pattern (random/sequential). For example, if the asymmetric I/O is considered, different probabilities can be allocated to read and write operations respectively, which can make one write operation equivalent to the effect of n read operations. Furthermore, we could also manage flash in FIFO manner as FaCE to transform flash space allocation into sequential pattern.

4   Parameter Tuning

The probabilities of transforming data among different memory levels are the crucial part in our stochastic page management policy. In this section, we provide study on how to automatically tune these probabilities based on the "cost" analysis. Table 2 facilitates fast check on the notations used in this section.

Definition 1. *For a cache management algorithm, we denote the place of the page to be evicted as the "evict position". (For example, the end of the LRU queue). N_{evict} is defined as the total hit number on the "evict position".*

To begin with, a definition is introduced. Note that we consider the hit number on a "position" instead of on a specific page in this definition. For example, if LRU algorithm is adopted in main memory, the N_{evict} stands for the number of accesses on the LRU end. If a read operation is hit on the least recently used

---

Algorithm 1. The HyPro algorithm

    Input: an operation request on page p
1 **if** *p exists in main memory* **then**
2   Perform read/write operation in main memory;
3 **else if** *p exists in Flash memory* **then**
4   **if** $rand() < p_{elevate}$ **then**
5   AcquireFreeMemPage();
6   write the evicted page to flash memory if any;
7   read from Flash memory, or just write to main memory;
8   **else**
9   Perform read/write operation towards Flash memory;
10   **end**
11 **else**
12   call OnMemFull algorithm when main memory is full;
13   AcquireFreeMemPage();
14   read from disk, or just write to main memory;
15 **end**

Algorithm 2. The OnMemFull algorithm

    Output: a free page in memory
1 **if** *memory is full* **then**
2   get a victim page from the buffer manager;
3   **if** $rand() < p_{sink}$ **then**
4   AcquireFreeFlashPage();
5   flush the page to Flash memory;
6   **end**
7   flush the page to disk if dirty;
8   return the page;
9 **end**
10 return a free page;

Table 2. Parameters used in this section

Parameters	Description
C_{dr}, C_{fr}	The read time cost of one page on disk, flash respectively
C_{dw}, C_{fw}	The write time cost of one page on disk, flash respectively
$p_{elevate}$	The probability to elevate
p_{sink}	The probability to sink
R_{fevict}, R_{mevict}	The number of read hit on the evict position of flash, main memory respectively
W_{fevict}, W_{mevict}	The number of write hit on the evict position of flash, main memory respectively

page P, N_{evict} increases. However, at this time page P is moved to LRU head and a new page named P' is moved to LRU end. Then next time we increase N_{evict} when the P' is accessed rather than P.

Assume that in a certain time period, a page P_f on flash is read R_f times and written W_f times. R_{mevict} and W_{mevict} stand for the read and write times on the "evict position" in the main memory during the same time period. The read and write costs of flash are denoted as C_{fr} and C_{fw}.

In HyPro, p_{sink} is adopted to balance the eviction to flash and disk. Hence, we discuss the tuning of p_{sink} by comparing the cost of the two cases: 1)evict the page to flash (Figure 4 (c)) and 2)evict the page to disk (Figure 4 (b)). The I/O costs of two cases are calculated as follows respectively.

Case 1. Evict page P_{mevict} to flash, and if flash is full evict the P_{fevict} to disk (when dirty). Consequently, P_{mevict} will be accessed from flash and P_{fevict} from disk, and thereby the corresponding I/O cost is:

$$C_{sinkf} = R_{mevict}C_{fr} + W_{mevict}C_{fw} + R_{fevict}C_{dr} + W_{fevict}C_{dw} + C_{fw} \quad (1)$$

Case 2. Evict page P_{mevict} to disk. In this case P_{mevict} will be accessed from disk while the P_{mevict} mentioned in Case 1 is still accessed from flash. Thus the I/O cost is:

$$C_{sinkd} = R_{mevict}C_{dr} + W_{mevict}C_{dw} + R_{fevict}C_{fr} + W_{fevict}C_{fw} \quad (2)$$

In the case of $C_{sinkf} < C_{sinkd}$, it is more I/O efficient to evict a page to flash, and hence, p_{sink} should be increased and vice versa. The above analysis only takes the I/O cost of page transferred, that is, page evicted and page elevated into consideration. Actually, the I/O costs of other pages will also be influenced which are not the primary cost and experiments show that the obtained C_{sinkf} and C_{sinkd} can deal with parameter adjusting effectively.

The C_{sinkf} and C_{sinkd} are very small and unstable for a single page on a short period of time. In practice, we accumulate C_{sinkf} and C_{sinkd} on all the evicted page in a certain window on the trace, so that the p_{sink} can be adjusted based on the comparison of C_{sinkf} and C_{sinkd} on each accumulation. Note that the calculation and parameter tuning described above need only $O(1)$ time for each access. The tuning will not significantly increase the overhead of whole strategy.

The tuning of $p_{elevate}$ can be performed in a similar way with p_{sink}, and thus is omitted here due to space limitation.

5   Performance Evaluation

In this section, we conduct a trace-driven simulation to evaluate the effectiveness of the proposed framework. The traces used include TPC-B, TATP [3] and making Linux kernel (MLK for short) to further evaluate the performance on various workloads. TAC and FaCE are chosen as the competitors. The simulation is developed in Visual Studio 2010 using C#. All experiments are run on a Windows 7 PC with a 2.4 GHz Intel Quad CPU and 2 GB of physical memory.

5.1   Experimental Setup

We use three traces mentioned above for performance evaluation. The benchmarks namely TPC-B [4], and TATP [3] are run on PostgreSQL 9.0.0 with default settings, e.g., the page size is 8KB. Dataset size of both TPC-B and TATP are 2GB. The MLK is a record of the page accesses of making Linux kernel 2.6.27.39. We use a tool named strace to monitor theses processes and obtain the disk access history. Specification on the traces was given in Table 3.

Table 3. Specification on the Traces

Filename	Page Number (10^3)	Reference Number (10^6)	Write Ratio
MLK	97.2	27.2	0.43%
TATP	135.1	2.5	4.59%
TPC-B	35.1	10.7	19.46%

In our experiment, the *total I/O time* is used as the primary metric to evaluate the performance. We employ The samsung SSD (64GB, 470 series) in our experiment. We obtain its access latency by testing. Access latency of hard disk is obtained from paper [15]. The parameters used in our experiments are listed in Table 4, in which *Flashsize/Pages* denotes the ratio between flash and dataset size. We fixed the main memory size to the 1% of the dataset size.

Table 4. Experimental Parameters

Parameter	Value
$C_r, C_w(\mu s)$	271, 803 (for SSD)
	12700, 13700 (for hard disk)
Flashsize/Pages	1.25%, 2.5%, 5%, 10%, 20%
$p_{elevate}$	0.001,0.01, 0.015,0.02,...,0.1
p_{sink}	0.01, 0.1,0.2,...,0.9

5.2   Parameter Tuning

To begin with, we inspect the effect of parameters on performance of algorithms. We illustrate how the performance of HyPro varies with p_{sink} in Figure 5 (a), along with the values of C_{sinkf} and C_{sinkd}. The HyPro achieves the best performance at around $p_{sink} = 0.15$. p_{sink} reflects the chance for a page evicted from main memory to be stored into flash. A low p_{sink} will result in a poor utility of flash memory. On the contrary, an excessive high p_{sink} would incurs great exchange cost between main memory and flash. Thus, our approach achieves the best performance with proper p_{sink}.

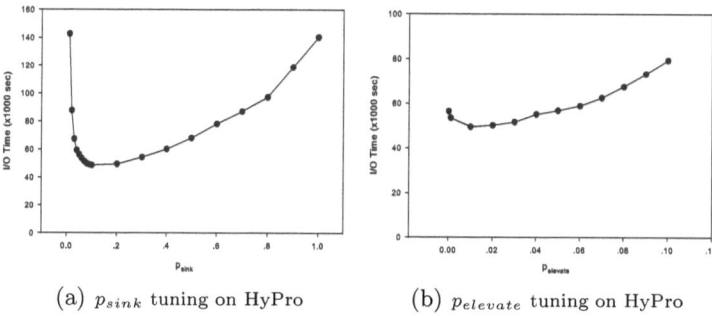

(a) p_{sink} tuning on HyPro   (b) $p_{elevate}$ tuning on HyPro

Fig. 5. Parameters Tuning

The performance of our approach also significantly varies with the $p_{elevate}$, as shown in Figure 5 (b) , and the minimum I/O time is reached when it locates between 0.01 and 0.02. As the $p_{elevate}$ controls whether a page on flash should be elevated, large $p_{elevate}$ will cause more exchanges between main memory and flash, which may deteriorates the whole performance, while no exchange will cause pages on flash has no opportunity to get into main memory (corresponding to the case $p_{elevate} = 0$). Other testings also show that the best parameter is often around 0.02 and 0.2, and thus, we adopt these values as the initial value and use dynamic parameter tuning in the following experiments.

5.3   Comparison with Other Approaches

In this part, we compare HyPro with FaCE and TAC. We test extensive configurations, but only shows some results here in Figure 6 due to the space limitation. In the following results, the main memory is 1% of the total workload size and the flash size varies from 1.25% to 20%. Our approach shows similar or better performance compared with FaCE and TAC approach. Our approach can reduce upto 50% of the total I/O time against other competitors. In MLK trace, FaCE performs the worst, since many pages in MLK are accessed only once, but the FaCE still cache these pages in flash memory. On TATP, HyPro and Face has similar performance and better than TAC. Because the TATP is not a very stable access pattern. FaCE and HyPro can adapt themselves to the workload changes quickly, but TAC needs a rather long time to learn this change. On TPC-B trace TAC has the best performance when the flash size is low, this is because the temperature-based hot detection can accurately discover the hottest pages, which has superiority when the cache size is low. However, When the flash size is large, the performance of TAC degrades. This phenomenon is partially because precisely hot page detection is not necessary for a larger cache size, and partially because of the write through cache design, as the write ratio is very high in TPC-B according to Table 3. The HyPro can adapt itself to the flash size enlargement and shows a good performance in all the cases.

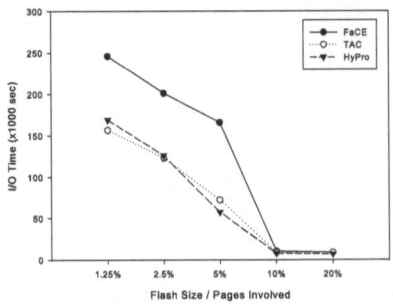

(a) Low-end SSD performance on MLK trace

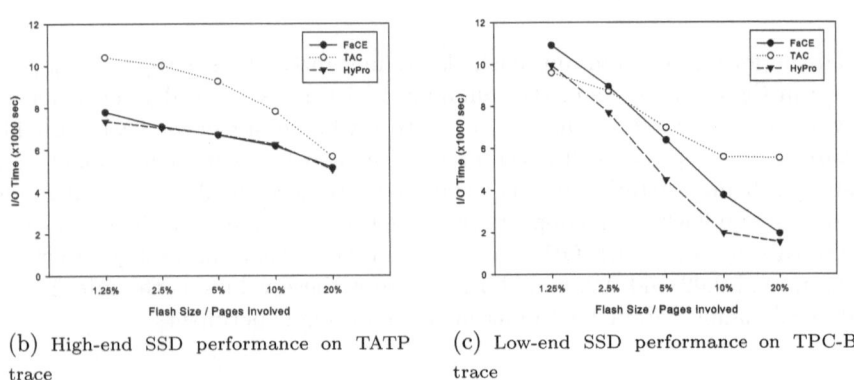

(b) High-end SSD performance on TATP trace

(c) Low-end SSD performance on TPC-B trace

Fig. 6. I/O performance comparison on benchmark traces

6   Conclusion

In this paper we propose a novel stochastic approach for flash based hybrid storage system management named HyPro. Different from the existing deterministic models, HyPro controls the data migration between devices using two probabilities. One probability describes the chance of which one page will be kept in main memory after accessed from flash. Another probability is adopted to determine the place where a page should be put after evicted from main memory. By doing this, Hypro can achieves lower hard disk access with a little exchange overhead increment on flash writes. We also developed an approach to determine the probabilities based on cost analysis. The experiments show that HyPro outperforms other competitors.

Acknowledgments. This research was supported by the grants of Natural Science Foundation of China (No. 61272155, 61073019) and MIIT grant 2010ZX01042-001-001-04.

References

1. http://windows.microsoft.com/en-US/windows-vista/products/features/performance
2. Momentus XT Solid State Hybrid Drives, http://www.seagate.com/www/en-us/products/laptops/laptop-hdd/
3. Telecom Application Transaction Processing Benchmark, http://tatpbenchmark.sourceforge.net/index.html
4. TPC Benchmark B (TPC-B), http://www.tpc.org/tpcb/
5. Bisson, T., Brandt, S.A., Long, D.D.E.: A hybrid disk-aware spin-down algorithm with I/O subsystem support. In: IPCCC, pp. 236–245 (2007)
6. Canim, M., Bhattacharjee, B., Mihaila, G.A., Lang, C.A., Ross, K.A.: An object placement advisor for DB2 using solid state storage. PVLDB 2(2), 1318–1329 (2009)
7. Canim, M., Mihaila, G.A., Bhattacharjee, B., Ross, K.A., Lang, C.A.: SSD buffer-pool extensions for database systems. PVLDB 3(2), 1435–1446 (2010)
8. Chen, S.: Flashlogging: exploiting flash devices for synchronous logging performance. In: SIGMOD Conference, pp. 73–86 (2009)
9. Debnath, B.K., Sengupta, S., Li, J.: Flashstore: High throughput persistent key-value store. PVLDB 3(2), 1414–1425 (2010)
10. Debnath, B.K., Sengupta, S., Li, J.: Skimpystash: RAM space skimpy key-value store on flash-based storage. In: SIGMOD Conference, pp. 25–36 (2011)
11. Do, J., Zhang, D., Patel, J.M., DeWitt, D.J., Naughton, J.F., Halverson, A.: Turbocharging DBMS buffer pool using SSDs. In: SIGMOD Conference, pp. 1113–1124 (2011)
12. Jiang, S., Zhang, X.: LIRS: an efficient low inter-reference recency set replacement policy to improve buffer cache performance. In: SIGMETRICS, pp. 31–42 (2002)
13. Kang, W.-H., Lee, S.-W., Moon, B.: Flash-based extended cache for higher throughput and faster recovery. Proc. VLDB Endow. 5(11), 1615–1626 (2012)
14. Koltsidas, I., Viglas, S.: Flashing up the storage layer. PVLDB 1(1), 514–525 (2008)
15. Lee, S.-W., Moon, B.: Design of flash-based DBMS: an in-page logging approach. In: SIGMOD Conference, pp. 55–66 (2007)
16. Luo, T., Lee, R., Mesnier, M.P., Chen, F., Zhang, X.: hStorage-DB: Heterogeneity-aware data management to exploit the full capability of hybrid storage systems. CoRR abs/1207.0147 (2012)
17. Megiddo, N., Modha, D.S.: ARC: A self-tuning, low overhead replacement cache. In: FAST (2003)
18. Ou, Y., Härder, T.: Trading memory for performance and energy. In: Xu, J., Yu, G., Zhou, S., Unland, R. (eds.) DASFAA Workshops 2011. LNCS, vol. 6637, pp. 241–253. Springer, Heidelberg (2011)
19. Wu, X., Reddy, A.L.N.: Managing storage space in a flash and disk hybrid storage system. In: MASCOTS, pp. 1–4 (2009)

An Influence Strength Measurement via Time-Aware Probabilistic Generative Model for Microblogs

Zhaoyun Ding[1,*], Yan Jia[1], Bin Zhou[1], Jianfeng Zhang[1], Yi Han[1], and Chunfeng Yu[2]

[1] School of Computer, National University of Defense Technology,
Changsha, 410073, China
{zyding,yanjia,binzhou,jfzhang,yihan}@nudt.edu.cn
[2] Naval Aeronautical Engineering Institute Qingdao Branch, Qingdao, Shandong, China
13953228232@163.com

Abstract. Micro-blogging such as Twitter provides users a platform to participate in the discussion of topics and find more interesting friends. While this large
number can generate notable diversity and not all influence strengths between
users are the same, it also makes measure the influence strength more accurately,
that not only rated as binary friendship relations, challenging and interesting. In
this work, we develop a time-aware probabilistic generative model to estimate the
influence strength by taking the time interval, relationship of following, and the
post content into consideration. In particular, the Gibbs sampling is employed to
perform approximate inference, and the interval of time and the multi-path influence propagation is incorporated to estimate the indirect influence strength more
microscopically according to the propagation of words. Comprehensive experiments has been conducted on a real data set from Twitter, which contains about
0.26 million users and 2.7 million tweets, to evaluate the performance of our proposed approach. As indicated, the experimental results validate the effectiveness
of our approach. Furthermore, we also observe that the influence strength ranking by our model is less correlative with the method which ranks the influence
strength according to the number of common friends.

Keywords: influence strength, probabilistic generative model, microblogs, social media, social network.

1 Introduction

Online social networking sites are becoming more popular each day, such as Facebook,
Twitter, and LinkedIn. Among all these sites, Twitter is the fastest growing one than any
other social network. Microblogs, such as Twitter, is a service that has emerged as a new
medium for communication recently. Unlike other social network services, the relationship of following between users can be unidirectional; a user is allowed to choose who
she wants to follow without seeking any permission. Twitter employs a social network
called "following" relationship; the user whose updates are being followed is called the

* This work was supported by NSFC (No.60933005, 91124002);863 (No. 012505,
2011AA010702, 2012AA01A401, 2012AA01A402);242 (No.2011A010);and NSTM
(No.2012BAH38B04, 2012BAH38B06).

Y. Ishikawa et al. (Eds.): APWeb 2013, LNCS 7808, pp. 372–383, 2013.

"friend", while the one who is following is called the "follower". In case the author is not protecting his tweets, they appear in the so-called public timeline and his followers will receive all messages from him.

It is well recognized that networks often contain both strong and weak ties, also influence strength between different users is usually different and it is unidirectional in microblogs, so treating all relationships as equal will increase the level of noise in the learned models and likely lead to degradation in performance. Recently, influence strength analysis has attracted considerable research interests. However, most of their works only considered the network structure or the user behaviors in social network, and the micro-level mechanisms of influence strength of a user on her followers at a specific topic, have been largely ignored.

Studies by Kwak et al [1] have shown that Twitter is more likely to be a news media. Users can not only find friends by microblogs, but also can publish messages by themselves, and microblogs transform people from content consumers into content producers. Usually, there are two aims for users to make use of microblog services, including finding friends in real life and finding friends according to the *homophily*. So, it is insufficient to measure the influence strength of two users only by the number of common friends.

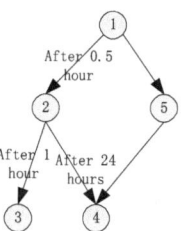

Fig. 1. The correlation of the influence strength and the words   **Fig. 2.** An example of multi-level hierarchial structure

Moreover, if a user is influenced by her friends, she will retweet (by making use of an informal convention such as "RT @user") or reintroduce (tweets are similar with the tweets previously posted by other users, but without acknowledgement of a source) the messages of her friends, resulting that some words in tweets are the same. The more the same words which are copied by a user are, the higher the influence strength on the user is. Figure 1 gives an example to illustrate the correlation of the influence strength and the number of the same words. In the topic of basketball, we can find the user A has more similar words with her friend B and we infer the user A copes more ideas from her friend B. In fact, figure 1 indicates two assumptions: 1) Users with similar interests have a stronger influence on each other, 2) Users whose actions frequently correlate have a stronger influence on each other, which have been validated in [2] [3].

Also, the interval of time is an important factor to measure the influence strength and it stands for the diffused rate of each word. If a user copes the ideas from her friend much faster for a topic, it will indicate that the friend has stronger influence on this user. The interval of time has been ignored in recent studies for influence strength analysis.

In addition, the network in microblogs is a multi-level hierarchial structure, implying that the influence is transitive. A small network of multi-level hierarchial structure is illustrated in the figure 2. Two key factors need be considered to measure the indirect influence strength. 1) The interval of time. Though the user *3* and the user *4* are influenced indirectly by the user *1*, the user *3* receives the idea indirectly from the user *1* faster than the user *4*. We can infer the user *1* has stronger indirect influence strength on the user *3*. 2) Multi-path influence propagation. In the figure 2, we can find that the user *4* is influenced indirectly by the user *1* through two users. How to combine the multi-path indirect influence strength more microscopically is unknow.

In order to measure the influence strength in microblogs, we propose a novel time-aware generative probabilistic model of a corpus to draw each word in tweet from the corresponding probability distribution, considering the time interval, the relationship of following, and the post content. The model is estimated by Gibbs sampling. Moreover, we take advantage of the interval of time and the multi-path influence propagation to measure the indirect influence strength more microscopically in microblogs.

Experiments are conducted on a real dataset from Twitter containing about 0.26 million users and 2.7 million tweets, and the results show that the model in this paper has better prediction performance than other methods. Furthermore, we find that the influence strength ranking of our model is less correlative with the method which ranks the influence strength according to the number of common friends. Results indicate that some users with higher influence strength by the number of common friends may be friends in real life and they form relationships because they know each other in real life, but they have little interaction in virtual life.

2   Related Work

Liu et al [2] [3] proposed a generative graphical model which leveraged both heterogeneous link information and textual content associated with each user in the network to mine topic-level influence strength. Based on the learned direct influence, they further studied the influence propagation and aggregation mechanisms. Guo et al [4] proposed a generative model for modeling the documents linked by the citations, called the Bernoulli Process Topic (BPT) model, to measure the influence strength in Citation Networks. Dietz et al [5] devised a probabilistic topic model that explained the generation of documents; the model incorporated the aspects of topical innovation and topical inheritance via citations to predict the citation influences.

Our work is similar to their works and we also propose a novel generative probabilistic model of a corpus which leverages both the network of following and post content associated with each user in the network to measure the influence strength in microblogs. However, the time is considered in our work. Usually, the influence strength between users is evolved with the variety of time. Moreover, the interval of time is an important factor to decide the direct and indirect strength of influence between users.

Xiang et al [6] developed an unsupervised model to estimate relationship strength from interaction activity (e.g., communication, tagging) and user similarity. More specifically, they formulated a link-based latent variable model, along with a coordinate ascent optimization procedure for the inference. However, influence strength analysis microscopically is ignored in their work.

Also, interaction data has been used to predict relationship strength [7] [8] but this work only considered two levels of relationship strength, namely weak and strong relationships.

3   Preliminary

Definition 1. *(Word) A word is the basic unit of discrete data, defined to be an item from a vocabulary indexed by* $\{1, ..., v\}$ *. Words are represented with unit-basis vectors that have a single component equal to one and all other components equal to zero. Thus, the vth word in the vocabulary is represented by a V-vector* w *such that* $w^v = 1$ *and* $w^u = 0$ *for* $u \neq v$.

Definition 2. *(Post/Tweet/Document) A post* **w** *is comprised of a bag of words from a vocabulary* $V = \{1, ..., v\}$, *and is represented with vector* $\mathbf{w} = (w_1, w_2, ..., w_n)$, *where* w_i *denotes the occurrence number of word* v_i *in document* **w**.

Definition 3. *(Topic) A semantic topic* θ *observed in a particular time period is defined as a multinomial distribution of words* $\{p(w|\theta)\}_{w \in V}$ *with the constrain* $\sum_{w \in V} p(w|\theta) = 1$.

Definition 4. *(Network) The network in microblogs is a directed graph* $G = (N, E)$. *The* N *stands for the set of users in microblogs. And the directed edge* $E \subseteq N \times N$ *stands for the relationships of following in microblogs. For* $\forall e_{uv} = (u, v) \in E$, *if there exists a directed edge between* u *and* v, $e_{uv} = 1$; *otherwise* $e_{uv} = 0$.

Definition 5. *(Direct and Indirect Influence) Given two user nodes* u, v *in the network* $G = (N, E)$, *we denote* $I_u(v) \in R$ *as the influence strength of user* v *on user* u. *Furthermore, if* $e_{uv} = 1$, *we call* $I_u(v)$ *the direct influence of user* v *on* u; *if* $e_{uv} = 0$, *we call* $I_u(v)$ *the indirect influence of user* v *on* u. *Because the relationship of following between users in microblogs can be unidirectional, the influence strength is asymmetric, i.e.,* $I_v(u) \neq I_u(v)$.

4   Probabilistic Generative Model

Similar to the existing topic models, each document is represented as a mixture over latent topics. In our model, the content of each document is a mixture of two sources: 1) the content of the given document, and 2) the content of other documents related to the given document through the network structure. This perspective actually reflects the process of writing a post (tweet) in microblogs: either depending on users' interests or influenced by one of their friends. when a user writes a post, she may create the idea innovationally or copy it from one of her friends.

In the model, we use a parameter s to control the influence situation. The s is generated from a Bernoulli distribution whose parameter is ρ. When s = 1, the behavior is generated based on users own interests. When s = 0, it means the behavior of the user is influenced by one of her friends. The parameter ρ of the coin is learned by the model, given an asymmetric beta prior $\alpha_\rho = (\alpha_{\rho_\theta}, \alpha_{\rho_\psi})$ which prefers the topic mixture θ of a

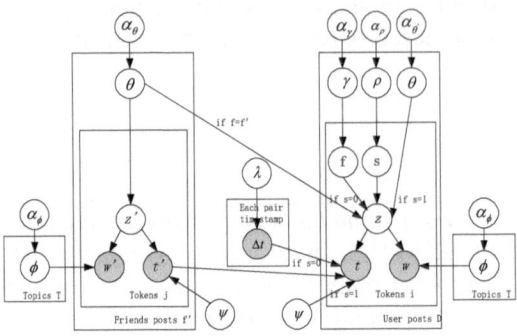

Fig. 3. Probabilistic generative model

friend document. Thus we need another parameter γ to select one influencing user v from friends set $L(u)$.

The key differences from the existing topic models are that the interval of time and the relationship of users are combined in our model. The mixture distribution over topics is influenced by both word co-occurrences and the document's time stamp. A per-topic Beta distribution generates the document's time stamp. The interval of two related documents Δt is drawn according to the negative exponential distribution with the parameter λ. Figure 3 shows the graphical structure of the model, which explains the process of generative each word and time stamp in posts. The key differences of the generative process are illustrated in the algorithm 1.

Algorithm 1. Probabilistic generative model

1: **for** all influenced users' documents $d_u \in D$ **do**
2:      Draw a friend mixture $\gamma_u = p(v|u)|_{L(u)} \sim dirichlet(\alpha_\gamma)$ restricted to v followed by u;
3:      Draw an innovation topic mixture $\theta_{d_u} = p(z|d_u) \sim dirichlet(\alpha_{\theta'})$;
4:      Draw the proportion between tokens associated with friends and those associated with
5:      the innovation topic mixture $\rho_{d_u} = p(s = 0|d_u) \sim beta(\alpha_{\rho_\partial}, \alpha_{\rho_\beta})$;
6:      **for** all tokens i **do**
7:          toss a coin $s_{d_u,i} \sim bernoulli(\rho_{d_u})$;
8:          **if** $s_{d_u,i} = 0$ **then**
9:              Draw a friend $f_{d_u,i} \sim multi(\gamma_u)$;
10:             Draw a topic $z_{d_u,i} \sim multi(\theta_{f_{d_u,i}})$ from the friend's document topic mixture;
11:             Draw a word $w_{d_u,i} \sim multi(\phi_{z_{d_u,i}})$ from $z_{d_u,i}$–specific word distribution;
12:             Draw an interval of time $\Delta t_{d_u,i} \sim Exponential(\lambda_{z_{d_u,i}})$;
13:             Get a time stamp $t_{d_u,i} = \Delta t_{d_u,i} + t_{f_{d_u,i}}$;
14:          **else**
15:             Draw the topic $z_{d_u,i} \sim multi(\theta_{d_u})$ from the innovation topic mixture;
16:             Draw a word $w_{d_u,i} \sim multi(\phi_{z_{d_u,i}})$ from $z_{d_u,i}$–specific word distribution;
17:             Draw a time stamp $t_{d_u,i} \sim Beta(\psi_{z_{d_u,i}})$;
18:          **end if**
19:      **end for**
20: **end for**

---

5 Parameter Estimation via Gibbs Sampling

We employ Gibbs sampling to perform approximate inference. We list the partial update equations for each variable as below which are same as [2] [3].

$$p(f_i|\vec{f}_{-i}, d_i, s_i = 0, z_i, \cdot) \propto \frac{N_{d,f,s}(d_i, f_i, 0) + \alpha_\gamma - 1}{N_{d,s}(d_i, 0) + L(d_i)\alpha_\gamma - 1} \cdot \frac{N_{f',z'}(f_i, z_i) + N_{f,z,s}(f_i, z_i, 0) + \alpha_\theta - 1}{N_{f'}(f_i) + N_{f,s}(f_i, 0) + T\alpha_\theta - 1}$$

(1)

$$p(s_i = 0|\vec{s}_{-i}, d_i, f_i, z_i, \cdot) \propto \frac{N_{f',z'}(f_i, z_i) + N_{f,z,s}(f_i, z_i, 0) + \alpha_\theta - 1}{N_{f'}(f_i) + f_{f,s}(f_i, 0) + T\alpha_\theta - 1} \cdot \frac{N_{d,s}(d_i, 0) + \alpha_{\rho_\theta} - 1}{N_d(d_i) + \alpha_{\rho_\theta} + \alpha_{\rho_\psi} - 1}$$

(2)

$$p(s_i = 1|\vec{s}_{-i}, d_i, z_i, \cdot) \propto \frac{N_{d,z,s}(d_i, z_i, 1) + \alpha_{\theta'} - 1}{N_{d,s}(d_i, 1) + T\alpha_{\theta'} - 1} \cdot \frac{N_{d,s}(d_i, 1) + \alpha_{\rho_\psi} - 1}{N_d(d_i) + \alpha_{\rho_\theta} + \alpha_{\rho_\psi} - 1}$$

(3)

The $\vec{f}_{-i}$ represents the friend assignments for all tokens except w_{di}. And the $\vec{s}_{-i}$ represents the coin assignments for all tokens except w_{di}. In the update equations, $N(*)$ is the function which counts occurrences of a configuration. For example, $N_{d,f,s}(1, 2, 0)$ denotes the number of tokens in user 1 that are assigned to friend 2, where the coin result s is 0.

Next, we need to calculate the $p(z_i|\vec{z}_{-i}, w_i, t_i, s_i = 0, f_i, \cdot)$ and $p(z_i|\vec{z}_{-i}, w_i, t_i, d_i, s_i = 1, f_i, \cdot)$, where $\vec{z}_{-i}$ represents the topic assignments for all tokens except w_{di}.

Based on the generation process, we can get the posterior probability of the whole data set by integrating out the distributions because the model uses conjugate priors.

$$
\begin{aligned}
&p(w, w', z, z', t, t', f, s|\alpha_\phi, \alpha_\theta, \alpha_{\theta'}, \alpha_\gamma, \alpha_\rho, \psi, \lambda, \cdot) = \\
&\int p(w, w'|z, z', \phi)p(\phi|\alpha_\phi)p(t, t'|\psi, z, z')p(\Delta t|\lambda, z') \cdot \\
&\int p(z, z'|s, f, \theta, \theta')p(\theta|\alpha_\theta)p(\theta'|\alpha_{\theta'}) \cdot \\
&\int p(f|\gamma, L)p(\gamma|\alpha_\gamma, L) \cdot \int p(s|\rho)p(\rho|\alpha_\rho)
\end{aligned}
$$

(4)

Using the chain rule, we can obtain the conditional probability conveniently. According to $p(z_i, z_{-i}) = p(z)$ and the bayesian inference. We can infer as follows.

$$
\begin{aligned}
p(z_i|\vec{z}_{-i}, w_i, t_i, s_i = 0, f_i, \cdot) &= \frac{p(w, w', z, z', t, t', f, s|\alpha_\phi, \alpha_\theta, \alpha_{\theta'}, \alpha_\gamma, \alpha_\rho, \psi, \lambda, \cdot)}{p(w, w', \vec{z}_{-i}, \vec{z'}_{-i}, t, t', f, s|\alpha_\phi, \alpha_\theta, \alpha_{\theta'}, \alpha_\gamma, \alpha_\rho, \psi, \lambda, \cdot)} \propto \\
&\frac{\int p(w, w'|z, z', \phi)p(\phi|\alpha_\phi)p(t, t'|\psi, z, z')}{\int p(w, w'|\vec{z}_{-i}, \vec{z'}_{-i}, \phi)p(\phi|\alpha_\phi)p(t, t'|\psi, \vec{z}_{-i}, \vec{z'}_{-i})} \cdot \frac{\int p(z, z'|s, f, \theta, \theta')p(\theta|\alpha_\theta)p(\theta'|\alpha_{\theta'})}{\int p(\vec{z}_{-i}, \vec{z'}_{-i}|s, f, \theta, \theta')p(\theta|\alpha_\theta)p(\theta'|\alpha_{\theta'})} \\
&\propto \frac{p(t|\psi, z')}{p(t|\psi, \vec{z}_{-i}')} \cdot \frac{\int p(w, w'|z, z', \phi)p(\phi|\alpha_\phi)}{\int p(w, w'|\vec{z}_{-i}, \vec{z'}_{-i}, \phi)p(\phi|\alpha_\phi)} \cdot \frac{\int p(z, z'|s, f, \theta, \theta')p(\theta|\alpha_\theta)p(\theta'|\alpha_{\theta'})}{\int p(\vec{z}_{-i}, \vec{z'}_{-i}|s, f, \theta, \theta')p(\theta|\alpha_\theta)p(\theta'|\alpha_{\theta'})} \\
&\propto \frac{(1-t_i)^{\psi_{z_i}^1 - 1} t_i^{\psi_{z_i}^2 - 1}}{B(\psi_{z_i}^1, \psi_{z_i}^2)} \cdot \frac{N_{w,z}(w_i, z_i) + N_{w',z'}(w_i, z_i) + \alpha_\phi - 1}{N_z(z_i) + N_{z'}(z_i) + V\alpha_\phi - 1} \cdot \frac{N_{f',z'}(f_i, z_i) + N_{f,z,s}(f_i, z_i, 0) + \alpha_\theta - 1}{N_{f'}(f_i) + N_{f,s}(f_i, 0) + T\alpha_\theta - 1}
\end{aligned}
$$

(5)

$p(z_i|\vec{z}_{-i}, w_i, t_i, d_i, s_i = 1, f_i, \cdot)$ is derived analogously.

$$
\begin{aligned}
&p(z_i|\vec{z}_{-i}, w_i, t_i, d_i, s_i = 1, f_i, \cdot) \propto \\
&\frac{(1-t_i)^{\psi_{z_i}^1 - 1} t_i^{\psi_{z_i}^2 - 1}}{B(\psi_{z_i}^1, \psi_{z_i}^2)} \cdot \frac{N_{w,z}(w_i, z_i) + N_{w',z'}(w_i, z_i) + \alpha_\phi - 1}{N_z(z_i) + N_{z'}(z_i) + V\alpha_\phi - 1} \cdot \frac{N_{d,z,s}(d_i, z_i, 1) + \alpha_{\theta'} - 1}{N_{d,s}(d_i, 1) + T\alpha_{\theta'} - 1}
\end{aligned}
$$

(6)

After the Gibbs process, we can estimate λ of the negative exponential distribution.

$$\lambda = 1 / \frac{1}{Y} \sum_{y=1}^{Y} \Delta t_{iy} \qquad (7)$$

The Y represents the number of all time intervals in the i topic. And the Δt_{iy} represents the yth time intervals in the i topic.

Moreover, we will also obtain the sampled coin s_i, friends f_i, and topic z_i for each word, and the influence strength can be then estimated by Eq.(8), which are averaged over the sampling chain after convergence. A denotes the length of the sampling chain.

$$I_u(v|z) = \gamma_d(f|z) = \frac{1}{A} \cdot \sum_{n=1}^{A} \frac{\sum_{i=1}^{N_{d,z,f,s}(d,z,f,0)^n} \lambda_z^n \exp(-\lambda_z^n \Delta t_i^n) + \alpha_\gamma}{N_{d,z,s}(d,z,0)^n + |L(d)|\alpha_\gamma} \qquad (8)$$

The Δt_i represents the shortest interval of two same tokens assigning to the same topic between the user u and the user v.

The equations reflect our assumptions in a statistical way. It indicates that if one user u copes more words of user v on topic z, then v has a stronger influence on u w.r.t. topic z. Moreover, if the user u copes ideas of user v faster (has a shorter interval of time) on topic z, then v has a stronger influence on u w.r.t. topic z. In addition, we can also illustrate the evolutionary trend of the influence strength for two pair users according to the time stamp of each user in microblogs.

6   Indirect Influence Strength

The above probabilistic model only measures the direct influence strength, but does not consider indirect influence strength. In reality, influence can propagate through the network. Liu et al [2] [3] and Guo et al [4] studied the propagation of influence macroscopically. They first measured the direct influence strength between users, then combined these influence strengths to measure indirect influence strength. However, the micro-level mechanisms of influence propagation have been ignored by them.

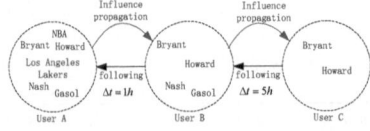

Fig. 4. Influence propagation through words

In fact, we can study the influence propagation more microscopically according to the propagation of words. Figure 4 gives an example of the propagation of words. User C copes two words from the user A through the user B in the topic of basketball. According to the statistical way in the above subsection, we can infer the indirect influence strength as follows.

$$I_u(f_1, f_2, \cdot | z) = \frac{1}{A} \sum_{n=1}^{A} \frac{N_{d,z,s,f_1,f_2,\cdot}(d, z, 0, f_1, f_2, \cdot)^n + \alpha_\gamma}{N_{d,z,s}(d, z, 0)^n + |L(d)|\alpha_\gamma} \tag{9}$$

$I_u(f_1, f_2, \cdot | z)$ represents the indirect influence of user f_2 on u through the user f_1. $N_{d,z,s,f_1,f_2,\cdot}(d, z, 0, f_1, f_2, \cdot)$ represents the number of the same tokens in user u that are assigned to both the user f_1 and the user f_2. For example, $N_{d,z,s,f_1,f_2,\cdot}(u, z, 0, f_1, f_2, \cdot) = 2$ in the figure 4. Analogously, we can get indirect influence strength for further hops.

Moreover, the influence of ideas decreases with the process of propagation. Fortunately, the interval of time can be used to illustrate the process of influence decrease. Usually, the influence of ideas decreases with the development of the time. If the interval of time is longer, the influence of ideas will decrease significantly. Otherwise, if the interval of time is shorter, the influence of ideas will decrease insignificantly. We take advantage of the function of negative exponential distribution to illustrate the process of influence decrease quantitatively.

$$I_u(f_1, f_2, \cdot | z) = \frac{1}{A} \cdot \sum_{n=1}^{A} \frac{\sum_{i=1}^{N_{d,z,s,f_1,f_2,\cdot}(d,z,0,f_1,f_2)^n} \lambda_z^n \exp(-\lambda_z^n \Delta t_i^n) + \alpha_\gamma}{N_{d,z,s}(d, z, 0)^n + |L(d)|\alpha_\gamma} \tag{10}$$

Also, the Δt_i represents the interval of two tokens between two indirect users. For example, $\Delta t_i(A, C) = 6h$ in the figure 4.

Next, we will analyze the multi-path influence propagation in microblogs. In the figure 2, we can find that the user 4 is influenced indirectly by the user 1 through two users (the user 2 and the user 5). In order to measure the indirect influence strength through multi-path, we give an aggregation function to combine multi-path indirect influence strength.

$$I_u(f_2|z) = \max(I_u(f_1^1, f_2, \cdot | z), I_u(f_1^2, f_2, \cdot | z)) \tag{11}$$

The u is influenced indirectly by the f_2 through two users f_1^1 and f_1^2. We refer the highest multi-path indirect influence strength as the final indirect influence strength.

7 Experiments

7.1 Dataset

For the purpose of this study, a set of Twitter data about Chinese-based twitters who have published at least one Chinese tweet from $2011 - 04 - 15$ to $2011 - 07 - 15$ was prepared as follows. About 0.26 million users and 2.7 million tweets were collected through the API of Twitter. We begin by describing the dataset we are trying to mine. As the figure 5 shows, the distributions of users' followers and tweets are approximately power-law, implying that the vast majority of users have a small quantity of followers and tweets, while a small fraction have a large number of followers and tweets.

In order to measure the influence strength with our model, we extracted tweets and users who published at least 10 tweets, then extracted their network relationships. Due to the power-law distributions of users' tweets, 3334 users and 168526 network relationships(edges) were extracted from the data set. Our model is trained with hyper parameters $\alpha_\phi = 0.01, \alpha_\theta = \alpha_{\theta'} = 0.1, \alpha_{\rho_\psi} = 0.1, \alpha_\gamma = 1.0$ and 50 topics.

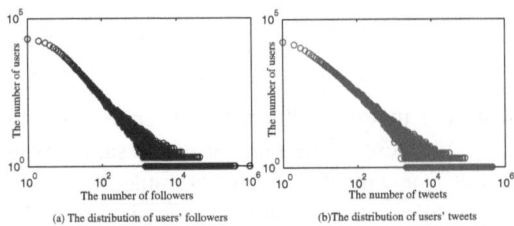

(a) The distribution of users' followers   (b) The distribution of users' tweets

Fig. 5. The distributions of users' followers and tweets

7.2 Evaluation

Predicting User Behaviors by Direct Influence Strength. We apply the derived influence strength to help predict user behaviors and compare the prediction performance with other methods. The behavior is defined as whether a user retweets a friend's microblog. To do this, we rank the pairs of users by influence strength and measure the area under the ROC curve (AUC) based on the feature values for the ranked pair (e.g., 1 if a user is retweeted by other user, 0 otherwise). We compare the rankings using influence strength to several alternative rankings (Named our model as TNIM-Time Network Influence Model):

1) The influence strength between users is calculated as the Kullback-Leibler (KL) divergence of user distributions over topics (Named as KLD).

$$I_u(v) = \frac{1}{KL(u,v)} = \frac{1}{\sum\limits_{0<i\leq T} z_u^i \log\frac{z_u^i}{z_v^i}} \tag{12}$$

2) The influence strength between users is calculated as the number of common friends (Named as NCF).

$$I_u(v) = \frac{|N(u) \cap N(v)|}{|N(u) \cup N(v)|} \tag{13}$$

3) The influence strength between users is calculated as the methods proposed by Liu et al [2] [3] (Named as NIM-Network Influence Model).

$$I_u(v) = \frac{1}{A} \sum_{n=1}^{A} \frac{N_{d,z,f,s}(d,z,f,0)^n + \alpha_\gamma}{N_{d,z,s}(d,z,0)^n + |L(d)|\alpha_\gamma} \tag{14}$$

In order to predict user behaviors, we divide the data set into two parts. We learn the influence strength by the data set from $2011-04-15$ to $2011-06-30$. And we predict user behaviors by the data set from $2011-07-01$ to $2011-07-15$. We select four users and their friends to evaluate our model. The ROC curve of different methods for four users are shown in the figure 6.

Then, we select 1000 users and their friends to compute the average AUC (Area Under Curve) of above four methods. The results are shown in the figure 7.

Experimental results show that the prediction performance of our model is the best. The average AUC (Area Under Curve) of our model is 0.0647 higher than the method of NIM, and is 0.0876 higher than the method of KLD. The prediction performance of NCF is the worst. We can infer it is not too accurate to predicate the retweets by the number of common friends and they do not always have common topics in virtual life.

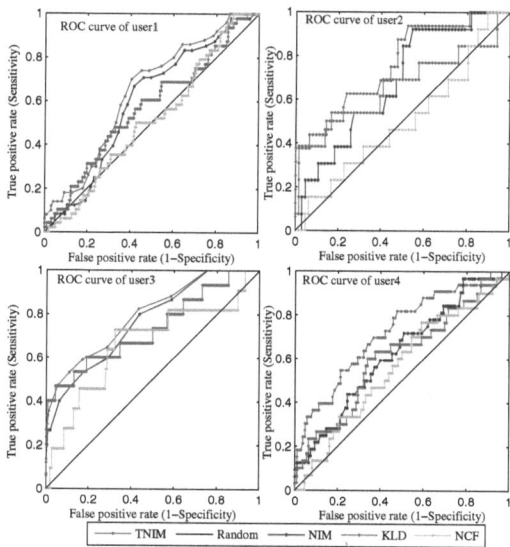

Fig. 6. The ROC curve of four users

Correlation Analysis by Direct Influence Strength. In this subsection, we will analyze the correlation of above three methods and our model by ranking the pairs of users according to the influence strength. *Spearman Rank Correlation* is used to analyze the correlation of two rankings.

$$\rho = 1 - \frac{6 \sum\limits_{0<i<n} \left(r_i^1 - r_i^2\right)^2}{n^3 - n} \tag{15}$$

We select 1000 users and rank their friends according to the influence strength. Then, we compute the correlation of our model and above three methods according to the *Spearman Rank Correlation*. The average correlations of 1000 users are shown in the figure 8.

Experimental results show that the influence strength ranking of our model (TNIM) is more correlative with NIM. However, our model is less correlative with NCF. Results indicate that some users with higher influence strength by the number of common friends may be friends in real life and they form relationships because they know each other in real life, but they have little interaction in virtual life.

Predicting User Behaviors by Indirect Influence Strength. In order to evaluate the prediction performance of indirect influence strength, the behavior is defined as whether a user retweets a user's microblog by multi-hops. For example, if a user A is influenced by the user B indirectly through the user C, we will estimate whether the relationship of retweets is as follows: "A RT @C RT @B...".

Because it is difficult to get indirect influence strength by the methods of the Kullback-Leibler (KL) divergence and the number of common friends, we only compare our model with the method proposed by Liu et al [2] [3]. Also, we select 1000 users and

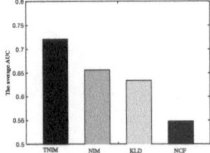
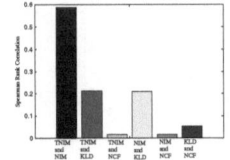
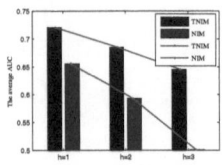

Fig. 7. The average AUC

Fig. 8. The correlation of our model and other methods

Fig. 9. The average AUC of indirect influence strength

their indirect friends to compute the average AUC (Area Under Curve). Experimental results are shown in the figure 9, where h denotes the hop of indirect influence.

Experimental results show that the prediction performance of our model is better than the method proposed by Liu et al [2] [3]. Moreover, we find the prediction performance of our model drops more slowly with the improvement of the hop. The indirect influence strength proposed by Liu et al [2] [3] is impacted by the hops of network. And all indirect influence strengths are computed by $I_u(w) \circ I_w(v)$, where $\circ$ is the concatenation function, e.g., multiplication and minimum value. However, the interval of time is ignored by them. Usually, the up-to-date tweets are recommended to followers, and followers are influenced more easily by these up-to-date tweets. It is not too accurate to compute the indirect influence strength, not considering the interval of time. In fact, Though the hop of network is farther, the interval of multi-hop may be shorter and the tweet is retweeted more easily by other users. So, the prediction performance of the method proposed by Liu et al [2] [3] drops faster with the improvement of the hop.

8   Conclusion

A novel time-aware probabilistic generative model is proposed to estimate influence strength, considering the time interval, the relationship of following, and the post content. Then, we take advantage of the interval of time and the multi-path influence propagation to to derive the indirect influence strength between nodes. Experimental results show that our model has better prediction performance than other methods, in despite of direct and indirect influence strength.

In the future work, we will take advantage of the influence strength to mine influencer more accurately in microblogs. Also, we can employ the influence strength to viral marketing and trigger a large cascade of influence more accurately.

References

1. Kwak, H., Lee, C., Park, H., Moon, S.: What is twitter, a social network or a news media? In: Proc. of the 19th International World Wide Web Conference, WWW 2010, Raleigh, USA, pp. 591–600 (April 2010)
2. Liu, L., Tang, J., Han, J., Jiang, M., Yang, S.: Mining topic-level influence in heterogeneous networks. In: Proc. of the 19th ACM Conference on Information and Knowledge Management, CIKM 2010, Toronto, Ontario, Canada, pp. 199–208 (October 2010)

3. Liu, L., Tang, J., Han, J., Yang, S.: Learning influence from heterogeneous social networks. Data Mining and Knowledge Discovery 25(3), 511–544 (2012)
4. Guo, Z., Zhang, Z., Zhu, S., Chi, Y., Gong, Y.: Knowledge discovery from citation networks. In: Proc. of the 9th IEEE International Conference on Data Mining, ICDM 2009, Miami, Florida, USA, pp. 800–805 (December 2009)
5. Dietz, L., Bickel, S., Scheffer, T.: Unsupervised prediction of citation influences. In: Proc. of the 24th International Conference on Machine Learning, ICML 2007, Corvallis, Oregon, USA, pp. 233–240 (June 2007)
6. Xiang, R., Neville, J., Rogati, M.: Modeling relationship strength in online social networks. In: Proc. of the 19th International World Wide Web Conference, WWW 2010, Raleigh, USA, pp. 981–990 (April 2010)
7. Gilbert, E., Karahalios, K.: Predicting tie strength with social media. In: Proc. of the 27th International Conference on Human Factors in Computing Systems, CHI 2009, Boston, USA, pp. 211–220 (April 2009)
8. Kahanda, I., Neville, J.: Using transactional information to predict link strength in online social networks. In: Proc. of the 3rd International AAAI Conference on Weblogs and Social Media, ICWSM 2009, San Jose, California, USA, pp. 74–81 (May 2009)

A New Similarity Measure Based on Preference Sequences for Collaborative Filtering

Tianfeng Shang[1,2], Qing He[1], Fuzhen Zhuang[1], and Zhongzhi Shi[1]

[1] The Key Laboratory of Intelligent Information Processing, Institute of Computing Technology, Chinese Academy of Sciences, Beijing, 100190, China
[2] Graduate School of Chinese Academy of Sciences, Beijing, 100049, China
{shangtf,heq,zhuangfz,shizz}@ics.ict.ac.cn

Abstract. Collaborative filtering is one of the most popular techniques in recommender systems, and the key point is to find similar users and items. There are already some similarity measures, such as vector cosine similarity and Pearson's correlation coefficient, and so on. However, in some cases, what recommender systems get are not the ratings, but preference sequences of users on a series of items. For this type of data, those traditional similarity measures may fail to meet the practical application requirements. In this paper, a similarity measure based on inversion is proposed for preference sequences naturally. Based on the Inversion similarity measure, some structural information of user preference sequences is analyzed. By merging average precision and weighted inversion into similarity computation, a new similarity measure based on preference sequences is proposed for collaborative filtering. Experimental results show that the proposed similarity measure based on preference sequences outperforms the common similarity measures on the datasets with continuous real numbers.

Keywords: Recommender System, Collaborative Filtering, Similarity Measure, Preference Sequences, Weighted Inversion, Average Precision.

1 Introduction

Recently, CF-based recommender systems have gradually become one of the most important recommender systems. Nowadays, they are widely used and make great success in many applications [2], especially in e-commerce web sites, such as www.amazon.com, www.netflix.com, www.ebay.com, et al. Besides, lots of researchers pay more and more attentions to CF techniques of these recommender systems. Generally speaking, CF techniques mainly include three categories [10]: memory-based algorithms [7], model-based algorithms [5], and hybrid CF algorithms [9].

In memory-based CF algorithms, they firstly compute similarity between users or items on the user-item rating matrix, and then generate predictions or recommendations for objective users. So similarity computation is one critical step of CF algorithms. At present, the most commonly used ones are vector cosine

Y. Ishikawa et al. (Eds.): APWeb 2013, LNCS 7808, pp. 384–391, 2013.

similarity and Pearson correlation coefficient [3]. Vector cosine similarity takes each user or item as a vector of ratings, and computes the cosine angle formed by the rating vectors. Pearson correlation coefficient measures the degree of two users or items linearly relate to each other. They have both made great success in many practical CF applications.

However, in some cases, what recommender systems get are not the ratings, but preference sequences of users on a series of items. For this type of data, those traditional similarity measures may fail to meet the practical application requirements. In this paper, from the point view of user behavior, a similarity measure based on inversion (Inversion) is proposed for preference sequences naturally. Based on the Inversion similarity measure, some structural information of user preference sequences is analyzed. By merging average precision (AP) and weighted inversion into similarity computation, a new similarity measure based on preference sequences is proposed for collaborative filtering. Finally some experimental results show that the proposed similarity measure based on preference sequences can get better performance than the common similarity measures in preference sequences datasets.

The rest of this paper is organized as follows. Section 2 introduces some preliminary knowledge, including overview of CF recommender systems and similarity measures. Section 3 gives out some analysis on structural information of user preference sequences, and a new similarity measure based on preference sequences is proposed for collaborative filtering. Some preliminary experimental results and evaluations are shown in Section 4. Finally, Section 5 presents conclusions and future work.

2 Preliminary Knowledge

2.1 Overview of CF Recommender Systems

In CF recommender systems, the fundamental assumption is that if users u and v have similar ratings on some items, they will also give similar ratings on other items [10]. Based on the above assumption, the basic form of CF recommender systems can be described as follows. In a typical CF recommender system, there are two lists: one is list of M users, $\{u_1, u_2, ..., u_M\}$, and the other is list of N items, $\{i_1, i_2, ..., i_N\}$. The users' ratings on items can either be explicit indications, such as real numbers or like/dislikes, or implicit indications, such as purchases or click-throughs. All the ratings constitute a user-item rating matrix, usually a very sparse matrix.

2.2 Similarity Measures in CF Recommender Systems

Similarity computation is one of the most important phases of CF recommender systems. For user based CF recommender systems, the co-rating items that the two users u and v have both rated should firstly be found out, and then select the appropriate method to compute the similarity between users u and v. It is similar for item based CF recommender systems. At present, there are many different methods for similarity computation [1], but the most commonly used are vector cosine similarity (Cosine) and Pearson's correlation coefficient (Pearson).

3   A New Similarity Measure Based on Preference Sequences

In this section, the ratings of one user are seen as a preference sequence and inversion is employed as a similarity measure of the preference sequences. By analyzing some problems of the similarity measure based on inversion and some structural information of preference sequences, a new similarity measure based on preference sequences for collaborative filtering is proposed by merging average precision and weighted inversion into similarity computation. Here the similarity measure based on inversion is firstly introduced as follows.

3.1   Similarity Measure Based on Inversion

From the point view of behavioral psychology, one's behavior on some items (such as high or low score, purchase or not, interested or disgusted, et al.) can be seen as a preference sequence, thus the similarity between two users can be measured by the two preference sequences of their behaviors [6,8]. In some sense, it may get better performance than the common vector cosine similarity and Pearson correlation coefficient.

Based on the above analysis, a similarity measure based on inversion can be proposed for collaborative filtering, and thus similarity computation between two users can be converted to inversion computation of their preference sequences. First, some definitions are given as follows.

Definition 1 (*Inversion*) [4]: Suppose that $P = (p_1, p_2, \cdots, p_n)$ is a sequence containing N different real numbers, $< p_i, p_j > (i, j = 1, 2, \cdots, n, i \neq j)$ is called a pair, the number of pairs at how many times $p_i > p_j$ occurs in sequence P is called its inversion, written as $t(P)$ and shown as Formula (1).

$$t(P) = \sum_{i=1}^{n} \sum_{j=i+1}^{n} f(p_i > p_j) \tag{1}$$

where, $f(\cdot)$ is an indicator function.

Definition 2 (*Inversion of two preference sequences*): Suppose that $P = (p_1, p_2, \cdots, p_n)$ and $Q = (q_1, q_2, \cdots, q_n)$ are two preference sequences containing N different real numbers respectively, and there is permutation relationship between them. The inversion of sequence Q can be calculated by taking P as benchmark. In detail, sort P in ascending order to get $P' = (p'_1, p'_2, \cdots, p'_n)$, meanwhile, according to the permutation relationship between P and Q, $Q' = (q'_1, q'_2, \cdots, q'_n)$ can be gotten. Thus, the inversion of sequence Q' is called the inversion from Q to P, written as $t(Q \rightarrow P)$.

According to the above definitions and theorems, similarity between two sequences can be calculated by normalizing the inversion between them, as shown in Formula (2).

$$Sim(P, Q) = 1 - \frac{2 \times t(Q \leftrightarrow P)}{|P| \times (|P| - 1)} \tag{2}$$

According to Formula (1) and Formula (2), the time complexity of similarity measure based on inversion is $O(n^2)$. However, according to merge-sort algorithm [4], similarity measure based on inversion can reduce its time complexity to $O(n \log n)$.

3.2 A New Similarity Measure Based on Preference Sequences

In the above section, similarity measure based on inversion tries to understand user behaviors by preference sequences. However, the similarity measure based on inversion may not give the best performance in some cases.

Generally speaking, for high efficiency and high accuracy of recommender systems, not all the items are considered, but just the items that users are more interested in should be considered in practical recommender systems, that is, items with higher preference should be taken more attention to. To address the above problems, we can take position information into account during similarity computation. According to our research, two kinds of position information should be considered as follows.

- In similarity measure based on inversion, inversion values on different positions are given the same weight. Obviously, this approach is not the most rational in practical recommender systems. One simple improved approach is that each inversion value can be weighted by divided by its position respectively. Similar with Formula (1), it can be written as Formula (3).

$$WInv(P) = \sum_{i=1}^{n} \frac{\sum_{j=i+1}^{n} f(p_i > p_j)}{i} \tag{3}$$

- Besides, there is another interesting phenomenon that some elements of preference sequences are equal to elements of benchmark sequence on the corresponding positions. Take preference sequence $P = (1, 2, 6, 5, 4, 3)$ and benchmark sequence $O = (1, 2, 3, 4, 5, 6)$ as example, the 1st and 2nd elements of P are equal to elements of O on the corresponding positions respectively, and this phenomenon will impact on similarity between P and O. However, in similarity measure based on inversion, this phenomenon is ignored. One way to improve the similarity computation can resort to average precision (AP) [3], which is an important measure for evaluating the performance of information retrieval systems, to merge position information into similarity computation. Specifically, construct an auxiliary sequence $\tilde{P} = (0, 0, 3, 1, -1, 3)$, and take element "0" as recall at its position, thus AP of $\tilde{P}$ can be calculated by Formula (4) to improve the similarity between objective sequence P and benchmark sequence O.

$$AP(P) = \frac{\sum_{i=1}^{N}(p(i) \times rel(i))}{N} \qquad (4)$$

where N is the length of sequence, $p(i)$ is the precision at cut-off i in the sequence $\tilde{P}$, and $rel(i)$ is an indicator function.

According to Formula (3) and Formula (4), two kinds of position information of each element in preference sequences can be taken into the similarity computation, and a new similarity measure based on preference sequences can be proposed for collaborative filtering, named similarity measure based on Weighted Inversion and Average Precision (WIAP), as shown in Formula (5).

$$Sim(P, O) = (1 - WInv(P)) \times AP(P) \qquad (5)$$

where O is the benchmark sequence, $WInv(P)$ and $AP(P)$ are calculated according to Formula (3) and Formula (4) respectively. Finally the proposed similarity measure WIAP can be described as Algorithm 1.

Algorithm 1. Similarity Measure by Merging Weighted Inversion and Average Precision (WIAP)

Input: Benchmark preference sequence O and objective preference sequence P, each sequence with N different nature numbers from 1 to N.
Output: $Sim(P, O)$, similarity between P and O.

1. Sort O in ascending order under the limit of keeping the permutation relationship between P and O;
2. Initialize $WI = 0$, $AP = 0$, $c = 0$;
3. For $i = 1$ to $N - 1$
4.   Calculate inv as inversion of $P(i)$;
5. $WI = WI + inv \ / \ i$;
6.   If $P(i) == i$
7. $c = c + 1$;
8. $AP = AP + c \ / \ i$;
9.   End If
10. End For
11. If $P(N) == N$
12. $c = c + 1$;
13. $AP = AP + c \ / \ N$;
14. End If
15. Normalize WI into $[0, 1]$;
16. Calculate $Sim(P, O) = (1 - WInv(P)) \times AP(P)$ and return it;

In Algorithm 1, Step (1-2) are some data preprocessing and initialization. Step (3-14) are to calculate of $WInv(P)$ and $AP(P)$ according to Formula (3) and Formula (4), where Step (4-5) are to calculate $WInv(P)$, Step (6-9) and Step (11-14) are to calculate of $AP(P)$. Step (15) is to normalize $WInv(P)$ into $[0,1]$. Step (16) is to calculate the final similarity measure WIAP according to Formula (5).

4   Experiments

In this section, some experiments are conducted on four similarity measures (Cosine, Pearson, Inversion and WIAP) to measure the performance of the proposed similarity measure based on preference sequences. The performance of all the similarity measures are measured by Recall, which is one of the most popular metrics in recommender systems. In information retrieval, recall is the fraction of the documents that are relevant to the query that are successfully retrieved, and it can be given by Formula (6).

$$Recall = \frac{|\{relevant\ documents\} \bigcap \{retrieved\ documents\}|}{|\{relevant\ documents\}|} \tag{6}$$

User-based top-N recommendation algorithm [10] is chosen for prediction and recommendation computation in the experiments. It firstly identifies the k most similar users to the objective user. After the k most similar users have been discovered, their corresponding rows in the user-item matrix are aggregated to identify a set of items, C, purchased by the group together with average of their weighted ratings. By sorting each row of set C according to their weights, user-based CF techniques then recommend the top-N items of set C to the objective user. In practice, about $10 \sim 20$ items (such as movies, music, books, et al) can shown in one web page, so in the experiments, we set N to 15, and set k to 10.

4.1   Experiments on Synthetic Dataset

In this section, some experiments are conducted on the synthetic dataset which is a 1,000 $\times$ 1,000 matrix generated by Matlab program. To verify the good suitability of the proposed similarity measure based on preference sequences, the synthetic matrix is randomly sampled to construct five matrices with different matrix filling ratios from 20% to 100%. The experimental results on the five matrices are shown in Fig.1.

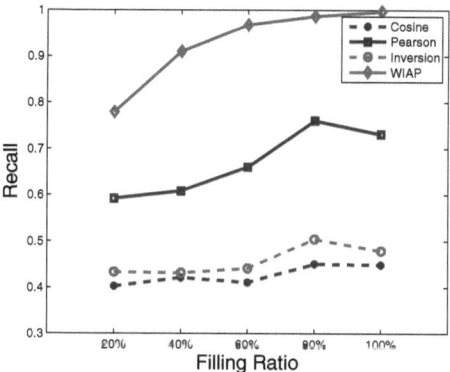

Fig. 1. Recall of Synthetic Dataset on Different Filling Ratios

From Fig.1, it can be seen that the proposed similarity measure WIAP outperforms all the other similarity measures Cosine, Pearson and Inversion on different matrix filling ratios. In particular, similarity measure WIAP almost recalls all the items on the full matrix. Besides, recall of WIAP can increase rapidly along with the rise of matrix filling ratio.

4.2   Experiments on Benchmark Datasets

In this section, some experiments are conducted on three benchmark datasets: Movielens dataset, Jester dataset and BookCrossing dataset. Similar with experiments on synthetic dataset, each benchmark dataset is randomly sampled to construct five matrices with different sampling ratios from 20% to 100%. The experimental results on three benchmark datasets are shown in Table 1.

Table 1. Recall of Benchmark Datasets on Different Sampling Ratios

Dataset	Measure	Sampling Ratio				
		20%	40%	60%	80%	100%
Movielens	Cosine	0.1444	0.2106	0.3400	0.3853	0.4790
	Pearson	0.3343	0.3723	0.4296	**0.4473**	0.4671
	Inversion	0.2893	0.2715	0.4195	0.3829	0.3942
	WIAP	**0.3949**	**0.3725**	**0.4311**	0.4469	**0.4987**
Jester	Cosine	0.5735	0.5059	0.4715	0.4621	0.4356
	Pearson	0.6235	0.5695	0.5805	0.5790	0.5644
	Inversion	0.6012	0.5108	0.4977	0.5037	0.4867
	WIAP	**0.7302**	**0.8138**	**0.9220**	**0.9647**	**0.9814**
BookCrossing	Cosine	0.9222	0.5204	0.2587	0.1566	0.1127
	Pearson	**0.9970**	0.9508	0.8618	**0.6873**	**0.5167**
	Inversion	**0.9970**	0.9529	0.8512	0.6564	0.4664
	WIAP	**0.9970**	**0.9534**	**0.8715**	0.6863	0.5128
Average[1]	Cosine	0.5467	0.4123	0.3567	0.3347	0.3424
	Pearson	0.6516	0.6309	0.6240	0.5712	0.5161
	Inversion	0.6292	0.5784	0.5895	0.5143	0.4491
	WIAP	**0.7074**	**0.7132**	**0.7415**	**0.6993**	**0.6643**

[1] "Average" represents average recall of each measure on three benchmark datasets.

In Table 1, it contains four blocks where the front three blocks are original recalls of three benchmark datasets and the last block is the average recalls of them. In all the 15 cases of the front three blocks, the proposed similarity measure WIAP can get much better performance than other similarity measures in 12 cases. In cases of 80% of Movielens dataset and 80% and 100% of BookCrossing dataset, Pearson is just a little bit better than WIAP. Actually, the values of Movielens dataset and BookCrossing dataset are just few discrete numbers which may be not suitable to be seen as preference sequences. Thus, WIAP may not get best performance in some cases of them. For the "Average" block of Table 1, it can be seen that WIAP can get much better performance than other similarity measures in all the cases. Besides, for BookCrossing dataset, there is an interesting phenomenon that on some sparser matrices, all the similarity measures get higher recalls. That is because the matrices are so sparse that there are not enough books to be recommended, and the similarity measures can recall the very few books with higher proportion. Generally speaking, it can be concluded that WIAP outperforms three other similarity measures when measuring preference sequences.

5   Conclusions

In this paper, some structural information of user preference sequences in recommender systems is analyzed, and a new similarity measure based on preference sequences, named WIAP, is proposed for collaborative filtering by merging AP and weighted inversion into similarity compuation. Experimental results show that the proposed similarity measure WIAP trends to get better performance on the datasets with continuous real numbers. However, there are still many further research issues. For example, how to improve the performance of WIAP on the datasets with few discrete numbers may be a valuable research point.

Acknowledgments. This work is supported by the National Natural Science Foundation of China (No.61175052, 60975039, 61203297, 60933004, 61035003), National High-tech R&D Program of China (863 Program) (No.2012AA011003), National Program on Key Basic Research Project (973 Program) (No.2013CB329502).

References

1. Ahn, H.: A new similarity measure for collaborative filtering to alleviate the new user cold-starting problem. Information Sciences 178(1), 37–51 (2008)
2. Alsaleh, S., Nayak, R., Xu, Y., Chen, L.: Improving matching process in social network using implicit and explicit user information. In: Du, X., Fan, W., Wang, J., Peng, Z., Sharaf, M.A. (eds.) APWeb 2011. LNCS, vol. 6612, pp. 313–320. Springer, Heidelberg (2011)
3. Chowdhury, G. Introduction to modern information retrieval. Facet Publishing (2010)
4. Cormen, T., Leiserson, C., Rivest, R., Stein, C.: Introduction to algorithms, 3rd edn. MIT Press and McGraw-Hill (2009)
5. Hofmann, T.: Latent semantic models for collaborative filtering. ACM Transactions on Information Systems (TOIS) 22(1), 89–115 (2004)
6. Jung, S., Hong, J., Kim, T.: A statistical model for user preference. IEEE Transactions on Knowledge and Data Engineering 17(6), 834–843 (2005)
7. Linden, G., Smith, B., York, J.: Amazon. com recommendations: Item-to-item collaborative filtering. IEEE Internet Computing 7(1), 76–80 (2003)
8. Park, M.-H., Hong, J.-H., Cho, S.-B.: Location-based recommendation system using bayesian user's preference model in mobile devices. In: Indulska, J., Ma, J., Yang, L.T., Ungerer, T., Cao, J. (eds.) UIC 2007. LNCS, vol. 4611, pp. 1130–1139. Springer, Heidelberg (2007)
9. Su, X., Greiner, R., Khoshgoftaar, T., Zhu, X.: Hybrid collaborative filtering algorithms using a mixture of experts. In: Proceedings of the IEEE/WIC/ACM International Conference on Web Intelligence, pp. 645–649. IEEE Computer Society (2007)
10. Su, X., Khoshgoftaar, T.: A survey of collaborative filtering techniques. In: Advances in Artificial Intelligence 2009, p. 4 (2009)

Visually Extracting Data Records
from Query Result Pages

Neil Anderson and Jun Hong

School of Electronics, Electrical Engineering and Computer Science
Queen's University Belfast, Belfast, UK, BT7 1NN
{nanderson423,j.hong}@qub.ac.uk

Abstract. Web databases are now pervasive. Query result pages are dynamically generated from these databases in response to user-submitted queries. Automatically extracting structured data from query result pages is a challenging problem, as the structure of the data is not explicitly represented. While humans have shown good intuition in visually understanding data records on a query result page as displayed by a web browser, no existing approach to data record extraction has made full use of this intuition. We propose a novel approach, in which we make use of the common sources of evidence that humans use to understand data records on a displayed query result page. These include structural regularity, and visual and content similarity between data records displayed on a query result page. Based on these observations we propose new techniques that can identify each data record individually, while ignoring noise items, such as navigation bars and adverts. We have implemented these techniques in a software prototype, *rExtractor*, and tested it using two datasets. Our experimental results show that our approach achieves significantly higher accuracy than previous approaches.

1 Introduction and Motivation

Web sites that rely on structured databases for their content are now ubiquitous. Users retrieve information from these databases by submitting HTML query forms. Query results are displayed on a web page, but in a proprietary presentation format, dictated by the web site designer. We call these pages, *query result pages*. Figure 1 shows a typical query result page from HMV.com. On this page each DVD is presented as a data record. In turn, each data record contains a collection of data items; including the title, description, release date, availability, price and customer rating, etc. as well as an image of the product. Any item on the page that does not belong to a data record is called a *noise item*, for example, a menu button or an advert. A query result page is designed for a human to read rather than a computer to process, thus there is no standard way to automatically extract structured data from the page.

Automatic data extraction is the process of extracting automatically a set of data records and the data items that they contain from a query result page. Such structured data can then be integrated with data from other data sources and presented to the user in a single cohesive view in response to their query. For instance, there is great commercial demand for comparison shopping search engines. A user may wish to buy a DVD; a comparison shopping search engine can extract data from many different online stores,

Y. Ishikawa et al. (Eds.): APWeb 2013, LNCS 7808, pp. 392–403, 2013.

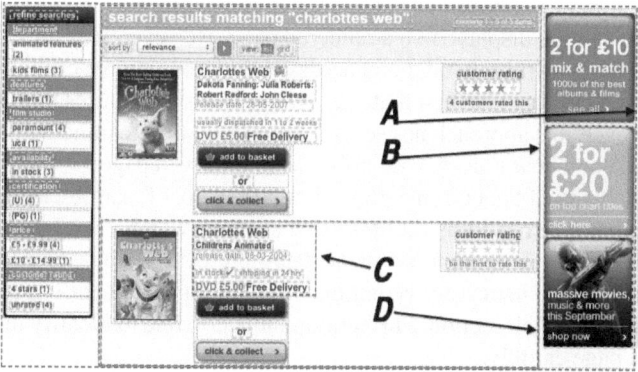

Fig. 1. Query Result Page from HMV.com with visual blocks highlighted

integrate the data and display it to the user. Other practical applications include flight and hotel booking sites, financial product comparisons, property sales and rentals.

In this paper, we focus on the problem of data record extraction, that is, to identify the groups of data items that make up each data record on a query result page.

Data records on a query result page display regularity in their content, structure and appearance. They exhibit structural and visual similarities: that is, they form visual patterns which are repeated on the page. This is because data records on the same site are often presented using the same template. The displayed data is in an underlying database, and as the data items for each record are retrieved from the database (in response to the user's query), the same template is used each time to present the record. Much of the existing work that deals with the extraction of data records is based on the theme of identifying repeated patterns. The common premise is to find and use the repeated patterns of data records. The main differences between the existing approaches are where they look for these patterns and how they use them in data extraction.

Early approaches [1,4] identify repeated patterns in the HTML source code of multiple training pages from a data source in order to infer a common structure or template, and use it to extract structured data from new pages from the same source. Other approaches [3,15] identify repeated patterns in the source code of a query result page in order to directly extract data records and align data items in them. However, these approaches are all limited by the rapidly increasing complexity of source code. For instance, the widespread use of Javascript libraries, such as jQuery, can make source code structurally much more complex; thus these approaches start to fail.

Later approaches [9,16,18,19] use the tag tree representation of a web page to identify repeated patterns in the subtrees of the tag tree. This representation is useful because it provides a hierarchal representation of the source code. However, the tag tree was designed for the browser to use when displaying the page, and unfortunately does not accurately resemble the structure of the data records on the displayed page. The use of scripts and other runtime features contribute further to the differences between the structure of the tag tree and the dynamically displayed web page.

The state-of-the-art approaches [11,14,17] identify visually-repeated patterns using additional information displayed on a rendered page. However, they all succumb to the same limitation: they still rely on direct access to either the source code or the tag tree.

The approach in [10] is the first that uses primarily visual features for data record extraction. However, this approach has several other limitations caused by the granularity of its input data and how it deals with noise items on a web page.

In addition, many of the current approaches [10,14,16,17] start by identifying a section of the page, referred to as the *data-rich section*, which contains all of the data records. However, correctly identifying the data-rich section can be problematic. There is a risk that some data items may be omitted from, or unwanted noise items incorrectly included in, the data-rich section. Furthermore, it is possible to identify the wrong subsection of the page entirely.

In this paper, we propose a novel visual approach to data record extraction, which shows a strong correlation with human intuition. Our approach is guided by how most humans expect query result pages to be visually presented. This also influences web site designers, who often present pages to meet human expectations.

A common convention is to place noise items, such as menus or adverts, around the periphery of the page, reserving the centre of the page for the data records. Furthermore, data items in each data record are displayed together, forming what is known as a locality constraint. As shown in Figure 1, query result pages follow these conventions.

To make it easy for a human reader to perceive data items in each data record as a group, designers display the data items repeatedly in the same relative position in each data record. Furthermore, designers also reserve the same horizontal space (or width) for each data record. Originally they did so because human readers disliked pages that required horizontal scrolling [7], therefore, the data records on the page must have a uniform width so they would not cause the page to exceed the screen width. Indeed this accepted design constraint contributes greatly to the regular visual appearance of data records. Technical constraints also have a part to play. Each data record on the page is displayed using the same template code, which, when rendered, reinforces a repeated visual appearance and therefore a similar composition for each data record.

A human reader uses the regularity of the visual-repeated pattern and structure that appears when the data records are displayed on the page to infer semantic relationships between data items and can therefore group the data items into data records.

In summary, our main contributions in this paper are as follows: First, we remove the requirement to identify the data-rich section of a query result page. We propose a novel visual approach which identifies, as a seed block, a single data item, which is a basic content block in a data record. We then implement our observations on visual and structural regularity to group together only the data items in each record. Second, our visual approach directly accesses a rendering engine to retrieve positional information and visual features of each item on the page, avoiding the need to interpret increasingly complex HTML source code and tag trees.

The rest of the paper is organised as follows. The fundamentals behind our proposed approach are presented in Section 2. Solutions for identifying a record block for each data record on the query result page are described in Section 3 and Section 4. Experimental results are reported in Section 5 and finally, Section 6 concludes the paper.

2   Visual Data Record Extraction

In this section, we first introduce the *visual block model* (VBM) to represent a rendered web page, which allows us to access the positional information and visual properties of each item on the page. Next, we present our method to measure the visual similarity between the visual blocks in the model, followed by our definitions of the spatial relationships between these visual blocks. Finally, we present an overview of our approach.

2.1   Visual Block Model

The visual block model of a query result page is a product of the tag tree and the Cascading Style Sheet (CSS) of the page. A layout engine generates visual blocks for each node in the tag tree, according to the instructions contained in the CSS. This process, called rendering, draws a rectangular box around the minimum boundary of each visible node on the page. We refer to each rectangular box as a *visual block*. The position of each visual block is represented by its four borders in the four directions on the two-dimensional plane. The plane has its origin at the top-left of the page, with the x-axis running from left to right and y-axis running from top to bottom.

Some of the visual blocks in the VBM, such as those outlined in blue in Figure 1, represent the structural components of the query result page. These blocks, which we call *container blocks*, can be thought of as the scaffolding that holds the structure of the page together. Each of these visual blocks contains at least one other visual block, each of which we call a *child block*. The larger visual blocks shown in Figure 1, such as those that appear to surround each data record, contain many other visual blocks.

The remaining visual blocks in the VBM, such as those shown in green in Figure 1, which we call *basic blocks*, represent the individual labels and data items displayed on the query result page. These blocks do not contain any other visual blocks. A container block for a data record typically contains a number of these basic blocks.

For each visual block, we obtain 23 visual properties that can be used to define the visual appearance of the visual block. We choose the properties that have historically enjoyed good cross-browser representation and are the most widely used by web site designers to define the visual appearance of query result pages, for example, *fontWeight*. Our approach uses these properties to determine the visual appearance of each block and is flexible; additional visual properties from the CSS specification can be incorporated very easily. We use the WebKit layout engine [8] to render query result pages, however, our approach is independent of any specific engine.

In contrast to the VIPS Algorithm [2], used in other visual approaches, our VBM makes no prior assumptions regarding the organisational hierarchy of a query result page, it simply provides the visual properties for each displayed item. Furthermore, our VBM has finer granularity of visual blocks than the VIPS algorithm. For instance, it can represent two consecutive visual blocks containing two visually distinct texts whereas the VIPS cannot segment the two texts into two visual blocks.

VBM vs. Tag Tree. While the VBM and the tag tree are related, they are not equivalent. The tag tree is a complex representation of the HTML code of the page, created for the browser to interpret and is only part of the information required to render the page as the designer intended. We choose to use the VBM in preference to the tag tree for data record extraction for a number of reasons. First, nodes that are close together on the tag tree may be spatially far apart on the displayed page and vice-versa. By considering only the visual blocks in the VBM, our approach can 'see' the result page in the same way that a human can. Crucially, this is how the page was designed: inferred relationships, such as a group of data items that form a data record, are much easier to identify in a visual context. Second, our approach is insulated from developments in coding practices and standards. Our approach relies on the rendering engine, accordingly we are at liberty to make use of the best engine without the need to adapt our VBM.

2.2 Similarity between Visual Blocks

Our approach decides if two visual blocks have visual, width or content similarity.

Definition 1. *Visual Similarity: Two visual blocks, A and B, are visually similar if all of the visual properties of both blocks are the same. Let $P_A = \{P_{a1}, P_{a2}, ..., P_{an}\}$ be a set of visual properties of A, and $P_B = \{P_{b1}, P_{b2}, ..., P_{bn}\}$ be a set of visual properties of B. The visual similarity between A and B, $Sim(A, B)$, is defined as follows:*

$$Sim(A, B) = \begin{cases} 1 & if P_{ai} = P_{bi}, i = 1, 2, ..., n; \\ 0 & Otherwise. \end{cases}$$

For example, as shown in green in Figure 1, the visual blocks that contain the DVD title in both records on the query result page have the same visual proprieties and are, therefore, visually similar.

Definition 2. *Width Similarity: Two visual blocks have similar widths if the width properties for both blocks are within a threshold of 5 pixels of each other.*

For example, as shown in blue in Figure 1, the visual block that contains all of the data items of the first record, is the same width as the visual block that contains all of the data items of the second record. We say these blocks have similar widths.

Our approach also needs to decide if two blocks have block content similarity, that is, they have similar sets of child blocks. Our observation is that two record blocks have a high degree of block content similarity. This is because both record blocks are created from the same template, repeated for each record on the page. For example, the two record blocks, as shown in Figure 1, contain a large number of visually similar blocks. In contrast, our observation is that there is little block content similarity between a record block and a noise block. For example, the container block for the 'sort by' options, which is the same width as, and directly above, the record blocks in Figure 1, does not share any visually similar child blocks with the record blocks.

We use a variant of the Jaccard index [13] to measure block content similarity between two visual container blocks. The index ranges between 0 and 1, where 1 means that the two blocks are identical, and 0 means they have nothing in common. In our approach, we consider that each container block contains a set of child blocks. We can then measure block content similarity between two container blocks by a similarity index between two corresponding sets of visual child blocks.

Definition 3. *Block Content Similarity: Two visual container blocks have similar block contents if they have a similarity index above a preset threshold.*

For example, as shown in blue in Figure 1, the visual block that contains all of the data items of the first record has a large number of child blocks that have the same visual appearance as those child blocks in the visual block that contains all of the data items of the second record. We say these blocks have block content similarity. In Section 4.2, we formalise our usage of the similarity index.

2.3 Spatial Relationships between Visual Blocks

We now define a number of spatial relationships between visual blocks. The structural regularity of data records on a query result page can be recognised by identifying these spatial relationships between data items in data records.

Definition 4. *Contains: A visual block contains another if all of the borders of the later are inside those of the former.*

For example, as shown in Figure 2, block F contains blocks A, B, C, D and E.

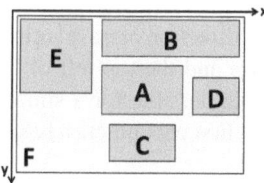

Fig. 2. Examples of Spatially Related Visual Blocks

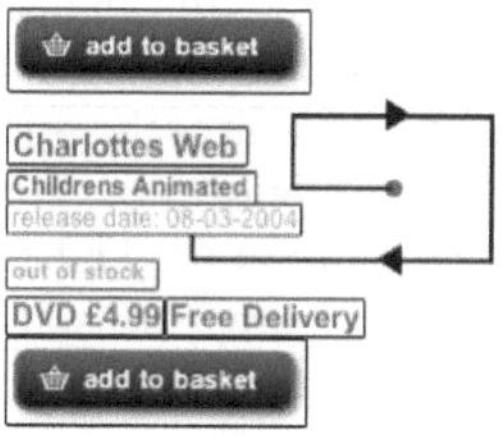

Fig. 3. A Clockwise Ulam Spiral Encountering a Basic Block

2.4 Data Record Extraction

Each data record on the page is represented by a visual block, which contains all of the contents of the data record and nothing else. We have completed a survey of 600 query result pages and found that in over 98% of cases there is a single visual block that exactly contains each data record. The goal of our approach is to identify this block for each data record on a query result page. We call such visual blocks, *record blocks*.

Our approach starts by identifying a single basic visual block that is very likely to be one of the basic blocks of a data record, we call this the *seed block*. The seed block is contained in a set of larger blocks, which we call the *candidate record blocks*, as only one of them is the record block for the data record that the seed block is in. Our approach must select the correct candidate record block as the record block.

3   Seed Block Selection

The goal of seed block selection is to identify a single basic visual block from the VBM which is part of a single data record. Let us look at the organisation of a query result page. Since the Western reading order is from top to bottom, and from left to right, it follows that the data records should start in the top left of the page. However, most web pages have common navigation menu and header structures that appear around the edges of the page. Thus, the starting point of the data records can be shifted down and to the right. As human readers expect this convention, they start looking for data records in this area of the page. A study on web usability by eye tracking [12] confirms that the highest priority area for content is between the centre and the top left of the page.

Our approach starts at the centre of the page, furthest from the noise blocks at the edges and closest to the highest priority area for data records. We trace a clockwise Ulam Spiral [5], as shown in Figure 3, which naturally grows from the centre towards the top left of the page. This is the area of the page most likely to contain data records.

The Ulam Spiral was specially selected as it covers the largest possible proportion of the highest priority area, before it reaches the edges of the page. A simple plane between the centre of the page and the top left corner of the page has, on the other hand, the potential to miss the basic blocks belonging to sparsely populated data records. Instead it could quickly reach the edge of the page and select as the seed a basic block belonging to a noisy feature such as a left menu.

The exponential growth of the spiral combined with its direction of travel (clockwise) ensures that it shows bias to the area between the centre and the top left of the page thereby covering more of the highest priority area than is possible for a simple plane cover. As shown in Figure 3, the spiral terminates when it first encounters a basic block. This block is taken as the seed block.

4   Data Record Selection

The goal of data record selection is to identify a set of container blocks from the VBM, one block for each of the data records on the query result page.

The seed block is contained inside a number of container blocks, each of which provides a structure on the page. Examples of these container blocks are shown in Figure 1, highlighted in blue. By isolating only the container blocks in which the seed block is actually contained, our approach identifies the set of candidate record blocks, as shown in Figure 1, highlighted in red. We observe that one of these candidate record blocks is the record block for the data record and furthermore, this visual block has the similar width to the record block for each of the records on the same query result page.

4.1   Getting Candidate Record Blocks and Clustering Container Blocks

The seed block is a basic block, which is contained inside one or more of the container blocks. As shown in Figure 1, the seed block, which was selected in the previous step, is contained inside four container blocks, highlighted in red. As one of these blocks is the record block, all four are taken as the candidate record blocks. Next, our approach filters the set of all container blocks on the page, discarding any block that is not the same width as one of the candidate record blocks. Our approach uses a one-pass algorithm to cluster the filtered container blocks into a strict partition based on block width. This step creates a number of clusters, one of which contains the record blocks. In our example, the algorithm would create four clusters, one for each of the candidate record blocks.

4.2 Measuring Block Content Similarity

Our approach uses a similarity measure to determine if two container blocks have similar block contents. Assume that block A contains a set of child blocks $\{a_1, a_2, ..., a_m\}$ and block B contains a set of child blocks $\{b_1, b_2, ..., b_n\}$.

It is reasonable to expect that a container block may contain more than one child block with the same visual properties. For example, a data record on a car sales web site may contain an individual child block for each feature of the car, such as the engine size, number of doors and fuel type. These child blocks could share the same visual properties. Accordingly, our approach uses a multi-set representation for each container block. This generalisation of the notion of a set, in which members may appear more than once, allows our approach to represent the child blocks of a container block.

Our approach uses a one-pass algorithm to cluster each of the child blocks contained in a set into a strict partition based on their visual similarity. The function Sim, defined in Definition 1, is used for this purpose. Two child blocks, a and a', are clustered together if $Sim(a, a') = 1$. So a set of child blocks, A, is clustered into a strict partition $A_c = \{A_1, A_2, ..., A_m\}$.

For instance, assume that container block A contains four child blocks, a_1, a_2, a_3 and a_4. Two child blocks, a_1 and a_2, are visually similar, while a_3 and a_4 are visually distinct. The corresponding multi-set, A, is clustered into a set of subsets, A_c, where each subset in A_c represents a single cluster:

$$A_c = \{\{a_1, a_2\}, \{a_3\}, \{a_4\}\} \tag{1}$$

Select one child block from each cluster in A_c as its representative, so we have a set of representative child blocks, A_x, for A_c:

$$A_x = \{x_1, x_2, ..., x_m\} \tag{2}$$

Given two sets of visual blocks, A and B, we use our similarity measure to find the block content similarity between A and B, defined as follows:

$$SimBlockContent(A, B) = \frac{|A \cap B|}{|A \cup B|} \tag{3}$$

We define an indicator function for A_x as follows:

$$1_{A_x}(x_i) = |A_i| \, x_i \in A_x \wedge A_i \in A_c, \, for \, i = 1, 2, ..., m \tag{4}$$

Assume that B is clustered into B_c and $B_y = \{y_1, y_2, ..., y_n\}$ is a set of representative child blocks for B_c. We define an indicator function for B_y as follows:

$$1_{B_y}(y_j) = |B_j| \, y_j \in B_y \wedge B_j \in B_c, \, for \, j = 1, 2, ..., n \tag{5}$$

We then have:

$$|A \cap B| = \sum_{i=1}^{m} \sum_{j=1}^{n} min\{1_A(x_i), 1_B(y_j)\}|Sim(x_i, y_j) = 1 \qquad (6)$$

$$|A \cup B| = \sum_{i=1}^{m} \sum_{j=1}^{n} max\{1_A(x_i), 1_B(y_j)\}|Sim(x_i, y_j) = 1 + |A - B| + |B - A| \quad (7)$$

where

$$|A - B| = \sum_{i=1}^{m} \sum_{j=1}^{n} 1_A(x_i)|Sim(x_i, y_j) = 0$$

and

$$|B - A| = \sum_{i=1}^{m} \sum_{j=1}^{n} 1_B(y_j)|Sim(x_i, y_j) = 0$$

If $SimBlockContent(A, B)$ is above a preset threshold, A and B are considered to have similar block contents.

4.3   Selecting Record Blocks

Only one of the candidate record blocks represents the container blocks that provide the structure for each data record on the page. The other candidate record blocks represent container blocks that are used to provide structure to other areas of the page. By selecting the candidate record block, which has content similarity to the maximum number of container blocks, our approach identifies the blocks for each data record.

However, our example demonstrates a scenario where two candidate record blocks both correspond to the maximum number of container blocks. In this case the candidate record blocks C and D in Figure 1 both have similar contents to the same number of container blocks, as both blocks are present for each displayed record. In the event of a tie for maximum blocks, our approach selects the candidate record block which represents the widest container block, as these container blocks provide the structure to the whole data record (for example, container block labelled D), rather than the internal structure provided by the smaller blocks (for example, container block labelled C).

5   Experimental Results

We have implemented our algorithms for data record extraction in a software prototype called *rExtractor*. In this section we first describe the two datasets that we use to evaluate the accuracy of our algorithms, and then discuss the performance metrics used to interpret the results of these experiments. We compare the performance of our prototype, *rExtractor*, with that of ViNTs [17], a state of the art extraction system available on the web (our experimental analysis makes use of the online prototype), which is based on both visual content features and HTML tag structures.

5.1   Datasets

In our experiments we use two datasets, *DS1* and *DS2*. DS1 is used to compare our method with ViNTs [17]. As these is no standard dataset for experimental analysis of data record extraction techniques, DS1 comprises of web sites taken from the third party list of web sites contained in *Dataset 3* presented in [17]. The web pages contained in Dataset 3 were downloaded from the Internet in 2004, as a result they are 'stale', that is, they are not representative of the layout and appearance of modern web pages. From a visual standpoint, modern pages are more vivid than older pages: they contain more images and much more noise in the form of adverts and menus for example. Therefore, for each web site from Dataset 3 we visited the same web site and downloaded five modern equivalent query result web pages. DS1 comprises 50 web pages from 50 web sites, one web page per web site. In total DS1, contains 752 data records.

DS2 contains a total of 500 web pages from 100 web sites, five per site. It is based on *Dataset 1* and *Dataset 2* in [17], as well as the list of web sites in [6]. As with the previous dataset, the source web pages used to compile DS2 are stale. Accordingly, for each web site included in DS2 from Dataset 1 and Dataset 2, we visited the same web site and downloaded a modern equivalent query result page. The web sites in [6] are listed by their query interface, for each web site included in DS2, we used the query interface to generate five modern query result pages. The web sites in our datasets are drawn from a number of domains, including Books, Music, Movies, Shopping, Properties, Jobs, Automobiles and Hotels. In total, DS2 contains 11,458 data records.

5.2   Performance Metrics

We use a number of performance metrics for our experiments.

Ground Truth. The set of data records from all of the web pages collected per web site.

True Positives. These are data records extracted correctly from all of the web pages per web site. Ideally, the true positives are the same as the ground truth for each web site.

False Positives. These are data records extracted incorrectly from all of the web pages per web site. Ideally, the number of the false positives should be zero.

Precision. This is the number of true positives divided by the number of both true and false positives per web site. The average precision across web sites is calculated by averaging the precisions of individual web sites.

Recall. This is the number of true positives divided by the number of data records in the ground truth per web site. The average recall is calculated by averaging the recalls of individual web sites.

5.3   Experimental Analysis

To undertake the experimental evaluation and comparison of *rExtractor* and ViNTs, we first extract manually the data records on each web page in DS1. Second, we run both *rExtractor* and ViNTs on all of the pages in DS1 and record the correctly extracted (true positives) and incorrectly extracted (false positives) data records on each web

Table 1. Results for DS1 and DS2

DataSet	Algorithm	Avg. Precision	Avg. Recall
1	*rExtractor*	95.5%	96.3%
1	ViNTs	46.3%	48.9%
2	*rExtractor*	97.0%	95.5%

page. Finally, we calculate the average precision and recall for all of the web sites in DS1 for both ViNTs and *rExtractor*. Next, we undertake the same procedure for DS2 and *rExtractor*, to determine how *rExtractor* preforms on a larger dataset. The results for both DS1 and DS2 are presented in Table 1.

5.4   Experimental Analysis

As we can see from Table 1, the performance of *rExtractor* is considerably better than that of ViNTs. Our analysis of the results for ViNTs on DS1 found that ViNTs identified a large number of false positive data records (427 in total). This contributed greatly to their low precision score. By inspecting these false positives, we discovered that ViNTs frequently selected the wrong sub-section of the page as the data-rich section section (DRS). For example, their approach often selected a centrally located, or near centrally located, page navigation menu (for instance, a large footer menu) and then extracted each menu item as a false positive record. In these cases, it was then impossible for their technique to extract the correct data records. Consequently, the number of true positive data records they identified was also small, which contributed to their low recall score. This is not a criticism of ViNTs. Their technique worked very well on the web pages that were available when ViNTs was developed; rather it serves to highlight that modern web pages are vastly different from older web pages.

Conversely, the seed block technique implemented by *rExtractor* selected correctly a child block belonging to a data record in all of the test cases. This demonstrates that our technique has no need to explicitly identify a DRS. Furthermore, as we can see from Tables 1 and 2, *rExtractor* preforms extremely well on both DS1 and DS2 (more than 12,000 data records in total). The experimental results show that the techniques implemented in *rExtractor* to identify candidate record blocks, measure block similarity and select record blocks are all robust and highly effective. Therefore, it is more worthwhile to investigate the cases where *rExtractor* failed to correctly extract data records.

First, a small number of data records (1.3% in total) were not displayed in record container blocks. *rExtractor* relies on these blocks to define the record boundaries. Close inspection of web pages containing these data records reveals they use out-dated page design conventions and are very much the exception rather than the norm.

Second, our content similarity technique prevented the extraction of a small number of data records. These records were determined to have not enough in common with other records on the same page. For instance, designers use a different structural layout and visual appearance to distinguish between a normal record and a 'featured' record. In the cases where *rExtactor* failed, it was often because a featured record was excluded. We see this as an opportunity to develop further our content similarity technique.

6 Conclusions

This paper presents a novel approach to the automatic extraction of data records from query result pages. Our approach first identifies a single visual seed block in a data record, and then discovers the candidate record blocks that contain the seed. Next the container blocks on the page are clustered and a similarity measure is used to identify which of these blocks have similar contents to each candidate record block. Finally, our approach selects the cluster of visual blocks that correspond to the data records. We plan to extend our approach so that the similarity threshold, used to determine if two blocks have similar contents, is set automatically by a machine learning technique.

References

1. Arasu, A., Garcia-Molina, H.: Extracting structured data from web pages. In: SIGMOD Conference, New York, NY, USA, pp. 337–348 (2003)
2. Cai, D., Yu, S., Wen, J., Ma, W.-Y.: Extracting content structure for web pages based on visual representation. In: Zhou, X., Zhang, Y., Orlowska, M.E. (eds.) APWeb 2003. LNCS, vol. 2642, pp. 406–417. Springer, Heidelberg (2003)
3. Chang, C.-H., Lui, S.-C.: Iepad: information extraction based on pattern discovery. In: WWW Conference, New York, NY, USA, pp. 681–688 (2001)
4. Crescenzi, V., Mecca, G., Merialdo, P.: Roadrunner: Towards automatic data extraction from large web sites. In: VLDB Conference, San Francisco, CA, USA, pp. 109–118 (2001)
5. Prime spiral (2012), http://mathworld.wolfram.com/PrimeSpiral.html
6. Tel-8 query interfaces (2004),
   http://metaquerier.cs.uiuc.edu/repository/datasets/tel8/
7. Jakob nielsen - usable i.t (2002),
   http://www.useit.com/alertbox/20021223.html
8. Webkit - layout engine, http://www.webkit.org/
9. Liu, B., Grossman, R., Zhai, Y.: Mining data records in web pages. In: SIGKDD Conference, New York, NY, USA, pp. 601–606 (2003)
10. Liu, W., Meng, X., Meng, W.: Vide: A vision-based approach for deep web data extraction. IEEE Transactions on Knowledge and Data Engineering 22, 447–460 (2010)
11. Miao, G., Tatemura, J., Hsiung, W.-P., Sawires, A., Moser, L.E.: Extracting data records from the web using tag path clustering. In: WWW Conference, pp. 981–990 (2008)
12. Nielsen, J., Pernice, K.: Eyetracking Web Usability, 1st edn., pp. 97–110. New Riders (2010)
13. Real, R., Vargas, J.M.: The probabilistic basis of jaccard's index of similarity. Systematic Biology 45, 380–385 (1996)
14. Simon, K., Lausen, G.: Viper: augmenting automatic information extraction with visual perceptions. In: CIKM Conference, New York, NY, USA, pp. 381–388 (2005)
15. Wang, J., Lochovsky, F.H.: Data extraction and label assignment for web databases. In: WWW Conference, New York, NY, USA, pp. 187–196 (2003)
16. Zhai, Y., Liu, B.: Web data extraction based on partial tree alignment. In: WWW Conference, New York, NY, USA, pp. 76–85 (2005)
17. Zhao, H., Meng, W., Wu, Z., Raghavan, V., Yu, C.: Fully automatic wrapper generation for search engines. In: WWW Conference, New York, NY, USA, pp. 66–75 (2005)
18. Zhao, H., Meng, W., Yu, C.: Automatic extraction of dynamic record sections from search engine result pages. In: VLDB Conference, pp. 989–1000 (2006)
19. Zhao, H., Meng, W., Yu, C.: Mining templates from search result records of search engines. In: SIGKDD Conference, New York, NY, USA, pp. 884–893 (2007)

Leveraging Visual Features and Hierarchical Dependencies for Conference Information Extraction

Yue You[1,2], Guandong Xu[3], Jian Cao[1],
Yanchun Zhang[2,4], and Guangyan Huang[2]

[1] Department of Computer Science and Engineering,
Shanghai Jiao Tong University, China
[2] Centre for Applied Informatics,
Victoria University, Australia
[3] Advanced Analytics Institute,
University of Technology Sydney, Australia
[4] University of Chinese Academy of Sciences, China
{yy71107103,cao-jian}@sjtu.edu.cn, guandong.xu@uts.edu.au,
{yanchun.zhang,guangyan.huang}@vu.edu.au

Abstract. Traditional information extraction methods mainly rely on visual feature assisted techniques; but without considering the hierarchical dependencies within the paragraph structure, some important information is missing. This paper proposes an integrated approach for extracting academic information from conference Web pages. Firstly, Web pages are segmented into text blocks by applying a new hybrid page segmentation algorithm which combines visual feature and DOM structure together. Then, these text blocks are labeled by a Tree-structured Random Fields model, and the block functions are differentiated using various features such as visual features, semantic features and hierarchical dependencies. Finally, an additional post-processing is introduced to tune the initial annotation results. Our experimental results on real-world data sets demonstrated that the proposed method is able to effectively and accurately extract the needed academic information from conference Web pages.

Keywords: Information Extraction, Visual Feature, DOM Structure, Tree-structured Conditional Random Fields.

1 Introduction

As we all know, thousands of conferences, symposiums and workshops are held all around the world every year. The conference Web site is a main and official platform to share and post related conference information which can be accessed by people everywhere. With the increase of conference Web pages, it becomes a cumbersome and time-consuming job for researchers to collect conference information and keep track of the hot research topics. Furthermore, people sometimes are more interested in discovering the future trends of the research field,

Y. Ishikawa et al. (Eds.): APWeb 2013, LNCS 7808, pp. 404–416, 2013.
© Springer-Verlag Berlin Heidelberg 2013

analyzing the social networks of scholars and the inner relationships among these conferences. Building a repository of conference information can satisfy all the above requirements and many value-added services and applications can be developed based on the repository. The ArnetMiner[1] system [1] is a good example of academic social networking based on the publication information repository. In order to automatically and effectively archive clean and high quality academic data, it is essential to extract useful academic information from the conference Web pages, which is more up-to-date, comprehensive and reliable than other sources.

Web information extraction is a classical problem, which aims at identifying interested information from unstructured or semi-structured data in Web pages, and translates it into a semantic clearer structure. During the past years, many techniques such as DOM structure analysis [2], visual based methods [3,4], machine learning methods [3,4] have been devised for Web information extraction. Likewise, this paper studies the problem of automatically extracting structured academic information from conference Web pages.

In the area of conference Web page extraction, most traditional approaches focused on the use of visual features in the Web page. When people design conference Web pages, they usually follow some implicit rules about how they structure certain types of information on the page. So, it is easy to understand that visual features are in some sense more stable than content features. VIsion-based Page Segmentation (VIPS) algorithm [5] was proposed by Deng Cai et al to segment the Web page into text blocks. But the segmentation results of VIPS on conference Web page sometimes are not satisfied due to the ignorance of the hierarchical dependencies (see the example as shown in Figure 3), thus we aim to address this segmentation problem by proposing a new hybrid approach combines these two kinds of features to archive better segmentation results. In order to better organize the segmented text blocks, classification or block labeling procedures are then employed, after which we can perform the refined information extraction on the annotated labels. The state-of-the-art models for sequence labeling problems include Support Vector Machine (SVM) [6] and Conditional Random Fields (CRFs) [7], which mainly rely on the sequence structure. Looking into the conference websites, we observe that apparent hierarchical relations exist between different information blocks. For example, the submission deadline information always appears directly after the phrase - "Important Dates" in a bold style. This observation gives us a hint that we should make full use of the hierarchical dependencies to archive a better annotation performance. Due to the fact that the hierarchical relation in conference Web page forms a tree structure, we therefore introduce a Tree-structured Conditional Random Fields model [8,9] to do the annotation work.

In summary, we have the following contributions: (1) we propose a hybrid page segmentation algorithm, which combines the vision-based segmentation algorithm with the DOM-based segmentation algorithm; (2) we introduce the Tree-structured Conditional Random Fields model [8,9] to annotate the text blocks, which making use of visual features, content features and hierarchical

---

[1] http://arnetminer.org/

dependencies concurrently; (3) we refine the annotation quality by introducing some heuristic post-processing rules. Our experimental results on the real world data sets verify that the proposed integrated approach is highly effective for extracting academic information from conference Web pages.

The rest of the paper is organized as follows. Related works are reviewed in Section 2. We overview our approach in Section 3. Page segmentation algorithm is discussed in Section 4. Text block annotation is described in Section 5. Experiment and evaluation results are reported in Section 6. Finally, in Section 7 we summarize our work and make some conclusions.

2   Related Work

In this section, we present a survey in two aspects: (1) Classification/sequence labeling problem and (2) Web information extraction.

First, our work is related to a classification or sequence labeling problem in some sense. Identifying the useful information from Web pages is sometimes equivalent to annotating the text blocks with different predefined labels. There have been a lot state-of-the-art approaches proposed for this problem, such as Bayes Network [10], SVM [6], Conditional Random Fields [7] and so on. But these approaches usually depend on individual features or linear-dependencies in a sequence of information, while in Web information extraction the information can be two-dimensional [4] or hierarchically depended [3].

Second, our work belongs to the area of Web information extraction, which receives a lot of attentions during past years. This work could be categorized into two types: (1) **Template level wrapper induction systems.** Several automatic or semi-automatic wrapper learning approaches based on the templates have been proposed. Typical representatives are RoadRunner [11], EXALG [12] and so on. (2) **Visual feature assisted techniques.** In contrast, the different display styles of different parts in a Web page provides an additional means for segmenting content blocks. For example, [13,3] reported that visual features, such as width, height, font, etc. are useful in page segmentation.

The limitations of existing Web information extraction methods are:

- Some rule-based Web information extraction techniques are not scalable for the heterogeneity of different conference Web pages.
- Previous studies may be effective for sequence of text blocks as a linear model. But they are not suitable for conference Web information extraction, where the information embedded in the page always follows a tree structure.
- Traditional methods can only extract information from a single Web page, but can not integrate the useful information of a conference that are located in multiple webpages.

3   Overview of Our Approach

The academic information of a specific conference is distributed within a set of pages of the conference Web site. In general, we are primarily concerned with three types of academic information: (1) **Information about conference**

events (e.g. conference name, time, location, submission deadline and submission URL. (2) **Information about conference topics** (e.g. call for papers and topics of interests). (3) **Information about related people and institutes** (e.g. chairs, program committee, authors, companies and universities).

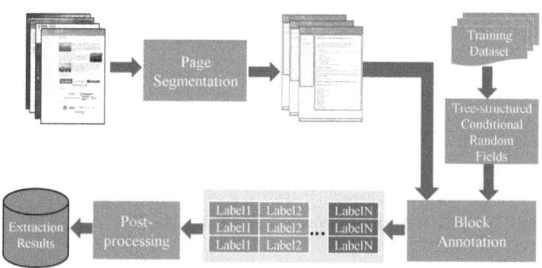

Fig. 1. Overview of our approach

This paper proposes and implements a Web information extraction system which aims at extracting academic information from conference Web pages. Figure 1 shows the framework of our approach. We first use a hybrid algorithm to segment the page into text blocks, then construct a Tree-structured Conditional Random Fields [8,9] to annotate each text block based on visual features and content features. Further processing including verification and some heuristic rules are used to improve the initial annotation results. Finally, we integrate the whole conference information based on the annotation results.

4   Page Segmentation Algorithm

The first task for web information extraction is to find a good representation for Web pages. A good representation can make information extraction easier and more accurate. Through a survey on a large amount of conference Web pages, we find that although different conference Web sites adopt different design styles and layouts, most pages are composed of some common function blocks. Therefore, we first use the VIPS algorithm [5] to segment web pages into text blocks. VIPS algorithm makes use of page layout features such as font, color, and size to construct a vision-tree for a page.

Figure 2 illustrates the layout structure and the vision-based content structure of the *Call for Papers* page of VLDB 2011. At the first level, the original web page has three objects, namely, visual blocks $VB1 - VB3$ and two separators $\varphi^1 - \varphi^2$, as specified in Fig.2(b). Then we can further construct a sub-content structure for each sub-web page. For example, $VB2$ has two off spring objects: $VB2\_1 - VB2\_2$ and one separator φ_2^1. It can be further analyzed recursively.

VIPS can obtain good segmentation results for most Web pages, but it may lose some important information when dealing with some conference Web pages. For example, the red blocks in Figure 3 shows parts of the segmentation results of the *Important Dates* page in VLDB 2011. We can find that it loses the blue blocks, which contains important submission deadline information. The block

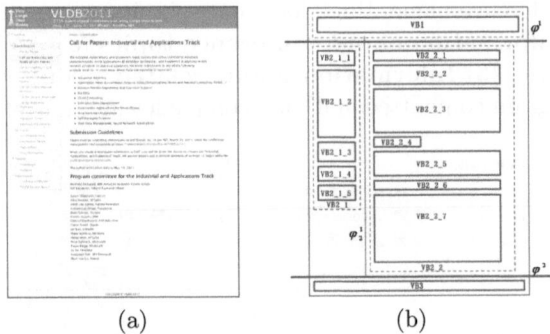

(a)   (b)

Fig. 2. The layout structure and vision-based structure of a conference Web page

lost happens because VIPS algorithm is dependent on visual features of the page elements too much. From Figure 3, we can find that the display style of the red blocks adjacent to the lost part is the same (in this case, they are all in bold), while the lost blocks are displayed in a different style (in this case, they are in non-bold). In other words, VIPS ignores the adjacent blocks, which are not displayed consistently. The lost of these blocks is intolerable.

Fig. 3. The segmentation result of *Important Dates* page in VLDB 2011.

After investigating lots of conference Web pages, we find that the most useful text blocks appear in tag <p>, <table>, <ul>, <h>, <a> or in text node, while these nodes are always located at the same level of the DOM tree and adjacent to each other. Therefore, we define these six types of HTML nodes as

Table 1. Rules for filtering content nodes

Rules	Description
Rule1:	for <p> tag, save node
Rule2:	for <h> tag, save node
Rule3:	for <ul> tag, if its sibling nodes belong to Content Nodes, save all child <li> nodes
Rule4:	for <table> tag, if it has no <table> children, and its sibling nodes belong to Content Nodes, save all child <tr> nodes
Rule5:	for text nodes, there are two cases: 1) if parent is <a> tag, and its sibling nodes belong to Content Nodes, save node 2) if parent is not tag define HTML display information (eg. <style>), save node
Rule6:	for tag, save the node
Rule7:	if the all the conditions above are not met, continue deep-first traversal

Table 2. Rules for redividing and reorganizing nodes

Rules	Description
Rule1:	for tag <h>, <li>, <tr>, self form a block
Rule2:	for tag <p>, redevise it into multiple blocks according to tag in the node
Rule3:	for tag <a>or text node, reorganize multiple continuous nodes into one block according to tag

Algorithm 1. Hybrid Page Segmentation Algorithm

Input: a conference Web page P.
Output: an array of segmented texts $ST[\,]$.
1: $ST_1[\,] \leftarrow VIPS(P)$ {Segment the pages by VIPS}
2: $ST_1[\,] \leftarrow RemoveNontextBlock(ST_1[\,])$
3: $ST_2[\,] \leftarrow DOMPS(P)$ {Analyzing the page by DOM structure}
4: $ST[\,] \leftarrow Combine(ST_1[\,], ST_2[\,])$ {Combine the results of two algorithms}
5: **Function** Combine($ST_1[\,], ST_2[\,]$)
6: **for** each $ST_1[i]$ in $ST_1[\,]$ **do**
7:    **if** $NotFindInST_2(ST_1[i])$ **then**
8:       continue
9:    **else if** LostBlocks($ST_1[i]$) **then**
10:      ProcessLost($ST_2[\,], ST_1[i]$)
11:    **else**
12:      $ST[\,].add(ST_1[i])$
13:    **end if**
14: **end for**

Content Nodes. A DOM-based page segmentation algorithm has two phases: (1) performing a depth-first traversal of the DOM tree, and saving useful Content

Nodes based on rules showed in Table 1. (2) re-dividing and reorganizing saved nodes into semantic blocks based on rules shown in Table 2.

According to the analysis above, we propose a hybrid page segmentation algorithm combining the VIPS and the DOM-based algorithm. Algorithm 1 shows the detail of the algorithm. First, it segments an input web page into visual blocks using the VIPS algorithm (line 1). Second, it removes the noise blocks (navigation blocks, copyright blocks, etc.) using some heuristic rules (line 2). Third, it segments the input web page using the DOM-based algorithm (line 3). Finally, it combines the text blocks produced by the VIPS and DOM-based algorithm to archive a more complete segmentation result (line 4).

5 Text Block Annotation

In this section, we use Tree-structured Conditional Random Fields (TCRFs) [8,9] to annotate the text blocks outputted by the segmentation algorithm described in Section 4.

5.1 Problem Definition

The input of our problem is a set of Web pages related to a conference, each Web page can be parsed into a sequence of text blocks, namely B. And the output is the structured academic information (as defined in Section 3) about the conference. In order to achieve effective Web information extraction, we first annotate the sequence of text blocks according to the display style and content information of text block. The label space L is defined in Table 3.

Table 3. Definition of Label Space L

Labels	Description
Title (TI):	Refers to the text blocks shown in bold and larger fonts.
Topic (TO):	Describes the topics of interest information of the conference.
List no text (LNT):	Text blocks displayed in list, but do not contain useful information.
Date Item (DI):	This label describes a time table of a conference process.
Date (D):	This label represents a specific time in the Web page.
Item (I):	It usually appears simultaneously with Date label, and it describes a process of the conference. Continuous Date category and Item category equal to a Date Item category in semantics, both describe the timing of a conference.
Committee Person (CP):	It refers to the text blocks describing information about organizers or program committee. This label always includes name, affiliation, country and so on.
Yes Text (YT):	It refers to the text blocks of natural language, and it may contain useful information.
No Text (NT):	It refers to the text blocks of natural language, but do not contain useful information.

Definition 1. *(Text block annotation): Given a sequence of text blocks b corresponding to a page, let $b = \{b_0, b_1, \cdots, b_n\}$ be the features of all the blocks and each b_i is a feature vector of one block, let $l = \{l_0, l_1, \cdots, l_n\}$ be one possible label assignment of the corresponding blocks and each $l_i \in L$ defined above. The goal of conference Web data extraction is to compute a maximum posteriori (MAP) probability of l and extract data from this assignment l^* :*

$$l^* = argmax \ p(l|b). \tag{1}$$

Based on the above definition, the main goal is to calculate $p(l|b)$. Here, we introduce TCRFs [8,9] to compute it, which will be described in the following.

5.2 Tree-Structured Conditional Random Fields

Conditional Random Fields (CRFs) [7] is a probabilistic model proposed by John Lafferty to segment and label sequence data. For understanding the formula, we define the following notations: X is a random variable over text blocks outputted by segmentation algorithm, and Y is a random variable over corresponding annotations. All components Y_i of Y are assumed to range over label space L. CRFs construct a conditional model $p(Y|X)$ based on a given set of features from text blocks and corresponding annotations. The definition of CRFs is given below, for more details please refer to the original paper [7].

Definition 2. *(Conditional Random Fields [7]): Let $G = (V, E)$ be a graph and Y is indexed by the vertices of G, i.e. $(Y_v)_{v \in V}$. Then (X, Y) is a conditional random field, when conditioned on X, Y_v obey the Markov property with respect to the graph: $p(Y_v|X, Y_w, w \neq v) = p(Y_v|X, Y_w, w \sim v)$, where $w \sim v$ means that there is an edge between w and v in G.*

Tree-structured Conditional Random Fields [8,9] is a particular case of CRFs, whose graphical structure is a tree. TCRFs can model the parent-child and sibling dependencies, while the simplest Linear-chain CRFs [7] can't do.

We define (y_p, y_c) as a parent-child dependency, (y_c, y_p) as a child-parent dependency, and (y_s, y_s) as a sibling dependency. A TCRFs model has the form:

$$p(y|x) = \frac{1}{Z(x)} \exp\left(\sum_{e \in \{E^{pc}, E^{cp}, E^{ss}\}, j} \lambda_j t_j(e, y|_e, x) + \sum_{v \in V, k} \mu_k s_k(v, y|_v, x) \right),$$

(2)

where x is a block sequence, y is annotation result, and $y|_e$ and $y|_v$ denotes y associated with edge e and vertex v in the tree; E_{pc} is set of (y_p, y_c), E_{cp} is set of (y_c, y_p), and E_{ss} is set of (y_s, y_s); t_j and s_k are feature functions; λ_j and μ_k are parameters corresponding to t_j and s_k respectively, and will be estimated from the training data; $Z(x)$ is the normalization factor.

5.3   Conference Web Page Annotation

Text Block Tree Representation. In order to apply TCRFs to conference Web page annotation, we construct a tree representation of the text blocks in a Web page. After investigating lots of typical top conference Web sites such as VLDB, CIKM, KDD etc. we find that almost all the Web pages follow the rule that the useful information displaying below titles are more obvious than other blocks. For example, the time of a conference event always follows an "Important Dates" block which belongs to Title (TI) label. Based on this observation, we can construct a text block tree to represent a Web page. The root of the tree represents the whole page. Each inner node represents the block displayed in a more highlighted style (usually in a larger font and in bold style). Each leaf node represents the block display below the inner node. Figure 4 shows the text block tree representation of the Web page corresponding to Figure 2(a). In addition to the parent-child dependencies, we also model the sibling dependencies in the text block tree. For example, topics of a conference are always located close to each other, and always in the same level of the text block tree.

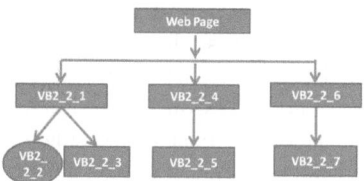

Fig. 4. The text block tree of the Web page showed in 2(a)

Features in Annotation Model. After studying characteristics of each label defined in label space L, two kinds of features are chosen to represent a text block, e.g. visual features and content features. The visual features include FontSize, FontWeight, FontColor, StartWithH and so on. The content features specify whether the text block contains month, country, institute, topic keyword and other keywords. These features belong to the vertex features. We also take edge features into account. We use $f(b_p, b_c)$, $f(b_c, b_p)$, and $f(b_s, b_s)$ to denote the current text block's parent-child dependency, child-parent dependency and sibling dependency, respectively.

Parameter Estimation. Parameter estimation is to determine the parameters $\Theta = \{\lambda_1, \lambda_2, \cdots ; \mu_k, \mu_{k+1}, \cdots \}$ from the training data $D = (x^{(i)}, y^{(i)})$ along with empirical distribution $\tilde{p}(x, y)$. Actually, we optimize the log-likelihood function of the conditional model $p(y|x, \Theta)$:

$$L_\Theta = \sum_i \tilde{p}(x^{(i)}, y^{(i)}) \log p_\Theta(y_{(i)}|x_{(i)})). \tag{3}$$

We use f to represent both edge feature function and vertex feature function; c to represent both edge and vertex; and λ to represent two types of parameters: λ and μ. Thus, the derivative of log-likelihood function with respect to parameter λ_j associated with clique index c is:

$$\frac{\partial L_\Theta}{\partial \lambda_j} = \sum_i \left[\sum_c f_j(c, y^{(i)}_{(c)}, x^{(i)}) - \sum_y \sum_c p(y_c|x^{(i)})f_j(c, y_{(c)}, x^{(i)}) \right], \tag{4}$$

where $y^{(i)}_{(c)}$ is the annotation result of clique c in $x^{(i)}$, and $y_{(c)}$ ranges over annotation results to the clique c. $p(y_{(c)}|x^{(i)})$ requires the computation of the marginal probabilities. We utilize the Tree Reparameterization (TRP) algorithm [14] to compute the approximate probabilities of the factors. The function L_Θ can be optimized by several techniques; but in this paper, we adopt gradient based Limited-memory BFGS method [15], which outperforms other optimization techniques for linear-chain CRFs [16].

Post-processing. The text blocks annotation results may directly affect the quality of the final conference academic information extraction result. In order to improve the quality of the annotation results, the post-processing is essential. Known from the conference Websites, the text blocks can be divided into five groups according to their display characteristics and contents, i.e., Title, Data Item/Date/Item, Topic/List no text, Committee Person, Yes Text/No Text. We summarize some heuristic rules for post-processing for different label groups, and repair text blocks with wrongly annotated labels from the following two aspects automatically. (1) A block has special features but cannot be annotated correctly. For example, given a text block Camera Ready Papers Due: Thursday, August 11, 2011, which have typical DI features. However, this block is classified as YT. For this situation, it can be repaired by identifying some typical combination of features. This method is suitable for text blocks with clear features, such as DI/D/I, TO and CP. (2) A text block is classified as a category but it does not contain corresponding features. For example, text blocks about submission instruction are classified as YT. Although it contains a lot of words, it usually does not contain any people name, date or location. Therefore, we can check its feature values ContainMonth, ContainURL and Country Keyword to determine whether it is a YT category. Due to the space limitation, we omitted the technical details here.

6   Experiments

6.1   System Implementation

The system is mainly implemented in Java, and the Web page segmentation module is implemented in C#. We also use MALLET[2], a Java-based package for machine learning applications. Our experiments are conducted on a PC with 2.93GHz CPU, 4GB RAM and Windows 7. Our system can process a conference Web page in average of 0.80 seconds, which contains 0.18 seconds for segmenting, 0.62 seconds for annotating and 0.002 seconds for post-processing.

6.2   Data Sets and Evaluation Measures

We collect our conference Web page data by meta-search. First, we prepare a list of seed top conference names in computer science field. Then we throw each name as a query to Google and get a list of urls. Using some heuristic rules, we choose the url which is most likely to be the home page of the conference Web site, then we request all the Web pages of that conference Web site. Finally, we collect 50 academic conference Web sites in computer science field, which has 283 web pages and 10,028 labeled text blocks. In order to evaluate our approach, 10 students are invited to manually annotate all the text blocks as the ground truth of labeling. We use the five-fold cross-validation method to evaluate our system, that is, one fold of conference Web sites are training data and the rest sites are test data. We use Precision, Recall and F-Measure as evaluation measures.

6.3   Experimental Results and Discussions

The object of the first experiment is to compare our hybrid segmentation algorithm with the baseline VIPS algorithm. To verify the effectiveness of our algorithm, we call these two algorithms for conference Web pages respectively, and take the manually segmented results as the ground truth. If a text block outputted by an algorithm is the same as the human segmented one, we call it a correct block. Table 4 shows the count number of correct text blocks of a conference Web site using the two segmentation algorithm respectively. From the results, we can find that our hybrid segmentation algorithm can get more correct text blocks than the VIPS segmentation algorithm. Especially for conference DASFAA-11, VIPS get a poor result because the visual separators between two text blocks is not so obvious, while our hybrid segmentation algorithm gets a satisfactory result.

---

[2] http://mallet.cs.umass.edu/

Table 4. Number of correct text blocks using two segmentation algorithms

Conference Name	VIPS	Hybrid Algorithm	Ground Truth	Δ
VLDB-11	185	226	230	+18%
AIED-11	180	195	195	+8%
CGI-11	190	198	201	+4%
CIKM-11	283	292	292	+3%
ICDE-11	220	237	241	+7%
ICDE-10	490	540	540	+9%
DASFAA-11	10	318	318	+97%
EURO-PAR-11	510	515	515	+1%
SIGMOD-11	145	170	174	+14%
SOSP-11	180	190	190	+5%

Table 5. Annotation Comparison of Different Models

Label	Bayes Network			SVM			CRFs			HY+TCRFs		
	Prec	Recal	F-ms	Prec	Recal	F-ms	Prec	Recal	F-ms	Prec	Recal	F-ms
TI	0.93	0.91	0.92	0.90	0.89	0.89	0.89	0.91	0.90	0.94	0.97	0.95
TO	0.76	0.65	0.70	0.87	0.80	0.83	0.91	0.83	0.87	0.93	0.92	0.92
DI	0.80	0.92	0.86	0.83	0.70	0.76	0.82	0.90	0.86	0.87	0.93	0.90
D	0.96	0.77	0.85	0.88	0.60	0.71	0.95	0.57	0.71	0.98	0.88	0.92
I	0.28	0.53	0.37	0.78	0.70	0.74	0.82	0.72	0.76	0.90	0.82	0.86
CP	0.96	0.83	0.89	0.96	0.80	0.87	0.97	0.88	0.93	0.97	0.90	0.93
YT	0.68	0.60	0.64	0.50	0.62	0.55	0.47	0.60	0.53	0.80	0.81	0.80
LNT	0.44	0.66	0.53	0.58	0.78	0.67	0.76	0.87	0.81	0.90	0.94	0.92
NT	0.75	0.78	0.76	0.75	0.88	0.81	0.77	0.85	0.80	0.85	0.87	0.86
Average	0.73	0.74	0.72	0.78	0.75	0.76	0.82	0.79	0.80	**0.90**	**0.89**	**0.90**

The aim of the second experiment is to demonstrate the effectiveness of combined TCRFs model. We have mentioned in Section 3, this work also approaches classification or sequence labeling problem, so we introduce Bayes Network [10], SVM [6], CRFs [7] as baselines for comparison. Table 5 shows the P, R, F measures of different labels in the label space L. We name our approach as HY+TCRFs, which indicates the combination of hybrid segmentation algorithm and TCRFs model. From the numeric results in Table 5, we find that HY+TCRFs always performs much better than the other models, because it takes both the visual and dependent features into account. It can also be found that by using HY+TCRFs, the TO/LNT, D/I/DI categories achieve a notable improvement than other labels. It is explainable because these labels have stronger dependencies in child-parent relation and sibling relation than other labels.

The last group of experiments as shown in Table 6 is the comparison between initial annotation results and the results after post-processing. It has demonstrated that post-processing generally improves the quality of the annotation results. Looking close to the results, we observe that for label TI/TO/DI/D/CP/LNT they all have more than 0.95 F-Measure, and NT has average 0.90 F-Measure. The explanation is that they have apparent and specific features on not only display style but also semantic level. The post-processing can improve the initial annotation easier than other labels. We also observe that YT/NT/I have a relatively low F-Measure for two reasons: (1) it is hard to effectively extract the features of YT/NT labels; (2) different conferences have different descriptions for an Item label, which has no constant format.

Table 6. Comparison of before and after post-processing

Label	Before			After		
	Prec	Recall	F-ms	Prec	Recall	F-ms
TI	0.94	0.97	0.95	0.98	0.99	0.98
TO	0.93	0.92	0.92	0.99	0.99	0.99
DI	0.87	0.93	0.90	0.96	0.99	0.97
D	0.98	0.88	0.92	0.98	0.94	0.96
I	0.90	0.82	0.86	0.94	0.87	0.90
CP	0.97	0.90	0.93	0.98	0.93	0.95
YT	0.80	0.81	0.80	0.82	0.80	0.81
LNT	0.90	0.94	0.92	0.96	0.97	0.96
NT	0.85	0.87	0.86	0.89	0.91	0.90
Average	0.90	0.89	0.90	**0.94**	**0.93**	**0.94**

7 Conclusions

This paper has proposed a hybrid approach to extract academic information from conference Web pages. Particularly, different from existing approaches, our method combined visual feature and hierarchical structure analysis. Meanwhile, we also developed heuristic-rule based post-processing techniques to further improve the annotation results. Our experimental results based on the real world data sets have shown that the proposed method is highly effective and accurate, and it is able to achieve average 94% precision and 93% recall.

Acknowledgments. This work is partially supported by China National Science Foundation (Granted Number 61073021, 61272438, 61272480), Research Funds of Science and Technology Commission of Shanghai Municipality (Granted Number 11511500102, 12511502704), Cross Research Fund of Biomedical Engineering of Shanghai Jiao Tong University (YG2011MS38).

References

1. Tang, J., Zhang, J., Zhang, D., Yao, L., Zhu, C., Li, J.Z.: Arnetminer: An expertise oriented search system for web community. In: Semantic Web Challenge. CEUR Workshop Proceedings, vol. 295 (2007)
2. Sun, F., Song, D., Liao, L.: Dom based content extraction via text density. In: SIGIR, pp. 245–254 (2011)
3. Zhu, J., Nie, Z., Wen, J.R., Zhang, B., Ma, W.Y.: Simultaneous record detection and attribute labeling in web data extraction. In: KDD, pp. 494–503 (2006)
4. Zhu, J., Nie, Z., Wen, J.R., Zhang, B., Ma, W.Y.: 2d conditional random fields for web information extraction. In: ICML. ACM International Conference Proceeding Series, vol. 119, pp. 1044–1051 (2005)
5. Cai, D., Yu, S., Wen, J.R., Ma, W.Y.: Block-based web search. In: SIGIR, pp. 456–463 (2004)
6. Duan, K.-B., Keerthi, S.S.: Which is the best multiclass SVM method? An empirical study. In: Oza, N.C., Polikar, R., Kittler, J., Roli, F. (eds.) MCS 2005. LNCS, vol. 3541, pp. 278–285. Springer, Heidelberg (2005)
7. Lafferty, J.D., McCallum, A., Pereira, F.C.N.: Conditional random fields: Probabilistic models for segmenting and labeling sequence data. In: ICML, pp. 282–289 (2001)

8. Bradley, J.K., Guestrin, C.: Learning tree conditional random fields. In: ICML, pp. 127–134 (2010)

9. Tang, J., Hong, M., Li, J., Liang, B.: Tree-structured conditional random fields for semantic annotation. In: Cruz, I., Decker, S., Allemang, D., Preist, C., Schwabe, D., Mika, P., Uschold, M., Aroyo, L.M. (eds.) ISWC 2006. LNCS, vol. 4273, pp. 640–653. Springer, Heidelberg (2006)

10. Heckerman, D.: A tutorial on learning with bayesian networks. In: Holmes, D.E., Jain, L.C. (eds.) Innovations in Bayesian Networks. SCI, vol. 156, pp. 33–82. Springer, Heidelberg (2008)

11. Crescenzi, V., Mecca, G., Merialdo, P.: Roadrunner: Towards automatic data extraction from large web sites. In: VLDB, pp. 109–118 (2001)

12. Arasu, A., Garcia-Molina, H.: Extracting structured data from web pages. In: SIGMOD Conference, pp. 337–348 (2003)

13. Song, R., Liu, H., Wen, J.R., Ma, W.Y.: Learning block importance models for web pages. In: WWW, pp. 203–211 (2004)

14. Wainwright, M.J., Jaakkola, T., Willsky, A.S.: Tree-based reparameterization for approximate inference on loopy graphs. In: NIPS, pp. 1001–1008 (2001)

15. Xiao, Y., Wei, Z., Wang, Z.: A limited memory bfgs-type method for large-scale unconstrained optimization. Computers & Mathematics with Applications 56(4), 1001–1009 (2008)

16. Sha, F., Pereira, F.C.N.: Shallow parsing with conditional random fields. In: HLT-NAACL (2003)

Aggregation-Based Probing
for Large-Scale Duplicate Image Detection

Ziming Feng, Jia Chen, Xian Wu, and Yong Yu

Computer Science Department, Shanghai Jiao Tong University
Dongchuan Road, Minhang District, Shanghai, China
{fengzm,chenjia,wuxian,yyu}@apex.sjtu.edu.cn

Abstract. Identifying visually duplicate images is a prerequisite for a broad range of tasks in image retrieval and mining, thus attracts heavy research interests. Many efficient and precise algorithms are proposed. However, compared to the performance duplicate text detection, the recall for duplicate image detection is relatively low, which means that many duplicate images are left undetected. In this paper, we focus on improving recall while preserving high precision. We exploit hash code representation of images and present a probing based algorithm to increase the recall. Different from state-of-the-art probing methods in image search, multiple probing sequences exist in duplicate image detection task. To merge multiple probing sequences, we design an unsupervised score-based aggregation algorithm. The experimental results on a large scale data set show that precision is preserved and the recall is increased. Furthermore, our algorithm on aggregating multiple probing sequences is proved to be stable.

1 Introduction

The explosion of web image scale inevitably brings overwhelming number of duplicate images. According to a recent study [18], more than 8.1% of web images have more than ten near duplicates and more than 20% of web images have duplicate images. Without proper processing, such duplicates can cause redundancies in web search results and reduces user experiences. Therefore, detecting duplicate images from a large image collection becomes a hot research topic.

As defined in work[17,9], the duplicate image detection task is to find all visually duplicate image groups from an image collection. According to the definition, it is a cluster style task rather than a search style task. The input is a gigantic collection of images and the output is a cleaned version of that image collection(all duplicate ones are grouped into one cluster).

To cope with large scale, efficiency is the core consideration in designing algorithms. State-of-the-art approaches usually exploit hash-code representation of images [9,17] and put images with the same hash-code together. Since duplicate images could be represented with different hash codes, such approaches achieve high precision but relatively low recall.

Y. Ishikawa et al. (Eds.): APWeb 2013, LNCS 7808, pp. 417–428, 2013.

In this paper, we propose to improve recall while preserving high precision. We design a probing style algorithm to look for more duplicate images. Different from single probing sequence in image search [13], multiple probing sequences exist in duplicate image detection. This raises a new challenge: how to merge multiple probing sequences. In this paper, we formulate this challenge into a rank aggregation problem. Under such formulation, each probing sequence is first represented as a partial ordered list and then aggregated into an unified probing sequence for execution.

In experiments, we compare the *Principal Components Analysis* (PCA) code and *Locality sensitive hashing* (LSH) code on millions of images in this task. It could reflect that PCA-code is more suitable in this task. Then we demonstrate the proposed approach with PCA-code. In developing the rank aggregation on probing sequences, we introduce three different distance measurements and compare them with the baseline. All these three measurements are proved to increase the recall while preserving high precision.

The rest of the paper is organized as follows. Section 2 discusses related work; Section 3 introduces the framework of the algorithm. Section 4 generates the PCA code; Section 5 proposes the aggregation-based probe algorithm. Section 6 provides the experimental results and section 7 concludes this paper.

2   Related Work

2.1   Duplicate Image Detection

Plenty of works [14,19,9,17,22] are devoted to duplicate image detection. According to different feature representations of an image, these works can be classified into two categories.

- Local Method: Typical works[3,9,22,16] are bases on SIFT[12]. The min-hash [9] is used to speed up the process. The advantage of adopting local features is that they are invariant to the scale, rotation, position and affine transform of images. While the drawback is burden on computational resources consumption, which makes it hard to build a system to handle the the billions of images.
- Global Method: [10,15,17] are based on global features. Typical global features include color histogram [20], gray value [17] and edge histogram [23]. PCA is useful in duplicate task[6]. Wang et al.[17] proposed a PCA-based hashing method depending on the global features. In their work, each image is divided into $k \times k$ regular grid elements and the corresponding mean gray values are extracted to get simple but effective visual features. However, it focuses on the precision but ignores the recall. In our work, we propose an efficient rank-based probe method to increase the low recall.

2.2   Rank Aggregation

A lot of relevant works on rank aggregation methods has been published [4,8,5,2,21,11]. In general, there are two main types: score-based aggregation and rank-based aggregation. Recent work focuses on rank-based methods [4,5], which means the order of the list is significant but the actual score of each element in the list is not. There are some common metrics for rank-based aggregation such as Canberra distance[7], Spearman footrule distance [4] and Kendall's tau distance [4].

In our task, the probing score to a given code represents the probability of the existence of duplicate images corresponding to it. Thus we use a score-based rank aggregation method to handle the problem. The existing score-based methods are often based on supervised learning [21,2]. Due to the large volume of web images, it's very expensive to acquire manual labeling. Therefore we adopt an unsupervised approach. Different from Klementiev[8] , which measures the agreement between the rankers, we design some distance measurements naturally from the existing rank-based measurement and directly optimize them.

3 Framework

In this section, we introduce the framework for the duplicate image detection task, which contains two steps:

At first, all images in the entire data set will be represented by gray value and indexed according to their PCA-code; Then we use the PCA codes to divide the image set into smaller groups, the images in each group share the same PCA code. We refer to each group as a *bucket* in the rest of this paper. In the generation of PCA-code, similar images may vary a lot in their PCA codes. The slight shift of the features may cause totally different code, thus put into different buckets. To improve the performance, the PCA-codes representing similar image features should be identified and related. The traditional similar metric is the hamming distance, which means the PCA-codes with small hamming distance will be grouped together. But it suffers a low efficiency to probe all images in the dataset.

To increase the efficiency, for each bucket, we generate a probing sequence. The probing sequence can be regarded as similarity measurement of the buckets. The later element in the sequence means the less probability that contains the duplicate image. If we can generate the probing sequence, it could effectively and efficiently find the duplicate image groups.

Recently, a multi-probe method [13] is proposed to generate the probing sequence for LSH. Since the Gaussian distribution assumption of PCA is equal to that of LSH in [13], we can apply [13] to the PCA-code. But [13] method relays on the query as the input parameter, however in duplicate image detection, there are more than one images in each bucket, therefore, we cannot apply [13] directly. We develop a two-step algorithm to handle the problem. First, for each image in the bucket, we generate the basic probing sequence. Second, we apply the rank aggregation algorithm to generate the merged probing sequence. Because the length of the candidate list is limited, it is a partial list rank aggregation.

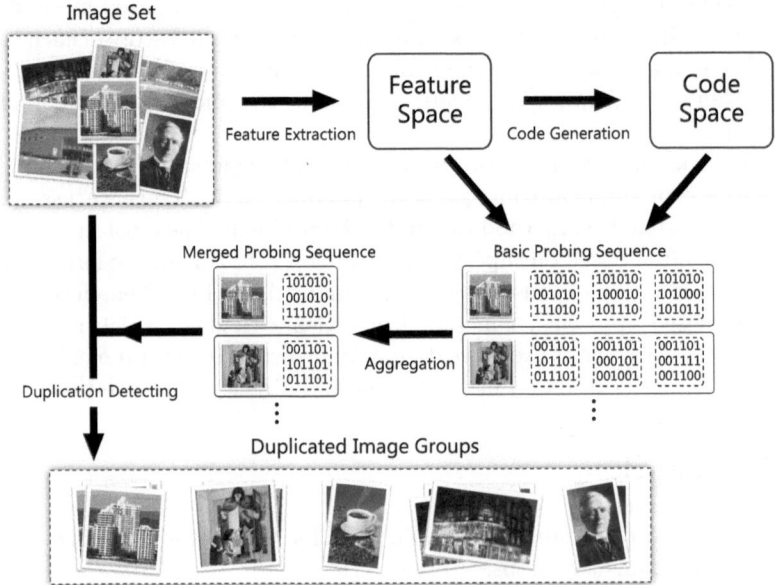

Fig. 1. Framework of our duplication detection system

Since we target to cope with large scale web images, it's hard and expensive to acquire the labeled data, thus it is hard to use supervised rank aggregation method. We turn to an unsupervised score-based method which is displayed in Figure 1.

4   PCA-Code Component

In this section, we introduce feature extraction and PCA hash-code generation. For feature extraction, an image is divided into a $k \times k$ grid, where k is set to 8 from previous work [17]. The mean gray value feature is extracted from each grid as a robust image representation for duplicate detection. It can tolerate most image transformations such as resizing, recoloring, compression, and a slight amount of cropping and rotation. Each feature element is calculated by the Equation (1)

$$F_i = \frac{1}{N_i} \times \sum gray_j^{(i)} i = 1, 2 \ldots k^2, \tag{1}$$

where $gray_j^{(i)}$ is the $j-th$ gray value in grid i and N_i is the number of pixels in grid i. So an image could be represented by a $k^2 \times 1$ vector.

Before code generation, the raw features should be compressed into a compact representation with noises reduced. The PCA hash-code algorithm uses a PCA-based hash function to map a high-dimensional feature space to a low-dimensional feature space. The PCA model is used to compress the features extracted from the image database. After the dimension reduction, visual features are transformed to a low-dimensional feature space. Then the hash code H for an image is built as follows: each dimension is transformed to one if its value in this dimension is greater than mean value, and zero otherwise. This is summarized in Equation (2).

$$H(i) = \begin{cases} 1 & v_i \geq mean_i \\ 0 & v_i < mean_i \end{cases} \tag{2}$$

where the $mean_i$ is the mean value of the dimension i. The flowchart of PCA-code component is shown in Figure 2.

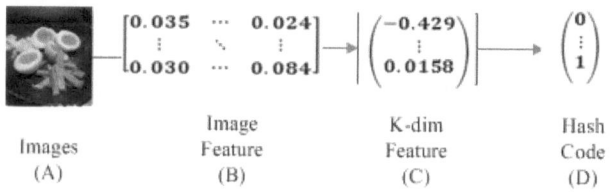

Images	Image Feature	K-dim Feature	Hash Code
(A)	(B)	(C)	(D)

Fig. 2. Illustration for hash-code generation component

5   Aggregation-Based Probe Algorithm

The input of our aggregation-based probe algorithm is a set of images in hash code and feature representation. The output is duplicate image groups(code groups). The algorithm traverses all the buckets. For a given bucket, it generates a merged probing sequence of buckets and then gathers them with the given bucket into a group. Then these buckets will not be probed again.

The merged probing sequence is generated in two steps.

Step 1: Basic Probing Sequence Generation. In this step, a probing sequence of buckets for each image in the starting bucket is generated. The top-K elements are selected into the sequence based on a score function. The score function measures the probability of containing duplicate images. A smaller score means larger probability of finding a duplicate image.

Step 2: Merged Probing Sequence Generation. In this step, the different basic probing sequences generated from the starting bucket are merged. The merging is formulated as a partial rank aggregation problem as shown in Table 1.

For elements in basic probing sequence, it is assigned with a probability value. We treat probability values as scores in our solution. Additionally, given that there is no labeled data available, we use an unsupervised approach in solving the aggregation problem.

Table 1. Terminology Mapping Table

Terms in duplicate detection	Terms in rank aggregation
Image	Ranker
Basic probing sequence	Partial ordered list
Merged probing sequence	Result list

5.1   Step 1: Basic Probing Sequence Generation

At the beginning, each image in collection are encoded by PCA-code. And all of them will generate a probing sequence by a score function. From [13], the score function defined for one image in starting bucket h_q to bucket h_j is as in Equation (3)

$$score_{LSH}(h_q, h_j) = \sum_{k \in \varphi} f_k^2 \tag{3}$$

where f is the compressed feature value of the query image and φ is the set of bits which are different between buckets h_q and h_j.

Then we apply this equation to PCA-code. Since we will compare the compressed feature values between different dimensions in PCA-code, we normalize them by $X' \equiv \frac{X-\mu}{\sigma}$. Considering the assumption that each dimension obeys the Gaussian distribution, the normalized value conforms to a standard Gaussian distribution. Then the score function of PCA is as in Equation (4)

$$score_{PCA}(h_q, h_j) = \sum_{k \in \varphi} (\frac{f_k - \mu_k}{\sigma_k})^2 \tag{4}$$

where the μ_k is the mean value of dimension k and σ_k means the standard deviation of dimension k. It is used to estimate the probability that images in h_j duplicate the query image. A lower score means that h_j is more likely to contain duplicate images. For each image in the starting bucket, we get a basic probing sequence based on $score_{PCA}$.

5.2   Step 2: Merged Probing Sequence Generation

Before modeling the score-based rank aggregation problem, we do some preprocessing to turn the partial ordered list into a full ordered list. We combine all the elements in partial ordered lists, to get a set T^*. Then we calculate scores for all the elements in T^* for each image in the starting bucket. It becomes a full ordered list problem.

Now we define some symbols to give a mathematical formulation for the score-based rank aggregation problem. We assume that the V^i represents the $i-th$ full ordered list and V^* is the result list. Finally, we cast the score-based rank aggregation problem into the optimization problem shown in Equation (5).

$$\min \sum_i dis(V^i, V^*) \tag{5}$$

The $dis(V^i, V^*)$ measures the distance between V^i and V^*.

In the formulation, different distance measurement $dis(V^i, V^*)$ leads to different result list V^*. The following three measurements are designed in this problem.

1. **Footrule distance**
   Motivated by the *Footrule distance*[4], the score-based distance measurement could be

$$dis(V^i, V^*) = \sum_j |V_j^i - V_j^*| \tag{6}$$

   It naturally describes the difference between the two rankers (The result list V^* also could be seen as a ranker).

2. **Canberra Distance**
   But in the **footrule distance**, we don't take into account that the large elements in the list will have an larger weight than the small elements (i.e. $|V_j^i - V_j^*|$ is not weighted). The weighted **footrule distance** is the Canberra Distance[7] .

$$dis(V^i, V^{/}*) = \sum_j \frac{|V_j^i - V_j^*|}{|V_j^i + V_j^*|} \tag{7}$$

   Adding the denominator to the equation prevents the unbalanced estimation problem.

3. **Discounted Canberra Distance**
   In our problem, if we take the fewer elements to achieve the same recall, it will be more efficient and effective. Then we can get a conclusion that the top-K elements in V^* are more important than others. So we add a rank adjustment factor to the Canberra Distance. This is the discounted Canberra Distance in the following equation,

$$dis(V^i, V^*) = \sum_j \frac{|V_j^i - V_j^*|}{|V_j^i + V_j^*|} \times \frac{1}{1 + \lg Pos_j^{V^*}} \tag{8}$$

   where the $Pos_j^{V^*}$ means the rank of element j in the V^*. The adjustment in this equation effectively increases the weight of the element at the top of the list.

Optimization Method. In this section, we show the details of optimization method for different distance measurements in the score-based rank aggregation.

1. In *footrule distance*, the optimal score of each element in T^* is the median value of the scores given by the rankers.

$$V_j^* = median(V_j^1 \ldots\ldots V_j^N) \tag{9}$$

Where N is the number of the rankers.

2. The problem of *Canberra Distance* can be solved by a general gradient descent algorithm. From Equation (8) we know that the elements in T^* are separated from each other. So we can calculate the optimal value of each element

independently. The formulation for the gradient descent step is listed in the following equation.

$$V_j^* = V_j^* - \alpha \times \sum_i \{I(V_j^i \geq V_j^*) \times \frac{-V_j^i \times 2}{(V_j^* + V_j^i)^2} + $$
$$I(V_j^i < V_j^*) \times \frac{V_j^i \times 2}{(V_j^* + V_j^i)^2}\} \tag{10}$$

Where $I(x)$ is the Indicator function and α is the learning rate(which is usually setting as 0.01). The smallest k elements in V^* are the result list.

3. Because the rank adjustment factor of each element in T^* has relativity, the problem of *Discounted Canberra Distance* cannot be handled by the previous algorithm. We add a restriction that the $V^* = \sum_i W_i \times V^i$. Then we follow the idea from the existing lambdarank optimization algorithm[1] (which approximately optimizes NDCG as its object) and this problem can be optimized directly by a approximate gradient descent algorithm. The gradient process is listed in the following equation.

$$W_k = W_K - \alpha \times \frac{1}{1 + \lg Pos_j^{V^*}} \times \sum_i \sum_j \{I(V_j^i \geq V_j^*) \times V_j^k \times $$
$$\frac{-2 \times V_j^i}{(V_j^* + V_j^i)^2} + I(V_j^i < V_j^*) \times V_j^k \times \frac{2 \times V_j^i}{(V_j^* + V_j^i)^2}\} \tag{11}$$

where α is the learning rate (which is usually setting to 0.01) and $I(x)$ is the Indicator function. The $Pos_j^{V^*}$ will be updated after each iteration process. Then we use the W to calculate the V^* and the smallest k elements in V^* are the result list.

6   Experiment

6.1   Experimental Setting

The dataset contains 1,003,440 images crawled from the web. It is split into two parts, the ground truth and the distractor set. The ground truth consists 239 groups of 1260 images. It is labeled manually. The distractor set is selected randomly from the web with images in the ground truth removed. The size of the distractor set is 1,002,180. All the images are resized to 160 pixels on their larger side. We use 100,000 images to train a stable PCA for dimension reduction.

6.2   Evaluation Measurements

To measure the performance of the methods, we use the same metrics as [17]:

$$Precision = \frac{(\#CorrectImagePairs)}{(\#DetectedImagePairs)} \times 100\% \tag{12}$$

$$Recall = \frac{(\#CorrectImagePairs)}{(\#GroundTruthImagePairs)} \times 100\% \qquad (13)$$

The *DetectedImagePair* denotes image pairs in a detected group. The *CorrectImagePairs* denotes the detected image pairs belong to the same group of the ground truth. The *GroundTruthImagePairs* represents the image pairs in the same group of the ground truth.

6.3   Comparison of PCA-Code and LSH

In this experiment, the PCA-code is compared to LSH on duplicate image detection tasks. To the best of our knowledge, the result of such comparison has not been shown before. The length of PCA-code and LSH are set as 40. Figure 3 shows the P-R curves of the PCA-code and LSH. The buckets in both codes are grouped by hamming distance. It can be seen that the curve of the LSH is under the curve of the PCA-code. It means that with the same precision, we can get a higher recall performance. So PCA-code is more effective than LSH. From another perspective, the curve of LSH declines rapidly. On the contrary, the curve of PCA declines slowly. It means that the buckets divided by PCA-code containing duplicate images are close to each other. And the buckets containing un-duplicate image are not close.

The reason is analyzed as follows. PCA model could be seen as a particular list of projections of the original feature space. Therefore it will perform better than LSH, whose projection is randomly generated.

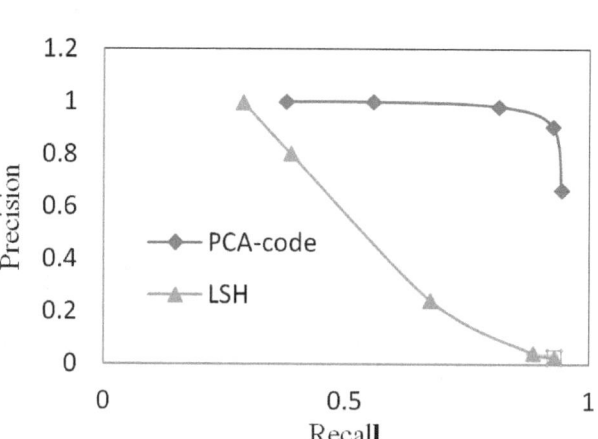

Fig. 3. Comparison on P-R curve of PCA-code and LSH

6.4   Comparison of Different Aggregation-Based Probe Method

In this experiment, we show the improvement of our aggregation-based probe method on three distances measurements. Then we evaluate the algorithm's stability on the partial ordered list length. The starting bucket is fixed for all experiments.

In evaluating the improvement of our aggregation-based probe method, we use two baselines. The first baseline is the PCA-code, which means the codes are divided by Hamming distance. Comparison with this baseline show the improvement of probing methods over non-probing ones. The second baseline is $MC4$ [4], a state-of-the-art *rank-based* rank aggregation method, which is the best method in [4]. Comparison with this baseline shows the improvement of our score-based rank aggregation methods to the rank-based rank aggregation methods.

Table 2 shows the performance of our aggregation-based probe method when the length of partial ordered list is fixed in probing. In this experiment, the length of result list is equal to the length of the partial ordered list.

Table 2. Comparison of Recall/Precision Performance of Various Methods. The *Length* means the length of the partial ordered list. $PCA, FDst, CaDst, DCaDst$ and $MC4$ means the PCA-code, the median value, Canberra distance OPT, Discounted Canberra distance OPT and Markov-4 aggregation method.

Length	PCA	MC4	Score-base Rank Aggregation		
			FDst	CaDst	DCaDst
16	NA	0.577/0.998	0.577/0.998	0.577/0.998	**0.750**/0.998
40	0.556/0.998	0.775/0.992	0.775/0.992	0.775/0.994	**0.813**/0.994
50	NA	0.783/0.991	0.783/0.991	0.784/0.994	**0.827**/0.993
100	NA	0.83/0.985	0.83/0.985	0.831/0.989	**0.894**/0.985
780	0.813/0.981	0.92/0.917	0.92/0.917	0.92/0.923	**0.93**/0.92

In the table, all the rank aggregation methods improve the recall and maintain the precision. The $FDst$ and $MC4$ methods show similar performance and the $CaDst$ is slightly better than either of those on precision performance.The improvement of $DCaDst$ on recall performance is higher than all the others.

When we fix the number of buckets at 40, *Recall/Precision* of *Baseline* is only 0.556/0.998 but $DCaDis$ reaches 0.813/0.994, improving the recall by 40%. The other methods are around 0.775/0.993, which is better than the *baseline* but lower than $DCaDst$.When we fix the recall score at 0.813, *Baseline* probes 780 buckets and $DCaDst$ probes only 40 buckets. $DCaDst$ probes far fewer buckets to reach the same recall. The other methods probe more than 50 buckets. This good performance of $DCaDst$ can be explained by its design principle: it focuses on the top-K items of the result list and optimizes them. This design is helpful for optimizing the top-k items of the result list. When the length of result list is shorter, the performance is more better than the others. So the $DCaDst$ could probe less buckets than the other methods to reach the same performance.

We conclude that $DCaDst$ is better than the other methods for this task. It effectively reduces the number of the buckets to be probed and increase the recall while preserving the precision performance. It also shows that the score function in Equation 3 is a reasonable inference, which could reflect the probability of the bucket containing the duplicate images.

Table 3 shows the stability of our aggregation-based probe algorithm on three distances when the partial list length is changed.

Table 3. The recall performance with respect to different settings. The LR means the length of the result list and the LP means the length of the partial ordered list.

LR	LP	FDst	CaDst	DCaDst
40	40	0.775	0.775	0.813
	80	0.775	0.775	0.813
50	50	0.783	0.784	0.827
	100	0.783	0.784	0.827

From Table 3, we can see that the recall performance is stable as the result list length fixed, which indicated that our proposed methods are robust as the length of partial ordered list varies. It is a exciting property in our task. So that we can set the length of the partial ordered list as length of the result list while maintaining the performance. It could reduce the computation and easy to apply to the next application.

To sum up, the experiments show that our aggregation-based probe algorithm can effectively increase the recall performance while maintaining the high precision with high robustness.

7 Conclusion and Future Work

In this paper, we propose a probing based method to detect duplicate images from a large image collection. Compared to previous works, the proposed approach reserves the advantages in high precision and at the same time improves the relatively low recall. To increase the recall, the proposed method generate the probing sequence to find the duplicate images more intelligently. Different from state-of-art probing methods for image search, multiple probing sequences exist in duplicate image detection. To unify multiple probing sequences, we present an unsupervised score-based rank aggregation algorithm. As shown in experimental results on millions of images, relatively low recall is increased and high precision is preserved.

In future work, we plan to construct a larger dataset and explore more image representations to analyze the content of image. Also we will discuss the convergence of the gradient descent and more rank aggregation methods.

References

1. Burges, C.J.C., Ragno, R., Le, Q.V.: Learning to rank with nonsmooth cost functions. In: NIPS, pp. 193–200 (2006)
2. Chen, S., Wang, F., Song, Y., Zhang, C.: Semi-supervised ranking aggregation. In: CIKM, pp. 1427–1428 (2008)

3. Chum, O., Philbin, J., Zisserman, A.: Near duplicate image detection: min-hash and tf-idf weighting. In: BMVC (2008)
4. Dwork, C., Kumar, R., Naor, M., Sivakumar, D.: Rank aggregation methods for the web. In: WWW, pp. 613–622 (2001)
5. Fagin, R., Kumar, R., Sivakumar, D.: Efficient similarity search and classification via rank aggregation. In: SIGMOD Conference, pp. 301–312 (2003)
6. Huang, Z., Shen, H.T., Shao, J., Zhou, X., Cui, B.: Bounded coordinate system indexing for real-time video clip search. ACM Trans. Inf. Syst., 27(3) (2009)
7. Jurman, G., Riccadonna, S., Visintainer, R., Furlanello, C.: Canberra distance on ranked lists. In: Ranking NIPS 2009 Workshop, pp. 22–27 (2009)
8. Klementiev, A., Roth, D., Small, K.: An unsupervised learning algorithm for rank aggregation. In: Kok, J.N., Koronacki, J., Lopez de Mantaras, R., Matwin, S., Mladenič, D., Skowron, A. (eds.) ECML 2007. LNCS (LNAI), vol. 4701, pp. 616–623. Springer, Heidelberg (2007)
9. Lee, D.C., Ke, Q., Isard, M.: Partition min-hash for partial duplicate image discovery. In: Daniilidis, K., Maragos, P., Paragios, N. (eds.) ECCV 2010, Part I. LNCS, vol. 6311, pp. 648–662. Springer, Heidelberg (2010)
10. Li, Y., Jin, J., Zhou, X.: Video matching using binary signature. In: Intelligent Signal Processing and Communication Systems, pp. 317–320 (December 2005)
11. Liu, Y., Liu, T.-Y., Qin, T., Ma, Z., Li, H.: Supervised rank aggregation. In: WWW, pp. 481–490 (2007)
12. Lowe, D.G.: Distinctive image features from scale-invariant keypoints. International Journal of Computer Vision 60(2), 91–110 (2004)
13. Lv, Q., Josephson, W., Wang, Z., Charikar, M., Li, K.: Multi-probe lsh: Efficient indexing for high-dimensional similarity search. In: VLDB, pp. 950–961 (2007)
14. Pönitz, T., Stöttinger, J.: Efficient and robust near-duplicate detection in large and growing image data-sets. In: ACM Multimedia, pp. 1517–1518 (2010)
15. Qamra, A., Meng, Y., Chang, E.Y.: Enhanced perceptual distance functions and indexing for image replica recognition. IEEE Trans. Pattern Anal. Mach. Intell. 27(3), 379–391 (2005)
16. Valle, E., Cord, M., Philipp-Foliguet, S.: High-dimensional descriptor indexing for large multimedia databases. In: CIKM, pp. 739–748 (2008)
17. Wang, B., Li, Z., Li, M., Ma, W.-Y.: Large-scale duplicate detection for web image search. In: ICME, pp. 353–356 (2006)
18. Wang, X.-J., Zhang, L., Liu, M., Li, Y., Ma, W.-Y.: Arista - image search to annotation on billions of web photos. In: CVPR, pp. 2987–2994 (2010)
19. Wang, Y., Hou, Z., Leman, K.: Keypoint-based near-duplicate images detection using affine invariant feature and color matching. In: ICASSP, pp. 1209–1212 (2011)
20. Zhang, D., Chang, S.-F.: Detecting image near-duplicate by stochastic attributed relational graph matching with learning. In: ACM Multimedia, pp. 877–884 (2004)
21. Zhao, X., Li, G., Wang, M., Yuan, J., Zha, Z.-J., Li, Z., Chua, T.-S.: Integrating rich information for video recommendation with multi-task rank aggregation. In: ACM Multimedia, pp. 1521–1524 (2011)
22. Zhou, W., Lu, Y., Li, H., Song, Y., Tian, Q.: Spatial coding for large scale partial-duplicate web image search. In: ACM Multimedia, pp. 511–520 (2010)
23. Zhu, J., Hoi, S.C.H., Lyu, M.R., Yan, S.: Near-duplicate keyframe retrieval by nonrigid image matching. In: ACM Multimedia, pp. 41–50 (2008)

User Interest Based Complex Web Information Visualization

Shibli Saleheen and Wei Lai

Faculty of Information and Communication Technologies,
Swinburne University of Technology,
Hawthorn, Victoria, Australia
{ssaleheen,wlai}@swin.edu.au

Abstract. Web graph allows end user to visualize web information and its connectivity. But due to its huge size, it is very challenging to visualize web information from the enormous cyberspace. As a result, web graph lacks simplicity while the user is searching for information. Clustering and filtering techniques have been employed to make the visualization concise but it is often not accurate to the expectation of user because they do not utilize user-centric information. To deliver the information concisely and precisely to the end users according to their interest, we introduce personalized clustering techniques for web information visualization. We propose a system architecture, which considers the user interests while clustering & filtering, to make the web information more meaningful and useful to the end users. By the implementation of the architecture, an experimental example is provided to reflect our approach.

Keywords: Clustering, Personalization, Visualization.

1 Introduction

It has become very difficult for end users to get their desired information quickly and accurately from the gigantic World Wide Web. Research attempts to help the users manage and control the information over the internet by presenting user-focused information [4] [6] [11]. However, the existing technology is not adequate enough to provide visual maps users expect to guide their web experience. As a consequence, the next research challenge is to provide focused information to the end users for quick and easy understanding of the web information and its pattern. To provide an easy way to explore information according to the need of user, techniques of information retrieval and visualization have been introduced [7]. Information visualization presents the abstract data to amplify human cognition and unfold the hidden relationships among the data. The purpose of information visualization is not only to create interesting pictures for the users but also to communicate information clearly and effectively [3] [7].

A web graph, which represents the web information with its internal connectivity revealed, is a very useful way to visualize various aspects of the web

Y. Ishikawa et al. (Eds.): APWeb 2013, LNCS 7808, pp. 429–436, 2013.
© Springer-Verlag Berlin Heidelberg 2013

information. Nevertheless, a huge web graph is less likely to create a good impression for the end user compared to a relatively smaller one with information relevant to the user's interests. Therefore, reducing the size of the web graph is very important from the end user viewpoint. Filtering and clustering techniques from different aspects such as structure, context, have been applied on the web graph to reduce the size. Besides the structure and content based clustering techniques, compromise the direct user's needs is expected for personalized visualization. First of all, users are different and their needs vary. Most of the work accomplished for personalization of web data is done for a group of people who can be addressed as similar minded [13]. But it is not obvious that every user of a specific group behaves similarly all the time. Secondly, short term interests or preferences of a particular user change over time or in different context.

To provide the above personalization in web information visualization, we propose to model every user separately. Instead of providing information directly in the traditional way, we introduce user profile driven clustering for web information personalization. We strongly believe that this can classify the web information into several categories and yield an improved visualization for the end user as the information is clustered according to the individual's interest.

The rest of the paper is organized as follows: Section 2 presents related works; Section 3 describes the proposed architecture for personalized visualization; Section 4 presents the experiment; and finally, Section 5 concludes the paper.

2   Related Work

Work has been done in different directions to make the representation of web information more effective to the end users. Some work deals with information searching through search engines [14] [1], i.e., on search results. For personalization, researchers consider the user search preference learned from the click history or browsing history [9]; history of similar minded users [13]; location metadata of the user for personalized search [2]; and other personalizing page ranking techniques [11]. In the case of navigating in the web space, clustering techniques plays important roles for reducing the complexity of the web graph. Structure based clustering can be found in the works of Huang et al. [6] and Rattigan et al. [12]. They provide structure based clustering techniques to reduce the web graph size. Graph clustering approaches like these only consider the structural aspect of the graph without considering the content of the documents. Besides structure, Gao [4] integrates content in the process of clustering. She first clusters using structure and then clusters using content only on the subclusters created in the first phase. As a consequence, this two-phased clustering heavily depends on the structure. Inclusion of structure in the visualization creates scopes for non-similar documents to be linked and hence put in the same cluster. Our work avoids this possibility by considering only content based similarities while relating documents. In contrast to previous research approaches, this work is personalized as we consider the user's point of view.

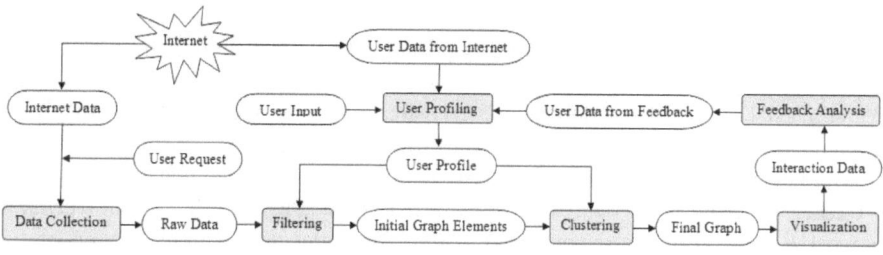

Fig. 1. System Architecture

3   System Architecture

Our proposed system to resolve the issue of personalized visualization uses data from user profile as preliminary knowledge-base to assist in filtering, clustering and hence final visualization. While the user uses the system, the system automatically updates the existing user profile or creates a new one where appropriate to reflect user's latest focusing trends. Finding appropriate scope for involving the user profiles in the system is very crucial for a personalized system. Scopes of the user profiles for personalized searching from electronic information space have been documented in [10]. In our system, after collecting information, we include the user profile while analysing the web information, i.e., in filtering & clustering, and finally we visualize the information.

Figure 1 depicts the architecture of the system for personalization of information visualization. The key modules of this model are:

i) **Result Analyser module** (composed of Data Collection, Filtering and Clustering components) collects the web data based on user's request, filters out irrelevant data, and finally generates a clustered web graph by using the user profile for a handy and compact visualization.

ii) **Visualization module** is responsible for visualizing the web graph to provide a good overview; to help the user to find out the expected information; and to create scopes for the user to interact with the information.

iii) **Feedback Analyser module** gets the user interaction data from the visualization module, then analyses the data, and lists possible updates.

iv) **Profiler module** gets the updates from Feedback Analyser module and consults with the existing user profile to keep user profile up-to-date to reflect user's latest interests in order to provide more relevant output afterwards. This module also gets direct inputs from the user where appropriate, to construct or modify the user profile.

3.1   Data Collection and Filtering

The first component of our architecture is the web crawler which collects information to represent URLs by their corresponding keywords for the initial web graph. To ensure quick processing and get adequate & useful information, our web crawler does not revisit the faulty, unreachable pages and pages containing very less information rather discards them. We filter out all other html tags, any data except text such as image, audio and video to keep page information simple. Once crawling is finished, we have the original text data for further processing. We convert the words to their base forms(for example: 'running' to 'run') to get the actual frequencies of words and remove stop words from the content.

After filtering the stop words, we have set of words, still big in size, to represent a URL. To reduce the size, we apply well known TF-IDF to get the terms of every URL weighted and then sort them. We take a fixed percentage of top weighted terms as keywords to make further steps time & cost effective.

Now we have a set of documents in the corpus to be visualized by the web graph. Every document is composed of a set of $(Keyword, Weight)$ pairs. All the distinct keywords of these documents form the total set of keywords. At this point, it is very easy to define the documents as vectors of weighted keywords. Hence, for the $i - th$ document, document vector d_i as $< t_{i1}, t_{i2}, t_{i3}, ..., t_{in} >$ where t_{ij} is the weight $j - th$ keyword where $1 <= j <= n$ and n is the number of total distinct keywords in the corpus.

We calculate the edges connecting the nodes for the initial graph based on the content similarity of the nodes. If we consider two documents, the more keywords they have common, the more they are connected. The similarity between them is computed by cosine similarity[1]. If the similarity score $CSim(d_y, d_z)$ is greater than a threshold value σ then we create an edge between the documents d_y and d_z. Finally, we have an undirected initial graph for next processing.

3.2   User Profiling

To optimize retrieval accuracy, we clearly need to model the user accurately and personalize the information retrieval according to each individual user, which is a difficult task. It is even harder for a user to precisely describe what the information need is. Unfortunately, in the real world applications, users are usually reluctant to spend time inputting additional personalized information which can provide relevant examples for feedback.

Representation of user profiles has been classified into three categories in [5]. The simplest one to represent and maintain is keyword-based user profile where the keywords are either automatically extracted from the web documents during a user's visit or through direct input by the user. Each keyword can represent a topic of interest and weight can be associated with it to represent the degree of interest. The other two categories are semantic network profile which deals to address the polysemy problem and concept based user profile which consists of concepts & reveals the relationship between the concepts.

In our system architecture, we adopt the keyword based user profiles for simplicity to represent a user. So the user profile consists of a set of weighted keywords, i.e., $(Keyword, Weight)$ pairs. User profiles are maintained in two phases by the 'Profiler' component. In the creation phase, the user data are collected directly from the user via a user interface. Here the user is allowed to answer

---

[1] http://en.wikipedia.org/wiki/Cosine_similarity

some predefined questions for the creation of basic user profile. After collecting the basic data from the user, the system will collect more user data from the internet according to the user's authorization. The maintaining or updating phase occurs mostly whenever the user uses the system. Profiler updates the preferences based on user interaction data passed from the Feedback Analyser.

3.3   Clustering

We want to cluster the total corpus into different clusters where the documents will be clustered by biased similarity values based on the user profile. We modify the document vectors computed in section 3.1 to represent both user interests and main topics of the document. For this we need to replace the existing document terms with the keywords of the user profile where appropriate.

We update the total number of keywords of the corpus by adding the user profile keywords in the keyword set. If the number of new distinct keywords is m, the total number of keywords becomes $(n+m)$. Now, we can write the modified document vector for d_i as: $< T_{i1}, T_{i2}, ..., T_{in}, T_{i(n+1)}, T_{i(n+2)}, ..., T_{i(n+m)} >$.

We calculate the $T_{i(n+j)}$; where $1 < j <= m$; in the following way. We measure the similarity value of c_j, where c_j is a keyword from user profile, with each keyword of the d_i. We use one of the wordNet similarity measures for this purpose. There are six similarity measures applied on wordNet, three measures including Lin [8] are based on information content of the least common subsumer(LCS). Lin yields slightly higher correlation with human judgements than others, which actuates us to use it instead of others.

If c_j is similar to document keyword k_{il}; where $1 < l <= n$; meaning that $(SIM(k_l, c_j) * T_{il}) > \tau$, we consider that document d_i is containing keyword c_j instead of the original document keyword. So, we remove the k_{il} by making the weight T_{il} as 0. We repeat it for every c_j of user profile for each d_i. As a particular c_j can be similar with multiple keywords of a document by different values, we take the maximum similarity value as the weight of that c_j.

Hence, the new vector to represent a document, d_i, reflects the user interests and the topics of the original content as well. We apply k-means algorithm for clustering, where the number k and initial means are calculated from the 'cosine similarity' values of the document vectors. We select d_i as an initial mean if the degree of d_i is greater than the sum of the average degree of the web graph and a predefined threshold.

3.4   Visualization and Feedback Analysis

For displaying the final view to the user, we have developed the visualization component based on popular JGraph[2] framework. The visualization interface has three panes. The pane on the left side of the interface produces tree view of the web graph clusters. In the middle of the interface the web graph is displayed. To distinguish between the cluster and the URL nodes we use rectangular and oval shapes, respectively. Beside navigating, we have also added browser support

---

[2] http://www.jgraph.com/jgraph.html

Table 1. Keywords representing URL nodes

URL		Keywords
1	-	justice, evacuation, screencasts, wireless, concept
2	-	centre, management, laboratory, college, group, resource
3	-	directory, bing, yahoo, google, hospitality
...	...	...
6	-	bachelor, flexibility, convenience, delivery
7	-	postgraduate, information, city, night, suite
...	...	...
19	-	command, wireless, sheet, layout

for the user to visit the web pages in the right pane. The user can- zoom in/out; re-arrange the web graph by drag and drop; delete an inappropriate node or cluster; and create new clusters.

As the user interacts with the visualization interface, it sends the browsing, click and arrangement information generated from the actions accomplished by the user to the feedback analyser component. Browsing feature is the primary contributing feature in profile updating. When user browses a page, the keywords with their weights along with the dwelling time are sent to the feedback analyser module, which updates the weights of existing interests and adds the new ones. While adding the new interests in the user profile, the interest and corresponding weight-value to be added are shown to the user for any amendment.

4   Experiment

In this section, we provide an experimental example. It begins with showing a simple web graph. For this example we analyse a university web site and set the crawl style property as BFS whose termination is set to 1. We get 19 nodes while crawling and name them according to their appearance number in the crawler. We get the sets of keywords for representing the nodes after applying techniques from 3.1, which are shown in Table 1. Figure 2a shows the original web graph. We get two clusters and eight other nodes which is shown in Figure 2b after applying content based clustering described in [4].

We also have two user profiles constructed earlier based on two different users. We consider one user profile, UP_a, represents an international student who looks for opportunity of postgraduate studies in computer science related fields whether the other, UP_b, is of a local student focusing on undergraduate studies in business, particularly interested in 'Hawthorn' campus. The corresponding user profiles are given in Table 2a as $(Interest, Weight)$ pairs.

First, we take the user profile UP_a and generate the clusters based on the methodology described in section 3.3. We get three clusters and eight unclustered nodes. Second, applying the same process on the initial graph for the user profile UP_b, we get four clusters and seven unclustered nodes as well. Table 2b shows how the nodes are assigned in the clusters for all three cases. With no user

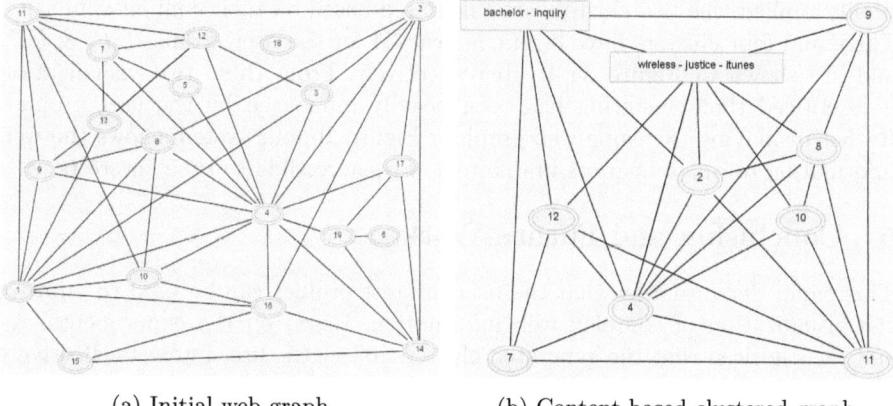

(a) Initial web graph   (b) Content based clustered graph

Fig. 2. Traditional Visualization

Table 2. User Profiles and Generated Clusters

(a) User Profiles

Profile	Interest	Weight
	Supervisor	0.5
	Scholarship	0.5
UP_a	Visa	0.3
	Computer	0.8
	International	0.9
	Business	0.8
UP_b	Bachelor	0.9
	Hawthorn	0.5
	Management	0.7

(b) Generated Clusters

Profile	Cluster Label	Nodes
-	bachelor-inquiry	3, 5, 6, 13, 17, 18
	wireless-justice-itunes	1, 14, 15, 16, 19
	computer	10, 14, 15
UP_a	supervisor-international	2, 4, 8, 16
	job-study	6, 7, 9, 13
	bachelor-business	6, 18
UP_b	management-hawthorn	2, 4, 7
	job-news	9, 11, 13
	wireless-justice-itunes	1, 14, 16, 19

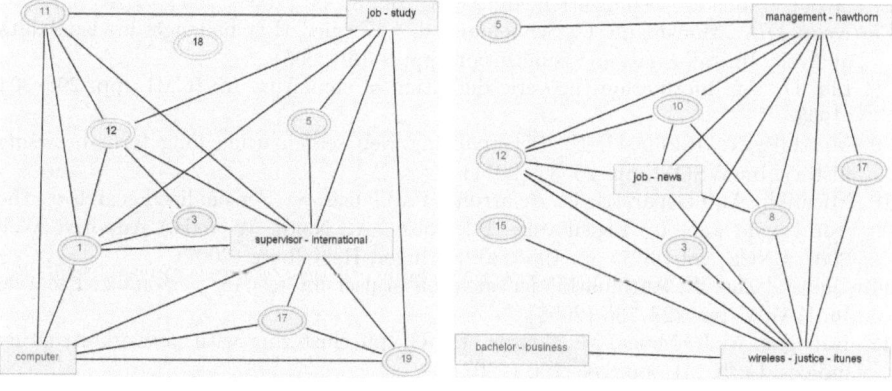

(a) Clustered graph using UP_a (b) Clustered graph using UP_b

Fig. 3. Personalized Visualization

profile applied, the two clusters are calculated based on their content similarity. Three and four clusters have been constructed for the user profiles UP_a & UP_b and are shown in Figure 3a & 3b respectively. From these two visualization it is noticed that clustering has been heavily influenced by the user profiles. Rather displaying the same web graph of Figure 2b, our system shows the web information to different users in different ways according to their interests.

5   Conclusion and Future Work

This paper demonstrates that the user interest profiles can be used to improve the visualization of complex web information. Based on the experimental result, it is noticed that the generated clusters reflect the user interests effectively meaning that the web graph is personalized. Further studies need to be done to design domain specific ontology to get the web information in a structured way and ontology for the user profile as well. Then a combination of both ontologies in the system architecture will be investigated.

References

1. Agrawal, R., Gollapudi, S., Halverson, A., Ieong, S.: Diversifying search results. In: WSDM, pp. 5–14 (2009)
2. Bennett, P.N., Radlinski, F., White, R.W., Yilmaz, E.: Inferring and using location metadata to personalize web search. In: SIGIR, pp. 135–144 (2011)
3. Card, S.: Information visualization. In: The Human-Computer Interaction Handbook: Fundamentals, Evolving Technologies, and Emerging Applications (2007)
4. Gao, J., Lai, W.: Visualizing blogsphere using content based clusters. In: Web Intelligence and Intelligent Agent Technology, pp. 832–835 (2008)
5. Gauch, S., Speretta, M., Chandramouli, A., Micarelli, A.: User profiles for personalized information access. In: Brusilovsky, P., Kobsa, A., Nejdl, W. (eds.) Adaptive Web 2007. LNCS, vol. 4321, pp. 54–89. Springer, Heidelberg (2007)
6. Huang, X., Eades, P., Lai, W.: A framework of filtering, clustering and dynamic layout graphs for visualization. In: ACSC, pp. 87–96 (2005)
7. Keim, D.A., Mansmann, F., Schneidewind, J., Ziegler, H.: Challenges in visual data analysis. In: Information Visualization, pp. 9–16 (2006)
8. Lin, D.: An information-theoretic definition of similarity. In: ICML, pp. 296–304 (1998)
9. Matthijs, N., Radlinski, F.: Personalizing web search using long term browsing history. In: WSDM, pp. 25–34 (2011)
10. Micarelli, A., Gasparetti, F., Sciarrone, F., Gauch, S.: Personalized search on the world wide web. In: Brusilovsky, P., Kobsa, A., Nejdl, W. (eds.) Adaptive Web 2007. LNCS, vol. 4321, pp. 195–230. Springer, Heidelberg (2007)
11. Qiu, F., Cho, J.: Automatic identification of user interest for personalized search. In: WWW, pp. 727–736 (2006)
12. Rattigan, M.J., Maier, M., Jensen, D.: Graph clustering with network structure indices. In: ICML, pp. 783–790 (2007)
13. Smyth, B., Balfe, E., Boydell, O., Bradley, K., Briggs, P., Coyle, M., Freyne, J.: A live-user evaluation of collaborative web search. In: IJCAI, pp. 1419–1424 (2005)
14. Wang, X., Zhai, C.: Learn from web search logs to organize search results. In: SIGIR, pp. 87–94 (2007)

FIMO: A Novel WiFi Localization Method

Yao Zhou, Leilei Jin, Cheqing Jin*, and Aoying Zhou

Shanghai Key Laboratory of Trust Worthy Computing,
Software Engineering Institute
East China Normal University, Shanghai 200062, China
sei_zy2006@126.com, lljin_ecnu@sina.com, {cqjin,ayzhou}@sei.ecnu.edu.cn

Abstract. With the development of technology and the proliferation of mobile computing devices, people's need for pervasive computing is rapidly growing. As a critical part of pervasive computing, Location Based Service (LBS) has drawn more and more attention. Localization techniques that report the real-time position of a moving object are key in this area. So far, the outdoor localization technologies (i.e, GPS) are relatively mature, while the indoor localization technologies are still under improvement. In this paper, we propose a novel WiFi localization method, called FIMO (FInd Me Out). In this method, we take the instability of signal strength and the movement of objects into consideration when determining the location based on Fingerprint. Experimental results show the proposed method is capable of estimating a moving object's location precisely.

Keywords: LBS, localization, indoor, WiFi.

1 Introduction

For a long time, people expect the electronic devices around them can have perception towards the environment.In early 2012, IDC (Internet Data Center) predicted that around 1.1 billion smart connected devices would be sold in 2012, and this number would even be doubled in 2016[1]. As one of the important applications of ubiquitous computing system, localization application has drawn more and more attention. In general, these systems often need to provide information or services according to user's current location. For example, a customer who is shopping in a new mall may need to know where his /her favorite shop is and how to reach it. Advertisements can be recommended to users based on their locations etc..

Nowadays, outdoor localization technologies, such as GPS, have become quite popular in daily life. Recent years also witness the increasing demand for indoor localization. Several indoor localization techniques have been developed in recent years, including Bluetooth [3], RFID [8] and WiFi [1]. Bluetooth and WiFi based approaches rely on Received Signal Strength (RSS) for localization, while RFID based approaches determine the location of the moving object by reading the

---

* Corresponding author.

[1] See http://www.199it.com/archives/29224.html

Y. Ishikawa et al. (Eds.): APWeb 2013, LNCS 7808, pp. 437–448, 2013.
© Springer-Verlag Berlin Heidelberg 2013

active RFID tag. WiFi localization has many advantages, such as ubiquitous coverage, scalability, no additional hardware required, extended range, no line of sight restrictions and free measurement etc.

We can see that WiFi is a good choice for localization in indoor space. However, the accuracy of basic approaches is low due to the instability of WiFi signal strength. Some projects combined WiFi with other devices such as RFID [5] [13], Bluetooth [2] [12] to improve the accuracy. Although the accuracy is improved, additional devices are required, and will cause extra burden to the setup process. Besides, very few approaches have taken advantage of continuous monitoring of users' locations and considered the features of WiFi signal strength itself.

In this paper, we propose a novel approach to estimate a moving object's location, by taking into consideration the instability of signal strength and the movement of objects. This method is based on the Fingerprint approach in estimating the location of a moving object.

The remainder of this paper is organized as follows. In Section 2, we survey related work in localization technologies. In Section 3, we discuss our research methodology. In Section 4, we present and analyze our approach. Section 5 reports a comprehensive experimental study. Finally, we present our conclusions and provide some future research directions in Section 6.

2   Related Work

The existing work on indoor localization can be divided into two main groups: signal propagation approaches and fingerprinting approaches.

2.1   Signal Propagation (SP) Approaches

Signal propagation approaches tend to as exponential attenuation models for WiFi signals and use the path loss[2] to determine location based upon distance from the Access Points (APs), whose locations are assumed known [10]. Although the signal propagation models may work well in open space, they cannot be used directly in indoor environment since there exists many kinds of obstructions. In other words, we need to consider additional path loss caused by the physical obstructions between the APs in indoor localization and make the models much more complex than in the open area. The model proposed by Seidel and Rappaport [11] is the most popular one till now. In this model, the walls made of different materials are assigned different attenuation factors for better precision. SP models can therefore be generalized to locations without available reference data. Furthermore, it only requires storage for the location of each AP and a simple description of its signal attenuation. However, since WiFi signals are unstable, attenuation is almost never radially symmetric. In addition, other complex factors, such as multipath, different construction materials, and various objects in the building, ever decrease the precision of the above methods.

As a conclusion, signal propagation approaches are easy to implement, but the performance is low.

---

[2] See http://en.wikipedia.org/wiki/Path_loss

2.2   Fingerprinting (FP) Approaches

FP approaches ignore attenuation and instead compute likelihoods from location-specific statistics compiled from reference data. FP needs to create a signal strength map in off-line manner in order to start localization. When a user submits a location request, his current signal strength values are compared with those in the database. The system then compute a most probable location for the user. Although FP techniques require reference data to setup, reference data is only needed to be collected once in a new building and the accuracy of FP is better than SP.

FP approach was developed by Microsoft Research RADAR [1]. In their paper, they proposed two ways to match signal strength values. One is to return the position with the nearest neighbor in signal space; the other is to return the average of the coordinates of k neighbors in signal space as an estimate. In general, it was found that WiFi signal strength at the same location changes with time, temperature and objects around it. Thus, it is hard to achieve good accuracy based on pure matching of signal strength.

Most existing methods estimate the user's location only by using of the current signal strengths. Ho et. al. proposed a method to retrain the system when the user finds that the estimation location is wrong [7]. N. Hernandez et. al. proposed a WiFi localization system based on fuzzy techniques. It is only suitable for the small-scale variations which happen when the user moves over a small distance (in the range of wavelength) [6]. Baniukevic et. al. proposed an indoor localization algorithm by combining WiFi with Bluebooth [2]. It uses WiFi signal strength to localization and adjusts the coordinates of the location using Bluetooth at reference points. The methods proposed by D. L. Lee and Q. X. Chen have considered the movement of objects, but these methods can only identify the room where the user is in [4] [9].

Our algorithm is a type of Fingerprint approach. Existing approaches determine the location according to signal strength received directly. However, the signal strength is unstable in the indoor environments. Meanwhile, few of them consider the continuous movement of the user. Our approach records the signal strength received at the mobile user continuously, eliminates the locations that are unlikely to be reachable from the previous location by considering the walk speed and then determines the user's location in the possible locations by matching the signal strength with weights (which are got by considering the instability of signal strength. To be discussed in Section 4.2).

3   Research Methodology

In this section, we describe the experimental testbed and data collection.

3.1   Experimental Testbed

Our experimental testbed is located at the ground floor of a 3-storey building. Figure 1 shows the layout of the floor (765.43 m^2). Five Access Points (APs) are

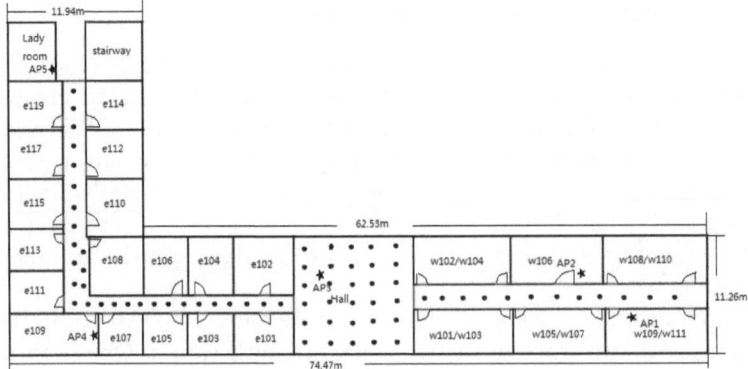

Fig. 1. Map of the floor where the experiments were conducted. The stars show the locations of the 5 access points. The black dots denote locations were empirical signal information was collected.

deployed at different locations, denoted as AP_1, AP_2, AP_3, AP_4 and AP_5. Each AP is a Totolink N300R Router. In our tests, a mobile host is a laptop computer equipped with Intel WiFi Link 5100 AGN running Windows 7. The APs provide overlapping coverage over the entire floor together.

3.2   Data Collection

Data collection phase is critical in this research methodology. We use Native WiFi API[3] to record the signal strength values. The reference data set and the testing data sets are created as follows:

- *reference data set*: We select 71 reference points, as shown in Figure 1. At each reference point, we repeatedly recorded the signal strength for 40 times, and used the average as the signal strength of this point.
- *testing data set 1*: This data set contains the signal strength values received by the user when he was walking *uniformly* from the location near AP_5 to the location near AP_1.
- *testing data set 2*: This data set contains the signal strength values received by the user when he was walking *with variable velocity* from the location near AP_5 to the location near AP_1.

In the first data set, each record is described as $(x, y, \overrightarrow{SS})$, where x and y represent the location, and $\overrightarrow{SS}$ is an array of signal strengths for five APs. In the last two sets, the form of the each record tuple is $(t, x, y, \overrightarrow{SS})$, where t represents the timestamp to record the signal strength, and the definition of the rest attributes are as same as the above definition.

---

[3] The Native Wifi API is supported by Windows platform. It automatically sets component configures, connects to, and disconnects from wireless networks. Native Wifi can store profiles on the networks it interacts with in the form of XML documents.

4   Our Algorithm

In this section, we describe our algorithm in detail. Ideally, we expect the signal strength of an AP remains almost stable with time going on. Figure 2 shows the signal strength levels at different distances. We can find that almost all of the signal strength values received 1m away from the AP are the same, while the stand deviation of the signal strength values received 10m away from the AP is around 3%. When the distance reaches 35m, the stand deviation can be around 6%. From this, we conclude that the signal strength keeps fluctuating all the time and the ones with higher value is more stable. We also test the stability of signal strengths received by a moving object, as shown in Figure 2. In this figure, a user walks from AP5 to AP1 slowly, we can observe that the signal strengths would have greater fluctuation when the user is moving. How to use these unstable signal strengths to localization is challenging to us.

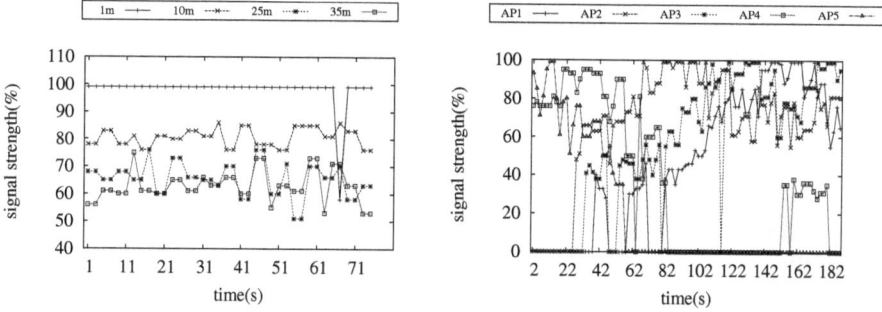

Fig. 2. Signal Level at different distances **Fig. 3.** The signal strength values received the user along work

Figure 2 and 3 show that the signal strength would reflect distance more accurately if the fluctuation of the signal strength could be smoothed. Besides, in Figure 2, it's easy to see that stronger signal strength are more stable. And the degree of fluctuation of the signal strength changes with time. So we can be convinced that the signal strength with different values and the signal strength received at different time should have different credibility.

Due to these particular features, we propose an approach, called FIMO (FInd Me Out), for indoor localization. FIMO continuously records signal strength values from each AP at the user's location at fixed intervals when the user is walking around. When the user submits a location request, FIMO will process the data and then determine the location of the user using Algorithm 1. FIMO backtracks the signal strengths received by the user which are recorded in the database and put them into *ssArray* (Line 2). It smoothes the signal strength values in *ssArray* chronologically and puts the Smoothed Values into *smoothArray* (Line 3), as discussed in Section 4.1. After calculating the weight of every signal strength value using the Original Value (The signal strength values that haven't been smoothed) (Line 5), as discussed in Section 4.2, FIMO returns the location to the user by matching the reference points (Line 7-14, Section 4.4) in the possible range (Line 6, Section 4.3).

Algorithm 1. FInd Me Out (FIMO)

Require: l: threshold of the records tracked back; n: the number of APs; $userID$:the ID of the moving object;
Ensure: location of $userID$: loc
1: Initialize $isFirst \leftarrow true$ and $minSim \leftarrow 0$;
2: Set $ssArray$ as the latest l records received by $userID$;
3: $smoothArray \leftarrow SMOOTH(ssArray)$;
4: // smooth the signal strength values in chronological order;
5: Set $w_1...w_n$ based on $SSArray$;
6: Set $rpArray$ as reference points which locate at the possible range;
7: **for** each $RP_c \in rpArray$ **do**
8:     $sim_c \leftarrow SIM(\vec{w}, smoothArray.get[0].\overrightarrow{SS}, RP_c.\overrightarrow{SS})$;
9:     // compute the similarity value between the signal strengths received right now and reference points;
10:    **if** $(isFirst == true || sim_c < minSim)$ **then**
11:       $minSim \leftarrow sim_c$;
12:       $loc \leftarrow RP_c$;
13:    **end if**
14: **end for**
15: **return** loc;

4.1 Smoothing the Signal Strength

Since the signal strength is unstable, we smooth the signal strength values before matching them. As shown in Figure 3, all of values have different degree of fluctuation. We need to smooth the fluctuation while keeping the trend of signal strength. A moving average[4] is commonly used with time series data to smooth short-term fluctuations and highlight longer-term trends or cycles. So we calculate the moving averages of all values to smooth their fluctuations. Figure 4(a) shows that this method is effective.

However, there is a special situation when the fluctuation is not continuous, just as shown in Figure 4(b) . If we just calculate the moving averages, the smoothed values may be inaccurate. Under the situation, we detect the outliers and deal with them through the following process before calculating the moving averages of all values.

1. **Detecting the outliers:** The system scans the signal strength values according to the time sequence. Assume that θ represents the threshold of normal value, and the system acquires four sequential values, a, b, c and d. If $(b - a > \theta$ and $b - c > \theta$ and $c - d \leq \theta)$ or $(a - b > \theta$ and $c - b > \theta$

---

[4] See http://en.wikipedia.org/wiki/Moving_average

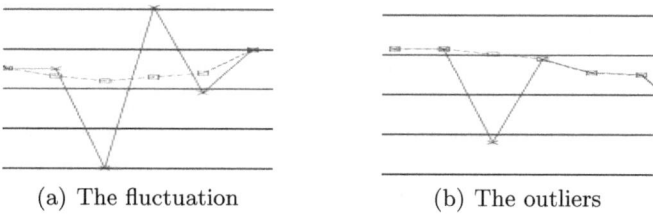

| (a) The fluctuation | (b) The outliers |

Fig. 4. Smoothing the signal strength

and $d - c \leq \theta$), b will be considered as an outlier. All of a, b, c and d are measured in percentage. As the threshold for the outlier, θ is set to be 20% in our experiments. The system will regard the value as an outlier (as shown in Figure 4(b)) and proceed to step 2.

2. **Dealing with the outliers.** The system replaces the outlier b with the average of a and c. Figure 4(b) shows the result of smoothing.

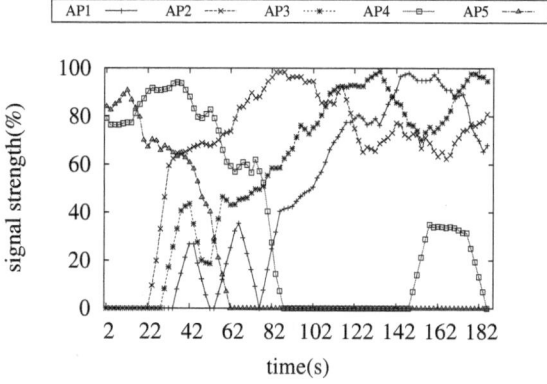

Fig. 5. The signal strength values received by the user along work after smoothing

Figure 5 shows the signal strengths after smoothing. Compared with Figure 3 which is before smoothing, it seems that the quality of the data has improved obviously after smoothing. We call the values after smoothing the Smoothed Value and the values having not been smoothed the Original Value.

4.2   Calculating the Weights

Figure 2 shows stronger signal strength tends to be more stable. To a certain extend, stronger signal strength has higher credibility. There are a lot of moving objects in indoor spaces, and the number and behaviors of them are constantly changing. These will cause the fluctuation of signal strength. Significant fluctuation will have material impact upon the credibility of the signal strength.

Therefore, even at the same location, different signal strength has different credibility, and we should give different weights to different signal strength when matching the signal strength in the database. Considering the two factors affecting the credibility of signal strength mentioned above, we propose a method to calculate the weights. The rule is to give the signal strength with the value which is higher and more stable higher weight. We can use the Original Value (o_i) of AP_i's signal strength to represent the credibility of AP_i's signal strength.

Then it comes to the question of how to measure the stability of signal strength. It is generally known that the sample standard deviation shows how much variation from the average (or expected value). A low sample standard deviation indicates that the data points tend to be very close to the mean, *vice verse*. However, the average value is not the expected value because the signal strength varies with distance, so we cannot simply use the sample standard deviation here. Although the trend of the signal strength should be described as curve function, we just focus on the stability of signal strength in a short time (such short time is deemed as a time window), and then it can be seen as linear. We regard the time period containing four continuous signal strength values as a time window in our experiments, and the length of the time window can be changed. If the length is too long, the items could not be fit with the linear function. And if it is too short, the linear function fitting out could not reasonably reflect the trend of the changes of signal strength. The method of least squares[5] is a standard approach to the approximate solution of overdetermined systems. Least Squares means that the overall solution minimizes the sum of the squares of the errors made in the results of every single equation. So we use Least Squares to find out the linear function which could describe the distribution of signal strength in a short time. Then we can get the expected signal strength value ($e_{i,j}$) received from AP_i at time t_j with the linear function.

Using the expected signal strength values, the sample standard deviation of several signal strength values (σ_i) in the same period can be calculated with Equation (1) below.

$$\sigma_i = \sqrt{\frac{1}{W-1}\sum_{j=1}^{W}(o_{i,j}-e_{i,j})^2} \tag{1}$$

where W indicates the length of the time window, and $o_{i,j}$ is the Original Value of AP_i at time t_j.

To sum up, the weight of every signal strength value of AP_i would be w_i coming from Equation (2).

$$w_i = \frac{o_i}{\sigma_i} \cdot \frac{1}{\max_k(o_k/\sigma_k)} \tag{2}$$

---

[5] See http://en.wikipedia.org/wiki/Least_squares

4.3   Considering the Walk Speed

The points on the circle centered at AP would have the same signal strength in free space. And the user might receive the same signal strength values at different locations. Therefore, the user's position might be located at a location which is far away from the actual one. In order to avoid this situation, we should consider the behavior of the user. In the indoor space, walking is the most common way of moving and the speed will has an upper limit. Thus, we could use the speed of the user and the last location of the user recorded to limit the possible candidate locations.

Specific studies have found pedestrian walking speed ranging from 1.25 m/s to 1.32 m/s for older individuals to 1.48 m/s to 1.51 m/s for younger individuals[6]. The highest normal pedestrian walk speed is about 1.5 m/s. Of course, with speed testing equipment, the accuracy of the speed can be improved. Here we use 1.5 m/s as the pedestrian walk speed to limit the candidates. We take the last location of the user as the center of circle and the maximum distance the user could walk from the last location as radius. The locations in the circle are the candidates. But the last location of the user recorded in database might not be the actual location which the user was, so we should extend the range of candidate locations properly because each record interval is the same. We use 2 times of the maximum distance as the radius of circle to limit the candidate locations. The process of calculation can be described as $r = 2vT$, where r is the radius of the circle containing candidate locations, v is the pedestrian walk speed (here we use 1.5m/s) and T is the time from the last record to this record.

The reference points located in the circle with the radius of r, which is centered at the last location of the user are the candidate locations of the user's current location. Thus, the location of the user must be one of the candidate locations.

4.4   Determining the Location

Similar with the base Fingerprint approach, we use the reference point with the most similar signal strength values as the user's location. We use the Equation (3) to compute the degree of similarity.

$$sim_c = \sum_{i=1}^{n} \left(w_i * (s_i - r_{c,i})^2 \right) \tag{3}$$

where sim_c is the similarity between the signal strength values received by the user and the signal strength values of the cth reference point which is a candidate location, n is the number of APs which can be detected by the user's device, w_i is the weight of AP_i's signal strength value, s_i is the Smooth Value of AP_i's signal strength value and $r_{c,i}$ is the AP_i's signal strength value of the cth reference point.

The reference point with the signal strength closest to that received by the user (sim_c is the minimum) would be returned to the user as the estimated location .

---

[6] See http://en.wikipedia.org/wiki/Walking

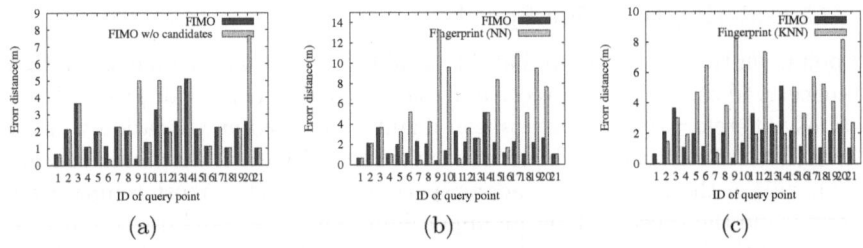

Fig. 6. The query points of the first user

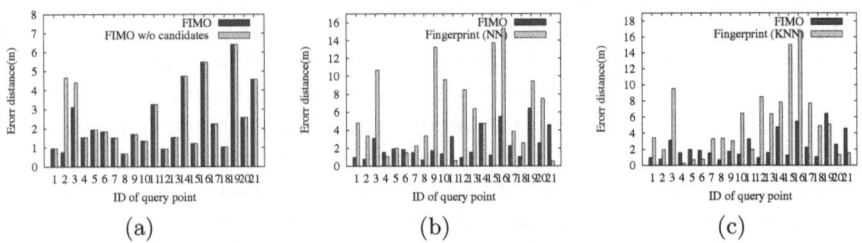

Fig. 7. The query points of the second user

5   Experiments

We have conducted series of experiments on the testbed described in Section 3.1. The testing data sets are also introduced in Section 3.2. We use four different approaches to locate the user. Two approaches are ours, one is FIMO introduced in Section 4, the other is FIMO without considering the walk speed. We use the Fingerprint using the nearest neighbor (Fingerprint(NN))and the Fingerprint using the multiple nearest neighbors (Fingerprint(kNN)) introduced in Section 2.2 as comparisons.

Figure 6 and 7 show the results of localization. Here, x-axis represents the ID of querying requirement, and the y-axis represents the error. In Figure 6(b)-(c) and 7(b)(c), we can observe that FIMO is much better than the Fingerprint(NN) and Fingerprint(kNN). In FIMO, considering the walk speed is the only step that might requires other device or estimation. In Figure 6(a) and 7(a), we can see the FIMO without considering the walk speed of the user already has good accuracy, but the FIMO considered the walk speed could generate better results.

Table 1 shows the sums of squares among the groups[7] (SSA) in Figure 6 and 7. The statistics suggests that the accuracy of FIMO has improved significantly comparing with the Fingerprint(NN) and Fingerprint(kNN). The FIMO is proved to have the best accuracy with respect to all the parameters (i.e. the average, medium and maximum values,as shown in Table 2). The errors in distances of all query points are less than 7m, and the points erred within 3m are more than 70%.

---

[7] SSA is a measure of how much the means differ from one another. Its conceptualized a little differently, because it is thought of as the variation of each mean from the mean of the total sample.

Table 1. Sum of Squares Among the Groups (SSA)

	FIMO vs. FIMO w/o Candidates	FIMO vs. Fingerprint (NN)	FIMO vs. Fingerprint (kNN)
1	3.79	78.71	43.69
2	0.64	136.69	86.33

Table 2. The average, median and maximum of the error distances

	FIMO	FIMO w/o Candidates	Fingerprint (NN)	Fingerprint (kNN)
average	2.17m	2.59m	5.34m	4.62m
median	1.95m	1.99m	4.04m	3.61m
maximum	6.42m	7.64m	15.33m	16.66m

6 Conclusions and Future Works

In this paper, we presented FIMO, a WiFi-based localization method. In this approach, the system records the signal strength values the user receives while he is moving, and determines the user's location by backtracking the signal strength values and the user's movement. Experimental reports show that the performance is significantly improved over existing methods.

In this paper, our work merely applies to specific environmental. We are considering to expand the application to more complicated situations, such as multiple floors or models of access points etc. Since FIMO is based on fingerprint approach, a lot of data is required to be collected for reference points in advance. If the signal strength values of some reference points could be estimated, the setup would be less time consuming. The curve fitting might be a practicable method. We believe that it is a direction worth exploring in the future.

Acknowledgments. The research of Cheqing Jin is supported by the National Basic Research Program of China (Grant No. 2012CB316200), the Key Program of National Natural Science Foundation of China (Grant No. 60933001), National Natural Science Foundation of China (Grant No.61070052). The research of Aoying Zhou is supported by National Science Foundation for Distinguished Young Scholars (Grant No. 60925008), and Natural Science Foundation of China (No. 61021004).

References

1. Bahl, P., Padmanabhan, V.N.: Radar: An in-building rf-based user location and tracking system. In: Proc. of INFOCOM, pp. 775–784 (2000)
2. Baniukevic, A., Sabonis, D., Jensen, C.S., Lu, H.: Improving wi-fi based indoor positioning using bluetooth add-ons. In: Proc. of MDM, pp. 246–255. IEEE Computer Society, Washington, DC (2011)

3. Bargh, M.S., de Groote, R.: Indoor localization based on response rate of bluetooth inquiries. In: Proc. of MELT, pp. 49–54. ACM, New York (2008)
4. Chen, Q., Lee, D.-L., Lee, W.-C.: Rule-based wifi localization methods. In: Proc. of EUC, pp. 252–258. IEEE Computer Society, Washington, DC (2008)
5. Chen, Y.-C., Chiang, J.-R., Chu, H.-H., Huang, P., Tsui, A.W.: Sensor-assisted wi-fi indoor location system for adapting to environmental dynamics. In: Proc. of MSWiM, pp. 118–125. ACM, New York (2005)
6. Hernández, N., Herranz, F., Ocaña, M., Bergasa, L.M., Alonso, J.M., Magdalena, L.: Wifi localization system based on fuzzy logic to deal with signal variations. In: Proc. of ETFA, pp. 317–322. IEEE Press, Piscataway (2009)
7. Ho, W., Smailagic, A., Siewiorek, D.P., Faloutsos, C.: An adaptive two-phase approach to wifi location sensing. In: Proc. of PERCOMW, pp. 452–456. IEEE Computer Society, Washington, DC (2006)
8. Jin, G.-Y., Lu, X.-Y., Park, M.-S.: An indoor localization mechanism using active rfid tag. In: Proc. of SUTC, pp. 40–43. IEEE Computer Society, Washington, DC (2006)
9. Lee, D.L., Chen, Q.: A model-based wifi localization method. In: Proc. of InfoScale, pp. 40:1–40:7. ICST (2007)
10. Letchner, J., Fox, D., Lamarca, A.: Large-scale localization from wireless signal strength. In: Proc. of the National Conference on Artificial Intelligence, pp. 15–20. The MIT Press (2005)
11. Seidel, S.Y., Rapport, T.S.: 914 MHz path loss prediction model for indoor wireless communications in multi-floored buildings. In: Proc. of Antennas Propagation. IEEE Trars (1992)
12. Wang, R., Zhao, F., Luo, H., Lu, B., Lu, T.: Fusion of wi-fi and bluetooth for indoor localization. In: Proc. of MLBS, pp. 63–66. ACM, New York (2011)
13. Yeh, L.-W., Hsu, M.-S., Lee, Y.-F., Tseng, Y.-C.: Indoor localization: Automatically constructing today's radio map by irobot and rfids. In: Proc. of Sensors, pp. 1463–1466 (2009)

An Algorithm for Outlier Detection on Uncertain Data Stream*

Keyan Cao, Donghong Han, Guoren Wang, Yachao Hu, and Ye Yuan

College of Information Science & Engineering, Northeastern University, China
Key Laboratory of Medical Image Computing (NEU), Ministry of Education
caokeyan@gmail.com

Abstract. Outlier detection plays an important role in fraud detection, sensor net, computer network management and many other areas. Now the flow property and uncertainty of data are more and more apparent, outlier detection on uncertain data stream has become a new research topic. Firstly, we propose a new outlier concept on uncertain data stream based on possible worlds. Then an outlier detection method on uncertain data stream is proposed to meet the demand of limited storage and real-time processing. Next, a dynamic storage structure is designed for outlier detection on uncertain data stream over sliding window, to meet the demands of limited storage and real-time response. Furthermore, an efficient range query method based on SM-tree(Statistics M-tree) is proposed to reduce some redundant calculation. Finally, the performance of our method is verified through a large number of simulation experiments. The experimental results show that our method is an effective way to solve the problem of outlier detection on uncertain data stream, and it could significantly reduce the execution time and storage space.

Keywords: Outlier detection, uncertain data stream, possible world.

1 Introduction

Uncertainty is inherent in data collected in various applications, such as sensor networks, marketing research, and social science [7]. Since the intrinsic differences between uncertain and deterministic data, existing data mining algorithms on deterministic data are not suitable for uncertain data. Such data present interesting research challenges for a variety of data set and data mining applications [1] [4] [6] [7] [9] [10] [11]. Recent years, more and more attention is paid to uncertain data mining.

Outlier detection is considered as an important data mining task, aiming at discovery of the elements that are significantly different from other elements,

---

* This research are supported by the NSFC (Grant No. 61173029, 61025007, 60933001, 75105487 and 61100024), National Basic Research Program of China (973, Grant No. 2011CB302200-G), National High Technology Research and Development 863 Program of China (Grant No. 2012AA011004) and the Fundamental Research Funds for the Central Universities (Grant No. N110404011).

Y. Ishikawa et al. (Eds.): APWeb 2013, LNCS 7808, pp. 449–460, 2013.

or does not conform to the expected normal behavior, or conforms well to a defined abnormal behavior [3] [5] [8]. Now the flow property and uncertainty of data more and more apparent, outlier detection on uncertain data stream has become a new research topic.

To the best of our knowledge, no existing algorithm considers distance-based outlier detection on uncertain data stream that uncertain tuple consist of a set of possible instances. In this paper, the algorithm is proposed for outlier detection on uncertain data stream over sliding window that uncertain data tuple consist of a set of instances. In summary, the major contributions of this work are as follows:

• The first algorithm for outlier detection over sliding window on uncertain data stream that are tuple level uncertainty is designed.

• A dynamic storage structure is designed for outlier detection on uncertain data stream over sliding window, to meet the demands of limited storage and real-time response.

• We propose an efficient range query method based on SM-tree(Statistics M-tree), it could reduce some of unnecessary calculations thereby improve the implementation efficiency of the algorithm.

• We also implement a series of experiment evaluations, showing that our methods are effective and efficient.

This paper is organized as follows. In the next section, we will introduce the related work. After that, we introduce semantic of outlier on uncertain data stream in section 3. Section 4 proposes the method for outlier detection on uncertain data stream. In section 5, we will present the experimental results. Section 6 concludes the paper.

2   Related Work

In this section, we illustrate related work in two fields: (1) outlier detection on uncertain data; (2) outlier detection on uncertain data stream.

2.1   Outlier Detection on Uncertain Data

The first method for outlier detection on uncertain data was proposed in [2]. Uncertain data object was represented by a probability density function. The algorithm used a probabilistic definition of outliers in conjunction with density estimation and sampling. However, their work only focused on detecting outlier objects without considering outlier instances.

Another outlier detection method for uncertain data was presented in [12]. If the existing probability of neighbor object was very low, then the centered object was a distance-based outlier. Their method only concerned tuple level uncertainty, i.e. probabilistic dimension call confidence, so it was not suitable for property level uncertainty.

Bin Jiang et al. [7] started with a comprehensive model considering both uncertain objects and their instances. They assumed that the uncertain objects with similar properties tended to have similar instances, the normal instances of each uncertain object using the instances of objects with similar properties.

The algorithms mentioned above work are all in static fashion. This means that the algorithm must be executed from scratch if there are changes in the underlying data objects, leading to performance degradation when operate on uncertain data stream.

2.2   Outlier Detection on Uncertain Data Stream

The first method for outlier detection on uncertain data stream was proposed in [13]. They studied the semantic of outlier detection on probabilistic data stream and proposed a new definition of distance-based outlier over sliding window. However, this method only concern the tuple level uncertainty, not the attribute level, so it is not suitable for the uncertain data model studied in this paper.

In this research, we propose a new algorithm for outlier detection on uncertain data stream which is described by a set of discrete instances over sliding window. Our techniques meet the demand of uncertain data stream for running time and storage requirements. Therefore, our study extends to this new research field.

3   Semantic of Outlier on Uncertain Data Stream

This section serves a two-fold purpose: firstly to introduce the uncertain outlier definition, and secondly to explain the meaning of the symbol. The traditional outlier definition on certain data is as follows:

Definition 1. (Neighbors) R *is a user specified threshold,* $R \geq 0$. x_i *and* x_j *are two data objects, if the distance between them is no longer than* R, *then* x_i *is a neighbor of* x_j. *Let* $n(x_i, R)$ *denotes the number of neighbors that data object* x_i *has.*

Definition 2. $((K, R)$-Outlier) *Let* $R \geq 0$, $K \geq 0$, *be a user-specified threshold. A data object* x_i *is marked as an outlier, if the number of neighbors that* x_i *has is less than* K, *i.e.* $n(x_i, R) < K$.

This definition is only suitable for determine data. For uncertain data set, we need to consider all possible worlds to determine whether the uncertain data tuple is an outlier in the current window.

It is assumed that uncertain data stream consist of uncertain tuples x_1, x_2, $\cdots$, x_i, $\cdots$, x_n, arriving at time stamp t_1, $t_2, \cdots$, t_i, $\cdots$, t_n, for any $i < n$, $t_i < t_n$. Each uncertain tuple has m possible instances and existence probability, i.e. $x_i = \{(x_i^1, p_{x_i^1}), (x_i^2, p_{x_i^2}), \cdots, (x_i^j, p_{x_i^j}), \cdots, (x_i^m, p_{x_i^m})\}$, $\sum_{j=1}^{m} p_{x_i^j} = 1$.

Definition 3. $((R, K)$-Outlier Possible World) *We define a possible world is a* (R, K)-*outlier possible world of uncertain tuple* x_i, *if the tuple* x_i *is an outlier in this possible world. Let* N_{x_i} *is a set of* (R, K)-*Outlier Possible World* that x_i has. $sum(x_i)$ is the probability summation of $((R, K)$-Outlier Possible World probability in N_{x_i}, $sum(x_i) = \sum_{w_i \in N_{x_i}} P_{w_i}$.

Definition 4. $((R, K, \lambda)$-**Outlier)** We define that an uncertain tuple x_i to be a (R, K, λ)-outlier, if $sum(x_i)$ is at lest λ.

The set of outliers is denoted by $O(R, K, \lambda)$, the set of the candidates is denoted by $C(R, K, \lambda)$, the candidates is non-outlier in current window, it is possible become outlier. The set of safe inliers (never the outliers) denoted by $I(R, K, \lambda)$. Outliers set and candidates set are stored, safe inliers set needn't be stored. Table 1 summarizes of the most frequently used symbols throughout the paper, along with interpretation.

4   Outlier Detection on Uncertain Data Stream

In this section, we provide a new method for outlier detection on uncertain stream. We start by describing the framework for outlier detection, and then introduce storage structure and incremental method. If we want to get the outlier on uncertain data stream, we can not use the outlier detection method on static database, because of the most straightforward scratch, towards updating the results. It is expected to be very computationally expensive. A wise choice is to design incremental algorithm, it processes only the changes based on previous result to produce the new result for current window. Next, we will introduce our methods.

Table 1. Frequently Used Symbols

symbol	Interpretation
K	the number of neighbors parameter
R	the distance parameter for the outlier detection
W	the windows size
S	the set of uncertain tuples in the current window
$slide$	the window slide length
x_i^j	the j-th instance of uncertain tuple x_i
$p_{x_i^j}$	the existing probability of instance x_i^j
$pp_{x_i^j}$	the probability that instance x_i^j is an outlier
$sum(x_i)$	the probability summation of possible worlds that x_i is an outlier in
$n(x_i^j, R)$	the number of neighbors that instance x_i^j has
$S_{x_i^j}$	the succeeding neighbors set of instance x_i^j has
$P_{x_i^j}$	the preceding neighbors set of instance x_i^j has
N_{x_i}	the set of (R,K)-outlier possible worlds that x_i has
$\tilde{x}_i$	the store structure of x_i after updated
$x_{arr.}$	the newly arrived tuple
$x_{exp.}$	the latest expired tuple

4.1   Framework

At first, we introduce the storage structure, in order to meet the demand for increment processing. We define the storage structure, which is designed to capture key information of outlier during the course of uncertain data stream. For example, we list partial uncertain tuple in uncertain data stream:

$$x_1\{[(1,2),(0.7)],[(1,1),(0.3)]\}, x_2\{[(2,1),(0.4)],[(3,1),(0.6)]\},$$
$$x_3\{[(2,2),(0.5)],[(2,3),(0.5)]\}, x_4\{[(4,4),(0.8)],[(3,3),(0.2)]\} \cdots$$

There are the value of the four tuples from x_1 to x_4. x_i^1 and x_i^2 are two instances of tuple x_i, $p_{x_i^1}$ and $p_{x_i^2}$ are exist probability of x_i^1 and x_i^2 respectively. let $R = 2$, $K = 2$, range query results for every instance in the current window is showed in Figure 1(range query is find neighbors for instance). We neaten the range query result on the two principles: (1) The instance of one tuple can not appear in the range query result of the same tuple's instance, i.e., in the range query result of x_1^1, x_1^2 can not exist, and delete x_1^2 from the range query results of x_1^1. (2) If there are at least two instances of the same uncertain tuple in the range queries result of a instance, then merge these instances existing probability of the same uncertain tuple, i.e., x_3^1 and x_3^2 are in the range queries result of x_1^1, the probability of x_3^1 and x_3^2 are added, $p_{x_3^1} + p_{x_3^2} = p_{x_3}$. The result after neatening is more conducive to analyze in figure 2.

Since the tuple is uncertain data, the probability of an uncertain tuple being an outlier is probability summation of its every instance being an outlier as Definition 5.

Definition 5. *(Uncertain Tuple Outlier Probability) We assume that the uncertain tuple x_i consist of m instances $((x_i^1, p_{x_i^1}), (x_i^2, p_{x_i^2}), \cdots, (x_i^j, p_{x_i^j}), \cdots, (x_i^m, p_{x_i^m})), 1 \le j \le m$, $P_{x_i^j}$ is the probability of instance x_i^j. The probability of uncertain tuple being an outlier is equal to the probability summation of each instance being an outlier, i.e. $pp_{x_i} = \sum_{j=1}^{m} pp_{x_i^j}$*

By the above Definition 5, we can calculate the probability of x_1 being an outlier, which is equal to the probability summation of x_1^1 and x_1^2 being outlier, i.e. $p_{x_1} = pp_{x_1^1} + pp_{x_1^2}$. It converts judging whether an uncertain tuple as an outlier to get the probability of each instance being an outlier.

Definition 6. *(Instance Outlier Probability) The probability of the instance x_i^j being an outlier is denoted by $pp_{x_i^j}$, $pp_{x_i^j} = p_{x_i^j} \times P_{n(x_i^j, R) < k}$, $P_{n(x_i^j, R) < k}$ denotes probability that the number of neighbor instance x_i^j has is less k, $0 \le P_{n(x_i^j, R) < k} \le 1$.*

In figure 2, the range query result of x_1^1 contain x_2^1 and x_3, the probability of x_1^1 being an outlier $pp_{x_1^1}$ is $0.7 \times (1 - 0.4) = 0.42$. The probability of x_1^2 being an outlier $pp_{x_1^2}$ is $0.3 \times (1 - 0.5) = 0.15$. Then we can get the probability of x_1 being an outlier is equal to $p_{x_1} = pp_{x_1^1} + pp_{x_1^2} = 0.42 + 0.15 = 0.57$.

Definition 7. *(Non-outlier Instance) Let s_{x_i} denote the set of instances that tuple x_i has. We determine the instance is a non-outlier instance, if the instance is not an outlier in any possible world. $s_{x_i}^y$ is a set of outlier instances that tuple*

$$x_1 \left\langle \begin{matrix} x_1^1(0.7)\{x_1^2(0.3), x_2^1(0.4), x_3^1(0.5), x_3^2(0.5)\} \\ x_1^2(0.3)\{x_1^1(0.7), x_2^1(0.4), x_2^2(0.6), x_3^1(0.5)\} \end{matrix} \right.$$

$$x_2 \left\langle \begin{matrix} x_2^1(0.4)\{x_1^1(0.7), x_1^2(0.3), x_2^2(0.6), x_3^1(0.5). x_3^2(0.5)\} \\ x_2^2(0.6)\{x_1^2(0.3), x_2^1(0.4), x_3^1(0.5), x_4^2(0.2)\} \end{matrix} \right.$$

$$x_3 \left\langle \begin{matrix} x_3^1(0.5)\{x_1^1(0.7), x_1^2(0.3), x_2^1(0.4), x_2^2(0.6). x_3^2(0.5), x_4^2(0.2)\} \\ x_3^2(0.5)\{x_1^1(0.7), x_2^1(0.4), x_3^1(0.5), x_4^2(0.2)\} \end{matrix} \right.$$

$$x_4 \left\langle \begin{matrix} x_4^1(0.8)\{x_4^2(0.2)\} \\ x_4^2(0.2)\{x_2^2(0.6), x_3^1(0.5), x_3^2(0.5), x_4^1(0.8)\} \end{matrix} \right.$$

$$x_1 \left\langle \begin{matrix} x_1^1(0.7)\{x_2^1(0.4), x_3(1)\} \\ x_1^2(0.3)\{x_2(1), x_3^1(0.5)\} \end{matrix} \right.$$

$$x_2 \left\langle \begin{matrix} x_2^1(0.4)\{x_1(1), x_3(1), \} \\ x_2^2(0.6)\{x_1^2(0.3), x_3^1(0.5), x_4^2(0.2)\} \end{matrix} \right.$$

$$x_3 \left\langle \begin{matrix} x_3^1(0.5)\{x_1(1), x_2(1), x_4^2(0.2)\} \\ x_3^2(0.5)\{x_1^1(0.7), x_2^1(0.4), x_4^2(0.2)\} \end{matrix} \right.$$

$$x_4 \left\langle \begin{matrix} x_4^1(0.8)\{\} \\ x_4^2(0.2)\{x_2^2(0.6), x_3(1)\} \end{matrix} \right.$$

Fig. 1. Result of range query without neaten

Fig. 2. Result of range query after neaten

x_i has, and $s_{x_i}^n$ is a set of non-outlier instance that tuple x_i has. Since these two sets do not overlap and cover the complete objects set, i.e., $s_{x_i}^y \bigcup s_{x_i}^n = s_{x_i}$, $s_{x_i}^y \bigcap s_{x_i}^n = \emptyset$. If the probability summation of non-outlier instance is greater than $1 - \lambda$, i.e. $\sum_{x_i^j \in s_{x_i}^n} P_{x_i^j} > 1 - \lambda$, then the tuple is non-outlier.

Through analysis of the results, we can get the outlier in the current window, and then we will design a storage structure. Since we are dealing with uncertain data in stream environment, the arrival time of uncertain data will affect the result. Firstly, we divide the range query results of every instance into two sets, preceding neighbors and succeeding neighbors set, denoted by $P_{x_i^j}$ and $S_{x_i^j}$ respectively.

Definition 8. *(Non-outlier Instance Probability) The non-outlier instance probability increase, along with the neighbor number $n(x_i^j, R)$ of an instance x_i^j increase. If the exist of a part neighbors, non-outlier instance probability is greater than $1 - \lambda$, then x_i^j is non-outlier instance.*

According to the above definition, we only need to store the neighbors that make the tuple become non-outlier(according to the time order), instead of storing all neighbors of each instance. The tuple in $P_{x_i^j}$ is eliminated before the tuple x_i. It may make tuple x_i become outlier from candidate set, therefore, we must consider the expires time of tuple in candidate set. The tuples in P_{x_i} is stored with time information and probability. The tuples in the S_{x_i} can not be expired before the tuple x_i was deleted, so we only store the probability information. For example, in figure 3, we depict an example in two dimensional, for $K = 3$, $\lambda = 0.7$. Let the subscripts denote the order of arrival of these tuples. We focus on tuple x_{18}, the range query result of instance x_{18}^1 is $\{x_6^1(0.4), x_{10}^2(0.6), x_{15}^1(0.5), x_{16}^2(0.6), x_{17}^1(0.7), x_{19}^2(0.8)\}$, the range query result of instance x_{18}^2 is $\{x_3^2(0.6), x_{16}^1(0.4), x_{20}^1(0.7), x_{22}^2(0.8)\}$. Range query result based on time, be divided into two sets to store, i.e., $S_{x_{18}^j}$ and $P_{x_{18}^j}$. In order to reduce the storage space to meet the demand of uncertain data stream, we only store the necessary information. According arrival time of each neighbor, to find the existence of the neighbors make tuple x_{18} is non-outlier, i.e., $p_{x_i(no)} > 1 - \lambda$.

When x_{22}, x_{20}, x_{19}, x_{17} and x_{16} exist, $p_{x_{18}(no)} = p_{x_{18}^1(no)} + p_{x_{18}^2(no)} = 0.3136 > 1 - \lambda$, then x_{18} is non-outlier. We only store structure of x_{18} as follows:

$$x_{18}^1(0.8)\{(16, 0.6), (17, 0.7)\}\{0.8\}$$

$$x_{18}^2(0.2)\{(16, 0.4)\}\{0.7, 0.8\}$$

4.2   Incremental Processing

In this paper, we use the count-based window which always maintains the W most recent uncertain tuples. The uncertain tuples maintained by the sliding window are termed active objects. When the tuple leaves the window we think that the tuple expires. Along with the arrival of the new uncertain tuples, the old tuples disappear from the window.

In the section, we introduce the incremental processing for outlier detection on uncertain data stream. Whenever the window slid, the new uncertain tuple arrives, and the old uncertain tuple is moved out from the window. In order to reduce calculation, we first delete the expired tuple, and then receive a new uncertain tuple. When the old tuple is deleted, it may make the tuple in candidate set become outlier. The impact that a new tuple arrives includes two classifications: (1) The impact to a new uncertain tuple from existing uncertain tuples. (2) The impact to exist uncertain tuples from a new uncertain tuple. It might make the outlier become candidate tuple, or make the candidate tuples become a safe inlier.

When a new tuple arrives, the range query is done according the probability descending of each instance. According to Definition 6, the important tuples are found and stored in the data structure. Range query could be done according the time of tuple. For instance, when tuple x_i arrivals, we first determine whether the instance of x_{i-1} is neighbor of instance of x_i, and then determine x_{i-2}, in turn forward according to the time of arrival, until finding the tuples that make x_i is non-outlier. These instances could be store in store structure of x_i. The pseudo code of these operations is given in Algorithm 1.

When the tuple is expired, it is removed from the window, if it is in $O(R, K, \lambda)$ or $C(R, K, \lambda)$. Tuple storage structure is updated, that the store structure contain expired tuple instance. If the tuple is in $O(R, K, \lambda)$, we only update the storage structure. If the tuple is in $C(R, K, \lambda)$, update the storage structure, and then determine whether the tuple become an outlier. The pseudo code of these operations is given in Algorithm 2.

In figure 3, we assume that a new tuple x_{28} arrives, instance x_{28}^1 is neighbor of x_{18}^1 and its existing probability is 0.9, then according Definition 6, when x_{28}, x_{22}, x_{20}, x_{19} and x_{17} exist, x_{18} is non-outlier but not a safe one. Then we update the store structure of x_{18} as follows:

$$x_{18}^1(0.8) : \{(17, 0.7)\}, \{0.8, 0.9\}$$

$$x_{18}^2(0.2) : \{\}, \{0.7, 0.8\}$$

Algorithm 1. Arrival (tuple x_{arr}, now)

Require: x_{now}: the arrival tuple, *now*: the current time window
1: make a range query for each instance of tuple x_{arr}, let $\Re$ the set of object returned;
2: **for** each instance $x_i^j \in \Re$ **do**
3:    add x_{arr} to S_{x_i}, update x_i storage structure
4:    **if** $x_i \in O(R, K, \lambda)$ **then**
5:       **if** $\widetilde{x}_i$ is candidate **then**
6:          add $\widetilde{x}_i$ to $C(R, K, \lambda)$
7:       **end if**
8:       **if** $\widetilde{x}_i$ is safe inlier **then**
9:          delete store structure of $\widetilde{x}_i$ from window
10:      **end if**
11:   **end if**
12:   **if** $x_i \in C(R, K, \eta)$ **then**
13:      **if** $\widetilde{x}_i$ is an safe inlier **then**
14:         add $\widetilde{x}_i$ to $I(R, K, \lambda)$
15:      **end if**
16:   **end if**
17: **end for**
18: construct x_{arr} storage structure
19: **if** x_{arr} is an outlier **then**
20:    add x_{arr} to $O(R, K, \lambda)$
       else
21:    add x_{arr} to $C(R, K, \lambda)$
22: **end if**

Some time later, when tuple x_{35} arrivals, instance x_{35}^2 is the neighbor of x_{18}^1, and its probability is 0.6, we can see, the existence of x_{35}, x_{28}, x_{22}, x_{20} and x_{19} make x_{18} become non-outlier, and they are all in set $S_{x_{18}}$, so x_{18} is a safe inlier. When x_{18} become a safe inlier along with window slid, we deleted store structure of x_{18}.

4.3 Improve the Efficiency of Range Queries

In the previous section, we introduce the method of range queries from back to front according arrival time. To be able to improve the efficiency of range query, we use the SM-tree(Statistics M-tree) to range query.

The formation of SM-tree that each instance in current window is a left node, and the radius of the leaf node is R is showed in figure 4. Parent node contains statistic information of child nodes, M^u and M^l are min and max time of M node respectively. When a new tuple arrives, the range query is done for each instance of the new tuple. We assume that the new instance in the range M_1 and M_2, because $M_1^u = t_4$, $M_1^l = t_3$, $M_2^u = t_1$, $M_2^l = t_1$, i.e., $M_2^u < M_1^l$, then first calculate the relationship between leaf nodes of M_1. If we get the all

---

Algorithm 2. Delete (tuple $x_i.exp$, now)

Require: x_{exp}: the departing tuple, *now*: the current time window
1: **if** x_{exp} in $C(R, K, \lambda)$ or $O(R, K, \lambda)$ **then**
2:    remove x_{exp} from $C(R, K, \lambda)$ or $O(R, K, \lambda)$
3: **end if**
4: **for** each tuple x_i which storage structure contain expired tuple x_{exp} **do**
5:    update x_i storage structure
6:    **if** $x_i \in C(R, K, \lambda)$ **then**
7:       **if** $\tilde{x}_i$ is an outlier **then**
8:          add $\tilde{x}_i$ to $O(R, K, \lambda)$
9:       **end if**
10:   **end if**
11: **end for**

---

instances that make the new tuple become non-outlier in the range of M_1, it isn't necessary to consider the leaf node of M_2. It could reduce the unnecessary calculations with SM-tree to improve efficiency of range queries.

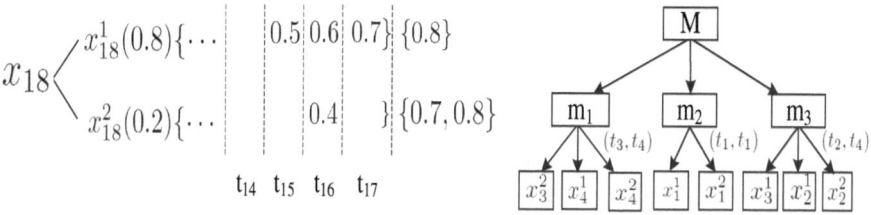

Fig. 3. Example of tuple store structure   **Fig. 4.** SM-tree

5   Performance Evaluation

We conducted a series of experiments on synthetic data set and real data sets to evaluate the performance of the algorithms proposed in this paper, namely Uncertain Outlier Detection on Uncertain Data Stream (denoted by UOD), and Uncertain Outlier Detection based on M-tree(denoted by UOD-M). For comparison, we implemented Naive Outlier Detection based on Examining Possible Worlds method named (denoted by NUOD).

All the algorithms are implemented in visual C 2010, and experiments are carried out on a PC with a Core i3-3.3 GHz CPU and 4G of main memory.

5.1   Running Time

We study the performance of the proposed methods by varying several of important parameters such as the window size W, the number of outliers, the number of required neighbors K, the distance R and the probability threshold λ.

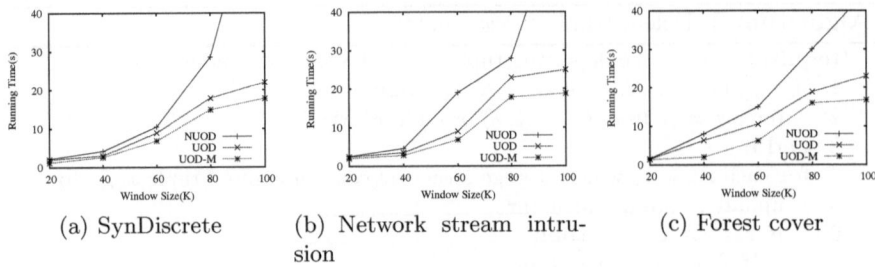

(a) SynDiscrete   (b) Network stream intrusion   (c) Forest cover

Fig. 5. Running time vs. window size

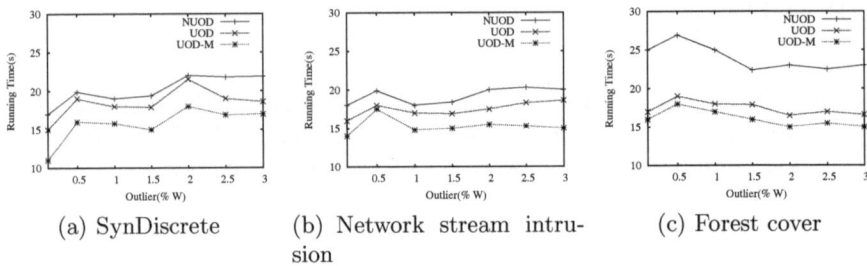

(a) SynDiscrete   (b) Network stream intrusion   (c) Forest cover

Fig. 6. Running time vs. number of outliers (%W)

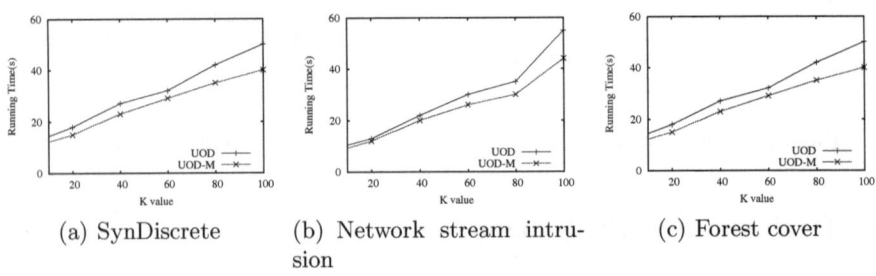

(a) SynDiscrete   (b) Network stream intrusion   (c) Forest cover

Fig. 7. Running time vs. k value

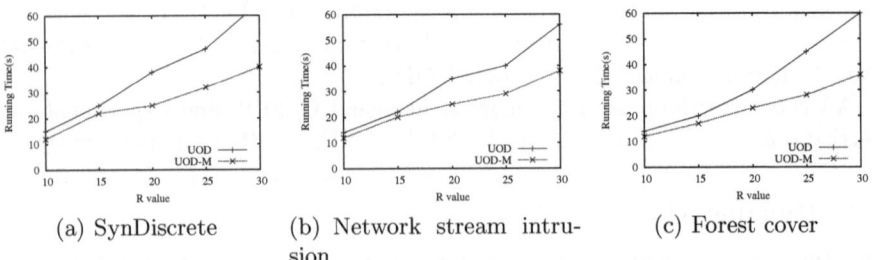

(a) SynDiscrete   (b) Network stream intrusion   (c) Forest cover

Fig. 8. Running time vs. R value

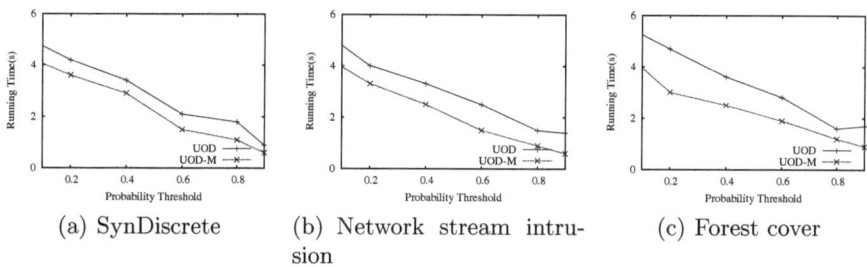

(a) SynDiscrete

(b) Network stream intrusion

(c) Forest cover

Fig. 9. Running time vs. λ value

First, we test the performance of the algorithm for varying values of W from 10^4 to 10^5. Figure 5 certify the running time of all algorithms increased when increasing the length of the sliding window. The performance of UOD-M and UOD algorithm performs better than NUOD.

Figure 6 depicts the results that number of outliers varying in range $[0.1\%, 3\%]$. As expected, UOD-M algorithm performs better than UOD algorithm.

In figure 7, we show that the performance of NUOD, UOD-SM and UOD algorithm. We vary K from 1 to 100. Figure 7 show that the running time of UOD-M and UOD decreased when increasing the value of K.

Figure 8 show that running time of UOD is much higher than UOD-M for varying R from $[10, 50]$. Figure 9 demonstrate that the running time when increasing the probability threshold λ. We vary λ value from 0.1 to 0.9. Figure 9 show that of both algorithms decreasing when increasing λ value.

5.2   Memory Consumption

Figure 10 show the memory consumption of three data sets. The consumed memory corresponds to the memory needed to store the information for uncertain tuples in current window. From the figure 10, we can see the required amount of memory is only a small fraction of the total memory available in modern machines.

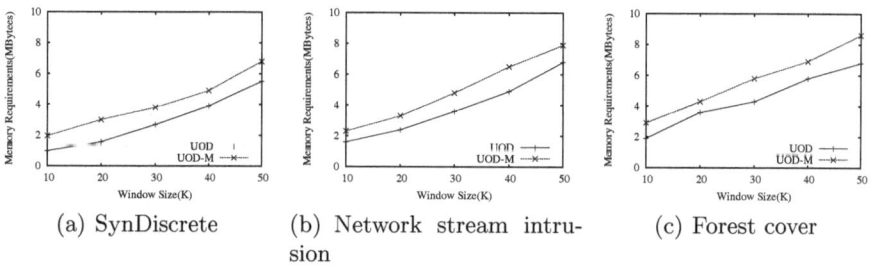

(a) SynDiscrete

(b) Network stream intrusion

(c) Forest cover

Fig. 10. Memory requirements vs. window size

6   Conclusion

In this paper we studied the problem of outlier detection on uncertain data stream, which the uncertain tuple consist of a set of instances and existing probability. We proposed outlier definition based on possible worlds, and then the method used to outlier detection on uncertain data stream is proposed. A dynamic storage structure was designed. We have conducted extensive experiments to evaluate the relative performance of the algorithms. The results showed that the algorithm can effectively detect outliers.

We will continue to study the problem based on outlier detection on uncertain data stream. Next steps of our work are taking more complex uncertain data, such as outlier detection on multi-dimensional uncertain data stream.

References

1. Aggarwal, C.C.: On density based transforms for uncertain data mining. In: ICDE, pp. 866–875 (2007)
2. Aggarwal, C.C., Yu, P.S.: Outlier detection with uncertain data. In: SDM, pp. 483–493 (2008)
3. Assent, I., Kranen, P., Baldauf, C., Seidl, T.: Anyout: Anytime outlier detection on streaming data. In: VLDB, pp. 228–242 (2012)
4. Burdick, D., Deshpande, P.M., Jayram, T.S., Ramakrishnan, R., Vaithyanathan, S.: Olap over uncertain and imprecise data. In: VLDB, pp. 970–981 (2005)
5. Chandola, V., Banerjee, A., Kumar, V.: Outlier detection: A survey. ACM Computing Surveys (2007) (to appear)
6. Cheng, R., Kalashnikov, D., Prabhakar, S.: Evaluating probabilistic queries over imprecise data. In: SIGMOD, pp. 551–562 (2003)
7. Jiang, B., Pei, J.: Outlier detection on uncertain data: Objects, instances, and inferences. In: ICDE, pp. 422–433 (2011)
8. Kontaki, M., Gounaris, A., Papadopoulos, A., Tsichlas, K., Manolopoulos, Y.: Continuous monitoring of distance-based outliers over data streams. In: ICDE, pp. 135–146 (2011)
9. Sarma, A.D., Benjelloun, O., Halevy, A., Widom, J.: Working models for uncertain data. In: ICDE, p. 7 (2006)
10. Singh, S., Mayfield, C., Prabhakar, S., Shah, R., Hambrusch, S.: Indexing uncertain categorical data. In: ICDE, pp. 616–625 (2007)
11. Tao, Y., Cheng, R., Xiao, X., Ngai, W.K., Kao, B., Prabhakar, S.: Indexing multi-dimensional uncertain data with arbitrary probability density functions. In: ICDE, pp. 922–933 (2005)
12. Wang, B., Xiao, G., Yu, H., Yang, X.: Distance-based outlier detection on uncertain data. CIT 1, 293–298 (2009)
13. Wang, B., Yang, X., Wang, G., Yu, G.: Outlier detection over sliding windows for probabilistic data streams. Journal of Computer Science and Technology 25(3), 389–400 (2010)

Improved Spatial Keyword Search
Based on IDF Approximation*

Xiaoling Zhou, Yifei Lu, Yifang Sun, and Muhammad Aamir Cheema

University of New South Wales, Australia
{xiaolingz,yifeil,yifangs,macheema}@cse.unsw.edu.au

Abstract. Spatial keyword search is a widely investigated topic along with the development of geo-positioning techniques. In this paper, we study the problem of top-k spatial keyword search which retrieves the top k objects that are most relevant to query in terms of joint spatial and textual relevance. Existing state-of-the-art methods index data objects in IR-tree which supports textual and spatial pruning simultaneously, and process query by traversing tree nodes and associated inverted files. However, these search methods suffer from vast number of times of accessing inverted files, which results in slow query time and large IO cost. In this paper, we propose a novel approximate IDF-based search algorithm that performs nearly twice better than existing method, which are shown through an extensive set of experiments.

1 Introduction

There has been a strong trend of incorporating spatial information into keyword-based search systems in recent years for the support of geographic queries. It is driven mainly by two reasons: highly user demands and rapid advances in Geo-positioning and Geo-coding technologies. User interests include finding the nearby restaurants, tourist attractions, entertainment services, public transport and so on, which renders the spatial keyword queries important. Studies [13] have shown that about 20% of web search queries are geographical and exhibit local intent. In addition, both the various advanced Geo-positioning technologies such as GPS, Wi-Fi and cellular geo-location services (offered by Google, Skyhook, and Spotigo), and different geo-coding technologies which enable annotating of web content with positions accelerate the evolution of spatial keyword search.

Spatial keyword queries take a set of keywords and a location as parameters and return the ranked documents that are most textually and spatially relevant to the query keywords and location. Usually, distance is used to measure spatial relevance and document similarity functions such as TF-IDF weighting [7] are used to measure textual relevance. A fair amount of investigations have been conducted aiming at this issue, such as the algorithm of solving kNNs (k-nearest neighbor queries) using R-tree [12], loose combinations of inverted file and R*-tree [2] [6] [15], and signature augmented IR²-tree [4]. Whereas limitations exist

---

* This research was supported by ARC DP130103401 and DP130103405.

Y. Ishikawa et al. (Eds.): APWeb 2013, LNCS 7808, pp. 461–472, 2013.

in those works, they either cannot support pruning simultaneously according to both keyword similarity and spatial information, or become infeasible when it comes to handling ranked queries.

One of the most promising and popular ones is IR-tree index based method. This particularly designed index structure enables simultaneous spatial pruning and textual filtering, and also an incremental searching and ranking process. Most importantly, it supports query specific IDF(Inverse Document Frequency) values computation for determining document textual scores, which is more reasonable than traditional IDF values used in purly textual search. Essentially, IDF value measures the selectivity(or rarity) of a query keyword. In purly textual search cases, for a word w, IDF_w is computed as the logrithm of *Collection Size* ($|D|$) divided by *Document Frequency* of $w(DF_w)$. While in spatial keyword search cases, it is inappropriate to keep using this tranditional IDF values as the word distribution among documents within query spatial range might be completely different with word distribution in whole documents set. Therefore, the query specific IDF value used in spatial keyword search is computed as the logrithm of *Total Number of Documents within Query Range* (D_s)divided by *Word DF within Query Range* (df_{w,D_s}). However, in order to determine IDF value of each query keyword for later calculating textual score of each object, IR-tree based searching method suffers from multiple times of accessing inverted files attached with leaf nodes of the index tree, which deteriorates searching performance substantially, espectially when their inverted files are stored on disk.

Based on the above observation, in this paper, we propose a new searching algorithm which does not access inverted files when computing IDFs thus results in an IDF range for each query keyword, and then retrieve top-k documents based on range scores computed using IDF ranges. Our approximate IDF-based method reduces query search time considerably to nearly half of the original search method while without compromising accuracy of final ranked results. We conduct a comprehensive set of experiments that clearly show the search performance gain of our proposed method in comparison with the original algorithm under a wide variation of various query or index parameters. Some essential preliminaries related to spatial keyword search are presented in Section 2. Section 3 details our proposed approximate IDF-based algorithm. Section 4 provides results of experiments over a wide range of parameter settings and the last section concludes the work.

2   Preliminaries

2.1   Problem Definition

We consider a database consists of N documents, each associated with a spatial location. A query Q also consists of a set of keywords (denoted as W_Q), and a location (denoted as $Q.loc$).

Given a query range S_Q and a scoring function $score(Q, d)$, a **top-k spatial keyword search** returns top k scoring documents from the database.

The scoring function should take into consideration both the textual relevance of the document to the query as well as the spatial proximity between the locations of the query and the document. Hence, the final function can be represented as $score(Q, d) = \alpha \cdot score_{\text{text}}(Q, d) + (1 - \alpha) \cdot score_{\text{spatial}}(Q, d)$, where $\alpha \in [0, 1]$ is a tuning parameter.

Spatial Relevance. We only need to consider documents within the search range as specified by S_Q (these doucments are also known as *spatially qualified*). For spatially qualified documents, we give its proximity score based on the euclidean distance between its location and the query location, i.e., $score_{\text{spatial}}(Q, d) = 1 - \frac{dist(d.loc, Q.loc)}{dist_{\max}}$, where $dist_{\max}$ is the maximum possible distance between the query and any spatially qualified document.

Textual Relevance. We adopt the popular Vector Space Model to measure the textual relevance between the document and the query. In this paper, we consider approaches using the query-specific IDF, which is defined as $idf_{w,D,S_Q} = \log \frac{D_s}{df_{w,D_s}}$. Note that D_s is number of spatially qualified documents rather than that of the entire collection of documents, and df_{w,D_s} refers to number of spatially qualified documents that contain w, therefore, idf_{w,D,S_Q} reveals query words distributions among spatially qualified documents rather than in the entire collection. Finally, the textual score is calculated as: $score_{\text{text}}(Q, d) = \sum_{w \in W_Q} (tf_{w,d} \cdot idf_{w,D,S_Q})$. We say a document d is *textually qualified* to a given query Q iff d contains at least one query keyword of Q.

Notations. We use the following notations in the rest of the paper. The root of an IR-tree is r. The number of documents within S_Q as D_s. Number of documents within S_Q that contain w as df_{w,D_s} and query specific idf value of w as idf_{w,D_s}. Similarly, for a tree node e, df_{w,D_e} refers to the number of spatially qualified documents stored underneath e that contain w. In later sections, unless explictly stated, when saying df_w or idf_w, we refer to df_{w,D_s} and idf_{w,D_s} respectively. Finally, the candidate set produced from IDF calculation step as B.

2.2   IR-Tree Index Structure

IR-tree [11] is one of the most effective and efficient query processing methods for top-k spatial keyword search. Its effectiveness largely comes from the query-specific IDF. In order to facilitate query processing, a novel index, IR-tree, is proposed, which is essentially a clever combination of R-tree [5] and inverted files [16]. Document locations are indexed in an R-tree enhanced as follows:

– each internal node of the R-tree is associated with textual summary information about documents in the subtree, including (1) the count of documents in the subtree, and (2) the max-TF and DF for each words appearing among documents in the sub-tree.

— each leaf node is linked to an inverted file indexing all the documents belonging to the nodes.

An example IR-tree is shown in Figure 1.

2.3   The Existing Search Algorithm for IR-Tree

The query processing algorithm in [11] consists of two steps, *IDF calculation* and *top-k document retrieval*.

Algorithm 1. IDFCalculation (r, W_Q, S_Q)

---

1 $T.push(r); D_s \leftarrow 0;$
2 **while** T *is not empty* **do**
3 $\quad$ $e \leftarrow T.pop();$
4 $\quad$ **if** $e.MBR$ is *fully contained in* S_Q **then**
5 $\quad\quad$ $D_s \leftarrow D_s + |D_e|;$
6 $\quad\quad$ **if** $\exists w \in W_Q$ *such that* $df_{w,D_e} > 0$ **then**
7 $\quad\quad\quad$ **forall** $w \in W_Q$ **do** $df_w \leftarrow df_w + df_{w,D_e};$
8 $\quad\quad\quad$ $B.add(e);$

9 $\quad$ **else if** e *is intersect with* S_Q **then**
10 $\quad\quad$ **if** e *is an internal node* **then**
11 $\quad\quad\quad$ **forall** *child node* c *of* e **do** $T.push(c);$
12 $\quad\quad$ **else** /*e is a leaf node */
13 $\quad\quad\quad$ $S_1 \leftarrow \emptyset; S_2 \leftarrow \emptyset;$
14 $\quad\quad\quad$ **forall** $d \in e$ *such that* $d.loc \in S_Q$ **do** $S_1 \leftarrow S_1 \cup \{d\};$
15 $\quad\quad\quad$ $D_s \leftarrow D_s + |S_1|;$
16 $\quad\quad\quad$ **if** $\exists w \in W_Q$ *such that* $df_{w,D_e} > 0$ **then**
17 $\quad\quad\quad\quad$ $I_e \leftarrow$ the inverted file for entry $e;$
18 $\quad\quad\quad\quad$ **forall** $w \in W_Q$ **do**
19 $\quad\quad\quad\quad\quad$ **forall** $d \in I_e[w] \cap S_1$ **do**
20 $\quad\quad\quad\quad\quad\quad$ $df_w \leftarrow df_w + 1;$
21 $\quad\quad\quad\quad\quad\quad$ $S_2 \leftarrow S_2 \cup \{d\};$

22 $\quad\quad\quad\quad$ **forall** $d \in S_2$ **do**
23 $\quad\quad\quad\quad\quad$ $B.add(d) ;$ /* Also keep $tf_{w,d}$ information about d */

24 **return** $\{\log \frac{D_s}{df_w}\}, \forall w \in W_Q$

IDF Calculation. We show the pseudo-code of IDF calculation in Algorithm 1. It is basically a top-down traversal of the IR-tree, and accumulates D_s and df_ws during the process.

Top-k Document Retrieval. This step performs best-first search on the IR-tree, and it terminates as soon as top-k objects have been determined. Initially, all the entries in B and their estimated maxscores are pushed into max-heap H.

The algorithm then iteratively pop an entry e in H until either H is empty or top-k results have been determined. If e is a document, it is added to R; if e is a non-leaf node, it pushes all its children and their scores to H; otherwise, e must be a leaf node, it pushes all its documents and their scores into H.

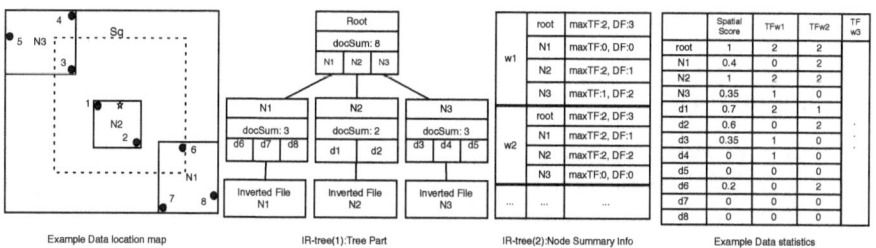

Fig. 1. Example Data statistics

Example 1. Consider the example IR-tree in Figure 1. Document statistics and spatial scores are given. Assume $k = 1$, $\alpha = 0.5$, $W_Q = (w1, w2)$. The algorithm works as follows:

- In the IDF calculation step, IR-tree is traversed from root, node N2 is contained in S_Q, so it is added into candidate set B, D_s is computed as 2, $df_{w1} = 1$ and $df_{w2} = 2$. Nodes N1 and N3 are leaf nodes that partially overlap with Q, therefore their associated inverted files are retrieved to calculate accurate $df_{w1} = 2$, $df_{w2} = 3$, and d3 and d6 are found to be both spatially and textually qualified and are added to B. At the end of this step, we have $B = \{N2, d3, d6\}$, $idf_{w1} = \log_{10} \frac{2}{2} = 0.301$, and $idf_{w2} = \log_{10} \frac{4}{3} = 0.125$.
- In the TopKRetrieval step, H is initialized as $\{(N2, 0.926), (d3, 0.326), (d6, 0.225)\}$ by adding all entries in B and their scores(for nodes, the scores calculated are their estimated maxscore; for docs, they are accurate score) to H. Then, node N2, the head entry in H is popped and expanded so that its children d1 and d2 are added to H. Now $H = \{(d1, 0.714), (d2, 0.425), (d3, 0.326), (d6, 0.225)\}$.
- Since d1's score is higher than the next highest score, d1 is returned as the top-1 answer, and the algorithm stops.

3   Approximate IDF-Based Searching Algorithm

One performance issue with the above query processing algorithm on IR-tree is that generally more than 70% of the search time is used for IDF calculation. For example, the shaded bars in Figures 2 indicate the IDF calculation time.

After analyzing the IDF calculation algorithm, Lines 17-23 of Algorithm 1 are usually the most time-consuming part. This is because the algorithm needs to access inverted files for each intersecting leaf node. These inverted files are typically proportional to the total size of document collection, and are much larger

than spatial information and other index overhead associated with each document. In addition, due to the sheer size of the inverted files, they are typically
stored on the disk, and many random I/Os are needed to retrieve them.

Therefore, our goal is *not* to access all of the inverted files while still retrieving
the top-k results. We instead approximate the IDF values by a range. We can
find conditions that guarantee the correctness of top-k answers even with such
approximate IDF values.

3.1   Framework

Our idea is to leave out Lines 17-23 of Algorithm 1, and compute a DF range
$[DF_w^{min}, DF_w^{max}]$ for each query word w without traversing inverted lists.

Consider the previous example. When node N1 is traversed, we find it is an
intersecting leaf node, instead of scanning its inverted files, we know that under
N1, there are $df_{w,D_{N1}}$ docs contain w, but not sure if all of them are spatially
qualified. Thus, we can bound DF_w as $[DF_w, DF_w + df_{w,D_{N1}}]$. Subsequently, we
can bound the IDF values for each query keyword as $[\log \frac{D_s}{DF_w+df_{w,D_{N1}}}, \log \frac{D_s}{DF_w}]$.

Once we have the IDF range for every query keyword, the min and max scores
of an IR-tree node or a document can be computed using the IDF ranges and
the scoring function presented in section 2. Note that spatial score of an IR-tree
node can also be bounded by computing the min and max distance between the
MBR of the node and the query location.

Denote minScore of d as $\bot(d)$ and maxScore of d as $\top(d)$. It is obvious that
if $\bot(d_1) > \top(d_2)$, then we know d_1's score is better than d_2 and we can safely
rank d_1 before d_2.

3.2   IDF Range Computation

The main difference of this algorithm with Algorithm 1 is the case when current
entry e is a leaf node(line 12). Instead of retrieving the inverted file of e, only
locations of child objects stored in e are scanned so as to update D_s(lines 14-15),
then e is added into candidate set simply if it is textually qualified, and replace
lines 17-23 by $DF_w^{max} \leftarrow DF_w^{max} + df_{w,D_e}$ to accumulate min and max DFs for
each query keyword.

A subtlety here is to additionally record intersecting leaf node if it is textually qualified, so that later when two objects have very similar score ranges and
cannot be ordered exactly, we need to expand these unchecked intersecting leafs
and update query word DFs to get tighter object score ranges (add a line under
line 16: $ILFs.add(e)$, where $ILFs$ denotes a list of intersecting leaf nodes).

3.3   Top-k Docs Retrieval

After getting candidate set, IDFs ranges, top-k documents are retrieved. As
there may be much more than k qualified documents in candidate set, we use
the incremental search algorithm which stops as long as k docs are identified.

Algorithm 2. TopKRetrieval $(W_Q, S_Q, k, \alpha, B, ILFs, minDFs, maxDFs)$

---

1 $newMinDFs \leftarrow minDFs$, $newMaxDFs \leftarrow maxDFs$; $R \leftarrow \emptyset$;
2 **for** $e \in B$ **do** $H.push(e, \top(e), \bot(e))$;
3 **while** $|H| > 0$ **and** $|R| < k$ **do**
4 | $e \leftarrow H.pop()$;
5 | **if** e is node **then**
6 | | ExpandTreeNode(e, H);
7 | **else**
8 | | **if** $|H| = 0$ **then**
9 | | | $R \leftarrow R \cup \{e\}$;
10 | | **else**
11 | | | $e' \leftarrow H.top()$;
12 | | | **if** $\bot(e) \geq \top(e')$ **then**
13 | | | | $R \leftarrow R \cup \{e\}$;
14 | | | **else**
15 | | | | **if** e' is node **then**
16 | | | | | $e' \leftarrow H.pop()$; $H.push(e)$;
17 | | | | | ExpandTreeNode(e', H);
18 | | | | **else**
19 | | | | | $H.push(e)$;
20 | | | | | expand a proportion(10-100%) of $ILFs$ to update DFs;
21 | | | | | compute $newMinIDFs, newMaxIDFs$;
22 | | | | | update entry scores in H;

23 **return** R

Algorithm 3 outlines top-k document retrieval process. Firstly, for each entry e in B, we estimate its $\bot(e)$ and $\top(e)$, and put entry $(e, \top(e) , \bot(e))$ into a max-heap H which sorts its entries in descending order of their maxScores.

Secondly, we iteratively check the head entry $(e, \top(e), \bot(e))$ of H to identify top-k documents until H is empty. If e is a node, expand this node by adding its children and their scores into H. If it is a document and H is not empty, the next top entry(n) in H needs to be examined and compared with e in order to determine whether e can be added into R. This leads to two cases: 1) Scores of e and n are comparable, i.e. $\bot(e)$ is greater than or equal to $\top(n)$, we can simply add e into R; 2) Their scores are not comparable, i.e. $\bot(e)$ is less than $\top(n)$, we cannot determine directly whether e is the most relevant document in H, thus extra process is required. If n is a node, push e back to H and expand node n, whereas if n is also a document, which means the current DFs ranges are too rough to determine exact top-k documents, more intersecting leaf nodes need to be expaned to update min and max DFs of query keywords, IDF values are recomputed and entry scores in H are also updated to a tighter bound for exact ranking.

ExpandTreeNode(n, H) process examines all the child nodes (cn) of the node to be expaned (n) and add $(cn, \top(cn), \bot(cn))$s into H for later processing if n

is an internal tree node. If n is a leaf node which is within query range, textually qualified documents (d) which is implied by that they exist in at least one of the inverted lists of query keywords associated with n are examined and $(d, \top(d), \bot(d))$s are added into H. However, if n is an intersecting leaf node($n.MBR$ is interect with S_Q), not only documents stored underneath n have to be checked whether they are textually qualified, they also have to be checked if spatially qualified so that they can be added into H and minDFs and maxDFs are also updated accordingly.

Lemma 1 (Correctness). *Algorithm 2 correctly retrieves the top-k result.*

Example 2. Using the same example in Figure 1, the proposed algorithm works as below:

- IDFRangeComputation procedure traverses IR-tree to find contained node N2, and intersecting leaf nodes N1 and N3, thus updates $D_s = 4$, $ILFs$=(N1, N3), B=(N1, N2, N3), and $df_{w1,D_s} = [1, 3]$, $df_{w2,D_s} = [2, 3]$, and computes $idf_{w1,D_s} = [0.125, 0.602]$, $idf_{w2,D_s} = [0.125, 0.301]$.
- TopKRetrieval first constructs initial $H = \{(N2, 1.403, 0.75)(N1, 0.501, 0.325)$ $(N3, 0.476, 0.238)\}$. Then node N2 is expanded to get $newH$=$\{(d_1, 1.103, 0.538)(d_2, 0.601, 0.425)(N1, 0.501, 0.325)(N3, 0.476, 0.238)\}$
- Two documents are found at the top of heap, but with incomparable scores. Therefore, more intersecting leaf nodes have to be expanded. Assume 50% of $|ILFs|$is expanded at one time, then in this case, node N1 is expanded. df_{w1,D_s} is not changed, $newdf_{w2,D_s} = [3, 3]$, so $newidf_{w2,D_s} = [0.125, 0.125]$. The new updated $H = \{(d_1, 1.015, 0.538)(N3, 0.476, 0.238)(d_2, 0.425, 0.425)(N1, 0.325, 0.325)\}$
- d_1 is returned as its $\bot(d_1) > \top(N3)$. Search stops.

4   Experiments

4.1   Experiments Setup

We select two publicly available datasets.

- **Trec** is from the TREC-9 Filtering Track[1]. We use TREC as text dataset.
- **POI-Italy** contains gas station and car parking locations of Italy[2]. for spatial data.

There are totally 239580 documents in Trec dataset. We randomly combine two documents to form a longer document, and 100000 of these combined documents are selected as test data set. We assign each combined document a unique spatial location selected randomly from POI-Italy dataset. All the x and y coordinates data in POI-Italy dataset are normalised into range [0, 40000]. Correspondingly, we randomly select 1 to 8 words from the vocabulary of the documents to form textual part of queries and assign random rectangle as the spatial part. We show some statistics in Table 1.

---

[1] http://trec.nist.gov/data/t9_filtering.html
[2] http://poi-osm.tucristal.es/

In the experiments, we investigate performance gains of proposed search algorithms and influences of 5 factors on search performance shown below in table 2. The underlined values are default settings if not specified.

- S_Q Query Spatial Range.
- α Score ratio, which measures the relative importance of textual relevance to spatial relevance.
- k The number of documents to be returned.
- $|W_Q|$ The number of query keywords.
- $|D|$ Test size, the total number of documents indexed.

The index structure implemented is IR-tree Index demonstrated in Section 2.2 using the public R-Tree Library at http://jsi.sourceforge.net. Two search algorithms are implemented, which are denoted as IRTree (described in section 2.3) and ApproIDF (our proposed algorithm).

4.2 Search Performance

In the experiments, we compare the search performance of two search algorithms over wide variation of different query and index factors.

Vary Query Word Number $|W_Q|$: In Figure 2(a), we compare the experimental results of the two search algorithms in terms of total query search time against query word numbers. The shaded area represents IDF computation time of the two algorithms. It is obvious that our proposed algorithm outperforms the original search algorithm, and consumes nearly half of the search time of original algorithm. This trend remains the same while the number of words are increasing. Another observation is that the IDF computation time of ApproIDF is reduced substantially as expected, which becomes less than one thirds of that of original search algorithm. Finally, the obvious trend shows that total search time increases linearly as query word number increases. This is due to the increase of textually qualified documents since our implementation supports OR semantic keyword search, which means a document is qualified as long as it contains at least one of the query keywords.

Table 1. Trec Dataset Statistics

Test Size	Total Distinct Words	Total Words	Average Document Size
100000	241297	35883968	358

Table 2. Experiment Parameter Settings

Parameters	Values
Query range	10, 100, 1000, 10000
Score ratio	0, 0.25, 0.5, 0.75, 1
Top k	1, 10, 20, 50, 100
Query word number	1, 2, 4, 8
Test size	10000, 20000, 40000, 60000, 80000, 100000

Vary Score Ratio α: Figure 2(c) presents search perfomance of varying weight ratio between textual and spatial relevance. It can be seen that search time stays steadily for the two algorithms when α varies from 0 to 1 except that when α is set to 0(only consider spatial score), approIDF performs particularly good than any other cases. The reason is that when α is 0, the ordering of documents only depends on spatial score, which renders IDF values useless no matter they are exact or approximate. Therefore, top k documents can be selected directly and quickly using accurate spatial score. In addition, the overall trend is the same as above, ApproIDF performs much better than original search algorithm in all ratio settings. Here,pre-checking of α can be added before search process starts as an optimization to deal with special cases such as α is 0 or 1.

Vary Top K k: Figure 2(d) demonstrates the impact of number of requested documents(k). It is shown that search time of the two algorithms increase slightly as k increases. As both of the search processes are incremental, which means they stop as soon as top-k docs have been retrieved without having to go through all candidate set, when k becomes larger, more scores of documents need to be computed and ranked. Besides, our proposed algorithm is still more efficient than IR-tree search algorithm.

Vary Query Range S_Q: Next, effects of query spatial range on search performance are evaluated. The log scale search time is summarized in Figure 2(b). When query range is small(10-1000), total search time for both algorithm is less than 0.04 millisecond, whereas when search range increases to 10000, total search time grows drastically to around 0.63ms for IR-tree and 0.33ms for ApproIDF. The is mainly due to increase of candidate size when query range grows. There are nearly 10% (9231) of documents are within query range when S_Q is set to 10000. Overall, ApproIDF outperforms IR-tree search algorithm, especially obvious when query range is large.

Vary Test Size $|D|$: As shown in Figure 2(e), search time grows up linearly for both algorithms along the increasing of test size (number of documents indexed). It is mainly due to the increase of candidate size when test size becomes larger as shown in Figure 2(f).

5   Related Works

Numerous works have been done in this field. Some works use inverted index for textual retrieval and separate index such as Quad-tree or Grid index for spatial filtering [2] [10]. Currently the most widely used index structure is hybrid index, and many variations exist besides the IR-tree structure [3] [11] we already mentioned, such as Naïve Hybrid Index [9], R*-tree [1] [8], KR*-tree [6], IR^2-tree [4], and W-IR-tree [14].

Naïve Hybrid Index [9] combines document location to every word in the document to be a new word, and build index on them. This method clearly wastes lots of space and does not support flexible spatial relevance computation.

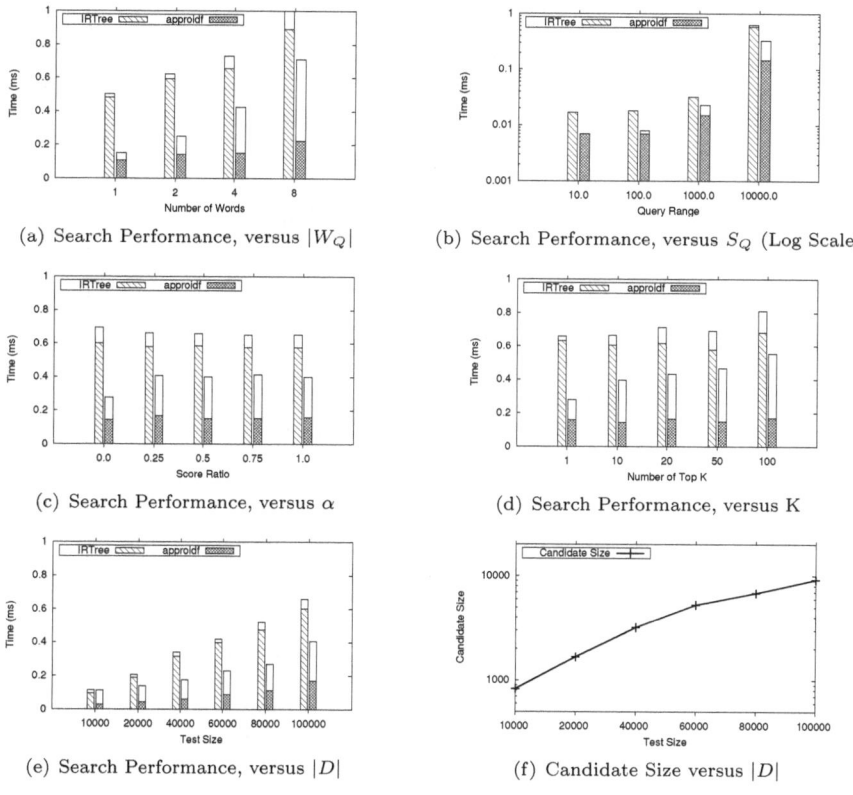

(a) Search Performance, versus $|W_Q|$

(b) Search Performance, versus S_Q (Log Scale)

(c) Search Performance, versus α

(d) Search Performance, versus K

(e) Search Performance, versus $|D|$

(f) Candidate Size versus $|D|$

Fig. 2. IDF Computation Time and Total Search Performance Comparison

R*-tree based method use loose combinations of an inverted file and an R*-tree [6] [15]. These works usually suffer from the inability to prune search space simultaneously according to keyword similarity and spatial information. **IR²-tree** [4] augments the R-tree with signatures enables keyword search on spatial data objects that each have a limited number of keywords. This technique suffers from substantial I/O cost since the signature files of all words need to be loaded into memory when a node is visited. Also, signature approach is useless and infeasible when it comes to handling ranked quires. **W-IR-tree** [14] is essentially an IR-Tree combined with word partitioning technique, which partitions keywords before constructing IR-tree. W-IBR-tree is simply a variation of W-IR-Tree by replacing inverted files attached to each node by inverted bitmap, which reduces the storage space and also saves I/O during query processing. This tree structure combines both keyword and location information, better matches the semantics of top-k spatial keyword queries and performs better than IR-Tree. However, it does not consider textual similarity as a ranking factor as it treats an object as either textual relevant (containing all the keyword queries) or not.

6   Conclusion and Future Work

In this paper, we introduce a new approximate IDF-based searching algorithm to deal with top-k spatial keyword search problem based on IR-tree index structure. We evaluate the performance of our proposed algorithm against the original one [11] via wide variation of experiment factors, and the experimental results demonstrate that our proposed search algorithm performs nearly twice better than the original one. The only drawback of IR-tree index structure is that it consumes substantial amount of storage space whose improvement can be a promising work in the future.

References

1. Beckmann, N., Kriegel, H.P., Schneider, R., Seeger, B.: The R*-Tree: An efficient and robust access method for points and rectangles. In: SIGMOD Conference, pp. 322–331 (1990)
2. Chen, Y.Y., Suel, T., Markowetz, A.: Efficient query processing in geographic web search engines. In: SIGMOD Conference, pp. 277–288 (2006)
3. Cong, G., Jensen, C.S., Wu, D.: Efficient retrieval of the top-k most relevant spatial web objects. In: PVLDB, vol. 2(1), pp. 337–348 (2009)
4. Felipe, I.D., Hristidis, V., Rishe, N.: Keyword search on spatial databases. In: ICDE, pp. 656–665 (2008)
5. Guttman, A.: R-trees: A dynamic index structure for spatial searching. In: SIGMOD Conference, pp. 47–57 (1984)
6. Hariharan, R., Hore, B., Li, C., Mehrotra, S.: Processing spatial-keyword (SK) queries in geographic information retrieval (GIR) systems. In: SSDBM, p. 16 (2007)
7. Hiemstra, D.: A probabilistic justification for using tf x idf term weighting in information retrieval. Int. J. on Digital Libraries 3(2), 131–139 (2000)
8. Hjaltason, G.R., Samet, H.: Distance browsing in spatial databases. ACM Trans. Database Syst. 24(2), 265–318 (1999)
9. Jones, C.B., Abdelmoty, A.I., Finch, D., Fu, G., Vaid, S.: The SPIRIT spatial search engine: Architecture, ontologies and spatial indexing. In: Egenhofer, M., Freksa, C., Miller, H.J. (eds.) GIScience 2004. LNCS, vol. 3234, pp. 125–139. Springer, Heidelberg (2004)
10. Lee, R., Shiina, H., Takakura, H., Kwon, Y.J., Kambayashi, Y.: Optimization of geographic area to a web page for two-dimensional range query processing. In: WISEW 2003, pp. 9–17. IEEE Computer Society, Washington, DC (2003)
11. Li, Z., Lee, K.C.K., Zheng, B., Lee, W.C., Lee, D.L., Wang, X.: IR-Tree: An efficient index for geographic document search. IEEE Trans. Knowl. Data Eng. 23(4), 585–599 (2011)
12. Roussopoulos, N., Kelley, S., Vincent, F.: Nearest neighbor queries. In: SIGMOD Conference, pp. 71–79 (1995)
13. Sanderson, M., Kohler, J.: Analyzing geographic queries. In: Workshop on Geographic Information Retrieval SIGIR (2004)
14. Wu, D., Yiu, M.L., Cong, G., Jensen, C.S.: Joint top-k spatial keyword query processing. IEEE Trans. Knowl. Data Eng. 24(10), 1889–1903 (2012)
15. Zhou, Y., Xie, X., Wang, C., Gong, Y., Ma, W.Y.: Hybrid index structures for location-based web search. In: CIKM, pp. 155–162 (2005)
16. Zobel, J., Moffat, A.: Inverted files for text search engines. ACM Comput. Surv. 38(2) (2006)

Efficient Location-Dependent Skyline Retrieval with Peer-to-Peer Sharing

Yingyuan Xiao[1], Yan Shen[1], Hongya Wang[2], and Xiaoye Wang[1]

[1] Tianjin Key Laboratory of Intelligence Computing and Novel Software Technology, Key Laboratory of Computer Vision and System, Tianjin University of Technology, 300384, China
[2] Donghua University, 201620, Shanghai, China

Abstract. Motivated by the fact that a great deal of information is closely related to spatial location, a new type of skyline queries, *location-dependent skyline query* (LDSQ), was recently presented. Different from the conventional skyline query, which focuses on non-spatial attributes of objects and does not consider location relationship (i.e., proximity), LDSQ considers both the proximity and non-spatial attributes of objects. In this paper, we explore the problem of efficient LDSQ processing in a mobile environment. We propose a novel LDSQ processing method based on peer-to-peer (P2P) sharing. Extensive experiments are conducted, and the experimental results demonstrate the effectiveness of our methods.

Keywords: location-based services, location-dependent skyline query, valid scope, peer-to-peer sharing.

1 Introduction

Owing to the ever growing popularity of mobile devices and rapid advent of wireless technology, information systems are expected to be able to handle a tremendous amount of service requests from the mobile devices. The majority of these mobile devices are equipped with positioning systems, which gave rise to a new class of mobile services, i.e., location-based services (LBS). LBS enable users of mobile devices to search for facilities such as restaurants, shops, and car-parks close to their route. In general, mobile clients send *location-dependent queries* to an LBS server from where the corresponding location-related information is returned as query results.

The skyline query is one very important query for users' decision making. The conventional skyline query focuses on non-spatial attributes of objects and does not consider proximity. Considering both the proximity and non-spatial attributes of objects, a new type of skyline queries, *location-dependent skyline query* (LDSQ), was recently presented [1, 2]. Unlike the conventional skyline query, whose result is based only on the actual query itself and irrelevant to the location of the mobile client, the query result of an LDSQ is closely relevant to the location of the mobile client, i.e., the location of the mobile client is a parameter of the LDSQ. To answer LDSQs, mobile clients need to connect to the corresponding LBS server through a

Y. Ishikawa et al. (Eds.): APWeb 2013, LNCS 7808, pp. 473–481, 2013.

GSM/3G/Wi-Fi service provider and the LBS server is responsible for processing each LDSQ. With the ever growing number of mobile devices, the sharply increasing query loads are already resulting in network congestion and the performance bottle of the LBS server. Aim to the above scalability limitations, in this paper, we explore to leverage P2P data sharing to answer LDSQs.

The remainder of this paper is organized as follows. We review the related work in Section 2. In Section 3, we first formulate the problem studied in this work, and then describe the reference infrastructure for processing LDSQ. Section 4 presents the proposed processing approach. We evaluate the proposed approaches through comprehensive experiments in Section 5. Finally, Section 6 concludes this paper.

2   Related Work

Skyline query has received considerable attention due to its importance in numerous applications. The skyline operator is first introduced into the database community by Borzsonyi et al. [3]. Since then a large number of methods have been proposed for processing skyline queries in a conventional set. These methods can be classified in two general categories depending on whether they use indexes or not. The popular indexes-based methods mainly include *index* [4], *nearest neighbor* [5], *branch-and-bound skyline* [6] and *ZBtree* [7]. The typical methods without indexes are *block nested loop* [3], *divide and conquer* [3], *sort first skyline* [8], *linear elimination sort for skyline* [9] and *sort and limit skyline algorithm* [10]. Furthermore, a number of interesting variants of the conversional skyline, like *multi-source skyline* [11], *distributed skyline* [12] and etc., have been addressed.

The abovementioned methods only focus on non-spatial attributes of objects and do not consider location relationship. Motivated by the popularity of LBS and the increased complexity of the queries issued by mobile clients, Zheng et al. [1] first address the problem of LDSQ which is the most relevant to our work. LDSQ enables clients to query on both spatial and non-spatial attributes of objects. In [1], the concept of *valid scope* is introduced to address the query answer validation issue and two query algorithms to answer LDSQ, namely *brute-forth* and *δ-Scanning*, are proposed. Furthermore, [2] explores how to efficiently answer LDSQ in road networks where the important measure is the *network distance*, rather than their *Euclidean distance*. The main disadvantage of the methods present in [1, 2] is lack of scalability and thus they cannot avoid network congestion and the performance bottle of the LBS server caused by the ever growing number of mobile devices.

3   Preliminary

In this section, we first formally define LDSQ and related concepts, and then describe the system infrastructure for supporting LDSQ.

Notations and Definitions. Let S be the set of spatial objects stored at the LBS server and every object $o \in S$ has both spatial location attributes denotes by $L(o)$ and a set of non-spatial preference attributes denoted by $P = \{p_1, p_2, ..., p_m\}$. We use $o. p_i$ ($1 \leq i \leq m$) to represent the o's value for the i-th non-spatial attribute and $d(L(o), L(q))$ to represent the distance between o and query point q where $d(.)$ denotes a distance metric obeying the triangle inequality. Let $t \prec u$ denote object t dominates object u with respect to non-spatial preference attributes. We use $SK(S)$ to denote the *skyline* over S, i.e., $SK(S) = \{o \in S | \nexists k \in S \text{ s.t. } k \prec o\}$. In the following, we formally define the LDSQ and related concepts.

Definition 1 ($\prec_q$). Given two spatial objects t, u and a query point q, we say $t \prec_q u$ with respect to q, if $t \prec u$ with respect to non-spatial preference attributes and $d(L(t), L(q)) < d(L(u), L(q))$, formally, $t \prec_q u \leftrightarrow t \prec u \wedge d(L(t), L(q)) < d(L(u), L(q))$.

Definition 2 (Location-dependent Skyline). Given a set of spatial objects S and a query point q, we call $LSK(S, q) = \{o \in S | \nexists k \in S \text{ s.t. } k \prec_q o\}$ the *location-dependent skyline* of q over S.

Definition 3 (LDSQ). An LDSQ is to find the location-independent skyline over a given set of spatial objects with respect to a given query point.

Definition 4 (Valid Scope). The valid scope of an LDSQ with respect to a query point q, denoted by $VS(S, q)$, is defined as a spatial region wherein all query points will receive an identical result, i.e., for any two query points q_1, q_2, if $L(q_1) \in VS(S, q)$ and $L(q_2) \in VS(S, q)$, then $LSK(S, q_1) = LSK(S, q_2) = LSK(S, q)$.

The introduction of valid scope provides the chance of sharing query results among neighboring mobile clients.

System Infrastructure. Our system infrastructure is a LBS system based on GSM or 3G and supports anytime, anywhere data access and processing using a portable size wireless mobile device. Mobile client in general communicates with a Mobile Base Station (MBS) to carry out any activities such as transaction and information query. MBS has a wireless interface to establish communication with mobile clients and can serve a large number of mobile clients. The power of a MBS defines its communication region, which we refer to as a *cell*. One or more MBSs are connected with a Base Station Controller (BSC), which coordinates the operations of MBSs when commanded by the Mobile Switching Center. The LBS server is connected to a BSC through a high-speed wired network and provides location based information services for mobile clients. Mobile clients equipping with GPSs can communicate with MBSs using wireless channels. A mobile client can freely move among cells and transparently accesses the spatial database residing at the LBS server. Benefiting from the increasing deployment of new P2P wireless communication technologies, there are short-range networks (MANETs) that allow ad hoc connections with neighboring mobile clients. This enables mobile clients to communicate with neighboring peers in an ad hoc manner for data sharing [13].

4   LDSQ Processing Approach

In this section, we propose the _sharing-based location-dependent skyline query approach_ (_SLSQ_) to improve the preceding δ-_Scanning_ algorithm present in [1]. Different form the existing LDSQ processing algorithms, in which mobile clients directly send the LDSQ requests and their current locations to the LBS server and the LBS server is responsible for answering each LDSQ, _SLSQ_ first tries to get the query result for each mobile client from the cached data by its neighboring mobile clients. Only when these cached data do not match the LDSQ request, are the corresponding LDSQ request and location information transmitted to the LBS server through the neighboring MBS. To leverage query results cached in the neighboring peers, _SLSQ_ needs each mobile client to cache its last LDSQ result and the corresponding valid scope.

Assume that R denotes the LDSQ request issued by a mobile client MC, $L(MC)$ denote the current location of MC, $peer_i$ for i=1, 2, ..., k denotes a single-hop mobile client of MC where k represents the number of its single-hop mobile clients, $LSK(S, peer_i)$ denotes the result of last LDSQ issued by $peer_i$, $LSK(S, MC)$ represents the result of last LDSQ issued by MC, $VS(S, peer_i)$ denotes the valid scope of the last LDSQ with respect to $peer_i$ and $VS(S, MC)$ denotes the valid scope of the last LDSQ with respect to MC. $LSK(S, peer_i)$ and $VS(S, peer_i)$ are cached in $peer_i$, and $LSK(S, MC)$ and $VS(S, MC)$ are cached in MC. Specifically, the processing strategy of _SLSQ_ is described as follows:

(1) MC checks whether $L(MC)$ falls within $VS(S, MC)$. If yes, MC returns $LSK(S, MC)$ and the algorithm ends; otherwise MC broadcasts the message <R, $L(MC)$> to all its single-hop $peer_i$ for i=1, 2, ..., k and waits the result from $peer_i$.

(2) For each $peer_i$ ($1 \leq i \leq k$), once receiving <R, $L(MC)$> from MC, it checks whether $L(MC)$ falls within $VS(S, peer_i)$. If yes, $peer_i$ sends the message "Yes" to MC; otherwise it sends the message "No" to MC.

(3) If MC receives "Yes" from a $peer_i$ within the specified deadline $T_{deadline}$, it sends the message "OK" to the $peer_i$ and waits the result from the $peer_i$; otherwise it forwards <R, $L(MC)$> to the LBS server and waits the result from the LBS Server.

Here, $T_{deadline}$ is calculated by the formula: $T_{deadline} = t_{issue} + 2 \times t_{delay} \times slack$, where t_{issue} is the time at which the message <R, $L(MC)$> is broadcast by MC, t_{delay} denotes the largest network delay between two neighboring mobile clients based on MANETs, and _slack_ represents the slack factor.

(4) Once receiving "OK" from MC, the $peer_i$ sends $LSK(S, peer_i)$ and $VS(S, peer_i)$ to MC; otherwise the LBS server calls the algorithm present in [1] to compute $LSK(S, MC)$ and $VS(S, MC)$ and sends them to MC, if it receives <R, $L(MC)$> from MC.

(5) MC caches the result received from the $peer_i$ or the LBS server and meanwhile returns $LSK(S, peer_i)$ (or $LSK(S, MC)$).

Based on the above processing strategy, we formalize _SLSQ_ in the following Algorithms 1-3. Algorithms 1-3 are respectively executed at mobile clients, those single-hop $peer_i$ and the LBS Server.

Algorithm 1. *SLSQ -MC (R, L(MC))*

Input: LDSQ request (*R*) and the location of *MC* (*L(MC)*)

Output: the result of the LDSQ

1: **If** *CacheData* ≠ ∅ ∧ *L(MC)* ∈ *CacheData.VS(S, MC)*

2: **return** *CacheData.LSK(S, MC)*

3: **else**

4: Broadcast the message *<R, L(MC)>* to all its single-hop mobile clients *peer$_i$*, set $T_{deadline}$ and wait response messages from them;

5: **if** receiving the message "Yes" from a *peer$_i$* within the specified deadline $T_{deadline}$ **then**

6: Send the message "OK" to the *peer$_i$* and wait the results from the *peer$_i$*;

7: Once receiving the result *<LSK(S, peer$_i$), VS(S, peer$_i$)>* from the *peer$_i$*, *CacheData.LSK(S, MC)* ←*LSK(S, peer$_i$)* and *CacheData.VS(S, MC)* ← *VS(S, peer$_i$)*;

8: **return** *CacheData. LSK(S, MC)*;

9: **else**

/ * all its single-hop peers send "No" or *MC* does not receive any "Yes" within $T_{deadline}$ */

10: Forward *<R, L(MC)>* to the LBS server through its MBS and wait the result from the LBS server;

11: Once receiving the result *<LSK(S, MC), VS(S, MC)>* from the LBS server, *CacheData.LSK(S, MC)*←*LSK(S, MC)* and *CacheData.VS(S, MC)* ←*VS(S, MC)*;

12: **return** *CacheData. LSK(S, MC)*;

In Algorithm 1 and the following Algorithm 2, *CacheData* denotes the cache capacity allocated by *MC* or *peer$_i$* for caching its last LDSQ result and the corresponding valid scope, *CacheData.LSK(S, MC)* represents the space of *CacheData* used to caching the last LDSQ result and *CacheData.VS(S, MC)* denotes the remaining space of *CacheData* used to caching the corresponding valid scope.

Algorithm 2. *SLSQ - peer$_i$ (R, L(MC))*

Input: LDSQ request (*R*) and the location of *MC* (*L(MC)*)

Output: *<LSK(S, peer$_i$), VS(S, peer$_i$)>* or null

1: **if** *CacheData* ≠ ∅ ∧ *L(MC)* ∈ *CacheData.VS(S, peer$_i$)*

2: Send "Yes" to *MC*, set $T_{overdue}$ and wait the message "OK" from *MC*;

3: **else**

4: Send "No" to *MC* and **return**;

5: **if** receiving the message "OK" from *MC* within $T_{overdue}$ **then**

6: Send *<LSK(S, peer$_i$), VS(S, peer$_i$)>* to *MC* and **return**;

7: **else**   /* time out */

8: **return**;

In Algorithm 2, *CacheData.VS(S, peer$_i$)* denotes the valid scope of the last LDSQ cached in *CacheData* and $T_{overdue}$ is calculated by the formula: $T_{overdue} = t_{current} + 2 \times t_{delay} \times slack$, where $t_{current}$ is the current system time of *peer$_i$*, t_{delay} denotes the largest network delay between two neighboring mobile clients based on MANETs, and *slack* represents the slack factor.

Algorithm 3. *SLSQ -LBS (S, L(MC))*

Input: the set of spatial objects (*S*) and the location of *MC* (*L(MC)*)
Output: *LSK(S, MC), VS(S, MC)*
1: *LSK(S, MC)* ← δ-*Scanning* (*S, L(MC)*);
2: *VS(S, MC)* ← δ-*Scanning VS* (*S, LSK(S, MC)*);
3: **return** < *LSK(S, MC), VS(S, MC)*>;

In Algorithm 3, functions δ-*Scanning*() and δ-*Scanning VS*() are responsible for respectively computing LDSQ and valid scope, whose detailed descriptions are present in [1].

5   Performance Evaluation

In this section, we first describe the experimental settings and then present the simulation results.

Experiment Settings. Our simulation experiments are based on the same data sets of spatial objects as [1], namely *School* and *NBA*. In the simulation, mobile clients have sufficient memory space to buffer their last query results. Like [1], we also assume the LBS server has sufficient memory space to load all the candidate objects. In order to reduce the randomness effect we average the results of the algorithms over 50 LDSQs. Table 1 lists the main simulation parameters.

Table 1. Main simulation parameters

Parameter	Description	Range / Default value
$S_{dataset}$	The size of spatial object set	48k or 17k
$|P|$	The number of non-spatial preference attributes	1~13 / 6
R	The P2P wireless transmission range of a mobile client based on MANETs	20~140(Meters) / 80
Pr	The probability of *MC* receives "Yes" from a *peer$_i$*	0.2~0.8 / 0.5
N_m	The number of single-hop peers of a *MC*	2~10 / 5

Our simulator consists of two main modules: the *MC* module and the BSC module. The *MC* module generates and controls the query requests of all *MC*s. Each *MC* is an independent object that decides its movement autonomously. The BSC module manages the spatial object set and meanwhile coordinates the operations of MBSs.

Simulation Results. First, we compare *SLSQ* with *brute-forth* and δ-*Scanning* in CPU time. Fig. 1 shows how CPU time of the three methods varies with the number of non-spatial preference attributes |P|, i.e., the performance of the three methods in terms of CPU time, as a function of |P|. As shown in Fig.1, *SLSQ* outperforms the other methods under two different data sets (*School* and *NBA*). This is because *SLSQ* enables a large number of LDSQs to be answered directly from the results cached in neighboring mobile clients and thus avoids a great majority of *dominance relationship tests* required by *brute-forth* and *δ-Scanning*. We can also see from Fig. 1 that *δ-Scanning* gets a distinct advantage over *brute-forth*. The reason is that *brute-forth* blindly checks all objects and thus incurs the significant overhead of *dominance relationship tests*. The another observation from Fig. 1 is the CPU cost under *School* and *NBA* data sets shows a little different trend. This is because *School* is an independent data set in non-spatial attributes while *NBA* follows ant-correlated distribution in non-spatial attributes, which causes the different overhead of *dominance relationship tests*.

Fig. 2 shows CPU time, as a function of the P2P wireless transmission range R. R reflects the range of P2P wireless transmission between neighboring mobile clients. The bigger R usually means a larger number of single-hop peers of a mobile client. As we expect, *SLSQ* achieves a better CPU performance than *brute-forth* and *δ-Scanning* and the performance gap becomes wider with the increase of R. This is not surprising because *SLSQ* avoids a great majority of *dominance relationship tests* by P2P sharing. Moreover, for *SLSQ*, the increase of R enables each mobile client to have a higher opportunity to fulfill its LDSQ by P2P sharing, but *brute-forth* and *δ-Scanning* are not influenced by R.

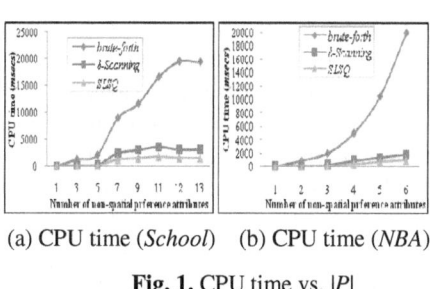
(a) CPU time (*School*)   (b) CPU time (*NBA*)

Fig. 1. CPU time vs. |P|

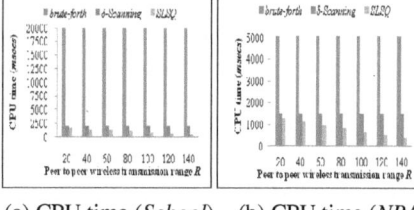

(a) CPU time (*School*)   (b) CPU time (*NBA*)

Fig. 2. CPU time vs. R

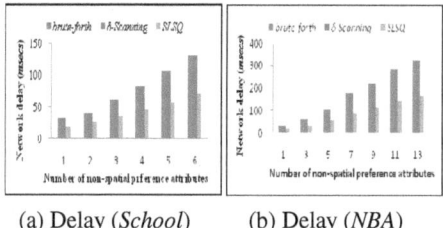
(a) Delay (*School*)   (b) Delay (*NBA*)

Fig. 3. Network delay vs. |P|

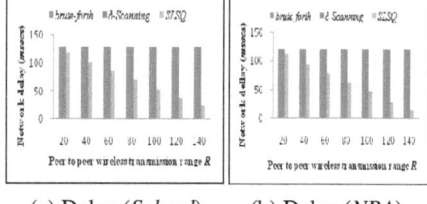
(a) Delay (*School*)   (b) Delay (*NBA*)

Fig. 4. Network delay vs. R

Fig. 3 shows the performance of the three methods in terms of network delay, as a function of |P|. As shown in Fig.3, *SLSQ* outperforms the other methods in network delay. This is because *SLSQ* tries to fetch the query result for each mobile client from the cached data by its neighboring peers and only when these cached data do not match the LDSQ request, does *SLSQ* ask for help from the remote LBS server. That is, in many cases *SLSQ* answers LDSQs by directly fetching data cached in the neighboring peers. This enables *SLSQ* to avoid the overhead of routing, message forwarding and handoff caused by connecting the remote LBS server. We can also see from

Fig. 3 that the network delay grows as attribute dimensionality increases. This is because the volume of query result, which needs to be transmitted, grows with the increase of attribute dimensionality.

Fig. 4 shows the performance of the three methods in terms of network delay, as a function of R. As we expect, *SLSQ* has a distinct advantage over *brute-forth* and *δ-Scanning* in network delay. The reason is that for many LDSQs, *SLSQ* avoids the overhead of routing, message forwarding and handoff caused by connecting the remote LBS server while the overhead is required by *brute-forth* and *δ-Scanning* for all LDSQs. We can also see from Fig. 4 that *brute-forth* and *δ-Scanning* are not influenced by R in network delay while the network delay of *SLSQ* decreases as R grows. The reason is that *brute-forth* and *δ-Scanning* directly ask the LBS server to compute LDSQs so do not involve the P2P wireless transmission between mobile clients, while *SLSQ* leverages the P2P sharing. Thus, the increase of R enables *SLSQ* to have a higher opportunity to fulfill its LDSQ by P2P sharing.

6   Conclusion

In this paper, we propose the *sharing-based location-dependent skyline query approach (SLSQ)*. *SLSQ* tries to get the query result for each mobile client from the cached data by its neighboring mobile clients. Only when these data cached do not match the LDSQ request, are the corresponding LDSQ request and location information transmitted to the LBS server. Extensive experiments are conducted, and the results demonstrate the effectiveness of our methods. A promising direction for future work is o consider on-air processing of LDSQ in road networks.

Acknowledgment. This work is supported by the Natural Science Foundation (NSF) of China under Grant No. 61170174 and the NSF of Tianjin under Grant No. 11JCYBJC26700.

References

1. Zhang, B., Lee, K.W.C., Lee, W.C.: Location-dependent skyline query. In: Proc. of MDM, pp. 148–155. IEEE Press, Beijing (2008)
2. Xiao, Y., Zhang, H., Wang, J.: Skyline computation for supporting location-based services in a road network. Information 15(5), 1937–1948 (2012)
3. Stephan, B., Donald, K., Konrad, S.: The skyline operator. In: Proc. of ICDE, pp. 421–430. IEEE Press, Heidelberg (2001)
4. Tan, K.L., Eng, P.K., Ooi, B.C.: Efficient progressive skyline computation. In: Proc. of VLDB, Roma, pp. 301–310 (2001)
5. Kossmann, D., Ramsak, F., et al.: Shooting stars in the sky: An online algorithm for skyline queries. In: Proc. of VLDB, pp. 275–286 (2002)
6. Papadias, D., Tao, Y., et al.: Progressive skyline computation in database systems. ACM Transactions on Database Systems 30(1), 41–82 (2005)
7. Lee, K., Zheng, B., Li, H., Lee, W.-C.: Approaching the skyline in Z order. In: Proc. of VLDB, pp. 279–290 (2007)

8. Chomicki, J., Godfrey, P., Gryz, J., et al.: Skyline with presorting. In: Proc. of ICDE, pp. 717–816 (2003)
9. Godfrey, P., Shipley, R., Gryz, J.: Maximal vector computation in large data sets. In: Proc. of VLDB, pp. 229–240 (2005)
10. Bartolini, I., Ciaccia, P., Patella, M.: Efficient sort-based skyline evaluation. ACM Transactions on Database Systems 33(4), 1–45 (2008)
11. Deng, K., Zhou, X., Shen, H.: Multi-source skyline query processing in road networks. In: Proc. of ICDE, Istanbul, Turkey, pp. 796–805 (2007)
12. Xiao, Y., Chen, Y.: Efficient distributed skyline queries for mobile applications. Journal of Computer Science and Technology 25(3), 523–536 (2010)
13. Ku, W.S., Zimmermann, R.: Location-Based Spatial Query Processing in Wireless Broadcast Environments. IEEE Transactions on Mobile Computing 7(6), 778–790 (2008)

What Can We Get from Learning Resource Comments on Engineering Pathway

Yunlu Zhang, Wei Yu*, and Shijun Li

School of Computer, Wuhan University, Wuhan, China
yuwei@whu.edu.cn

Abstract. K-Gray Engineering Pathway (EP) is a digital library website that could help users share their learning resources. Users can catalog, and comment news, blogs, videos, books and papers after login as registered users, and search without needing to login. We have already had the ability to search over comments for all resources, and the "most commented" resources are accessible on the K-12, higher education and disciplinary pages, but for now the ranking is only based on the comments number of each learning resources or the average rating.

To help users to evaluate the learning resources not only based on their comments' quantity, but also on their quality, we introduce the Analytic Hierarchy Process (AHP). For AHP couldn't completely suit to our situation, to make it work well, we established a three layers model with objective layer, criteria layer and alternative layer. Subjective expert and objective quantitative information are used to build the judgement matrices. The attributes of the comment,including total number, ratings, length of its content, poster, posting time, are used to defined the absolute importance, and further used to calculate the relative importance and judgement matrices. High dimensional calculation of weightings and consistence are avoided by adopting our estimation method. Experimental results and the comparisons with existing methods validate the effectiveness of our proposed methods.

Keywords: learning resources, Analytic Hierarchy Process, digital library, guide user.

1 Introduction

K-Gray Engineering Pathway (EP),website on http://www.engineeringpathway.com:8080/engpath/ep/Home, is an online digital library for multiple disciplines. The photographic images on our homepage have been selected to capture the excitement and diversity of engineering, computing education and practice. Different taglines are used for the different audiences within EP. The homepage and general tagline is to *Turn Ideas into RealityLearn, Connect and Create*. The K-12 pages communicate aspirations for the future *Shape the FutureDream, Design*

---

* Corresponding author.

Y. Ishikawa et al. (Eds.): APWeb 2013, LNCS 7808, pp. 482–493, 2013.

and Do. The broadening participation tagline motivates professional development and personal goals: *Make a World of DifferenceDream Big, Love What You Do.*

We have already done some data mining research based on our new EP in the last two years, such as [1,2], in this work we want to try another method to mine more useful information from our EP. For we can get search results by inputting query keywords in our EP homepage, as shown in Figure 2, we input "data mining" as search keyword, and we can get 26 learning resource objects. From the results page, we can get a summary of each item, such as: title, resource, discipline, and the source where they come from. If we click the link under the first learning resource title, we get more details about it and add comment about it.

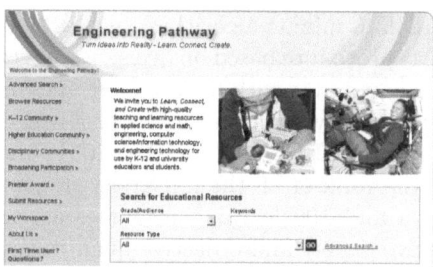 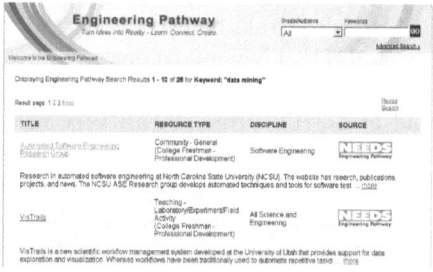

Fig. 1. Screen image of the Engineering Pathway

Fig. 2. Search results by keyword "data mining "

The "100 Most Commented" is what we used in this paper. The contents of comment should contain rating, comment title, comment description, author and post time. How to use these dataset to answer the questions like *"Which learning resource object should we choose"* or *"Which one is the best search result and which one is exactly what I want"* ? That is the reason why we focus on using these comments to evaluate the learning resources, and we can achieve this by selecting the best alternative that matches all of the digital library users criteria.

Learning recourses evaluation by their comments is such a multiple, diverse criteria decision-making problem with high complexity, so we draw lessons from Analytic Hierarchy Process (AHP), a classical method presented by Saaty[3] and be cited more than 1,000 times. We make some modifications on this method and will give more details on the following sections.

2   Related Work

AHP can be used in many applications besides our web data mining area, such as in ecosystem [4], emergency management [5], plant control [6] and so on so forth.

From those research works, it is clear that AHP has been widely applied in many of the data management and decision making areas. To address the problem encountered in our web data mining area, that is, digital learning resource mining for EP, we summarize some related former research works as follows:

Some researchers choose a combination of fuzzy theory and AHP, such as Tao et al. [8] , Uzoka et al. [9] and Feng et al. [10]. Tao proposed a decision model by the application of AFS (axiomatic fuzzy set) theory and AHP method to get the ranking order. They also provided the definitely semantic interpretations for the decision results by their theory. Uzoka used fuzzy logic and AHP for medical decision support systems. It is interpreting idea to introduce the semantic infor- mation proposed by Tao, thus we try to add them in our EP comments mining this time. However it causes high computational complexity, we will try this in our future work in this aspect.

AHP also could be used with traditional data mining algorithm, such as Svo- ray et al [11]. They used a data mining (DM) procedure based on decision trees to identify areas of gully initiation risk, and compared it with AHP expert system. Rad et al. [13] proposed an very inspiring idea by considering the problem of clustering and ranking university majors in Iran. By eight different criteria and 177 existing university majors, they give the rank of all the university majors. In our paper, we use five different criteria and 100 learning resources as ex- perimental dataset, however, both subjective expert and objective quantitative information are used to build the judgement matrices.

As for educational digital library applications, we focus on how to provide both teachers and students enriching and motivational educational resources as the former researchers do in [14] and [15] . Jing and Qingjun [16] provided teachers and students a virtual knowledge environment where students and teachers enjoy a high rate of participation. Alias et al. [17] described an implementation of digital libraries that integrated semantic research. Mobile ad hoc networks are becoming an important part of the digital library ecology . Social navigation can be used to replicate some social features of traditional libraries, to extend digital library portals and to guide portal users to valuable resources in [13].

3   Learning Recourses Automatic Evaluation

3.1   Learning Recourses Comments Hierarchy

There are three lays on our system, that is ,the objective layer, criteria layer, and alternatives layer. The framework of the comments hierarchy and the content of each layer is as shown in Figure 7.

- Objective layer: Select the most valuable learning resource by comments;
- Criteria layer: rating, comment title, comment description/comment content, comment author type , posting time;
- Alternatives layer: the top 100 most commented resources in one certain discipline.

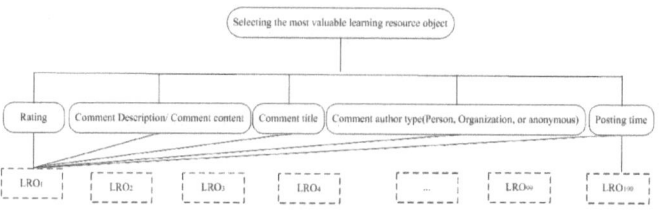

Fig. 3. comments hierarchy

3.2   Judgement Matrices

There are one judgement matrix in criteria layer and five judgement matrices in alternatives layer to be determined. We will establish them based on the expert information and quantitative information respectively.

Judgement Matrices in Criteria Layer. Let matrix $M \in R^{5 \times 5}$ denote the judgement matrix in criteria layer and C_i, $i = 1, \ldots, 5$, denote the five criteria (factors) in the criteria layer respectively. In matrix M, its element $M_{i,j}$ means the quantitative relative importance judgement of pairwise factors C_i and C_j, that is, $M_{j,i} = 1/M_{i,j}$. The standard relative importance scale is employed, as shown in Table.1.

Table 1. The relative importance of the element C_i to C_j $(M_{i,j})$

The scale	The meaning
1	C_i and C_j have the same importance
3	C_i is slightly more important than C_j
5	C_i is obviously more important than C_j
7	C_i is strongly more important than C_j
9	C_i is extremely more important than C_j
$2, 4, 6, 8$	Compromise between the above judge

Judgement Matrices in Alternative Layer. Let matrices $M^i \in R^{100 \times 100}$, $i = 1, \ldots, 5$, denote the judgement matrix for criteria C_i in the alternative layer and A_i, $i = 1, \ldots, 100$, denote the top 100 most commented resources respectively. To establish each matrix M^i, we use the following two steps:

- Calculate the Absolute Importance (AI^i)
  Based on the related information, such as the rating value and number of comments for rating C^1, calculate the absolute importance value for each alternative under criterion C_i;

– Scale and calculate the Relative Importance (RI^i)
  Scale the absolute importance value to 1 to 9, and calculate the relative importance to get the matrix M^i;

Specifically, we present the following methods to determine the judgement matrices in the alternative layer

1. Rating matrix M^1
   For the criterion of rating, the average rating value, varying from 1 to 5, and the number of comments are used to determine the absolute importance of the rating for each resource $j = 1, \ldots, 100$ as follows:

$$AI_j^1 = \frac{1}{N_j} \sum_{n=1}^{N_j} r_{j,n} + \alpha^1 N_j$$

where N_j is the number of comments for resource j, $r_{j,n}$ is the rating value given by comment n and α^1 is a properly selected weighting for N_j. Suppose AI_{max}^1 and AI_{min}^1 are the maximum and minimum absolute importance of the top 100 most commented resources respectively, thus the relative importance value of resource j is given by the following linear scaling method:

$$RI_j^i = \frac{8}{AI_{max}^i - AI_{min}^i} (AI_j^i - AI_{min}^i) + 1$$

We assume that $AI_{max}^i > AI_{min}^i$. The element in matrix M^i can be given by $M_{k,j}^i = \frac{RI_k^i}{RI_j^i}$. Note that we will use the same linear scaling method and definition of RI_j^i for the following matrices, $i = 2, \ldots, 5$, and omit them.

2. Comment title matrix M^2
   For the criterion of comment title, we assume the longer the comment title is, the more important it is, and also the number of comments are key factor for the importance of resource. Thus the absolute importance of the comment title for resource $j = 1, \ldots, 100$ is defined by

$$AI_j^2 = \frac{1}{N_j} \sum_{n=1}^{N_j} LT_{j,n} + \alpha^2 N_j$$

where N_j is the number of comments for resource j, $LT_{j,n}$ is the length of the comment title given by comment n and α^2 is a properly selected weighting for N_j. The relative importance RI^2 and comment title matrix M^2 can be obtained by the same method in rating matrix.

3. Comment description matrix M^3
   Similar to the previous analysis, we use the length of the comment and the number of comments to define the absolute importance of the comment description for each resource $j = 1, \ldots, 100$ as follows:

$$AI_j^3 = \frac{1}{N_j} \sum_{n=1}^{N_j} LC_{j,n} + \alpha^3 N_j$$

where N_j is the number of comments for resource j, $LC_{j,n}$ is the length of the comment given by comment n and α^3 is a properly selected weighting for N_j.

4. Author type matrix M^4

   Since there are three types of authors, that is, anonymous , person and organization. Here we assign different author type value to each of them: $V_{anonymous} = 1$, $V_{person} = 2$ and $V_{organization} = 3$. Thus the absolute importance of the comment author type for resource $j = 1, \ldots, 100$ is determined as follows:

$$AI_j^4 = \frac{1}{N_j} \sum_{n=1}^{N_j} V_{j,n} + \alpha^4 N_j$$

where N_j is the number of comments for resource j, $V_{j,n}$ is author type value of comment n and α^4 is a properly selected weighting for N_j.

5. Posting time matrix M^5

   We assume the more greater the average posting time is, the more important the resource is. Thus the absolute importance of the posting time for each resource $j = 1, \ldots, 100$ is defined as follows:

$$AI_j^5 = \frac{1}{N_j} \sum_{n=1}^{N_j} (T_{j,n} - T_c) + \alpha^5 N_j$$

where N_j is the number of comments for resource j, $T_{j,n}$ is posting time (day) of comment n, T_c is the current day and α^5 is a properly selected weighting for N_j.

3.3 Weighting

After getting the judgement matrix M, we obtain the weightings of the five criteria in the second layer by calculating its normalized eigenvector w_{max} corresponding to the maximum eigenvalue λ_{max}, that is,

$$M w_{max} - \lambda_{max} w_{max}, \quad \sum_{i=1}^{5} w_{max,i} = 1.$$

Similarly, we get weightings of different alternative under criteria C_i, $i = 1, \ldots, 5$, as the normalized eigenvectors w_{max}^i of M^i as follows:

$$M^i w_{max}^i = \lambda_{max}^i w_{max}^i, \quad \sum_{j=1}^{100} w_{max,j}^i = 1.$$

Note that for M^i, $i = 1, \ldots, 5$, they are 100×100 dimension matrices, thus it is time-consuming job to exactly calculate the eigenvalues and eigenvectors. To obtain a high quality estimation in short time, we use the following approach :

1. Normalize each column in M, that is, $M'_{i,j} = \frac{M_{i,j}}{\sum_{i=1}^{n} M_{i,j}}$;
2. Sum $M'_{i,j}$ in each row, that is, $M'_i = \sum_{j=1}^{n} M'_{i,j}$;
3. Estimate the desired eigenvector w by normalize M^i, that is, $w = \{w_j\}_{j=1,\ldots,n}$, where $w_j \approx \frac{M'_j}{\sum_{i=1}^{n} M'_i}$;
4. The maximum eigenvalue λ can be estimated by $\frac{1}{n} \sum_{i=1}^{n} \frac{(Mw)_i}{w_i}$.

3.4   Consistency

A well-defined judgement matrix should be transitive (that is, if A is better than B and B is better than C, then we have A is also better than C) and numerically consistent (that is, if $A = 2B$ and $B = 2C$, then we have $A = 4C$). To verify the consistence of the judgement matrix $M \in R^{n \times n}$, Saaty [3] suggests to calculating the following Consistency Ratio (CR):

$$CR_M = \frac{CI_M}{RI_n}$$

where CI_M is so-called Consistence Index of matrix M, defined as, $CI_M = \frac{\lambda_{max} - n}{n - 1}$ and RI_n is the so-called Random Index for $n \times n$ matrix

Saaty Rule says that **if $CR_M < 0.1$, then matrix M is consistent; otherwise, we need to adjust M until the condition satisfies.**

However, for matrices $M^i, i = 1, \ldots, 5$, we assert that they are always consistent because

$$M^i_{k,j} M^i_{j,l} = \frac{RI^i_k}{RI^i_j} \frac{RI^i_j}{RI^i_l} = M^i_{k,l}$$

thus $\lambda^i_{max} = n$ and $CI_{M^i} = 0$. Furthermore, the total consistency is also holds because

$$\frac{w_{max,1} CI_{M^1} + w_{max,2} CI_{M^2} + \cdots + w_{max,5} CI_{M^5}}{w_{max,1} RI_{100} + w_{max,2} RI_{100} + \cdots + w_{max,5} RI_{100}} = 0 < 0.1.$$

3.5   Implementation of AHP

In this subsection, we summarize the implementation of AHP for our problem.

1. Criteria Evaluation
   Get judgment matrix M based on expert information, calculate its maximum eigenvalue and corresponding eigenvector w_{max}, and check its consistence by Saaty Rule (adjust it if necessary);

2. Alternative Weightings Calculation:
   For each criteria C_i, $i = 1, \ldots, 5$, calculate its judgement matrix based on method in subsection 3.2, and further calculate its maximum eigenvalue w_{max}^i and corresponding eigenvector based on method in subsection 3.3;

3. Alternative Evaluation:
   The final evaluation v_j for alternative j is given as

$$v_j = \sum_{i=1}^{5} w_{max,i} w_{max,j}^i.$$

Then the top 10 learning resources will be returned to users.

4   Experiment

In this section, we will evaluate the proposed AHP method on the K-Gray Engineering-Pathway library data. The proposed method will be experimentally compared with other general methods, such average rating, average comment number, to validate its effectiveness.

4.1   Experimental Set Up

The top 100 commented learning resources on computer science from EP homepage are selected as the the test data. We collect the comment information, that is, comment title, comment description, rating, author and posting time for each learning resource.

To implement the proposed AHP method, first the following judgement matrix M in criteria layer is obtained based on the expert information from the questionnaire.

$$M = \begin{bmatrix} 1 & 4 & 1 & 5 & 6 \\ \frac{1}{4} & 1 & \frac{1}{4} & 1 & \frac{4}{5} \\ 1 & 4 & 1 & 6 & 5 \\ \frac{1}{5} & 1 & \frac{1}{6} & 1 & \frac{1}{2} \\ \frac{1}{6} & \frac{5}{4} & \frac{1}{5} & 2 & 1 \end{bmatrix}$$

The maximum eigenvalue λ_{max} of M is 5.0859, the $CI = \frac{\lambda_{max}-5}{5-1} = 0.0215$ and $CR = \frac{CI}{RI} = 0.0192 < 0.1$, thus it is consistent. The corresponding normalized eigenvector is $w_{max} = (0.381, 0.081, 0.379, 0.067, 0.092)$, which is also the weightings of the five criteria.

To properly set parameters α^i, $i = 1, \ldots, 5$, we calculate the average number of comments, rating, comment title length, comment length, author value and posting time value $(4.31, 3.51, 38.3, 481.4, 2.51, 0.24)$. Note that α^i can be viewed as a scale factor for the number of comments, so we set α^i according to the above average value, as shown in Table. 2. Then we can obtain the eigenvectors of M^i, $i = 1, \ldots, 5$, by the estimation method in subsection 3.3, and further get the final evaluation of the top 100 commented resources.

Table 2. Parameters setting for AHP

i	1	2	3	4	5
α^i	0.8	8	10	0.5	0.05

4.2   Comparison and Discussion

The proposed AHP method is compared with the general average methods, such as average rating, average number of comments methods.

We normalized the evaluation of the 100 resources by average rating (AR), average comments length (ACL), average comment title length(ACTL), average number of comments (ANC), average author value (AAV) and average posting time value (APTV). Figure 4 - Figure 9 show the first 50 resource evaluation value by AHP and the above average methods. From Figure 4-Figure 9, we can see that the evaluation results obtained by AHP and ACL own the highest similarity. Furthermore, AHP can be viewed as a mixed method of these average method.

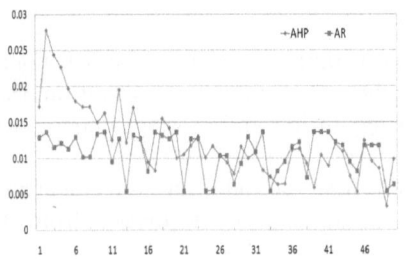

Fig. 4. Comparison with AR   **Fig. 5.** Comparison with ACL

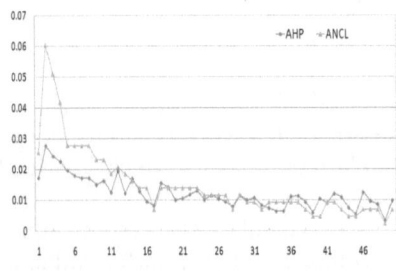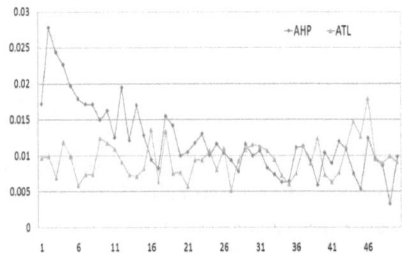

Fig. 6. Comparison with ANCL   **Fig. 7.** Comparison with ATL

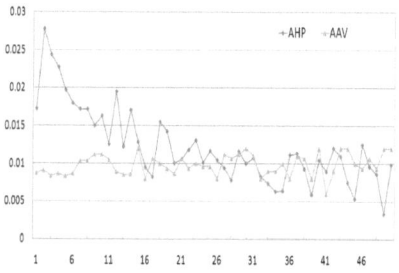 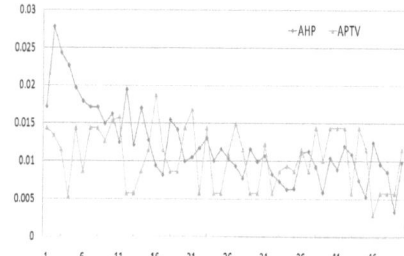

Fig. 8. Comparison with AAV   **Fig. 9.** Comparison with APTV

To make it clear, we define the standard deviation between AHP and method $*$ as

$$SD_* = \left(\sum_{j=1}^{100} (v_j^{AHP} - v_j^*) \right)^2$$

where v_j^{AHP}, v_j^* are the normalized evaluation value for resource j obtained by AHP and method $*$. Table. 3 reports the standard deviation for different methods. ACL and AR have the smaller SD values compared with other methods. It is consistent with the expert information and our proposed AHP method from the judgement matrix M, we see that criteria rating and comment length have the bigger weightings. Finally, we report the top 10 recommened resources by AHP, ACL and AR based on the 100 most commented resources on computing sciences.

Table 3. The squared deviation of different methods with AHP

	ACL	AR	AAV	ATL	APTV	ANCL
SD	0.034	0.041	0.050	0.057	0.059	0.060

Table 4. The top 10 recommended resources by different methods

	AHP	ACL	AR
1	Alice 2.0	The Black Swan	Alice 2.0
2	iWoz	ACM K-12	Web-CAT
3	First Barbie	Pair Programming	Jeliot 3
4	Pair Programming	Educating Engineers	Award Courseware
5	The Black Swan	Computer Museum	RVLS
6	Computing	ThinkCycle	Greenfoot
7	Computational Geometry	Initiatives	World Without Oil
8	Language	Media	The Enigma
9	Achieving Dreams	Scratch	Girl Geeks
10	ACM K-12	Children Website	Award Courseware

5 Conclusion and Future Work

The AHP opens up a wide variety of novel applications. So after introducing the problem of automatic learning resources evaluation, we make an attempt to solve it by using this method for reference. However the traditional method is not satisfactory for our situation, the primary ones are: the ranking criteria contains multiple variables, each criteria in each Hierarchy may has their own sub-criteria, and the vector matrix for selecting items is huge. So we made some improvement on AHP to solve our problem. By establishing three layers model with objective layer, criteria layer and alternative layer, we can provide our users a more comprehensive ranking of our learning resources. By using subjective expert and objective quantitative information, we can build the judgement matrices, then we calculate the absolute importance, relative importance and finally the judgement matrices with the quantified attributes of comments, such as the number of comments, rating, the length of content, the author type and the posting time.

Acknowledgments. This work is supported by the National Natural Science Foundation of China (NSFC-61272109). Thanks to Alice Merner Agogino.

References

1. Zhang, Y., Agogino, A.M., Li, S.: Lessons Learned from Developing and Evaluating a Comprehensive Digital Library for Engineering Education. In: JCDL 2012 Proceedings of the 12th ACM/IEEE-CS Joint Conference on Digital Libraries, pp. 393–394 (2012)
2. Zhang, Y., Zhou, G., Zhang, J., Xie, M., Yu, W., Li, S.: Engineering pathway for user personal knowledge recommendation. In: Gao, H., Lim, L., Wang, W., Li, C., Chen, L. (eds.) WAIM 2012. LNCS, vol. 7418, pp. 459–470. Springer, Heidelberg (2012)
3. Saaty, T.L.: Fundamentals of Decision Making and Priority Theory with the AHP, 2nd edn. RWS Publications, Pittsburgh (2000)
4. Koschke, L., Fuerst, C., Frank, S.: A multi-criteria approach for an integrated land-cover-based assessment of ecosystem services provision to support landscape planning. Ecological Indicators 21, 54–66 (2012)
5. Ergu, D., Kou, G., Peng, Y.: Data Consistency in Emergency Management. International Journal of Computers Communications & Control 7(3), 450–458 (2012)
6. Forsyth, G.G., Le Maitre, D.C., O'Farrell, P.J.: The prioritization of invasive alien plant control projects using a multi-criteria decision model informed by stakeholder input and spatial data. Journal of Environmental Management 103, 51–57 (2012)
7. Tao, L., Chen, Y., Liu, X.: An integrated multiple criteria decision making model applying axiomatic fuzzy set theory
8. Uzoka, F.-M.E., Obot, O., Barker, K.: An experimental comparison of fuzzy logic and analytic hierarchy process for medical decision support systems. Computer Methods and Programs in Biomedicine 103(1), 10–27 (2011)
9. Feng, Z., Wang, Q.: Research on health evaluation system of liquid-propellant rocket engine ground-testing bed based on fuzzy theory. Acta Astronautica 61(10), 840–853 (2007)

10. Svoray, T., Michailov, E., Cohen, A.: Predicting gully initiation: comparing data mining techniques, analytical hierarchy processes and the topographic threshold. Earth Surface Processes and Landforms 37(6), 607–619 (2012)

11. Amin, G.R., Gattoufi, S., Seraji, E.R.: A maximum discrimination DEA method for ranking association rules in data mining. International Journal of Computer Mathematics 88(11), 2233–2245 (2011)

12. Rad, A., Naderi, B., Soltani, M.: Clustering and ranking university majors using data mining and AHP algorithms: A case study in Iran. Expert Systems with Applications 38(11), 755–763 (2011)

13. Brusilovsky, P., Cassel, L., Delcambre, L., Fox, E., Furuta, R., Garcia, D.D., Shipman III, F.M., Bogen, P., Yudelson, M.: Enhancing digital libraries with social navigation: The case of ensemble. In: Lalmas, M., Jose, J., Rauber, A., Sebastiani, F., Frommholz, I. (eds.) ECDL 2010. LNCS, vol. 6273, pp. 116–123. Springer, Heidelberg (2010)

14. Fernandez-Villavicencio, N.G.: Helping students become literate in a digital, networking-based society: a literature review and discussion. International Information and Library Review 42(2), 124–136 (2010)

15. Hsu, K.-K., Tsai, D.-R.: Mobile Ad Hoc Network Applications In The Library. In: Proceedings of the 2010 Sixth International Conference on Intelligent Information Hiding and Multimedia Signal Processing (IIHMSP 2010), pp. 700–703 (2010)

16. Jing, H., Qingjun, G.: Prospect Application in the Library of SNS. In: 2011 Third Pacific-Asia Conference on Circuits, Communications and System (PACCS), Wuhan, China, July 17-18 (2011)

17. Alias, N.A.R., Noah, S.A., Abdullah, Z., et al.: Application of semantic technology in digital library. In: Proceedings of 2010 International Symposium on Information Technology (ITSim 2010), pp. 1514–1518 (2010)

Tuned X-HYBRIDJOIN
for Near-Real-Time Data Warehousing

M. Asif Naeem

School of Computing and Mathematical Sciences, Auckland University of Technology,
Private Bag 92006, Auckland 1142, New Zealand
mnaeem@aut.ac.nz

Abstract. Near-real-time data warehousing defines how updates from
data sources are combined and transformed for storage in a data warehouse as soon as the updates occur. Since these updates are not in
warehouse format, they need to be transformed and a join operator is
usually required to implement this transformation. A stream-based algorithm called X-HYBRIDJOIN (Extended Hybrid Join), with a favor-
able asymptotic runtime behavior, was previously proposed. However,
X-HYBRIDJOIN does not tune its components under limited available
memory resources and without assigning an optimal division of memory
to each join component the performance of the algorithm can be suboptimal. This paper presents a variant of X-HYBRIDJOIN called Tuned X-
HYBRIDJOIN. The paper shows that after proper tuning the algorithm
performs significantly better than that of the previous X-HYBRIDJOIN,
and also better as other join operators proposed for this application found
in the literature. The tuning approach has been presented, based on measurement techniques and a revised cost model. The experimental results
demonstrate the superior performance of Tuned X-HYBRIDJOIN.

Keywords: Data warehousing, Tuning and performance optimization,
Data transformation, Stream-based joins.

1   Introduction

Near real-time data warehousing exploits the concepts of data freshness in traditional static data repositories in order to meet the required decision support
capabilities. The tools and techniques for promoting these concepts are rapidly
evolving. Most data warehouses have already switched from a full refresh [7] to
an incremental refresh policy [4]. Further the batch-oriented, incremental refresh
approach is moving towards a continuous, incremental refresh approach.

One important research area in the field of data warehousing is data transformation, since the updates coming from the data sources are not in the format
required for the data warehouse. Furthermore, a join operator is required to
implement the data transformation.

In traditional data warehousing the update tuples are buffered in memory and
joined when resources become available [6]. Whereas, in real-time data warehousing these update tuples are joined when they are generated in the data sources.

Y. Ishikawa et al. (Eds.): APWeb 2013, LNCS 7808, pp. 494–505, 2013.

One important factor related to the join is that both inputs of the join come from different sources with different arrival rates. The input from the data sources is in the form of an update stream which is fast, while the access rate of the lookup table is comparatively slow due to disk I/O cost. It creates a bottleneck in processing of update stream and the research challenge here is to minimize this bottleneck by optimizing the performance of the join operator.

To overcome these challenges a novel stream-based join algorithm called X-HYBRIDJOIN (Extended Hybrid Join) [1] was proposed recently by the author. This algorithm not only addressed the issues described above but also was designed to take into account typical market characteristics, commonly known as the 80/20 sale Rule [2]. According to this rule 80 percent of sale focus on only 20 percent of the products, i.e., one observes a Zipfian distribution. To achieve this objective, one component of the algorithm called disk-buffer is divided into two equal parts. The contents of one part of the disk buffer (called the non-swappable part) are kept fixed while the contents of the other part (called the swappable part) are exchanged for each iteration of the algorithm. As the non-swappable part of the disk buffer always contains the most frequently used disk pages, most stream tuples can be processed without invoking the disk. Although the author presented an adaptive algorithm to adopt the typical market characteristics, the components of the algorithm are not tuned to make efficient use of available memory resources. Further details about this issue are provided in Section 3.

On the basis of these observations, a revised X-HYBRIDJOIN is proposed with name Tuned X-HYBRIDJOIN. The cost model of existing X-HYBRIDJOIN is revised and the components of the proposed algorithm are tuned based on that cost model. As a result the available memory is distributed among all components optimally and consequently it improves the performance of the algorithm significantly.

The rest of the paper is structured as follows. The related work to proposed algorithm is presented in Section 2. Section 3 describes problem statement about the current approach. The proposed solution for the stated problem is presented in Section 4. Section 5 presents the tuning of the proposed algorithm based on the revised cost model. The experimental study is discussed in Section 6 and finally Section 7 concludes the paper.

2   Related Work

Some techniques have already been introduced to process the join queries over continuous streaming data [5]. This section presents only those approaches which are directly related to the stated problem domain.

A stream-based algorithm called Mesh Join (MESHJOIN) [8] was designed specifically for joining a continuous stream with disk-based data in an active data warehouse. This is an adaptive approach but there are some research issues related to inefficient memory distribution among join components due to unnecessary constraints and an inefficient strategy for accessing the disk based relation.

R-MESHJOIN (reduced Mesh Join) [9] is a revised version of MESHJOIN that focuses on the optimal distribution of memory among the join components. The R-MESHJOIN algorithm introduces the new strategy for memory distribution among the join components by implementing real constraints. However, the mechanism used for accessing the disk based relation is similar to MESHJOIN.

One approach to improve MESHJOIN has been a partition-based join algorithm [10]. It uses a two-level hash table in order to attempt to join stream tuples as soon as they arrive, and uses a partition-bases waiting area for other stream tuples. In this approach the author keeps focus about the analysis of the stream buffer in terms of back log tuples rather than analysing the performance of the algorithm.

A recent piece of work introduces a novel stream-based join called HYBRID-JOIN (Hybrid Join) [11]. The key objective of this effort is to minimize the disk overhead using an index-based approach and to deal with intermittency in the stream. Although both issues are successfully addressed in this approach, it is not optimal with respect to stream data with Zipfian distribution.

3   Preliminaries and Problem Definition

This section presents a working overview of X-HYBRIDJOIN along with the research issue. In the field of real-time data warehousing X-HYBRIDJOIN is an adaptive algorithm for joining the bursty data stream with disk-based master data. Although the typical characteristics of market data are considered, optimal settings for the available limited memory resources are not considered. Before describing the problem it is first necessary to explain the major components of X-HYBRIDJOIN and the role of each component. Figure 1 presents an abstract level working overview of X-HYBRIDJOIN where m is the number of partitions in the queue to store stream tuples and n is the number of pages in disk-based master data R. Moreover, R is assumed to be sorted with respect to the access frequency. The stream tuples are stored in the hash table while the join attribute values are stored in the queue. The queue is implemented using a doubly linked-list data structure to allow the random deletion of matching tuples. The disk buffer is used to load the disk pages into memory. To make efficient use of R by minimizing the disk access cost, the disk buffer is divided into two equal parts. One is called the non-swappable part which stores a small but most frequently used portion of R into memory on a permanent basis. The other part of the disk buffer is swappable and for each iteration it loads the disk page p_i from R into memory.

Before the join process starts X-HYBRIDJOIN loads the most frequently used page of R into the non-swappable part of the disk buffer. During the join process, for each iteration the algorithm dequeues the oldest join attribute value from the queue and using this value as an index it loads the relevant disk page into the swappable part of the disk buffer. After loading the disk page into memory the algorithm matches each of the disk tuples available in both the swappable and non-swappable parts of the disk buffer with the stream tuples in the hash

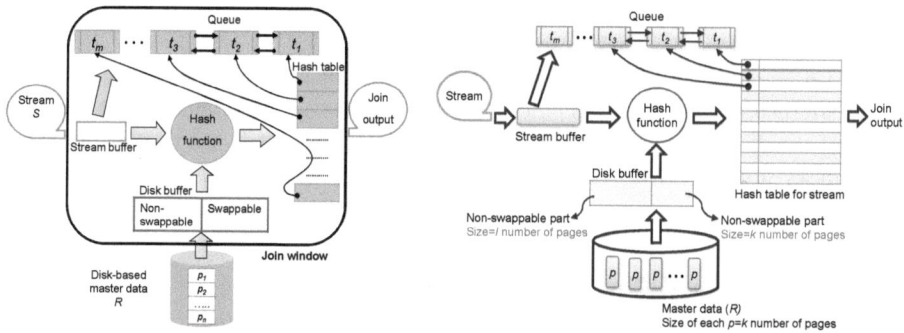

Fig. 1. X-HYBRIDJOIN working overview

Fig. 2. Memory architecture for Tuned X-HYBRIDJOIN

table. If the tuple match, the algorithm generates the resulting tuple as output and deletes the stream tuple from the hash table along with its join attribute value from the queue. In the next iteration the algorithm again dequeues the oldest element from the queue, loads the relevant disk page into the swappable part of the disk buffer and repeats the procedure.

X-HYBRIDJOIN minimizes the disk access cost and improves performance significantly by introducing the non-swappable part of the disk buffer. But in X-HYBRIDJOIN the memory assigned to the swappable part of the disk buffer is equal to the size of disk buffer in HYBRIDJOIN [11] and the same amount of memory is allocated to the non-swappable part of the disk buffer. In the following it will be shown that this is not optimal. The problem considered in this paper is to tune the size of both parts of the disk buffer so that the memory distribution among these two components is optimal. Once these two components acquire optimal settings, based on that memory can be assigned to the rest of join components.

4   Proposed Solution

As a solution for the above stated problem, a revised version of X-HYBRIDJOIN called Tuned X-HYBRIDJOIN is proposed. This section presents the memory architecture and a revised cost model for the proposed algorithm. Most importantly, the tuning for the proposed algorithm is presented while the tuning procedure is based on both a measurement strategy and the cost model.

4.1   Memory Architecture

The memory architecture that Tuned X-HYBRIDJOIN uses is shown in Figure 2. From the figure, Tuned X-HYBRIDJOIN includes the same number of components as X-HYBRIDJOIN however, the memory size for each component is different to that in X-HYBRIDJOIN. In Tuned X-HYBRIDJOIN since memory is allocated to the each component after executing the tuning module and therefore, each component is assigned an optimal size of memory. Particulary, X-HYBRIDJOIN uses same memory for the both swappable and non-swappable parts of the disk buffer but after tuning it has been explored that the optimal size of memory for the both components is different. The reason for it is presented in Section 5.

4.2   Cost Calculation

This section revises the cost formulas derived in X-HYBRIDJOIN. The reason for revising the cost model is that X-HYBRIDJOIN uses equal memory for both the swappable and non-swappable parts of the disk buffer, and therefore the formulas do not apply for other relative sizes. Following the style of cost modeling used for MESHJOIN, the cost for any algorithm is expressed in terms of memory and processing time. Equation 1 describes the total memory used to implement the algorithm while Equation 2 calculates the processing cost for w tuples.

Memory Cost. Since the optimal values for the sizes of both the swappable part and non-swappable part can be different, it is assumed k number of pages for the swappable part and l number of pages for the non-swappable part. Overall the largest portion of the total memory is used for the hash table while a much smaller amount is used for each of the disk buffer and the queue. The memory for each component can be calculated as given below:

Memory for the swappable part of disk buffer (bytes)= $k \cdot v_P$ (where v_P is the size of each disk page in bytes).
Memory for the non-swappable part of disk buffer (bytes)= $l \cdot v_P$.
Memory for the hash table (bytes)= $\alpha[M - (k + l)v_P]$ (where M is the total allocated memory and α is memory weight for the hash table).
Memory for the queue (bytes)= $(1 - \alpha)[M - (k+l)v_P]$ (where $(1 - \alpha)$ is memory weight for the queue).
The total memory used by the algorithm can be determined by aggregating the above.

$$M = (k + l)v_P + \alpha[M - (k + l)v_P] + (1 - \alpha)[M - (k + l)v_P] \qquad (1)$$

Currently the memory reserved for the stream buffer is not included because it is small (0.05 MB was sufficient in all experiments presented in this paper).

Processing Cost. This section presents the processing cost for the proposed approach. The cost for one iteration of the algorithm is denoted by c_{loop} and express it as the sum of the costs for the individual operations. Therefore the processing cost for each component is first calculated separately.

Cost to read the non-swappable part of disk buffer (nanoseconds) $= c_{I/O}(l \cdot v_P)$.
Cost to read the swappable part of disk buffer (nanoseconds)= $c_{I/O}(k \cdot v_P)$.

Cost to look-up the non-swappable part of disk buffer in the hash table (nanoseconds) $= d_N c_H$ (where $d_N = l\frac{v_P}{v_R}$ is the size of the non-swappable part of disk buffer in terms of tuples, v_P is size of disk page in bytes, v_R is size of disk tuple in bytes, and c_H is look-up cost for one disk tuple in the hash table).

Cost to look-up the swappable part of disk buffer in the hash table (nanoseconds)$=$ $d_S c_H$ (where $d_S = k\frac{v_P}{v_R}$ is the size of the swappable part of disk buffer in terms of tuples).

Cost to generate the output for w matching tuples (nanoseconds) $= w \cdot c_O$ (where c_O is cost to generate one tuple as an output).

Cost to delete w tuples from the hash table and the queue (nanoseconds)$= w \cdot c_E$ (where c_E is cost to remove one tuple from the hash table and the queue).

Cost to read w tuples from stream S into the stream buffer (nanoseconds)$= w \cdot c_S$ (where c_S is cost to read one stream tuple into the stream buffer).

Cost to append w tuples into the hash table and the queue (nanoseconds)$= w \cdot c_A$ (where c_A is cost to append one stream tuple in the hash table and the queue). As the non-swappable part of the disk buffer is read only once before execution starts, it is excluded. The total cost for one loop iteration is:

$$c_{loop}(\text{secs}) = 10^{-9}[c_{I/O}(k \cdot v_P) + (d_N + d_S)c_H + w(c_O + c_E + c_S + c_A)] \qquad (2)$$

Since in every c_{loop} seconds the algorithm processes w tuples of stream S, the performance or service rate μ can be calculated by dividing w by the cost for one loop iteration.

$$\mu = \frac{w}{c_{loop}} \qquad (3)$$

5   Tuning

The stream-based join operators normally execute within limited memory and therefore tuning of join components is important to make efficient use of the available memory. For each component in isolation, more memory would be better but assuming a fixed memory allocation there is a trade-off in the distribution of memory. Assigning more memory to one component means less memory for other components. Therefore it needs to find the optimal distribution of memory among all components in order to attain maximum performance. A very important component is the disk buffer because reading data from disk to memory is expensive.

In the proposed approach tuning is first performed through performance measurements by considering a series of values for the sizes of the swappable and non-swappable parts of the disk buffer. Later a mathematical model for tuning is also derived from the cost model. Finally, the tuning results of both approaches are compared to validate the cost model. The details about the experimental setup are presented in Table 1.

5.1   Tuning through Measurements

This section presents the tuning of the key components of the algorithm through measurements. In the measurement approach the performance is tested on

Table 1. Data specification

Parameter	value
Disk-based data	
Size of R	0.5 *million* to 8 *million tuples*
Size of each tuple v_R	120 *bytes*
Stream data	
Size of each tuple v_S	20 *bytes*
Size of each node in queue	12 *bytes*
Benchmark	
Based on	Zipf's law
Characteristics	Bursty and self-similar

particular memory settings for swappable and non-swappable parts rather than on every contiguous value.

The measurement approach assumes the size of total memory and the size of R are fixed. The sizes for the swappable and non-swappable parts vary in such a way that for each size of the swappable part the performance is measured against a range of sizes for the non-swappable part. By changing the sizes for both parts of the disk buffer the memory sizes for the hash table and the queue are also affected.

The performance measurements for varying the sizes of both swappable and non-swappable parts are shown in Figure 3. The figure shows that the performance increases rapidly by increasing the size for the non-swappable part. After reaching a particular value for the size of non-swappable part the performance starts decreasing. The plausible reason behind this behavior is that in the beginning when the size for the non-swappable part increases, the probability of matching stream tuples with disk tuples also increases and that improves the performance. But when the size for the non-swappable part is increased further it does not make a significant difference in stream matching probability. On the other hand, due to higher look-up cost and the fact that less memory is available for the hash table the performance decreases gradually. A similar behavior is seen when the performance against the swappable part is tested. In this case, after attaining the maximum performance it decreases rapidly because of an increase in the I/O cost for loading the growing swappable part. From the measurements shown in the figure it is possible to approximate the optimal settings for both the swappable and non-swappable parts by finding the maximum on the two-dimensional surface.

5.2   Tuning Based on Cost Model

A mathematical model for the tuning is also derived based on the cost model presented in Section 4.2. From Equation 3 it is clear that the service rate depends on the size of w and the cost c_{loop}. To determine the optimal settings it is first necessary to calculate the size of w. The main components on which the value

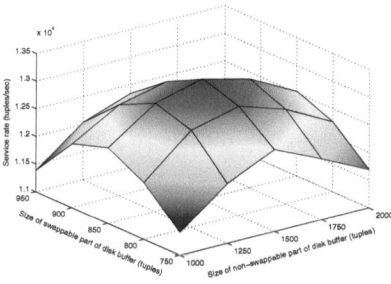

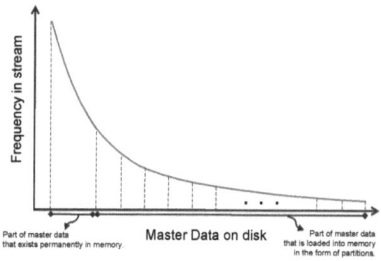

Fig. 3. Tuning using measurement approach

Fig. 4. A sketch of matching probability of R in stream

of w depends are:- size of the non-swappable part (d_N), size of the swappable part (d_S), size of the master data (R_t), and size of the hash table (h_S).

Typically the stream of updates can be approximated through Zipf's law with a certain exponent value. Therefore, a significant part of the stream is joined with the non-swappable part of the disk buffer. Hence, if the size of the non-swappable part (i.e. d_N) is increased, more stream tuples will match as a result. But the probability of matching does not increase at the same rate as increasing d_N because, according to Zipfian distribution, the matching probability for the second tuple in R is half of that for the first tuple and similarly the matching probability for the third tuple is one third of that for the first tuple and so on [2]. Due to this property, the size of R (denoted by R_t) also affects the matching probability. The swappable part of the disk buffer deals with the rest of the master data denoted by R' (where $R' = R_t - d_N$), which is less frequent in the stream than that part which exists permanently in memory. The algorithm reads R' in partitions, where the size of each partition is equal to the size of the swappable part of the disk buffer d_S. In each iteration the algorithm reads one partition of R' using an index on join attribute and loads it into memory through a swappable part of the disk buffer. In the next iteration the current partition in memory is replaced by a new partition, and so on. As mentioned earlier, using the Zipfian distribution the matching probability for every next tuple is less than the previous one. Therefore, the total number of matches against each partition is not the same. This is explained further in Figure 4, where n total partitions are considered in R'. From the figure it can be seen the matching probability for each disk partition decreases continuously as one moves toward the end position in R.

The size of the hash table is another component that affects w. The reason is simple: if there are more stream tuples in memory, the number of matches will be greater and vice versa. Before driving the formula to calculate w it is first necessary to understand the working strategy of Tuned X-HYBRIDJOIN. Consider for a moment that the queue contains stream tuples instead of just join attribute values. Tuned X-HYBRIDJOIN uses two independent inner loops under one outer loop. After the end of the first inner loop, which means after finishing the processing of the non-swappable part, the queue only contains those stream tuples which are related to only the swappable part of R, denoted by R'. For the next outer iteration of the algorithm these stream tuples in the queue are considered to be an old part of the queue. In that next outer iteration the algorithm loads some new stream tuples into the queue and these new stream tuples are considered to be a new part of the queue. The reason for dividing the queue into two parts is that the matching probability for both parts of the queue is different. The matching probability for the old part of the queue is denoted by p_{old} and it is only based on the size of the swappable part of R i.e. R'. On the other hand, the matching probability for the new part of the queue, known as p_{new}, depends on both the non-swappable as well as the swappable parts of R. Therefore, to calculate w it is first needed to calculate both these probabilities.

Therefore, if the stream of updates S obeys Zipf's law, then the matching probability for any swappable partition k with the old part of the queue can be determined mathematically as shown below.

$$p_k = \frac{\sum\limits_{x=d_N+(k-1)d_S+1}^{d_N+kd_S} \frac{1}{x}}{\sum\limits_{x=d_N+1}^{R_t} \frac{1}{x}}$$

Each summation in the above equation generates a harmonic series, which can be summed up using the formula $\sum\limits_{x=1}^{k} \frac{1}{x} = \ln k + \gamma + \varepsilon_k$, where γ is a Euler's constant whose value is approximately equal to 0.5772156649 and ε_k is another constant which is $\approx \frac{1}{2k}$. The value of ε_k approaches 0 as k goes to ∞ [3]. In this paper the value of $\frac{1}{2k}$ is small and therefore, it is ignored.

If there are n partitions in R', then the average probability of an arbitrary partition of R' matching the old part of the queue can be determined using Equation 4.

$$\bar{p}_{old} = \frac{\sum\limits_{k=1}^{n} p_k}{n} = \frac{1}{n} \tag{4}$$

Now the probability of matching is determined for the new part of the queue. Since the new input stream tuple can match either the non-swappable or the swappable part of R, the average matching probability of the new part of the queue with both parts of the disk buffer can be calculated using Equation 5.

$$\bar{p}_{new} = p_N + \frac{1}{n} p_S \tag{5}$$

where p_N and p_S are the probabilities of matching for a stream tuple with the non-swappable part and the swappable part of the disk buffer respectively. The values of p_N and p_S can be calculated as below.

$$p_N = \frac{\sum_{x=1}^{d_N} \frac{1}{x}}{\sum_{x=1}^{R_t} \frac{1}{x}} \qquad \text{and} \qquad p_S = \frac{\sum_{x=d_N+1}^{R_t} \frac{1}{x}}{\sum_{x=1}^{R_t} \frac{1}{x}}$$

Assume that w are the new stream tuples that the algorithm will load into the queue in the next outer iteration. Therefore,

The size of the new part of the queue (tuples)$=w$

The size of the old part of the queue (tuples)$=(h_S - w)$

If w are the average number of matches per outer iteration with both the swappable and non-swappable parts, then w can be calculated by applying the binomial probability distribution on Equations 4 and 5 as given below.

$$w = (h_S - w)\bar{p}_{old}(1 - \bar{p}_{old}) + w\bar{p}_{new}(1 - \bar{p}_{new})$$

After simplification the final formula to calculate w is described in Equation 6.

$$w = \frac{h_S\bar{p}_{old}(1 - \bar{p}_{old})}{1 + \bar{p}_{old}(1 - \bar{p}_{old}) - \bar{p}_{new}(1 - \bar{p}_{new})} \qquad (6)$$

By using the values of w and c_{loop} in Equation 3 the algorithm can be tuned.

5.3   Comparisons of Both Approaches

To validate the cost model the tuning results based on the measurement approach are compared with those that are achieved through cost model.

Swappable Part: This experiment compares the tuning results for the swappable part of the disk buffer using both the measurement and cost model approaches. The tuning results of each approach (with 95% confidence interval in case of measurement approach) are shown in Figure 5 (a). From the figure it is evident that at every position the results in both cases are similar, with only 0.5% deviation.

Non-swappable Part: Similar to before, the tuning results of both approaches for the non-swappable part of the disk buffer are also compared. The results are shown in Figure 5 (b). Again, it can be seen from the figure, the results in both cases are nearly equal with a deviation of only 0.6%. This proves the correctness of the tuning module.

6   Experimental Study

To strengthen the arguments an experimental evaluation of proposed Tuned X-HYBRIDJOIN is performed using the synthetic datasets. Normally, in Tuned X-HYBRIDJOIN kinds of algorithms, the total memory and the size of R are

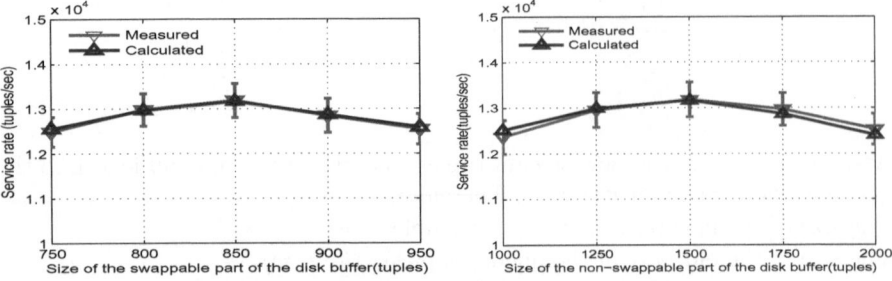

(a) Tuning Comparison for swappable part: based on measurements Vs based on cost model

(b) Tuning Comparison for non-swappable part: based on measurements Vs based on cost model

Fig. 5. Comparisons of tuning results

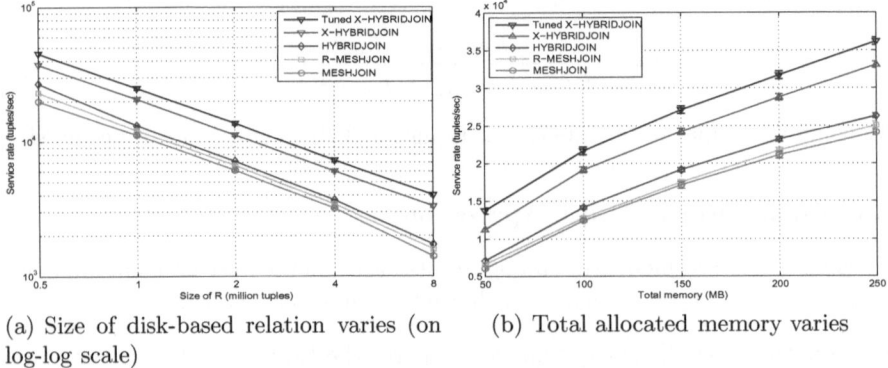

(a) Size of disk-based relation varies (on log-log scale)

(b) Total allocated memory varies

Fig. 6. Performance comparisons

the common parameters that vary frequently. Therefore, the experiments presented here compare the performance by varying both parameters individually.

Performance Comparisons for Different Sizes of R: This experiment compares the performance of Tuned X-HYBRIDJOIN with the other related algorithms for different sizes of R. Therefore it is assumed that the size of R varies exponentially while the total memory budget remains fixed (50MB) for all values of R. For each value of R the performance is measured separately. The performance results of this experiment are shown in Figure 6 (a). From the figure it is clear that for all settings of R the Tuned X-HYBRIDJOIN performs significantly better than other approaches.

Performance Comparisons for Different Memory Budgets: Second experiment analyses the performance of Tuned X-HYBRIDJOIN using different memory budgets while the size of R is fixed (2 million tuples). Figure 6 (b) depicts the performance results. From the figure it can be observed that for all memory budgets the Tuned X-HYBRIDJOIN again performs significantly better than all approaches. This improvement increases gradually as the total memory budget increases.

7   Conclusions

This paper investigates a well known stream-based join algorithm called X-HYBRIDJOIN. Main observation about X-HYBRIDJOIN is that the tuning factor is not considered but it is necessary, particularly when limited memory resources are available to execute the join operation. By omitting the tuning factor, the available memory cannot be distributed optimally among the join components and consequently the algorithm cannot perform optimally. This paper presents a variant version of X-HYBRIDJOIN called Tuned X-HYBRIDJOIN. The cost model presented in X-HYBRIDJOIN is revised and the proposed algorithm is tuned based on that revised cost model. To strengthen the arguments a prototype of Tuned X-HYBRIDJOIN is implemented and the performance with existing approaches is compared.

References

1. Naeem, M.A., Dobbie, G., Weber, G.: X-HYBRIDJOIN for Near-Real-Time Data Warehousing. In: Fernandes, A.A.A., Gray, A.J.G., Belhajjame, K. (eds.) BNCOD 2011. LNCS, vol. 7051, pp. 33–47. Springer, Heidelberg (2011)
2. Anderson, C.: The Long Tail: Why the Future of Business is Selling Less of More. Hyperion (2006)
3. Milton, A., Irene, A.S.: Handbook of Mathematical Functions with Formulas, Graphs, and Mathematical Tables. Ninth Dover printing, Tenth GPO printing, New York (1964)
4. Labio, W.J., Wiener, J.L., Garcia-Molina, H., Gorelik, V.: Efficient resumption of interrupted warehouse loads. SIGMOD Rec. 29(2), 46–57 (2000)
5. Golab, L., Tamer Özsu, M.: Processing Sliding Window Multi-Joins in Continuous Queries over Data Streams. In: VLDB 2003, Berlin, Germany, pp. 500–511 (2003)
6. Wilschut, A.N., Apers, P.M.G.: Dataflow query execution in a parallel main-memory environment. Distrib. Parallel Databases 1(1), 103–128 (1993)
7. Gupta, A., Mumick, I.S.: Maintenance of Materialized Views: Problems, Techniques, and Applications. IEEE Data Engineering Bulletin 18, 3–18 (2000)
8. Polyzotis, N., Skiadopoulos, S., Vassiliadis, P., Simitsis, A., Frantzell, N.: Meshing Streaming Updates with Persistent Data in an Active Data Warehouse. IEEE Trans. on Knowl. and Data Eng. 20(7), 976–991 (2008)
9. Naeem, M.A., Dobbie, G., Weber, G.: R-MESHJOIN for Near-real-time Data Warehousing. In: DOLAP 2010: Proceedings of the ACM 13th International Workshop on Data Warehousing and OLAP. ACM, Toronto (2010)
10. Chakraborty, A., Singh, A.: A partition-based approach to support streaming updates over persistent data in an active datawarehouse. In: IPDPS 2009: Proceedings of the 2009 IEEE International Symposium on Parallel & Distributed Processing, pp. 1–11. IEEE Computer Society, Washington, DC (2009)
11. Naeem, M.A., Dobbie, G., Weber, G.: HYBRIDJOIN for Near-real-time Data Warehousing. International Journal of Data Warehousing and Mining (IJDWM) 7(4) (2011)

Exploiting Interaction Features
in User Intent Understanding

Vincenzo Deufemia, Massimiliano Giordano,
Giuseppe Polese, and Luigi Marco Simonetti

Università di Salerno, Via Ponte don Melillo, Fisciano(SA), Italy
{deufemia,mgiordano,gpolese}@unisa.it, l.simonetti@studenti.unisa.it

Abstract. Understanding user intent during a web navigation session is a challenging topic, which is drawing the attention of many researchers in this area. The achievements of such research goals would have a great impact on many internet-based applications. For instance, if a search engine had the capability of capturing user intents, it could better suite the order of search results to user needs. In this context, the research has mainly focused on the analysis of user interactions with Search Engine Result Pages (SERPs) resulting from a web query, but most methods ignore the behavior of the user during the exploration of web pages associated to the links of the SERP s/he decides to visit. In this paper we propose a novel model that analyzes user interactions on such pages, in addition to the information considered by other mentioned approaches. In particular, captured user interactions are translated into features that are part of the input of a classification algorithm aiming to determine user informational, navigational, and transactional intents. Experimental results highlight the effectiveness of the proposed model, showing how the additional analysis it performs on visited web pages contributes to enhance user intent understanding.

Keywords: user intent understanding, interaction features, web search.

1 Introduction

The success of internet applications is bound to the capability of search engines to provide users with information meeting their expectations. This cannot be guaranteed by merely analyzing the web structure as several existing search engines do. Thus, nowadays, search engines need to incorporate the capability of predicting user intents in order to adapt the order of search results so as to meet their expectation. For this reason, user intention understanding (UIU) has recently become an important research area.

Some approaches to user behavior analysis in web navigation have highlighted the importance of analyzing user interactions with web pages in order to infer their interest and satisfaction with respect to the visited contents [1–3]. Other studies have investigated how user interactions with *Search Engine Result Pages* (SERPs) can be exploited to infer user intent [4–9]. However, such methods

Y. Ishikawa et al. (Eds.): APWeb 2013, LNCS 7808, pp. 506–517, 2013.

for UIU limit their analysis to the results contained in a SERP, ignoring many important interactions and contents visited from such results. On the contrary, the aim of this paper is to show that UIU can be considerably improved by performing additional analysis of user interactions on the web pages of a SERP that the user decides to visit. In particular, we define a new model for UIU that incorporates interactions analyzed in the context of SERP results and of the web pages visited starting from them. The interaction features considered in the model are not global page-level statistics, rather they are finer-grained and refer to portions of web pages. This is motivated on the basis of the results of literature studies, performed with eye-trackers, revealing that although a user might focus on many sections composing a web page, s/he will tend to overlook portions of low interest [10]. Thus, capturing interaction features on specific portions of web pages potentially conveys a better accuracy in the evaluation of the user actions. These and other features are evaluated in our approach by means of a classification algorithm to understand user intents. In particular, to simplify the classification process, we use a common taxonomy that defines three types of queries: *informational*, *navigational*, and *transational* [11].

Experimental results highlight the efficiency of the proposed model in query classification, showing how the interaction features extracted from visited web pages contribute to enhance user understanding. In particular, the proposed set of features has been evaluated with three different classification algorithms, namely Support Vector Machine (SVM) [12], Conditional Random Fields (CRF) [13], and Latent Dynamic Conditional Random Fields (LDCRF) [14]. The achieved results have been compared with those achieved with other subsets of features and demonstrate that the proposed model outperforms them in terms of query classification. Moreover, a further analysis highlights the effectiveness of the proposed features for the classification of transactional queries.

The rest of this paper is organized as follows. In Section 2, we provide a short review of related work. Then, we present the model exploiting interaction features for UIU in Section 3. Section 4 describes experimental results. Finally, conclusions are given in Section 5.

2   Related Work

A pioneer study by Carmel *et al.* [15] in the early 90s focusses on the analysis of hypertext. In particular, the study highlights three navigation strategies: *scan browsing*, aiming to inquire and evaluate information contained in the hypertext based on interest, *review browsing*, aiming to integrate information contained in the user mental context, and *search oriented browsing*, aiming to look for new information based on specific goals, and integrating them with information collected in successive scans.

A further interesting study by Morrison *et al.* [16] aimed to classify types of queries based on search intents. Such taxonomies are defined as formalizations of three basic questions that the user asks him/herself before starting a search session: *why, how,* and *what* to search. The taxonomy that is closest to user intent understanding is the one used in the context of a research called (*method taxonomy*), which aims to detect the following types of search activities: *explore, monitor, find,* and *collect.* By monitoring user search activities, Sellen *et al.* extend previously defined taxonomies, introducing transactional search types [17], defined as *transacting*, and finalized to search purposes and to the use of web services on which it is possible to better exploit user experiences while performing search activities. In this context, the goal of the search is not to find information, rather services of information manipulation: hence, the search engine becomes a means to migrate towards a web service. Successively, new approaches to user intent understanding have highlighted the necessity to introduce a clear classification of search queries based on user intents. To this end, one of the former studies by Broder *et al.* [11] refers to a simple taxonomy of search queries based on three distinct categories: *navigational queries* to search for a particular web site, *informational queries* to collect information from one or more web pages, and *transactional queries* to reach a web site on which further interactions are expected.

In the last decade, all existing approaches have aimed at applying taxonomies in a recognition and automatic classification context. One of the former methodologies by Kang *et al.* [4] aims at two types of search activities: *topic relevance,* that is, searching documents for a given topic, of informational type, and *homepage finding*, aiming to search main pages of several types of navigational web sites. Starting from common information used by Information Retrieval (IR) systems, such as web page content, hyperlinks, and URLs, the model proposes methods to classify queries based on the two categories mentioned above.

The analysis and the comprehension of user interaction models during web navigation is the basis of a predictive model on real case studies, proposed by Agichtein *et al.* [6]. The model tries to elicit and understand user navigation behaviors by analyzing several activities, such as clicks, scrolls, and dwell times, aiming to predict user intention during web page navigation. The features this study proposes to analyze are used to characterize the complex system of interactions following the click on a result. Successively, a study by Guo *et al.* [9] starts from the hypothesis that such interactions can help accurately inferring two particular tightly correlated intents: search and purchase of products, a two phase activity defined as *search-purchase.*

Starting from experimental studies on real user navigation strategies, Lee *et al.* propose a model for the automatic identification of search goals based on features, and restricted to navigational and informational queries [5]. These studies have primarily revealed that most queries can be effectively associated to one of two categories defined within the taxonomy, making it possible the construction of an automatic identification system. Moreover, most queries that are not effectively associable to a category are related to few topics, such as proper nouns or names of software systems. This makes it possible to use *ad-hoc* systems of features for all those unpredictable cases. The model proposes two features: *past user-click behavior* that infers user intent from the way s/he has interacted with the results in the past, and *anchor-link distribution*, which uses possible targets of links having the same text of the query.

The strategies for user intent understanding described so far aim to classify search queries exclusively using features modeled to describe several aspects of search queries. A study by Tamine *et al.* proposes to analyze search activities that the user has previously performed in the same context, aiming to derive data useful for inferring the type of the current query [8]. The set of queries already performed represents the *query profile*.

3 The Proposed Model

In this section we describe the proposed model and the features used for the classification process.

3.1 Search Model: Session, Search, Interaction

Several approaches and models have been proposed to provide solutions to the user intention understanding problem, but all of them mainly focus on the interaction between users and the SERP. Additional interactions originating from SERP's contents, such as browsing, reading, and multimedia content fruition are not considered by the research community.

The proposed model aims to extend existing models, by analyzing user interactions between users and web pages during a search session. Our aim is to analyze not only the interactions between users and SERP, but also between users and web pages reached by clicking on SERP's results. We believe that data about interactions between users and web pages may be very useful to clarify the intent of the user, because these interactions are driven by the same motivation behind the initial search query. All user interactions with web pages could be reduced to three main categories, which are the same we consider in our model: *session*, *search*, and *interaction*.

Session. A *session* is a sequence of search activities aimed at achieving a goal. When the first query does not provide the desired result, the user tries to gradually approach the target, refining or changing search terms and keywords. All these research activities constitute a session.

Search. A *search* activity is the combination of the following user actions: submission of a query to a search engine, analysis of search results, navigation on one or more web pages inside them. During a search activity a user has a specific goal, generally described by the query itself. This goal is classifiable in a taxonomy, as defined by previous studies [11, 18].

Interaction. *Interaction* is the navigation of a web page using a wide range of interactions that include mouse clicks, page scrolling, pointer movements, and text selection. Starting from these interactions, combined with features such as dwell time, reading rate, and scrolling rate, it is possible to derive an implicit feedback of users about web pages [2]. Moreover, several studies have proven the usefulness of user interactions to assess the relevance of web pages [1, 3, 19, 20], to classify queries, and to determine the intent of search sessions [9, 21].

The proposed model introduces a new methodology for tracing interactions between users and web pages. In particular, user interaction analysis is restricted to a smaller portion of a web page: the *subpages*. Indeed, a web page consists of several blocks of text, graphics, multimedia, and can have a variable length. As shown by studies using eye-tracking [10], the user frequently adopts a *E or F reading pattern* during browsing, excluding areas of little interest. Compared to a global analysis of the entire page, the introduction of *subpage-level* analysis gives higher accuracy in the assessment of the interaction.

3.2   Features

Interaction data extracted from user web navigation have been encoded into features that characterize user behavior. We organize the set of features into the following categories: *query, search, interaction*, and *context*.

Query. These features are derived from characteristics of a search query such as keywords, the number of keywords, the semantic relations between them, and other characteristics of a search or an interaction.

Search. These features act on the data from search activities such as: results, time spent on SERP, and number of results considered by the user. The *DwellTime* is measured from the start of the search session until the end of the last interaction originated by the same search session. The reaction time, *TimeToFirstInteraction,* is the time elapsed from the start of the search session and the complete loading of the first selected page. Other features dedicated to interactions with the results are *ClicksCount*, which is the number of visited results, and *FirstResultClickedRank*, determining the position of the first clicked result.

Interaction. These features act on the data collected from interactions with web pages and subpages, taking into account the absolute dwell time, the effective dwell time, all the scrolling activities, search and reading activities. The *DwellRate* measures the effectiveness of the permanence of a user on a web page, while the reading rate *ReadingRate*, measures the amount of reading of a web page [2]. Additional interactional features are: *ViewedWords*, the number of words considered during the browsing, *UrlContainsTransactionalTerms*, which verifies if the URL of the page contains transactional terms (download, software, video, watch, pics, images, audio, etc.), *AjaxRequestsCount*, which represents the number of AJAX requests originated during browsing.

Context. These features act on the relationship between the search activities performed in a session, such as the position of a query in the sequence of search requests for a session.

4   Experiments

In this section we describe the dataset constructed for evaluating the proposed approach and the results achieved with different classification algorithms.

In particular, after an overview on the evaluation metrics used and on the subsets of features taken into consideration, experimental results are presented.

4.1 Experiments Configuration

The goal of this experimentation is to evaluate the effectiveness of the proposed model for inferring user intention in web searches. In particular, we analyze the performances of the model for classifying queries, based on the query taxonomy: informational, navigational, and transactional.

Search activities have been accomplished by a pool of thirteen subjects (seven men and six women), whose age ranges from 24 to 37 years old. Sixth subjects had a master degree (5 in computer science and 1 in linguistics), 3 a bachelor degree (1 in computer science, 1 in political science, and 1 in chemistry), and 4 a high school degree in accounting. The average experience in web browsing was of 9.6 years.

Dataset. The dataset has been constructed by the thirteen participants to the test. In particular, all the subjects were requested to perform a series of search activities related to various topics and with different intentions. Each subject was asked to determine his/her own goal in advance. After starting the session, subjects performed a series of searches using the Google search engine, without any limitation as to time or the results to visit. By following this protocol, we had 129 sessions and 353 web searches, which were subsequently manually classified by relying on the intent of the user. Starting from web searches, 490 web pages and 2136 sub pages were visited. All interactions were detected by the YAR plug-in for Google Chrome/Chromium [2].

Evaluation Metrics. In order to evaluate the effectiveness of the proposed model, we adopted the classical evaluation metrics of Information Retrieval: *precision*, *recall*, and *F1-measure*. They are defined as follows:

$$\textbf{Precision} = \sum_{Category(i)} \frac{\#correctly\ classified\ queries}{\#classified\ queries} \times \frac{\#category\ queries}{\#total\ queries}$$

$$\textbf{Recall} = \frac{\#correctly\ classified\ queries}{\#total\ queries}$$

$$\textbf{F1} - \textbf{measure} = 2 \times \frac{Precision \times Recall}{Precision + Recall}$$

Feature Subsets. In order to verify the effectiveness of the proposed features in different combinations, we have grouped them into several subsets:

- **All**: subset of all the proposed features *query, search, interaction,* and *context*;

- **Query**: subset of all the features related to *queries*;
- **Search**: subset of all the features related to *search* and *context*;
- **Interaction**: subset of all the features related to *interactions*;
- **Query+Search**: subset of the features derived as union from *Query* and *Search*. The goal is to evaluate the effectiveness of query classification by using the features considered in other studies [3, 6, 9];
- **Transactional**: subset of all the features related to interactions over transactional queries *ViewWords*, *AjaxRequestsCount*, *ScrollingDistance*, *ScrollingCount*, and *UrlContainsTransactionalTerms*. The goal here is to evaluate the classification of transactional queries by adopting more specific features;
- **All−Transactional**: subset derived by the exclusion of the transactional features from the set *All*. The goal here is to evaluate the effectiveness of the classification of transactional queries by comparing results achieved with all features to those achieved by excluding transactional features.

Classifiers. We considered three classifiers to evaluate the proposed feature model: SVM [12], CRF[13], and LDCRF[14]. Support Vector Machine (SVM) is a discriminative model, based on the structural risk minimization principle from statistical learning theory [12], widely used in binary classification problems. In particular, SVM maps the input data in a higher dimensional feature space, and splits the data into two classes by computing a hyperplane that separates them. CRF is a discriminative probabilistic model used for labeling sequential data [13]. The model utilizes inter-dependent features to select the most probable label sequence for one observation LDCRF extends CRF by incorporating hidden state variables which model the sub-structure of input sequences [14]. It is able to learn the transitions between input elements by modeling a continuous stream of class labels, and the internal sub-structure by using intermediate hidden states.

In the context of query classification, SVM assumes that the queries in a user session are independent, Conditional Random Field (CRF) considers the sequential information between the queries, and LatentDynamic Conditional Random Fields (LDCRF) models the sub-structure of user sessions by assigning a disjoint set of hidden state variables to each class label. They have been configured as follows:

1. **SVM.** We used MSVMpack [22] as the SVM toolbox for model training and testing. The SVM model is trained using a linear kernel and the parameter C has been determined by cross-validation.
2. **CRF.** We used the HCRF library[1] as the tool to train and test the CRF model. For the experiments we used a single chain structured model and the regularization term for the CRF model was validated with values 10^k with $k = -1 \ldots 3$.
3. **LDCRF.** We used the HCRF library for training and testing LDCRF model. In particular, the model was trained with 3 hidden states per label, and the regularization term was determined by cross-validation to achieve the best performance.

---

[1] http://sourceforge.net/projects/hcrf/

4.2   Results

In order to simulate an operating environment, the set of queries made by users was separated into two subsets, which included 60% and 40% of web searches, and were used for training and testing the classifiers, respectively.

Taking into account informational queries (see Figure 1), it could be found that best classifications were achieved by using the CRF classifier with the *All* set of features, whereas worst performances were achieved by considering the subsets of features *Query* and *Search*. Similar considerations hold for *Search+Query*. These initial observations demonstrate the importance of the subset of interactional features, which alone are able to achieve classification performances close to those achieved by the *All* set of all features.

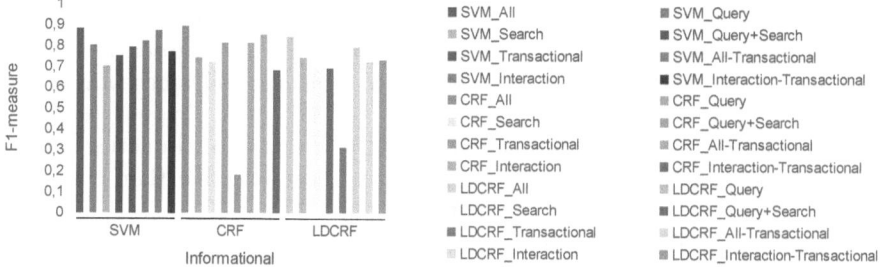

Fig. 1. Performances of the considered feature subsets for informational queries

Among the interactional set of features proposed above, there is a subset, named *Transactional*, which was introduced to improve the classification of transactional queries. The experimental results highlight that these features also improve the classification of informational queries. In fact, the *Search+Query* subset achieves performances that are closer to the subset *All−Transactional*, as shown in Figure 1. This means that the subset of interactional features, which are not part of *Transactional*, are not effective for the classification of informational queries. The relevance of transactional features in this type of query is confirmed by two other important experimental results. The first one regards the application of the single subset *Transactional*, which is able to achieve performances equivalent to the *Search+Query* and *All−Transactional* subsets. The second one is the performance degradation of the *Interaction−Transactional* features, which achieve results by far lower than the ones achieved through the application of all the interactional features, and any other subset of features considered individually.

Turning to navigational query results shown in Figure 2, there may be some surprising findings. First of all, the set *All* of all the features does not provide the best performances, which are instead provided by the subset *All−Transactional*. This highlights the problem of classifying navigational and transactional queries due to their similarity. Even if slightly, performances of the model improve by excluding transactional features. The inefficiency of these features can also be

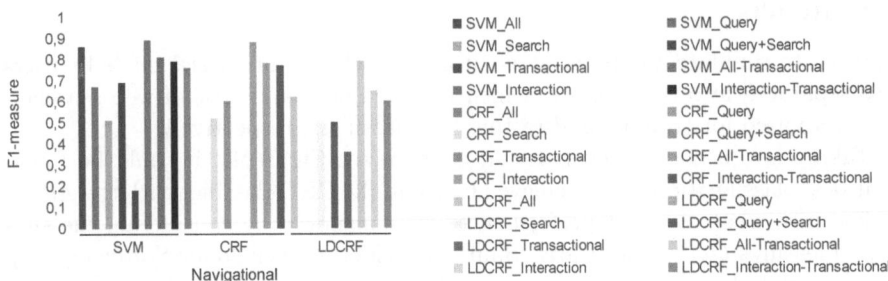

Fig. 2. Performances of the considered feature subsets for navigational queries

seen by considering the subset *Transactional*, which provides performances worse than with other subsets. Poor results were also achieved from subsets of features considered individually, except for *Interaction*, which provides acceptable performances.

Finally, let us analyze performances related to transactional queries (Figure 3). In this context, the set *All* confirms to be more effective than the other subsets. What we want to highlight here is the efficacy of transactional features. The subset *Transactional*, in terms of performances, is second only to the subset of all features, but it is close to it. By excluding such features in the subset *All−Transactional*, performances degrade providing results comparable to *Query+Search*. The inefficiency of pure interactional features with respect to transactional ones also arises by a direct comparison: The subset *Interaction* achieves performances comparable to *Transactional*, but far better than *Interaction−Transactional*, which excludes transactional features from Interaction.

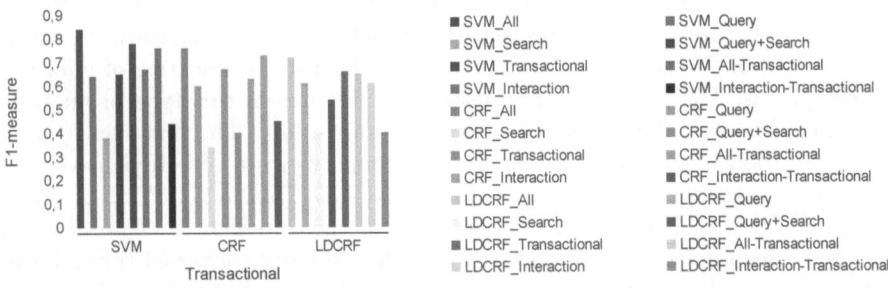

Fig. 3. Performances of the considered feature subsets for transactional queries

An Analysis of Transactional Features. A substantial limit of traditional approaches described in the literature is represented by the classification of *transactional queries*. In this paper we propose some features that aim to characterize transactional queries by analyzing scrolling behaviors, ajax requests, read text, and presence of transactional terms in queries or URLs. The goal here is to compare the effectiveness of query classification by applying all the features (*All*) with respect to applying a subset of them that excludes the transactional features (*All−Transactional*). Referring to transactional queries, Table 1 shows that there is a conspicuous improvement of the set *All*, especially when using SVM. To be noted the precision value of 0.93, by far higher than the precision measured through all the classifiers for the subset *All−Transactional*.

Table 1. Precision, Recall, and F1-measure of the classification of transactional queries with the feature sets *All* and *All−Transactional*

Class_Model	Precision	Recall	F1-measure
SVM_All	0.93	0.76	0.84
SVM_All−Transactional	0.72	0.62	0.67
CRF_All	0.78	0.74	0.76
CRF_All−Transactional	0.85	0.50	0.63
LDCRF_All	0.77	0.68	0.72
LDCRF_All−Transactional	0.65	0.65	0.65

Moreover, transactional features are also useful to improve the quality of the classification process of informational queries (Table 2), whereas they have no effect on navigational ones (Table 3).

Table 2. Precision, Recall, and F1-measure of the classification of informational queries with the feature sets *All* and *All−Transactional*

Class_Model	Precision	Recall	F1-measure
SVM_All	0.83	0.90	0.88
SVM_All−Transactional	0.80	0.84	0.82
CRF_All	0.86	0.92	0.89
CRF_All−Transactional	0.76	0.87	0.81
LDCRF_All	0.84	0.83	0.84
LDCRF_All−Transactional	0.80	0.78	0.79

Table 3. Precision, Recall, and F1-measure of the classification of navigational queries with the feature sets *All* and *All−Transactional*

Class_Model	Precision	Recall	F1-measure
SVM_All	0.82	0.90	0.86
SVM_All−Transactional	0.87	0.90	0.89
CRF_All	0.84	0.70	0.76
CRF_All−Transactional	0.90	0.87	0.88
LDCRF_All	0.59	0.67	0.62
LDCRF_All−Transactional	0.77	0.80	0.79

5 Conclusions

In this paper we have proposed a new model for user intent understanding in web search. Assuming that the interactions of the user with the web pages returned by a search engine in response to a query can be highly useful, in this research we aimed at defining a new model, based on the results returned by the search engine, on the interactions of the user with them, and with the web pages visited by exploring them. By examining these interactions, we have produced a set of features that are suitable for determining the intent of the user. Each feature involves a different level of interaction between the user and the "system": query, search, and web pages (interaction). To simplify and make the process of classification more efficient, we also adopt a simple one level taxonomy: *informational, navigational,* and *transactional.* In addition to the set of all the proposed features, during the testing phase we also considered some subsets corresponding to features of individual interactional contexts and to their union or difference, in order to evaluate the effectiveness of the classification with respect to traditional models, and to interactional, or transactional features.

Experimental results have highlighted the effectiveness of query classification when applying both features representing interactions on web pages and those representing interactions in the context of queries and results. This also arise in the classification of transactional queries, further highlighting the effectiveness of interactional features, and more important, of transactional features.

References

1. Agichtein, E., Brill, E., Dumais, S.: Improving web search ranking by incorporating user behavior information. In: Proceedings of the International Conference on Research and Development in Information Retrieval, SIGIR 2006, pp. 19–26. ACM (2006)
2. Deufemia, V., Giordano, M., Polese, G., Tortora, G.: Inferring web page relevance from human-computer interaction logging. In: Proceedings of the International Conference on Web Information Systems and Technologies, WEBIST 2012, pp. 653–662 (2012)
3. Guo, Q., Agichtein, E.: Beyond dwell time: estimating document relevance from cursor movements and other post-click searcher behavior. In: Proceedings of the International Conference on World Wide Web, WWW 2012, pp. 569–578. ACM (2012)
4. Kang, I., Kim, G.: Query type classification for web document retrieval. In: Proceedings of the Conference on Research and Development in Informaion Retrieval, SIGIR 2003, pp. 64–71. ACM (2003)
5. Lee, U., Liu, Z., Cho, J.: Automatic identification of user goals in web search. In: Proceedings of the International Conference on World Wide Web, WWW 2005, pp. 391–400. ACM (2005)
6. Agichtein, E., Brill, E., Dumais, S., Ragno, R.: Learning user interaction models for predicting web search result preferences. In: Proceedings of the International Conference on Research and Development in Information Retrieval, SIGIR 2006, pp. 3–10. ACM (2006)

7. Jansen, B.J., Booth, D.L., Spink, A.: Determining the user intent of web search engine queries. In: Proceedings of the International Conference on World Wide Web, WWW 2007, pp. 1149–1150. ACM (2007)

8. Tamine, L., Daoud, M., Dinh, B., Boughanem, M.: Contextual query classification in web search. In: Proceedings of International Workshop on Information Retrieval Learning, Knowledge and Adaptability, LWA 2008, pp. 65–68 (2008)

9. Guo, Q., Agichtein, E.: Ready to buy or just browsing? detecting web searcher goals from interaction data. In: Proceedings of the International Conference on Research and Development in Information Retrieval, SIGIR 2010, pp. 130–137. ACM (2010)

10. Nielsen, J.: F-shaped pattern for reading web content (2006),
http://www.useit.com/articles/f-shaped-pattern-reading-web-content/

11. Broder, A.: A taxonomy of web search. SIGIR Forum 36, 3–10 (2002)

12. Cortes, C., Vapnik, V.: Support-vector networks. Mach. Learn. 20, 273–297 (1995)

13. Lafferty, J.D., McCallum, A., Pereira, F.C.N.: Conditional random fields: Probabilistic models for segmenting and labeling sequence data. In: Proceedings of the International Conference on Machine Learning, ICML 2001, pp. 282–289 (2001)

14. Morency, L.P., Quattoni, A., Darrell, T.: Latent-dynamic discriminative models for continuous gesture recognition. In: Proceedings of IEEE Conference Computer Vision and Pattern Recognition, CVPR 2007, pp. 1–8 (2007)

15. Carmel, E., Crawford, S., Chen, H.: Browsing in hypertext: a cognitive study. IEEE Transactions on Systems, Man and Cybernetics, Part A: Systems and Humans 22, 865–884 (1992)

16. Morrison, J., Pirolli, P., Card, S.K.: A taxonomic analysis of what world wide web activities significantly impact people's decisions and actions. In: Extended Abstracts on Human Factors in Computing Systems, CHI EA 2001. pp. 163–164. ACM (2001)

17. Sellen, A., Murphy, R., Shaw, K.: How knowledge workers use the web. In: Extended Abstracts on Human Factors in Computing Systems, CHI 2002, pp. 227–234. ACM (2002)

18. Rose, D., Levinson, D.: Understanding user goals in web search. In: Proceedings of the International Conference on World Wide Web, WWW 2004, pp. 13–19. ACM (2004)

19. Guo, Q., Agichtein, E.: Towards predicting web searcher gaze position from mouse movements. In: Proceedings of the International Conference on Human Factors in Computing Systems, CHI EA 2010, pp. 3601–3606. ACM (2010)

20. Kelly, D., Teevan, J.: Implicit feedback for inferring user preference: a bibliography. SIGIR Forum 37, 18–28 (2003)

21. Guo, Q., Agichtein, E.: Exploring mouse movements for inferring query intent. In: Proceedings of the International Conference on Research and Development in Information Retrieval, pp. 707–708. ACM (2008)

22. Lauer, F., Guermeur, Y.: MSVMpack: a multi-class support vector machine package. Journal of Machine Learning Research 12, 2269–2272 (2011)

Identifying Semantic-Related Search Tasks in Query Log

Shuai Gong[1,3], Jinhua Xiong[1], Cheng Zhang[1], and Zhiyong Liu[1,2]

[1] Institute of Computing Technology, Chinese Academy of Sciences
[2] State Key Laboratory of Computer Architecture
[3] University of Chinese Academy of Sciences
{gongshuai,zhangcheng2}@software.ict.ac.cn, {xjh,zyliu}@ict.ac.cn

Abstract. Users often submit multiple related queries in order to accomplish one search task. Identifying search tasks faces two challenges: 1) Search tasks are often intertwined and may span from seconds to days. 2) Queries triggered by semantic-related search tasks may share few common terms or clicked documents. To address the challenges, we exploit semantic features of named entities to improve semantic-related search tasks identification. A novel approach to learning the semantic-related distance function between pair-wise queries is proposed. The approach uses categories of named entities as regularization, which reinforces that queries containing entities from the same category more probably belong to one search task. Finally, semantic-related search tasks are identified by the hierarchical agglomerative clustering algorithm with the learned distance function. Experiments show significant improvement of our approach over corresponding state-of-the-art ones.

Keywords: Query log analysis, User intent, Search task, Named entity.

1 Introduction

Detecting user intents in queries is crucial for search engines to better satisfy users' information need. Usually users may issue multiple related queries to accomplish one search task [7], e.g. multiple queries "magnolia junior high" and "chino high school" may belong to the same search task for looking for "schools in the City of Chino". Identifying search tasks could help search engines to better understand user intents. Many works have been proposed to identify search tasks, but long-term and semantic-related search task identification has not been well studied. Search tasks are often intertwined [11], and may span from seconds to days [2]. For example, the search task "trip planning" can span many days. During this period the user may issue many queries of other search tasks. To our best knowledge, existing technologies do not work well to identify such long-term search tasks [3,9,8]. Because of the sparsity of clicks and lexical features of query log, queries triggered by the same semantic-related search task may share few common terms or clicked documents. Identifying long-term and semantic-related search tasks is difficult. Firstly, it is not straightforward how to exploit semantic

Y. Ishikawa et al. (Eds.): APWeb 2013, LNCS 7808, pp. 518–525, 2013.

information. It is also difficult to combine properly both semantic and lexical features between pair-wise queries. Secondly, search tasks may span widely, so annotators have to check all the queries issued by the same user, in order to label sufficient semantic-related queries for learning a distance function. Labeling large-scale queries is expensive, therefore the distance function between pair-wise queries has to be learned from a small set of labeled data. It is prone to suffer the over-fitting problem.

To address the above challenges, this paper exploits semantic information properly and learns an optimal distance function between pair-wise queries. Firstly, this paper extends respectively named entities and their contexts in user queries with semantic information. In most cases, queries containing named entities of the same category tend to belong to the same search task, this paper exploits the category of named entities to regularize the learning of the distance function. Semantic-related search tasks, i.e. query clusters, are identified by the hierarchical agglomerative clustering algorithm with the learned distance function.

The rest of the paper is structured as follows. Related work is discussed in Section 2. Then we formalize basic concepts of identifying semantic-related search tasks in Section 3. In Section 4, we introduce our approach to identifying semantic-related search tasks. In Section 5, we present our experiments on commercial search log. Finally, we conclude our work in Section 6.

2   Related Work

The related work can be organized into the following two categories:

Query intent analysis has attracted substantial attention. User queries are often classified into three high-level intents: navigational, informational and transactional intents [4]. Recently, many works [6,12] attempted to exploit named entities to better understand user intent. Guo et al. [6] exploited the category of named entities to classify user queries. Yin et al. [12] built a hierarchical taxonomy of the generic search intents based on named entities. Recently the research on query intent of multiple queries such as query refinements [10] and search tasks [9] have also attracted much attention [5,1,8,2].

Search tasks identification was proposed by Jones et al. [9] in the setting of queries sessions detection [3]. Jones et al. [9] determined the boundary of search tasks between chronological queries with a decision tree. Donato et al. [5] determined whether pairs of chronologically issued queries targeting the same search task. This approach could not deal with intertwined search tasks well. Aiello et al. [1] clustered the identified search tasks of different users into more general topics by the approach proposed in [5]. But this paper aims to discover more semantic-related search tasks accomplished by the same user. Lucchese et al. [8] clustered user queries in a time-split session. They exploited both content features and semantic features in a straightforward way. The different importance of terms and dependency among terms in queries is ignored. The semantic of queries might deform. Sadikov et al. [10] clustered refinements of a query according to different user intents. But the intertwined search tasks could not be effectively identified. Boldi *et al.* [3] introduced the query-flow graph to represent query log. This model was used to separate the chronological queries into sets of related queries, named query chain, with three categories of features. Kotov *et al.* [2] modeled the user behaviors of cross-sessions search tasks. This work also showed the importance of identifying long-term search tasks.

3   Problem Formalization

This section introduces the distance function and semantic-related features used for search task identification.

3.1   Distance Function

Q denotes all the queries issued by a user. Search task T is a subset of Q. E denotes the named entities set collected from collaborative knowledge bases, such as Wikipedia. Cat denotes the categories set of named entities. The category of a named entity could be denoted as $cat(e)$. A user query $q_i \in Q$ can be formalized as a bi-tuple as $q_i =< e_i, c_i >$, where e_i denotes the entity contained in the query q_i and c_i denotes the remaining terms of that query. $\tilde{q} =< q_i, q_j >$ denotes any pairs of queries issued by a user. The function $f : Q \times Q \to [0, 1]$ denotes semantic-related distance between pair-wise queries. Given labeled semantic-related user query pairs, finding out the distance function f becomes the following optimal problem:

$$f^* = \arg \min_{f \in H_K} [\frac{1}{l} \sum_{i=1}^{l} V(f, x_i, y_i) + \lambda \|f\|^2] \tag{1}$$

where V is the loss function over the labeled data, such as squared loss for Regularized Least Squares, H_k is an Reproducing Kernel Hilbert space of functions with kernel function K, l is the size of labeled queries pairs, and λ is the coefficient of regularization on the ambient space. This paper defines the distance function as a linear function about feature vector $x_{\tilde{q}}$ between pair-wise queries. The solution of the minimization problem will be discussed in Section 4.

3.2   Entity Feature

The features between pair-wise queries can be split into two parts: entity features and context features. The entity features used in the distance function are:

Semantic Relevance of Named Entity. To get the semantic relevance feature between pair-wise named entities, it is necessary to use some external semantic sources, such as Wikipedia. Therefore we crawled and indexed the web pages from Wikipedia, built a search engine. The relevance between a named entity and all indexed documents constitutes the semantic feature vector for that entity.

The semantic relevance between pair-wise named entities is cosine similarity between the semantic feature vectors.

Context Cluster Relevance of Named Entity. Named entities occurring in analogous contexts tend to target relevant tasks. But considering spelling errors in user queries, directly using contexts of named entities will make the feature vector space very sparse. So the contexts of entities are clustered. Each element of a context cluster feature vector for each named entity is $P(clu_i|e)$. The context clusters relevance between pair-wise named entities is cosine similarity of context cluster feature vector of each named entity.

3.3   Context Feature

The context of a query is the remaining part of the query except the named entity. The semantic and lexical features of context are introduced as follows:

Context Semantic Relevance. This paper also uses the vector model to build semantic feature vector for each context. Different with [8], this paper submits each context as a whole to the search engine. The semantic relevance between pair-wise contexts is cosine similarity between the semantic feature vectors.

Context Lexical Distance. As illustrated in [9,8], the queries containing common terms may target relevant search tasks. This paper uses both Jaccard index on tri-grams and normalized Levenstein distance to represent lexical distance between pair-wise contexts respectively, as suggested by the works [9,3].

4   Identifying Semantic-Related Search Tasks

This section introduces our approach for long-term **S**emantic-related search **T**asks **I**dentification (*STI*). The *STI* approach consists of two major steps:

1) Learning Distance Function: In the first step, *STI* learns an optimal distance function. The distance function will be used to measure similarity between pair-wise queries when clustering user queries in the second step.

2) Clustering Semantic-Related Queries: At the second step, *STI* uses the hierarchical agglomerative clustering algorithm with the learned distance function to cluster user queries. The resulting query clusters are semantic-related search tasks.

4.1   Learning Distance Function

As discussed in Section 3, learning a distance function can be transformed into solving an optimal problem (Eq. 1). Because it is difficult to label large scale search tasks, learning the distance function will suffer the over-fitting problem. Therefore, the entity category regularization is proposed.

Entity Category Regularization. In most cases, queries issued by a user containing named entities of the same categories tend to be in the same search task, such as queries "Toyota radios" and "Ford parts" probably constitute the search task for maintaining a car. So regularizing a distance function with categories of named entities does make sense. That is to say the semantic distance between pair-wise queries containing named entities of the same category may be close. The regularization of categories of named entities can be formalized as follows:

$$\frac{1}{N}\sum_{i=1}^{N}(f(\tilde{q}_i)^2 g(\tilde{q}_i)) < \epsilon \tag{2}$$

where $g(\tilde{q})$ denotes whether the named entities contained in the query pair $\tilde{q}$ belong to the same category. N is the number of pair-wise queries.

Distance Function Optimization. When minimizing the objection function (Eq. 1), the constraint of Eq. 2 must be satisfied. By leveraging the Lagrange multiplier and using sum of squared loss as the loss function, the object function can be reformulated as follows:

$$f^* = \arg\min_{f\in H_K}[\frac{1}{l}\sum_{i=1}^{l}(y_i - f(\tilde{q}_i))^2 + \lambda\|f\|^2 + \frac{\epsilon}{N}\sum_{i=1}^{N}(f(\tilde{q}_i)^2 g(\tilde{q}_i))] \tag{3}$$

where l is the number of labeled pair-wise queries, N is the total number of pair-wise queries, y_i denotes the labeled value of semantic-related distance function, λ is the coefficient of regularization in ambient space, and ϵ is the coefficient of same categories regularization. As above mentioned, the distance function is a linear combination of different features. So Eq. 1 can be reformulated as follows:

$$w^* = \arg\min_{w}[\frac{1}{l}\sum_{i=1}^{l}(y_i - w^T x_i)^2 + \lambda w^T w + \frac{\epsilon}{N}w^T \hat{X}^T \hat{X} w] \tag{4}$$

where w is the parameters of the distance function. $\hat{X}$ is a $N \times D$ matrix, the i-th row of $\hat{X}$ denotes the feature vector between i-th pair-wise queries which containing named entities of the same category. D is the dimension of feature vector between pair-wise queries. Setting the gradient of Eq. 4 to zero gives

$$(X^T X + \lambda l I + \frac{\epsilon l}{N}\hat{X}^T \hat{X})w = X^T Y \tag{5}$$

where X is a $l \times D$ matrix, which is usually named as design matrix. The i-th row of X is the feature vector between the i-th labeled pair-wise user queries. I is the identity matrix. Solving the Eq. 5 will get the optimal w^*.

4.2   Clustering Semantic-Related Queries

The hierarchical agglomerative clustering algorithm is used to cluster semantic-related queries with the leaned distance function. As illustrated in [10], the complete-link similarity of two sets of queries works best in query clustering setting. Therefore this paper uses the complete link similarity to merge different query clusters. The algorithm works by picking each pair of query sets that have the highest value of complete-link similarity. If the similarity is greater than η, then merges the selected query sets. This proceeds until no more similar query sets are left. The resulting query clusters are semantic-related search tasks.

5   Experiments

5.1   Experiment Setup

This section discusses our experiment data set, two extended state-of-the-art approaches to be compared with our approach STI and evaluation measures.

Data Set. The 2006 AOL query log is used for experiment, in order to compared with related approaches tested by the same log. This query log consists of about 20 million of web queries issued by more than 650,000 users over 3 months. In order to learn the distance function, 200 pairs of queries random selected are labeled as semantic-related and 200 pairs of queries random selected are labeled as not-semantic-related. Named entities are collected from Wikipedia with the "list of *" query such as "list of states in U.S.".

Extended State-of-the-Art Approaches. 1) Query Chain Identification Boldi *et al.* [3] proposed an approach to identify the chains of queries, which are similar to search tasks. They used the logistic regression to learn the distance function between pair-wise queries. To compare with STI, we modify the approach, named as $mQCI$. **2) Search Task Discovery in Sessions** The approach proposed in [8] was used to address task-based session discovery problem ($TSDP$) in time-split sessions. It can not be used to identify long-term and semantic-related search tasks directly. Therefore we modified the approach to enable that all the queries issued by the same user will be included in for search tasks identification, named as $mTSDP$.

Search Task Analysis. To compare different approaches, we use User Annotation and three measures (Time Span, Cohesion and Coverage) for evaluation.

5.2   Evaluation

The user annotation results are presented in Table 1. Each evaluator is required to rate each identified search task as 1(Poor), 2(Bad), 3(Acceptable), 4(Good) or 5(Perfect) and indicates optionally whether an identified search task is under clustered or over clustered. According to the result of user rating score, the differences among the three approached are statistically significant, and STI performs the best. The column of the ratio of under clustered search tasks shows that STI yields the least one. As showed by Fig. 1, the coverage of identified search tasks by STI is much higher than both $mQCI$ and $mTSDP$. The column of the ratio of over clustered search tasks shows that STI yields the largest one.

We now look at the reason that is likely to underlie the observed results. Both *mTSDP* and *mQCI* perform better to aggregate the misspelling queries, such as **"800flowers"** and **"1-800flower"**. But *STI* tends to yield long-term and more semantic-related search tasks than others, such as **"magnolia junior high"** and **"chino high school"**, which are not in the results of both *mQCI* and *mTSDP*. This is consistent with what Fig.1 and Fig. 2 shows.

Table 1. User Annotations Result

Method	Average Rating	Under Clustered	Over Clustered
STI	4.4	7%	10%
mQCI	2.6	18.9%	5%
mTSDP	3.8	17%	7%

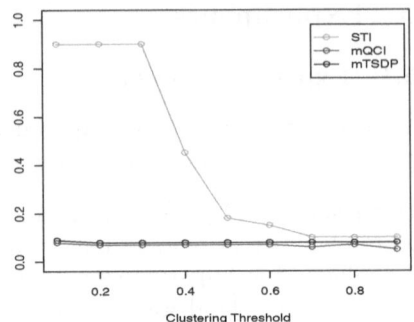

Fig. 1. Average Coverage

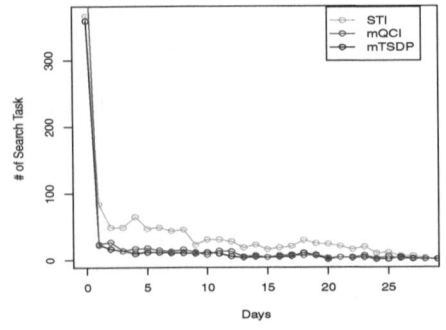

Fig. 2. Time Span Distribution

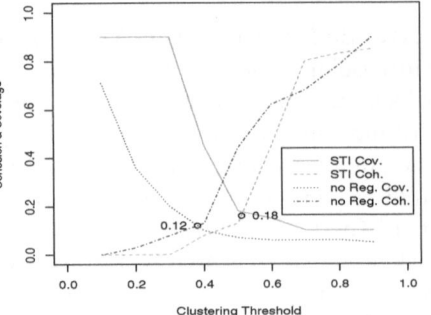

Fig. 3. Category Regularization Impact

Fig. 3 shows the impact of the entity category regularization, which has been discussed in Section 4.1. When learned the distance function with entity category regularization, the balance point of coverage and cohesion curves is 50% higher than that of without regularization. With regularization, the coverage of the identified search tasks is always higher. And when $\eta > 0.6$, the coverage and cohesion of identified search tasks with regularization are both higher than those of without regularization. The category regularization helps to overcome the over-fitting problem and reinforces that the queries containing named entities of the same category probably belong to the same search task.

6   Conclusion

In this paper, we propose an approach to identifying long-term and semantic-related search tasks in query log. A distance function between pair-wise queries from the semantic-related point of view is learned with the entity category regularization. Then, the hierarchical agglomerative algorithm is used to cluster semantic-related queries with the learned function. We compared our approach with two state-of-the-art ones. The result shows our approach outperforms the others significantly.

Our work can be further studied in two aspects. Firstly, how to represent a search task is important. A better representation is very helpful for users to understand the search task. Secondly, how to use the identified search tasks to improve user search experience can also be studied, e.g. to recommend more semantic-related queries to the user.

Acknowledgment. This work was supported by Foundation for Innovative Research Groups of the NSFC(Grant No. 60921002) and the National Key Technology R&D Program(Grant No. 2011BAH11B02).

References

1. Aiello, L.M., Donato, D., Ozertem, U., et al.: Behavior-driven Clustering of queries into topics. In: Proc. of the 20th CIKM, pp. 1373–1382 (2011)
2. Kotov, A., Bennett, P.N., White, R.W., et al.: Modeling and analysis of cross-session search tasks. In: Proc. of the 34th SIGIR (2011)
3. Boldi, P., Bonchi, F., Castillo, C., et al.: The query-flow graph: model and applications. In: Proc. of the 17th CIKM, pp. 609–618 (2008)
4. Broder, A.: A taxonomy of web search. SIGIR Forum 36, 3–10 (2002)
5. Donato, D., Bonchi, F., Chi, T., et al.: Do you want to take notes?: identifying research missions in Yahoo! search pad. In: Proc. of the 19th WWW (2010)
6. Guo, J., Xu, G., Cheng, X., et al.: Named entity recognition in query. In: Proc. of the 32nd SIGIR, pp. 267–274 (2009)
7. Ji, M., Yan, J., Gu, S., et al.: Learning search tasks in queries and web pages via graph regularization. In: Proc. of the 34th SIGIR, pp. 55–64 (2011)
8. Lucchese, C., Orlando, S., Perego, R., et al.: Identifying task-based sessions in search engine query logs. In: Proc. of the 4th WSDM, pp. 277–286 (2011)
9. Jones, R., Klinkner, K.L.: Beyond the session timeout: automatic hierarchical segmentation of search topics in query logs. In: Proc. of the 17th CIKM (2008)
10. Sadikov, E., Madhavan, J., Wang, L., et al.: Clustering query refinements by user intent. In: Proc. of the 19th WWW, pp. 841–850 (2010)
11. Spink, A., Park, M., Jansen, B.J., et al.: Multitasking during web search sessions. Information Processing and Management 42(1), 264–275 (2006)
12. Yin, X., Shah, S.: Building taxonomy of web search intents for name entity queries. In: Proc. of the 19th WWW, pp. 1001–1010 (2010)

Multi-verifier:
A Novel Method for Fact Statement Verification

Teng Wang[1,2], Qing Zhu[1,2], and Shan Wang[1,2]

[1] Key Laboratory of the Ministry of Education for Data Engineering and Knowledge
Engineering, Renmin University of China, Beijing, China
[2] School of Information, Renmin University of China, Beijing, China
{wangteng,zq,swang}@ruc.edu.cn

Abstract. Untruthful information spreads on the web, which may mislead users
or have a negative impact on user experience. In this paper, we propose a method
called *Multi − verifier* to determine the truthfulness of *a fact statement*. The basic
idea is that whether a fact statement is truthful or not depends on the information
related to the fact statement. We utilize a popular search engine to collect the
top-n search results that are most related to the target fact statement. We then
propose *support score* to measure the extent to which a search result supports the
fact statement, and we make use of *credibility ranking* to rank the search results.
At last, we combine the *support score* and *credibility ranking* value of a search
result to evaluate its *contribution* to a fact statement determination. Based on the
contributions of the search results, we determine the fact statement. Our proposals
are evaluated by experiments and results show availability and high precision of
the proposed method.

Keywords: Fact statements, Credibility ranking, Support score.

1 Introduction

The Web has become one of the most important information sources in people's daily
life, however, much untruthful information spreads on the web. Untruthful information
may mislead users or affect user experiences, therefore, it is urgent to determine whether
a piece of information is trustful or not. Information is mainly loaded by sentences.
Sentences that state *facts*, rather than *opinions*, are called fact statements[8]. Generally
speaking, fact statements are in two categories: positive fact statements and negative
fact statements. In this paper, we mainly focus on positive fact statements, since for
any negative fact statement, there is a positive one corresponding to it, and truthfulness
determination of a negative fact statement can be inferred through determination of the
positive one. In this paper, when we say a fact statement, we mean a positive one.

A fact statement is either trustful or untruthful, depending on wether the contents
it carries is a objective fact or not. When a user determine the trustfulness of a fact
statement, he/she will specify some part(s) of the fact statement he/she is not sure about.
The part(s) are called the doubt unit(s) [8]. After the doubt unit(s) is/are specified, the
fact statement can be regarded as an answer to a question. Based on the number of
correct answers of the corresponding questions, fact statements can be divided into two

Y. Ishikawa et al. (Eds.): APWeb 2013, LNCS 7808, pp. 526–537, 2013.

categories: unique-answer fact statement and multi-answer fact statement. For example, *Obama is the American President* is a trustful fact statement. If the doubt unit is *Obama*, it is an answer of *who is American President*. Since the question has only one correct answer, the fact statement is an unique-answer fact statement.

Most of the existing works utilize a popular search engine to collect the search results of the target fact statement, and the truthfulness of the target fact statement is determined by analyzing the search results, e.g., the amount and the sentiments of the search results are considered as the dominant factors in some works. However, this is not desirable for the following reasons: (i). The amount of the search results is not always reliable in trustfulness determination. Since not all search results support the target fact statement. (ii).The sentiments of the search results do not always represent their sentiments on the target fact statement, since not every search results' theme happens to be the fact statement. Fig.1 shows a search result about the fact statement *Warren Moon was born in 1956*. Obviously, the theme is not the fact statement. In other works, the trustful fact statement is identified from the alternative fact statements of a target fact statement. Here, the alternative statements are the fact statements which are answers to the question corresponding to the target fact statement. There are three limitations in these works: (i). The doubt unit must be specified, otherwise the alternative fact statements can't be found. (ii). If the doubt unit includes much information, it is difficult, sometimes even impossible to find the proper alternative fact statements. (iii). This method is not fit for the trustfulness determination of multi-answer fact statements, since only one fact statement is picked out as the trustful fact statement.

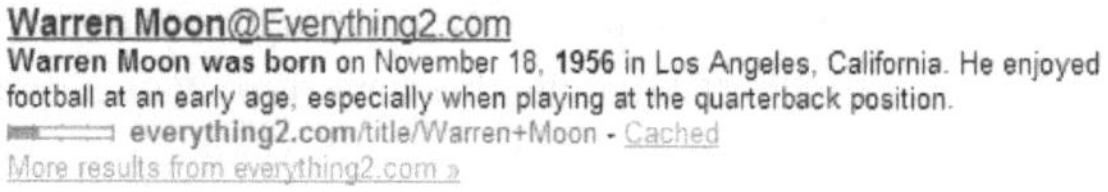

Warren Moon@Everything2.com
Warren Moon was born on November 18, **1956** in Los Angeles, California. He enjoyed football at an early age, especially when playing at the quarterback position.
everything2.com/title/Warren+Moon - Cached
More results from everything2.com »

Fig. 1. A example of search result

In order to solve these problems, we propose a method, called Multi-verifier, to determine the truthfulness of a fact statement, which is suitable for the truthfulness determination of multi-answer fact statements. In summary, the contributions of this paper are summarized as follows.

- We propose a novel method, called Multi-verifier, to determine the truthfulness of a fact statement. Multi-verifier can be used for the truthfulness determination of multi-answer fact statements.
- We measure the support score of a search result to the target fact statement. We propose a way based on grammatical relationships between words to extract the words, which make sense for the truthfulness determination, from the search result and measure the support score of the search result to the fact statement based on the words.
- We rank the search results on the target fact statement based on credibility. We consider the importance and the popularity of the sources from which the search results are derived and get the importance ranking and popularity ranking of the search results. Through merging the two rankings, the credibility ranking of the search results is captured.

- We run a set of experiments and evaluate the method in terms of reasonability and precision. The results show availability and high precision of our proposals.

For the rest of this paper, we first summarize the related works. Then, we describe the method in details. Experiments and analysis are shown in section 4. At last, we conclude this paper.

2   Related Works

We conduct a brief review of existing works on the following aspects: web page credibility and statement trustfulness.

Generally speaking, credible sources are likely to present trustful information. Some researchers study credibility of web pages, and they utilize web page features to determine whether a web page is credible or not, e.g., page keywords, page tittle, page style, etc[1][2][3]. However, the trustfulness of web page content hasn't been considered in these works. Some web pages which meet these features may present incorrect facts. Other studies focus on spam web pages detection based on link analysis and content for filtering out low quality pages[4][5]. However spam web pages are not equal to untruthful web pages and non-spam web pages may present incorrect facts. These studies can't be used in fact statement determination directly.

Some researchers have started discussing the truthfulness of individual statements in very recent years. [6] introduces system Honto?search1.0, which doesn't determine an uncertain fact directly, but provides users with additional data on which users can determine the uncertain fact. Authors believe that if an uncertain fact is trustful, the sentiments of the search results on the uncertain fact should be consistent. Honto?search2.0[7] is the improved version of Honto?search1.0. This system has two function: comparative fact finding and aspect extraction. Given an uncertain fact, the system collects comparative facts and estimates the validity of each fact. Just for checking the credibility of these facts in details, necessary aspects about the uncertain fact are extracted. Scoring the facts and the aspects, users can get more information and determine the uncertain fact. Verify[8] can determine the truthfulness of the fact statement directly. Given a target fact statement, Verify finds the alternative fact statements of the fact statement, ranks these fact statements, and chooses the fact statement at the highest position as the trustful fact statement. In [7][8], doubt unit(s) of the fact statement should be specified. In addition, when the doubt unit includes much information, it is difficult, sometimes even impossible to find the proper alternative(comparative) fact statements. Especially, for multi-answer fact statements, they can't work well.

3   Proposed Solution

The goal of this paper is to determine whether a fact statement is trustful or not. Firstly, we submit the target fact statement to a popular search engine, and get the top-n search results; secondly, the *support scores* of the search results of the fact statement are measured; thirdly, the credibility ranking of the search results is captured; at last, based on the support scores and credibility ranking, the truthfulness of the fact statement is determined.

3.1 Notation

We use fs to denote the target fact statement. After deleting stop words in fs, the rest are the keywords of fs, which is represented by K. R denotes the collection of the search results of fs. We set $R = \{r_1, \ldots, r_n\}(n \geq 2)$ and $r_i = < t_i, u_i, s_i > (1 \leq i \leq n)$. Suppose that r_i is extracted from web page p_i, t_i and u_i denote the *title* and *url* of p_i respectively. And s_i is the snippet related to fs in p_i. K_{r_i} is the collection of words which belong to both K and r_i. c_i is used to denotes the shortest consecutive sentences including K_{r_i} in r_i. $sup(r_i, fs) \in [0, 1]$ is used to describe the support score of r_i to fs.

3.2 Support Score Measurement

Not all words and sentences in a search result make sense for the trustfulness determination of the target fact statement. The words and sentences in the search result, which make sense for the truthfulness determination of the fact statement, are called *necessary words* and *necessary sentences* respectively. Necessary words are extracted from necessary sentences, but not every word in necessary sentences is necessary. Fig.1 shows a search result on *Moon was born in 1956*. Obviously, *Born Harold Warren Moon, November 18, 1956...* is necessary. In the sentence, *Los Angeles, CA* are not necessary words.

In order to measure the support score of a search result more accurately, we extract the necessary words from the search result, and compute the corresponding support score based on the necessary words. The process to measure the support score is composed of two stages: necessary words detection and support score computation.

Necessary Words Detection. Given a search result r_i, the sentences in c_i are necessary sentences, since c_i is the shortest consecutive sentences including K_{r_i} in r_i. We analyze the relationships of words in c_i and find the necessary words of r_i. In this paper, we use Stanford Parser to find the grammatical relationships of words in c_i.

Stanford Parser[1] is used to find the grammatical relationships of words in a sentence[10]. The grammatical relationship format is $d.name(d.dependent, d.governor)$. For example, $amod(great, president)$ is a grammatical relationship. Here, *great* is a dependent and *president* play a role of governor. *amod* is the name of the relation. The relation means *great* is the modifier of *president*. There are 52 grammatical relationships in Stanford typed dependencies. We divide the grammatical relationships provided by Stanford Parser into two categories based on their importance to the skeleton of a sentence: essential grammatical relationships, represented by R_e and unessential grammatical relationship, represented by R_o.

N_i is used to denote the collection of necessary words in r_i. $D_i = \{d_{i1}, \ldots, d_{im}\}(1 < m)$ denotes the collection of the grammatical relationships between words in c_i. We take the following steps to find N_i.

- Capture the shortest consecutive sentence c_i in r_i;
- Get the collection of the grammatical relationships between the words in c_i by Stanford parser and set N_i to K_{r_i}.
- Use heuristic rules on the collection of grammatical relationships to find N_i.

---

[1] http://nlp.stanford.edu/software/stanford-dependencies.shtml

The heuristic rules are as follows: (i). If $d_{ij}(1 \leq j \leq m)$ is essential, $d_{ij}.dependent$ and $d_{ij}.governor$ are put into N_i. (ii). If $d_{ij}(1 \leq j \leq m)$ isn't essential and $d_{ij}.dependent \in N_i$, $d_{ij}.governor$ are put into N_i.

Algorithm 1 shows the procedure of finding the necessary word collection of a search result. The input is the search result $r_i =< t_i, u_i, s_i >$, the set of words which belong to the keyword set of fs and the search result, represented by K_{r_i}, the essential grammatical relationship collection R_e and the unessential grammatical relationship collection R_o. While the output is the necessary word collection N_i. Firstly, c_i is captured. If t_i contains the all words of K_{r_i}, c_i is set to t_i; if not, c_i is the shortest consecutive sentences containing K_{r_i} in s_i (Line1-7). Secondly, the collection of grammatical relationships between words in c_i, called D_i, is found out(Line 8). Thirdly, N_i is set to K_{r_i}(Line 9). Fourthly, retrieving D_i and applying the heuristic rules, new necessary words are put into N_i (Line 12-17). The algorithm repeat the fourth step, until no new necessary word is found(Line 10-18).

Algorithm 1. Finding necessary words

    Input : r_i, R_e, R_o, K_{r_i}
    Output: N_i
1 $c_i = s_i$;
2 **if** t_i *includes the all words in* K_{r_i} **then**
3      $c_i = t_i$;
4 **else**
5   **for** *each consecutive sentences* s_{ij} *in* s_i *and* s_{ij} *includes the all words in* K_{r_i} **do**
6   **if** $len(s_{ij}) \lesssim len(c_i)$ **then**
7            $c_i = s_{ij}$;
8 D_i=Stanford Parser(c_i);
9 $N_i = K_{r_i}$;
10 **while** $LastN_i \neq N_i$ **do**
11     $LastN_i = N_i$;
12   **for** *each* $d_{ij} \in D_i$ **do**
13   **if** $d_{ij}.name \in R_e$ **then**
14          $N_i \leftarrow d_{ij}.governor$; $N_i \leftarrow d_{ij}.dependent$;
15   **if** $d_{ij}.name \in R_o$ **then**
16   **if** $d_{ij}.dependent \in N_i$ **then**
17            $N_i \leftarrow d_{ij}.governor$;
18 **return** N_i;

Support Score Computation. We compute $sup(r_i, fs)$ based on the distance between r_i and fs, represented by $dis(r_i, fs)$. $dis(r_i, fs)$ can be computed based on the distance of N_i and K. Inspired by edit distance, we define the distance between two sets.

Definition 1 (Set distance). *The distance of two sets is the number of insertion or deletion operations required to transform a set into another.*

For example, $S_1 = \{a,b,c\}$ and $S_2 = \{b,c,d\}$ are two set. If a is deleted from S_1 and d is inserted in S_1, $S_1 = S_2$. So the set distance of S_1 and S_2 is 2.

We use $Sdis(N_i, K)$ to denote the set distance of N_i and K. Here, K is the collection of keywords of fs. In order to get accurate $Sdis(N_i, K)$, the words in K or N_i are stemmed before computing $Sdis(N_i, K)$. However, $Sdis(N_i, K)$ can't be used to measure $dis(r_i, fs)$ directly. The reason is that fs can be express in a different way. Given a sentence fs_1, which is different from fs but express the meaning of fs. fs can be transformed into fs_1 by word change. We suppose there are some words in fs can't be changed from fs to $fs1$. And if these words changes, the meaning fs can't be kept. Here we use Du to denote the collection of these unchangeable words in fs. Obviously, $Du \in K$. Since word matching is a factor considered in returning search results, Du is composed of the nouns and numbers in fs. Based on K and Du, we give out an equation to compute $dis(r_i, fs)$.

$$dis(r_i, fs) = \begin{cases} \frac{Sdis(N_i, K) + |Du| - |K|}{|K|} & \text{if } Sdis(N_i, K) + |Du| - |K| > 0 \\ 0 & \text{if } Sdis(N_i, K) + |Du| - |K| <= 0 \end{cases} \quad (1)$$

Inspired by inverse hyperbolic function, we use the following equation to compute $sup(r_i, fs)$. The value of α is determined in our experiments.

$$sup(r_i, fs) = \begin{cases} \frac{\alpha^{-2*dis(r_i, fs)} - 1}{\alpha^{-2*dis(r_i, fs)} + 1} & \text{if } dis(r_i, fs) > 0 \\ 1 & \text{if } dis(r_i, fs) = 0 \end{cases} \quad (2)$$

In support score computation, we do not consider whether the search result oppose the fact statement or not. Since the focus is the positive fact statements in this paper and there are few search results explicitly opposing the target positive fact statement. In addition, for the search results implicitly opposing the fact statement, corresponding support scores may be smaller based on the equations or the search results will be considered neutral.

3.3 Credibility Ranking Capture

If a fact statement is supported by more credible search results, the fact statement is more likely to be trustful. Thus, the credibility of the search results is an important factor in fact statement determination. Usually, credible search results are likely to be presented by credible web sites/pages. The credibility of search results can be measured by the credibility of the web sites/pages presenting them. Moreover, reputable web sites/pages are likely to be credible[1][2][9]. If a web site/page is popular and important, it could be reputable. The credibility of the web site/page can be measured based on its importance and popularity.

Importance Ranking. The search results of a given fact statement display in a certain order. We consider the order as the importance ranking of the search results. The order is brought by the search engine which is used to find the search results. Besides, the importance of the sources of the search results are considered by the search engine. For a fact statement fs and a search result r_i on fs, $Srank$ is used to denote the importance ranking of the search results on fs, and $Srank_i$ is the position of r_i in $Srank$. Given another search result r_j on fs, if $Srank_i < Srank_j$, r_i is more important than r_j. Note that, we do not adopt the popular Pagerank for the following reasons: (i). The Pagerank values of web sites/pages can't be derived. (ii). Although the Pagerank levels are available, the range of Pagerank level is too small ([0, 10]) to satisfy our requirements.

Popularity Ranking. We use $Arank$ to denote the popularity ranking, $Arank_i$ represents the $Arank$ value of r_i. We capture the popularity ranking of search results based on Alexa[2] ranking. Alexa ranking is a traffic ranking of each site based on reaches and page views. It can be used to measure popularity of web sites. Given a fact statement fs and a collection of search results R on fs. $Alexa_i$ is used to denote Alexa ranking value of the site from which r_i is derived. Based on $Alexa_i (1 \leq i \leq n)$, we can get the popularity ranking of the search results in R. Based on Alexa ranking, $Arank_i$ can be derived by the following equations.

$$IntervalAmid = \min((\max(Alexa_i) - Alexa_m),$$
$$(Alexa_m - \min(Alexa_i))) * 2/(n-1) \tag{3}$$

$$Amid_i = 1 + [\frac{Alexa_i - \min(Alexa_i)}{IntervalAmid}] \tag{4}$$

$$IntervalArank = (\max(Amid_i) - 1)/(n-1) \tag{5}$$

$$Arank_i = 1 + [\frac{Amid_i - 1}{IntervalArank}] \tag{6}$$

Since Alexa ranking is for all web sites, given $Alexa_i$ and $Alexa_j$, the gap between them may be very large. We use the above four equations to map $Alexa_i$ into range [1, n]. We first introduce the median of $Alexa_i$ represented by $Alexa_m$ and get the interval ranking $Amid$ by Equation.3 and Equation.4, just for keeping the distribution of $Alexa_i (1 \leq i \leq n)$ unchanged. Then, we linearly map $Amid_i$ into [1, n] and get $Arank_i$ by Equation.5 and Equation.6.

Credibility Ranking. We use $Crank$ to denote the credibility ranking of the search results on the same fact statement. $Crank_i$ is the position of r_i in $Crank$. Based on $Srank_i$ and $Arank_i$, $Crank_i$ can be computed. The follow equation describes how to compute $Crank_i$ based on $Srank_i$ and $Arank_i$.

$$Crank_i = \lceil w_1 * Srank_i + w_2 * Arank_i \rceil \tag{7}$$

Here, $Srank$ and $Arank$ play the same important roles in getting $Crank$. We set w_i and w_2 to 0.5.

---

[2] https://www.alexa.com

3.4 Fact Statement Determination

The search results on a given fact statement are in two categories: positive search results and neutral search results. S_{pos} denotes the collection of the positive search results and S_{neu} is the collection of the neutral search results. Given a search result r_i, if $K_{r_i} \supseteq Du$ (Du is the unchangeable word collection of fs), $r_i \in S_{pos}$; otherwise, $r_i \in S_{neu}$. We define the contributions of r_i as the ratio of $sup(r_i, fs)$ and $Crank_i$. The $\triangle_{Pos}$ is the sum of contributions of the search results in S_{pos}. Similarly, $\triangle_{Neu}$ is the sum of contributions of the search results in S_{Neu}.

We believe if a fact statement is trustful, $\triangle_{Pos}$ should be larger than $\triangle_{Neu}$. Here, $\triangle_{Pos}/\triangle_{Neu}$ is the key factor in fact statement determination. We make use of a threshold δ. If $\triangle_{Pos}/\triangle_{Neu}$ is larger than or equal to δ, fs is trustful; otherwise, fs is untrustful.

Algorithm 2. Fact statement determination

    Input : $R = \{r_1, \ldots, r_n\}$, $Crank$, $sup(r_i, fs)$, $K_{r_i}(i = 1, \ldots, n)$, Du, δ
    Output: trustful or untruthful
1 $\triangle_{Neu} = \triangle_{Pos} = 0$;
2 **for** *each* $r_i \in R$ **do**
3   **if** $K_{r_i} \supseteq Du$ **then**
4        $\triangle_{Pos} += \frac{sup(r_i, fs)}{CRank_i}$;
5   **else**
6        $\triangle_{Neu} += \frac{sup(r_i, fs)}{CRank_i}$;
7 **if** $\frac{\triangle_{Pos}}{\triangle_{Neu}} \geq \delta$ **then**
8   **return** trustful;
9 **else**
10   **return** untruthful;

Algorithm 2 shows the procedure of determining a fact statement. The input of the algorithm is the collection of search results on the target fact statement R, the credibility ranking of the search results $Crank$, the support scores of the search results to the fact statement $sup(r_i, fs)(1 \leq i \leq n)$, the set of words which belong to the keyword set of fs and the search result, represented by K_{r_i}, the unchangeable word collection of the fact statement Du, and the threshold δ. While the output is the result of determination (*trustful* or *untruthful*). First, the contributions form positive search results and neutral search results are worked out respectively (Line 1-6). Secondly, the ratio between positive contributions and neutral contributions are computed. At last, the algorithm can decide whether the fact statement is truthful or not (Line 7-10).

4 Experiments

We run a set of experiments to evaluate Multi-verifier in terms of precision and availability. First, we detect the distribution of the search results including the target fact statements; secondly, we evaluate the availability of the method to capture credibility ranking; thirdly, we evaluate the availability of the method to compute support score; at last, we measure precision of Multi-verifier while varying threshold δ and the number of adopted search results n.

We make use of a synthetic dataset, which can be generated by the method introduced in [8]. The dataset is composed of 50 trustful fact statements and 50 untruthful fact statements. These fact statements are fetched from Trec2007[3]. In the trustful fact statements, 30 are unique-answer ones and the rest are multi-answer ones. Yahoo boss 2.0[4] is used to collect search results. Here, for each fact statement, top-150 search results are collected. 11 experienced users, who are graduate students and experienced Internet users, help to mark the search results. If the search result includes the meaning of the corresponding fact statement, it is marked as 0, otherwise, it is marked as 1. The experiments run on an Intel Core 2 Quad 2.66Hz, windows 7 machine with 2GB main memory.

4.1 Distribution of Search Results

In this experiment, we detect the average percentage of the search results which include the meanings of the target fact statements. For untruthful fact statements, there are few search results including them. Thus, this experiment is about trustful fact statements. We use P_f to represent the average percentage of the search results including the target fact statements. fs_u denotes the trustful unique-answer fact statements, fs_m represents the trustful multi-answer fact statements. Fig.2 shows P_f values when n changes. Here, n is the number of the search results adopted for a fact statement. It can be seen from the figure, both P_f for fs_u and P_f for fs_m decrease with the increase of n. For a multi-answer fact statement, there are more than one correct answers to the corresponding question. Thus P_f for fs_u is always larger than the P_f for fs_m. In addition, when n increases, more search results which don't include the target fact statements come out.

4.2 Evaluation of Credibility Ranking

By computing the *spearman correlation* value between importance ranking($Srank$) and popularity ranking($Arank$), we discuss whether the two rankings can replace each other or not. Fig.3 shows the average correlative value of $Srank$ and $Arank$, when n changes. It can be seen that, the average correlative value is always smaller than 0.63; when $n = 40$, the average correlative value is at the peak; when $n \gtrsim 40$, the average correlation value decreases with the increase of n. Therefore, $Srank$ and $Arank$ can not replace each other.

In the experiments, distributions of credibility ranking ($Crank$) and popularity ranking ($Arank$) are detected. Fig.4 shows distributions of credibility ranking and popularity ranking. The horizontal axis represents the positions of search results in importance ranking($Srank$), and the vertical axis represents the average credibility rank values or popularity rank values of the search results at corresponding positions . It can be seen in Fig.4, the average $Arank$ values increase with the increase of $Srank$. However, changes

---

[3] http://trec.nist.gov
[4] http://boss.yahoo.com

of average *Crank* values are not so sharply. It means that the *Arank* or *Srank* can not replace *Crank*, and some search results at higher positions in *Srank* or *Arank* may be at lower positions in *Crank*. It is consistent with our observation.

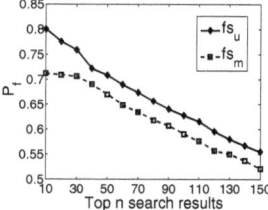
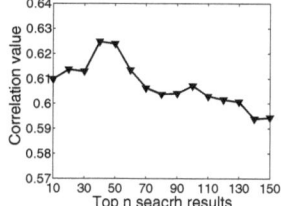
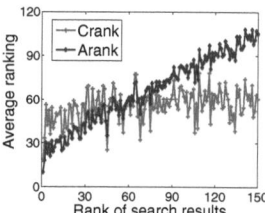

Fig. 2. Average percent on n **Fig. 3.** Correlation of *Srank* and *Arank*   **Fig. 4.** Distribution of *Crank* and *Rank*

4.3 Evaluation of Support Score

We evaluate the availability of the way to measure the distance of a search result and the target fact statement and ascertain the optimal values of α by experiments.

Fig.5(a) shows the average distance for trustful unique-answer fact statements, when n changes. rd_{su} represents the average distance for these search results which do not included the meaning of the corresponding fact statement. rd_{si} denotes the average distance for these search results including the meaning of the target fact statements. It can be seen from the figure, rd_{si} is smooth. With n increases, rd_{su} increases. rd_{su} is always greater than rd_{si}. The difference between rd_{su} and rd_{si} is larger than 0.3. Fig.5(b) shows the average distance for trustful multi-answer fact statements, when n changes. rd_{su} and rd_{si} represent the same as them in Fig.5(a). In this figure, rd_{su} is always above rd_{si} and the smallest difference is larger than 0.4. Fig.5(c) shows the average distance for un-truthful fact statements, when n varies. In Fig.5(c), rd_{su} is always above rd_{si} and the smallest difference is greater than 0.8. From the experiments, we can know whether a search result includes the target fact statement can be reflected by the distance.

In computation of support scores, α plays an important role (Equation.2). We believe if the value of α makes the difference between the average support scores of the search results including target fact statements and that of the other search results reach a maximum, the α value is optimal. Fig.6 shows the difference on $n = 150$, when α changes. From this figure, when $\alpha = 1.3$, the difference reach the peak(0.2). It means 1.3 is the optimal value of α.

4.4 Evaluation of Truthfulness Determination

During the truthfulness determination of fact statements, the number of the search results adopted for a fact statement n and the threshold δ influence the precision of the truthfulness determination. In the experiments, we measure the precision of Multi-verifier, based on various n and δ values. Since δ is more than one, the range of δ is from 1 to 5.

Fig.7 shows the precision on n when $\delta = 1.5$. From this figure, when the $n = 30$, the precision reaches the peak(0.9). When $n \lesssim 30$, the precision is enhanced with the

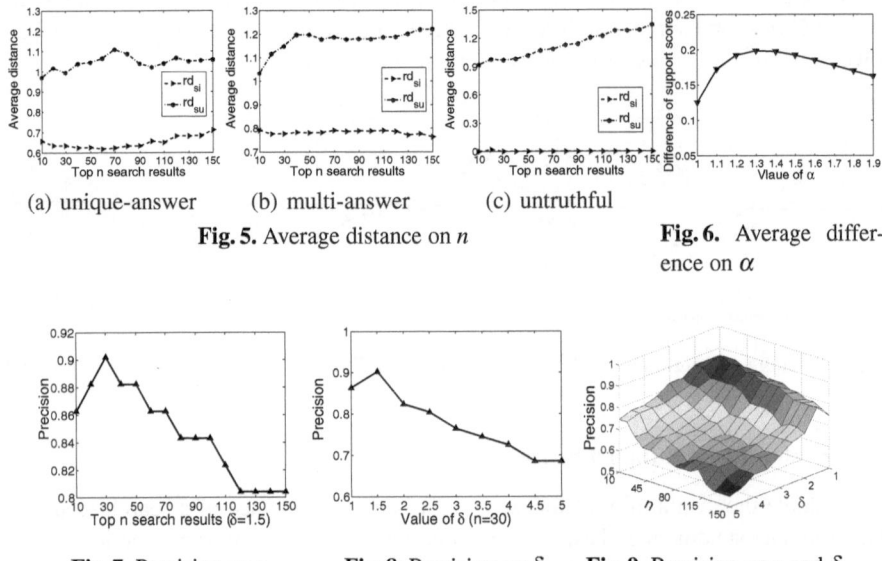

(a) unique-answer   (b) multi-answer   (c) untruthful

Fig. 5. Average distance on n

Fig. 6. Average difference on α

Fig. 7. Precision on n **Fig. 8.** Precision on δ **Fig. 9.** Precision on n and δ

increase of n. The reason is that more search results are positive in top 30 search results, and more positive search results are considered with the increase of n. When $n \gtrsim 30$, the precision decreases with the increase of n. The reason is that more neutral search results are considered with increase of n when $n \geq 30$. Fig.8 shows the precision on δ when $n = 30$. From this figure, When $\delta = 1.5$, the precision reaches the greatest value(0.9). When $\delta \leq 1.5$, the precision is enhanced with the increase of δ. Since, when δ is smaller, some untruthful fact statements are regarded as trustful ones. However, when δ is greater, some trustful fact statements are regarded as untruthful ones. Thus, when $\delta \gtrsim 1.5$, the precision decreases with the increase of δ. Fig.9 shows the precision on δ and n. From this figure, the overall trend is that the precision increases and then decreases as the increase of n and δ. Especially, when $\delta = 1.5$ and $n = 30$, the precision reach the peak(0.9).

The experiments show the method is available and accurate. Since the dataset includes many multi-answer fact statements, we can see that the method is also available for multi-answer fact statements determination.

Since Honto?search can not determine a fact statement directly, it is not necessary to run experiments to compare it with our method. Although Verify can directly determine a fact statement, it is based on alternative fact statements and not suitable for multi-answer fact statements. Thus, it is not necessary to compare it with our method.

5   Conclusion and Future Work

In this paper, we propose a method called $Multi-verifier$ to determine the truthfulness of a fact statement. In $Multi-verifier$, the search results on the fact statement,whose truthfulness is needed to be determined, are found by a popular search engine. And the support scores and the credibility ranking are considered in determining the truthfulness of the fact statement. The experiments shows the method is powerful. However, our focus is domain-independent fact statements, and the domain knowledge isn't used. In the future, we will focus on the domain-dependent fact statements, we think that usage of domain knowledge can bring higher precision. In addition, the quality of related information is important for the truthfulness determination. We will try to find a method to enhance the quality of related information on a fact statement.

Acknowledgments. This work is partly supported by the Important National Science & Technology Specific Projects of China (Grant No.2010ZX01042-001-002), the National Natural Science Foundation of China (Grant No.61070053), the Graduates Science Foundation of Renmin University of China(Grant No.12XNH177) and the Key Lab Found (07dz2230) of High Trusted Computing in Shanghai.

References

1. McKnight, D.H., Kacmar, J.: Factors and effects of information credibility. In: 9th International Conference on Electronic Commerce, pp. 423–432. ACM Press, NewYork (2007)
2. Schwarz, J., Ringel Morris, M.: Augmenting web pages and search results to scport credibility assessment. In: International Conference on Human Factors in Computing Systems, pp. 1245–1254. ACM Press, NewYork (2011)
3. Lucassen, T., Schraagen, J.M.: Trust in Wikipedia: How Users Trust Information from an Unknown Source. In: 4th ACM Workshop on Information Credibility on the Web, pp. 19–26. ACM Press, New York (2010)
4. Gyongyi, Z., Garcia-Molina, H., Pedersen, J.: Combating Web Spam with TrustRank. In: 30th International Conference on Very Large Data Bases, pp. 576–587. Morgan Kaufmann, San Fransisco (2004)
5. Ntoulas, A., Najork, M., Manasse, M., Fetterly, D.: Detecting spam web pages through content analysis. In: 15th International Conference on World Wide Web, pp. 83–92. ACM Press, NewYork (2006)
6. Yamamoto, Y., Tezuka, T., Jatowt, A., Tanaka, K.: Supporting Judgment of Fact Trustworthiness Considering Temporal and Sentimental Aspects. In: Bailey, J., Maier, D., Schewe, K.-D., Thalheim, B., Wang, X.S. (eds.) WISE 2008. LNCS, vol. 5175, pp. 206–220. Springer, Heidelberg (2008)
7. Yamamoto, Y., Tanaka, K.: Finding Comparative Facts and Aspects for Judging the Credibility of Uncertain Facts. In: Vossen, G., Long, D.D.E., Yu, J.X. (eds.) WISE 2009. LNCS, vol. 5802, pp. 291–305. Springer, Heidelberg (2009)
8. Li, X., Meng, W., Yu, C.: T-verifier: Verifying Truthfulness of Fact Statements. In: 27th International Conference on Data Engineering, pp. 63–74. IEEE Press, New York (2011)
9. Fogg, B.J.: Persuasive Technology: Using Computers to Change What We Think and Do. Morgan Kaufmann, San Francisco (2002)
10. Marneffe, M., Maccartney, B., Manning, C.: Generating typed dependency parsers from pharse structure parses. In: The International Conference on Language Resources and Evaluation, ELRA, Luxembourg (2006)

An Efficient Privacy-Preserving RFID Ownership Transfer Protocol

Wei Xin, Zhi Guan, Tao Yang*, Huiping Sun, and Zhong Chen

Institute of Software, EECS, Peking University, Beijing, China
MoE Key Lab of High Confidence Software Technologies (PKU)
MoE Key Lab of Network and Software Security Assurance (PKU)
{xinwei,guanzhi,ytao,sunhp,chen}@infosec.pku.edu.cn

Abstract. RFID technology is increasingly become popular in supply chain management. When passing tags on to the next partner in the supply chain, ownership of the old partner is transferred to the new partner. In this paper, we first introduce some existing RFID tag ownership transfer protocols, then give the security and privacy requirements for such kind of protocols, finally, we propose a novel RFID tag ownership transfer protocol which supports constant-time authentication, and effectively protects the privacy of the old tag owner and the new tag owner.

Keywords: RFID, Ownership transfer, security, privacy, supply chain.

1 Introduction

Radio Frequency Identification (RFID) technology represents a fundamental change in the information technology infrastructure. RFID has many applications for both business and private individuals. Several of these applications will include items that change owners at least once in its lifetime. The swapping and resale of items is a practice that is likely to be popular in the future, and so any item that depends on RFID for function or convenience should be equipped to deal with change of ownership. Ownership transfer presents its own set of threats, and therefore demands the attention of security researchers.

Generally speaking, a secure ownership transfer protocol should follow the following assumptions: The old owner should not be able to access the tag after the ownership transfer has taken place; The new owner should be able to perform mutual authentication with the tag after the ownership transfer has taken place.

In this paper, we concentrate on designing a RFID ownership transfer protocol with high efficiency and robust security and privacy properties. Our protocol consists of three sub-protocols: an authentication protocol, an ownership transfer protocol, and a secret update protocol. The rest of the paper is organized as follows. In Section 2, we introduce a brief overview of some ownership transfer protocols in RFID systems. Section 3 presents the security and threat model. Section 4 describes the proposed ownership transfer protocol. Sections 5 demonstrates security and privacy analysis. And finally, Section 6 concludes.

---

* Corresponding author.

Y. Ishikawa et al. (Eds.): APWeb 2013, LNCS 7808, pp. 538–549, 2013.
© Springer-Verlag Berlin Heidelberg 2013

2   Related Work

One of the earliest protocols addressing ownership transfer with Trusted Third Party (TTP) was proposed by Saito et al.[1]. The goal of the protocol is to transfer ownership from one entity R_1 to another R_2. The protocol steps are summarized in Fig. 1. An adversary can commit de-synchronization attacks by blocking the message from R_2 to the tag. TTP and R_2 have the new key, while the tag keep the previous one. This can be prevented by TTP and R_2 storing the previous key. Unfortunately, in this case, since R_2 knows s_1, backward security is violated.

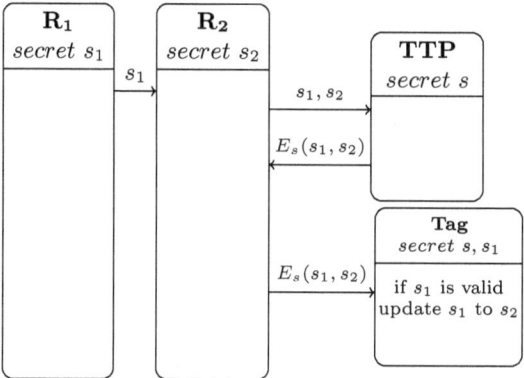

Fig. 1. Saito's three party model

Kapoor and Piramuthu proposed a lightweight and secure ownership transfer protocol with TTP [2]which was shown in Fig. 2. In order to provide backward unlinkability, the new owner R_2 does not have access to the key of the previous R_1. In addition, R_2 does not update the shared key unless he receives an acknowledgment from T in order to solve de-synchronization issues. However, as the scheme relies on symmetric primitives, it suffers from linear time authentication and is not suited for resource limited tags.

Song put forward an ownership transfer protocol without TTP [3] in 2008. The protocol consists of two sub-protocols: an ownership transfer protocol P_1, and a secret update protocol P_2. P_1 transfers ownership from one owner R_1 to another R_2 and the second stage R_2 updates the secret. The protocol P_1 was shown in Fig. 3. The protocol suffers from linear complexity problems that tag authentication is linear in the number of tags. Another vulnerability is de-synchronization problems when an adversary blocks the last communication $M3$ between R_2 and the tag.

Elkhiyaoui et al. proposed an ownership transfer protocol without TTP [4](see Fig. 4) achieving constant time authentication and issuer verification. Elkhiyaoui's protocol violates tag owner privacy that any legitimate owner which has been the owner of a specific tag is able to trace it by computing pairing of $e(v_i, g_2)$.

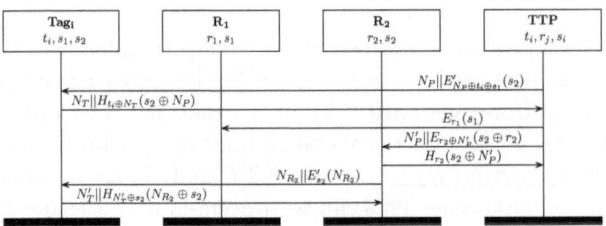

Fig. 2. Protocol of Kapoor and Piramuthu

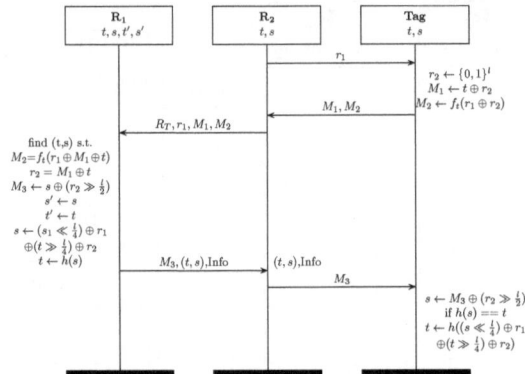

Fig. 3. Protocol of Song

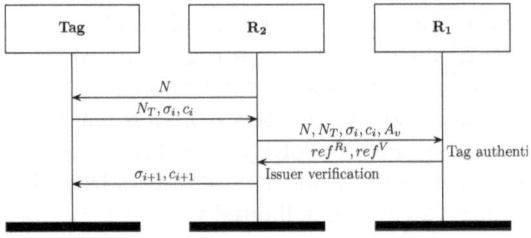

Fig. 4. Protocol of Elkhiyaoui

3   Security and Threat Model

3.1   Privacy Issues

RFID privacy is one of the areas which are most discussed in recent years. There are some common notions of privacy in RFID system:

1. *Anonymity*: an adversary $\mathcal{A}$ should not be able to disclose the identity of tags he reads from or writes into.

2. *Untraceability*: an adversary $\mathcal{A}$ should not be able to trace or track the person(product) attached with tags by a fake reader. *Untraceability* contains a special case called *backward untraceability*, which implies that an adversary $\mathcal{A}$ should not be able to to trace past transactions between an owner and a tag, even capturing the tag's secret information.

A great number of privacy-preserving RFID protocols have been proposed in the literature. Tag *anonymity* corresponds to anonymization protocol, one trivial approach is proposed by Sarma named hash-lock[5], using meta-ID instead of real ID. *Untraceability* implies that the tag response differently upon receiving the same request. The approach to reach *untraceability* is either based on the psedudorandom numbers or the random noise [6,7,8]. The Ohkubo-Suzuki-Kinoshita protocol (OSK) [9,10] made backward privacy possible, the scheme relies on the use of two one-way hash functions. Variants of the OSK scheme proposed in [11], making it resistant to replay attacks. An further improvement to the OSK [12] was proposed by Berbain using a pseudorandom number generator and a universal family of hash functions, moreover, they provided a provable secure proof under standard model. Recently, Ma [13] gave a simpler one of [12] using tags only equipped with pseududorandom generator.

In addition to the common notions of privacy, there are two extra privacy issues dedicated for ownership transfer protocols listed as follows:

1. *New owner privacy*: Once ownership of a tag has been transferred to a new owner, only the new owner should be able to identify and control the tag. The previous owner of the tag should no longer be able to identify or trace the tag.
2. *Old owner privacy*: When ownership of a tag has been transferred to a new owner, the new owner of a tag should not be able to trace past interactions between the tag and its previous owner.

3.2   Attacks on Ownership Transfer Protocols

An adversary $\mathcal{A}$ of the system is a malicious entity whose aim is to perform some attacks, either through the wireless communications between readers and tags (e.g., eavesdropping), or on the RFID devices themselves (e.g., corruption a device and getting all the information stored on it). We mainly consider the following attacks:

1. *Replay attacks*: Attackers intercept a valid response emitted by a tag and retransmit it to a legitimate reader.
2. *Man-in-the-middle attacks*: In this case, an attacker is able to insert or modify messages exchanged by legitimate principals without detection.
3. *Eavesdropping*: Attackers listen passively to messages exchanged by legitimate principals and are able to decode them.
4. *Denial-of-service (DoS)*: The attacker disturbs or impedes communications between principals.

5. *De-synchronization*: Attackers makes the owner and the tag no longer be able to successfully communicate with each other because they are using different keys.

4   RFID Protocol for Tag Ownership Transfer

Our protocol works under the following assumptions. A TTP maintains a secure database of information for all the readers and the tags that it owns, a reader have significantly greater processing ability than a tag. Each tag has a rewritable memory that may not be tamper-resistant, can generate pseudorandom numbers.

We use Song's protocol and Elkhiyaoui's protocol as our main references, Song's scheme suffers from linear complexity problems that tag authentication is linear in the number of tags, we address it by using Elgamal re-encryption mechanisms introduced by Elkhiyaoui. However, Elkhiyaoui's protocol have the vulnerabilities that any legitimate owner which has been the owner of a specific tag is able to trace it. We solve this problem by updating the index I in ownership transfer process. In addition, we use only pseudorandom number generator at the tag side which is well suited for resource limited passive RFID tags. To sum up, we try to design a ownership transfer protocol satisfying the following properties:

1. Constant-time for tag ownership transfer.
2. Old owner privacy, new owner privacy.
3. Issuer verification.
4. Resist common RFID attacks such as replay attacks, de-synchronization etc.

4.1   Pseudorandom Number Generators

A pseudorandom generator is a deterministic algorithm that receives a short truly random seed and stretches it into a long string that is pseudorandom. Stated differently, a pseudorandom generator uses a small amount of true randomness in order to generate a large amount of pseudorandom.

Let $g : \{0,1\}^\kappa \rightarrow \{0,1\}^{2\kappa}$ and $g = (g_1, g_2)$. g_1 and g_2 respectively map the input of g to the first n output bits of g and the last n output bits of g. We take g as the pseudorandom generator in the tag.

4.2   Protocol Description

We introduce a novel RFID ownership transfer protocol with high efficiency and robust security and privacy properties. Our protocol consists of three sub-protocols: an authentication protocol, an ownership transfer protocol, and a secret update protocol. Before introducing the authentication process and ownership transfer process, we first do some initiation works including setup and tag Initialization.

Setup: The issuer I outputs $(q, \mathbb{G}_1, \mathbb{G}_2, \mathbb{G}_T, g_1, g_2, e)$, where $\mathbb{G}_1$, $\mathbb{G}_2$, $\mathbb{G}_T$ are subgroups of prime order q, g_1 and g_2 are random generators of $\mathbb{G}_1$ and $\mathbb{G}_2$

respectively, and $e : \mathbb{G}_1 \times \mathbb{G}_2 \to \mathbb{G}_T$ is a bilinear pairing. I chooses $x \in Z_q^*$ and computes g_2^x . I's secret key is $sk = x$ and his public key is g_2^x. The system supplies each owner with his secret key $sk = \lambda_i$ and his public key $pk = g_1^{\lambda_i}$.

Tag Initialization: A tag T is initialized by the issuer I. I picks a random number $t \in \mathbb{F}_q$. Using a cryptographic hash function $H : \mathbb{F}_q \to \mathbb{G}_1$, I computes $h = H(t)$, then computes $u_0 = 1$ and $v_0 = h^x$, and stores $I_e^0 = (u_0, v_0)$ into the tag, I also choose a random $s \in_R \{0, 1\}^\kappa$ as the initialized state for pseudorandom number generator in the tag. Before accepting the tag, the reader checks the equation: $e(h, g_2^x) = e(I, g_2)$, in which $I = h^x$. If true, it implies that T is actually issued by I.

Authentication Protocol: Our authentication protocol is based on Ma's proposal [13] using only pseudorandom number generator as its cryptographic primitive, we extend the protocol from one way tag authentication protocol to mutual authentication protocol. According to the proof of Ma, the authentication protocol satisfies backward security. We also use idea of re-encryption in Elkhiyaoui's protocol. The whole process is summarized in Fig. 5.

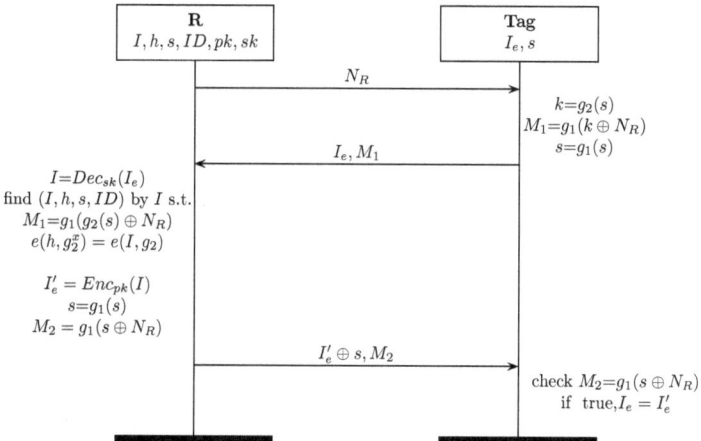

Fig. 5. Authentication Process

1. The reader R initiates the protocol by sending a random number to the tag.
2. Upon receiving a challenge N_R from R, the tag T derives two values, a new secret $k = g_2(s)$ and a new state $s = g_1(s)$. After updating the state to $g_1(s)$, T sends I_e and $M_1 = g_1(k \oplus N_R)$ to R.
3. On receipt of the messages, R first decrypts I_e using his secret key as $I = Dec_{sk}(I_e)$, then find (I, h, s, ID) in database using index I and satisfying both $M_1 = g_1(g_2(s) \oplus N_R)$ and $e(h, g_2^x) = e(I, g_2)$. If $M_1 = g_1(g_2(s) \oplus N_R)$ is not true, for $j = 1$ to ω, R judges whether M_1 equals $g_1(g_2(g_1^j(s)) \oplus N_R)$. If true, update s to $g_1^j(s)$, otherwise, reject. If $e(h, g_2^x) = e(I, g_2)$ is not true, reject as well. After that, R encrypts I by his public key as $I_e' = Enc_{pk}(I)$ and updates the state of T to $s = g_1(s)$. In our proposal, we use Elgamal encryption

primitive to encrypt I, so I'_e is different each time according to the security properties of the scheme. Finally R sends $I'_e \oplus s$ and $M_2 = g_1(s \oplus N_R)$ to T.

4. When T receives $I'_e \oplus s$ and M_2, T first verifies whether $M_2 = g_1(s \oplus N_R)$, if true, then update I_e to $I'_e \oplus s \oplus s$. Otherwise, T do not update I_e.

Ownership Transfer Protocol: In this protocol, the new owner of a tag first requests the TTP for ownership of the tag. If the request is valid, the TTP then transfers all the information related to the tag to the new owner via a secure channel. The ownership transfer protocol is based on our authentication protocol, since the authentication protocol satisfies backward privacy, the ownership transfer protocol follows old owner privacy innately. However, we have to update the tag state to protect new owner privacy. In addition, the index I must be changed in order to achieve untraceability. The protocol steps are summarized in Fig. 6 and described as follows:

1. At first, the old owner R_1 sends ID_{R_1} to the new owner R_2 through a secure communication channel(omitted in Fig. 6 for better representing).
2. The new owner R_2 initiates the protocol by sending a random number N_R to the tag.
3. Upon receiving a challenge N_R from R_2, the tag T derives two values, a new secret $k = g_2(s)$ and a new state $s = g_1(s)$. After updating the state to $g_1(s)$, T sends I_e and $M_1 = g_1(k \oplus N_R)$ to R_2.
4. R_2 then forwards I_e, $M_1 = g_1(k \oplus N_R)$, N_R, ID_{R_1} and ID_{R_2} to TTP. In our protocol, we use issuer as our TTP.
5. On receipt of the messages, TTP first decrypts I_e using R_1's secret key as $I = Dec_{sk_{R_1}}(I_e)$, then find (I, h, s, ID) in database using index I and satisfying both $M_1 = g_1(g_2(s) \oplus N_R)$ and $e(h, g_2^x) = e(I, g_2)$. If $M_1 = g_1(g_2(s) \oplus N_R)$ is not true, TTP tries to find $g_1^j(s)$ that M_1 equals $g_1(g_2(g_1^j(s)) \oplus N_R)$ for $j = 1$ to ω and update s to $g_1^j(s)$. After that, TTP picks a new random number $t' \in \mathbb{F}_q$ and computes $h' = H(t')$, $u_1 = g_1^{r_1}$, $v_1 = h'^x g_1^{\lambda r_1}$ (λ is the secret key of R_2) and let $I' = h'^x$, $I'_e = (u_1, v_1)$. Then, TTP updates the state of T to $s = g_1(s)$. Finally TTP sends $info(ID), s, I', h', I'_e, M_2$ to R_2.
6. R_2 stores $info(ID), s, I', h'$ and forwards $I'_e \oplus s$ and M_2 to T.
7. When T receives $I'_e \oplus s$ and M_2, T first verifies whether $M_2 = g_1(s \oplus N_R)$, if true, then update I_e to $I'_e \oplus s \oplus s$. Otherwise, T do not update I_e.

Secret Update Protocol: The main purpose of secret update protocol is to establish new secrets to protect the new owner's privacy. More specifically, we need to update the state s of the tag, our protocol is also based on authentication protocol, the whole process is summarized in Fig. 7 and described as follows:

1. The reader R initiates the protocol by sending N_R and $M_1 = g_1(g_2(s) \oplus N_R)$ to the tag T.
2. Upon receiving a challenge N_R and M_1 from R, the tag T computes $k = g_2(s)$ and verify $M_1 = g_1(k \oplus N_R)$. If true, computes $M_2 = g_2(k \oplus N_R)$ and update the tag state to $s = g_1(s \oplus N_R)$, then sends $I_e \oplus k$ and M_2 to R.

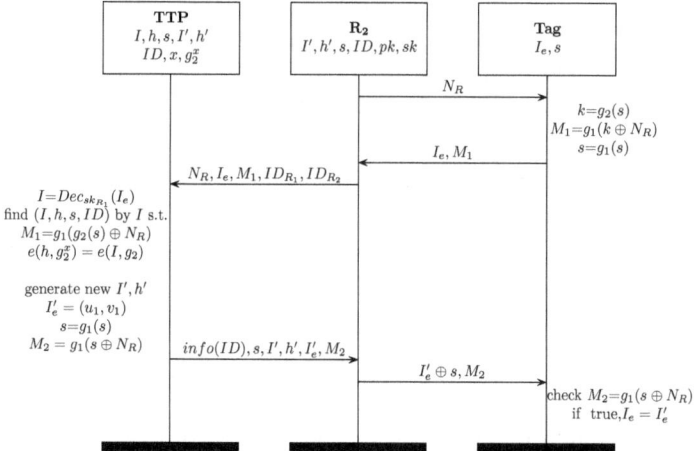

Fig. 6. Ownership transfer Protocol

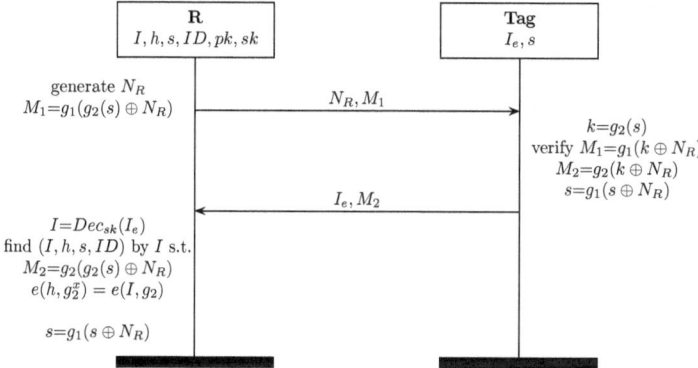

Fig. 7. Secret Update Protocol

3. When R receives $I_e \oplus k$ and M_2, R first get I_e by computing $I_e \oplus k \oplus g_2(s)$, then decrypts I_e using his secret key as $I=Dec_{sk}(I_e)$, then find (I, h, s, ID) in database using index I and satisfying $M_2=g_2(k\oplus N_R)$ and $e(h, g_2^x) = e(I, g_2)$. If true, update the tag state s to $s=g_1(s \oplus N_R)$.

5   Security Analysis

5.1   Security Issues

In this section, we mainly discuss some attacks that are commonly used by adversaries against RFID systems and present how our authentication protocol to resist these attacks, including Replay (RA) attacks, man-in-the-middle (MITM) attacks, backward traceability (BT) attacks, Denial of Service (DoS) attacks. In addition, we also do the analysis on how to protect new owner privacy and old owner privacy in our ownership transfer protocol. The security analysis of secret update protocol is just as authentication protocol, so we do not analyze it here due to repetitiveness.

Replay Attack. All communicating messages between the reader and the tag always contain a dynamically generated fresh random number. Adversaries are unable to pass the authentication scheme simply by replaying messages.

Man-in-the-Middle Attack(MITM). Since adversaries are unable to launch replay attacks and they do not have the tag state s, so they cannot pass our authentication scheme by masquerading as a reader or tag in an MITM attack.

Backward Traceability. Backward security refers to the scenario that even if the secret of T is compromised, the adversary cannot trace past transactions. In our protocol, backward security is guaranteed because the state s of T is updated each time using PRNG, according to the conclusion of Berbain et al. [14], the previous states of T are not computable.

DoS/Synchronization Problem. We address the DoS problem in the following way: Suppose that an adversary prevents message M_1 from reaching R. Then T will update its state s, R will not. However, R knows the previous value of s, and can use $g_1^j(s)$ to recover synchronization with T, j is limited to chains of length ω. Our protocol is more effective than song's proposal, because we use index I for locating the tuple in stead of searching it in the whole database. Note that the secret update protocol is also vulnerable to de-synchronization attacks if the adversary prevent M_2 from reaching R. However, R can compute $g_1(s \oplus N_R)$ itself to recover synchronization with T.

Old Owner Privacy. Old owner privacy implies that the new owner should not be able to trace past interactions between the tag and its previous owners. That property is guaranteed by backward traceability of the authentication protocol that the state s of T is updated in each session, and the previous states of T are not computable..

New Owner Privacy. New owner privacy implies that the old owner cannot trace the tag. We use Elgamal re-encryption mechanisms introduced by Elkhiyaoui to encrypt the tag index I in each session, However, in Elkhiyaoui's scheme, I keeps unchanged which can be used for adversary to trace the tag. Therefore, in our proposal, the TTP changed I when ownership transfer happens. In addition, we use secret update protocol to change the state of the tag into a new one, which cannot be inferred by the old owner if he does not obtain the values of N_R queried by the new owner. Under such assumption, the privacy of new owner is protected.

We compare our proposed scheme with other ownership transfer schemes in terms of the aspects as mentioned above, which is shown in Table 1.

Table 1. Security features comparison

Schemes \\ Security	RA	MITM	DoS	BT	NP	OP
Saito [1]	X	O	O	X	O	X
Kapoor [2]	O	O	O	O	O	O
Kulseng [15]	O	X	X	X	*	X
Osaka [16]	O	O	X	X	*	X
Fouladgar [17]	X	X	O	X	O	X
Song [3]	X	O	X	O	*	O
Elkhiyaoui [4]	O	O	O	X	X	X
ours	O	O	O	O	*	O

O: success
X: failure
*: success under an assumption

Table 2. Performance comparison

Schemes	Tag	Reader	TTP/Database
Saito	T_E	0	T_E
Kapoor	$2T_E+2T_H+2T_{RNG}$	$3T_E+2T_H+T_{RNG}$	$3T_E+2T_H+2T_{RNG}$
Kulseng	$4T_{RNG}+2T_H$	$3T_H+3T_{RNG}$	T_H+5T_{RNG}
Osaka	$2T_H$	T_{RNG}	$3T_E+T_H$
Fouladgar	$2T_H+T_{RNG}$	0	$2T_H$
Song	$3T_H+T_{RNG}$	T_{RNG}	$2T_H$
Elkhiyaoui	$T_{RNG}+3T_H$	$2T_E+3T_H+T_{RNG}$ $+7T_{PO}+2T_{exp}$	0
ours	$4T_{RNG}$	T_{RNG}	$2T_E+4T_{RNG}+2T_{PO}$

5.2 Performance

In this section, we analyze our protocol in computation costs and proves our scheme is viable for lightweight passive RFID tags. We will mainly focus on the performance characteristics of ownership transfer protocol. T_E denotes the time for one encryption or decryption; T_{RNG} is the time to generate a random number; T_H is the time for one hash function; T_{PO} stands for the time of pairing operation; and T_{exp} denotes the time of exponentiation operation. Table 2 compares the performance characteristics of our ownership transfer protocol with the other proposed schemes.

To notice that, we do not consider database searching process in the Table 2. Actually, Osaka's protocol, Fouladgar's protocol and Song's protocol do not support index, so the computation costs are much more than what we have listed in table 2. Although Kulseng's protocol supports index, the index may cause de-synchronization problem (index updates asynchronously between the reader and the tag), the reader still has to search the whole database when it happens. We adopt Elkhiyaoui's scheme to use re-encryption primitive to encrypt the same index in each session. It eliminates de-synchronization problem, meanwhile, increases the computation loads on the reader. Saito's protocol and Kapoor's protocol use symmetric encryption/decryption method at the tag side which is not suitable for resource limited RFID tags.

6   Conclusion

In this paper, we presented a novel RFID ownership transfer protocol to address security and privacy issues related to RFID ownership transfer in supply chains. Our proposal provides following advantages: First, we use Elgamal re-encryption mechanisms to solve linear complexity problem that authentication takes a linear time in the number of tags. Second, at the tag side, we use only pseudorandom number generator as its cryptographic element which is well suited for lightweight passive RFID tags. Third, we protect old owner privacy by adopting backward security authentication protocol and preserve new owner privacy by using secret update protocol. Our analysis show that our proposed protocol provides stronger security robustness and higher performance efficiency in comparison with existing solutions.

Acknowledgments. This work is partially supported by the HGJ National Significant Science and Technology Projects under Grant No. 2012ZX01039-004-009, Key Lab of Information Network Security, Ministry of Public Security under Grant No.C11606, the National Natural Science Foundation of China under Grant No. 61170263.

References

1. Saito, J., Imamoto, K., Sakurai, K.: Reassignment scheme of an RFID tag's key for owner transfer. In: Enokido, T., Yan, L., Xiao, B., Kim, D.Y., Dai, Y.-S., Yang, L.T. (eds.) EUC Workshops 2005. LNCS, vol. 3823, pp. 1303–1312. Springer, Heidelberg (2005)
2. Kapoor, G., Piramuthu, S.: Single RFID Tag Ownership Transfer Protocols. IEEE Transactions on Systems, Man, and Cybernetics, 1–10 (2011)
3. Song, B., Mitchell, C.J.: Rfid authentication protocol for low-cost tags. In: WISEC, pp. 140–147 (2008)
4. Elkhiyaoui, K., Blass, E.-O., Molva, R.: ROTIV: RFID Ownership Transfer with Issuer Verification. In: Juels, A., Paar, C. (eds.) RFIDSec 2011. LNCS, vol. 7055, pp. 163–182. Springer, Heidelberg (2012)
5. Sarma, S., Weis, S., Engels, D.: Radio-Frequency Identification: Security Risks and Challenges. Cryptobytes, RSA Laboratories 6(1), 2–9 (2003)
6. Hopper, N.J., Blum, M.: Secure human identification protocols. In: Boyd, C. (ed.) ASIACRYPT 2001. LNCS, vol. 2248, pp. 52–66. Springer, Heidelberg (2001)
7. Juels, A., Weis, S.A.: Authenticating Pervasive Devices with Human Protocols. In: Shoup, V. (ed.) CRYPTO 2005. LNCS, vol. 3621, pp. 293–308. Springer, Heidelberg (2005)

8. Bringer, J., Chabanne, H., Emmanuelle, D.: HB^{++}: a Lightweight Authentication Protocol Secure against Some Attacks. In: IEEE International Conference on Pervasive Services, Workshop on Security, Privacy and Trust in Pervasive and Ubiquitous Computing – SecPerU 2006, Lyon, France. IEEE Computer Society (June 2006)

9. Ohkubo, M., Suzuki, K., Kinoshita, S.: Cryptographic Approach to "Privacy-Friendly" Tags. In: RFID Privacy Workshop. MIT, Massachusetts (2003)

10. Ohkubo, Suzuki, K., Kinoshita, S.: Efficient Hash-Chain Based RFID Privacy Protection Scheme. In: International Conference on Ubiquitous Computing – Ubicomp, Workshop Privacy: Current Status and Future Directions, Nottingham, England (September 2004)

11. Avoine, G., Dysli, E., Oechslin, P.: Reducing time complexity in RFID systems. In: Preneel, B., Tavares, S. (eds.) SAC 2005. LNCS, vol. 3897, pp. 291–306. Springer, Heidelberg (2006)

12. Berbain, C., Billet, O., Etrog, J., Gilbert, H.: An efficient forward private rfid protocol. In: ACM Conference on Computer and Communications Security, pp. 43–53 (2009)

13. Chang-She, M.: Low cost rfid authentication protocol with forward privacy. Chinese Journal of Computers 34(8) (2011)

14. Berbain, C., Billet, O., Etrog, J., Gilbert, H.: An Efficient Forward Private RFID Protocol. In: Al-Shaer, E., Jha, S., Keromytis, A.D. (eds.) Conference on Computer and Communications Security – ACM CCS 2009, Chicago, Illinois, USA, pp. 43–53. ACM Press (November 2009)

15. Kulseng, L., Yu, Z., Wei, Y., Guan, Y.: Lightweight mutual authentication and ownership transfer for rfid systems. In: INFOCOM, pp. 251–255 (2010)

16. Osaka, K., Takagi, T., Yamazaki, K., Takahashi, O.: An efficient and secure RFID security method with ownership transfer. In: Wang, Y., Cheung, Y.-m., Liu, H. (eds.) CIS 2006. LNCS (LNAI), vol. 4456, pp. 778–787. Springer, Heidelberg (2007)

17. Fouladgar, S., Afifi, H.: A simple privacy protecting scheme enabling delegation and ownership transfer for rfid tags. JCM 2(6), 6–13 (2007)

Fractal Based Anomaly Detection over Data Streams

Xueqing Gong[1], Weining Qian[1], Shouke Qin[2], and Aoying Zhou[1]

[1] Software Engineering Institute, East China Normal University, Shanghai, China
[2] Baidu Inc., Beijing, China
{xqgong,wnqian,ayzhou}@sei.ecnu.edu.cn, qinshouke@baidu.com

Abstract. Robust and efficient approaches are needed in real-time monitoring of data streams. In this paper, we focus on anomaly detection on data streams. Existing methods on anomaly detection suffer three problems. 1) A large volume of false positive results are generated. 2) The training data are needed, and the time window of appropriate size along with corresponding threshold has to be determined empirically. 3) Both time and space overhead is usually very high. We propose a novel self-similarity-based anomaly detection algorithm based on piecewise fractal model. This algorithm consumes only limited amount of memory and does not require training process. Theoretical analysis of the algorithm are presented. The experimental results on the real data sets indicate that, compared with existing anomaly detection methods, our algorithm can achieve higher precision with reduced space and time complexity.

Keywords: Data Streams, Anomaly Detection, Fractal.

1 Introduction

Real-time monitoring over data streams have been attracting much attention. Anomaly detection aims to detect the points which are significantly different from others, or those which cause dramatic change in the distribution of underlying data stream. It is a great challenge to accurately monitor and detect the anomalies in real time because of the uncertainty of properties of the anomaly.

In general, existing anomaly detection methods detect abnormality based on a model derived from normal historical behavior of a data stream. They try to reveal the differences over short-term or long-term behaviors that are inconsistent with the derived model. However, the detected results largely depend on the model defined, as well as the data distribution and time granularity of the underlying data stream. For the data stream whose distribution changes constantly, the existing methods will result in a large volume of false positives. On the other hand, due to the application-dependent definition of abnormal behavior, each existing method can only detect certain types of abnormal behaviors. In real situation, it is meaningful to define abnormal behavior in a more general sense, and is highly desirable to have an efficient, accurate, and scalable mechanism for detecting abnormal behaviors over dynamic data streams.

Y. Ishikawa et al. (Eds.): APWeb 2013, LNCS 7808, pp. 550–562, 2013.

Self-similarity is ubiquitous both in natural phenomena and social events. It represents the intrinsic nature of the data. Abnormal behavior defined based on the change of self-similarity naturally captures the essence of the event. Fractals, which are built upon the statistical self-similarity, have been found with wide variety of applications [14]. In this paper, we focus on modeling data using fractals based on recurrent iterated function systems (RIFS).

Most existing data stream algorithms are proposed for specific monitoring tasks [10,16,19,5,15].

In [20], the authors use a Shifted Wavelet Tree (SWT) to detect bursts, which is an important kind of anomalies, on multiple windows. A more efficient aggregate monitoring method is put forward in [4]. However, they can only monitor monotonic aggregates, and both the maximum window size and the number of monitored windows are limited. The moving average, or Kalman-like filter [9] are also important measures to be monitored for anomaly detection. An adaptive change detecting method for such measurements, based on Inverted Histogram, is proposed in [19]. However, no any overall bound for the error is given.

Monitoring abnormal behaviors in the network traffic systems has been studied in [7,11,10]. They can monitor the changes of the behavior on only one granularity. The detection of trends and surprise in time series database is studied in [17]. In [6], the exception of trend is defined and detected.

Fractal geometry has been known to be a powerful tool to describe complex shapes in nature [12]. It has been successfully applied to modeling and compressing the data sets [18,13,14]. However, it is difficult to apply them directly in the data stream scenario, since the time complexity of constructing the fractal model is polynomial, while multi-pass scan of the whole data is required.

In this paper, we propose a fractal-based novel approach to detect anomaly in data streams. Our contributions are summarized below.

- We propose a general-purpose and adaptable definition for abnormal behavior based on the change of self-similarity. It captures the intrinsic property of the data streams.
- We introduce the fractal analysis based on self-similarity into anomaly detection of the data stream. The piecewise fractal model is presented to process data streams in guaranteed space cost and error.
- Both the theoretical analyses and experimental results illustrate that the proposed method can monitor data streams accurately and efficiently with limited resource consumption.

2   Preliminaries of Fractals

If a quantitative property q is measured on a time scale s, then the value of q depends on s according to the $q = ps^r$ scaling relationship [12]. This is called *power law scaling relationship* of q with respect to s. p is a factor of proportionality and r is the scaling exponent. The value of r can be determined by the slope of the linear least squares fitting to the graph of data pairs $(\log s, \log q)$:

$\log q = \log p + r \log s$. For discrete time series data, the power-law scaling relationship can be transformed to $q(st) = s^r q(t)$, where t is the basic time unit and $s(s \in \mathbb{R})$ is the scaling measurement [3]. r can be determined by $r = \frac{\log(q(st)/q(t))}{\log s}$.

The power-law scaling relationship is also retained when q is a first order aggregate function (e.g. *sum*, *count*) or some second order statistic variables (e.g. *variance*).

Intuitively, objects satisfying self-similarity property can be mapped (by a series of maps) to part of itself. This mapping process can be conducted iteratively until a *fixed point* called *attractor*, is found. Given the attractor. by iteratively applying the inverse of the mapping, we can approximate the original objects. The mapping process introduced above is mathematically called the *iterated function system* (IFS).

More precisely, an IFS consists of a finite collection of contraction maps M_i ($i = 1, 2, 3, ..., m$). Each M_i can map a compact metric space onto itself, provided that the metric space is linear. A point (x, y) can be mapped/transformed by $M_i[x, y]$ with the form $M_i \begin{bmatrix} x \\ y \end{bmatrix} = \begin{bmatrix} a_{11}^i & a_{12}^i \\ a_{21}^i & a_{22}^i \end{bmatrix} \begin{bmatrix} x \\ y \end{bmatrix} + \begin{bmatrix} b_1^i \\ b_2^i \end{bmatrix}$.

Assume $\mathbf{A}$ is the attractor, then $\mathbf{A} = \bigcup_{i=1}^m M_i(\mathbf{A})$. By modeling the data with an IFS, we can use the attractor of the IFS to approximate the data. When a_{12}^i is zero, the mapping is called *shear transformation* [14]. Then, $a_{22}^i < 1$ is called the contraction factor for map M_i.

IFS can only be used to manipulate *linear* fractals, which have a fixed contraction factor for all objects in all scales. However, the natural objects may be more complex. Thus recurrent iterated function systems (RIFS) are introduced. RIFS can be generalized for arbitrary metric spaces [2]. Intuitively, different from IFS, in RIFS each map can be applied on part of the objects. The ranges of objects the maps can be applied on may overlap. Thus, different objects may be mapped to different parts and with different contraction scale.

In RIFS, given the contraction maps M_i, the attractor $A \subset R^d$ is $\bigcup_{i=1}^N A_i$ where $A_i \subset R^d$ ($i = 1, ..., N$) are possibly overlapping partitions. Furthermore, $M_j(\mathbf{A}) = A_j = \bigcup_{<i,j>\in G} M_j(A_i)$. Here G is a directed graph, and $\mathbf{A} = (A_1, ..., A_N) \in (R^d)^N$. Each edge $< i, j >\in G$ indicates that the mapping composition $M_j \diamond M_i$ is allowed. The attractor can also be represented by $A = \mathbf{M}(A)$, in which $\mathbf{M} : (R^d)^N \to (R^d)^N$ is defined as $(M_1, M_2, ...M_N)$. Thus, given a set vector $\mathbf{D} = (D_1, D_2, ..., D_N)$ with $D_i \subset R^d$, $\mathbf{A} = \lim_{i \to \infty} \mathbf{M}_i(\mathbf{D})$. Similar to IFS, attractor A can be used to approximate the data. When the graph G is complete, the RIFS degrades to an IFS [8].

IFS and RIFS are often used to summarize and compress fractal data. The problem lies in the determination of the attractor. Since the attractor can be computed given the mapping, the computation of the parameters of the mapping is usually called the *inverse problem* of IFS or RIFS [8].

Given the mapping $\mathbf{M}$, $\mathbf{D}' = \mathbf{M}(\mathbf{D})$, we consider the condition that $\mathbf{D}' \subset \mathbf{D}$ in this paper. Under this condition, the *open set property* holds, so that the *Collage Theorem* can be applied to guarantee the precision of approximating the original data using the attractor [14,2]. This theorem provides a way to measure the goodness of fit of the attractor associated with an RIFS and a given function.

Collage Theorem: Let (X, d) be a complete metric space where d is a distance measure. Let $\mathbf{D}$ be a given function and $\epsilon > 0$ be given. Choose an RIFS, with contractility factor $s = max\{s_i : i = 1, 2, 3, ..., N\}$ so that $d(\mathbf{D}, \bigcup_{i=1}^{N} M_i(\mathbf{D})) \leq \epsilon$. Then $d(\mathbf{D}, A) \leq \frac{\epsilon}{1-s}$ where $\mathbf{A}$ is the attractor of the RIFS.

Since no assumption that the data streams to be monitored are linear fractals can be made, we consider RIFS in the rest of this paper.

3 Problem Statement

A data stream X is considered as a sequence of points $x_1, ..., x_n$. Each element in the stream can be a nonnegative value, which is denoted as a (*timestamp* : i, *value* : y_i) pair or i for simplicity. Function F being monitored can be not only the monotonic aggregates with respect to the window size, for example *sum* and *count*, but also the non-monotonic ones such as *average* and *variance*. One common property of the above functions is that they all retain the power-law scaling relationship [16].

Monitoring anomalies in data stream is to detect the change of power-law scaling relationship, i.e., self-similarity, on the sequences with continuous incoming new value x_i. Under such a data stream model, the problem about anomaly detection can be described as follows.

PROBLEM STATEMENT: A data stream X is represented as a sequence of points $x_1, ..., x_n$. An anomaly is detected if the self-similarity of X changes (i.e., the historic one is violated) when new value x_n comes.

Our basic idea is to use the attractor to represent the data stream, and monitor events that change the attractor, which mean anomalies. Thus, the problem lies in *"How to efficiently estimate the attractor in a streaming data environment?"*, *"How good is it to use estimated attractor to approximate the original data stream?"*, and *"How to modeling the anomaly based on the attractor?"*.

4 Piecewise Fractal Model for Data Streams

Constructing a piecewise fractal model is equivalent to solving a inverse problem in data stream scenario. We investigate two approaches. One is L_2 error optimal, whose complexity is $O(n^2)$ in time and $O(n)$ in space. The other is an approximate method with at most $(1 + \epsilon)$ times the error for the optimal solution. It uses $O(n)$ time and $O(m)$ space, where m is the number of pieces needed, and $m < \log n$. For ease of understanding, we list the notations used in Figure 1.

4.1 Inverse Problem

RIFS maps a partition D_i of the whole metric space D to another partition D_i' under two constraints. The first constraint is: $D_i' = M_i(D_i), D_i' \subset D_i$.

x_i	i-th data point in stream X
m	number of pieces/contraction maps
A_i	distance
B_i	distance
α	error constraint of piecewise fractal model
M_i	i-th contraction map
D_i	i-th partition of whole metric space D
D_i'	the partition mapped from D_i, $D_i' \subset D_i$
P_i	i-th piece of fractal model
P_i'	the piece mapped from P_i by M_i, $P_i' \subset P_i$

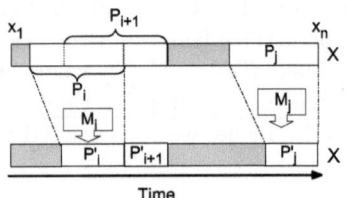

Fig. 1. List of notations

Fig. 2. Contraction mapping on stream X in our piecewise fractal model

The second constraint which needs to be maintained in RIFS is that every contraction mapping maps the two endpoints $(x_{(i,0)}, y_{(i,0)})$ and $(x_{(i,1)}, y_{(i,1)})$ of D_i to two endpoints $(x'_{(i,0)}, y'_{(i,0)})$ and $(x'_{(i,1)}, y'_{(i,1)})$ of D_i' (Figure 2). M_i transforms a larger interval D_i of stream X to a smaller interval D_i' of the same stream. Thus the endpoints of D_i' are also the endpoints of the attractor of M_i,

that is, $M_i \begin{bmatrix} x'_{(i,0)} \\ y'_{(i,0)} \end{bmatrix} = \begin{bmatrix} x_{(i,0)} \\ y_{(i,0)} \end{bmatrix}, M_i \begin{bmatrix} x'_{(i,1)} \\ y'_{(i,1)} \end{bmatrix} = \begin{bmatrix} x_{(i,1)} \\ y_{(i,1)} \end{bmatrix}$.

Once the contraction factor a_{22}^i for each mapping is determined, the remaining parameters can be obtained using the endpoint constraint (4.1) as follows.

$$a_{11}^i = \frac{x'_{(i,1)} - x'_{(i,0)}}{x_{(i,1)} - x_{(i,0)}}, \qquad b_1^i = \frac{x_{(i,1)} x'_{(i,0)} - x_{(i,0)} x'_{(i,1)}}{x_{(i,1)} - x_{(i,0)}} \tag{1}$$

$$a_{21}^i = \frac{y'_{(i,1)} - y'_{(i,0)}}{x_{(i,1)} - x_{(i,0)}} - a_{22}^i \frac{y_{(i,1)} - y_{(i,0)}}{x_{(i,1)} - x_{(i,0)}} \tag{2}$$

$$b_2^i = \frac{x_{(i,1)} y'_{(i,0)} - x_{(i,0)} y'_{(i,1)}}{x_{(i,1)} - x_{(i,0)}} - a_{22}^i \frac{x_{(i,1)} y_{(i,0)} - x_{(i,0)} y_{(i,1)}}{x_{(i,1)} - x_{(i,0)}} \tag{3}$$

The contraction mapping M_i denotes the similarity between D_i and D_i'. Provided that D_i' is a subset of D_i, the M_i denotes the *self-similarity* between D_i and D_i'. With the self-similarity/M_i, we can reconstruct D_i from D_i'. More importantly, we alarm anomaly when the self-similarity/M_i of the latest partition D_i is violated by newly arrived data point. Piecewise fractal model is used to maintain such contraction mappings (self-similarity) over data streams.

Now the problem of constructing the piecewise fractal model for data streams is reduced to determining the appropriate value of contraction factor a_{22}^i.

4.2 Piecewise Fractal Model

In a piecewise fractal model, the goal of the global optimal solution is to minimize the sum of error of each piece.

Algorithm 1 OptimalModel(x_n)	**Algorithm 2** ApproximateModel(x_n)
1: compute $\Sigma_{i=s}^e A_i B_i$;	1: compute $\Sigma_{i=s}^e B_i^2$ incrementally;
2: compute $\Sigma_{i=s}^e A_i^2$ incrementally;	2: compute $\Sigma_{i=s}^e A_i^2$ incrementally;
3: **if** $\Sigma_{i=s}^e A_i B_i / \Sigma_{i=s}^e A_i^2 < 1$ **then**	3: **if** $\Sigma_{i=s}^e B_i^2 / \Sigma_{i=s}^e A_i^2 < 1$ **then**
4: add x_n to current piece;	4: add x_n to current piece;
5: **else**	5: **else**
6: create a new piece for x_n;	6: create a new piece for x_n;
7: $pieceNumber + +$;	7: $pieceNumber + +$;
8: **if** $pieceNumber > m$ **then**	8: **if** $pieceNumber > m$ **then**
9: delete the oldest piece;	9: delete the oldest piece;
10: **end if**	10: **end if**
11: **end if**	11: **end if**
12: add x_n to all pieces;	12: add x_n to all pieces;

Optimal Model. Let the start point of a piece P of stream X be (s, y_s), and the end point of P be (e, y_e), $e > s$. A contraction mapping M can be defined as $M(i, y_i) = (int(i \cdot a_{11} + b_1), i \cdot a_{21} + y_i a_{22} + b_2)$. The data points in a larger piece of a data stream are mapped by M from $P\{(s, y_s), (e, y_e)\}$ to a smaller piece of the same stream $P'\{(s', y_{s'}), (e', y_{e'})\}$, where $P' \subset P$ and $e - s > e' - s'$. Every two adjacent P's meet only at the endpoints and do not overlap. For all pieces of the whole data stream, however, there might be some overlaps, that is, $\bigcup P_i = \bigcup P_i' = X$. Suppose the L_2 error between the data points in mapping $M(P)$ and those in the original piece P' is $E(M) = (M(P) - P')^2$, then the error of a mapping M is $E(M) = \Sigma_{i=s}^e ((i \cdot a_{21} + y_i a_{22} + b_2) - y_j)^2$, where $j = \lfloor i \cdot a_{11} + b_1 \rfloor$. Given P and P', the optimal mapping M_{opt} is the one such that the minimum value of $E(M)$ is reached.

For data stream X, it is ensured that each contraction mapping M maps the start point and end point of piece P to corresponding ones of piece P', namely, $(s', y_{s'}) = M(s, y_s)$ and $(e', y_{e'}) = M(e, y_e)$. a_{21} and b_2 can be obtained by equations 2 and 3. Therefore, we have $E(M) = \Sigma_{i=s}^e (A_i a_{22} - B_i)^2$, in which $A_i = y_i - (\frac{e-i}{e-s} y_s + \frac{i-s}{e-s} y_e)$, and $B_i = y_j - (\frac{e-i}{e-s} y_{s'} + \frac{i-s}{e-s} y_{e'})$.

The coefficient a_{22} of the optimal mapping M_{opt} can be computed using $\frac{\partial E(M)}{\partial a_{22}} = 2\Sigma_{i=s}^e (A_i a_{22} - B_i) A_i = 0$, then $a_{22} = \Sigma_{i=s}^e A_i B_i / \Sigma_{i=s}^e A_i^2$. Consequently, the optimal contraction mapping M_{opt} can be obtained by maintaining $\Sigma_{i=s}^e A_i B_i$ and $\Sigma_{i=s}^e A_i^2$ over the data streams. Algorithm 1 illustrates the method for constructing the optimal piecewise fractal model for a data stream.

Theorem 1. *Assume the length of a data stream is n. Algorithm 1 maintains optimal piecewise fractal model with $O(n + m)$ space in $O(n^2)$ time.*

Approximate Model. It is impossible to store the whole data stream in memory. Thus we cannot maintain the value of $\Sigma_{i=s}^e A_i B_i$ in the data stream scenario, as Algorithm 1 requires. To address this problem, we construct a approximate piecewise fractal model to compute the approximate value of $\Sigma_{i=s}^e A_i B_i$.

Denote the approximate value of a_{22} to be a_{22}'. We have 1) $C_i = y_i - (\frac{e'-i}{e'-s'} y_s' + \frac{i-s'}{e'-s'} y_e')$, 2) $\Sigma_{i=s}^e B_i^2 = \frac{e-s}{e'-s'} \Sigma_{i=s'}^{e'} C_i^2$, and 3) $(a_{22}')^2 = \frac{\Sigma_{i=s}^e B_i^2}{\Sigma_{i=s}^e A_i^2}$. Hence we only need to maintain $\Sigma_{i=s'}^{e'} C_i^2$ and $\Sigma_{i=s}^e A_i^2$ incrementally for current pair of P and P'

on the data stream. In this way, the a'_{22} and corresponding approximate results M_{app} can be computed.

The approximate piecewise fractal model consists of m pieces which can be used to reconstruct the original data stream. The error of reconstruction is bounded by Collage Theorem. Each piece P_i corresponds to a contraction mapping M_i and a pair of partition P and P'. One value needed to be stored for P_i is the suffix sum $F(n - si')$ starting from $(si', y_{si'})$, where a suffix sum of data stream is defined as $F(w_i) = \Sigma_{j=n-wi+1}^{n} x_j$. The other value needed to be stored for P_i is the number of points included in this suffix sum. We first try to add each data point of the evolving stream into the current piece. If failed, then a new piece is created for this point. Because the end point of piece P'_i is the start point of piece P'_{i+1}, it is not necessary to store the end point. The method of constructing an approximate piecewise fractal model for a data stream is described in Algorithm 2.

Theorem 2. *Algorithm 2 can maintain the approximate piecewise fractal model for a data stream with $O(m)$ space in $O(n)$ time.*

4.3   Error Bound

Theorem 3. *Assume Algorithm 2 produces an approximate solution M_{app} which maps the partition P to P' with error $E(M_{app})$, and the optimal method maps the partition P to P' with error $E(M_{opt})$, then $E(M_{app}) \leq 2E(M_{opt})$.*

The error bound of Algorithm 2 to the optimal result is shown in Theorem 3. The proof is omitted due to space limitation. For some applications in which the bounded absolute error is wanted, we provide a parameter α to control the error. Then the error of mapping M_{app} on partitions P and P' can be guaranteed: $E(M_{app}) \leq \alpha \cdot E(P'_{LSF})$, where $E(P'_{LSF})$ is the Least Linear Fitting error on partition P'. Therefore, the error produced by approximate piecewise fractal model on partition P' is less than the smaller value of $\alpha E(P'_{LSF})$ and $2E(M_{opt})$, which is $E(M_{app}) \leq min\{\alpha E(P'_{LSF}), 2E(M_{opt})\}$.

To control the parameter α, we only need to check whether $E(M_{app}) \leq \alpha E(P'_{LSF})$ holds or not when a new x_n comes. If $E(M_{app}) \leq \alpha E(P'_{LSF})$ holds, x_n is put into the current partition; otherwise, a new partition is created for x_n. The rest is the same as that of Algorithm 2.

5   Anomaly Detection Mechanism

5.1   History-Based Model

The *history-based model* trains a model incrementally, and compares the new coming data against the model. The data that deviate from the model are reported as anomalies. Existing burst detection [20,4,16] and change detection [10] methods conform to this model. For instance, the threshold of bursty behaviors of aggregate function F on i-th window is set as follows. At first, the moving

$F(w_i)$ is computed over some training data which could be the foremost 1% of the data set. The training data forms another time series data set, say, Y. Then the threshold of anomaly is set to be $T(w_i) = \mu_y + \lambda\sigma_y$, where μ_y and σ_y are the mean and standard deviation, respectively. The threshold can be tuned by varying the coefficient λ of standard deviation. These methods then look for the significant difference in short-term or long-term behaviors which are inconsistent with the model.

To maintain the model, piecewise fractal model is used. For each piece, the line connecting end points of a piece is used to approximate the data within this piece. The error is denoted by $E(M)$ and is guaranteed. The data generated by all pieces are summed up to approximate the original data stream. Thus, the piecewise fractal model is used as a synopsis data structure.

5.2 Self-similarity-Based Model

It is natural for us to investigate the another mechanism for detecting abnormal behaviors in the data streams. The behaviors observed by the mechanism on the data streams can be depicted with the equation: $Y = X + e$, where Y is either an aggregate function(e.g., $sum, count, average$), measurement of trend and deviant (say, $variance$), or distance of two distributions on the data stream. X is the result of Y in normal condition. e, the effect caused by abnormal behaviors on Y, can be seen as a noise. We assume that X and e have their own distributions and the uncertainty noise e is white Gaussian [9].

In the scenario of discrete data points, Gaussian white noise can only be modelled and processed in time series data. The constraints imposed on the data stream processing include limited memory consumption, sequentially linear scanning, and on-line accurate result reporting over evolving data streams. Therefore, we build piecewise fractal model on the data streams to model the change of self-similarity induced by white Gaussian noise.

The occurrence of abnormal behaviors is rooted in the change of self-similarity in the data stream. The piecewise fractal model can find the pieces with self-similarity in a data stream. That a new piece is created means there is a change in current self-similarity. When a new piece is created for the incoming x_n, an alarm is reported on the anomaly of x_n. If the bounded absolute error is wanted, a parameter α can be used in the algorithm to control the error.

With this model, we can evaluate both the signal and noise in the data stream. Since the anomaly detection only focuses on the prominent noise which causes the change of self-similarity, the evaluation of anomaly on e is not affected by the evolving of normal signal X.

6   Performance Evaluation

6.1   Experimental Setup

The experiments are conducted on a platform with a 2.4GHz CPU, 512MB main memory on Windows 2000. We applied our algorithms to a variety of data sets.

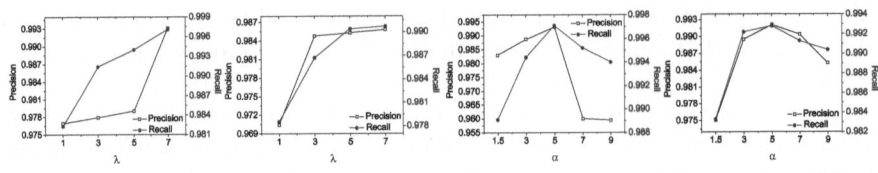

a) Varying λ (D1)   b) Varying λ (D3)   (a) Varying α (D1)   (b) Varying α (D3)

Fig. 3. The accuracy when varying λ **Fig. 4.** The accuracy when varying α

Due to the space limitation, only results on three real-life datasets are reported: 1) *Stocks Exchange (D1)*: This data set contains one year tick-by-tick trading records in 1998 from a corporation on Hong Kong main board. 2) *Network Traffic (D2)*: This data set BC-Oct89Ext4 is a network traffic tracing data set obtained from the Internet Traffic Archive [1,7]. 3) *Web Site Requests (D3)*: It consists of all requests made to the 1998 World Cup Web site between April 26, 1998 and July 26, 1998 [1].

Two accuracy metrics, recall and precision, are considered. Recall is the ratio of true alarms raised to the total true alarms which should be raised; precision is the ratio of true alarms raised to the total alarms raised.

6.2   Experimental Results

Detecting Anomalies in History-Based Model. The piecewise fractal model can be applied to reconstruct data and detect history-based anomalies defined by existing methods [20,4,16]. Here we present the results of comparison tests with these methods.

To set the threshold for $F(w_i)$ on i-th window w_i, we compute moving $F(w_i)$ over some training data. The training data is the foremost 1% of each data set. It forms another time series data set, called Y. The absolute threshold is set to be $T(w_i) = \mu_y + \lambda\sigma_y$. The length of windows is $4 * NW$ time units, where NW is the number of windows ranging from 50 to 1100. Similar testing results can be got from all the data sets.

Given $NW = 70$, and $m = 8$, α being 7 (D1) or 9 (D3), Figure 3 shows that under any setting of λ the precision and recall of our method are always above 97%. Larger λ means larger threshold. For our method can model such significant differences more accurate with self-similar pieces, it can be concluded that with the increasing value of λ, our method can guarantee better accuracy for detecting history-based anomalies.

While fixing the threshold parameter λ of piecewise fractal model (7 on D1 and 5 on D3), Figure 4 shows that when the value of α increases, the precision and recall also increase in certain domain. When α exceeds certain value, both precision and recall decrease. α cannot be too small, because smaller α means a piece can only store less data points, hence the length of the overlapping pieces for the piecewise fractal models becomes shorter when the number of pieces m is fixed. Thus the estimated aggregate value of larger windows is more error prone,

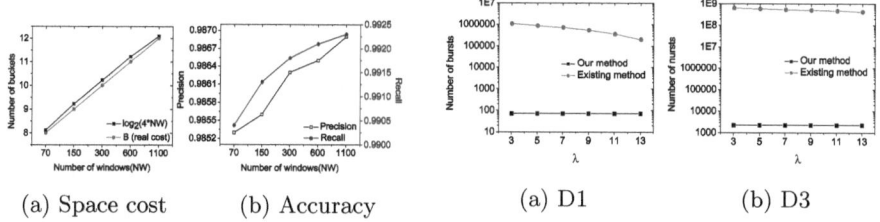

(a) Space cost   (b) Accuracy   (a) D1   (b) D3

Fig. 5. The efficiency of anomaly detection on D3 when varying number of windows

Fig. 6. Comparing number of anomalies, with $NW = 100$, $\lambda = 7$ for SWT (history-based method)

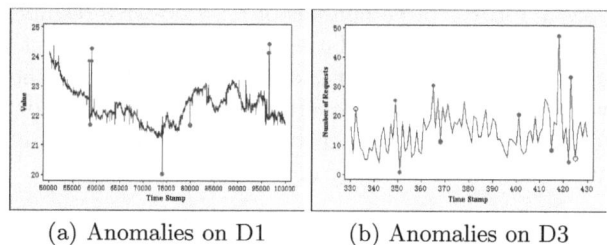

(a) Anomalies on D1   (b) Anomalies on D3

Fig. 7. Anomalies detected on D1 and D3

which leads to the increase of the error of anomaly detection. On the other hand, it is easy to understand that larger α results in longer pieces, therefore larger reconstruction error.

Given $\lambda = 5$ and $\alpha = 9$, with the increase of number of monitored windows, the number of used pieces also increases. But the number of consumed pieces can still be guaranteed by $m < \log_2 4 * NW$. It can be seen from Figure 5 that with only a few pieces the precision and recall are still high. The accuracy becomes better when NW increases. This is because the average size of monitored windows is larger with the increase of NW. The fractal approximation is more accurate over large time scales. So the error decreases with the increasing value of NW. This illustrate our algorithm's ability to monitor large amounts of windows with high accuracy over streaming data.

Anomaly Detection in Self-similarity-Based Model. In Figure 7 ($m = 1, \alpha = \infty$) (a) and (b), the solid red points are anomalies identified by our algorithm on two pieces of data streams D1 and D3. Both the precision and recall are 100%. Such significant difference depicts the abnormal behavior in stock exchanging and web site requesting.

In this experiment, we also compare the self-similarity-based method with existing history-based anomaly detection methods [20,4,16]. Since they have the same definition on anomalies, we choose [20] to represent the rest. When detecting anomaly on data streams, existing methods will return large number of false positive results. In the experiment, we set $NW = 100$. It can be seen from

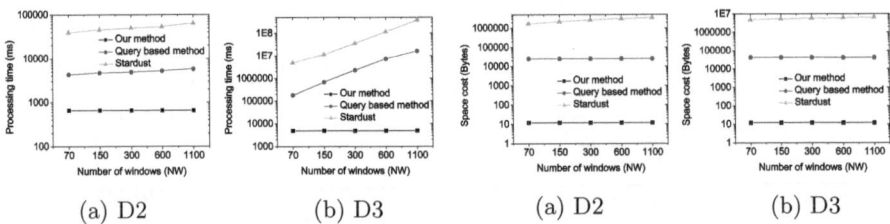

(a) D2   (b) D3   (a) D2   (b) D3

Fig. 8. Time consumption comparison on **Fig. 9.** Space consumption comparison on different data sets by varying the number different data sets by varying the number of monitored windows of monitored windows

Figure 6 that SWT returns large number of false positive results in every case. For example, in Figure 6.(b), 10^8 alarms are returned, which is definitely not informative to the users. Moreover, it is difficult to set the threshold for anomalies. A slightly smaller or larger λ can either boost false alarms or false positive, respectively. Therefore, our self-similarity-based method is more adaptable for providing accurate alarms in anomaly detection.

Space and Time Efficiency. To show the time and space efficiency of our algorithm, we compare our method with *Stardust* [4] and the query-based method [16]. It can be observed from Figures 8 and 9 that our method uses much less memory, runs more than tens of times faster than other methods. We use the code provided by the authors of [4] and [16] to do the comparing test. The rest of setting is the same as before, except for the special setting for Stardust with box capacity $c = 2$.

Figure 8 shows the time efficiency testing results. With the increasing of the number of monitored windows, the processing time saved by using self-similarity-based algorithm is getting larger and larger. Our method can process $400 * 10^7$ tuples in 10 seconds. This means that it is capable of processing traffic rates on 100Mbs links, and with some work than 1Gbps or even higher are within reach.

Figure 9 shows the space efficiency testing results. The space cost of our piecewise fractal model is only affected by the number of pieces maintained. However, Stardust has to maintain all the monitored data points and the index structure at the same time. The query-based method has to maintain an inverted histogram (IH) in memory. It can be seen that the space saved by the piecewise fractal model is considerable. Thus, our method is more suitable to detect anomalies over streaming data.

7   Concluding Remarks

In this paper, we incorporate the fractal analysis technique into anomaly detection. Based on the piecewise fractal model, we propose a novel method for detecting anomalies over a data stream. Fractal analysis is employed to model the self-similar pieces in the original data stream. We show that this fractal-based method can be used to detect not only the bursts defined before, but more general anomalies that cause the fractal characteristics change. Piecewise fractal model is proved to provide accurate monitoring results, while consuming only limited storage space and computation time. Both theoretical and empirical results show the high accuracy and efficiency of the proposed method. Furthermore, this approach can also be used to reconstruct the original data stream with the error guaranteed.

Acknowledgments. This work is partially supported by National Science Foundation of China under grant numbers 60925008 and 61170086.

References

1. Internet traffic archive, `http://ita.ee.lbl.gov/`
2. Barnsley, M., Elton, J., Hardin, D.: Recurrent iterated function systems. Constructive Approximation 5(1), 3–31 (1989)
3. Borgnat, P., Flandrin, P., Amblard, P.: Stochastic discrete scale invariance. IEEE Signal Processing Letters 9(6), 181–184 (2002)
4. Bulut, A., Singh, A.: A unified framework for monitoring data streams in real time. In: Proceedings of the 21st International Conference on Data Engineering, ICDE 2005, pp. 44–55. IEEE (2005)
5. Chandola, V., Banerjee, A., Kumar, V.: Anomaly detection: A survey. ACM Computing Surveys (CSUR) 41(3), 15 (2009)
6. Chen, Y., Dong, G., Han, J., Wah, B., Wang, J.: Multi-dimensional regression analysis of time-series data streams. In: Proceedings of the 28th International Conference on Very Large Data Bases, pp. 323–334. VLDB Endowment (2002)
7. Cormode, G., Muthukrishnan, S.: What's new: Finding significant differences in network data streams. In: Twenty-third Annual Joint Conference of the IEEE Computer and Communications Societies, INFOCOM 2004, vol. 3, pp. 1534–1545. IEEE (2004)
8. Hart, J.: Fractal image compression and recurrent iterated function systems. IEEE Computer Graphics and Applications 16(4), 25–33 (1996)
9. Jain, A., Chang, E., Wang, Y.: Adaptive stream resource management using kalman filters. In: Proceedings of the 2004 ACM SIGMOD International Conference on Management of Data, pp. 11–22. ACM (2004)
10. Kifer, D., Ben-David, S., Gehrke, J.: Detecting change in data streams. In: Proceedings of the Thirtieth International Conference on Very Large Data Bases, vol. 30, pp. 180–191. VLDB Endowment (2004)
11. Krishnamurthy, B., Sen, S., Zhang, Y., Chen, Y.: Sketch-based change detection: methods, evaluation, and applications. In: Proceedings of the 3rd ACM SIGCOMM Conference on Internet Measurement, pp. 234–247. ACM (2003)
12. Mandelbrot, B.: The fractal geometry of nature. Times Books (1982)
13. Mazel, D., Hayes, M.: Using iterated function systems to model discrete sequences. IEEE Transactions on Signal Processing 40(7), 1724–1734 (1992)
14. Michael, F.: Fractals everywhere. Academic Press, San Diego (1988)
15. Patcha, A., Park, J.: An overview of anomaly detection techniques: Existing solutions and latest technological trends. Computer Networks 51(12), 3448–3470 (2007)
16. Qin, S., Qian, W., Zhou, A.: Approximately processing multi-granularity aggregate queries over data streams. In: Proceedings of the 22nd International Conference on Data Engineering, ICDE 2006, p. 67. IEEE (2006)

17. Shahabi, C., Tian, X., Zhao, W.: Tsa-tree: A wavelet-based approach to improve the efficiency of multi-level surprise and trend queries on time-series data. In: Proceedings of the 12th International Conference on Scientific and Statistical Database Management, pp. 55–68. IEEE (2000)
18. Wu, X., Barbará, D.: Using fractals to compress real data sets: Is it feasible. In: Proc. of SIGKDD (2003)
19. Zhou, A., Qin, S., Qian, W.: Adaptively detecting aggregation bursts in data streams. In: Zhou, L., Ooi, B.-C., Meng, X. (eds.) DASFAA 2005. LNCS, vol. 3453, pp. 435–446. Springer, Heidelberg (2005)
20. Zhu, Y., Shasha, D.: Efficient elastic burst detection in data streams. In: Proceedings of the Ninth ACM SIGKDD International Conference on Knowledge Discovery and Data Mining, pp. 336–345. ACM (2003)

Preservation of Proximity Privacy
in Publishing Categorical Sensitive Data

Yujia Li[1], Xianmang He[2,*], Wei Wang[1], Huahui Chen[2], and Zhihui Wang[1]

[1] School of Computer Science and Technology, Fudan University,
No.220, Handan Road, Shanghai, 200433, P.R. China
{071021056,weiwang1,zhhwang}@fudan.edu.cn
[2] School of Information Science and Technology, NingBo University
No.818, Fenghua Road, Ning Bo, 315122, P.R. China
{hexianmang,chenhuahui}@nbu.edu.cn

Abstract. In this paper, we address the problem of proximity privacy in publishing categorical sensitive data. When using traditional approach to anonymize the data and generalize the QI-groups, some sensitive attribute values with semantical proximity may exist in the same QI-group, and thus lead to privacy leakage. To solve this issue, the concept of *m-Color* constraint is introduced and a method based on the *m-Color* constraint is proposed to prevent this kind of privacy leakage. The properties of *m-Color* constraint and related generalization algorithm are given, which reduce the loss of information greatly. The experiment results are provided to explain practicality and efficiency of the algorithm proposed in this paper.

Keywords: privacy preservation, *k*-anonymity, proximity privacy, *m-Color*.

1   Introduction

The information age has witnessed a tremendous growth of personal data that can be collected and analyzed, including a great deal of released microdata (i.e. data in raw, non-aggregated format) [12].These released microdata offer significant advantages in terms of information availability, which is particularly suitable for ad hoc analysis in a variety of domains, such as medical research.

However, the publication of microdata leads to the concerns of individual private information. For example, the medical institute and its cooperative organization need a large amount of private data on patients for disease research. Table 1 is such an example. One straightforward approach to achieve this is to exclude unique identifier attributes, such as Name from Table 1, which is not sufficient for protecting privacy leakage under linking-attack [1-3]. Suppose the attacker has known about Andy with attribute values <age=5, sex=M, zip=12k>, then according to Table 1, s/he can deduce Andy's disease, which is pneumonia. Such combination of Age, Zipcode and Sex can

---

[*] Corresponding author.

Y. Ishikawa et al. (Eds.): APWeb 2013, LNCS 7808, pp. 563–570, 2013.

be potentially used to identify an individual in Table 1 and is called a quasi-identifier (QI for short) in literatures.

To avoid linking attack, numerous approaches have been proposed. Generalization is applied on the quasi-identifiers and replaces QI-values with "the less-specific but semantically consistent values" Achieving k-anonymity privacy protection using the method of L. Sweeney [1]. Consequently, more records will have the same set of QI-values, in which are indistinguishable from each other. For example, Table 2 is 3-anonymized by generalizing QI-values in Table 1. Age=5, Zipcode=12k and Sex =M of Andy have been replaced with intervals [5-9], [12k-18k] and M respectively. Clearly, even if an adversary has the exact QI-values of Andy, s/he can not infer exactly which tuple in the first group is Andy. In general, we refer to the set with the same projection on the quasi-identifier as QI-Group.

Table 1. Microdata

Name	Age	Sex	Zip	Disease
Andy	5	M	12k	Pneumonia
Bill	9	M	14K	Pulmonary embolism
Ken	6	M	18K	Flu
Nash	8	M	19K	Bronchitis
Joe	12	M	22K	Colotis
Sam	11	M	20K	Colon cancer
Linda	21	F	58k	Stomach cancer
Jame	26	F	36k	Gastric ulcer
Sarah	28	F	37k	Gastritis
Mary	56	F	33k	Pulmonary edema

Table 2. Generalization Table

ID	Age	Sex	Zip	Disease
	[5-9]	M	[12k-18k]	Pneumonia
1	[5-9]	M	[12k-18k]	Pulmonary embolism
	[5-9]	M	[12k-18k]	Flu
	[8-12]	M	[19K-22k]	Bronchitis
2	[8-12]	M	[19K-22k]	Colotis
	[8-12]	M	[19K-22k]	Colon cancer
	[21-56]	F	[33k-58k]	Stomach cancer
3	[21-56]	F	[33k-58k]	Gastric ulcer
	[21-56]	F	[33k-58k]	Gastritis
	[21-56]	F	[33k-58k]	Pulmonary edema

1.1   Motivation

Recently, there are many works to address the problem of privacy preserving data publishing. However, most of them ignore the case that there may exist semantically proximity for sensitive categorical data. For example, an adversary possessing the QI-values of Andy is able to find out that Andy's record is in the first QI-group of Table 2. Without further information, assume that each tuple in the group has an equal chance of being owned by an adversary, s/he can conclude that Andy suffers from some respiratory infection disease with 2/3 probability.

The existing anonymization principles fall short of solving such proximity privacy leakage in our setting, which is a privacy threat specific to categorical sensitive attribute (i.e. pneumonia in Table 2). To address the issue, this paper proposes a new anonymization principle termed as *m-Color* constraint.

The rest of the paper is organized as follows. The previously related work is reviewed in Section 2. Section 3 clarifies the basic concepts and problem definition. Section 4 discusses the properties of *m-Color* constraint, and Section 5 elaborates the generalization algorithm. In Section 6, we experimentally evaluate the efficiency and effectiveness of our techniques. Finally, the paper is concluded in Section 7.

2   Related Work

In this section, previous related work will be surveyed. All the privacy-preserving transformation of the microdata is referred to as recoding, which can be classified into two classes of models: global recoding and local recoding[11,16]. In global recoding, a particular detailed value must be mapped to the same generalized value in all records. Local recoding allows the same detailed values being mapped to different generalized values of different QI-groups. Obviously, global recoding is a more special case of local recoding. Efficient greedy solutions following certain heuristics have been proposed [7],[9],[11],[15] to obtain a near optimal solution. Incognito [8] provides a practical framework for implementing full-domain generalization, borrowing ideas from frequent item set mining, while Mondrian [9] takes a partitioning approach reminiscent of kd-trees. To achieve k-anonymity, Ghinita [16] presents a framework mapping the multi-dimensional quasi-identifiers to 1-Dimensional(1-D) space. It is discovered that k-anonymity a data set is strikingly similar to building a spatial index over the data set, so that classical spatial indexing techniques can be used for anonymization [17]. The idea of non-homogeneous generalization was first introduced in k-anonymization revisited [18], which studies techniques with a guarantee that an adversary cannot associate a generalized tuple to less than K individuals, but suffer additional types of attack. Authors of paper [19] proposed a randomization method that prevents such type of attack and showed that k-anonymity is not compromised by the attack, but its partitioning algorithm is only a special of the top-down algorithm presented in [11]. However, in our algorithm, we mainly discuss privacy preservation of categorical sensitive attributes over m-Color constraint.

3   Problem Statement

Let T be a microdata table that contains the private information of a set of individuals. T has d QI-attributes $A_1,....,A_d$, and a sensitive attribute $(SA)S$. A partition P consists of several subsets G_i $(1\leq i\leq k)$ of T, such that each tuple in T belongs to exactly one subset and $T = \cup_i^k G_i$. We refer to each subset G_i as a QI-group or a bucket.

Definition 1. (Normalized Certainty Penalty (NCP) [11]). Suppose a table T is anonymized to T^*. In the domain of each attribute in T, suppose there exists a global order on all possible values in the domain. If a tuple t in T^* has range $[x_i, y_i]$ on attribute $A_i(1\leq i\leq d)$, then the normalized certainty penalty in t on A_i is $NCP_{A_i}(t) = \frac{|y_i - x_i|}{|A_i|}$, where $|A_i|$ is the domain of the attribute A_i. For tuple t, the normalized certainty penalty in t is $NCP(t) = \sum_i^d NCP_{A_i}(t)$. The normalized certainty penalty in T is $\sum_{t\in T} NCP(t)$.

   In reality, an attacker with sufficient background knowledge [10] may be able to deduce a concrete QI group. Assume that adversary deduce the sensitive attribute values of victims with the same probability. For example, after realizing that Andy's record appears in the first QI-group of Table 2, an adversary derives $P[X = s] = 1/3$, where X models the disease of Andy, and s can be any disease (i.e., Pneumonia, Pulmonary embolism, Flu) in that group.

3.1   Problem Definition

In our anonymization framework, we present a new anonymization method, which eliminates proximity privacy leakage in publishing categorical sensitive attributes. In general, we classify each tuple in table T based on a given taxonomy tree. For instance, as shown Fig 1, there is a taxonomy tree of respiratory and digestive system diseases. In the tree, ten types of diseases are described and belong to different sub-trees. We simply divide these diseases into four classes, and color each class with red, blue, green and black in Fig 2 respectively. According to the Fig 2, then we can get the Table 3, where these tuples are labeled by four class-tags, red, blue, green and black.

Table 3. Color Table

Name	Age	Sex	Zip	Disease	Color
Andy	5	M	12k	Pneumonia	red
Bill	9	M	14K	Plu embolism	blue
Ken	6	M	18K	Flu	red
Nash	8	M	19K	Bronchitis	red
Joe	12	M	22K	Colotis	black
Sam	11	M	20K	Colon cancer	black
Linda	21	F	58k	Stomach cancer	green
Jame	26	F	36k	Gastric ulcer	green
Sarah	28	F	37k	Gastritis	green
Mary	56	F	33k	Pul edema	blue

Table 4. *3*-Color Constraint

ID	Nmae	Age	Sex	Zip	Disease	Color
	Andy	5	M	12k	Pneumonia	red
1	Bill	9	M	14k	Pul embolism	blue
	Linda	21	F	58k	Stomach cancer	green
	Ken	6	M	18k	Flu	red
2	Joe	12	M	22k	Colotis	black
	Jame	26	F	36k	Gastric ulcer	green
	Nash	8	M	19k	Bronchitis	red
3	Sarah	28	F	37k	Gastritis	green
	Mary	56	F	33k	Pul edema	blue
	Sam	11	M	20k	Colon cancer	black

While generalizing the table T, we do not change the sensitive attribute values of disease but take colors as the sensitive values. From the Table 3, we can observe that the generalization algorithm will not cause any information loss in sensitive attribute values while the personality privacy [5] was not. We may label the tuples whose sensitive values are "similar" with the same color class-tags, and assign the tuples with the same color class-tags into different QI-groups to prevent the problem of proximity privacy leakage.   Hence, *m-Color* constraint is introduced.

Definition 2. A QI-group satisfies *m-Color* constraint, if the QI-group has at least m tuples, and each tuple of this QI-group owns different color class-tags. Then a table fulfills *m-Color* constraint, if in this table, each QI-group satisfies *m-Color* constraint.

Without considering the information loss, we can provide a possible partition of Table 1 again (see the Table 4). Based on the definition 2, the partition fulfills *3-Color* constraint.

In our paper, *m-Color* constraint is used to solve the categorical attribute proximity privacy problem, which limits proximity privacy leakage effectively.

4   The Properties of *m-Color*

Given table T and m, after each tuple was labeled different color class-tags, we wonder the following question: Given a table T and an integer m, can we find a generalized T^* that fulfills *m-Color* constraint? This question is equivalent to the following definition 5. Hence, the results are given as follows.

Definition 3. Given table T and m, if there exists a partition that each QI-group has at least m tuples, and all the sensitive attribute values of tuples in each QI-group are unique. We refer to such table as *m-eligible*.

Theorem 1. T is a *m-eligible*, if and only if the number of tuples with the same sensitive values is no more than $1/m$ the total number of tuples in the table T.

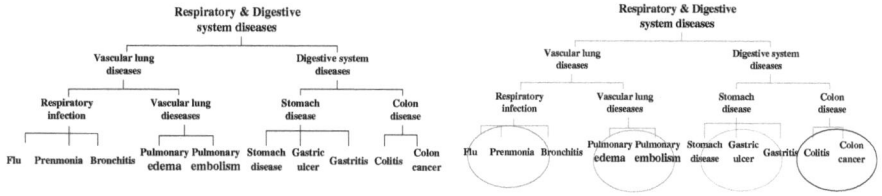

Fig. 1. Taxonomy Tree   **Fig. 2.** Color Taxonomy Tree

5   Generalization Algorithm

The principle of generalization algorithm is to partition the microdata into QI-groups and to reduce the information loss of table as much as possible. Here, the whole algorithm consists of three steps: partition, assign and split. The first step, according to top-down algorithm [11], we can reduce the scale of questions continuously. Secondly, we partition each sub-table, which comes from the first step, into buckets. The last step, the buckets were partitioned into QI-groups.

5.1   Top-Down Partitioning Algorithm

The basic idea of the algorithm is to partition the table iteratively. A set of tuples is partitioned into subsets if each subset is more local and satisfies *m-Color* constraint. The algorithm framework is shown in ALG_1. To keep the algorithm simple, we consider that in each round, we partition a set of tuples into two subsets, and to check whether the both subsets satisfy the *m-Color* constraint. Otherwise, after trying k times in the first step 1-5, we use the Assign algorithm to create the buckets.

```
ALG₁: Partition Algorithm
Input : dyed T, m, K
Output : the set of sub-tables S
Method:
Let S₁, S₂ initial value empty set and K partition times.
FOR I=1 TO K
1.   S₁ ={t₁}, S₂={t₂}
2.   FOR each w∈ T-{t₁,t₂},
        Compute □₁=NCP(w∪S₁)-NCP(S₁), □₂=NCP(w∪S₂)-NCP(S₂);
3.   IF (□₁>□₂), S₂ =S₂∪{w};
     ELES S₁ =S₁∪{w};
```

4.   Find the max value a_1, b_1;
5.   Check if S_1 and S_2 are *m-eligible*.

$$a_1 \leq \frac{1}{m}(a_1 + a_2 + \cdots + a_\lambda) \ , \quad b_1 \leq \frac{1}{m}(b_1 + b_2 + \cdots + b_\lambda) \ ;$$

6.   Random Shuffler the set T;

Step1: We can employ the following strategy to select t_1, t_2 :   select a point u randomly, then to choose t_1 to minimize NCP(t_1, u) (see the definition 3). We can scan all tuples once to find tuple t_2 that maximizes NCP(t_1, t_2).

Step 2: For each tuple w in table T, we compute the $\square_1$= NCP($w \cup S_1$)-NCP(S_1), and $\square_2$=NCP($w \cup S_2$)-NCP(S_2), and distribute w to the group that leads to a   lower information loss.

Step 3: If $\square_1$=$\square_2$, then w was put to the group with less cardinality.

Step 4-5 is to check whether S_1, S_2 satisfy *m-eligible*. If both are generalizable, we remove T from S, and add S_1, S_2 to S; otherwise T is retained in S. The attempts to partition T are tried K times and tuples of G are randomly shuffled for each time. In the second step, we use assign algorithm [3] to create the set of buckets.

5.2   Split Algorithm

In this section, we use buckets passed from the assign algorithm to generate QI-groups. Let the bucket's signature be $\{v_1, v_2, ..., v_\beta\}$ ($\beta \geq m$) and each color tag corresponds to α tuples.   If α=1, the bucket is a QI-group; if α>1, we divide the bucket into two disjoint buk$_1$ and buk$_2$, and each bucket has the same signature. The process to split the bucket until each bucket contains exactly β tuples, each color tag corresponds to one tuple.

   The idea underlying the split algorithm is given as follows: sort all the tuples in the bucket according to attribute A_1. Then put the first α/2 tuples from the bucket into buk$_1$ and the rest tuples into buk$_2$. In this way, we can obtain a partition of the bucket. For the attribute A_2, A_3,...., we can get another d-1 choices, where d is the dimension of QI-attributes. Among all the d partitions, we pick the one that minimizes the sum of NCP(buk$_1$) and NCP(buk$_2$)   as the final partition.

6   Experiments

In this section, we experimentally evaluate the effectiveness and efficiency of the proposed algorithm. In our paper, the experiments run on AMD Dual Core CPU 1.9GHZ, RAM 1GB, Window 2003 Server machine. Our experimentation deploys a real database INCOME from http://ipums.org. We treat the first five columns as QI-attributes, and Income as the sensitive attribute. We compare our algorithm (denoted by *m-Color*) with perturbation technique [21] (perturbation in short).

We measure utility by answering aggregate queries on published anonymous data, since such queries are the basic operations for many database tasks. Specifically, we use queries of the following form: select count(*) from INC where $A_1 \in b_1$ And $A_2 \in b_2$ ...And $A_v \in b_v$

In the experiment, we compare our algorithm with perturbation technique [21]. We fixed s as 0.1 and the dimensionality of queries is two. The technique of perturbation from the experiment results in the first set is relatively poor. With the increase of m, from about 40% to about 280%, the results of our paper are not to exceed 10%, seeing the Fig 3.

Fig 4 illustrates the cost of our anonymous algorithm, while the perturbation technique is finished in less than 1 second (omitted from the Fig 4). The executing time of generalization algorithm probably decreases from 350s to 110s when m from 4 to 30.

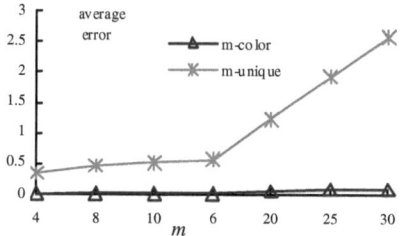
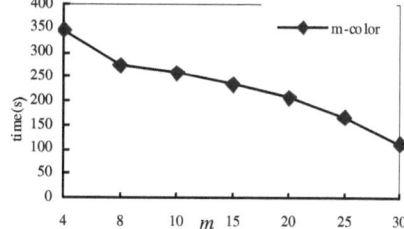

Fig. 3.   Query Accuracy vs. m **Fig. 4.** Running Time

7   Conclusion

The weakness of k-anonymity motivates us to develop a novel anonymization principle called categorical sensitive attributes *m-Color* constraint, which provides better privacy preservation in order to prevent the same categorical sensitive attributes proximity privacy leakage. We systematically analyze the theoretic properties of this new technique and propose a corresponding algorithm. As verified by extensive experiments, our algorithm supports more effective query and data analysis, and simultaneously providing enough privacy preservation.

Acknowledgement. This work was supported in part by the National Natural Science Foundation of China (NO. 61033010, 61170006, 61202007), the Project of Education Department of Zhejiang Province(Y201224678).

References

[1]   Sweeney, L.: k-anonymity: A model for protecting privacy. Journal on Uncertainty, Fuzziness, and Knowlege-Based Systems 10(5), 557–570 (2002)
[2]   Machanavajjhala, A., Gehrke, J., Kifer, D.: l-diversity: Privacy beyond k-anonymity. In: ICDE (2006)
[3]   Xiao, X., Tao, Y.: m-invariance: Towards privacy preserving re-publication of dynamic datasets. In: SIGMOD 2007, pp. 689–700 (2007)

[4]   Li, N., Li, T.: t-closeness: Privacy beyond k-anonymity and l-diversity. In: ICDE (2007)

[5]   Xiao, X., Tao, Y.: Personalized privacy preservation. In: SIGMOD 2006, pp. 229–240 (2006)

[6]   Bayardo, R., Agrawal, R.: Data privacy through optimal k-anonymization. In: ICDE 2005, pp. 217–228 (2005)

[7]   Fung, B.C.M., Wang, K., Yu, P.S.: Top-down specialization for information and privacy preservation. In: ICDE 2005, pp. 205–216 (2005)

[8]   LeFevre, K., DeWitt, D.J., Ramakrishnan, R.: Incognito: Efficient full-domain k-anonymity. In: Sigmod 2005, pp. 49–60 (2005)

[9]   LeFevre, K., DeWitt, D.J., et al.: Mondrian multidimensional k-anonymity. In: ICDE, pp. 277–286 (2006)

[10] Li, J., Tao, Y., Xiao, X.: Preservation of Proximity Privacy in Publishing Numerical Sensitive Data. In: Sigmod 2008 (2008)

[11] Xu, J., Wang, W., Pei, J., et al.: Utility-based anonymization using local recoding. In: KDD 2006, pp. 785–790 (2006)

[12] Willenborg, L., de Waal, T.: Elements of Statistical Disclosure Control. Lecture Notes in Statistics. Springer (2000)

[13] Samarati, P., Sweeney, L.: Protecting Privacy when Disclosing Information: k-Anonymity and Its Enforcement through Generalization and Suppression (1998)

[14] Samarati, P.: Protecting Respondents'Identities in Microdata Release. IEEE Trans.on Konwl. and Data Eng. 13(6), 1010–1027 (2001); Control. Lecture Notes in Statistics. Springer (2000)

[15] Wong, W.K., Mamoulis, N., Cheung, D.W.L.: Non-homogeneous generalization in privacy preserving data publishing. In: SIGMOD 2010: Proceedings of the 2010 International Conference on Management of Data, pp. 747–758. ACM, New York (2010)

[16] Ghinita, G., Karras, P., Kalnis, P., Mamoulis, N.: Fast data anonymization with low information loss. In: VLDB 2007: Proceedings of the 33rd International Conference on Very Large Data Bases, pp. 758–769. VLDB Endowment (2007)

[17] Iwuchukwu, T., Naughton, J.F.: K-anonymization as spatial indexing: toward scalable and incremental anonymization. In: VLDB 2007: Proceedings of the 33rd International Conference on Very Large Data Bases, pp. 746–757. VLDB Endowment (2007)

[18] Gionis, A., Mazza, A., Tassa, T.: k-anonymization revisited. In: ICDE 2008: Proceedings of the 2008 IEEE 24th International Conference on Data Engineering, pp. 744–753. IEEE Computer Society, Washington, DC (2008)

[19] Wong, W.K., Mamoulis, N., Cheung, D.W.L.: Non-homogeneous generalization in privacy preserving data publishing. In: SIGMOD 2010: Proceedings of the 2010 International Conference on Management of Data, pp. 747–758. ACM, New York (2010)

[20] Zhang, Q., Koudas, N., Srivastava, D., Yu, T.: Aggregate query answering on anonymized tables. In: ICDE 2007: Proceedings of the 23nd International International Conference on Data Engineering, vol. 1, pp. 116–125 (2007)

[21] Agrawal, R., Srikant, R.: Privacy-preserving data mining. SIGMOD Rec. 29(2), 439–450 (2000)

S2MART: Smart Sql to Map-Reduce Translators

Narayan Gowraj, Prasanna Venkatesh Ravi, Mouniga V, and M.R. Sumalatha

Department of Information Technology, Anna University,
Chennai, India
{ngowraj,prasanvenkyr,meek.mounik,
sumalatha.ramachandran}@gmail.com

Abstract. With the advent of rapid increase in the size of data in large cluster systems and the transformation of data into big data in major data intensive organizations and applications, it is very necessary to build efficient and flexible Sql to Map-Reduce translators that would make Tera to Peta bytes of data easy to access and retrieve because conventional Sql-based data processing has limited scalability in these cases. In this paper we propose a Smart Sql to Map-Reduce Translator (S2MART), which transforms the Sql queries into Map-Reduce jobs with the inclusion of intra-query correlation for minimizing redundant operation, sub-query generation and spiral modeled database for reducing data transfer cost and network transfer cost. S2MART also applies the concept of views in database to perform parallelization of big data easy and streamlined. This paper gives a comprehensive study about the various features of S2MART and we compare the performance and correctness of our system with two widely used Sql to Map-Reduce translators' hive and pig.

Keywords: Sql to Map-Reduce Translators, big data, intra-query relationship and sub-query generation.

1 Introduction

According to a recent presentation from the vice-president of Facebook Jeff Rothschild at the UC San Diego, Facebook has 30,000 servers supporting its operations which were 10,000 servers on April 2008. This drastic increase in the amount of data in large online organizations had led to the development of extensive data processing applications like the Sql to Map-Reduce translators [1] [2] [3]. Due to the ever increasing size of data in these large clusters, traditional relational databases can be expensive; hence many companies have maneuvered their focus towards NoSql database framework such as BigTable, HadoopDB, Simple DB, Microsoft's SDS cloud database, MongoDB, Apache CouchDB etc [4] [5] [6] [7]. The importance and uniqueness of cloud infrastructure is closely coupled with the development of the Map-Reduce paradigm [8] [9]. In summary the major contributions of this paper are: 1) Support intra-query correlations by building a Sql relationship tree to minimize redundant operations and computations.2) Build a spiral modeled database to store and retrieve the recently

Y. Ishikawa et al. (Eds.): APWeb 2013, LNCS 7808, pp. 571–582, 2013.

used query results to minimize the data transfer cost and network cost and increase the time efficiency.3) The spiral modeled database is adaptive to dynamic workload patterns.4) Sub-query generation to increase the throughput of the system.5) Use the concept of Views in database to store intermediate results which reduces the amount of records to be checked for every computation.6) Experiment our system with legacy systems like Hive, pig etc.

2   Literature Survey

The popularity of a hybrid system transforming Sql queries to Map-Reduce jobs has increased due to the emergence of big data in many data centric organizations. S2MART uses nuoDB [10] [11] as its cloud storage as nuoDB provides full indexing support, easy and flexible Sql querying with DB replication, provides drivers for ODBC, JDBC and Hibernate, it can be deployed anywhere in the cloud, provides scalability upto 50 nodes and the most important thing is that it provides support for Hadoop HDFS. In this paper we focus on three important legacy Sql to Map-Reduce translators namely Pig, Hive and Scope. Companies and organizations which deal with big data in the form of click streams, logs and web crawls have focused their attention towards developing an open source high-level dataflow system. This has led yahoo develop an open source incarnation called Pig to support parallelization of data retrieval in the cloud environment [12]. Pig supports a simple language called Pig Latin [13] to support query formulations and data manipulation. Pig Latin is written using java class libraries and provides support for simple operations like filtering, grouping, projection and selection. The major issue concerned with Pig is that it provides limited flexibility towards join operations and nested operators. Hive developed by Facebook adds Sql like functions to Map-Reduce but follows syntax different from the conventional Sql and it is called as HiveQl [14] [15]. HiveQl transforms the Sql query into Map-Reduce jobs that are executed in Hadoop (an open source incarnation of Map-Reduce). Hive provides the functionalities of a metastore that supports query compilations and optimizations [16]. Even though Hive follows a structured approach towards database concepts, it provides limited functionality towards join predicate. The Sql to Map-Reduce translator developed by Microsoft was called as Scope [17]. The compiler and optimizer of Scope generates a new scripting language for managing and retrieving data from large data repositories. The compiler and optimizer are responsible for automatic job generation and parallelization of job execution. However, Scope provides no support towards user defined types and metastore which reduces the performance of the system. Hence Scope doesn't provide a rich data model when compared to Hive and pig as it doesn't include partitioning of tables. The other translators developed by Microsoft are Cosmos [18] and Dyrad [19].

3   Proposed System

The proposed system consists of two linchpin modules which are:

3.1   System Architecture

The full-fledged system architecture is shown in fig.1 which is divided into three
major layers. The architecture of S2MART uses the classic master/worker pattern
of the Map-Reduce framework where the master node is referred as the query
distributor and the worker nodes are referred as query analyzers. The input
to the system is the Sql query from the user. This Sql query is given to the
Sql query Parser which in turn is given to the IQ-RT (Intra-query relationship
tree) parser which constructs the tree based on the input query. IQ-RT parser
generates the tree structure which we call as the relationship tree for the input
query from the Sql query parser. The tree is then given as an input to the IQ-RI
(Intra-query relationship identifier). After identifying the existing relationships,
the operations similar to each other are consolidated so that the number of scans
of a particular database reduces drastically. The resultant IQ-RT is subjected
to the Sub-Query generator engine which disintegrates the different operations
in the IQ-RT into sub-queries rather than a single large query. Now each query
are distinctly separated and every query is compared with the queries stored in
the Spiral database which uses the FQR (Frequent Query Retrieval) algorithm
to check whether the input query matches any one of the entries in the Spiral
database. Now there exists two possibilities, the first scenario is the case where
the input query and the results of the input query are cached from the Spiral
database and the other scenario is that the query and the results are not found
in the Spiral database. In the first case the results of the individual sub-query
are cached from the Spiral database [20] and are given as an input to the query
optimizer which optimizes the query results and gives the query results back to
the user. In the other case, each individual query is given to the Map-Reduce
framework which is built inside a Hadoop infrastructure [21][22]. Here in this case

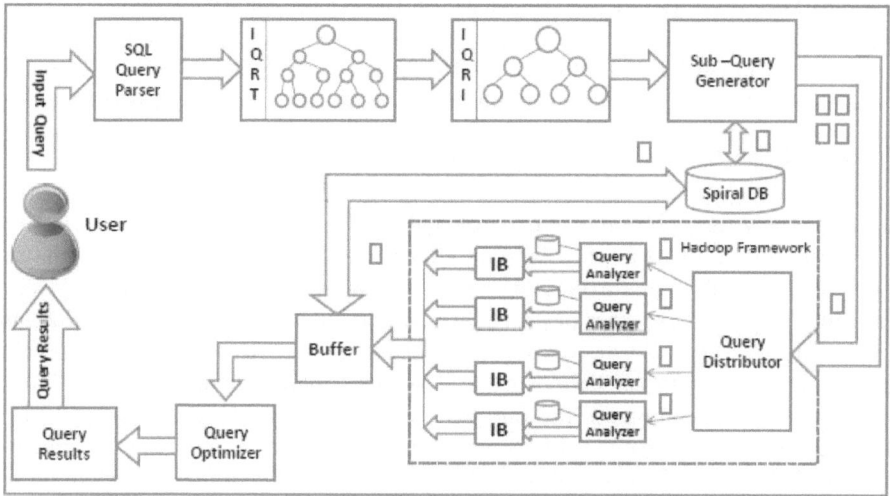

Fig. 1. System Architecture

a single node alone is not subjected to retrieve data from the database nodes hence avoiding a bottleneck situation and no single node does the entire mapping job of a single query hence increasing the efficiency of data retrieval from the database nodes without changing the underlying structure of the Map-Reduce framework. Data transfer cost is reduced in this method by the use of spiral DB. The frequently used queries are stored, thus reducing the need to transfer data unnecessarily. Network transfer cost is reduced by the use of buffers after the map-reduce step.

3.2   Methodology

The entire working of the S2MART basically depends on the following modules:

3.2.1 SQL Query Parser
The Sql Query Parser is a module which accepts Sql queries as input from the user, converts them into a format which is accepted by the S2MART and outputs the query to the IQ-RT module to generate the relationship tree for the input query.

3.2.2 Intra-query Relationship Tree
The relationship tree is constructed based on a standard prototype where every operation (selection, aggregation, ordering and sorting) is represented within an oval and every database node is represented with a square. Let us consider an example where a large input Sql query shown in fig.2 gets converted into the relationship tree. The query is used to retrieve the department name (dept_name), department id (dept_id) and employee name (empname) where the employee is of Indian origin with a salary greater than 25000 who belongs to the IT department and lives in the state Mumbai (MUM). Now given the input query, we construct the relationship tree based on the IQ-RT algorithm shown in fig.6. Fig.3 represents the relationship tree for the given input query where D, E, C represents the tuples which represent the Department, Employee and Country tables respectively. Here the total number of scans are 7 and requires 7 selection operations and 4 join operation to get the required query results.Each node in the fig.3 represents the individual database.

3.2.3 Intra-query Relationship Identifier
The first step in IQ-RI module is to find out the intra-query relationship existing between the various operations performed by the queries. The intra-query relationships are based on a set of procedures which are: Procedure 1: The selection operations on the same database can be grouped so as to eliminate redundant scans on the same table. Procedure 2: Two join operations involving the same partition key are combined into one join operation either using 'OR' or 'AND'. Procedure 3: Aggregation operations involving sum, max, min, avg etc on the same database can be grouped into one. Procedure 4: Combine individual outer join and inner join into a single query. Based on these procedures certain operations are grouped together to avoid redundant scans on the same table and to

Select d.dept_name,d.dept_id (Select e.empname from
emp e where (d.empno=e.empno)) from (dept d, emp e)
where e.nationality='indian' and e.salary>25000 and
d.dept_name='IT' and c.state_name='MUM' and
c.empno=e.empno;

Fig. 2. Input Query to SQL Parser

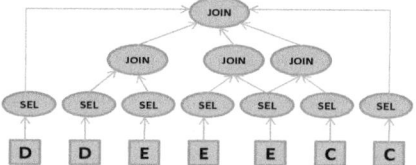

Fig. 3. Relationship Tree for Input Query

Select d.dept_name, d.dept_id, e.empname from
dept d, emp e, country c where d.empno=e.empno
and e.empno=c.empno and e.nationality='indian'
and e.salary>25000 and d.dept_name='IT' and
c.state_name='MUM';

Fig. 4. Resultant Query after IQ-RI module

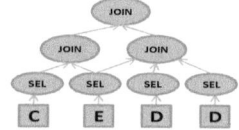

Fig. 5. Resultant Tree after IQ-RI module

optimize the query efficiency. Now the relationship tree is generated again based
on all the groupings. The query which has been subjected to the IQ-RI module
is shown in fig.4. Now, the IQ-RT is again generated for the query obtained
which is shown in fig.5. The IQ-RT obtained represents 4 selection operations
and 3 join operations totally, hence shows less number of operations compared
to the IQ-RT obtained in fig.3. Hence the number of table scans and operations
that has to be performed against the database has decreased which increases
the throughput and reduces execution time. The algorithm used to find the
intra-query relations and combine similar operations are shown in fig.7.

3.2.4 Sub-query Generator Engine
The input to the sub-query generator engine is the relationship tree which has
been subjected to the IQ-RI module. Now the consolidated query obtained from
the relationship tree is given a sequence number which is used as a reference.
This query is checked with the queries present in the Spiral Database to find

IQRT Generator

Input : query, a string

create empty tree IQRT;
while *query parsing till the end* **do**
  make *Join* node;
  assign *Join* as root node of IQRT;
  if *select operation is found in query* **then**
    make *select* node;
    assign the input *table_id* to *select* node;
    do point node to *Join* root node of query;
return *IQRT*;

Fig. 6. IQ-RT Algorithm

IQRI Generator

Input: *IQRT*, a tree

create empty tree *IQRT_new*;
if *no_of_table_ids is greater than six* **then**
  if *join.sel.table_ids are repeating* **then**
    group selection operations on same *table_id*;
  if *join.partitionkey are same* **then**
    use 'OR' or 'AND' operation into one join;
  if *aggregate operations nodes on same table_id* **then**
    group aggregate operations on same *table_id*;
  combine individual outer join and inner join into a -
  single query, *IQ*;
  IQRT_new=IQRT(*IQ*);
  GenQI(*IQRT_new*);
else
  GenQI(*IQRT*)

Fig. 7. IQ-RI Algorithm

any match. If there is any match then the results of the query is directly given to the buffer or else the query is divided into a number of sub-queries which is done based on the GenQL algorithm. In order to bring concurrency between the queries to get the proper results we use the concept of Views in database. After the sub-queries are generated, each sub-query represents a unique independent operation performed by the relationship tree. The number of views that has to be created is equal to the number of sub-queries generated for the same. The query which has been subjected to the IQ-RI module shown in the fig.4 can be divided into 4 sub-queries based on the GenQL algorithm shown in fig. 12. Since the number of sub-queries is 4 the number of views required is also 4. Every individual query is subjected to a view creation which reduces the amount of records or data that has to be accessed or referred for further operations on the table and the view creation can group operations that scans the same tables, minimizing the number of operations. The sub-queries generated for the query shown in fig.4 are shown in fig 8, 9, 10, 11.

Select d.dept_name, e.empno, e.empname
from dept d, emp e, country c where
d.empno=e.empno and e.empno=c.empno;

Fig. 8. Sub-Query 1

Select v.empno, v.dept_name, v.empname from
view1 v, emp e where .empno=e.empno and
e.salary>25000 and e.nationality='indian';

Fig. 9. Sub-Query 2

Select v.empno, v.dept_name, v.empname
from view2 v, dept d where
v.empno=d.empno and d.dept_name='IT';

Fig. 10. Sub-Query 3

Select v.empno, v.dept_name, v.empname
from view3 v, country c where
v.empno=c.empno and c.state='MUM';

Fig. 11. Sub-Query 4

3.2.5 Spiral Database

The spiral database acts as the cache storage for the S2MART system which is used to directly retrieve the most frequently used Sql query and its results. Eventually, this reduces the data and network transfer cost for the overall system. The name Spiral database comes from the fact that the Sql queries are stored dynamically in the database based on the frequency count. In this case we store the six frequently referred queries along with its results so as to have a less space complexity for the system. If the number of queries is more than six then we use the FQR (Frequent Query Retrieval) to perform deletion of unused queries. Thus we perform an operation which involves dynamic inclusion and exclusion of queries and its results and hence the name Spiral Database. The input to the Spiral database is from the buffer which stores the results of the query executed. The results along with the sequence number of the query are stored in the database for caching purposes. The FQR (Frequent Query Retrieval) algorithm is implemented within the spiral database to dynamically store the different queries based on the frequency counts. In the FQR algorithm implemented, we have assumed six as the maximum number of queries that can be stored in the spiral DB. The number of queries that are stored in the spiral DB can be changed depending on our needs. The work flow of the Spiral Database has already been discussed in Section 3. The complete work procedure of the FQR algorithm is depicted in fig. 13.

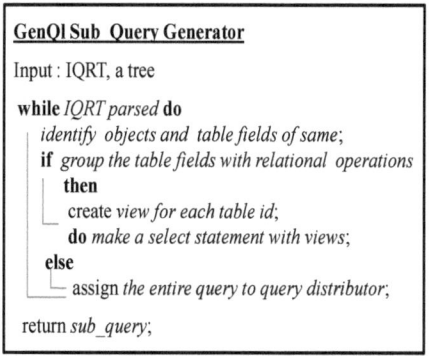

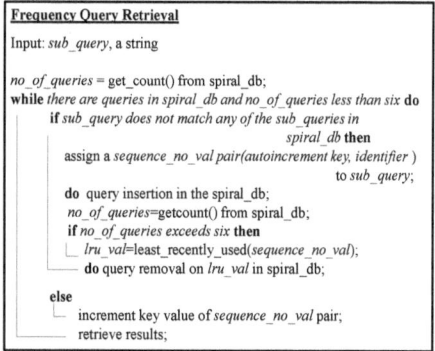

Fig. 12. GenQl Algorithm   **Fig. 13.** FQR Algorithm

3.2.6 Query Distributor and Analyzer

The query distributor and the query analyzer are set up inside the Hadoop framework where the query distributor represents the master node of the Map-Reduce framework and the query analyzers represent the worker nodes. The master node sends the individual sub-query to the query analyzers which are connected to the various multifaceted database nodes. The general format for the execution of every individual query is shown in fig.14. The final records are sent to the intermediate buffer once all the sub-queries have been executed, to undergo the reduce function. The point of using the intermediate buffers is to increase the fault tolerance of the entire system in situations where the buffer is subjected to single point of failure. In case of a bottleneck situation on the buffer, the query results can be retrieved back from the intermediate buffer which eliminates the requirement of performing the entire process once again. This increases the robustness of the system. A copy of the results are sent back to the Spiral Database to associate with the sequence number of the query.

3.2.7 Query Optimizer

The query results from the buffer in the Map-Reduce framework or results from the Spiral database are the two possibilities of query input to the query optimizer. The main function of the query optimizer is to check whether all the sub-queries which are generated from the relationship tree are executed successfully and the query results are optimized and are ordered in a presentable manner to be presented to the user.

3.2.8 Query Results

The final step in the methodology of the S2MART is the query results which provide the user interface where the query results are presented back to the user.

4   Experimental Results

We have conducted various experiments to verify the sanity of S2MART. We have conducted our experiments by comparing our novel system with our benchmarks Hive and Pig [23]. A cluster with 40 nodes where each node uses a 4GB ram, 250 GB hard disk with 2.1GHz speed is used for implementation of various queries [24] [25]. The size of the database used is 1GB with 7 attributes, 3 databases, and 10,000 tuples. Comprehensive evaluation of our novel system is done by using various paradigms of queries such as [26] [27] [28]:

4.1   Lightweight Query without Intra-query Relations and Sub-query Generation

In this case we consider a simple query which cannot be subjected to any of the modules of the S2MART. The query shown in fig.15 is used to find the employee name (emp_name) from the Employee table where employee id (emp_id) is 101.

create view *\<view name\>* as (*\< Individual sub-query is given here \>*);	Select empname from employee where emp_id=101;

Fig. 14. General Format for sub-queries   Fig. 15. Simple Lightweight query

The comparison between the various execution time of the query in S2MART, Hive and Pig is shown in fig.19 where x-axis represents the number of nodes and y-axis represents the execution time in seconds. The graph shows that there is no drastic decrease in the execution time for simple light weighted queries.

4.2   Lightweight Query with Intra-query Relationships But No Sub-query Generation

In this case we consider a query that can be subjected to the IQ-RT and the IQ-RI modules but it cannot be divided into sub-queries as it is atomic. The query shown in fig.16 is used to retrieve the Department name (dept_name), Department id (dept_id), Employee name (empno) from the tables Employee, Department and Country where the join operation is performed between the tables using the primary key of each table. Now the query which has been subjected to the IQ-RT and IQ-RI modules is shown in fig.17.

Fig.20 shows the comparison between S2MART and the benchmarks. The graph where x-axis represents the number of nodes and y-axis represents the execution time in seconds shows that there is gradual decrease in the execution time of the query when compared to the other systems after 25 nodes.

Select d.dept_name,d.dept_id (Select e.empname from emp e where (d.empno=e.empno)) from (dept d, emp e) where e.empno=c.empno;	Select d.dept_name,d.dept_id, e.empname from employee e , department d and country c where d.empno=e.empno and e.empno=c.empno;

Fig. 16. Not Subjected to IQ-RT & IQ-RI   **Fig. 17.** Subjected to IQ-RT & IQ-RI

4.3   Lightweight Query with Sub-query Generation and No Intra-query Relations

In this case we consider a consolidated query which can be divided into a number of Sub-queries but there are no intra-query relationships existing within the queries. The query shown in fig.18 is used to find the names of the Indian employees whose Department id (dept_id) is 101 from the Employee and the Department table. This query is divided into two sub-queries which are executed in two different views.

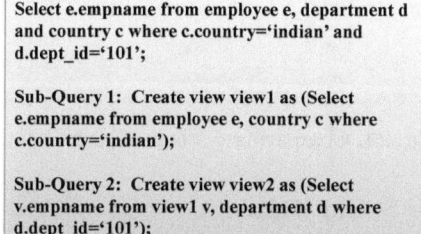
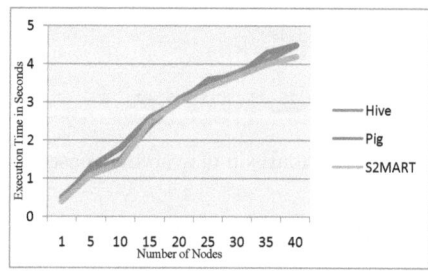

Fig. 18. Subject to Sub-queryGeneration   **Fig. 19.** Comparison of systems - case 4.1

Fig.21 shows the experimental comparison between the systems. The x-axis of the graph represents the amount of data in thousands and y-axis represents the execution time of the query. The graph shows that the execution time drastically decreases after 20,000 records and thus it can be inferred that S2MART can be very useful when we deal with large amount of data in the data repositories.

4.4   Heavyweight Query with Both Intra-query Relationships and Sub-query Generation

In this case we consider the same query which is shown in fig.4. The query is subjected to all the modules of the S2MART which includes IQ-RT, IQ-RI and Sub-Query Generation. Fig.22 shows the experimental comparison between the various systems where the x-axis represents the number of nodes and y-axis represents the execution time in seconds. The graph shows that the execution time gradually decreases as the number of nodes or the amount of data increases. We can infer that S2MART can be more efficient than its benchmarks when we have a complex heavyweight Sql query as the input.

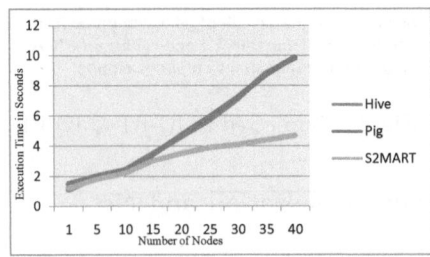
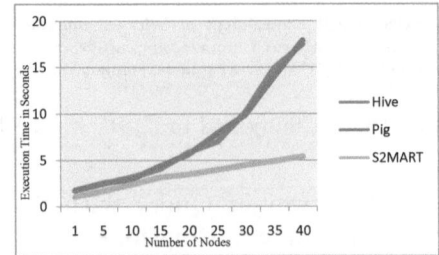

Fig. 20. Comparison of systems - case 4.2   **Fig. 21.** Comparison of systems - case 4.3

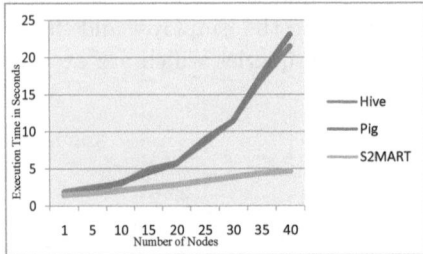
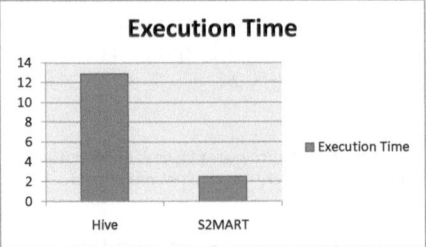

Fig. 22. Comparison of systems - case 4.4   **Fig. 23.** Comparing systems of cached data

4.5   Query Results Cached from the Spiral Database

In this case we consider the result of the query to be cached from the Spiral Database. In this case there is no query execution time spent for the query because the FQR algorithm finds whether a match is present before the execution of the query. Fig.23 shows the comparison between the execution time of the query for Hive and S2MART. We don't consider Pig for the comparison as it doesn't provide the functionalities of a metastore or caching of data. The graph shows that the execution time in hive[29] is very much greater when compared to S2MART.

5   Conclusion

Execution of complex queries involving join operations between multiple tables, aggregation operations are very much necessary for data intensive applications, thus we developed a system that could support the concept of big data in data intensive applications and ease data retrieval to provide useful information. We implement many novel modules such as Intra-Query relationships (IQ-RT and IQ-RI), Sub-Query generation and Spiral Database in the S2MART system that makes the complete system a novel one. We verified the correctness of S2MART system by performing various experimental results using different paradigms of queries and we achieved efficiency in all the experimental results. The intended applications for S2MART are all data intensive applications which use big data analytics. Hence in the future we can build the system with more efficient query optimization for a particular application rather than being generic and enable secure cloud storage [30] for confidentiality and integrity of data.

References

1. Jiang, D., Tung, A.K.H., Chen, G.: MAP-JOIN-REDUCE: Toward Scalable and Efficient Data Analysis on Large Clusters. IEEE Transactions on Knowledge and Data Engineering 23(9) (September 2011)
2. Warneke, D., Kao, O.: Exploiting Dynamic Resource Allocation for Efficient Parallel Data Processing in the Cloud. IEEE Transactions on Parallel and Distributed Systems 22(6) (June 2011)
3. Jiang, W., Agrawal, G.: MATE-CG: A MapReduce-Like Framework for Accelerating Data-Intensive Computations on Heterogeneous Clusters. In: 2012 IEEE 26th International Parallel and Distributed Processing Symposium (2012)
4. Sakr, S., Liu, A., Batista, D.M., Alomari, M.: A Survey of Large Scale Data Management Approaches in Cloud Environments. IEEE Communications Surveys and Tutorials 13(3) (Third Quarter 2011)
5. Han, J., Song, M., Song, J.: A Novel Solution of Distributed Memory NoSQL Database for Cloud Computing. In: 10th IEEE/ACIS International Conference on Computer and Information Science (2011)
6. Bisdikian, C.: Challenges for Mobile Data Management in the Era of Cloud and Social Computing. In: 2011 12th IEEE International Conference on Mobile Data Management (2011)
7. Nicolae, B., Moise, D., Antoniu, G., Bougé, L., Dorier, M.: BlobSeer: Bringing High Throughput under Heavy Concurrency to Hadoop Map-Reduce Applications. In: International IEEE Conference (2010)
8. Zhang, J., Wu, X.: A 2-Tier Clustering Algorithm with Map-Reduce. In: The Fifth Annual ChinaGrid Conference (2010)
9. Pallickara, S., Ekanayake, J., Fox, G.: Granules: A Lightweight, Streaming Runtime for Cloud Computing With Support for Map-Reduce. In: IEEE International Conference (2009)
10. Starkey, J.: Presentation of nuoDB,
    http://www.siia.net/presentations/software/AATC2012/NextGen_NuoDB.pdf
11. Starkey, J.: Presentation of nuoDB,
    http://www.cs.brown.edu/courses/cs227/slides/dtxn/nuodb.pdf
12. Zhang, Z., Cherkasova, L., Vermam, A., Loo, B.T.: Optimizing Completion Time and Resource Provisioning of Pig Programs. In: 2012 12th IEEE/ACM International Symposium on Cluster, Cloud and Grid Computing (2012)
13. Olsten, C., Reed, B., Srivastava, U., Kumar, R., Tomkins, A.: Christopher Olsten, Benjamin Reed, Utkarsh Srivastava, Ravi Kumar, Andrew Tomkins. Presented by Welch, D.
14. Thusoo, A., Sen Sarma, J., Jain, N., Shao, Z., Chakka, P., Zhang, N., Antony, S., Liu, H., Murthy, R.: Hive – A Petabyte Scale Data Warehouse Using Hadoop. In: IEEE International Conference (2010)
15. Michel, S., Theobald, M.: HadoopDB: An Architectural Hybrid of MapReduce and DBMS Technologies for Analytical Workloads. Presented by Raber, F.
16. Thusoo, A., Sen Sarma, J., Jain, N., Shao, Z., Chakka, P., Anthony, S., Liu, H., Wyckoff, P., Murthy, R.: Hive A Warehousing Solution Over a MapReduce Framework. In: VLDB 2009, Lyon, France (August 2009)

17. Chaiken, R., Jenkins, B., Larson, P.-Å., Ramsey, B., Shakib, D., Weaver, S., Zhou, J.: SCOPE: Easy and Efficient Parallel Processing of Massive Data Sets. In: VLDB 2008, Auckland, New Zealand, August 24-30 (2008)
18. Yang, C.: Osprey: Implementing MapReduce-Style Fault Tolerance in a Shared-Nothing Distributed Database. In: ICDE Conference (2010)
19. Isard, Y.Y.M.: DryadLINQ: A System for General-Purpose Distributed Data-Parallel Computing Using a High-Level Language. Microsoft research labs
20. He, Y., Lee, R.: RCFile: A Fast and Space-efficient Data Placement Structure in MapReduce-based Warehouse Systems. In: ICDE Conference (2011)
21. Foley, M.: High Availability HDFS, http://storageconference.org/2012/Presentations/M07.Foley.pdf
22. Chandar, J.: Join Algorithms using Map/Reduce. University of Edinburgh (2010)
23. Chen, T., Taura, K.: ParaLite: Supporting Collective Queries in Database System to Parallelize User-Defined Executable. In: 2012 12th IEEE/ACM International Symposium on Cluster, Cloud and Grid Computing (2012)
24. Leu, J.-S., Yee, Y.-S., Chen, W.-L.: Comparison of Map-Reduce and SQL on Large-scale Data Processing. In: International Symposium on Parallel and Distributed Processing with Applications (2010)
25. Hsieh, M.-J.: SQLMR: A Scalable Database Management System for Cloud Computing. In: 2011 International Conference on Parallel Processing (2011)
26. Zhu, M., Risch, T.: Querying Combined Cloud-Based and Relational Databases. In: 2011 International Conference on Cloud and Service Computing (2011)
27. Husain, M.F.: Scalable Complex Query Processing Over Large Semantic Web Data Using Cloud. In: 2011 IEEE 4th International Conference on Cloud Computing (2011)
28. Hu, W.: A Hybrid Join Algorithm on Top of Map Reduce. In: 2011 Seventh International Conference on Semantics, Knowledge and Grids (2011)
29. Introduction to Hive by Cloudera 2009, http://www.cloudera.com/wp-content/uploads/2010/01/6-IntroToHive.pdf
30. Wallom, D.: myTrustedCloud: Trusted Cloud Infrastructure for Security-critical Computation and Data Management. In: 2011 Third IEEE International Conference on Cloud Computing Technology and Science (2011)

MisDis: An Efficent Misbehavior Discovering Method Based on Accountability and State Machine in VANET

Tao Yang[1,2,3], Wei Xin[1,2,3,*], Liangwen Yu[1,2,3], Yong Yang[4],
Jianbin Hu[1,2,3], and Zhong Chen[1,2,3]

[1] MoE Key Lab of High Confidence Software Technologies, Peking University,
100871 Beijing, China
[2] MoE Key Lab of Computer Networks and Information Security, Peking University,
100871 Beijing, China
[3] School of Electronics Engineering and Computer Science, Peking University,
100871 Beijing, China
[4] School of Software Engineering, Beijing Jiaotong University,
100044 Beijing, China
{ytao,xinwei,yulw,hujb,chen}@infosec.pku.edu.cn,
yycola@263.net.cn

Abstract. VANET has some prominent features such as channel-opening, large-scale, fast, dynamic-changing, so there is some misbehavior vehicles. Due to the large number of vehicles, even if there is only 1% percent of misbehavior vehicles, there can be thousands of misbehavior vehicles. How to identify these vehicles quickly, and then use appropriate cancellation and isolation measures to reduce system losses to its minimum and controllable range. In this paper, we propose MisDis, a method that can identify misbehavior vehicle by implementing state automata and supervision, as well as a special security log to record the behavioral characteristics of the target vehicle. If the host vehicle violates the established security policy, then the system need to record the behavioral and try to dump to a safer roadside unit or department in charge.

Keywords: VANET, secure log, secure record, accountability, misbehavior detection.

1   Introduction

VANET (Vehicular ad hoc network) are gaining significant prominence from both academia and industry in recent years. VANET aims to enhance the safety and efficiency of road traffic, and have attractive prospects for development. VANET can improve People's livelihood and has a good prospect. We can get better driving experience, ease the traffic pressure and reduce traffic accidents through the establishment of VANET with which we constituent network infrastructure environment of "smart car". Due to the large number of vehicles, even if there is only 1% percent of

---

* Corresponding author.

Y. Ishikawa et al. (Eds.): APWeb 2013, LNCS 7808, pp. 583–594, 2013.

misbehavior vehicles, there can be thousands of misbehavior vehicles. It also faces serious security threats because of its huge scale of vehicle number, its variable node velocity, and its openness. VANET has been strictly constrained by security because users would not accept or participate it with fear for their safety or personal privacy leakage. As a result, the VANET security issue becomes a key, urgent, fundamental and challenging problem, and the research about that is becoming a hot spot.

Vehicles play a basic role in the VANET because they are the original location of messages' generation and transmission. Some types of messages are exchanged among vehicles such as traffic information, emergency incident notifications, and road conditions. When something goes wrong with vehicles, such as message leakage or misbehaviors, the whole VANET will be problematic, and its correctness and effectiveness cannot be guaranteed. Malicious vehicles can start malfunctioning due to some internal failures or propagate false identity, position or time information to solve their selfish motives. For example, an attacker node may create an illusion of traffic congestion by pretending multiple vehicles simultaneously and launch DoS attack by impairing the normal data dissemination operation.

There are number of security threats[1-3], which may degrade the performance of VANET and put the lives of passengers in danger. Misbehaving vehicles perform in several ways and have different objectives such as attackers eavesdrop the communication, drop, and change or inject packets into the network. Therefore, one solution is that vehicles should cooperate together to enhance the security performance of a network. Therefore, security mechanisms, facilities, and protocols are needed to diminish and to eliminate the attacker's effect. In this paper, the terms attacker, illegitimate behavior, malicious vehicles are synonymously used. However, current research mainly focuses on the integrity and confidentiality of the communication in the VANET and seldom cares about the possibility of OBUs' abnormal behaviors.

In this paper, we propose MisDis, a method focus on the misbehavior vehicles detection. In general, the contributions of MisDis are:

1. MisDis can introduce accountability to the vehicles' behavior by record evidence, and confirm the property both inside and outside the vehicle.
2. MisDis can identify misbehavior vehicle by implementing state automata and supervision.
3. MisDis designs a special security log to record the behavioral characteristics of the target vehicle. And if the host vehicle violates the established security policy, then the system need to record the behavioral and try to dump to a safer roadside unit or department in charge.

Organization: The rest of this paper is organized as follows: in Section 2, we present the related work. Section 3, introduces the preliminaries such as system model and security objectives. Section 4 describes the design rationale and details of our scheme. Then, the security analysis result is discussed in Section 5. Finally, we will give conclusion in Section 6.

2   Related Work

There are only few misbehavior detection schemes suggested for finding attacker.

Golle et al. [4] propose a scheme to detect malicious data by finding outlying data. If there is data that does not fit to the current view of the neighborhood, it will be marked as malicious. The current view, in turn, is established cooperatively. Vehicles share sensor data with each other. An adversary model helps to find explanations for inconsistencies. The main goal is to detect multiple adversaries or a single entity carrying out a sybil attack. This work, however, does not provide concrete means to detect misbehavior but only mentions a "sensor-driven detection".

Raya et al. [5] introduced a solution of immediate revocation of certificates of a "misbehaving" vehicle and formulated a detection system for misbehavior. It is assumed that the PKI is not omnipresent and hence the need for an infrastructure assisted solution. One component of their work is a local misbehavior detection system running at each node. This system relies on gathering information from other nodes, and then comparing the behavior of a vehicle with some model of average behavior of the other vehicles built on the fly. It is based on the deviation of attacker node from normal behavior and used timestamp, assigned messages and trusted components (hardware and software). Honest and illegitimate behavior is differentiated by using *clustering techniques*. The basic idea behind the autonomous solution is to evaluate deviation from normal behavior of vehicles, while they always assume an honest majority. Once, misbehavior is detected, a revocation of a certificate is indicated over a base station, a vehicle connects to. This revocation is then distributed to other vehicles and the Certification Authority itself. But, the assumption of an honest majority of vehicles does not hold as identities may be generated arbitrarily.

Ghosh et al [6,7] presented and evaluated a misbehavior detection scheme (MDS) for post crash notification (PCN) application. Vehicles can determine the presence of bogus safety message by analyzing the response of driver at the occurrence of an event. The PCN alert raised will be received by multiple vehicles in the vicinity, and more than one of these receiving vehicles may use an MDS in their OBUs to detect misbehavior. The basic cause-tree approach is illustrated and used effectively to jointly achieve misbehavior detection as well as identification of its root-cause.

Schmidt et al [8] introduced a framework (VEBAS) based on reputation models. By combining the output of multiple behavior analysis modules, a vehicle calculates the trustworthiness value in other vehicles in its near vicinity from the various sensor values obtained. The value is then used to identify a misbehaving vehicle and build up reputation. VEBAS requires continuous exchange of information between the vehicles that is used by a vehicle to analyze the behavior of nearby vehicles. Based on this information, vehicles are classified into trustworthy, untrustworthy or neutral. Applications may then take this trust rating into consideration in order to react appropriately on incoming information. However, the paper only describes the scheme, no evaluation for any specific application is presented.

A message filtering model is introduced by Kim et al [9] that uses multiple complementary sources of information. According to this model, drivers are sent alert messages after the existence of certain event is proved. The model leverages multiple complementary sources of information to construct a multi-source detection model such that drivers are only alerted after some fraction of sources agree. The model is based on two main components: a threshold curve and a Certainty of Event (CoE) curve. A threshold curve implies the importance of an event to a driver according to the relative position, and a CoE curve represents the confidence level of the received messages. An alert is triggered when the event certainty surpasses a threshold.

PeerReview[10], presented by Haeberlen et al., is a system that provides accountability in distributed systems. It's useful in accounting for the behavior of every peer and identifying those peers that break the protocol. It ensures that Byzantine faults whose effects are observed by a correct node are eventually detected and irrefutably linked to a faulty node. It also ensures that a correct node can always defend itself against false accusations. These guarantees are particularly important for systems that span multiple administrative domains, which may not trust each other. PeerReview's key idea is maintaining a per-node secure log which records the messages a node has sent and received, and the inputs and outputs of the application. Any node i (as a given reference implementation) can request the log of another node j and independently determine whether j has deviated from its expected behavior. PeerReview is widely applicable: it only requires that a correct node's actions are deterministic, that nodes can sign messages, and that each node is periodically checked by a correct node. To do this, i replays j's log using a reference implementation that defines j's expected behavior. By comparing the results of the replayed execution with those recorded in the log, PeerReview can detect Byzantine faults without requiring a formal specification of the system.

3   Preliminaries

3.1   Definition

First, we define two concepts:

- **Malicious vehicles**: it means that vehicles drop or duplicate packets in network in order to mislead other vehicles or abolish important messages based on their personal aims.
- **Honest (normal) vehicles**: A vehicle that has a normal behavior, where a vehicle with normal behavior forwards messages correctly or generates right messages.

3.2   Tamper Resistant Hardware

A basic approach that aims to prevent attacking over V2V and V2I is the usage of tamper-resistant hardware. Tamper resistant in general means that something is resistant against tampering, independent of the type of access a person might have to the system. With respect to hardware, it means the device in question is difficult to be manipulated or exchanged by another device, without the system would take notice. The implementation of tamper resistance ranges from complicating access to device internals up to self-destruction of the device upon the detection of tampering attempts.

Closely related are the following two terms, tamper proof and tamper evident. Tamper-proof is a more strict definition of tamper resistant, which claims to be 100% secure against tampering, whereas tamper evident means that one is able to detect that a device has been tampered with. Tamper resistance means the vehicle's devices have to be secured, and communication buses have to be protected. Furthermore, tamper resistant devices can provide secure storage for keys and certificates, and maybe even for a history of recent messages sent over the external communication system, like an event data recorder (EDR), which could be used for legal purposes. When properly applied, tamper-proof and tamper-resistant hardware enable secured communication and will prevent most attacks on the active safety communication system from inside of the vehicle.

3.3   Time Verification

Time is the most important element in a secure log. Timing information based verification correlates the time data fields in secure log against the vehicle's internal clock (synchronized and updated using information provided by the GPS system). Accurate time service can support the upper modules, like Detector and State Machine. Verifications with the secure log in order to detect misbehavior are possible with regard to the following aspects. A first step consists in comparing the dump log time to the creation log time stamp. Time verification with several logs originating from a single node can provide additional insight on the nodes behavior (or on the fact another node is trying to impersonate this node). In combination with position information, time based plausibility checks for single nodes also leads towards vehicle speed related plausibility checking.

3.4   Secure Log

Security log is the basic data set for vehicle misbehavior detection. Each OBU maintains a security log with the following security features: it can only add normal permission, only entities with high privileges, such as RSU and DoT, can dump   or delete the security log.

The secure log entry is defined as follows:

$$le_n = (\alpha_n, \beta_n, \gamma_n, \theta_n)$$

where:

α_n: a strictly increasing sequence number;

β_n: either SEND or RECV which is the category of the log entry;

γ_n: some type-specific content;

θ_n: a recursively defined hash value which is defined as $\theta_n = H(\theta_{n-1} \| \alpha_n \| \beta_n \| H(\gamma_n))$; $\|$ stands for concatenation and the base hash θ_{n-1} is a predefined value. The hash function $H(.)$ is pre-image resistant, second pre-image resistant, and collision resistant.

The secure log is an append-only list and is secure controlled by DoT. If required, the log can be dump to the DoT by some security channels. In this paper, we suppose the secure log about a node vehicle can take enough behavior evidence for judging.

3.5   OBU Model

The OBU model architecture is as shown in Fig. 1.

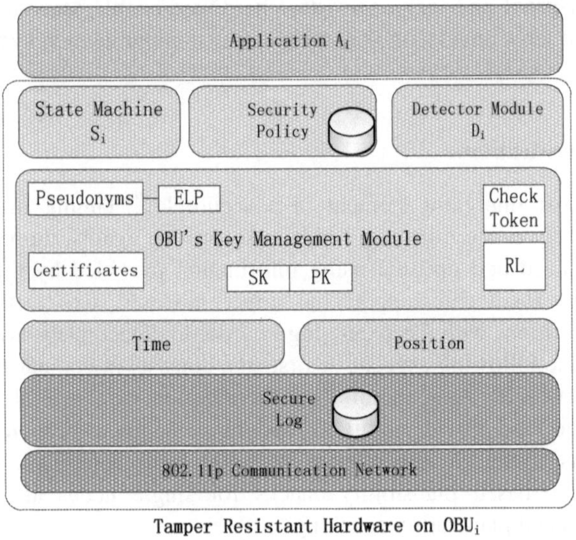

Fig. 1. OBU components

In general, the OBU i is modeled as a deterministic state machine S_i, a detector module D_i, and an application A_i. We use the state machine to denote all the functionality that should be checked, and use the application to denote other functions which are unnecessary to check (e.g., sending messages). The detector module D_i, which observes all inputs and outputs of S_i, implements MisDis, and it can communicate with other vehicles' detector modules. A normal vehicle i implements S_i and D_i as specified; on the other hand, a misbehavior vehicle may behave arbitrarily. D_i sends the failure indications about other vehicles to its A_i. Informally, *exposed(j)* is set when vehicle i has obtained enough proof of vehicle j's misbehavior; *suspected(j)* indicates that i suspects that j does not send a message that it should send; *trusted(j)* is raised otherwise. The model have two important database in the different level. One is the security policy which can directly instruct the OBU's misbehavior checking task, and another is the secure log which keep the behavior evidence of the host vehicle.

In the core level, there is the OBU's key management module which include the secret and public key of the OBU, the pseudonyms (related to the vehicle's electronic license plate), the digital certificates, the check token given by RSU and the Revocation List (RL for short) issued by DoT.

In the bottom level, the 802.11p communication module guarantees the interactive with the other OBUs and RSUs.

3.6   Secure Objectives

We define the security requirements and will show the fulfillment of these requirements after presenting the design details.

— **Accountability:** The requirement states that the misbehavior can be detectable by the evidence kept by the system. We consider the use of accountability to detect and expose node faults in distributed systems [11]. An accountable system maintains a tamper evident record that provides nonrepudiable evidence of all nodes' actions. Non-repudiation also requires that violators or misbehaving users cannot deny the fact that they have misbehaved. Based on this record, a faulty node whose observable behavior deviates from that of a correct node can be detected eventually. At the same time, a correct node can defend itself against any false accusations.

— **Identificability:** An accountable system requires strong node identities. Otherwise, an exposed faulty node could escape responsibility through assuming a different identity. In this paper, we assume that each node is in possession of a cryptographic key pair that can be linked to a unique node identifier. And we also assume that correct actions are deterministic and that communication between correct nodes eventually succeeds.

4   Design of MisDis

We expire by Timeweave[12] and employ some ideas of the scheme in PeerReview[11]. The basic idea of MisDis is to set a security log for each vehicle to record its behavioral state（determined behavioral state machines） strictly, and introduce a moderate size of monitoring OBUs which have check-token authorized by DoT. Then we use this collection to monitor the security log of the target vehicle real-time and on-site. If we find any exceptions, we should report to DoT through RSU for future investigation and isolation.

First, Table 1 shows the notation table in our scheme. Next, we elaborate on the technical design of the proposed scheme as following.

Table 1. Notation Table

Notation	Description
V_i	i^{th} Vehicle
RSU_j	j^{th} RSU
RID_i, PID_i	V_i's true and pseudo ID
s, P_{pub}	DoT's master and public key
$K_{RSU_j}^-$, $K_{RSU_j}^+$	RSU_j's secret and public key
K_i^-, K_i^+	V_i's secret and public key

Table 1. (*Continued*)

Notation	Description
$H(.)$	Hash function $H:\{0,1\}^* \rightarrow \mathbb{Z}_q^*$
$Sign(.)$	Signature function
$Enc(.)$	symmetric encryption function
T	The total database in DoT
T_1	The mid-database in RSU_1
T_2	The mid-database in RSU_2

4.1 System Initialization

In the system initialization, the DoT initializes the system parameters in offline manner, and registers vehicles and *RSU*s.

- 1) **Setup:** Given the security parameter k, the bilinear parameters $(q, G_1, G_2, \hat{e}, P, Q)$ are first chosen, where $|q| = k$. Then DoT randomly selects a master secret $s \in_R \mathbb{Z}_q^*$, and computes the corresponding public key $P_{pub} = sP$. In addition, the DoT chooses a secure symmetric encryption algorithm $Enc()$, a cryprographic hash function $H:\{0,1\}^* \rightarrow \mathbb{Z}_q^*$, and a secure signature algorithm $sign()$. In the end, DoT sets the public system parameters as $(q, G_1, G_2, \hat{e}, P, P_{pub}, H, HMAC, Enc, Sign)$.

- 2) **RSU-KeyGen:** DoT randomly select $K_{RSU}^- \in_R \mathbb{Z}_q^*$ for every RSU, and computes the corresponding public key $K_{RSU}^+ = K_{RSU}^- \cdot P$. DoT sends the pair (K_{RSU}^+, K_{RSU}^-) to the target *RSU* by secure channel.

- 3) **OBU-KeyGen:** Suppose every vehicle has a unique real identity, denote RID. Vehicle V_i first randomly chooses $K_i^- \in_R \mathbb{Z}_q^*$ as its secret key and computes the corresponding public key $K_i^+ = K_i^- \cdot P$. Then V_i submits a four-tuple (K_i^+, RID_i, a_i, b_i) to DoT, where $a_i = H(t_i P \| RID_i), b_i = (t_i - K_i^- \cdot a_i), t_i \in_R \mathbb{Z}_q^*$. DoT verifies the tuple's validity by judge $a_i = H((b_i P + (K_i^+)^{a_i}) \| RID_i)$. If pass, DoT assigns a PID for V_i, issues a anonymous certification binding K_i^+ to V_i.

- 4) **OBU-TokenGen:** According to the DoT's history record and the choice policy, DoT authorizes RSU to issue Check-Token to some OBUs. Each token has a limit valid time.

4.2 Behavior Logging

Suppose a vehicle i wants to send a message *msg* to a vehicle j, the process of message transmission is as follows:

1) i creates a log entry $(\alpha_n, SEND, j, msg)$, attaches θ_{n-1}, α_n and $\sigma_i(\alpha_n \| \theta_n)$ to msg, and sends the result to j.

2) j has enough information to calculate θ_n and to extract u_n^i: If the signature in u_n^i is not valid, j discards msg; otherwise, j creates its own log entry $(\alpha_l, RECV, i, \alpha_n, msg)$ and returns an ack with θ_{l-1}, α_l, and $\sigma_j(\alpha_l \| \theta_l)$ to i.

3) i extracts and verifies u_l^j: If the signature in u_l^j is valid, transmission completes; otherwise, i sends a challenge to j's monitors.

From the process above, MisDis ensures that the sender (or the receiver) of each message msg obtains verifiable evidence that the receiver (or the sender) of msg has logged the transmission. We focus on two categories of vehicles (i.e., misbehavior and correct vehicles) and their situations. For misbehavior vehicles, we detect and identify them; for correct vehicles, we prevent them from false accusations.

4.3   Secure Log Checking

Fig. 2 shows the secure log checking procedure (suppose V_j/V_k have check-token and meet V_i on the road).

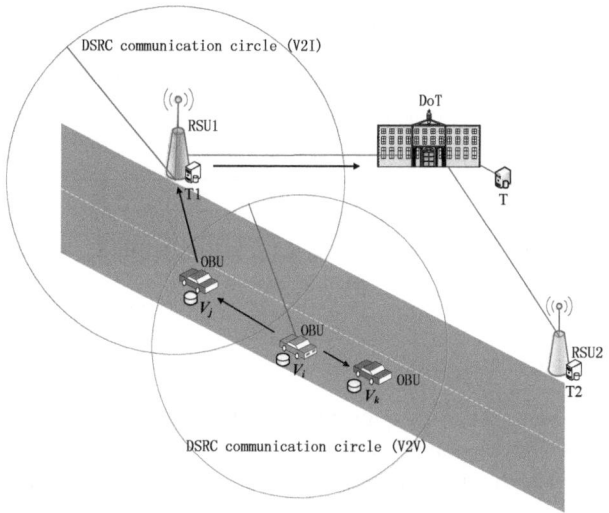

Fig. 2. OBUs and their monitor OBUs

In MisDis, an *authenticator* is defined as $u_n^i = \sigma_i(\alpha_n, \theta_n)$, a signed statement by vehicle i that its log entry le_n has a hash value θ_n; $\sigma_i()$ represents that the argument is signed with vehicle i's private key. Through sending u_n^i to vehicle j, a vehicle i commits to having logged entry le_n and to the contents of its log before le_n.

If vehicle i subsequently cannot produce a prefix of its log that matches the hash value in u_n^i, then vehicle j has verifiable evidence that i has tampered with its log and is faulty. In addition, u_n^i can be used by vehicle j as the verifiable evidence to persuade other vehicles or RSUs an entry le_n exists in vehicle i's log. Any OBU or RSU can also use u_n^i to inspect le_n and the entries before it in i's log.

The secure log checking includes the following 3 procedures:

- 1) **Hello:** OBU_i broadcasts Hello-Message periodically to show its existence for the near vehicle. Hello-Message is $[K_i^+]$.
- 2) **Check-Req:** After received the Hello-Message from V_i, V_j sends back a Check-Req-Message which is $[K_i^+ \| check - token \| Log \| T]$'s encryption version by a session key between V_i and V_j. The Check-Req Algorithm is shown as following.

Table 2. Check-Req Algorithm

Algorithm 1: Check-Req Algorithm
1. V_j randomly selects a secret $r \in_R Z_q^*$;
2. V_j computes a session key $\phi = r(K_i^+)$;
3. V_j computes a hint $\psi = rP$;
4.   Let $M = [K_i^+ \| check - token \| Log \| T]$ where Log denotes Log check part, T denotes Timestamp;
5.   Let $M' = [\psi, Enc_\phi(M)]$, and send it to V_i.

- 3) **Check-Confirm:** After received the Check-Req from V_j, V_i sends a message include the target log back.

4.4   Log Dumping

The token V_j can dump the secure log part into the RSU_l's T_l database through the same way of Check-Req Algorithm. Furthermore, RSU_l can forward the T_l to DoT's T database for the thorough misbehavior analysis by secure channel such as SSL.

5   Security Analysis

— **Accountability:** The OBU's secure log keeps the evidence of the vehicle node, and can be further dump onto the DoT's total database. OBU's core components are tamper-resistant hardware. Based on this outsourcing record, a misbehavior car whose observable behavior deviates from that of a correct node can be detected eventually.

— **Identificability**: we assume that each vehicle is in possession of a cryptographic key pair that can be linked to a unique vehicle identifier. And more, the correct vehicle' actions are deterministic and that communication between correct nodes eventually succeeds.

6   Conclusion

In VANET, each car on the road is equipped with one OBU. In this paper, we introduce MisDis, a method that can identify misbehavior vehicle by state automata and supervision, as well as a special security log to record the behavioral characteristics of the target vehicle. If the host vehicle violates the established security policy, then the system need to record the behavioral and try to dump to a safer roadside unit or department in charge.

Acknowledgments. I would like to thank Yonggang Wang from Peking University and Le Cai from UIBE for their helpful advice and kind encouragement. This work was supported in part by the NSFC under grant No. 61170263 and No. 61003230.

References

1. Aijaz, A., Bochow, B., Dtzer, F., Festag, A., Gerlach, M., Kroh, R., Leinmller, T.: Attacks on Inter Vehicle Communication Systems - an Analysis. In: Proceedings of WIT, pp. 189–194 (2006)
2. Raya, M., Hubaux, J.P.: Securing Vehicular Ad Hoc Networks. Journal of Computer Security, Special Issue on Security of Ad Hoc and Sensor Networks 15(1), 39–68 (2007)
3. Grover, J., Prajapati, N.K., Laxmi, V., Gaur, M.S.: Machine Learning Approach for Multiple Misbehavior Detection in VANET. In: Abraham, A., Mauri, J.L., Buford, J.F., Suzuki, J., Thampi, S.M. (eds.) ACC 2011, Part III. CCIS, vol. 192, pp. 644–653. Springer, Heidelberg (2011)
4. Golle, P., Greene, D., Staddon, J.: Detecting and Correcting Malicious Data in VANETs. In: Proceedings of the First ACM Workshop on Vehicular Ad Hoc Networks (VANET), pp. 139–151. ACM Press, Philadelphia (2004)
5. Raya, M., Papadimitratos, P., Gligor, V.D., Hubaux, J.P.: On data centric trust establishment in ephemeral ad hoc networks. In: 27th IEEE Conference on Computer Communications (INFOCOM 2008), pp. 1238–1246 (2008)
6. Raya, M., Papadimitratos, P., Aad, I., Jungels, D., Hubaux, J.P.: Eviction of Misbehaving and Faulty Nodes in Vehicular Networks. IEEE Journal on Selected Areas in Communications, Special Issue on Vehicular Networks 25(8), 1557–1568 (2007)
7. Ghosh, M., Varghese, A., Kherani, A.A., Gupta, A.: Distributed Misbehavior Detection in VANETs. In: 2009 IEEE Conference on Wireless Communications and Networking Conference (WCNC 2009), pp. 2909–2914. IEEE Press, Piscataway (2009)
8. Ghosh, M., Varghese, A., Kherani, A.A., Gupta, A., Muthaiah, S.N.: Detecting Misbehaviors in VANET with Integrated Root-cause Analysis. In: Ad Hoc Networks, vol. 8, pp. 778–790. Elsevier Press, Amsterdam (2010)

9. Schmidt, R.K., Leinmuller, T., Schoch, E., Held, A., Schafer, G.: Vehicle Behavior Analysis to Enhance Security in VANETs. In: 4th Workshop on Vehicle to Vehicle Communications (V2VCOM 2008), pp. 168–176 (2008)

10. Kim, T.H., Studer, H., Dubey, R., Zhang, X., Perrig, A., Bai, F., Bellur, B., Iyer, A.: VANET Alert Endorsement Using Multi-Source Filters. In: Proceedings of the Seventh ACM International Workshop on Vehicular Internetworking, pp. 51–60. ACM Press, New York (2010)

11. Haeberlen, A., Kouznetsov, P., Druschel, P.: PeerReview: practical accountability for distributed systems. In: 21st ACM Symposium on Operating Systems Principles (SOSP 2007), pp. 175–188. ACM Press, New York (2007)

12. Maniatis, P., Baker, M.: Secure History Preservation Through Timeline Entanglement, pp. 297–312. USENIX Association, Berkeley (2002)

13. NCTUns 5.0, Network Simulator and Emulator,
http://NSL.csie.nctu.edu.tw/nctuns.html

A Scalable Approach
for LRT Computation in GPGPU Environments

Linsey Xiaolin Pang[1,2], Sanjay Chawla[1], Bernhard Scholz[1], and Georgina Wilcox[1]

[1] School of Information Technologies, University of Sydney, Australia
[2] NICTA, Sydney, Australia
{qlinsey,scholz}@it.usyd.edu.au,
{sanjay.chawla,georgina.wilcox}@sydney.edu.au

Abstract. In this paper we propose new algorithmic techniques for massively data parallel computation of the Likelihood Ratio Test (LRT) on a large spatial data grid. LRT is the state-of-the-art method for identifying hotspots or anomalous regions in spatially referenced data. LRT is highly adaptable permitting the use of a large class of statistical distributions to model the data. However, standard sequential implementations of LRT may take several days on modern machines to identify anomalous regions even for moderately sized spatial grids.

This work claims three novel contributions. First, we devise a dynamic program with a pre-processing step of $\mathcal{O}(n^2)$ that allows us to compute the statistic for any given region in $\mathcal{O}(1)$, where n is the length of the grid. Second, we propose a scheme to accelerate the likelihood computation of a *complement* region using a bounding technique. Third, we provide a parallelization strategy for the LRT computation on GPGPUs. In concert all three contributions result in a speed up of nearly four hundred times reducing the LRT computation time of large spatial grids from several days to minutes.

Keywords: Spatial outlier, GPGPUs, LRT, 1EXP, upper-bounding.

1 Introduction

With the widespread availability of GPS-equipped smartphones and mobile sensors, there has been an urgent need to perform large scale spatial data analysis. For example, by carrying out a geographic projection of Twitter feeds, researchers are able to narrow down "hotspot" regions where a particular type of activity is attracting a disproportionate amount of attention. In neuroscience, high resolution MRIs facilitate the precise detection and localization of regions of the brain which may indicate mental disorder. The statistical method of choice for identifying hotspots or anomalous regions is the Likelihood Ratio Test (LRT) statistic. Informally, the LRT of a spatial region compares the likelihood of the given spatial region with its complement, and hence can be used to identify hotspot regions. In [10], it was known that the LRT value always follows a χ^2 distribution, independent of the distribution of the underlying data.

For a $n \times n$ spatial grid, the worst case execution time for identifying the most anomalous region is $\mathcal{O}(cn^4)$, where c is the execution time of computing the LRT over a single region. As noted by Wu et al. [8], a naive implementation of LRT for a moderate

Y. Ishikawa et al. (Eds.): APWeb 2013, LNCS 7808, pp. 595–608, 2013.

64×64 spatial grid may take nearly six hundred days.[1] Wu et al. [8] proposed a method which reduces the computation time to eleven days. However as noted previously in [9], this approach will not scale for larger data sets and the biggest spatial grid reported in [8] was 64×64.

The nature of LRT permits the computation of regions independently of each other, which facilitates parallelization to some degree. However, new algorithmic techniques and parallelization strategies are required in order to fully harvest the computational power of GPGPUs. We have identified the following **challenges** that need to be addressed for achieving a speed up of several orders of magnitude with GPGPUs:

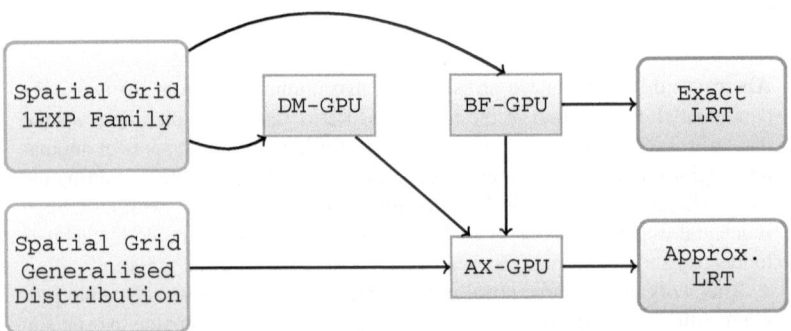

Fig. 1. Workflow: We distinguish between distributions which belong to the one-parameter exponential family (1EXP) and distributions outside of 1EXP. We then design two algorithms for 1EXP which are used to efficiently and exactly estimate the LRT. For general data distributions, we design the AX-GPGPU algorithm which uses the exact algorithms as subroutines.

- Computing the LRT of a given region R requires the computation of the likelihood of $\bar{R}$, i.e. the complement of R. While R has a rectangular shape, $\bar{R}$ is irregularly shaped making the parallelization more difficult [11,5,6]. The experiments of [9] showed that enumerating the complement regions is a bottleneck for efficient computation.
- The LRT computation on sequential computers is intractable for large spatial grids. GPGPUs can perform massively data parallel computations, which is well-suited to LRT computation. However, the LRT computation must be adapted to the GPGPU programming model that exhibits a complex memory hierarchy consisting of a grid, blocks and threads. To minimize the communication between CPU and GPGPU memory, it is advantageous to perform the majority of the computation on GPGPUs using appropriate schemes for dividing work between threads and allocating threads to blocks.

Our strategy to overcome the challenges noted above is shown in Figure 1. First, in our framework, we distinguish between data distributions following the one-parameter exponential family (1EXP) and those that are outside of 1EXP. When the data belongs

---

[1] The experiment results were reported in 2009.

to a 1EXP distribution, we can estimate the LRT of a given region R by omitting the computation of $\bar{R}$ (c.f. [3]). When the data is assumed to be generated by an arbitrary statistical distribution, we provide a solution to eliminate the bottleneck associated with the computation of $\bar{R}$: we use an "upper-bounding technique" to replace the computation of $\bar{R}$ with that of its constituent regions [8]. Second, we provide three GPGPU algorithms named as: Brute-Force GPGPU (BF-GPGPU), De Morgan GPGPU (DM-GPGPU) and Approximate GPGPU (AX-GPGPU). The BF-GPGPU and DM-GPGPU implementations provide exact solutions for 1EXP. The AX-GPGPU algorithm provides approximate solution for any underlying data distribution and it uses the BF-GPGPU or DM-GPGPU as a sub-routine. We designed our "blocking" scheme for partitioning the work by dividing the spatial grid into overlapping sub-grids and mapping these regions onto blocks of threads [4]. The majority of the computation is performed on the GPGPUs and we utilize shared memory for each block by pre-loading the data that will be used by multiple threads.

The rest of the paper is structured as follows. In Section 2, we provide background materials on LRT computation and upper-bounding technique. Related work is given in Section 3. In Section 4, we explain how we use De Morgan's law and dynamic programming to speed up the enumeration and processing of regions measurements. The description and details of three designed algorithms (i.e. BF-GPGPU, DM-GPGPU and AX-GPGPU) are presented in Section 5. In Section 6, we evaluate the algorithms on both synthetic and real data sets consisting of Magnetic Resonance Imaging (MRI) scans from patients suffering from dementia. We give our conclusions in Section 7.

2 Background

2.1 The Likelihood Ratio Test (LRT)

We provide a brief but self-contained introduction for using LRT to find anomalous regions in a spatial setting. The regions are mapped onto a spatial grid. Given a data set X, an assumed model distribution $f(X, \theta)$, a null hypothesis $H_0 : \theta \in \Theta_0$ and an alternate hypothesis $H_1 : \theta \in \Theta - \Theta_0$, LRT is the ratio

$$\lambda = \frac{\sup_{\Theta_0}\{L(\theta|X)|H_0\}}{\sup_{\Theta}\{L(\theta|X)|H_1\}}$$

where $L()$ is the likelihood function and θ is a set of parameters for the distribution [8]. In a spatial setting, the null hypothesis is that the data in a region R (that is currently being tested) and its complement (denoted as $\bar{R}$) are governed by the same parameters. Thus if a region R is anomalous then the alternate hypothesis will most likely be a better fit and the denominator of λ will have a higher value for the maximum likelihood estimator of θ. A remarkable fact about λ is that under mild regularity conditions, the asymptotic distribution of $\Lambda \equiv -2\log\lambda$ follows a χ_k^2 distribution with k degrees of freedom, where k is the number of free parameters[2]. Thus regions whose Λ value falls in the tail of the χ^2 distribution are likely to be anomalous [8].

---

[2] If the χ^2 distribution is not applicable then Monte Carlo simulation can be used to ascertain the p-value.

For example, if we assume the counts $m(R)$ in a region R follow a Poisson distribution with baseline b and intensity λ, then a random variable $x \sim Poisson(\lambda\mu)$ is a member of 1EXP with $T(x) = x/\mu$, $\Phi = 1/\mu$, $a(\Phi) = \Phi$, $\eta = log(\lambda)$, $B_e\eta = exp(\eta)$, $g_e(x) = log(x)$. For any regions R and $\bar{R}$, $m(R)$ and $m(\bar{R})$ are independently Poisson distributed with mean $\{exp(\eta_R)b(R)\}$ and $\{exp(\eta_{\bar{R}})b(\bar{R})\}$ respectively. Then $b_R = \frac{b(R)}{b(R)+b(\bar{R})}$ and $m_R = \frac{m(R)}{m(R)+m(\bar{R})}$. The log-likelihood ratio is calculated by: $c(m_R log(\frac{m_R}{b_R}) + (1-m_R)log(\frac{1-m_R}{1-b_R}))$ (c.f. [3]). The closed-form formula for LRT for Poisson distribution generalizes to the 1EXP family of distributions [3].

2.2   The Upper-Bounding Technique

The upper-bounding technique for LRT was introduced by Wu et al. [8] to reuse the likelihood computation, i.e., the likelihood of a region could be upper-bounded in terms of its sub-regions. The basic observation is that the likelihood value of any given region R under the complete parameter space is not greater than the product of the likelihood value of all its non-overlapping sub-regions under the null parameter model. For instance, if a region R is composed of two non-overlapping sub-regions R_1 and R_2, then $L(\theta_R|X_R) \leq L(\theta'_{R_1}|X_{R_1}) \times L(\theta'_{R_2}|X_{R_2})$. This is equivalent to $logL(\theta_R|X_R) \leq logL(\theta'_{R_1}|X_{R_1}) + logL(\theta'_{R_2}|X_{R_2})$. Here θ_R, θ'_{R_1} and θ'_{R_2} are the maximum likelihood estimators (MLEs). The θ_R is computed under the complete parameter space, and θ_{R_1} and θ'_R are computed under the null parameter space.

The upper-bounding technique can be used to prune non-outliers as follows: *if we replace the likelihood of a region R by the product of the likelihoods of its sub-regions and the new LRT is below the anomalous threshold (i.e. confidence level α), then R cannot be anomalous.* As noted earlier, in this paper we use the upper-bounding technique to speed up the computation of the likelihood of the complement of a region $\bar{R}$ when the data is assumed to follow a general statistical distribution.

3   Related Work

Previous attempts to parallelize LRT computation have only achieved limited success. For example, the Spatial Scan Statistic (SSS), which is a special case of LRT for Poisson data, is available as a program under the name SatScan [2]. It has been parallelized for multi-core CPU environments and its extension for a GPGPU hardware [7] has achieved improved speed up of two over the multi-core baseline. The GPGPU implementation in [7] has proposed loading parts of the data into shared memory but has achieved only a modest speed up. The other attempt of [14] applied their own implementation of a spatial scan statistic program on the GPU to the epidemic disease dataset. The number of blocks was decided by both the number of the exploring villages and the number of diseases to take full advantage of stream multiprocessors on GPGPUs. This solution is only applicable to its special disease scenario. In each of these cases, we believe there is further room for optimising the algorithms for the GPGPU by carefully exploiting the thread and block architecture and further utilising shared memory.

Furthermore, all the existing parallel solutions perform simplified Poisson LRT tests in a circular or cylindrical way, not in a grid-based scenario. Our parallel GPGPU solution is different and provides a fully paralleled template for a general LRT computation in a grid, based on the work of [8].

4 The De Morgan Model

For the 1EXP family, we can estimate the LRT of a region R by just computing the statistic of the region. We do not need to separately compute the statistic of $\bar{R}$. For a more general statistical family, we can upper-bound the likelihood of $\bar{R}$ by estimating the likelihood of the four sub-regions that are anchored at the corners of the grid.

We now present a novel and unique approach, based on De Morgan's law and dynamic programming to accumulate the statistic of a region R. We will build a table in time $O(n^2)$ and then we will be able to compute the statistic of any region R in constant ($O(1)$) time.

Consider Figure 2(a) that shows a rectangular area R embedded in a grid G. Instead of counting the number of elements in R directly, we express set R as set intersections of two sets A and B as shown in Figure 2(b). Set A is a rectangular region that starts in the upper left corner of the grid and ends at the lower right corner of R. Set B is a rectangular region that starts at the upper left corner of R and ends at the lower right corner of the grid. Hence, $R = A \cap B$. We denote the region from the lower left corner of G to the lower left corner of R by X and the region from the upper right corner of R to the upper right corner of G by Y. By applying De Morgan's law and inclusion/exclusion principle we obtain following relationship:

$$|R| = |A \cap B| \tag{1}$$
$$= |G| - (|\bar{A}| + |\bar{B}| - |\bar{A} \cap \bar{B}|) \tag{2}$$
$$= |G| - (2|G| - |A| - |B| - (|X| + |Y|)) \tag{3}$$
$$= |A| + |B| + |X| + |Y| - |G| \tag{4}$$

To obtain a query time of $\mathcal{O}(1)$, we need to pre-compute sets A, B, X, and Y for all possible regions in G. Since one of the corner is fixed we can pre-compute the cardinalities of these sets in tables of size $\mathcal{O}(n^2)$. Hence, the query of counting the

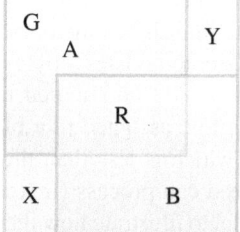

(a) Region R in Grid G   (b) R= $A \cap B$, $\bar{A} \cap \bar{B} = X \cup Y$

Fig. 2. Set Relations for Region Problem

number of elements of region $R(x_1, y_1, x_2, y_2)$ where (x_1, y_1) is the upper left corner and (x_2, y_2) is the lower right corner is expressed by:

$$|R(x_1, y_1, x_2, y_2)| = |A(x_2, y_2)| + |B(x_1, y_1)| + |X(x_1, y_2)| \\ + |Y(x_2, y_1)| - |G| \tag{5}$$

To obtain the tables for A, B, X, and Y, we employ dynamic programming. For example the table for A can be computed using the following recurrence relationship:

$$|A(i, j)| = |A(i, j - 1)| + |A(i - 1, j)| - |A(i - 1, j - 1)| \\ + |R(i, j, i, j)| \tag{6}$$

where $|R(i, j, i, j)|$ counts whether there is an element in the cell location (i, j). The first element and the first column and row need to be populated (initialized) so that all cardinalities of A can be computed.

5 Three GPGPU Algorithms

We present three GPGPU-based algorithms and provide a complete parallel solution for LRT computation. The Brute-Force GPGPU (BF-GPGPU) and De Morgan GPGPU (DM-GPGPU) algorithms provide an exact solution of LRT when data is assumed to be generated from 1EXP. For general data distributions the Approximate GPGPU (AX-GPGPU) uses the BF-GPGPU and DM-GPGPU as a subroutine to upper bound the likelihood of the complement region $\bar{R}$.

In the GPGPU memory hierarchy architecture, global memory is large and shared memory is small. Access to shared memory is nearly two orders of magnitude faster compared to global memory. A common strategy to exploit this mismatch is to partition the data into subsets called *tiles* which can fit into shared memory [13,12]. We now describe our block and thread schemes which underpins the three algorithms.

5.1 Exact Solution for 1EXP

5.1.1 BF-GPGPU

Smart Block and Thread Scheme: We assume that spatial grid[3] G $(n \times n)$ is small enough so that it can be loaded into GPGPU global memory but too big to be loaded into shared memory directly. The entire spatial grid G is partitioned into tiles which can be separately read into shared memory. Assume that the size of cell data is f bytes and shared memory accommodates s bytes. There are s/f cells or equivalently $w \times h$ sub-grids that can be stored into shared memory, where $w \times h = s/f$. Each such rectangular region will be assigned to a block of threads. If the number of threads in a block is $tw \times th$, which is less than the maximum number of threads allowed per block, then each thread can process $(w/tw) \times (h/th)$ cells. We use an example with the Poisson data model to illustrate how the computation is done, but our algorithm also works with more general distributions.

---

[3] To differentiate our grid in outlier detection with the grid in CUDA, we call the grid in our CUDA program "CUDA grid" and the grid in the outlier detection context grid the "spatial grid". We assume the grid size is power of 2.

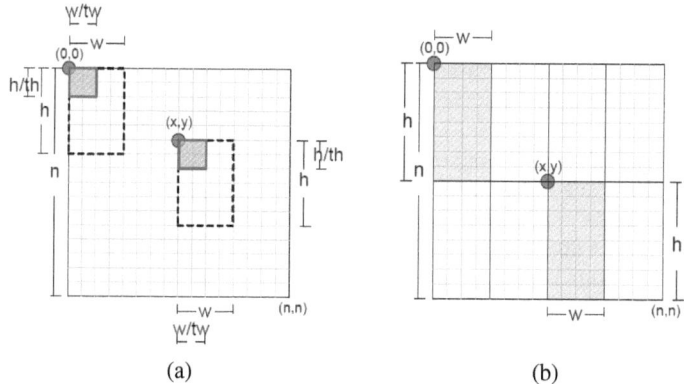

(a)   (b)

Fig. 3. Block and thread schemes: (a) is applicable for exact computation and pre-computation of R in AX-GPGPU. Each (x, y) point corresponds to a block coordinate. Each thread is responsible for processing $(w/tw, h/th)$ regions and each thread processes $(2, 2)$ regions in our implementation; Figure (b) is applicable to DM-GPGPU and pre-computation of $\bar{R}$. The grid is divided into equal disjoints parts. Each part is associated with a block.

In a Poisson model, each cell has a population count and a success count, which takes 8 bytes. If the size of shared memory is 16KB, a maximum of 2048 cells can be loaded into the shared memory. This resembles a rectangular region of 45 x 45 cells. To use threads efficiently, the rectangular region is set to 44×44 cells and associated with a block. A block with coordinates (x,y) in the CUDA grid is associated to the region with $(x, y) - (x+43, y+43)$ in the spatial grid. In our implementation, the maximum number of threads per block is limited by the hardware to 512. Therefore, 22×22 threads can be assigned to process 44×44 cells for maximizing the usage of shared memory. So each thread is responsible to process four sub-regions: (x_1, y_1, x_2, y_2), $(x_1, y_1, x_2 + 1, y_2)$, $(x_1, y_1, x_2, y_2 + 1)$, $(x_1, y_1, x_2 + 1, y_2 + 1)$, where (x_1, y_1) is the upper left cell and (x_2, y_2) is the lower right cell. See Figure 3a and algorithm 1.

A two-pass scheme is used to obtain the likelihood ratio of each region:

(a) In the first pass, all rectangular regions of fixed sizes up to 44×44 are enumerated in the grid. The data for each block is copied to its shared memory. After the block data is stored into shared memory, each thread computes the score function of its four rectangular regions and stores it back into global memory.

(b) The second pass enumerates all rectangular regions that do not fit into a single block on the GPGPU and uses the intermediate results from the first pass to calculate the top outlier over all regions.

5.1.2 DM-GPGPU

We now illustrate the GPGPU pre-computation step based on De Morgan's Law. Rest of the computation is identical as BF-GPGPU. In our De Morgan model, four datasets are pre-computed for all possible rectangular regions. From above, we already know that the pre-computed datasets come from four groups of regions, each group of which shares one corner with the grid G. The computation of a $n \times n$ grid requires the pre-computation of $4 \times n \times n$ regions.

Algorithm 1. Exact GPGPUs solution (all regions with size up to (w, h))

Input: grid G (grid_width, grid_height), block size (w, h) ,thread size (tw, th)
Output: subset of scores, m_R, b_R

1: //Step 1: Load sub-grid into shared memory of each block
2: $x1 \leftarrow blockIdx.x$, $y1 \leftarrow blockIdx.y$
3: **for** $i \leftarrow 0$ to tw **do**
4:    **for** $j \leftarrow 0$ to th **do**
5:       $x2 \leftarrow blockIdx.x+tw \times threadIdx.x+i$
       $y2 \leftarrow blockIdx.y+th \times threadIdx.y+j$
6:       **if** $(x2 < grid\_width$ and $y2 < grid\_height)$ **then**
7:          //Copy region (x1,y1, x2,y2) data from grid G to shared_array (SM)
8:          $SM[(tw \times threadIdx.y+j) \times th \times blockDim.x + th \times threadIdx.x+i] \leftarrow grid[y \times$ grid_width+ x2]
9: _synchronize threads();
10: //Step 2: process all the regions within each block
11: **for** $i \leftarrow 0$ to tw **do**
12:    **for** $j \leftarrow 0$ to th **do**
13:       //Retrieve region (x1,y1, x2,y2) data from SM
14:       $x2 \leftarrow blockIdx.x +tw \times threadIdx.x+i$
       $y2 \leftarrow blockIdx.y+th \times threadIdx.y+j$
15:       **for** $tmpx \leftarrow 0$ to $x2-x1$ **do**
16:          **for** $tmpy \leftarrow 0$ to $y2-y1$ **do**
17:             $m\_R \leftarrow accumulate\_mr()$; $b\_R \leftarrow accumulate\_br()$;
18:       $score \leftarrow lrt(m\_R, b\_R)$
19:       $save(score)$; $save(m\_R)$; $save(b\_R)$;

To parallelize the pre-computation, we split the whole spatial grid G into equal disjoint parts. For a grid of size $n \times n$, if the shared memory can load region with maximum size of $w \times h$, we create $(n/w) \times (n/h)$ non-overlapping parts. Each such disjoint part is associated to a block. If the thread number in a block is $tw \times th$, each thread can be assigned to process $(w/tw) \times (h/th)$ regions. Each thread is only responsible for the regions that share one of the four corners of its correspondent disjoint sub-grid. The approach for pre-computing $\bar{R}$ in the next section will be along similar lines. See Figure 3b.

An additional second pass is carried out to merge the results from each sub-grid to get the full pre-computed set. In our implementation, each block holds 32×16 cells to fully utilize the threads capability.

5.2   AX-GPGPU

5.2.1   Block and Thread Scheme
When we cannot assume that data follows 1EXP, the likelihood of $\bar{R}$ cannot be estimated from the statistic of R. Since $\bar{R}$ has an irregular shape, a naive implementation will not be able to leverage the strengths of a GPGPU environment. Experiments in

a CPU environment [7] have identified that the LRT computational bottleneck comes from the enumeration of the complement region for each given region.

However we can leverage the strategy proposed by Wu et al. [8] for pruning non-outliers, to provide an efficient parallelization of $\bar{R}$. The likelihood of $\bar{R}$ is upper-bounded by the sum of the likelihood of four regions. These four regions share one of the corner of the grid separately and they constitute $\bar{R}$. We use an approach similar as DM-GPGPU for $\bar{R}$ computation and BF-GPGPU for R computation.

Each point (x,y) in the spatial grid is the coordinate of a block. To satisfy the tight-bound criteria and fully utilize the thread capability, our block size is defined as 32×16. Therefore, 32×16 regions are stored into shared memory for processing R. There are 8568 sub-regions to be enumerated in each block. So each thread is responsible for 2 sub-regions. Figure 3b shows the block scheme in pre-computing $\bar{R}$. It is different from Figure 3a. The pre-computation of $\bar{R}$ only involves the group of regions that share one of the four corners [8]. Each point (x, y) does not need to be associated to a block, the entire sub grid is divided into several equal disjoint parts and each part corresponds to a block. We choose 32×32 regions to be block size for pre-computation of $\bar{R}$ and there are $(n/32) \times (n/32)$ blocks.

5.2.2   Concurrent Execution of Different Kernels

For the general distribution model we have to first compute an bounds for R, $\bar{R}$ and then use them to get an upper bound for the LRT. We can break the code up into three distinct kernels: $pre\_compute\_r$, $pre\_compute\_\bar{r}$ and $search$. Since the two kernels $pre\_compute\_r$, $pre\_compute\_\bar{r}$ are independent of each other, we can use streaming operations to run them concurrently.

After each kernel finishes execution, the results are transferred back to the CPU where the merging is carried out for bigger regions. Instead of loading the cell values, kernel $search$ loads the scores of pre-computed sets from the previous two kernels into shared memory. In our implementation, we can only load a sub-grid of maximum size 32×8 into shared memory. After the final score of each region is obtained, one more kernel for parallel sorting is executed and the top outlier is obtained.

5.3   Processing Larger Data Grid

All our algorithms assume that the entire spatial grid can be loaded into GPGPU global memory. This may not be a realistic scenario. For example, in our environment, the maximum size of the spatial grid that can be processed by a single GPGPU is 128×128. We provide several solutions to overcome this constraint.

Split the Grid into Small Sub-grids. We can split the spatial grid into several sub-grids and load each of them in a sequential fashion onto a GPGPU. The block and thread scheme is applied to each sub-grid and the final computation is merged on the CPU. This may involve multiple iterations of the same kernel function.

Multiple GPGPUs Approach. We combine multiple GPGPU cards into a computational node. For example, in our case, a single GPGPU implementation can handle a maximum grid size of 128×128 grid due to the 4GB global memory limitation. For a 256×256 grid, we can split it into 4 equal 128×128 sub grids and send the grids to the two GPGPUs , separately. Merging is carried out on the cpu after the results are obtained from the two GPGPUs.

Multi-tasked Streaming. CUDA offers APIs for asynchronous memory transfer and streaming. With these capabilities it is possible to design algorithm that allows the computation to proceed on both CPU and GPGPUs, while memory transfer is in progress. In our implementation, we use streaming operation to support asynchronous memory transfer while the kernel is being executed.

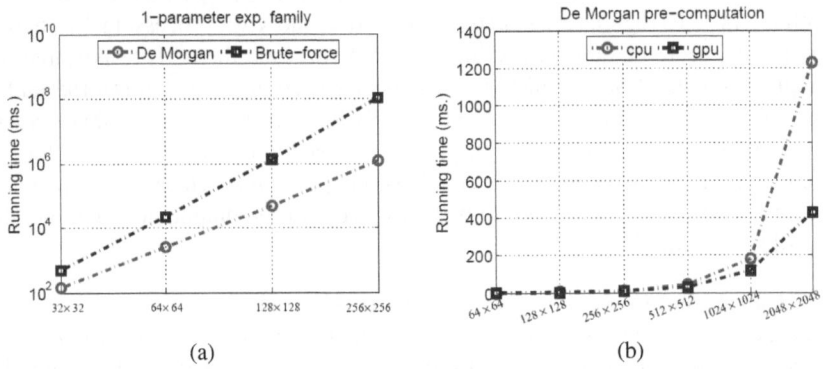

(a)   (b)

Fig. 4. Figure (a) The running time of CPU implementation of De Morgan vs. brute-force searching. Figure (b) pre-computation time for De Morgan: CPU vs. GPGPUs implementation.

6   Experiments and Analysis

We have designed and implemented a set of experiments to answer the following questions.

- What are the performance gains of the brute-force versus the De Morgan technique on a sequential CPU?
- What are the performance gains of sequential versus parallel outlier detection?
- What are the performance gains of sequential versus parallel pruning using our upper-bounding technique?
- How does our technique perform on real-world data sets?

The experiments were conducted on an 8-core E5520 Intel server that is equipped with two GPGPU $Tesla\ C1060$ cards supporting CUDA 4.0. Each GPGPU card has $4GB$ global memory, $16KB$ shared memory, 240 cores and 30 multiprocessors. The experiments are performed on a Poisson distribution model and a randomly generated anomalous region was planted for verification in synthetic data sets. To ensure correctness of the parallel approaches, the results were verified by implementing a sequential version of all the algorithms which ran on a single CPU machine. For the real data, we used MRI images of people suffering from dementia and used MRIs of normal subjects as the baseline.

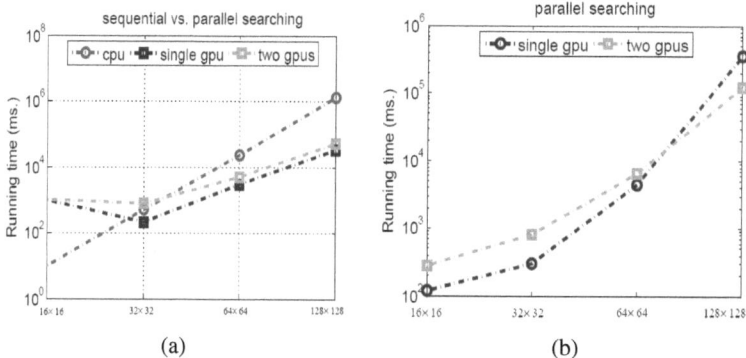

(a)   (b)

Fig. 5. (a) The running time of exact outlier detection algorithm on a CPU, and single GPGPU. It shows that BF-GPGPU is faster than the BF-CPU. BF-GPGPU becomes nearly 40 times faster than BF-CPU on the grid of 128×128. (b) The running time of exact searching on single GPGPU and two GPGPUs.

6.1 Performance on Synthetic Data

6.1.1 De Morgan vs. Brute-Force Processing on 1EXP Family

Figure 4a illustrates the performance of the brute-force method versus the De Morgan method executed on a single CPU core. The x-axis represents the size of the input instances and the y-axis the runtime in milliseconds. It is clear that the De Morgan approach consistently out performs the brute-force and for the largest grid size, the difference is almost 90 times more.

We compare the overheads of De Morgan pre-computation in a CPU vs. a GPGPU environment. The results are shown in Figure 4b. For smaller grid sizes, the CPU implementation has a smaller overhead compared to the GPGPU implementation. However, as the size of the spatial grid increases, the GPGPU implementation is three times faster than its CPU counterpart.

6.1.2 Sequential vs. Parallel Outlier Detection

To compare the performance of the brute-force approach on a CPU and a GPGPU environment for outlier detection, we implemented the BF-CPU and BF-GPGPU algorithms. Experiments were conducted on a on single-core CPU, one GPGPU, and two GPGPUs. We have applied parallel sorting provided by the CUDA Thrust API. Figure 5a illustrates the results. The BF-CPU performs well on small data sets but is clearly outperformed by parallel GPGPU searching for larger dataset. BF-GPGPU is six times faster than BF-CPU on the grid of size 64×64 and around forty times faster than on a grid of size 128×128. With the current memory limitation of $4GB$, single GPGPU implementation cannot process very large data sets. For larger data sets, multiple GPGPUs become a viable option. Figure 5b depicts the running time of exact outlier detection algorithm on a single GPGPU versus two GPGPUs without performing parallel sorting. The gains obtained from using two GPGPUs is around a factor of three compared to a single GPGPU.

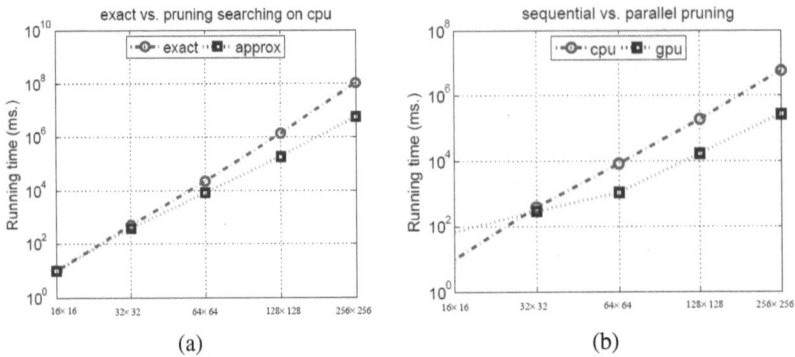

Fig. 6. (a) The running time of AX-CPU is nearly 10 times faster than BF-CPU. (b) The running time of AX-GPGPU is 10 times faster than AX-CPU and almost 400 times than BF-CPU.

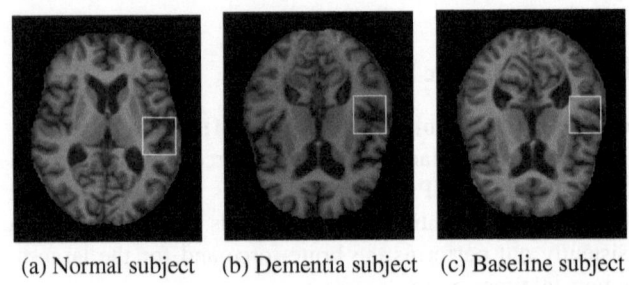

(a) Normal subject   (b) Dementia subject   (c) Baseline subject

Fig. 7. MRI Images: the LRT value of each region in (a) and (b) is compared to a corresponding region in the baseline image (c). The region in the yellow box in (b) has a LRT value significantly higher than the baseline (7562), so it is a potential outlier. The corresponding region in (a) has a low LRT value (2134) so it is normal.

6.1.3   Sequential Pruning vs. Parallel Pruning

Figure 6a and Figure 6b depict the performance gains obtained by using exact pruning versus approximate pruning on a GPGPU and CPU implementation respectively. As shown in the figures, the pruning approach accelerates the execution. Furthermore, the parallelized pruning approach shows a vast improvement with a running time of at least 20 times faster than AX-CPU. Moreover, it is almost 400 times faster than BF-CPU. The generalized GPGPU pruning approach provides us with a suitable template for fast computing of the likelihood ratio under any statistical model.

6.2   Outlier Detection on MRI Image

To illustrate a real-world application, the GPGPU implementation was tested on a MRI data set (OASIS) made available by Washington University's Alzheimer's Disease Research Center [1]. It consists of a cross-section collection of subjects including individuals with early-stage Alzheimer's Disease (AD). MRIs of a subject suffering from dementia and two normal subjects were selected as shown in Figure 7. The resolution of the MRI data is $176 \times 176 \times 208$. Figure 7 shows one slice of each subject's data, which is 176×208. A large number of voxels are black on the first and last 20 rows, and the first 40 and the last 40 columns. The area tested has dimensions of 128×128.

The application of LRT using both AX-GPGPU and BF-GPGPU identifies regions where the subjects suffering from dementia have a high likelihood ratio value for the affected regions in the brain. The same areas of normal subjects have low likelihood ratio values. The AX-GPGPU found the anomalous region in $24s$ compared to $1min6s$ by BF-GPGPU. The BF-CPU on the other hand required $17min55s$ to find the anomalous region.

7   Conclusion

The Likelihood Ratio Test Statistic (LRT) is the state-of-the-art method for identifying hotspots or anomalous regions in large spatial settings. To speed up the LRT computation, this paper proposed three novel contributions: (i) a fast dynamic program to enumerate all the possible regions in a spatial grid. Compared to a brute force approach, the dynamic programming method is nearly a hundred times faster. (ii) a novel way to use an upper bounding technique, initially designed for pruning non-outliers, to accelerate the likelihood computation of a complement region. (iii) a systematic way of mapping the LRT computation onto the GPGPU programming model. In concert the three contributions yield a speed up of nearly four hundred times compared to their sequential counterpart. Until now, existing GPGPU implementations for specialized versions of LRT such as Spatial Scan Statistic had given only modest gains. Moving the computation of the LRT statistics to the GPGPU enables the use of this sophisticated method of outlier detection for larger spatial grids than ever before.

Acknowledgment. We thank Dr. Jinman Kim at the University of Sydney for providing us with the MRI image resources and helping us analyze the data.

References

1. http://www.oasis-brains.org/
2. SatScan, http://www.SatScan.org
3. Agarwal, D., Phillips, J.M., Venkatasubramanian, S.: The hunting of the bump: On maximizing statistical discrepancy. In: SODA, pp. 1137–1146 (2006)
4. Beutel, A., Mølhave, T., Agarwal, P.K.: Natural neighbor interpolation based grid dem construction using a gpu. In: GIS 2010, pp. 172–181. ACM, New York (2010)
5. Gregerson, A.: Implementing fast mri gridding on gpus via. cuda. Nvidia Tech. Report on Medical Imaging using CUDA (2008)
6. Hong, S., Kim, S.K., Oguntebiv, T., Olukotun, K.: Efficient parallel graph exploration on multi-core cpu and gpu. In: Proceedings of the 16th ACM Symposium on Principles and Practice of Parallel Programming, PPoPP 2011 (2011)
7. Larew, S.G., Maciejewski, R., Woo, I., Ebert, D.S.: Spatial scan statistics on the gpgpu. In: Proceedings of the Visual Analytics in Healthcare Workshop at the IEEE Visualization Conference (2010)

8. Wu, M., Song, X., Jermaine, C., Ranka, S., Gums, J.: A LRT Framework for Fast Spatial Anomaly Detection. In: Proceedings of the 15th ACM SIGKDD International Conference on Knowledge Discovery and Data Mining (KDD 2009), pp. 887–896 (2009)

9. Pang, L.X., Chawla, S., Liu, W., Zheng, Y.: On mining anomalous patterns in road traffic streams. In: Tang, J., King, I., Chen, L., Wang, J. (eds.) ADMA 2011, Part II. LNCS, vol. 7121, pp. 237–251. Springer, Heidelberg (2011)

10. Wilks, S.S.: The large sample distribution of the likelihood ratio for testing composite hypotheses. Annals of Mathematical Statistics (9), 60–62 (1938)

11. Vuduc, R., Chandramowlishwaranv, A., Choi, J., Guney, M., Shringarpure, A.: On the limits of gpu acceleration. In: HotPar 2010 Proceedings of the 2nd USENIX Conference on Hot Topics in Parallelism, pp. 237–251 (2010)

12. Wu, R., Zhang, B., Hsu, M.: Gpu-accelerated large scale analytics. In: IACM UCHPC 2009: Second Workshop on UnConventional High Performance Computing (2009)

13. Wu, R., Zhang, B., Hsu, M.C.: Clustering billions of data points using gpus. In: IACM UCHPC 2009: Second Workshop on UnConventional High Performance Computing (2009)

14. Zhao, S.S., Zhou, C.: Accelerating spatial clustering detection of epidemic disease with graphics processing unit. In: Proceedings of Geoinformatics, pp. 1–6 (2010)

ASAWA: An Automatic Partition Key Selection Strategy

Xiaoyan Wang[1,2,3], Jinchuan Chen[3], and Xiaoyong Du[1,3,*]

[1] School of Information, Renmin University of China
[2] School of Information and Electrical Engineering, Ludong University
[3] Key Laboratory of Data Engineering and Knowledge Engineering, MOE
{wxy,jcchen,duyong}@ruc.edu.cn

Abstract. With the rapid increase of data volume, more and more applications have to be implemented in a distributed environment. In order to obtain high performance, we need to carefully divide the whole dataset into multiple partitions and put them into distributed data nodes. During this process, the selection of partition key would greatly affect the overall performance. Nevertheless, there are few works addressing this topic. Most previous projects on data partitioning either utilize a simple strategy, or rely on a commercial database system, to choose partition keys. In this work, we present an automatic partition key selection strategy called ASAWA. It chooses partition keys according to the analysis on both dataset and workload schemas. In this way, intimate tuples, i.e. co-appearing in queries frequently, would be probably put into the same partition. Hence the cross-node joins could be greatly reduced and the system performance could be improved. We conduct a series of experiments over the TPC-H datasets to illustrate the effectiveness of the ASAWA strategy.

Keywords: partition key selection, data partitioning.

1 Introduction

Nowadays, more and more applications need to manage huge volume of data and serve large amount of requests concurrently. As it is too hard for a single machine to support the whole system operations, these applications are mostly installed on clusters containing multiple data nodes. Thus, the whole data in an application needs to be partitioned and then stored in different machines. The goal of a partitioning algorithm is to achieve low response time and high throughput by minimizing inter-node communications.

During data partitioning, partition key selection, which would select attributes for tables as the baseline in partitioning, is the most important decision made regarding the scalability of database [1]. Data in different partitions are serviced by different processing nodes, and concurrent Read and Write operations over multiple partitions will be a bottleneck for applications. Poor partition key selection can then greatly affect query performance [2]. For example, an application, which has partitioned data

---

[*] Corresponding author.

Y. Ishikawa et al. (Eds.): APWeb 2013, LNCS 7808, pp. 609–620, 2013.
© Springer-Verlag Berlin Heidelberg 2013

by the *Location* attribute but accesses data with queries containing many joins on the *Date* column, would undoubtedly has an expensive communication cost.

Previously, there have been many efforts on the problem of data partitioning [1, 2, 5-8]. However, few of them address the problem of partition key selection. Instead, most of them utilize some simple strategy, while some others rely on a commercial database system for this selection.

In this work, we discuss how to select partition keys automatically for tables in applications and propose ASAWA (**A**utomatic **S**election based on **A**ttribute **W**eight **A**nalysis) strategy. Its general idea is based on the observation that if an attribute is found to be accessed in many queries, it would probably act as keys in join operations. Thus it would be better to partition datasets according to the attributes in search condition defined in query statements. The basic process of ASAWA would first compute an aggregated score for each attribute by identifying the data dependency among tables and finding out associated attributes, and then select partition keys with top scores evaluated by grouping features in workload schema. Combined both data and workload information predefined, it is expected to reduce communication cost among data nodes and improve system performance.

The contributions of this paper are summarized as follows:

1. We propose an automatic partition key selection strategy.
2. We implement the strategy within a novel algorithm for choosing partition keys based on data and workload information.
3. We conduct a series of experiments on TPC-H [3] to show both the efficiency and effectiveness of our proposed approaches.

The remainder of this paper is organized as follows. Section 2 presents related works in data partitioning with particular focus on partition key selection. Next, Section 3 illustrates the automatic partition key selection method ASAWA. Using three kinds of partition key selection strategies, Section 4 describes the performance comparison to show the efficiency of ASAWA. Finally, we have briefly summarized the contributions in this work and discussed works in the future in Section 5.

2   Related Works

Data partitioning [4] is known as one of the most famous technologies for its efficiency in large-scale application scalability. Traditionally, there are 3 kinds of partitioning strategy: horizontal partitioning [5], vertical partitioning [6] and hybrid approaches [7], any of which needs to select a partition key for each relation as the basis of partitioning algorithms. In recent years, there have emerged some new researches, which do not need to select partition keys. Schism [8] uses Metis [9] to divide mapping graph of relations in data into fine-grained partitioning results. In [2], it has proposed a method to have a skew-aware partitioning solution based on large-scale neighborhood search. It would like to select the most frequently accessed column in the workload as the partition key for each table originally, and change it when a better one founded during the process of relaxation. As followed by the partitioning strategies, when there are too many columns in a large amount of tables, the search space would be fairly huge. Designed in a bottom-up view, both of them need to be recalculated and partitioned if there is incremental data. Consequently, this kind of solutions is much more suitable for the applications with fixed data and workloads.

The ideal solution is supposed to partition data without manual assignment, while the performance is as well as the best-known. In most research works, they assume [10] that the proper partitioning keys have been chosen, and concentrate on partitioning algorithms with less aware on partition key selection. Although many traditional commercial database systems (e.g. Oracle [11], DB2 [12], SQL Server [13], etc.) have mentioned the importance of partition key selection and have their own mechanisms for recommendation, most of them are not open sources and combined deeply with underlying database optimizers. Therefore, they could not be applied on others directly. Based on some guidelines [14, 15], it is mostly operated manually in cases that the data management architecture is new or open source.

Basically, there are four kinds of partition key selection strategies:

- **Primary Key Selection:** This is the most widely used partition key selection strategy as default in commercial database systems. One of its variations is to use foreign key or the first column in a table as its partition key.
- **Random Selection:** This occurs when partition keys are specified without sufficient database management experiences or by systems without any constrains.
- **Exhaustive Selection:** This would like to try all possible partition key groups to have the one with best performance. It needs to use a sample data and workload set to simulate as in the real world, which is hard to predict the cost correctly.
- **Analytical Selection:** This is based on data and workload schema analysis, which is usually applied manually, or recommended by optimizing advisors integrated within expensive commercial database systems.

As data amount increasing rapidly and workload requests variously, this work would concentrate on automatically partition key selection from a top-down view that suits the requirements from data and workload. It prompts ASAWA method for automatic partition key selection, which not only takes the data and workload schemas into consideration, but also combined with design and execution information.

3   Partition Key Selection

To illustrate the problem of partition key selection, consider the following motivating scenario: *Given a data center with many query workloads, it needs to build a data management configuration that consists of many storage nodes and to partition data to suitable nodes so that the workloads could be executed in parallel. As the first step in data partitioning, a critical problem is how to select partition keys for relationships in the data center to minimize communication costs as much as possible.*

3.1   Problem Description

For a well-designed application, the basic information of the system, like data schema, workload pattern and their utility features, is given before data partitioning. In this work, we focus on applications with well defined data schema and static workload pattern which could be obtained with access frequencies for queries in it.

The partition key selection problem can be generally formulated as follows:

Definition. **Partition Key Selection**: Given a database $D = \{R_1, R_2, ..., R_d\}$, a query workload $W = \{Q_1, Q_2, ..., Q_w\}$, and a storage bound $S = \{B_1, B_2, ..., B_s\}$, select an attribute PKS_x $(1 \leq x \leq d)$ for each table as the partition key. Here,

- R_x $(1 \leq x \leq d) = <A_{x1}, A_{x2}, ..., A_{xt}>$ is a predefined table schema with a scale factor coefficient C_x in database D. A_{xi} $(1 \leq i \leq t_x)$ is one of the attributes in R_x.
- Q_y $(1 \leq y \leq w)$ is a predefined query pattern with an access frequency AF_y in workload W,
- B_z $(1 \leq z \leq s)$ is the storage bound of node z.

The total execution time T_{total} consists of two parts: data loading time T_{load} and the whole workload execution time T_{exec}. In this work, we assume all the data nodes are the same in physical features and use the widely used Hash function $h1:S \rightarrow [0,1]$ to distribute datasets into data nodes uniformly. The best partition key selection solution should partition data to nodes and complete queries within the least response time.

3.2   ASAWA Method

The basic process of ASAWA would first compute an aggregated score for each attribute based on identifying the dependency among tables and finding out the corresponding associated attributes, and then select partition keys with top scores optimized by grouping features in workload schema.

Database Analysis

Data Schema
As mentioned in Section 3.1, we assume that the data schema is well defined to guarantee data integrity for the database and all of the basic information (including attribute name, data type, constraints in the table, etc.) could be obtained.

To guarantee the efficiency of ASAWA, we would like to select the attributes with data integrity declarations as partition key candidates, including *Primary Key, Foreign Key* and columns declared *Unique* or *Not Null*. The weight of affect in dimension of data schema for an attribute could then be assigned equally as the values of $w_{DS}(PK)$, $w_{DS}(FK)$, $w_{DS}(U)$ and $w_{DS}(NN)$ respectively for each of declarations mentioned above. The values of weights could be the same or different, which depend on specific applications. The existences of the 4 kinds of statements are present as $e_{DS}(PK)$, $e_{DS}(FK)$, $e_{DS}(U)$ and $e_{DS}(NN)$ respectively, whose value is 1 or 0 based on the integrity statement on the relative attribute.

For a given attribute in the data schema, its weight in dimension of data schema is shown as $w_{DS}(GA)$ in Formula (1), which is the product of the existences that have defined integrity statements on it and the values of weights in a relative sequence:

$$w_{DS}(GA) = \sum(w_{DS}(PK) \times e_{DS}(GA_{PK}) + w_{DS}(FK) \times e_{DS}(GA_{FK})$$

$$+ w_{DS}(U) \times e_{DS}(GA_U) + w_{DS}(NN)) \times e_{DS}(GA_{NN})) \qquad (1)$$

Scale Factor.

A scale factor is the number used as a multiplier in scaling. This is mainly used in data generation. In ASAWA, the weight of affect for a given attribute in dimension of scale factor is as in Formula (2):

$$w_{SF}(GA) = 1 + log(SF_x / SF_{MIN}) \qquad (2)$$

in which $SF_{MIN} = min\{ SF_1, SF_2,..., SF_d \}$. Obviously, $min(w_{SF}(GA))=1$.

In some applications, each table's volume might increase in the same growth rate, which means that the scale factor for each of them is 1. Thus, w_{SF} of each table is *1*. In most of real world applications, tables' volumes are not increased in the same pace. For example, in TPC-H, the scale factor coefficient of table Lineitem is 6000000, while Orders's is 15000000 and Supplier's is 10000. According to Formula (2), $w_{SF}(A_{MIN}) = w_{SF}(A_{SUPPLIER}) = 1$, $w_{SF}(A_{LINEITEM}) \approx 3.78$, and $w_{SF}(A_{ORDER}) \approx 4.18$.

An extreme case is that, there are tables that too tiny to be worth partitioning. For example, in an OLAP application with Star Schema, the volume of fact tables increases fast while dimension tables always stay the same. These tiny tables are considered do not have scale factor. Instead of partitioning, they should be replicated for considerations on system performance.

Thus, the total weight of a given attribute in database is present in Formula (3):

$$w_{DATABASE}(GA) = w_{DS}(GA) \times w_{SF}(GA) \qquad (3)$$

Workload Analysis

Query Pattern

Query pattern is a piece of code that will make a reasonably normal query expression compile. As mentioned in Section 3.1, we assume that each query would be executed using the same structure with variable parameters. This ensures that the attributes touched in each query is the same in every execution of it. As mentioned in [10], The most relevant operations that present the attributes touched in SQL statements include equi-joins, Group By operations, duplicate elimination, and selections. The weights of them in dimension of query pattern are as $w_{QP}(Join)$, $w_{QP}(GB)$, $w_{QP}(DR)$, $w_{QP}(SelectConstant)$ and $w_{QP}(SelectHostVariable)$ relatively, while the existences as $e_{QP}(Join)$, $e_{QP}(GB)$, $e_{QP}(DR)$, $e_{QP}(SelectConstant)$ and $e_{QP}(SelectHostVariable)$ respectively, which are the counts of their appearance in the query.

For a given attribute in a given query, the weight of it in dimension of data schema is shown as $w_{QP}(GA_{GQ})$ in Formula (4):

$$w_{QP}(GA_{GQ}) = \sum(w_{QP}(Join) \times e_{QP}(Join) + w_{QP}(GB) \times e_{QP}(GB)$$

$$+ \ w_{QP}(DR) \times e_{QP}(DR) + w_{QP}(SelectConstant) \times e_{QP}(SelectConstant)$$

$$+ \ w_{QP}(SelectHostVariable) \times e_{QP}(SelectHostVariable)) \tag{4}$$

Access Frequency

Access frequency represents how many times an application runs in a given period. Its measurement is decided by the designer, e.g. an hour, a day, et al. In ASAWA, we assume the access frequency is steady for queries in the workload and define that the weight of affect in dimension of access frequency is as in Formula (5):

$$w_{AF}(Q_y) = 1 + log\ (\ AF_y\ /AF_{MIN}\ ) \tag{5}$$

in which $AF_{MIN} = min\{\ AF_1,\ AF_2,...,\ AF_w\ \}$. Obviously, $min(w_{AF}(Q_y))=1$.

For example, we assume that the access frequency of Q_1 and Q_{11} is 7000 and 150 respectively during an hour in TPC-H, while the minimized access frequency is 3 on Q_{21}. According to Formula (6), $w_{AF}(Q_{MIN}) = w_{AF}(Q_{21}) = 1$, $w_{AF}(Q_1) \approx 3.37$ and $w_{AF}(Q_{11}) \approx 2.70$.

Thus, the total weight of a given attribute in workload is present in Formula (6):

$$W_{WORKLOAD}(GA) = \sum_{y=1}^{w}(w_{QP}(GA_{Q_y}) \times w_{AF}(Q_y)) \tag{6}$$

According to the above analysis, the general weight of a given attribute w(a) could be calculated as presented in Formula (7):

$$w(a) = w_{DATABASE}(a) \times w_{WORKLOAD}(a)$$

$$= w_{DS}(a) \times w_{SF}(a) \times \sum_{y=1}^{w}(w_{QP}(a_{Q_y}) \times w_{AF}(Q_y)) \tag{7}$$

ASAWA Algorithm

Although there have been many partition key selection strategies for manual operation tips, how to combine these tips for automatic partition key selection is a problem. Here we propose ASAWA (**A**utomatic **S**election based on **A**ttribute **W**eight **A**nalysis) method to solve this issue.

The whole ASAWA algorithm is listed in Fig.1. Combining both data and workload information predefined, it takes attributes with data integrity constrains into consideration to reduce the complexity in practice and is expected to reduce communication cost among data nodes and benefit for system performance.

4   Experimental Study

In this section, we have shown results of our experiments using different partition key selection strategies and compared with the results with analysis.

Input:
    1. Data schema of the designed database
    2. Scale Factors of tables defined in the data schema
    3. Query Pattern of the designed workload
    4. Access Frequency of each query in the workload

Output:
    Solution $PKSD = \{PKS_1, PKS_2, ..., PKS_x\}$ $(1 \leq x \leq d)$, in which PKS_j $(1 \leq j \leq x)$ is an attribute in table $T_x \in D$.

Procedure ASAWA:
    1. Eliminate attributes without impact and obtain partition key candidates with its weight $w(a)$ for each.
        a. Extract the attributes with data integrity constrains from the data schema.
        b. Assign the weight of each attribute in dimension of data schema, scale factor, Query Pattern and access frequency as $w(a)$ using Formulate (8).
    2. Exhaustively consider all combinations of attributes that are not eliminated:
        a. Call optimizer with candidate set and grouping to estimate execution times of queries.
        b. Estimate aggregate Response time of all queries, and update {best partition key, relation grouping} to the current one to choose the best.
    3. Output $PKSD = \{ PKS_1, PKS_2, ..., PKS_x\}$ $(1 \bullet x \bullet d)$ over D as the partition key for the given data schema.

Fig. 1. The ASAWA Algorithm

4.1   Environment and Configuration

In the experiment, we have setup 3 data nodes, each of which is equipped with 2.66GHz * 8 cores, more than 300G disk, and CentOS 6.1 x86-64 with Open SSH 5.2 inside. As the scenario mentioned in Section 3, we select Greenplum Database (GPDB for short) 4.2.0.0 [26] as the platform. The dataset and queries used are generated from TPC-H 2.14.2, which has well-defined data schema and query patters to illustrate a system that examine large volumes of data, execute queries with a high degree of complexity, and give answers to critical business questions. The node with 32G main memories is chosen as the master node, while the other two nodes have 16G main memory and are set as slave nodes. Each of them has 8 segments.

4.2   Experimental Procedure

A complete experimental procedure is shown in Fig. 2. It would be executed in every round of each strategy on a generated dataset with a specific scale factor. To be fair, this procedure should be repeated in each round of every case.

```
Input:
    1. A specific scale of datasets generated from TPC-H
    2. Queries defined in the benchmark
Output:
    1. Data partitioning time
    2. Executing time of each query
    3. Total time of the whole process
Procedure:
    1. Analysis the partition key for each tables based on
    applied strategy within the round.
    2. Generate a new TPC-H dataset with the specific
    scale factor
    3. Partition data to data nodes based on keys se-
    lected, use Hash function h1:S•[0,1] to distribute it
    to nodes and record this time period as data loading
    time T_load.
    4. Execute specific query set for this round and
    record this time period as query execution time T_exec.
    5. Record the whole process executing time as T_total
    when the last query finishes its execution.
```

Fig. 2. One Round of Experimental Procedure in Each Case

It needs to select partition keys based on different strategies in the first step. In this work, we implement three partition key selection strategies for comparison, which are based on Primary Key, Random Assigned and ASAWA respectively. The detail descriptions of the first two strategies could be found in Section 2. For ASAWA, we set the weights in dimension of query pattern as: $w_{QP}(Join) = 1.0$, $w_{QP}(GB) = 0.1$, $w_{QP}(DR) = 0.08$, $w_{QP}(SelectConstant) = -0.05$ and $w_{QP}(SelectHostVariable) = 0.05$ as in [10]. In order to present the underlying structure information in the schema, we assign $w_{DS}(PK) = w_{DS}(FK) = 1$ and $w_{DS}(U) = w_{DS}(NN) = 0.5$ respectively for $w_{DS}(GA)$ in this work. Then we could run Procedure ASAWA shown in Fig. 1 for partition key selection for case ASAWA with the specific scale factor in this round.

In the TPC-H Schema, there are two tables named Nation and Region respectively that are not affected by scale factors. For the size for both of them are fixed and quite limited, instead of partitioned, both of them are copied to each data node in every selected strategy case for performance consideration.

4.3   Results and Analysis

As a matter of fact that the servers may have many different networks and application status, the running result is most probably different from each other, more or less.

In order to get a steady result and have a much more reliable result analysis, each strategy with relative dataset and workload in a specific scale should be executed multi round in the same physical environment. Due to the limitation of current hardware environment, the scale factor of dataset is set to be 1, 10 and 100 respectively. We run query set 5 times within the newly partitioned dataset in every round.

Overall Results

As a decision support oriented benchmark, it is hard to predict the access frequency of queries in TPC-H. We first all of the 22 queries in the data set of scale factor =1, 10 and 100 respectively. The results are shown in Figure 3.

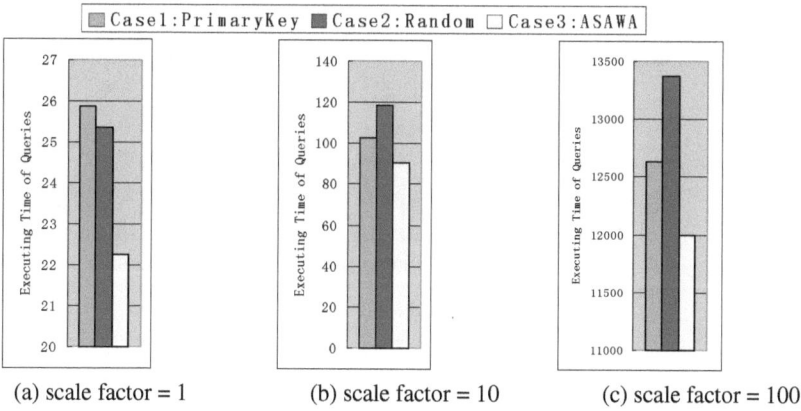

(a) scale factor = 1   (b) scale factor = 10   (c) scale factor = 100

Fig. 3. Executing Time of 22 Queries

From the series figures shown in Fig. 3, we could see that in each scale factor round, Case 2: Random Strategy always runs the longest time and shows the worst performance, Case 1: Primary Key Strategy is faster than Case 2 and shows fine performance, while Case 3: ASAWA is even better than Case 1 and always shows the best performance in the 3 scales of datasets. Besides the results shown in Figure 3, we have also recorded data loading time in each round of every case. In these records, Case 2 shows obviously longer data loading time than the others.

The result verifies again that the performance in queries is affected by the key selection switch. Also, we found that compared with Case 1, the advantages of Case 3 in Figure 3(c) are not so distinct than it in Figure 3(a) and 3(b). This is because GPDB has implemented a mechanism named Motion to do live migration and eliminate partition differences during running time, which means the long time running, the less distance. But this would add extra workload to data nodes and are not common implemented in every product.

Specific Query Results

The series of figures shown in Fig. 4 indicate the average executing time of each query in the data set of scale factor =1, 10 and 100 respectively. From these figures, we could see that for most TPC-H queries in each scale factor round, Case 2 runs the longest time and shows the worst performance, Case 1 shows fine performance than Case 2 especially in large scales, while Case 3 is shows the best. This indicates that the partition strategy should take the actual workloads of applications into consideration to get much better performance. Due to identify tuples specifically but having not considered the business semantical information, Primary Keys strategy always obtains reasonable but limited performance. For improper partition keys selection, Random strategy leads to bad system performance. Having taken data and workload features into consideration, ASAWA shows the best performance in all of these 3 strategies.

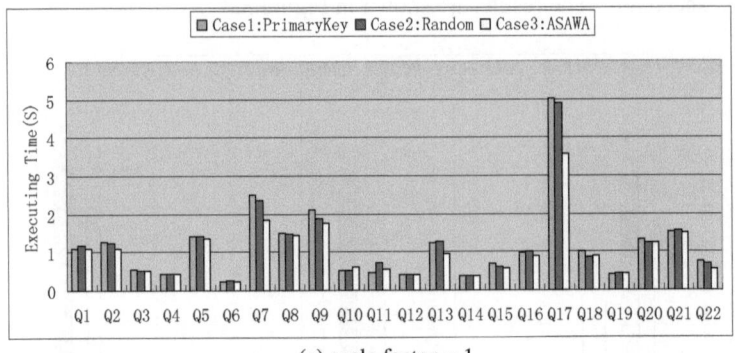

(a) scale factor = 1

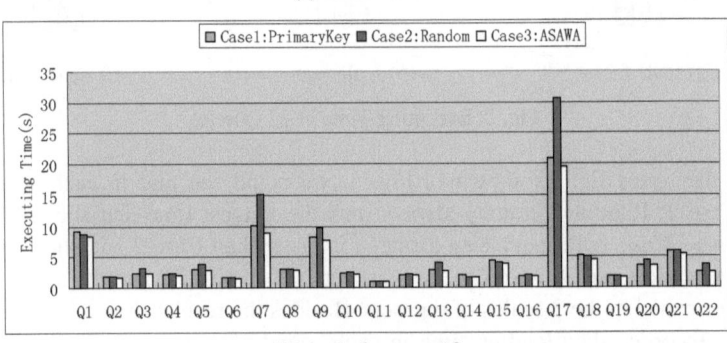

(b) scale factor = 10

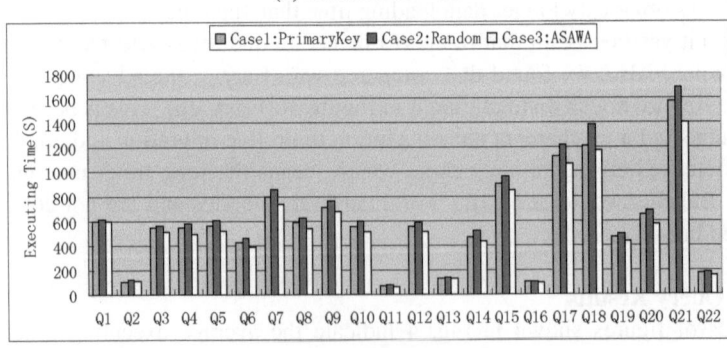

(c) scale factor = 100

Fig. 4. Executing Time of Each Query

Besides, we have intentionally selected Q1, Q5 and Q13 from the 22 queries of TPC-H, in which Q1 is represented for queries on a single large table, Q5 for queries on multi tables and Q13 for queries only on small tables, to take a continuous running for 7 rounds respectively on the dataset with the sale factor set as 100. The result shown in Fig.5 indicates that, for the queries that data distributed originally and memory could not cached, the executing time does not change too much, like Q1 and Q5, while for the queries that memory cacheable as Q13, the executing time is the most at the first time, less in the second time, and lest and maintaining in the following 5 times. It also shows the downtrend slightly in Q1 and Q5. Due to the automatic motion operation (like live migration) in GPDB, Round 3-7 show steady results. Obviously, if similar queries could be scheduled closer, which our strategy want to achieve, executing performance would be improved a lot.

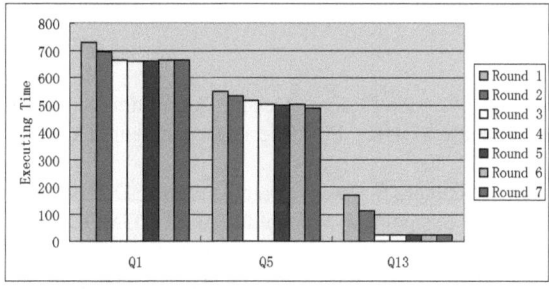

Fig. 5. Continuous Running of Single Query (scale factor = 100)

5   Conclusion

As more and more applications need to manage data in huge volume and response large amount of requests concurrently, the whole dataset for an application need to be partitioned into multiple parts and put into different machines. During this process, the selection of partition key would greatly affect the partitioning results. Nevertheless, there are few works addressing this topic. In this work, we present an automatic partition key selection strategy, called ASAWA. It would first compute an aggregated score for each attribute based on identifying the dependency among tables and finding out the corresponding associated attributes, and then select partition keys with top scores optimized by grouping features in workload schema. In this way, close tuples, i.e. co-appearing in queries frequently, would be probably put into the same partition. Hence the inter-node joins could be greatly reduced and the system performance could be improved. We then conduct a series of experiments over TPC-H queries with datasets generated in different scale factors. The experiment results illustrate the effectiveness of ASAWA method.

For most applications with huge data volume and great concurrent requests, partitioning strategy should be well planed before data storage and would be difficult to change once it adopted. If it could be adjusted automatically according to the actual applications, this would be undoubtedly more meaningful and useful for the widely use of database system products and release the burden of database administrators or developers. Partition key selection is the first step in data partitioning, and we only use empirical weight values assigned in ASAWA in the experiments. In the future work, we would like to adjust the values by learning methods to get a more practical weight set. Of course, there must be some differences among application features and visions. We would like to explore an automatic partitioning method combined with both ASAWA and proper partitioning algorithms in further research.

Acknowledgements. This work is supported by State Key Laboratory of Software Development Environment Open Fund under Grant No.SKLSDE-2012KF-09, the National Science Foundation of China under Grant No. 61003086, and the Graduate Student Scientific Research Foundation of Renmin University of China under Grant No. 42306176. The authors would like to thank the anonymous reviewers for their helpful comments and valuable suggestions.

References

1. Zilio, D.C.: Physical Database Design Decision Algorithms and Concurrent Reorganization for Parallel Database Systems. PhD Thesis, Department of Computer Science, University of Toronto (1998)
2. Pavlo, A., Curino, C., Zdonik, S.: Skew-aware Automatic Database Partitioning in Shared-Nothing Parallel OLTP Systems. In: Proc. of the ACM SIGMOD, pp. 61–72 (2012)
3. TPC Benchmark™ H, http://www.tpc.org/tpch/
4. Stonebraker, M., Cattell, R.: 10 Rules for Scalable Performance in 'Simple Operation' Datastores. Communications of the ACM 54, 72–80 (2011)
5. Ceri, S., Negri, M., Pelagatti, G.: Horizontal Data Partitioning in Database Design. In: Proc. of the ACM SIGMOD, pp. 128–136 (1982)
6. Navathe, S., Ceri, G., Wiederhold, G., Dou, J.: Vertical Partitioning Algorithms for Database Systems. ACM Transactions on Database Systems 9(4), 680–710 (1984)
7. Agrawal, S., Narasayya, V., Yang, B.: Integrating Vertical and Horizontal Partitioning into Automated Physical Database Design. In: Proc. of the ACM SIGMOD, pp. 359–370 (2004)
8. Curino, C., Jones, E., Zhang, Y., Madden, S.: Schism: a Workload-Driven Approach to Database Replication and Partitioning. Proc. of the VLDB Endowment 3, 48–57 (2010)
9. Metis, http://glaros.dtc.umn.edu/gkhome/views/metis/index.html
10. Zilio, D.C., Jhingran, A., Padmanabhan, S.: Partition Key Selection for a Shared-nothing Parallel Database System. Technical Report RC 19820(87739) 11/10/94, IBM T. J. Watson Research Center (1994)
11. Eadon, G., Chong, E.I., Shankar, S., Raghavan, A., Srinivasan, J., Das, S.: Supporting Table Partitioning by Reference in Oracle. In: Proc. of the ACM SIGMOD, pp. 1111–1122 (2008)
12. Zilio, D.C., Rao, J., Lightstone, S., Lohman, G., et al.: DB2 Design Advisor: Integrated Automatic Physical Database Design. In: Proceedings of the VLDB, pp. 1087–1097 (2004)
13. Nehme, R., Bruno, N.: Automated Partitioning Design in Parallel Database Systems. In: Proc. of the ACM SIGMOD, pp. 1137–1148 (2011)
14. Özsu, M.T., Valduriez, P.: Principles of Distributed Database Systems, 3rd edn. Springer, New York (2011)
15. Rahimi, S., Haug, F.S.: Distributed Database Management Systems: A Practical Approach. IEEE Computer Society, Hoboken (2010)
16. Greenplum Database, http://www.greenplum.com/products/greenplum-database

An Active Service Reselection Triggering Mechanism

Ying Yin, Tiancheng Zhang, Bin Zhang, Gang Sheng, Yuhai Zhao, and Ming Li

College of Information Science and Engineering, Northeastern University
Shengyang, 110004, China
yinying@ise.neu.edu.cn

Abstract. Actively identifying service faults and actively trigger service reselection are important management methods for efficiently promoting system reliability. Composite Web services, however, are often long-running, loosely coupled and cross-organizational applications. They always run in a highly dynamic and changing environment(The Web) which imposes many uncertainties, such as server unavailable or network interruption or temporarily interrupt and so on. Under these uncontrollable circumstances, it is impractical to monitor the changes in Quality of Service parameters for each and every service in order to timely trigger service reselection, due to high computational costs associated with the process. In order to overcome the above problem, this paper proposes an efficient reselection mechanism by mining early patterns in advance. The system will trigger service reselection once potential execution failure is detected by matching these early patterns with the current situation. The process is real time and low-cost, and as such, the proposed mechanism will improve system reliability.

Keywords: Early Pattern, Prediction, Service Reliability.

1 Introduction

The advantage of composite Web services is that it realizes a complex application by connecting multiple component services seamlessly. However, in real applications, Web service lives in a highly dynamic environment, both the network condition and the operational status of each of the constituent Web Services(WSs) may change during the life-time of a business process itself. The instability brought on by various uncertain factors often makes the composite services to be interrupted or interrupted temporally. Therefore, it is very important to ensure the normal execution of the composite service applications and to provide a reliable software system[11].

Reselection mechanisms for failed services are important technologies which ensures the reliability of composite services. On the other hand, reselection of services triggered by interruptions may not be suitable as the pre-computing of the Quality of Service (QoS) is costly both in time and financial cost. The system can identify the unreliable service by predicting potential interrupting

Y. Ishikawa et al. (Eds.): APWeb 2013, LNCS 7808, pp. 621–628, 2013.

factors in advance and then avoids the exceptional service interruption. That is, prediction has gradually become a valid method to trigger service reselection [9]. Prediction is realized by some algorithms which monitor and evaluate the quality of composite services. There are several prediction methods, such as ML-based methods known in literature[2], QoS-aware based methods[3,4] and collaborative filtering-based methods[5,9] and so on.

Despite the existence of previous work, prediction methods still three major challenges remain, as defined below. First of all, most of the traditional prediction models, such as Support Vector Machines (SVM) and Artificial Neural Networks (ANN), can only take input data as vectors of features [12]. However, there are no explicitly features in sequential data. Secondly, prediction models based on QoS monitoring, such as Naive Bayes and Markov model[10], which assume that sequences in a class are generated by an underlying model M and the probability distributions are described by a set of parameters. These parameters however, are obtained by predicting the QoS during the whole lifecycle of the services and will therefore lead to high overhead costs. Another method is prediction based on sequence distance, such as collaborative filtering[9], this type of methods require the definition of to define a distance function to measure the similarity between a pair of sequences. However, how to select an optimal similarity function is far from trivial, as it will introduce numerous parameters and measures for distances can be subjective.

In order to maintain the reliability of composite services, an important thing for us is actively identify exception patterns and trigger service reselection algorithm quickly. In this paper, we proposes an efficient service reselection mechanism by mining previous patterns in advance. The system will trigger service reselection once the currently executing sequence matches any mined historical sequence indicating imminent failure.

The remainder of this paper is organized as follows: Section 2 discusses the system architecture. In Section 3, we present the process of extracting execution instances from execution logs while in Section 4 the basic related definitions are given, and an efficient method for identifying early patterns based on an interest measure is proposed, and methods for implementation are presented. We summarize our research and discuss some future work directions in Section 5.

2   System Architecture

Below we describe the system architecture of our service reselection triggering mechanism. After accepting a user's query specification, our BPEL Engine should be able to provide a composite service execution sequence. Once some latent exception is identified during execution, our system will trigger the reselection process.

There are four phases in the trigger process: (1) recording the composite service execution log information to generate the execution service sequence repository; (2) cleaning collected data and converting service log into service execution sequences; (3) mining early prediction patterns from converted execution sequences and deducing predictive rules; (4) triggering the service reselection process as soon as an executing service sequence matches one of the early prediction patterns. Figure 1 shows the whole reselection triggering process and below each of the steps will be briefly discussed.

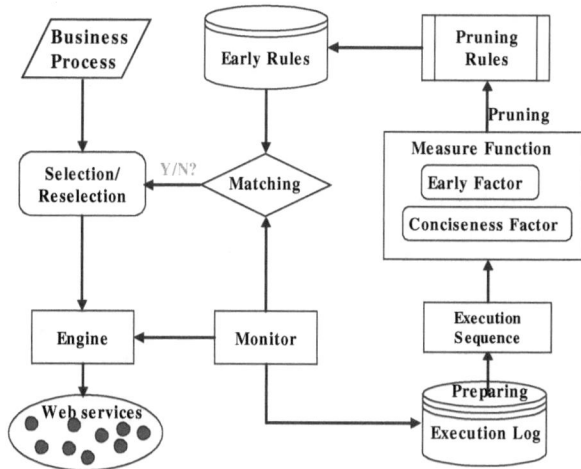

Fig. 1. The framework

Recording of composite service execution log. During this phase, the system needs to collect large amounts of failed composite service execution information, aiming to mine useful knowledge and to analyze the causes of failure reason so as to trigger service reselection as quickly as possible. The generated composite service execution log repository forms the basis for early prediction.

Extracting of the service execution sequence. the original service log includes a large variety of information, such as Web Service Description Language (WSDL) information, keywords, IP address, QoS and so on. The data needs to be cleaned and QoS information needs to be transferred into sequences so as to mine early pattern effectively. For the functionally similar Web services, even for each Web service itself, perceptive QoS values may differ at different time points or different network environments thus presumably indicating the difference in quality. That means that the quality of Web service might different in different situation. Therefore, extracting descriptive information from the service log and converting it into forms that can readily be mined are the key part of our methods. The details of this step will be given in Section 3.

Mining of early prediction pattern. Various behavioral patterns in different environment can be lead to composite service execution failure. However, both the time when the fault occurs and the length of the behavioral pattern influence when the service reselection will be triggered. Under the various environments, mining triggering patterns with minimal cost and triggering the reselection as earlier as possible are the important issues in our work. The contents of this part will be described in Section 4.

Triggering of Reselection. These mined execution sequence patterns can be used to timely detect whether the composite service deviates from normal execution. This kinds of knowledge should be applied to improve the reliability of the composite service. Based on these mined rules, how to construct an efficient search tree aiming to trigger the reselection process becomes an more important issues in promoting the reliability. The details of this part will be described in the further work.

3   Preliminary

Once a service-oriented application or a composite service is deployed in a runtime execution environment, the application can be executed in many execution instances. Each execution instance is uniquely identified with an identifier (i.e., id). In each execution instance, a set of service sequences can be triggered. Various uncertain factors may make a execution instance fail. Reason for this phenomenon is that there exists a large number of potential exception status information under different uncertain factors. We record triggered events in a log which contains different types of information. We are interested in failed QoS information. It is helpful for service quality management by extracting the execution status information from execution log to predict the service reliability.

Table 1. Composite QoS status Information

A	$CompositeQoS$	$Description$
S^0	$< av^0, exe^0 >$	Server unavailable, runtime delay
S^1	$< av^{0.5}, exe^0 >$	Server available intermittently, runtime delay
S^2	$< av^1, exe^0 >$	Server available, runtime delay
$\times$	$< av^0, exe^1 >$	Server unavailable, normal execution
S^3	$< av^{0.5}, exe^1 >$	Server available intermittently, normal execution
S^4	$< av^1, exe^1 >$	Server available, normal execution

In order to depict various exception status information clearly, for each component service, we only consider two QoS attributes as an example, such as the availability attribute (AV) and the execution time attribute(EXE). There are three possible states for AV, namely inaccessible, intermittently accessible and accessible, which are denoted by av^0, $av^{0.5}$ and av^1 respectively. EXE on the other hand can only signify two states, namely exe^0 and exe^1 representing delayed execution and normal execution respectively. Under these situations, we can obtain five possible groups of service execution statuses information which are shown in Table 1: $< av^0, exe^0 >$, denoted by S^0, represents server unavailable but runtime delay, $< av^{0.5}, exe^0 >$, denoted by S^1, represents server available intermittently but runtime delay, $< av^1, exe^0 >$, denoted by S^2, represents server available but runtime delay, $< av^{0.5}, exe^1 >$, denoted by S^3, represents server available intermittently but normal execution and $< av^1, exe^1 >$, denoted by S^4, represents server available but normal execution. However, for the status information $< av^0, exe^1 >$, denoted by S^0, represents server unavailable but normal execution. We do not consider this status, as it cannot occur in practice.

Table 2. An example of service execution instances

ID	$ExecuteLogEntry$	$Count$	$Class$
1	S_1^0	3	$failed$
2	S_1^1, S_3^1	3	$failed$
3	S_1^2, S_2^2, S_3^2	2	$failed$
4	S_1^1, S_2^1, S_3^1	3	$failed$
5	$S_1^4, S_2^2, S_3^2, S_4^2$	2	$failed$
6	S_1^4, S_2^2, S_3^2	2	$failed$

Based on the description assumption above large collections of composite service execution records D which are stored as sequential data as shown in Table 2. Let $S=\{S_1, S_2,...,S_n\}$, then each record consists of one or more component services from S. Note that every component service will have different QoS expression values, which can evaluate describe the execution status of that component service. For simplify, we let the capital letters with subscript, such as S_1, S_2, S_3, S_4, denotes different abstract service class and the superscript, denotes different QoS statuses of the component service. Table 2 only shows a service execution data with 15 failed users(not including successful users) and four abstract service class. Bellow are some definitions related.

4   Early Prediction Pattern Mining

In order to classify the each service sequence status into different fault types, we need to extract features from execution service log. Due to the unsupervised nature of the service log data, the first step is to mine those patterns that are effective for predicting services to be failed.

4.1   Basic Definition

In this section, we introduce some of the basic concepts for mining statuses and give the problem definition.

Definition 1. Pattern. *Let $\mathcal{D}$ be an execute service log with service set $\mathcal{S}$, $\mathcal{S}=\{S_1, S_2,...,S_n\}$. Let $P = \{S_1^i S_2^j...S_l^k\} \subseteq S\,(l-1,2,...,n)$ be a component service or a subset of the execution sequence with different status information, s_i^k denotes the kth status information of the ith service, where $\{i, j, k\} \in [0, 1, 2, 3, 4]$. A component service with status information or a set of component pairs is called a Pattern. .*

Given a pattern p=$s_1^i s_2^j, ...p_m^k$, we say pattern p appears in a sequence s if there exists an index $1 \le i_0 \le n - m + 1$ such that $a_{i_0}=s_1^i$, $a_{i_0+1}=s_2^j$,..., $a_{i_0+m-1}=s_m^k$.

For example, $p = \{acd\}$ appears in sequence $s = \{aacdfgacde\}$. However, a pattern p may appear multiple times in a sequence. Below, we give the minimal prefix length definition.

Definition 2. *Earliest Matching Pattern.* *Given Pattern p and sequence s, where $p \in s$. The* Earliest Matching Pattern*(EMP) is the distance from initial position of s to the first occurrence of p in s, and is denoted by $EMP(P, S)$.*

Definition 3. *Earliest Support.* *Let p be a pattern. The* Earliest Support*(ES) of pattern p is the ratio of the sum of the reciprocals of the earliest position of p to the number of instances in the data set, i.e.:*

$$ES(p) = \frac{\sum_{p \in S} \frac{1}{EMP(p,S)}}{\| S \|}$$

Problem Statement. For a given failed service log entry, a **Pattern** p should satisfy: (1) Pattern p should frequently be present in failed type; (2) Pattern p should be as short as possible and should have low prediction cost; All patterns satisfy the above conditions called *Early Patterns*. the optimal rules are deduced called *Early and Concise (EC) Rule Sets.*

4.2   Early Pattern Mining Algorithm

Previous existing pattern mining methods require the definition of various thresholds as quality measures such as support, confidence , such that only those relatively frequent rules are selected for classifier construction. However, these mining methods are ineffective in practice and the set of mined rules are still very large. It is very hard for a user to seek out appropriate rules. In our case, we consider the earliness and conciseness properties to design a utility measure and only preserve those early, concise patterns passing the measure from the training database.

The mining process is conducted on a prefix tree. Limited by space, we omit the description of constructing such a prefix tree structure. A complete pseudocode for mining optimal EP sets is presented in Algorithm 1.

Finding interesting contrast sequence rules by checking all subsequence along the paths is still a tough task. Various pruning rules are introduced below to prevent unnecessary rule generation and only preserve interesting rules. Reducing the search space are extremely important for efficiency of EP. In algorithm 1, we use two efficient pruning rules to improve efficiency.

Pruning Rule 1. *Pruning by ES.*
Given ES (Definition 3), Pattern pa denotes pattern p and all its possible proper supersets. If $0 \leqslant EP(p) \leqslant \gamma$, then pattern p and its supersets will not be an optimal rule.

Algorithm 1. Mining the EP
Input: data set D
Output: Pattern sets EC

  1: Set EC = ϕ
  2: Count numbers of 1-patterns
  3: Generate 1-pattern set
  4: Calculate ES of 1-pattern
  5: Select 1-pattern respectively and add them to EC
  6: new pattern set ← Generate (2-pattern set)
  7: **while** new pattern set is not empty **do**
  8:    Calculate ES(p) of candidates in new pattern set
  9:    For each pattern p in the (l+1)-pattern set
10:    **Apply pruning rule 1:** IF EP(p)$< \gamma$
11:    remove pattern EP;
12:    Else if there is a superpattern pa in l-pattern set
13:    **Apply pruning rule 2:** that EP(pa)= EP(p)
14:    Then remove pattern P;
15:    Select EP to EC;
16:    **ENDIF**
17: **end while**
18: new pattern set ← Generate(next level pattern sets)
19: Return R;

Note: The presentation of pruning rule 1 describes the early support of a pattern, not the general support. Because in a exception execution log, pattern p may emerge several times. We only record the first position so as to trigger service reselection procedures as soon as possible. This is difference with traditional association rules.

Pruning Rule 2. *Pruning by concise.*
If pattern p satisfy $EP(p)= EP(pa)$, where pa denotes its proper superset, then pattern pa and all its possible proper supersets will not be useful for EP.

Limited by space, we omit the proof. The above pruning rules are very efficient since it only generates a subset of frequent patterns with maximal interestingness instead of all ones.

5   Conclusion

In this paper, we discuss the mining of early prediction patterns which are important for predicting service reliability, and propose two interest measures in order to decide whether the rules are concise and as short as possible in length for quickly identifying exception patterns. Based on the proposed interest measures, we propose a new algorithm with efficient pruning rules to mine all optimal EP rules.

Acknowledgments. This work was supported by a grant from the National Natural Science Foundation of China under grants (No.61100028, 61272182, 61272180, 61073062); the New Century Excellent Talents in University Award (NCET-11-0085); the Ph.D. Programs Foundation of Ministry of Education of China(young teacher)(No.20110042120034); the Fundamental Research Funds for the Central Universities under grants (No. N110404017, N110404005); China Postdoctoral Science Foundation (No. 20100481204, 2011M500568, 2102T50263);

References

1. Yu, Q., Liu, X.M., Bouguettaya, A., Medjahed, B.: Deploying and managing Web services: issues, solutions, and directions. The VLDB Journal 17, 537–572 (2008)
2. Wang, F., Liu, L., Dou, C.: Stock Market Volatility Prediction: A Service-Oriented Multi-kernel Learning Approach. In: IEEE SCC, pp. 49–56 (2012)
3. Goldman, A., Ngoko, Y.: On Graph Reduction for QoS Prediction of Very Large Web Service Compositions. In: IEEE SCC, pp. 258–265 (2012)
4. Tao, Q., Chang, H., Gu, C., Yi, Y.: A novel prediction approach for trustworthy QoS of web services. Expert Syst. Appl. (ESWA) 39(3), 3676–3681 (2012)
5. Lo, W., Yin, J., Deng, S., Li, Y., Wu, Z.: Collaborative Web Service QoS Prediction with Location-Based Regularization. In: ICWS 2012, pp. 464–471 (2012)
6. Yu, T., Zhang, Y., Lin, K.J.: Efficient Algorithms for Web Services Selection with End-to-End QoS Constraints. ACM Trans. Web 1(1), 1–25 (2007)
7. Leitner, P., Michlmayr, A., Rosenberg, F., Dustdar, S.: Monitoring, Prediction and Prevention of SLA Violations in Composite Services. In: ICWS, pp. 369–376 (2010)
8. Zheng, Z., Ma, H., Lyu, M.R., King, I.: QoS-Aware Web Service Recommendation by Collaborative Filtering. IEEE Transactions on Service Computing (TSC) 4(2), 140–152 (2011)
9. Chen, L., Feng, Y., Wu, J., Zheng, Z.: An Enhanced QoS Prediction Approach for Service Selction. In: IEEE Conference on Web Service, ICWS (2011)
10. Park, J.S., Yu, H.-C., Chung, K.-S., Lee, E.: Markov Chain Based Monitoring Service for Fault Tolerance in Mobile Cloud Computing. In: AINA Workshops, pp. 520–525 (2011)
11. Zhang, L., Zhang, J., Hong, C.: Services Computing. TSingHua University Press, Beijing (2007)
12. Han, J., Kamber, M.: Data Mining: Concepts and Techniques, 2nd edn. The Morgan Kaufmann Series in Data Management Systems. Morgan Kaufmann Publishers (March 2006)

Linked Data Informativeness

Rouzbeh Meymandpour and Joseph G. Davis

School of Information Technologies
The University of Sydney
{rouzbeh,jdavis}@it.usyd.edu.au

Abstract. By leveraging Semantic Web technologies, Linked Open Data provides an extensive amount of structured information in a wide variety of domains. Principles of Liked Data facilitate access and re-use of semantic information, both for human and machine consumption. However, information overload due to the availability of a large amount of semantic data, as well as the need for automatic interpretation and analysis of the Web of Data require systematic approaches to manage the quality of the published information. Identifying the value of information provided by Linked Data enables a general understanding about the significance of semantic resources. This can lead to better information filtering functionalities in Semantic Web-based applications. The aim of this paper is to propose a novel approach, derived from information theory, to measure the informativeness in the context of the Web of Data. We experiment with the metric and present its applications in a variety of areas, including Linked Data quality analysis, faceted browsing, and ranking.

Keywords: Semantic Web, Linked Data, the Web of Data, DBpedia, Semantic Network, Partitioned Information Content, Information Filtering.

1   Introduction

The Semantic Web is a major evolution to the current World Wide Web (WWW) that provides machine-readable and meaningful representation of Web content. Formal knowledge representation languages such as RDF (Resource Description Framework) [1] and OWL (Ontology Web Language) [2], and query languages such as SPARQL [3] are key enablers of semantic applications. These technologies enable the Web of Data, a giant graph of interconnected information resources, known as Linked Data.

Based on the principles of the *Linking Open Data (LOD)* project, an extensive amount of structured data is now published and available for public use [4]. This includes interlinked datasets in a wide variety of domains that range from encyclopedic knowledge bases such as *DBpedia*[1], which is the Semantic Web-based representation of Wikipedia[2] [5], to domain-specific datasets such as Geo Names[3] and Bio2RDF[4].

---

[1] http://dbpedia.org
[2] http://wikipedia.org
[3] http://www.geonames.org
[4] http://bio2rdf.org

Y. Ishikawa et al. (Eds.): APWeb 2013, LNCS 7808, pp. 629–637, 2013.

According to the Linking Open Data cloud[5], more than 295 datasets in various domains, including Geography, Publications, Life Sciences, and so on are now interconnected together. This valuable source of semantic knowledge can be exploited in practical and innovative applications. It has also introduced several research opportunities.

Due to the fact that most of Linked Datasets have been created by automatic or semi-automatic extraction of structured data from Web pages, assessing the quality of the Web of Data has become a critical challenge for semantic data consumption. Moreover, not all available Linked Data is valuable for semantic software agents or human understanding. For example, it is likely that among all relations available in the Web of Data relating to a particular movie, its director and actors are more interesting to most users than its language or runtime. Likewise, for a software agent that compares two movies, having a same director conveys more information than sharing a same language. Understanding the informativeness of resources and structured data associated with them facilitates browsing as well as automatic consumption and interpretation of Linked Data. This paper, introduces two novel metrics to evaluate the amount of information that a resource and its relations carry. After proposing the metrics, we present a set of experiments conducted in various domains and applications in order to demonstrate their characteristics.

2   Measuring Informativeness in Linked Data

In this section, we first introduce the concept of informativeness measurement according to its mathematical theory. Then, by formalizing the definition of Linked Data, we propose metrics to assess the value of information in the Web of Data.

2.1   Measuring the Information Content

Information Theory describes the mathematical foundation of communication for transmitting an information source over a communication channel by means of codding schemes [6, 7]. Shannon [8] proposed *entropy* as a measure of information contained in a process. He showed that entropy, also known as *Information Content (IC)* or *Self-Information*, is the amount of binary symbols or bits that the receiver needs in order to recreate the original transmitted process. Derived from the probability theory paradigm, Information Content is defined as the engendered amount of surprise [9]:

$$IC(a) = -\log(\pi(a)) \tag{1}$$

where $\pi(a)$ is the probability of presence of term or concept a in a given corpora. The logarithm is often calculated to the base 2. Thus, the unit of information is *bits*. For the base 10 and natural logarithms, the unit is bans and nats, respectively.

By looking closely at this metric, we can describe its important characteristics. The amount of surprise evoked by occurrence of a particular term has a negative correlation with its probability. More popular terms, with a higher chance of presence, are more probable and therefore, there is less surprise in their occurrence. In other words, less frequent entities are more specific and informative than common terms.

---

[5] http://lod-cloud.net

2.2   Formal Description of Linked Data

In order to be able to present a systematic approach for Linked Data analysis, we mathematically explain and formalize the Web of Data based on a combination of graph and set theories. Linked Data is a labeled directed graph consists of resources as nodes and relations as edges that are identifiable by their URIs (Uniform Resource Identifiers). Thus, nodes, edges, and the directions of edges have to be considered for a proper representation of Linked Data and its resources:

Definition 1 (Linked Data Graph): Linked Data is a labeled directed graph LD, defined as $\langle R, L \rangle$, *where* $R = \{r_1, r_2, ..., r_{|R|}\}$ *is a set of resources, and* $L = \{l_1, l_2, ..., l_{|L|}\}$ *is a set of links, in which* l_i *is defined as* $\langle a, l, b \rangle$, *where* l *is the link that connects resource* a *to resource* b.

Therefore, incoming and outgoing edges can be used to describe a resource:

Definition 2 (A Resource in Linked Data): A resource $a \in R$ *in Linked Data LD is represented as its set of features A, which is defined as:*

$$A = \{\forall \langle l_i, t_i \rangle, l_i \in L, t_i \in R \mid \exists t_i \in R, \exists \langle a, l_i, t_i \rangle \in LD \lor \exists \langle t_i, l_i, a \rangle \in LD\} \qquad (2)$$

2.3   Informativeness of Resources in Linked Data

By deriving from the formal definition of Linked Data and the principles of information theory, we propose a metric to measure the value of information associated with resources in the Web of Data. According to the Definition 2, each resource is represented as its set of characteristics, which is a collection of its incoming and outgoing edges. Based on the basis of the probability theory, we define the surprise value associated with a resource in Linked Data:

Definition 3 (Probability of a Resource in Linked Data): For a resource such as $a \in R$, *represented as a set of features* $A = \{a_1, a_2, ..., a_{|A|}\}$, *the probability of A is*

$$\pi(A) = \pi(a_1)\,\pi(a_2)\cdots\pi(a_{|A|}) \qquad (3)$$

Thus, we have

$$IC(A) = -\log(\pi(A)) = -\log\left(\pi(a_1)\,\pi(a_2)\cdots\pi(a_{|A|})\right) \qquad (4)$$

Based on this, we define the *Partitioned Information Content*:

Definition 4 (Partitioned Information Content for Linked Data): Information Content of a resource is defined as the sum of the Information Content values of its features:

$$PIC(A) = \sum_{\forall a_i \in A} IC(a_i) \qquad (5)$$

where IC of each feature is calculated using the Equation (1):

$$IC_{\forall a_i \in A}(a_i) = -\log(\pi(a_i)) = -\log\left(\frac{\varphi(a_i)}{N}\right) \tag{6}$$

such that $\varphi(a_i)$ is the frequency of the feature a_i and N is the frequency of the most common feature in a given Linked Dataset.

In this definition, the amount of information conveyed by a given resource is calculated according to the probability of its features. Popular features carry less information while specific ones are more informative. This metric also allows us to study the Web of Data in terms of the value of information contained in each resource and domain. It highlights the most informative and influential information sources among a large number of available resources in a particular domain.

2.4   Informativeness of Relations in Linked Data

Another key component of Linked Data is relations, also known as properties. Understanding the importance and informativeness of relations allows better automatic interpretation and enhanced filtering of Linked Data. From the users' perspective, it supports information browsing and visualization by identifying the most significant properties in a particular domain. The informativeness of a given property is indicated by measuring the value of information produced by the relation in a given context:

Definition 5 (Generated Information Content of Relations): Let $C_i \subset C$ and $L_i \subset L$ be the set of all features and all links in the domain i, respectively, and let $C_{l_i} \subset C_i$ be the subset of all features that have $l_i \in L_i$ as their link:

$$GIC_{\forall l_i \in L_i}(l_i) = \sum_{\forall c_{l_i} \in C_{l_i}} IC(c_{l_i})$$

$$C_{l_i} = \{\forall \langle l_i, t_j \rangle \in C_i, t_j \in R, l_i \in L_i\} \tag{7}$$

The domain could be the whole Linked Data graph or a subset of it that is filtered by a search query. This metric enables us to indicate the most important relations that convey the highest amount of information about resources inside a particular domain.

3   Experiments

3.1   Experimental Context

The main Linked Dataset used in our experiments was DBpedia. We downloaded the main DBpedia Ontology as well as five fundamental datasets in English, including Ontology Infobox Properties, Ontology Infobox Types, Article Categories, SKOS Categories, and YAGO Links. Finally, by simultaneous executing of a number of SPARQL queries, we extracted instances and their features from the DBpedia dataset in three domains: Music, Actors, and Films (see Table 1).

At the time of writing this paper (October 2012), the latest version of DBpedia was 3.8, released on August 2012. In our final DBpedia dataset, the parameter N in the Equation (6) was set as the number of *(rdf:type, owl:thing)* relations (2,350,906).

Table 1. Different domains and their size characteristics

	Films	**Music**		**Actors**
rdf:type [a]	dbo:Film [b]	dbo:MusicalArtist	dbo:Band	yago:Actor [c]
Resources	71,715	34,246	27,061	31,482
Unique Features	237,429	640,157		236,184
Total Triples	1,729,718	2,302,008		1,002,901

[a] 'rdf:' is the prefix for the namespace http://www.w3.org/1999/02/22-rdf-syntax-ns#
[b] 'dbo:' is the prefix for the namespace http://dbpedia.org/ontology/
[c] The actual property value is http://dbpedia.org/class/yago/Actor109765278

3.2   Distribution of Information in Different Domains

The first experiment was performed using the Partitioned Information Content-based metric in order to compare information distribution across several domains in DBpedia. It is observable that information in the Music domain is more distributed than the others, with standard deviation of 590.2 in comparison to 148.2 for Films and 437.6 for Actors (see Fig. 1 (a) and Fig. 2). Moreover, there is more distinctive information available for musical artists and actors than films (Fig. 1 (b)). This may be explained by the fact that each artist has an increasing number of outputs during their career.

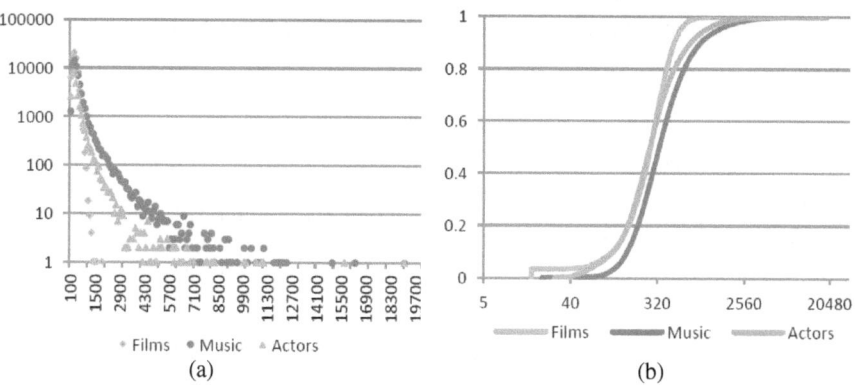

(a)   (b)

Fig. 1. (a) Distribution of IC and (b) Cumulative distribution of IC among DBpedia resources

Further experiments were performed in order to compare the informativeness of semantic data provided by various Linked Datasets. LinkedMDB[6] (Linked Movie Database) is a specialized open Linked Data for movies [10]. It provides structured information for 85,600 movies described using more than 6 million triples[7]. The experiments illustrate a wider distribution of information in LinkedMDB in comparison to

---

[6] http://www.linkedmdb.org
[7] According to the latest downloadable version of LinkedMDB, released on February 2012.

the Films dataset extracted from DBpedia (see Fig. 2). Comparing these figures underline the difference between the qualities of structured information provided by DBpedia, which is a general purpose Linked Data, against LinkedMDB.

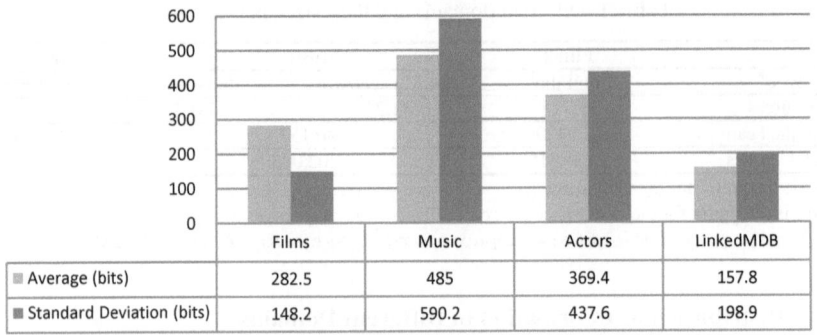

	Films	Music	Actors	LinkedMDB
■ Average (bits)	282.5	485	369.4	157.8
■ Standard Deviation (bits)	148.2	590.2	437.6	198.9

Fig. 2. Information distribution among instances in various domains

3.3   Applications to Faceted Browsing and Information Filtering

Information filtering techniques help users to overcome data overload by facilitating browsing and exploring the knowledge base. Facets improve semantic data browsers by providing suggestions according to the currently viewed item. We applied the proposed metrics to discover the most relevant resources to a given instance. Table 2 lists the top 10 recommended artists and bands when the user browses a specific domain.

Table 2. Applications of PIC for faceted browsing[a]

rdf:type dbo:MusicalArtist		dbo:genre db:Country_music[b]		dc:subject db:RockandRoll_HF[cde]	
Artist Name	**PIC (bits)**	**Artist Name**	**PIC (bits)**	**Artist Name**	**PIC (bits)**
David Bowie	19,008.9	Bob Dylan	14,901.7	Bob Dylan	14,901.7
Prince	16,220.9	Van Morrison	11,652.0	Van Morrison	11,652.0
Bob Dylan	14,901.7	Elvis Presley	11,096.9	Elvis Presley	11,096.9
The Beatles	12,337.7	Dolly Parton	8,536.0	Johnny Cash	8,091.1
Kanye West	12,176.2	Johnny Cash	8,091.1	Chet Atkins	6,346.4
Paul McCartney	11,920.7	Tony Brown	7,986.3	Ray Charles	4,959.8
Ilaiyaraaja	11,887.7	Reba McEntire	7,888.2	Bob Seger	3,936.0
Van Morrison	11,652.0	George Strait	7,858.5	Jackson Browne	3,421.4
Elvis Presley	11,096.9	Willie Nelson	7,758.0	Tom Petty	3,370.6
Elton John	11,005.2	Rick Rubin	7,186.3	Neil Diamond	3,245.9

[a] For a clear presentation, artist names are shown instead of DBpedia URIs.
[b] The applied filters also include *rdf:type dbo:MusicalArtist*
[c] 'dc:' is the prefix for the namespace `http://purl.org/dc/terms/subject`
[d] The actual applied filter is *dc:subject db:Rock_and_Roll_Hall_of_Fame_inductees*
[e] The applied filters also include *rdf:type dbo:MusicalArtist* and *dbo:genre db:Country_music*

Another application of the metrics is to identify the most important relations in each domain. Table 3 shows the top 10 list of properties with the highest amount of

Generated Information Content. These properties hold the most valuable and distinctive information about instances. It is useful especially for Linked Data browsers and search engines to allow users easily navigate through a large collection of results.

Table 3. Applications of GIC for faceted browsing

Films		Music		Actors	
Relation	**GIC (bits)**	**Relation**	**GIC (bits)**	**Relation**	**GIC (bits)**
dbo:starring	1,175,262.8	dbo:artist	2,166,102.0	dbo:starring	991,276.0
owl:sameAs[a]	937,578.8	dbo:producer	1,389,607.9	owl:sameAs	666,309.8
dbo:writer	643,112.6	dbo:associatedBand	1,230,862.1	dbo:occupation	461,851.4
dbo:producer	467,498.6	dbo:associatedMusicalArtist	1,230,862.1	dbo:director	418,187.0
dbo:director	404,065.1	owl:sameAs	998,533.4	dbo:writer	335,585.8
dc:subject	287,922.5	dbo:musicalBand	846,536.5	dc:subject	309,823.7
dbo:musicComposer	217,413.4	dbo:musicalArtist	846,536.5	dbo:producer	286,722.6
dbo:editing	150,631.0	foaf:homepage	626,596.5	rdf:type	210,529.5
dbo:cinematography	142,578.4	dbo:bandMember	603,403.8	dbo:spouse	143,141.6
rdf:type	123,186.2	dbo:writer	558,319.8	dbo:birthPlace	137,186.4

[a] 'owl:' is the prefix for the namespace http://www.w3.org/2002/07/owl#

4   Related Work

Since being introduced by Resnik [11], Information Content-based metrics have been widely exploited for semantic similarity measurement between concepts in lexical taxonomies [12-15]. These metrics demonstrated better performance than conventional edge-counting and feature-based methods. However, the notion of informativeness has been rarely used in Linked Data. Cheng et al. [16] embed the mutual information of resources into a random surfer model and present an approach for centrality-based summarization of entities in Linked Data.

A number of metrics have been developed to analyze the Web of Data from various aspects. Ell et al. [17] propose metrics to evaluate labels in Linked Data in terms of completeness, unambiguity, and multilinguality. Guéret et al. [18] apply network analysis techniques to measure and improve the robustness of Linked Data graph.

Another series of research has been carried out for entity ranking in Linked Data. Although traditional graph-based ranking methods, such as PageRank [19], SimRank [20], and HITS [21], can also be applied to the link structure of the Web of Data, its semantics, represented using various relation types, will not be taken into account. Numerous studies have attempted to extend the existing algorithms to consider link types in the rating methodology [22-27]. ObjectRank [22] adds link weights to PageRank for ranking in a directed labeled graph. Bamba and Mukherjea [23] extend PageRank to arrange the results of Semantic Web queries. TripleRank [26] is a generalization of the authority (importance) and connectivity (hub) scores of HITS algorithm for Linked Data, with applications in faceted browsing. These methods rate the importance of resources according to the link structure of the Web of Data graph while our approach is based on the amount of useful information that each individual resource and relation conveys. Nevertheless, further research needs to be done to assess the performance of different algorithms in terms of ranking functionality.

5   Conclusion and Future Work

In this paper, we addressed the problem of Linked Data quality assessment, which has a significant impact on the performance of Semantic Web applications. By relying on the principles of information theory, we have developed two metrics, namely Partitioned Information Content (PIC) and Generated Information Content (GIC), for measuring the informativeness of resources and relations in Linked Data. We analyzed the Web of Data using the metrics in several domains. As it was demonstrated in the experiments, assessing the quality of information contained in Linked Data offers the potential for its deployment in a variety of applications including information filtering, multi-faceted browsing, search engines, and entity ranking. Moreover, it can also be exploited for Linked Data clustering and visualization to highlight the most important information sources and relations. Currently, we are verifying and evaluating the proposed metrics in a range of application areas. We are also devising them in order to enable further analysis and comparison of resources in the Web of Data.

References

1. Lassila, O.: Resource Description Framework (RDF) model and syntax specification. World Wide Web Consortium, W3C (1999)
2. McGuinness, D.L., Harmelen, F.V.O.: Web Ontology Language: Overview. World Wide Web Consortium, W3C (2004)
3. Prud'hommeaux, E., Seaborne, A.: SPARQL Query Language for RDF. World Wide Web Consortium, W3C (2008)
4. Bizer, C., Heath, T., Berners-Lee, T.: Linked Data - The Story So Far. International Journal on Semantic Web and Information Systems 5(3), 1–22 (2009)
5. Auer, S., Bizer, C., Kobilarov, G., Lehmann, J., Cyganiak, R., Ives, Z.: DBpedia: A Nucleus for a Web of Open Data. Springer, Heidelberg (2007)
6. Gray, R.M.: Entropy and Information Theory. Springer, New York (2009)
7. Edwards, S.: Elements of information theory, 2nd edition. Information Processing & Management 44(1), 400–401 (2008)
8. Shannon, C.E.: A Mathematical Theory of Communication. Bell System Technical Journal 27, 379–423, 623–656 (1948)
9. Ross, S.M.: A First Course in Probability. Prentice Hall (2002)
10. Hassanzadeh, O., Consens, M.P.: Linked Movie Data Base (2009)
11. Resnik, P.: Using information content to evaluate semantic similarity in a taxonomy. In: Proceedings of the Fourteenth International Joint Conference on Artificial Intelligence, pp. 448–453 (1995)
12. Jiang, J.J., Conrath, D.W.: Semantic Similarity Based on Corpus Statistics and Lexical Taxonomy. In: Proceedings of the International Conference Research on Computational Linguistics (ROCLING X), Taiwan (1997)
13. Lin, D.: An information-theoretic definition of similarity. Morgan Kaufmann, San Francisco (1998)
14. Resnik, P.: Semantic Similarity in a Taxonomy: An Information-Based Measure and its Application to Problems of Ambiguity in Natural Language. Journal of Artificial Intelligence Research 11, 95–130 (1999)

15. Zili, Z., Yanna, W., Junzhong, G.: A New Model of Information Content for Semantic Similarity in WordNet (2008)
16. Cheng, G., Tran, T., Qu, Y.: RELIN: Relatedness and Informativeness-Based Centrality for Entity Summarization. Springer, Heidelberg (2011)
17. Ell, B., Vrandečić, D., Simperl, E.: Labels in the Web of Data. Springer, Heidelberg (2011)
18. Guéret, C., Groth, P., van Harmelen, F., Schlobach, S.: Finding the Achilles Heel of the Web of Data: Using Network Analysis for Link-Recommendation. Springer, Heidelberg (2010)
19. Brin, S., Page, L.: The anatomy of a large-scale hypertextual Web search engine. In: Proceedings of the Seventh International Conference on World Wide Web 7, Brisbane, Australia. Elsevier Science Publishers B. V. (1998)
20. Jeh, G., Widom, J.: SimRank: a measure of structural-context similarity. In: Proceedings of the Eighth ACM SIGKDD International Conference on Knowledge Discovery and Data Mining, Edmonton, Alberta, Canada. ACM (2002)
21. Kleinberg, J.M.: Authoritative sources in a hyperlinked environment. J. ACM 46(5), 604–632 (1999)
22. Balmin, A., Hristidis, V., Papakonstantinou, Y.: Objectrank: authority-based keyword search in databases. In: Proceedings of the Proceedings of the Thirtieth International Conference on Very Large Data Bases - Volume 30, Toronto, Canada. VLDB Endowment (2004)
23. Bamba, B., Mukherjea, S.: Utilizing Resource Importance for Ranking Semantic Web Query Results. Springer, Heidelberg (2005)
24. Ding, L., Pan, R., Finin, T., Joshi, A., Peng, Y., Kolari, P.: Finding and Ranking Knowledge on the Semantic Web. Springer, Heidelberg (2005)
25. Hogan, A., Harth, A., Decker, S.: ReConRank: A Scalable Ranking Method for Semantic Web Data with Context. In: Proceedings of Second International Workshop on Scalable Semantic Web Knowledge Base Systems (SSWS 2006), in conjunction with International Semantic Web Conference (ISWC 2006) (2006)
26. Franz, T., Schultz, A., Sizov, S., Staab, S.: TripleRank: Ranking Semantic Web Data by Tensor Decomposition. Springer, Heidelberg (2009)
27. Delbru, R., Toupikov, N., Catasta, M., Tummarello, G., Decker, S.: Hierarchical Link Analysis for Ranking Web Data. Springer, Heidelberg (2010)

Harnessing the Wisdom of Crowds
for Corpus Annotation through CAPTCHA*

Yini Cao and Xuan Zhou

DEKE Lab, Renmin University of China
Beijing, China 100872
{novgddhy,xzhou}@ruc.edu.cn

Abstract. CAPTCHA is a type of widely applied security measures to prevent abuses of online services. Our research is focused on designing a new type of CAPTCHA called AnnoCAPTCHA, which can harness the intellectual power of Web users to perform corpus annotation – an expensive step for computational linguistic analysis. We analyzed the applicability and the efficiency of AnnoCAPTCHA. We also conducted user study to demonstrate its accessibility.

Keywords: CAPTCHA, corpus annotation, human computation.

1   Introduction

The term "CAPTCHA" (Completely Automated Public Turing test to tell Computers and Humans Apart) was created by Luis von Ahn in 2000 [1], to name a simple Turing test which requires users to type letters from distorted images. Since its creation, CAPTCHA has become a widely applied security measure on the Web, to prevent large scale abuses of online services. According to Ahn's estimates [2], humans around the world type more than 100 million CAPTCHAs every day. Suppose it takes 10 seconds to enter a CAPTCHA. In accordance with the use of 100 million times a day, it will be huge labor resources of 150,000 hours. It is desirable if we can turn them into useful services. In recent years, there have been several proposals for better utilization the labor resources on CAPTCHA. A typical example is ReCAPTCHA, a special type of CAPTCHA that can be used to digitize printed material.

In this work, we explore the possibility of utilizing the labor resources on CAPTCHA to annotate textual corpus. We propose a new type of CAPTCHA, named AnnoCAPTCHA, which not only provides the functionality of CAPTCHA but also works a platform to perform corpus annotation. In each Turing test issued by Anno-CAPTCHA, a user is presented with a set of short textual items and is asked to select the right annotation for each item. Among the textual items, some items' annotations are already known, so they can be used to verify if the user is a human being. The other items' annotations are unknown, such that the user's selections become new

---

* This work is supported by the Basic Research fund in Renmin University of China from the central government-No.12XNLJ01.

Y. Ishikawa et al. (Eds.): APWeb 2013, LNCS 7808, pp. 638–645, 2013.

annotations to the target corpus. If AnnoCAPTCHA can be applied broadly, it can help us construct large scale annotated corpus and there after enhance the performance of NLP. In the paper, we conduct analytical study to evaluate its effectiveness and efficiency. In addition, we report a user study to demonstrate its accessibility to human users.

The rest of this paper is organized as follows. We review some related work in Section 2. In Section 3, we introduce our system and analyze its effectiveness and efficiency. In Section 4, we present a user study for evaluating the accessibility of AnnoCAPTCHA. Finally, Section 5 provides a conclusion.

2   Related Work

2.1   Human Computation

Our work can be regarded as an application of human computation to the field of NLP. The concept of human computation was first coined by Luis Von Ahn. Guided by its philosophy, Ahn invented ReCAPTCHA [2] and the well known ESP Game [3]. Game is the most common area to implement human computation. There have already been a number of online games [3,4] that can exploit players' intelligence to label the contents of images. CAPTCHA is another typical area to apply human computation. Except ReCAPTCHA, there are a number of proposals for generating values out of CAPTCHA. For instance, ASIRRA [5] has shown that CAPTCHA can be used for humanitarian purpose. TagCaptcha [6] uses CAPTCHA to perform image annotation on the Web. Besides, SolveMedia[1], captcha.tw[2] and other ads applications use CAPTCHAs to display ads. To the best of our knowledge, we are the first to use CAPTCHA to perform corpus annotation.

2.2   Corpus Annotation

Corpus annotation is an expensive preparing step in natural language processing and text mining. According to our study, the existing methods for large-scale corpus annotation usually fall in the following three categories:

1. Traditional method. The traditional method mainly counts on domain experts to perform manual annotation. The well known Brown Corpus was generated by this method. Using this method, large-scale annotation tasks will require considerable human efforts and thus are very costly.
2. WWW-based non-experts method. The main idea is to call up non-expert volunteers on the Web to annotate corpuses. The most famous example is the Open Mind Commonsense project[3]. The efficiency of this method is also questionable, as there is a lack of incentive for Web users to perform annotation for free.

---

[1] For more information, go to http://www.solvemedia.com/
[2] For more information, go to http://captcha.tw/
[3] For more information, go to http://openmind.media.mit.edu/

3. Game-based method. This method applies human computation. It exploits human intelligence in playing games to perform corpus annotation. For instance, by using competitive games, the Learner project [7] creates 1.3 million labels over a large image repository. However, the existing methods only deal with multimedia data such as images. There is no game designed to annotate textual corpus.

In summary, corpus annotation is a labor intensive task. There is a dearth of human computation mechanisms that exploit the free labor on the Web to perform textual corpus annotation. Our work in this paper aims to provide such a mechanism.

3   The AnnoCAPTCHA System

To illustrate the functionality of AnnoCAPTCHA, we use sentiment annotation as an example. Sentimentally annotated corpus is the base for sentiment analysis, a hot topic in the areas of NLP and Web mining. As the creation of such a corpus is very expensive, it is an ideal application scenario for AnnoCAPTCHA.

3.1   User Interface

The user interface of AnnoCAPTCHA is illustrated in Fig.1. In each test issued by AnnoCAPTCHA, a user is presented with four context-adjective phrase. The user is asked to judge their sentiment polarity (either *positive* or *negative*). Among the four phrase items, two's sentiment polarities are already known to the system and the other two are waiting to be annotated. As the four context-adjective phrases are randomly placed, the user cannot distinguish between the known ones and the unknown ones. The items with known sentiment polarities are used to verify if the user is a human. If the verification succeeds, i.e. the user's judgment on the known items is correct, we regard that the user's judgment on the unknown items is very likely to be correct too. Then, the user' judgment on each unknown item is recorded and counted as a vote for that item's sentiment polarity. As more and more users use AnnoCAPTCHA, the number of polarity votes for each unknown item increases. When the overall judgment on an unknown item reaches a certain statistical significance, such that we are confident about its real sentiment polarity (e.g. 9 out of 10 vote for positive polarity), we label the item's sentiment polarity and turn it into a known item. Then, the new known item becomes a member of the target corpus. It can subsequently be used to perform user verification.

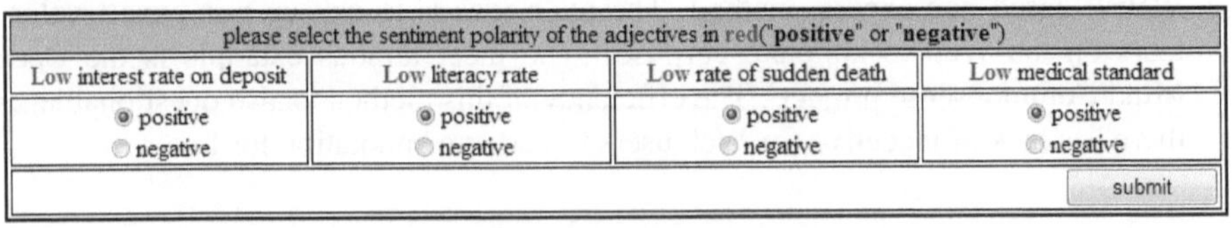

please select the sentiment polarity of the adjectives in red("**positive**" or "**negative**")			
Low interest rate on deposit	Low literacy rate	Low rate of sudden death	Low medical standard
◉ positive ○ negative	◉ positive ○ negative	◉ positive ○ negative	◉ positive ○ negative
			submit

Fig. 1. The User interface

3.2   Effectiveness as CAPTCHA

AnnoCAPTCHA can provide the functionality of CAPTCHA. It performs human verification using the two context-adjective phrases whose sentiment polarities are already known. Usually, a human's judgment on sentiment polarity is much more accurate than a machine's. Suppose a human's accuracy can achieve at least 95%. Then, a human user's chance of passing the test is around 90% ($=0.95^2$). The test is much harder to a computer. Suppose a computer can only perform random guess. Its chance of passing the test is thus only about 25% ($=0.5^2$). Even though a user may fail the test occasionally, the user can try more tests. Eventually, the verification will succeed. In contrast, to perform attacks, such as email spamming, an adversary has to rely on a machine to obtain a large number of accounts in a short period of time. Thus, it has to be tested by AnnoCAPTCHA repeatedly. A machine may pass some tests by chance. According to the theory of probability, as the number of tests increases, the machine's success rate will quickly approach its expected accuracy. As a consequence, a machine can be easily recognized by a small number of tests, before it can perform any effective attack. Upon detection, the website can immediately cut off the connection with the attacker to prevent it from taking further action.

3.3   Efficiency of Annotation Generation

In the following, we analyze how fast sentiment annotations can be generated by AnnoCAPTCHA.

The Model.

- We assume that each CAPTCHA contains $k+2$ items, in which 2 items' annotations are known and the other k items' annotations are unknown.
- A user of AnnoCAPTCHA is either a human or a machine. Let $s\%$ be the accuracy of a human's judgment. Let $t\%$ be the accuracy of a machine's judgment. Obviously, $s\%$ is much greater than $t\%$.
- As mentioned previously, once a machine is detected by AnnoCAPTCHA, its connection will be cut off. This effectively restricts the number of annotations by machines. Let the proportion of CAPTCHA tests finished by humans be $m\%$ and that by machines/attackers be $1-m\%$. Thus, $m\%$ should be far greater than $1-m\%$.
- After a user passes the verification based on the 2 known items, the system will record user's votes on the k unknown items in a database. The vote record of each unknown item is in the form $i : j$, where i represents the number of users voting for positive polarity and j represents the number of users voting for negative polarity. When the difference between i and j reaches a certain statistical significance, that is, when i (or j) is significantly larger than j (or i), we turn the unknown item into a known item. We use error probability to measure the significance level, which is calculated as:

$$P_{ij} = \begin{cases} C_{i+j}^i P_0^i (1 - P_0)^j, & i > j \\ C_{i+j}^i P_0^j (1 - P_0)^i, & i < j \end{cases} \tag{1}$$

where P_0 denotes the probability that a vote is incorrect, that is

$$P_0 = t\% \times m\% - m\% \times s\% - t\% + 1$$

In principle, we regard the sentiment polarity of an unknown item positive if $i > j$ and negative if $i < j$. The error probability P_{ij} measures the probability that we make a wrong judgment. AnnoCAPTCHA uses a predefined threshold of the error probability, e.g. 1%, to represent the maximum error rate it can accept. Once the P_{ij} of an unknown item falls below the threshold, AnnoCAPTCHA finalizes its sentiment polarity and turns it into a known annotation.

Analysis of Efficiency

When an item receives i positive votes and j negative votes, if the error probability P_{ij} falls below the threshold, the annotation process for this item will terminate. Thus, $i+j$ measures the cost of the annotation, i.e. the number of votes required to determine the sentiment polarity of a context-adjective phrase. The higher the number, the longer the duration we need to generate an annotation. We are interested in the expected annotation cost, which we use to depict the efficiency of AnnoCAPTCHA.

Let Pt denote the threshold of error probability. Given an arbitrary pair of i and j, let Q_{ij} denote the probability that the process does not terminate before it receives i positive votes and j negative votes. It can be inferred that the value of Q_{ij} depends on that of $Q_{(i-1)j}$ and $Q_{i(j-1)}$, because the vote record $i : j$ must be evolved from either $i-1 : j$ or $i : j-1$. Thus, Q_{ij} can be calculated recursively as follows:

$$Q_{ij} = \begin{cases} Q_{(i-1)j} \times P_0, & i > 0, j = 0, P_{(i-1)j} > Pt \\ Q_{i(j-1)} \times (1 - P_0), & i = 0, j > 0, P_{i(j-1)} > Pt \\ Q_{i(j-1)} \times (1 - P_0) + Q_{(i-1)j} \times P_0, & i > 0, j > 0, P_{i(j-1)}, P_{(i-1)j} > Pt \\ Q_{(i-1)j} \times P_0, & i > 0, j > 0, P_{i(j-1)} \leq Pt, P_{(i-1)j} > Pt \\ Q_{i(j-1)} \times (1 - P_0), & i > 0, j > 0, P_{(i-1)j} \leq Pt, P_{i(j-1)} > Pt \\ 0, & otherwise \end{cases} \qquad (2)$$

Let O_{ij} denote the probability that the process terminates when it receives i positive votes and j negative votes. Then O_{ij} can be computed by:

$$O_{ij} = \begin{cases} Q_{ij}, & P_{ij} \leq Pt \\ 0, & otherwise \end{cases} \qquad (3)$$

Finally, the expected annotation cost can be calculated as:

$$\text{Cost} = \Sigma_{i,j \geq 0, i \neq j}(i + j)(O_{ij} + O_{ji}) \qquad (4)$$

We did not manage to integrate Equations (1), (2), (3) and (4) into a single formula to measure the expected annotation cost. However, it does not prevent us from evaluating the relationship between the expect cost and the various factors in our model. As suggested by our analysis, the main influence factors of the expected annotation cost include the accuracies of the annotations, i.e. $s\%$ and $t\%$, and the proportion of attackers' votes, i.e. $1-m\%$.

We created a Java program to compute the expect annotation cost based on Equations (1)~(4). To evaluate the influence of human's accuracy ($s\%$) on the expected cost, we set $m\%$ to 80% and $t\%$ to 50% (machines only make random guess), and vary $s\%$ from 55% to 100%. Then, the relationship between the annotation cost and human's accuracy ($s\%$) is plotted in Fig. 2. Obviously, the higher the human accuracy, the smaller of the expected cost. As we can see, to guarantee the efficiency of the annotation process, we need to ensure that the accuracy of humans be significantly higher than that of computers. For example, to ensure that the average cost is below 10 votes, the human accuracy must be above 75%. Fortunately, this condition can be satisfied for most corpus annotation tasks.

To evaluate the influence of attacks, we vary m % from 50% to 100%, and fix $s\%$ to 90% and $t\%$ to 50%, and. Then, the relationship between the annotation cost and the proportion of human's votes (m %) is plotted in Fig.3. As we can see, the higher the proportion of human's votes, the smaller the average cost. The proportion of human's votes depends on how effectively a website detects and blocks attackers. When the proportion is more than 70%, the average cost can be controlled below 5 votes.

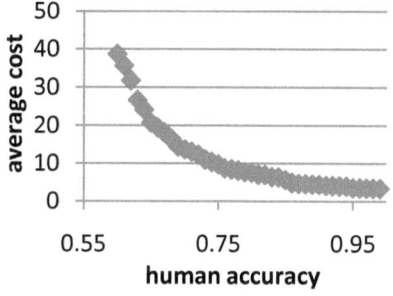

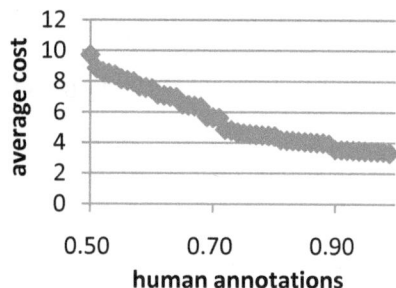

Fig. 2. Annotation cost vs human accuracy (s %)

Fig. 3. Annotation cost vs proportion of human annotations (m %)

4   Evaluation of Accessibility

We carried out a user study to evaluate AnnoCAPTCHA's accessibility. In the study, we asked 9 volunteers to do the same 10 CAPTCHAs. The volunteers are all university students. Every CAPTCHA has 4 items, which are all Chinese sentiment phrases selected from the SemEval-2010 test set [4]. Each phrase contains an adjective and a context. We recorded the users' votes and the time each user took to finish each CAPTCHA. In addition, we let the machine to annotate the same 10 CAPTCHAs by random guess, to study the impact of attacks to the annotation efficiency.

---

[4] For more information, go to
http://semeval2.fbk.eu/semeval2.php?location=tasks#T.

4.1   User Response Time

User response time measures the time a human user needs to complete a CAPTCHA test. Fig.4 shows the response times of the 9 volunteers on each CAPTCHA. As we can see, there is a learning curve for this new type of CAPTCHA. When a volunteer is presented with the first CAPTCHA, she / he needs a certain time to understand its requirements and get used to the test environment. This time usually takes around 25 seconds, and can occasionally go up to 1 minute. After the second CAPTCHA, the volunteers become familiar with the exercises. Most of the response times fall in the interval between 5 seconds to 25 seconds. The average response time is below 10 seconds. Overall, no volunteer encountered difficulty when doing the CAPTCHAs. The accessibility of AnnoCAPTCHA appears acceptable to most users.

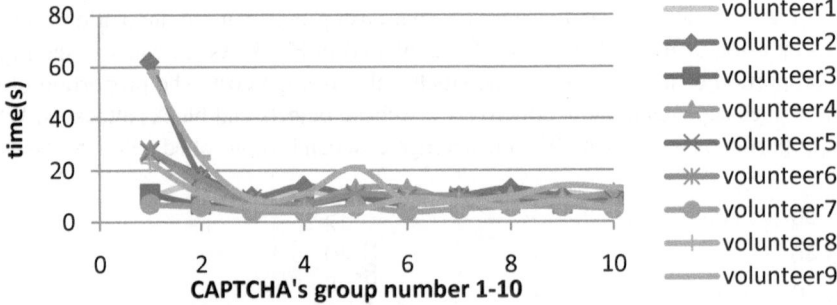

Fig. 4. User Respond Time

4.2   User Accuracy

The accuracy of the 9 volunteers is shown in Fig.5. As we can see, all volunteers can provide correct answers to more than 90% of the test questions. The average accuracy is 96.67%. This also indicates that AnnoCAPTCHA is not a difficult exercise to literate users.

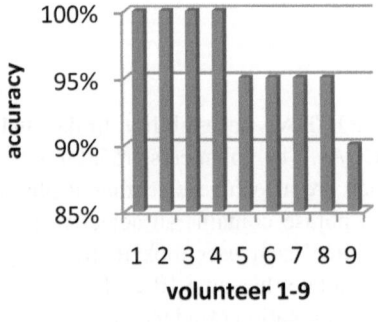

Fig. 5. User Accuracy

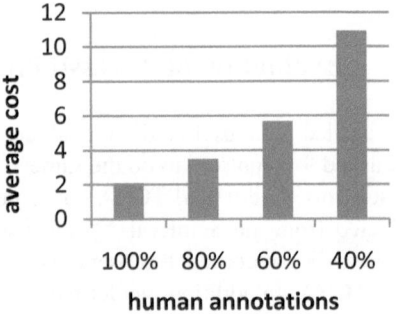

Fig. 6. Annotation Cost vs Proportion of Human Annotations

4.3   Efficiency

Using the annotations collected from the user study, we simulated the process of annotation generation by AnnoCAPTCHA. We increased the proportions of random annotations made by machines to evaluate the impact of attacks on the annotation efficiency. We assume that human's accuracy is 96.67%, the exact value we measured in the user study. And we set the error probability threshold to 1%. The average cost for annotating each context-adjective phrase (i.e. the average number of users we need to annotate each phrase) is shown in Fig. 6.   As we can see, with the accuracy of 96.67%, the annotation is very efficient. Without the disturbance of machines, the annotation cost is only about 2 votes. As more and more test results of machines are added, the annotation cost goes up. However, the overall cost still appears acceptable.

5   Conclusions

The main contribution of this paper is to provide a new type of CAPTCHA that can utilize crowd intelligence to do corpus annotation. We used the sentiment corpus annotation as example and discussed the effectiveness and efficiency of our system. We clarified the relationship between the efficiency of the system and the accuracy of human annotation as well as the influence of attacks. Finally we conducted a user study to prove our system is easy to access and efficient in annotation generation.

References

1. von Ahn, L., Blum, M., Hopper, N.J., Langford, J.: CAPTCHA: Telling Humans and Computers Apart. In: Biham, E. (ed.) EUROCRYPT 2003. LNCS, vol. 2656, Springer, Heidelberg (2003)
2. von Ahn, L., Maurer, B., McMillen, C., Abraham, D., Blum, M.: reCAPTCHA:human-Based Character Recognition via Web Security Measures,
   http://www.sciencemag.org/content/321/5895/1465.full
3. von Ahn, L., Dabbish, L.: Labeling Images with a Computer Game,
   http://www.cs.cmu.edu/~biglou/ESP.pdf
4. von Ahn, L., Liu, R., Blum, M.: Peekaboom: A Game for locating Objects in Images,
   http://www.cs.cmu.edu/~biglou/Peekaboom.pdf
5. Elson, J., Douceur, J.R., Howell, J., Saul, J.: Asirra: A CAPTCHA that Exploits Interest-Aligned Manual Image Categorization. In: Proceedings of the 14th ACM Conference on Computer and Communications Security. Association for Computing Machinery, New York (2007)
6. Morrison, D., Marchand-Maillet, S., Bruno, E.: Tagcaptcha: annotating images withcaptchas. In: Proceedings of the ACM SIGKDD Workshop on Human Computation, HCOMP 2009, pp. 44–45. ACM, New York (2009)
7. Chklovski, T.: Improving the design of intelligent acquisition interfaces for collecting world knowledge from web contributions. In: Proceedings of the 3rd International Conference on Knowledge Capture, pp. 35–42 (2005)

A Framework for OLAP in Column-Store Database: One-Pass Join and Pushing the Materialization to the End

Yuean Zhu[1,2], Yansong Zhang[1,2,3], Xuan Zhou[1,2], and Shan Wang[1,2]

[1] DEKE Lab, Renmin University of China, Beijing 100872, China
[2] School of Information, Renmin University of China, Beijing 100872, China
[3] National Survey Research Center at Renmin University of China, Beijing 100872, China
{iwillgoon,swang}@ruc.edu.cn

Abstract. In data warehouse modeled with the star schema, data are usually retrieved by performing a join operation between the fact table and dimension table(s) followed by a selection and project operation, while join operator is the most expensive operator in RDBMS. In column-store database, there are two ways to do join. The first way is *early materialization* join (EM join); the other way is *late materialization* join (LM join). In EM join, the columns involved in the query are glued together firstly, then the glued rows are sent to join operator. Whereas, in LM join, only the attributes participated in the join operator are accessed. The problem that access to inner table is out-of-order can't be ignored for LM join. Otherwise, the naïve LM join is usually slower than EM join [9]. Since the *late materialization* is good for memory bandwidth and CPU efficiency, the LM join attracts more attention in academic research community. The state-of-art LM joins in column-store such as radix-cluster hash join [8] in MonetDB, *invisible join* [10] in C-Store all try to avoid accessing table randomly. In this paper, we devised a framework for OLAP called CDDTA-MMDB where a new join algorithm called CDDTA-LWMJoin (we contract it to LWMJoin in the following) is introduced. The LWMJoin is on the basis of our prior work: CDDTA-Join [7]. We equip the CDDTA-Join with light-weight *materialization* (LWM) which is designed to cut down the memory access and reduce production of intermediate data structure. Experiments show that CDDTA-MMDB is efficient and can be 2x faster than MonetDB and 4x faster than *invisible join* in the context of data warehouse modeled with star schema.

Keywords: OLAP, in-memory column-store database, join, *materialization*.

1 Introduction

Data warehousing (DW) and On-line analytical processing (OLAP) consist of essential elements of decision support, increasingly becoming a focus of the database industry. Deviating from the queries to transactional databases, the nature of the queries to DWs is more complex, longer lasting and read-intensive, but less attributes to touch. Since column-store database only retrieves attributes required in the query, it can optimize cache performance. Thus columnar storage is more suitable for DW in this sense, and now the Microsoft also adds columnar storage in SQL server 2012 to

Y. Ishikawa et al. (Eds.): APWeb 2013, LNCS 7808, pp. 646–653, 2013.

improve performance on data warehousing queries [5]. With the increment of RAM density and reduction of its price there are a number of in-memory column-oriented database systems such as MonetDB [1, 2] and C-Store [3] emerging for better performance. These systems offer order-of-magnitude gains against traditional, row-oriented databases on certain workloads, particularly on read-intensive analytical processing workloads, such as those encountered in DWs. In this paper, we propose a framework for OLAP called CDDTA-MMDB (Columnar Direct Dimensional Tuple Access-Main Memory Database) to efficiently process the query on DWs modeled with star scheme. This work is on the basis of our prior work to develop one-pass join algorithm [7] which is variant of invisible join [10]. The following is our focus innovations:

- **The Storage Model.** In CDDTA-MMDB we employ the decomposed storage model (DSM) [4] which vertically partitions the tables and maintain a collection of triplets *<surrogate, key, value>* tables. In addition to make use of CPU cache line more efficiently, this representation is one of crucial measures for our goal to gain light-weight *materialization* and LWMJoin algorithm.
- *Materialization* **Strategy in Column-Store Database.** We propose a new materialization strategy called light-weight *materialization* (LWM) which adopts notion of *late materialization* (LM) to push the materialization as late as possible. Compared to the LM, LWM tries to avoid producing intermediate data structure such as bitmap and reduce random memory access during the whole query plan. Experiments show that CDDTA-Join [7] equipping with LWM (CDDTA-LWMJoin, in the following we contract it to LWMJoin) outperforms the state-of-art join algorithm such as radix-cluster join [8], invisible join [10] in the context of DWs modeled with star schema. To the best of our knowledge, this is first time to propose such materialization for column-store database.

The rest of this paper is organized as follows: In section 2, prior work on columnar storage, *materialization* and join algorithms is described. Section 3 gives the light-weight *materialization* techniques. In Section 4, system overview is presented. Section 5 shows representative experimental results. Finally, conclusion is drawn in section 6.

2   Related Works

Storage. The decomposition storage Model (DSM) [4] reduces each n-attribute relation to n separate relations. DSM organizes values from same attribute close in physical. Supposed the record reconstruction cost is low, cache performance of main-memory database systems can be improved by DSM. As mentioned to the detail of representation of DSM, MonetDB proposed the Binary Units (or BUNs) [2] to construct the decomposed table in the form of *<OID, value >* pairs.

***Materialization* and Join Algorithm in Column-Store.** Column-store modifies the physical data organization in database. In logical, the applications treat column-store the same as row-store. At some point in a query plan, a column-oriented DBMS must glue multiple attributes from the same logical tuple into a physical tuple. The strategies of materializing are divided into two categories: *early materialization* (EM) and *late materialization* (LM). A recent work on *materialization* in column-store is in [14]. It used a dynamic and cache-friendly data structure, called *cracker maps* which provide direct mapping between the pairs of attributes in the query, to reduce the random data access. In [6], trade-offs between different strategies are systematically explored and advices are given for choosing a strategy for a particular query.

The naïve join algorithm using LM strategy (naïve LM join) in column-store database is about 2x slower than the join algorithm using EM strategy (EM join) [9]. The main reason is that in naïve LM join the access to inner table is out-of-order, causing the high cache miss. The state-of-art LM join algorithms in column store such as radix hash join [8], MCJoin[13] and invisible join [10] all try to avoid the random memory access to inner table. Radix-clustered hash join [8] partitions the input relation into H clusters. By controlling the number of clustering bits each pass, the H clusters generated can be kept under the count of Translation Lookaside Buffer (TLB) entries and cache lines available on that CPU. The random access only happens in storage block. Invisible join [10] uses bitmap to discard tuples that don't satisfy the prediction and can reduce the random memory accesses. LWMJoin tries to avoid producing the intermediate data structure and further reduces memory access compared to invisible join [10] and CDDTA-Join [7]. Experiments show that LWMJoin runs about 4x faster than invisible join and 2x faster than radix-cluster join on average.

3   Light-Weight *Materialization*

Column-store is physical modification to database. In terms of logical and view level it is identical to row-store. The applications involved with database, row-oriented or column-oriented, treat the interface as row-oriented. So a column-oriented DBMS must glue multiple attributes, at some time, from the same logical tuple into a physical tuple, which is called *materialization*. There are two ways to materialize: *early materialization* (EM) and *late materialization* (LM). At each point when a column is needed, it is read into CPU and added to intermediate tuple representation. This policy to reconstruct the tuple is called *early materialization*. It is not surprising that EM is not always the best strategy for a column-store [9], because some CPU works are useless. Consider this scenario: a column-store DBMS place all columns of a table in separate files and sort them in the same order; a query has two selection operators σ_1 and σ_2 over two columns (R.a, R.b), a GROUP BY operator on R.a, where σ_1 has higher selective than σ_2. Under the EM policy, the operations are as follows: read in a block of R.a and a block of R.b; glue them together to form the rows and apply σ_1 and σ_2 on the rows in turn. Then, send the tuples for aggregation. A more efficient way for this process is that: keep two intermediate position lists; scan the block of R.a and R.b, meanwhile apply σ_1 and σ_2 on them respectively; use the lists to record whether attributes in the corresponding position in column satisfy the prediction; then use position-wise AND to intersect the two position lists. According to intersected result, the block of *R*.a is re-accessed for aggregation. The latter strategy is called *late materialization*. Light-weight *materialization* (LWM) adopts notion of LM to push the materialization as far as possible in query plan, but try to reduce the memory access and avoid producing intermediate data structure. Instead of re-accessing the block of *R*.a to get record value for GROUP BY, LWM just lets the *key* of triplet produced by scanning (*R*.a, *R*.b) go through whole query plan. Furthermore, LWM doesn't need to be assisted by the intermediate data structure except triplets (in some case such as the attributes that are not used any more but just serve as filters, the triplets can be replaced by bitmap to save space). Considering the above scenario again, we don't need to keep two intermediate position lists and use position-wise AND to intersect the two positions to produce intersected result, but just utilize *key* in triplet and let it go through query plan. So, the LWM can accelerate the system speed, and it greatly facilitates aggregation operation.

4   System Design

The sub-queries on fact table and dimension tables are handled separately in CDDTA-MMDB. When a query is received, it is broken off. Then, sub-queries on dimension table(s) are rewritten in order to produce triplets, while the sub-query on fact table is parsed and information is stored for later use. The *Query Parser* is responsible for those tasks. Based on the triplets and information from parsing sub-query on fact table, the result set can be obtained by scanning fact table one pass. Figure 1 illustrates the overview of the system.

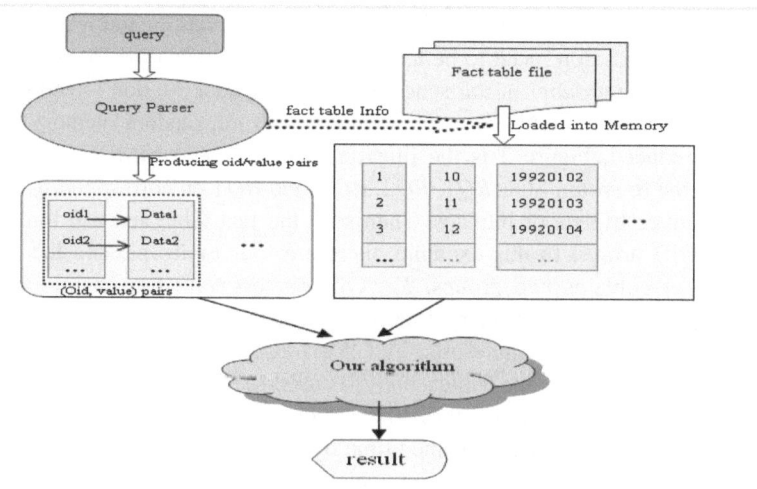

Fig. 1. System overview

4.1   Details of CDDTA-LWMJoin Algorithm

CDDTA-LWMJoin is derived from invisible join [10]. Invisible join is also divided into three phases. First, apply the prediction on the dimension table and use key of rows that pass the prediction to build hash table. Second, scan the fact table; meanwhile fetch the foreign key to probe the hash table. The bitmaps on dimension tables are built up in this step. Then intersect the bitmaps to product final bitmap. At last, according to final bitmap scan the fact table again to construct result relation. If the i-th value is set to *true* in the final bitmap, the i-th record in the fact table is retrieved. From above analysis, the time t used by invisible join can be expressed as $t_0 + 2t_1 + t_2$, where t_0 is spent on producing hash table; t_1 on scanning the fact table; t_2 on producing the final bitmap. Since size of the final bitmap is equal to size of fact table, we can assume that t_1 is equal t_2. So, the total time t spent by invisible join is $t_0 + 3 t_1$.

Compared with invisible join, LWMJoin algorithm just utilizes the *key* information in triplet which was produced in first phase and thus avoids scanning fact table twice. The following is the algorithm procedure. First, decomposed queries involved with dimension table are rewritten; then we get the result sets from dimension tables. The last step of this phase is to produce the *<surrogate, key, value>* triplets. For those subqueries such as *RQ*1, bitmap is enough to represent them; for the other case, the *<surrogate, key, value>* triplet is to be produced. This representation forms the base of LWMJoin. Second, Fact table is scanned sequentially; foreign keys are used to probe the *key* vectors producing in first phase, deciding whether the row satisfies the prediction. If keys equal to zero is greater than one, this row should be discarded. Otherwise, a physical tuple is constructed. Now, the operators such as GROUP BY can be applied on rows passing qualification.The benefits brought by LWM are obvious: dynamically allocating the memory to build the temporary position lists is no need; it accelerate the aggregation especially when the fact table is big and result sets are big because the records don' need to be fetched again for aggregation. In general, the *key* of triplet serves as *grouper* in star schema and in such case we don't need to re-access attributes in dimension table for aggregation. Therefore, random memory access can be further reduced. Figure 2 is the illustration of CDDTA-MMDB handling Q4.1. First apply the rewritten SQL *RQ*1, *RQ*2, *RQ*3, and *RQ*4 on corresponding dimension table to produce triplets or bitmaps. Then scan the fact table and use foreign key to probe *key*/OID arrays. In this example, there are four tuples passing the prediction, and result set contains two groups. *Key*/OID vectors with broken line are used in GROUP BY (if it is end of query plan, the program will fetch value by *key* to construct tuple directly). The time t used by LWMJoin is expressed as: $t_0 + t_1$, where the time t_0 is used to produce the triplets and t_1 is used to scan the fact table. So, the time spent by *invisible join* is almost three times more than LWMJoin. The following experiments will show that the consumed time by invisible join is more than three times compared with time used by LWMJoin, which accords with above analysis.

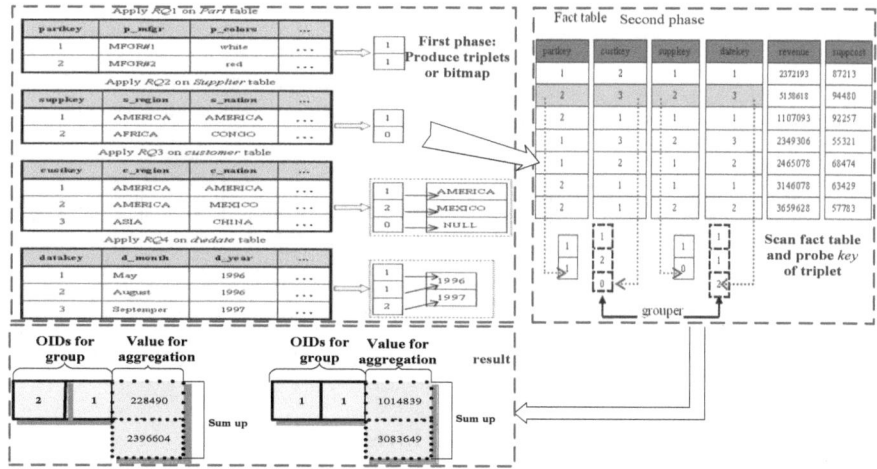

Fig. 2. Demonstration of CDDTA-LWMJoin

5   Experiments

In this section, we will evaluate CDDTA-MMDB using SSBM. We generated 100G data set of SSBM using data generator with *SF* = 100. Only attributes of fact table participating in experiment are loaded into memory, and all dimension tables are resident in memory. The experiment environment is as following: 48GB memory, two Intel® xeon® processors (each runs at 2.4GHz and has six superscalar processor cores that support simultaneous multi-threading with two hardware threads per core), 1TB disk space and 64-bit Ubuntu 10.10 OS (linux kernel version 3.0.0-19, gcc version 4.6.1). We focus the benefits LWM brings, the performance of CDDTA-MMDB and the scalability of the system.

5.1   Aggregation Time Comparison

Query 1.x doesn't have GROUP BY, so we omit them in the experiment. We simulate state-of-art LM join algorithms such as in [7, 10] that after join phase the data is retrieved for aggregation (LMJoin). Then we compare aggregation time used by LMJoin with that used by LWMJoin. To our surprise, as is shown in Figure 3 time consumed by aggregation in LMJoin even exceeds the time by join phase on Q3.1. On Q3.1 the time took up by aggregation is up to 61% of total time, followed by 35% on Q4.1, 21% on Q2.1 and 17% in Q4.1. Whereas, in LWMJoin the time took by aggregation is reduced to 28%, 13%, 8% and 5% respectively. The difference of aggregation time between LMJoin and LWMJoin lies in the random memory access. After join phase, LMJoin fetches the tuple for aggregation. This out-of-order memory access will incur cache miss. So the aggregation time in LMJoin is much more than that in LWMJoin, especially when the selectivity is high.

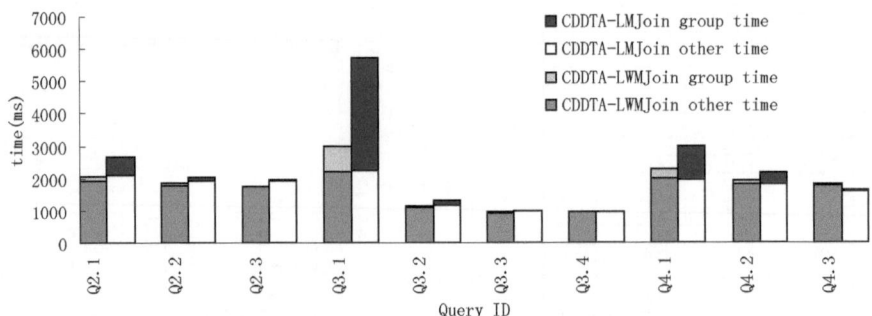

Fig. 3. Comparison of aggregation time by CDDTA-LWMJoin and by CDDTA-LMJoin

5.2   CDDTA-MMDB vs. MonetDB and Invisible Join

CDDTA-MMDB and *invisible join* implemented by us can automatically detect parameters of the host CPU, and then horizontally partitions the fact table evenly according to CPU cores. Figure 4 gives the performance comparison of CDDTA-MMDB, MonetDB (v11.7.9) and *invisible join*. We implement *invisible join* using the C++ *Boost* Library 1.48. All systems run on the thirteen queries several times such that the experiment is done on "hot" data set. CDDTA-MMDB has best performance among them. The time consumed by *invisible join* is almost four times more than the time by CDDTA-MMDB, which accords with the above analysis. Our system has worse performance than MonetDB on Q.2.2 and Q2.3, but it outperforms MonetDB greatly on Q1.x and Q4.x. On Q3.x, our system is slightly better than MonetDB. Overall, the total time to execute all queries by our system is about 22532ms, and total time by MonetDB is about 50311ms.

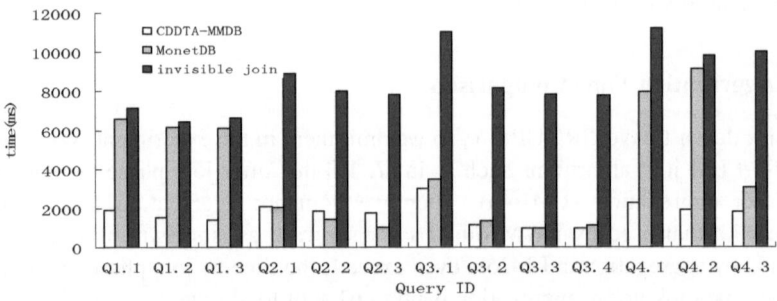

Fig. 4. Time used by CDDTA-MMDB vs. MonetDB vs. *invisible join*

6   Conclusion

The main contribution of this paper is that we propose a framework which introduces LWMJoin to efficiently deal with the queries on DW modeled with star scheme.

LWMJoin is the combination of CDDTA-Join [7] and LWM. Compared to the state-of-art *materialization* in columnar database, LWM reduces production of intermediate data structure and can cut down memory access greatly. Experiments show that CDDTA-MMDB has good scalability and can be 2x faster than MonetDB and 4x faster than *invisible join*. In future, we will add triplet representation and LWM strategy into open source columnar database such as MonetDB.

Acknowledgement. This work is supported by the Important National Science & Technology Specific Projects of China ("HGJ" Projects, Grant No.2010ZX01042-001-002), the National Natural Science Foundation of China (Grant No.61070054).

References

1. Boncz, P.A., et al.: MonetDB/X100: Hyper-pipelining query execution. In: CIDR, pp. 225–237 (2005)
2. Boncz, P.A., et al.: MIL primitives for querying a fragmented world. VLDB Journal 8(2), 101–119 (1999)
3. Stonebraker, M., et al.: C-Store: A Column-Oriented DBMS. In: VLDB, pp. 553–564 (2005)
4. Copeland, G.P., et al.: a decomposition storage model. In: ACM SIGMOD, pp. 268–279 (1985)
5. Larson, P.-A., et al.: Columnar Storage in SQL Server 2012. In: ICDE, pp. 15–20 (2012)
6. Abadi, D.J., et al.: Integrating compression and execution in column-oriented database systems. In: ACM SIGMOD, pp. 671–682 (2006)
7. Yan-Song, Z., et al.: One-size-fits-all OLAP technique for big data analysis. Chinese Journal of Computers 34(10), 1936–1946 (2011)
8. Boncz, P.A., et al.: Breaking the Memory Wall in MonetDB. Communications of the ACM 51(21), 77–85 (2008)
9. Abadi, D.J., et al.: Materialization strategies in a column-oriented DBMS. In: ICDE, pp. 466–475 (2007)
10. Abadi, D.J., et al.: Column-Stores vs. Row-Stores: How Different Are They Really? In: ACM SIGMOD, pp. 967–980 (2008)
11. O'Nei, P.E., et al.: The Star Schema Benchmark (SSB), http://www.cs.umb.edu/~poneil/StarSchemaB.PDF
12. O'Neil, P.E., et al.: Adjoined Dimension Column Index (ADC Index) to Improve Star Schema Query Performance. In: ICDE, pp. 1409–1411 (2008)
13. Begley, S., et al.: MCJoin: A Memory-Constrained Join for Column-Store Main-Memory Databases. In: ACM SIGMOD, pp. 121–132 (2012)
14. Idreos, S., et al.: Self-organizing Tuple Reconstruction in Column-stores. In: ACM SIGMOD, pp. 1–12 (2009)

A Self-healing Framework for QoS-Aware Web Service Composition via Case-Based Reasoning

Guoqiang Li[1,2], Lejian Liao[1], Dandan Song[1,*], Jingang Wang[1], Fuzhen Sun[1], and Guangcheng Liang[1]

[1] Beijing Engineering Research Centre of High Volume Language Information Processing & Cloud Computing Applications, Beijing Key Laboratory of Intelligent Information Technology, School of Computer Science, Beijing Institute of Technology, Beijing, China
[2] School of informatics, Linyi University, Linyi, China
{lgqsj,sdd,liaolj,bitwjg,10907023,195812}@bit.edu.cn

Abstract. The self-healing ability is very important for service-oriented systems in a highly dynamic environment. Case-based reasoning is adopted to cope with the component services failing to meet with the functional and nonfunctional requirements in this paper. We take previous failure instances as cases which are stored in a case base. When a new fault occurs, its symptoms are extracted and matched against the case base to look for the most similar case. A case representation and a similarity function are proposed. Meanwhile, a novel reuse approach is designed to find the solutions satisfying the symptoms. Moreover we present a self-healing framework and conduct experiments with real QoS dataset. Experimental results show that case-based reasoning improves the self-healing ability encouragingly.

Keywords: Quality of Service, case-based reasoning, self-healing framework, web service.

1 Introduction

Service-oriented architecture has become a mainstream technology for business process integration. However, in complex network environments, composite services may encounter unexpected risks, e.g. network breaking down, or a service unavailability. Therefore, service-oriented systems should be equipped with self-healing ability in order to run without interruption.

The main idea of our approach is that each failure instance can be taken as a failure case. We engage in using previous experiences to understand and solve new problems, which is the core ideology of the case-based reasoning (CBR). Its advantages include: (1) proposing solutions to problems quickly, (2) allowing us to make assumptions and predictions based on the past, (3) remembering previous experiences to avoid repeating past mistakes. (4) adapting to changes

---

* Corresponding author.

Y. Ishikawa et al. (Eds.): APWeb 2013, LNCS 7808, pp. 654–661, 2013.

of the environment [1]. To the best of our knowledge, our work is the first one to use CBR in the service adaptive domain.

The remainder of this paper is structed as follows. In Section 2 we present the related work. The self-healing framework is described in Section 3. Section 4 sets forth a case representation, a retrieval algorithm and a reuse approach in detail. In Section 5 we present experiments of our work. At last, in Section 6 we discuss conclusion and future work.

2   Related Work

Although WSBPEL[1] specifies business processes behavior as a standard, its fault-handling mechanism is limited as it is difficult for users to consider the dynamic information of services and their running environment in advance. Many works extended BPEL engine to enhance their fault-handling capabilities. A multi-layered control loop is proposed in [2], which focuses on the framework structure. VieDAME [3] monitors BPEL process according to Quality of Service (QoS) and replaces existing partner services based on various strategies. In [4], AOP (aspect-oriented programming) is used to extend the engine ActiveBPEL by using ECA (event-condition-action) rule paradigm. In [5], Web service Recovery Language is defined inspired by the ECA rule paradigm with emphasis on the process recovery. Similarly, in [6], ECA rules are designed to change the component's states according to the fault events. In addition, vital and non-vital components are distinguished to express the importance of activities in the workflow. In Dynamo [7] and DISC [8], JBOSS rule and EC (event calculus) model are used respectively. But in the above mentioned works, it is difficult for designers to design the complete set of rules and event models before the process execution. Our approach can automatically generate the failure cases on the basis of the fault information. With this advantage, our approach is easier to be accepted by the users.

When QoS constraints are violated during the execution, the problem of selecting supplementary service is formalized as a mixed integer linear programming problem in [9], where negotiation techniques are exploited to find a feasible solution. In [10], a staged approach generates multiple workflows for one template (abstract workflow), and some workflows are selected as an IP optimization problem by considering QoS parameters. In [11], a reconfigurable middleware is proposed, covering the whole cycle of adaptation management including monitoring, QoS analysis and substitution-based reconfiguration. In [12], failures are classified into several categories from the perspective of a service composition. Then process reorganization, substitution and retrial/data mediation are proposed to react to the failures. In [13], a region-based reconfiguration algorithm is designed to minimize the reconfiguration cost. Supplementary services are selected as a backup source for recovery by using the linear programming. In [14], DSOL re-planning technique is leveraged by Q-DSOL to optimize the orchestration supporting QoS in presence of faults. Different from previous works, our

---

[1] http://docs.oasis-open.org/wsbpel/2.0/wsbpel-v2.0.html

chief consideration is on the selection of adaptive action according to the typical fault information considering QoS constraints. We carry on the candidate chain search based on the similarity using previous experiences as much as possible.

3   Self-healing Framework

There are three parts in our framework. The first part is the business process which achieves user's requirements. The framework is shown in Figure 1. The second part extractor gets the fault information (FIN), services' functional information (FI) and non-functional information (NFI) which are passed to the third part case-based reasoner. After reasoning, a *case with solution* (CWS) is returned to the business process and the satisfied solution keeps the process running forward rightly. A *case without solution* (CWOS) is viewed as a new fault. We assume that fault information has been caught using some method(e.g. AOP). The reasoning pseudo code is presented in Algorithm 1.

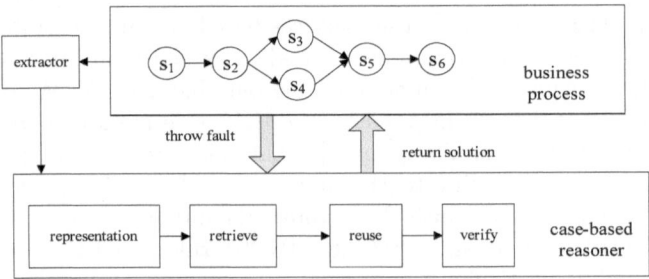

Fig. 1. Self-healing framework

Algorithm 1. Implementation of CBR

1: symbols: fault information data base(FIDB), case base(CB)
2: input: FIN, NFI
3: output: solution (sn)
4: Begin:
5: interacting with FIDB , CWOS= generate(FIN, NFI)
6: searching CWS in CB with the similarity function, CWS = retrieve(CWOS)
7: extracting the solution of CWS, denoted as sn
8: if sn is *replace*
9:     verify(sn) goto 12.
10: else sn is *repeat*
11:     return sn. goto 15.
12: if sn satisfies QoSC
13:     return sn. goto 15.
14: else goto 6.
15: updating CB with sn and CWOS.
16: end

---

4   Detailed Implementation

Suppose there is a case base CB $= (C_1, C_2, C_3)$ and a services composition SC $=$ $\{f(S_1, S_2, S_3,\dots), \text{QoSC}\}$, where $f(S_1,S_2,S_3,\dots)$ is the composition connecting all services by using sequence, condition and loop structures. QoSC denotes the user's constraints about QoS. So, our task is to find a case C_i whose solution can deal with a failed service S_j and satisfy QoSC simultaneously.

4.1   Case Representation

In our approach, FI and NFI information can be easily extracted from WSDL files. FIN is represented by using the HTTP status codes as in [15]. The case symptoms (CS) can be described as a triple: CS $=$ (FI, NFI, FIN), where FI $=$ (input, output), NFI $=$ (availability, cost, successability, response time), FIN $=$ (fri, nfri). Their definitions refer to WSBPEL and [15]. Here we only give one part as follows:

1. functional information(FI)
   (a) input: abstract message format for the solicited request.
   (b) output: abstract message format for the solicited response.
2. nonfunctional information(NFI)
   (a) availability: the percentage of time is operating during a time interval
   (b) throughput: number of successful invocations/total invocations
3. fault information(FIN)
   (a) fri: function related information
   (b) nfri: non-function related information

We classify adaptive actions into two categories *repeat* and *replace* which are commonly used in previous works, e.g.[12]. For example, the code 404 denotes that service is not found, so *replace* should be chosen. There are 20 standard faults defined in WSBPEL specification where they are distinguished according to whether they are related to input data. For example, the fault "invalidExpressionValue" is data related, and *repeat* can be used. Analysis shows only "missingReply" and "missingRequest" can be thought as data unrelated.

We construct a case representation with solution and solution templates shown in Table 1 and 2 respectively. Codes with the prefix "d" can be defined by the users in the first column in Table 2. Codes with prefix "b" come from WSBPEL and the prefix is added by us. The remaining codes come from [15]. The underlined tasks in template column will be replaced with a concrete service in Table 2. A new problem is a case without the solutions part of Table 1.

4.2   Retrieval Algorithm

In light of our initial study, the case base is stored in a linear list. In real word, we can index the cases according to attributes of the service to improve the retrieval efficiency. Due to the fact that attributes of QoS vary in units, they

Table 1. Case in the case base

Problem(symptoms)	Solutions
(1)problem: fault description	
(2)service location:	
(3)service input:	
(4)service output:	(1)Diagnosis: the used data is faulty
(5)availability:89	(2)Repair: repeat invoking the service with
(6)throughput:7.1	correct data
(7)successability:90	
(8)response time:302.75	

Table 2. Solution templates

code	explanation	templates
403	forbidden	Diagnosis: the network doesn't work well.
405	method not allowed	Repair: replace this service with another one
b901	ambiguousReceive	Diagnosis: the service doesn't work well with incorrect data.
411	Length required	Repair: repeat this service with correct input data.
b906	MissingReply	Diagnosis: the service violates the QoS constraints
d801	service's availability is lower	Repair: replace this service with another one, reset the weight factors of the QoS.

should be normalized to be [0,1]. A QoS attribute matrix is shown in Eq. (1) same as the one in [16].

$$A = \begin{bmatrix} q_{11} & q_{12} & \cdots & q_{1n} \\ q_{21} & q_{22} & \cdots & q_{2n} \\ \cdots\cdots\cdots\cdots\cdots \\ q_{m1} & q_{m2} & \cdots & q_{mn} \end{bmatrix} \tag{1}$$

where q_{ij} is in the range [0,1]. So, we use the Euclidean distance to compute two cases' similarity:

$$d_{ij} = \sqrt{\sum_{k=1}^{N} (C_{ik} - C_{jk})^2 \lambda_k} \tag{2}$$

The factor λ_k is the weight denoting the user's preference for an attribute of QoS and sum of these factors is 1. C_{ik} comes from the matrix A of C_i .

4.3   Reuse Approach

After retrieving top N cases, we use the solutions to achieve adaption. Two conditions are considered.

- Condition 1: if the adaptive action is *repeat*, the failed service can be re-invoked directly.
- Condition 2: if the adaptive action is *replace*, the solution is verified to check whether QoS attributes deviate from user's constraints.

Our approach automatically generates two alternatives for one new case. Assume S_1, S_2 and S_3 occur in the case base corresponding to C_1, C_2 and C_3. C_2 is similar to C_1 and its solution is: replace S_2 with S_3. So, C_3 is also a candidate of C_1. Even though their similarity may be bigger than the similarity threshold, we think that our method has a great success rate as long as the threshold is as small as possible.

Finally, we transform the validation of solution against QoS requirements into a constraint satisfaction problem. In our paper, it is assumed that the candidate service of the failed service can be found in the case base. The reason is that the business process will return to a correct state by using the compensation handler in WSBPEL. How to implement this compensation mechanism with CBR is a future research topic.

5   Experiment

The QWS dataset[2] is used in our experiments including 2,507 Web services. We use five attributes: response time, availability, throughput, successability, reliability. We assume all services are functionally equivalent and the services composition uses sequential structure including 20 services. We record the comparison times(CPT) about the failed services which are generated randomly.

We build the case base only using the first 2400 services. In the first experiment, we do not limit the search step (SS). The results in Figure 2(a) show that CPT is not proportional to the number of failed services. We can see that the CPT is very high at point 8 and 10, where the CPT of point 10 is 938.2 because there are 30% of found services coming from the remaining 107 services. So, it is very important to build a case base as completely as possible.

In the second experiment, two conditions are compared: step-limited and step-unlimited. The latter is our reuse approach and the value of SS is equal to 5. In Figure 2(b), we can see that the CPT of step-unlimited is lower than that of step-limited. In particular, when there are 9 failed services, the comparison times reduce 26 times. So, our method takes less time to achieve the adaption.

At last, we make a further study on the influence on CPT by different search steps . In Figure 2(c), the results show that the CPT can be reduced significantly with the increase of search steps. But, it does not mean that the bigger steps are better. For example, when SS is equal to 4 and 5, the comparison times are very close. So, in practice, we should determine a proper value of the search steps.

---

[2] http://www.uoguelph.ca/∼ qmahmoud/qws/index.html#Quality_of_Web_Service (QWS)_ This dataset has more QoS attributes than that on the web site http://www.zibinzheng.com/

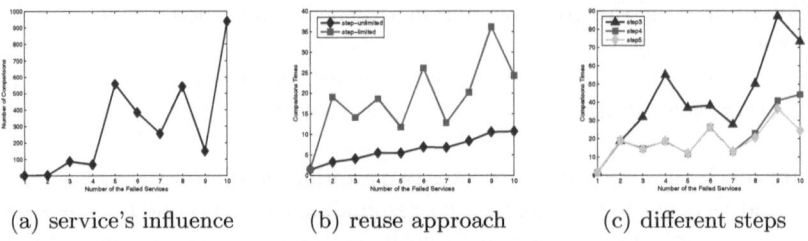

(a) service's influence   (b) reuse approach   (c) different steps

Fig. 2. Comparison times under different conditions

6   Conclusion and Future Work

We presented a framework by means of the case-based reasoning. This framework can dynamically adapt to the failure of composite services efficiently. According to the principle of the case-based reasoning, we focus on the developments of the case representation, the retrieval algorithm and the reuse approach. Experiments are conducted to show the effectiveness of our approach with real dataset.

In practice, the case base may become very large. So, how to maintain the base will be a challenge to use this method in the future. An interesting research direction is to determine how the efficiency of our approach is affected by complex services composition structures.

Acknowledgment. This work is funded by the National Program on Key Basic Research Project(973 Program, Grant No.2013CB329605), Natural Science Foundation of China (NSFC, Grant Nos. 60873237 and 61003168).

We are very grateful to Qusay H. Mahmoud for providing the QWS Dataset.

References

1. Kolodner, J.L.: An introduction to case-based reasoning. Artif. Intell. Rev. 6(1), 3–34 (1992)
2. Guinea, S., Kecskemeti, G., Marconi, A., Wetzstein, B.: Multi-layered monitoring and adaptation. In: Kappel, G., Maamar, Z., Motahari-Nezhad, H.R. (eds.) ICSOC 2011. LNCS, vol. 7084, pp. 359–373. Springer, Heidelberg (2011)
3. Moser, O., Rosenberg, F., Dustdar, S.: Non-intrusive monitoring and service adaptation for ws-bpel. In: Proceedings of the 17th International Conference on World Wide Web, WWW 2008, pp. 815–824. ACM, New York (2008)
4. Liu, A., Li, Q., Huang, L., Xiao, M.: Facts: A framework for fault-tolerant composition of transactional web services. IEEE Transactions on Services Computing 3(1), 46–59 (2010)
5. Baresi, L., Guinea, S.: A dynamic and reactive approach to the supervision of bpel processes. In: ISEC 2008, pp. 39–48 (2008)
6. Ali, M.S., Reiff-Marganiec, S.: Autonomous failure-handling mechanism for wf long running transactions. In: Proceedings of the 2012 IEEE Ninth International Conference on Services Computing, SCC 2012, pp. 562–569. IEEE Computer Society, Washington, DC (2012)

7. Baresi, L., Guinea, S., Pasquale, L.: Self-healing bpel processes with dynamo and the jboss rule engine. In: International Workshop on Engineering of Software Services for Pervasive Environments: in Conjunction with the 6th ESEC/FSE Joint Meeting, ESSPE 2007, pp. 11–20. ACM, New York (2007)

8. Zahoor, E., Perrin, O., Godart, C.: Disc: A declarative framework for self-healing web services composition. In: IEEE International Conference on Web Services, ICWS 2010, Miami, Florida, USA, July 5-10, pp. 25–33. IEEE Computer Society (2010)

9. Ardagna, D., Pernici, B.: Adaptive service composition in flexible processes. IEEE Transactions on Software Engineering 33(6), 369–384 (2007)

10. Chafle, G., Dasgupta, K., Kumar, A., Mittal, S., Srivastava, B.: Adaptation inweb service composition and execution. In: International Conference on Web Services, ICWS 2006, pp. 549–557 (September 2006)

11. Ben Halima, R., Drira, K., Jmaiel, M.: A qos-oriented reconfigurable middleware for self-healing web services. In: IEEE International Conference on Web Services, ICWS 2008, pp. 104–111 (September 2008)

12. Subramanian, S., Thiran, P., Narendra, N.C., Mostefaoui, G.K., Maamar, Z.: On the enhancement of bpel engines for self-healing composite web services. In: Proceedings of the 2008 International Symposium on Applications and the Internet, SAINT 2008, pp. 33–39. IEEE Computer Society, Washington, DC (2008)

13. Li, J., Ma, D., Mei, X., Sun, H., Zheng, Z.: Adaptive qos-aware service process reconfiguration. In: Proceedings of the 2011 IEEE International Conference on Services Computing, SCC 2011, pp. 282–289. IEEE Computer Society, Washington, DC (2011)

14. Cugola, G., Pinto, L., Tamburrelli, G.: Qos-aware adaptive service orchestrations. In: IEEE 19th International Conference on Web Services (ICWS), pp. 440–447 (June 2012)

15. Al-Masri, E., Mahmoud, Q.H.: Investigating web services on the world wide web. In: Proceedings of the 17th International Conference on World Wide Web, WWW 2008, pp. 795–804. ACM, New York (2008)

16. Al-Masri, E., Mahmoud, Q.: Qos-based discovery and ranking of web services. In: Proceedings of 16th International Conference on Computer Communications and Networks, ICCCN 2007, pp. 529–534 (August 2007)

Workload-Aware Cache for Social Media Data

Jinxian Wei, Fan Xia, Chaofeng Sha, Chen Xu,
Xiaofeng He, and Aoying Zhou

Institute of Massive Computing, Software Engineering Institute,
East China Normal University
{jinxianwei,fanxia,chenxu}@ecnu.cn, cfsha@fudan.edu.cn,
{xfhe,ayzhou}@sei.ecnu.edu.cn

Abstract. The success of social network services has brought up many interesting Web 2.0 applications, while posed great challenges for human-real-time data management for huge volume of data with unstructured nature. Timeline query is a specific type of queries that are widely used in social network services and analysis. A workload-aware cache scheme for efficient evaluation of home timeline queries in human-real-time manner is proposed in this paper. It utilizes the communities within followship network and considers the high skew of access frequency across users to generate cache units. Thus, timeline queries are transformed to a process of merging cache units. Empirical studies show the superiority of overlapping cache strategy over other three existing strategies.

1 Introduction

The success of social network services (SNS), such as Facebook and Twitter, has brought up many interesting Web 2.0 applications, while posed great challenges for human-real-time data management for huge volume of data with unstructured nature.

Many SNS tasks, including those for both SNS providers and end-users, can be represented by a specific kind of queries, i.e. *timeline query (TQ)*. Intuitively, a timeline query is a top-k or windowed query for records satisfying specific conditions, in reverse order of timestamps associated with those records. *Home timeline query (HTQ)*, as a kind of typical timeline queries, is to retrieve a timeline with records created by a given user's followees.

Fig. 1, for example, shows a global timeline within the period between timestamps t_0 and t_j of six users, i.e. u_1, u_2, u_3 and v_1, v_2, v_3, whose following relationships are also given. Here, $u_{i,j}$ means the jth message posted by user u_i. The top-5 results of home timeline queries for u_1, u_2, u_3 at timestamp t_j are also listed. Here, the query condition is that the message queried should belong to a querying user himself or his followees. Thus, $v_{3,1}$ is included in the results of $HTQ(u_1, t_j, 5)$ and $HTQ(u_2, t_j, 5)$, but not of $HTQ(u_3, t_j, 5)$, since user u_3 is not following v_3.

However, evaluation of home timeline queries in a human-real-time manner is non-trivial. On the one side, the data should be cached in memory with orders or

Y. Ishikawa et al. (Eds.): APWeb 2013, LNCS 7808, pp. 662–673, 2013.

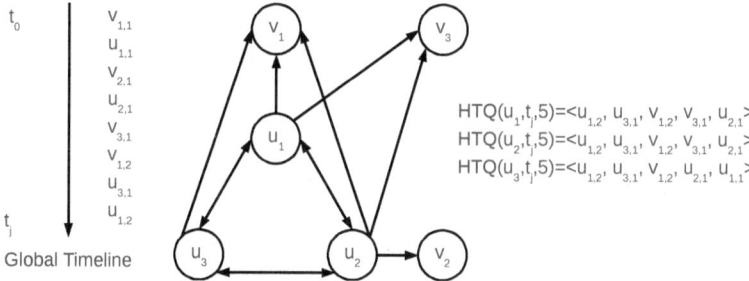

Fig. 1. An example of home timeline queries over a global timeline of six users

indices on timestamps, so that the top-k or time-window part of the query can be efficiently evaluated. On the other side, a specific query may only retrieve a very small portion of the data distributed in the whole timeline, while different queries may extract different set of records with overlaps. The sparsity of the results makes the ordering of the global timeline in the system meaningless.

The report in [1] shows three caching schemes to facilitate the evalation of home timeline queries over social media data. The *pull*-base method caches the messages of each user separately by putting any item to the cache unit of its corresponding user simply. This strategy requires to merge the messages to construct home timeline so that it suffers the problem of high-cost in cache merge leading to high latency. The *push*-base approach directly caches home timeline query result by delivering user generated message to his followers eagerly. Clearly, it consumes too much memory for duplicated items and imposes heavy load to CPU for vast deliveries. To strike a balance between pull-base and push-base approach, a *hybrid* strategy distinguishes *popular* users, such as celebrities who have a large volume of followers, from the ordinary ones. The messages generated by popular users are cached following pull-based method, while messages from the other users are processed as push-based approach. Consequently, active users are most likely to have worse user experience in the view of retrieval latency since they are always following a certain number of popular ones.

A workload-aware caching scheme named *overlapping caching* is proposed in this paper with the observation that users of SNS tend to form communities with homogenization and access frequency has high skew among users in SNS. It distinguishes itself from previous work in the following aspects.

- A cache unit is shared by a group of users with close relationships in the followship network, while user groups may have overlap with each other. The evaluation of timeline queries is transformed to an efficient process of merging cache units.
- User's access frequency is taken into consideration ensuring that our method outperforms other native caching schemes in general case.
- Empirical studies upon data sets simulated on the basis of real-life social media data show the superiority of overlapping cache strategy over other three existing strategies in the view of both system cost and performance.

Table 1. Symbols used in this paper

Notation	Description	Notation	Description
u, u_x, u_i	User identifiers	t_y, t_j	Timestamps
U	A set of user identifiers	T_G	Global timeline
ϕ_i	Update frequency of u_i	φ_i	Request frequency of u_i
F_i	Set of users followed by u_i	S	Set of all shared timelines
Ts_l	a shared timeline	C_l	Set of users covered by Ts_l
P_l	Set of users who share Ts_l	X_i	Set of shared timelines owned by u_i

The rest of the paper is organized as follows. In Section 2, we state the problem of home timeline queries formally. The workload-sensitive caching scheme is illustrated in Section 3 and the empirical study results are shown in Section 4. Section 5 lists the related work followed by the conclusion of this paper in Section 6.

2   Problem Statement

A *timeline* is a sequence of items (e.g., messages) that are ordered chronologically. Each item is a triple $< u_x, t_y, m_{x,y} >$, in which u_x is the user identifier, t_y denotes the timestamp, and $m_{x,y}$ is the id of the item created by u_x at t_y. A *followship network* is a directed graph $G : (V, E)$, in which V is the set of users, and E is a subset of $V \times V$. $(u, u_i) \in E$ means that user u is following user u_i. In many real-life SNS, every user in default subscribes his own messages, i.e. $\forall u, (u, u) \in E$. We denote all u_i's followees as F_i, i.e. $F_i = \{u | (u_i, u) \in E\}$.

The raw data that to be queried is a *global timeline*, denoted as T_G, in which items from different users are interlaced. A *timeline query* $TQ(U, t_j, k)$ is a query at t_j to retrieve last k items $< u_x, t_y, m_{x,y} >$ in T_G satisfying both $t_y \leq t_j$ and $u_x \in U$, where U is a set of user identifiers.

Home timeline queries, as a kind of typical timeline queries, are notoriously common in SNS applications. Every time a user logs in or refreshes his/her web page, home timeline query is triggered. It retrieves u_i's *home timeline*, denoted as T_i, which is the timeline with all items from this user's followees. Thus, the home timeline query of user u_i at time t_j can be denoted as $HTQ(u_i, t_j, k) = TQ(F_i, t_j, k)$.

The main problem in this work is how to achieve low latency and high throughput with limited system resources. Here, latency is the response time of home timeline query, whereas throughput means the operations consisting of creating items and retrieving home timelines per time unit (e.g., seconds). In addition, we take RAM and CPU as system resources into consideration, since most of the data is stored in RAM for performance reasons in many real applications.

Formally, let us define $Cost_{RAM}()$ as the total consumption of memory and $Cost_{CPU}()$ as the aggregate CPU utilization. Then, the *home timeline query problem* is to minimize both $Cost_{RAM}()$ and $Cost_{CPU}()$, while ensuring the

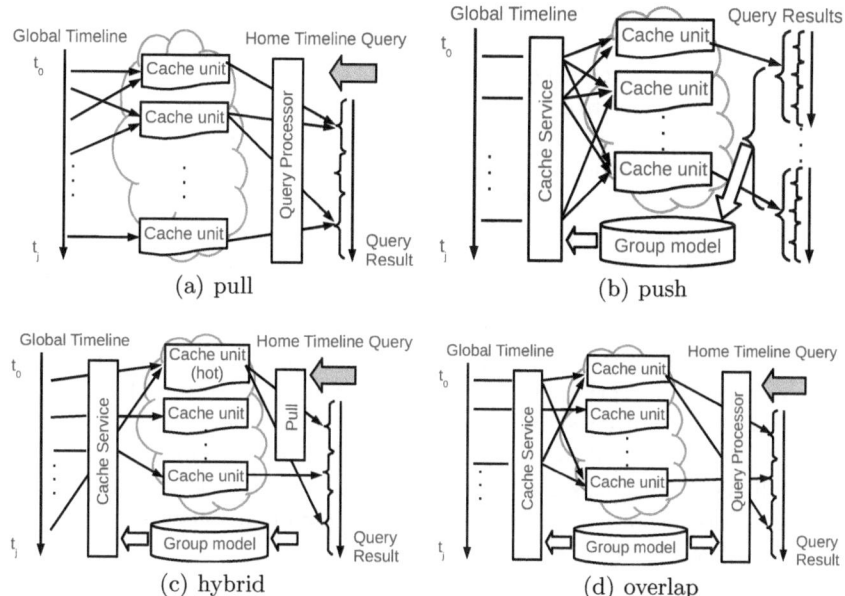

Fig. 2. An overview of caching for timeline query processing

average query latency is less than M time units (e.g., milliseconds) and the average throughput is higher than N operations per time unit in the premise that home timelines are provided correctly.

3   Workload-Sensitive Caching

3.1   Overview of Four Caching Strategies

Two trivial solutions for home timeline query processing are *pull*-based method and *push*-based method, which are illustrated in Fig. 2(a) and Fig. 2(b) respectively. Under *pull* strategy, any message item $< u_x, t_y, m_{x,y} >$ is cached only once in u_x's *user timeline* where all messages posted by u are organized chronologically. Thus, while processing a home timeline query $HTQ(u_i, t_j, k)$, the problem is to retrieve user timelines of u_i's followees, i.e, F_i. Differently, the push-based method assumes that all home timeline queries are known in advance and results of them are pre-computed and cached. While the query is triggered, the system just retrieve the result from cache and deliver it to the user. Clearly, to support *push* strategy, each item $< u_x, t_y, m_{x,y} >$ is cached as many times as the number of u_x's followers. However, neither pull-based nor push-based method is ideal for home timeline query processing. The former suffers the problem of high-cost in cache merge leading to high latency, whereas the later consumes too much memory for duplicated items and imposes heavy load to CPU.

As a trade-off between the two strategies aforementioned, a *hybrid* approach is reported that combines pull-based approach with push-based one [1], which is shown in Fig. 2(c). In this method, messages from *popular* users, i.e., the users with many followers, are cached once in their own user timelines as pull-based method, while messages from the other users are delivered to home timelines of their followers as push-based approach. Generally, the hybrid strategy decreases CPU cost and requires less memory space in comparison with the push one, and reduces the latency in cache merge in contrary to pull one. However, it is oblivious to the *active* user who have higher request rates and tend to have a higher possibility to follow more popular users. Unfortunately, those active users, who play crucial roles in applications, will have worse user experience in the view of home timeline query latency under *hybrid* strategy.

Those drawbacks above drive us to study the properties of workload from an alternative aspect rather than simply distinguishing the popular users from the other ones. It is observed that users of SNS tend to form communities with homogenization and users within the same community are most likely to have similar followships such as many common followers. Intuitively, the messages from a group of users can be shared by their common followers so that it is unnecessary to cache these messages in duplicate.

Shared timeline, denoted as Ts_l, is a timeline generated by a corresponding shared group of users $C_l \subseteq V$. In other words, each item $< u_x, t_y, m_{x,y} >$ in Ts_l satisfies that $u_x \in C_l$. The overlapping caching strategy we propose is illustrated in Fig. 2(d), where each message $< u_x, t_y, m_{x,y} >$ is delivered to any Ts_l satisfying $u_x \in C_l$ and each Ts_l is used to answer home timeline query $HTQ(u_i, t_j, k)$ if $C_l \subseteq F_i$.

Besides exploiting the community characteristic in SNS as above, this workload-sensitive caching method also takes users' access frequency into consideration due to the fact that both read (query a home timeline) frequency and write (post a message) frequency have high skew across users. On one hand, write frequencies of users in one shared group are relevant to the size of the corresponding shared timeline, which influences system cost including RAM usage and CPU utilization. On the other hand, decreasing latency for users with high read frequency results in low aggregated read latency, in which way, system throughput is improved. A natural question is how to determine shared groups while taking both social networks and user access frequency into account followed by answering home timeline queries via shared timelines based on those shared groups.

3.2   Shared Groups Determination

The result shared groups are expected to resolve the problem stated in Section 2. That is how to achieve low latency and high throughput with limited system resources. Thus, the determination of shared groups is forced to facilitate the evaluation of home timeline queries to consume RAM and utilize CPU in an economical manner while satisfactory performance is guaranteed.

We first formulate the *shared groups determination* problem, then present our greedy solution to this problem followed by a linear community detection algorithm that making use of user profiles, which serves as the basis of our greedy solution.

Problem Formulation. To formulate the *shared groups determination* problem, we additionally introduce some notations. Let P_l denote the set of users who share Ts_l covering updates from users in C_l. Clearly, followees of users in P_l are supersets of C_l. We use X_i to denote the set of shared timelines that user u_i participating to share. Given the update (write) and request (read) frequency of user u_i, denoted as ϕ_i and φ_i respectively, we now define the shared groups determination problem, with a setting threshold τ, as follows:

$$\min \sum_{Ts_l \in S} \sum_{u_i \in C_l} \phi_i$$

s.t.

$$\bigcup_{u_i \in P_l} C_l = F_i, \forall i \tag{1}$$

$$|X_i| \leq \tau/\varphi_i, \forall i \tag{2}$$

Constraint 1 ensures that one user's home timeline can be generated by merging all his/her shared timelines. Constraint 2 limits the number of shared timelines each user owns, which is bound to request frequencies of users' home timelines.

The problem of determination of the shared timelines is non-trivial. We simplify the original problem by setting all $\phi_i = 1$ and formally define the simplified problem as *Volume Bounded Set Basis Problem*, which can be reduced to the classic Set Basis problem [2].

Theorem 1. *The Volume Bounded Set Basis Problem is NP-hard.*

Details of the definition of *Volume Bounded Set Basis Problem* and its complexity analysis along with the proof of Theorem 1, are attached in Appendix.

Community-Based Greedy Identification of Groups. Although a greedy algorithm upon the entire network gives a good approximation to the problem of determining shared groups, it is expensive for large social networks like Sina weibo who has a user base of hundreds of millions. The main idea of our solution is to divide the network into overlapped communities, and adopt greedy algorithm for selecting shared groups in communities for the purpose of efficiency. The calculated shared groups, on the other side, can be reused by users across communities.

Suppose that we already divide the network into communities(to be discussed in the following subsection).We now present our greedy algorithm to determine shared groups as Algorithm 1, where the gain of C_t is calculated as:

$$gain(C_t) = \sum_{u \in C_t, C_t \subseteq F_u} \phi_u$$

Algorithm 1. Greedy algorithm for determination of shared groups

---
Input: The followship network: $G(V, E)$; community: C
Output: A set of user groups: $R = \{R_i | R_i \subseteq V\}$
  1: **for all** $v_i \in V$ **do**
  2:     Compute the number of shared timeline w_i for v_i; $w_i' \leftarrow w_i$;
  3:     $F_i \leftarrow \{u | (v, u) \in E\}$; $F_i' \leftarrow F_i$;
  4: **end for**
  5: **while** there are v_i's that cannot be satisfied by Ts_j's for $R_j \in R$ **do**
  6:     **for all** v_i whose $w_i' = 1$ **do**
  7:         Create a new $R_k \leftarrow F_i'$; $R \leftarrow R \bigcup \{R_k\}$; $w_i' \leftarrow 0$;
  8:     **end for**
  9:     **for all** v_i with $w_i' > 0$ in ascendant order of w_i' **do**
 10:         Create a new $R_k \leftarrow \arg\max_{gain(C_t)} C_t : \{u | u \in F_i, (u, t) \in C, (v_i, t) \in C\}$;
 11:         $F_i' \leftarrow F_i' - R_k$; $R \leftarrow R \bigcup \{R_k\}$; $w_i' \leftarrow w_i' - 1$;
 12:         **for all** u with $(u, t) \in C$ and $F_u' \cap R_k \neq \emptyset$ **do**
 13:             $F_u' \leftarrow F_u' - R_k$; $w_u' \leftarrow w_u' - 1$;
 14:         **end for**
 15:     **end for**
 16: **end while**

Profile-Based Community Detection. Intuitively, users of one community tend to share a considerable portion of common followers and followees. Derived from that, we propose a community detection algorithm to filter users in the determination of shared groups in a more efficient manner. Graph cluster algorithms can be used to find out the communities, but their high complexity make it impractical to apply them to a real life social network[3]. We here explore the profiles of users and assign initial communities to each user based on their profiles. Then Label Propagation Algorithm(LPA)[4] is used to decide the final communities for them. In LPA, the communties of users are voted by their followees. First, LPA will propagate the communities of a user to each of its followers. Then, it chooses topK voted communities for each user. We provide a map reduce implementation of this algorithm. In this implementation, one job is used to propagate the communities to the followers. The other job aggregates the votes and decides the top communities for each user. Clearly, labels of a user are sent to all his followers, so the network traffic of these two jobs is $O(|E|C)$, where $|E|$ stands for the number of edges and C is an constant constraining the maximum number of communities one user belongs to.

3.3   Timeline Query Processing in Overlapping Caching

Once shared groups are determined, each shared timeline, which organizes messages from users in corresponding shared group, can be generated. Under overlapping caching strategy, home timeline query can be answered by merging all related shared timelines. The merging phrase is to remove duplicates and return top items among all the related shared timelines, whose algorithm can be found in Algorithm 2.

Algorithm 2. Merge Algorithm

Input: all T_{s_i} for u, return count l;
Output: top l items stored in set Q,
 1: create priority queue P
 2: enqueue all T_{s_i} into P according to the first element of T_{s_i}
 3: **while** $P.size > 0$ and $Q.size < l$ **do**
 4:    $tl \leftarrow P.poll$
 5:    $Q \leftarrow Q \bigcup \{tl[0]\}$
 6:    $tl \leftarrow tl - \{tl[0]\}$
 7:    **if** $tl \neq \emptyset$ **then**
 8:        enqueue tl into P
 9:    **end if**
10: **end while**
11: **return** Q;

4   Empirical Study

We now examine our experimental results. The four strategies(i.e., *pull*, *push*, *hybrid* and *overlap*) described in Section 3 are implemented to handle the message load obtained from a real-life social media(Sina Weibo). We show a comparison of the above four caching strategies from the aspects of system cost and performance.

Our system cost metric is aggregate memory usage and CPU utilization across all servers. We put a limit on memory capacity in order to evaluate performance of the strategies when some cache misses happen, which is the common case in many real-life applications. CPU utilization is measured by Unix sar command. Performance is evaluated in terms of latency and throughput. We measure the latency as response time observed from the client, for retrieving one user's home timeline. Throughput is measured as the number of operations performed per second. Here, an operation means posting a message or retrieving a home timeline.

In summary, our results show: The overlapping caching strategy strikes best balance between system load and performance: it utilizes least CPU, consumes second least memories, and can achieve lowest query latency and highest throughput at most time under the memory constraint in our experiment.

4.1   Experimental Data and Setup

We construct our data set on the basis of a real-life data set crawled from Sina Weibo, the Chinese edition of Twitter. The real data set consisting of 1,289,986 users and all the messages they post from in 2011. We got the number of their followers and the users they are following. Unfortunately, the social network we crawled is incomplete and the retrieval log of home timeline is unavailable. Therefore, as a substitute we reconstruct their followships among the 1,289,986 users following the distribution of the number of their followers, which is about 100 on average, and generate a retrieval log in which users has bigger post frequencies tend to retrieve more often. Also, we rank the users by the number of

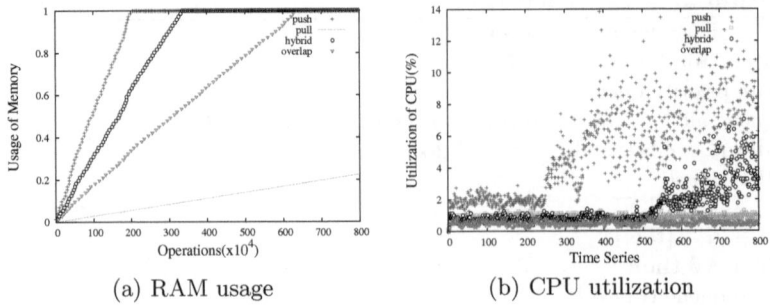

(a) RAM usage   (b) CPU utilization

Fig. 3. A comparison of system cost for the four strategies

their followers and regard the top 7,509 users as popular ones, who has more than 1,000 followers. All the popular users are treated differently in *hybrid* strategy.

To make a fair comparison of system cost and performance among the four caching strategies, each of them is run against the same workload. That is, the same sequence of messages posted in real-life application and the same retrieval log we simulate as mentioned are used in our experiments. The ratio of posts and retrievals is 1:4. We also launch the same number of threads for each strategy.

We run our experiments on server-class machines(8 core E5606@2.13Hz Intel Xeon CPU, 16GB of RAM, 1TB of disk and gigabit ethernet). In particular, we use twelve Redis storage servers(six for meta data) and one node to launch and process requests. Memory capacity for each the Redis server is limited to 100MB, under which a portion of cache misses encounter under all the strategies except the *pull* one.

4.2 System Cost

We start by examining the memory usage and CPU utilization of the six servers under four strategies. Fig. 3 indicates: The *push* strategy reaches the maximum capacity of memory first and imposes heaviest load to system CPU with increasing number of posted messages; The *pull* based strategy uses the memory most economically; Our proposal utilizes least CPU and consumes the second least memory, confirming our hypothesis.

4.3 Performance

Next, we evaluate the performance of the four strategies. As for the latency and throughput reported in Fig. 4, we use average values. Our experiment results show that: The *push* based strategy brings lowest query latency and highest throughput before reaching the maximum memory capacity and results in a sharp drop in performance since it consumes most memory and meets the limit of memory capacity first; Performance of the *pull* method is quite stable yet trivial; The *overlap* based strategy achieves lowest query latency and highest throughput at most time.

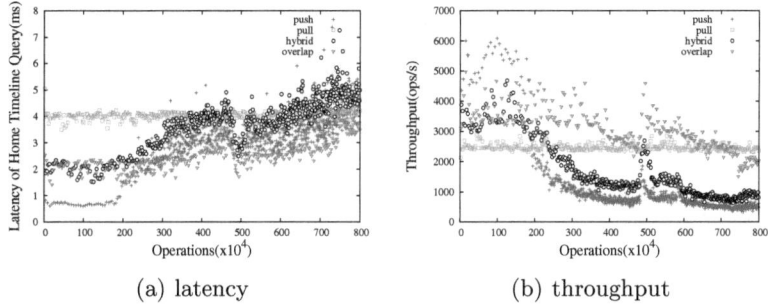

(a) latency   (b) throughput

Fig. 4. A comparison of performance for the four strategies

5   Related Work

Both industry and research community have paid great effort to develop approaches to support home timeline query, which is a key feature in SNS. There are three strategies(i.e pull,push and hybrid) that were proposed previously to generate the home timelines for users . Facebook is one of the most giant SNS providers in the world. According to its report[5], pull based strategy was implemented to support its service and push based one was constructed on the fly. On the other hand,Twitter[1] reported recently a hybrid strategy: push messages from the majority of ordinary users and pull messages from popular ones.

A.Silberstein.et.al[6] have introduced a formal solution to hybrid strategy from the aspect of materialization. They proposed a cost based heuristic method to decide whether the tweets should be materialized for each pair of followees and followers. Besides, some arguments can be tuned to make a tradeoff between the latency and throughput. Some other researchers[7][8] try to approach this problem from the distributed management of graph data. Each user in the social network is represented by a node in the graph and home timeline is generated by issuing read requests to the neighbors. They try their best effort to minimize the communications between sites. In order to sustain the local semantic of query, [7]replicates all the nodes that need to be accessed on remote sites to local. However, this approach causes a great of storage redundancy and may result in great efforts to maintain the replications. [8] introduces a more elaborate approach. The nodes are partitioned randomly and clustered into multiple clusters in each partition.The replication decision is made for each pair of clusters and partitions in according to their access behaviors.The experiments illustrate that their method is better than [7]. Besides, there are several decentralized social network systems[9][10][11] proposed. [10] incorporates a concept similar to hybrid strategy. Users in its system are divided into social users and media users. Media users typically have to broadcast messages to a lot of users. As a result, messages of media users are disseminated by gossiping.

Our work focuses on the microbloging service and proposes a new strategy. [4][3] all aim to detecting overlapping communities in the social network. [4] provides an algorithm that is linear to the number of edges in the network. We provide a MapReduce implementation to preprocess the dataset and help us understand the data.

6   Conclusion

In this paper, to minimize memory usage and CPU utilization while ensuring high throughput and low latency, we propose a novel workload-aware cache scheme, i.e., overlap-base strategy, to evaluate the home timeline query. This strategy makes full use of workload information including social communities and user access frequency in SNS. Our experimental studies show overlap-based approach outperforms all other three strategies including pull-based, push-based and hybrid strategies in both system cost and performance.

Acknowledgement. This work is partially supported by National Major Projects on Science and Technology under grant number 2010ZX01042-002-001-01.

References

1. Asemanfar, A.: Timelines @ Twitter. In: International Software Development Conference (QCon London) (2012)
2. Stockmeyer, L.: Set Basis Problem is NP-Complete. Technical report, Report RC-5431, IBM T.J. Watson Research Center, Yorktown Heights, NY (1976)
3. Wei, F., Qian, W., Wang, C., Zhou, A.: Detecting overlapping community structures in networks. World Wide Web 12(2), 235–261 (2009)
4. Xie, J., Szymanski, B.K.: Towards linear time overlapping community detection in social networks. In: PAKDD (2), pp. 25–36 (2012)
5. Kumar, S.: Social networking at scale. In: HPCA. IEEE (2012)
6. Silberstein, A., Terrace, J., Cooper, B.F., Ramakrishnan, R.: Feeding frenzy: selectively materializing users' event feeds. In: SIGMOD Conference, pp. 831–842 (2010)
7. Pujol, J.M., Erramilli, V., Siganos, G., Yang, X., Laoutaris, N., Chhabra, P., Rodriguez, P.: The little engine(s) that could: scaling online social networks. In: SIGCOMM, pp. 375–386 (2010)
8. Mondal, J., Deshpande, A.: Managing large dynamic graphs efficiently. In: SIGMOD Conference, pp. 145–156 (2012)
9. Xu, T., Chen, Y., Jiao, L., Zhao, B.Y., Hui, P., Fu, X.: Scaling microblogging services with divergent traffic demands. In: Kon, F., Kermarrec, A.-M. (eds.) Middleware 2011. LNCS, vol. 7049, pp. 20–40. Springer, Heidelberg (2011)
10. Sandler, D., Wallach, D.S.: Birds of a fethr: open, decentralized micropublishing. In: IPTPS, p. 1 (2009)
11. Buchegger, S., Schiöberg, D., Vu, L.H., Datta, A.: Peerson: P2p social networking: early experiences and insights. In: SNS, pp. 46–52 (2009)

Appendix: Volume Bounded Set Basis Problem

Definition 1. *Volume Bounded Set Basis Problem*
  Instance: Collection F of subsets of a finite set U, and positive integers m, W such that $m \leq W$.
  Question: Is there a collection B of subsets of U and $\sum_{b \in B} |b| \leq W$ such that, for each $f \in F$ there is a subcollection of B with size not greater than m whose union is exactly f?

To analyze the complexity of Volume Bounded Set Basis Problem, we introduced the classic Set Basis problem [2] first.

Definition 2. *Set Basis Problem [2]*
  Instance: Collection C of subsets of a finite set S, postive interger $K \leq |C|$.
  Question: Is there a collection B of subsets of S with $|B| = K$ such that, for each $c \in C$, there is a subcollection of B whose union is exactly c?

Lemma 1. *The Set Basis Problem is NP-hard. [2]*

We prove Theorem 1 that the Volume Bounded Set Basis Problem is NP-hard by transforming set basis problem to the Volume Bounded Set Basis Problem.

Proof. Let collection C of subsets of a finite set S, postive interger $K <= |C|$ be an instance of the set basis problem. We construct the instance of Volume Bounded Set Basis Problem as follows: collection $F = C$ of subsets of a finite set $U = S$, postive interger $m = K, W = \sum_{c \in C} |c|$.
  Let B be subsets of S with $|B| = K$ such that, for each $c \in C$, there is a subcollection of B whose union is exactly c. Therefore for each $f \in F = C$, there is a subcollection of B whose union is exactly c which has size not greater than $|B| = k = m$. And the volume of all basis sets in B is upper bouned by W, i.e. $\sum_{b \in B} |b| \leq \sum_{c \in C} |c| = W$, which is also due to the fact that each $c \in C$ is an union of a subcollection of B. Then B is a solution to the instance of Volume Bounded Set Basis Problem.
  The another direction is oblivious.

Shortening the Tour-Length
of a Mobile Data Collector in the WSN
by the Method of Linear Shortcut

Md. Shaifur Rahman and Mahmuda Naznin

Bangladesh University of Engineering & Technology, Dhaka, Bangladesh
shaifur.at.buet@gmail.com, mahmudanaznin@cse.buet.ac.bd

Abstract. In this work, we present a path-planning method for the
Mobile Data Collector (MDC) in a wireless sensor network (WSN). In
our method, a tour for the MDC is generated such that the latency
of delivering data to the sink is reduced. We show that the TSP tour
covers all the nodes of the WSN. However, the latency for the TSP tour
may be prohibitively high for delay-sensitive real-time WSNs. We reduce
the latency by shortening the given TSP tour using a method called
Linear Shortcut. We observe that the MDC need not visit the exact
location of the node. It only need to be in the proximity of the node
as required by the transmission radius. We present an algorithm that
iteratively shortens tour-length and hence, reduces the latency. We term
the resulting tour as Tight Label Covering (TLC) tour. Experimental
results show that TLC tour reduces latency by a significant margin.

Keywords: Wireless Sensor Network, Path-planning, Mobile Data Collector, TSP tour.

1 Introduction

Wireless Sensor Network (WSN) is widely used for tracking, monitoring and
other purposes. The problem of collecting data packets from the sensor nodes
and depositing those to the sink node is known as *Data Gathering Problem* [1,2].
Using mobile elements for data gathering in the WSN is a recent trend [3]. It
has many advantages. For example, it increases connectivity, reduces cost of
deployment of a dense WSN and increases the lifetime of the WSN. However,
it has the disadvantage of high latency as these mobile elements have limited
speed (compared to the high speed data gathering by routing). The dedicated
mobile element in the WSN which collects data packets and brings those to
the static sink is called *Mobile Data Collector* (*MDC*). A convenient way to
control the latency in the case of data collection by the MDC is to carefully
plan its path so that the path is as short as possible. In this work, we present a
path-planning method that shortens the path of the MDC iteratively. We test
our path-planning method in a simulation which is run on a realistic testbed.
Experimental results show that the shortening of the path indeed translates into
decreased latency and other improvements.

Y. Ishikawa et al. (Eds.): APWeb 2013, LNCS 7808, pp. 674–685, 2013.
© Springer-Verlag Berlin Heidelberg 2013

Rest of the sections are organized as follows. Related works are discussed in Section 2. The problem is formulated in Section 3. Our method is presented in Section 4. Experimental results are presented in Section 5. The prospect of future work is presented in Section 6.

2   Related Works

A complete survey on using mobile elements for data collection in the WSN is presented in [3]. Earlier works on mobile elements can be found in [4], [5], [6] and [7]. However, random motion of the mobile elements in these works is not suitable for optimization. Mobile sinks have been considered in [8] and [9]. Mobile relay based approaches for opportunistic networks have been surveyed in [10]. However, these methods are not suitable for the WSN because of its difference with such networks. In [11], an energy-efficient data gathering mechanism for large-scale multi-hop network has been proposed. The inter-cluster tour proposed in this work is NP-hard. Latency issue is not addressed in it. One of the heuristics used in this work produces edges which are not connected with any nodes of the WSN. Latency is considered while planning path for the mobile collector in [12] and [13]. Methods presented in these works produce a shorter tour termed *Label Covering* tour from a *TSP-tour*. However, the transmission range of the sensor nodes is ignored in the shortening process. In [14], authors propose an approximation algorithm which is based on the *TSP-tour* of the MDC. Although, the computation time $(O(n))$ is impressive, the solution is applicable to only certain kind of *TSP-tour* (tours for which the centroid of the tour-polygon lies within that polygon). If a condition regarding the concavity of the given *TSP-tour* is not met, the problem of finding the optimal solution becomes NP-hard. In [15], authors address the problem of planning paths of multiple robots so as to collect the data from all sensors in the least amount of time. The method presented here exploits earlier work on *TSP-tour with neighborhood* problem. However, this work does not utilize the available location information of the sensor nodes to the fullest as it allows the traversal of the full boundary of the transmission region of a node. In sparse network, one or more sensor nodes have no neighbors at all. As a result, the traversal of the boundary those nodes is futile and adds up to the tour-length.

3   Preliminaries

We represent a WSN with n nodes by a complete graph K_n where the graph-nodes represent the sensors and the sink. The edges in this graph represent the Euclidian distances between two nodes. We adopt the *disk model* of the given transmission range TXR. A circle with radius TXR centered at a node represent the area of radio transmission of that node. We assume that there are one sink and one MDC in the WSN and that both the sink and the sensor nodes are static. A *tour* or *cycle* for the MDC is a closed path in the graph K_n which starts and ends at the sink node.

Definition 1. *A TSP-tour is a tour in which the MDC visits the exact location of all the nodes in the WSN exactly once. The min-cost TSP tour[1] is a TSP-tour in which the MDC covers the minimum Euclidian distance.*

Definition 2. *A tour T by the MDC is complete if each sensor node of the WSN can send data packets to the visiting MDC directly or via a neighboring which forwards packets. Otherwise, the tour is incomplete.*

By definition, a *TSP-tour* is complete. However, it is not a good choice for a delay-sensitive WSN as explained in the following section.

Definition 3. *Data Delivery Latency (DDL) of a packet is the time-difference between the generation and delivery of the packet.*

Let a packet i be generated at t_g time after the MDC has set out from the sink node. The MDC completes the current tour in t_T time according to some tour plan T. DDL for a packet i is given by:

$$t_l(i) = t_T - t_g(i) \tag{1}$$

If n packets are collected in the current tour by the MDC, average packet delivery latency t_{avg} is given by:

$$t_{avg} = \frac{\sum_{i=1}^{n}[t_T - t_g(i)]}{n}$$
$$= t_T - \frac{\sum_{i=1}^{n} t_g(i)}{n} \tag{2}$$

The term $\frac{\sum_{i=1}^{n} t_g(i)}{n}$ in Equation 2 is the *average packet generation time*. This parameter is not controllable as it depends on the sampling rate of the sensor nodes and the event frequency. However, we may improve both per packet DDL (t_l) and average DDL (t_{avg}) by decreasing the tour-time t_T (See Equation 1 and 2). The tour-time of the *TSP-tour* i.e. t_{TSP} has two components: the fraction of tour-time t_h when the MDC halts and collects data from the nearby nodes and the fraction of tour-time t_m when the MDC travels between the node positions.

$$t_{TSP} = t_h + t_m \tag{3}$$

When the number of nodes is very high and/or the network is sparse, $t_h \ll t_m$ and thus t_m dominates tour-time t_{TSP}. This assumption is logical for practical scenario where the speed of a commercially available robotic car used as MDC usually is $5\ ms^{-1}$ whereas packet transfer from a sensor node to the MDC happens in the order of miliseconds [16]. Thus, decreasing the motion time t_m contributes to improving latency. If the speed of the MDC is v_{MDC}, and if we assume that it accelerates to this speed instantly and stops instantly, then

$$t_m = \frac{|t_{TSP}|}{v_{MDC}} \tag{4}$$

---

[1] We use *TSP-tour* to denote the minimum cost *TSP-tour* in this work.

where $|t_{TSP}|$ is the path-length of the *TSP-tour*. Given a particular MDC, v_{MDC} is fixed. The only way to decrease tour-time is decreasing the length of the tour i.e. $|t_{TSP}|$ (See Equation 4). However, decreasing the tour-length arbitrarily has the risk of making the resulting tour incomplete. Therefore, we address the issue carefully so that, the resulting tour is complete and shorter than the *TSP-tour*.

Problem Statement

Given a TSP-tour of the MDC in a WSN, find a tour T_d that is complete and shorter than the TSP-tour.

4   Improving Latency by Means of Linear Shortcut

4.1   Linear Shortcut of a Tour

Definition 4. *A linear shortcut of given a tour is derived by choosing 0, 1 or 2 points (called Anchor Points) from each tour-edge according to some strategy and connecting those points by straight lines in the order of visiting those edges. It is called linear as only new straight lines are introduced in the resulting tour instead of any curves.*

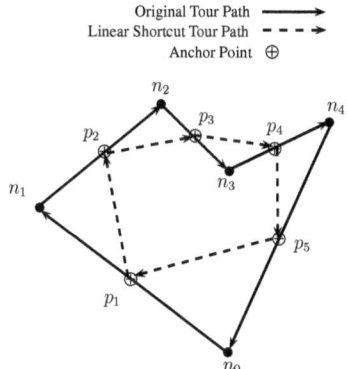

Fig. 1. Example of *Linear Shortcut* of a tour

An example of linear shortcut of a tour is shown in Figure 1. Five Anchor Points p_1, p_2, p_3, p_4 and p_5 are chosen from five tour-edges. Those are connected in the order of their visiting in the given tour to produce the linear shortcut. The first and the last points are also connected to make the path a cycle. The tour $< p_1, p_2, p_3, p_4, p_5, p_1 >$ is a linear shortcut but $< p_1, p_3, p_2, p_4, p_5, p_1 >$ is not. Using principle of triangle inequality, Lemma 1 can be easily proved.

Lemma 1. *If at least one anchor point is not coincident with the endpoint of the tour-edge, the linear shortcut is shorter than the given tour. (Proved using Triangle Inequality)*

4.2   Linear Shortcut of the *TSP-tour*

In [12] and [13], a tour known as *Label Covering* or *LC* tour is derived from the *TSP-tour*. As shown in Figure 2, the tour-edges between nodes 1 and 2 are also within the range of Node 3 and 5. Therefore, the label set of this edge is $\{1, 2, 3, 5\}$. Similarly, the sets of labels of all other edges are determined. the minimum length label covering tour is determined by making shortcut of zero or more edges of the given *TSP-tour*. For the graph shown in Figure 2, the *TSP-tour* is $< 1, 2, 3, 4, 5, 1 >$ and the *LC*-tour derived from it is $< 1, 2, 3, 1 >$.

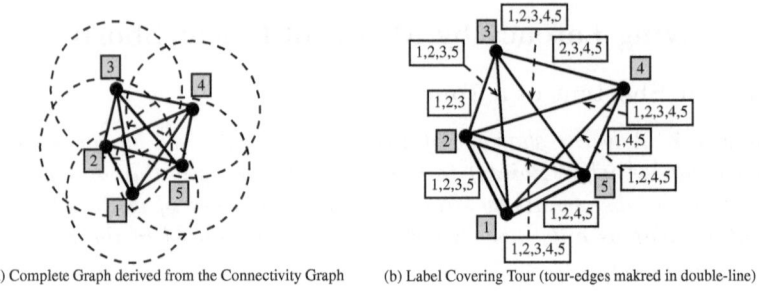

(a) Complete Graph derived from the Connectivity Graph   (b) Label Covering Tour (tour-edges makred in double-line)

Fig. 2. Label Covering tour in a cluster with five nodes

4.3   Linear Shortcut of the *LC*-tour

Definition 5. *The segment of the tour-edge which is within the circle representing the transmission area of a node is called the Contact Interval (CI) of that node on that edge.*

As shown in Figure 3, four nodes are covered by an edge connecting Node n_i and n_j. Each of their CI's is represented by two points on the edge i.e the l (which is encountered first by the MDC on this edge) and the r points. For example, the CI of Node n_{i+1} is given by (ln_{i+1}, rn_{i+1}). If the intersection of

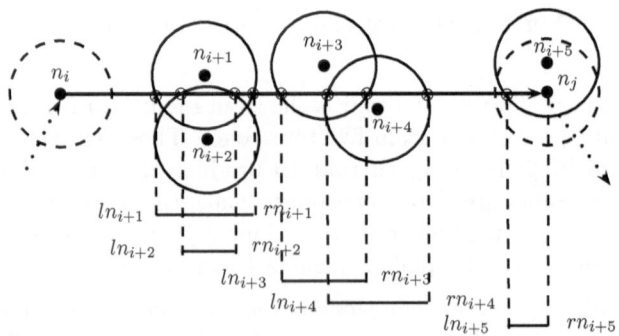

Fig. 3. *Contact Intervals of an edge*

the circle and the straight line is beyond the edge, the end-point of the CI is the nearest end-point of the edge. For example, in Figure 3, r-point of Node n_{i+5} is the location of Node n_j.

Definition 6. *Given a list of contact intervals CI_e of a tour-edge e, Critical Contact Interval or CCI is the interval of the minimum length which has at least one point from each contact interval.*

Critical Contact Interval or CCI of i-th edge is represented by two points- $lcci$ and $rcci$ on that edge. These points can be determined as follows:

$$lcci \leftarrow r\text{-point closest to the first end-point along the tour-edge}$$
$$rcci \leftarrow l\text{-point closest to the last end-point along the tour-edge}$$

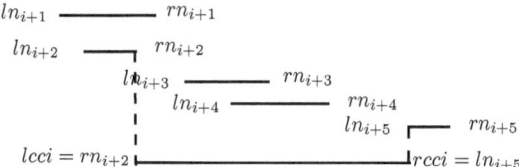

Fig. 4. *Critical Contact Interval for a given list of intervals*

For example, the CCI of the edge shown in Figure 3 is determined in Figure 4.

After the CCI's have been computed, we can connect r point of the CCI of an edge with the l point of the CCI of the next edge. However, the nodes visited in the given tour may be missed as shown in Figure 5(a). Therefore, to cover those nodes, we apply the following method for a Node n_i:

1. If both the edges have non-Null CCI's, i.e. there are intermediate nodes on both the edges, then we add the r point of the incoming edge with the l point of the outgoing edge. We call this line segment r-l line segment.
    (a) If r-l line segment intersects circle of Node n_i, then CCI's of both of the adjacent edges are kept unchanged (See Figure 5(b)).
    (b) If r-l line segment does not intersect the circle of Node n_i, then we draw a straight line that is parallel to the r-l line segment and tangent to that circle. Let this line intersects the incoming and outgoing edges at points p_i and p_o respectively. (see Figure 5(b) for Node n_{11}). We update the r point of the incoming edge and the l point of the outgoing edge as p_i and p_o respectively.
2. If the incoming edge does not have any intermediate node with overlapping circle or (in case it has) its r point is farther from point n_i by at least TXR, then we compute the point p_i as the intersection between the incoming edge and the circle centered at n_i. If the incoming edge has non-null CCI, then we update its r point as p_i. Otherwise, we set the incoming edge's r and l point as p_i. This case applies for Nodes n_1, n_2 and n_{13} as shown in Figure 5(b).

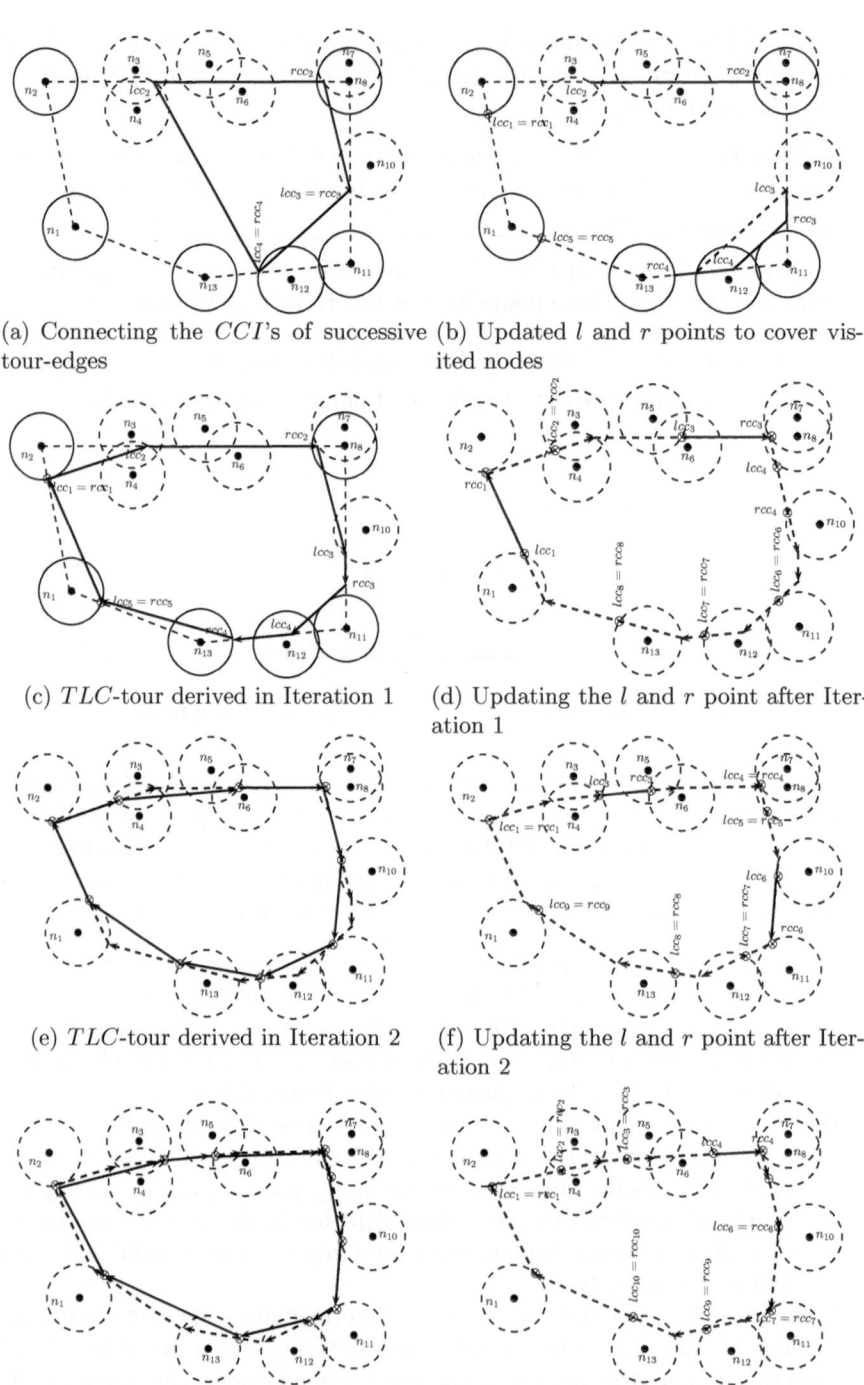

(a) Connecting the *CCI*'s of successive tour-edges

(b) Updated *l* and *r* points to cover visited nodes

(c) *TLC*-tour derived in Iteration 1

(d) Updating the *l* and *r* point after Iteration 1

(e) *TLC*-tour derived in Iteration 2

(f) Updating the *l* and *r* point after Iteration 2

(g) *TLC*-tour derived in Iteration 3

(h) Updating the *l* and *r* point after Iteration 3

Fig. 5. Generating *TLC-tour* using Linear Shortcut

Now, we join the r point of the previous edge with the l point of the next edge. The final edges are shown as bold straight lines in Figure 5(c). We term this shortening as *tightening* of the given tour by the linear shortcut method. We term the shorter tour derived from the LC-tour as *Tight Label Covering* tour or TLC-tour.

4.4   Iterative Improvement of the TLC-tour

The path found in Figure 5(c) can be further shortened using the linear shortcut method. We divide each iteration of improvement into 2 steps:

1. Connect the r point rcc_i of i-th edge with l point lcc_j of the next edge (j-th edge such that $j > i$) with non-Null CCI and include the edge connecting lcc_j and rcc_j in the edge set.
2. *Re-associate* the intermediate circles with the resulting edges and *recompute* the CCI's for each edge.

Now, we use the following policy to *re-associate* the circles when existing tour-edges *break* into shorter ones and new edges are *added*:

1. For each circle adjacent with two edges, the CI's for the both the edges are calculated. Then the circle is associated with the edge on which the circle has longer CI.
2. If the CI's for both the edges are equal, the circle is associated with the incoming edge.

As shown in Figure 5(d), there are eight edges. Circle of Node n_1 has overlaps with both the outgoing and the incoming edges. However, unlike the incoming edge, the circle has a CI of non-zero length with the outgoing edge. Therefore, we associate Node n_1 with this edge. Node n_2 has CI's of zero length with both the incoming and the outgoing eges. Therefore, we associate it with the incoming edge.The CCI of the edge connecting n_1 and n_2 is updated after this *re-association*. In similar ways, we determine the l and r points of the CCI's of the remaining edges. After this round of re-associating of circles and computation of the CCI's of respective edges, we join the r point of an edge with the l point of the next edge. Thus, the tour as shown in Figure 5(e) is derived. According to the Lemma 1, it is shorter than the tour derived in the previous iteration.

We can continue this way in to more iterations to tighten the given tour. The steps are illustrated in Figure 5(f), 5(g) and 5(h).

The given LC-tour and the TLC-tour derived after Iteration 4 has been imposed on each other for comparison in Figure 6. The derived TLC-tour is a significantly shorter than the LC-tour.

Condition for the Termination of Iterations: We can define *path gain* $g_i(t_{TLC})$ for a TLC-tour derived in iteration i as follows:

$$g_i(t_{TLC}) = \frac{|t_{TLC}|_{i-1} - |t_{TLC}|_i}{|t_{TLC}|_i} \tag{5}$$

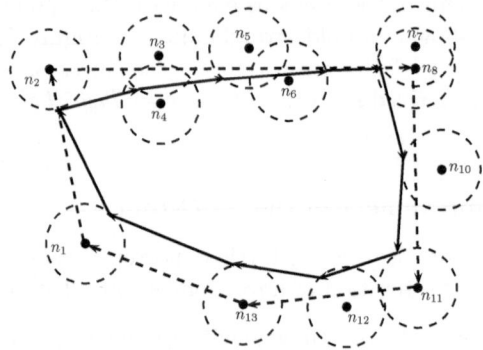

Fig. 6. Comparison between input *LC-tour* (doted path) and *TLC-tour* (solid path) derived in Iteration 4

Here $|t_{TLC}|_i$ is the length of the TLC-tour derived in iteration i. We stop the iterations for path-shortening as soon as the *path gain* is below a threshold like 1%, 5% etc.

Computation Complexity: Our method of generating *TLC-tour* is formally presented in Algorithm 1. First, we generate the CI's for all the circles in $O(n)$ time. Then, we sort the CI's in the non-decreasing order of the distance from the first endpoint of the edge in $O(n \log n)$ time. Therefore, we generate the CCI for an edge in $O(n + n \log n) = O(n)$ time. For a given tour, there are $O(n)$ edges. Therefore, we generate CCI's of all the edges in $O(n^2)$ time. Then, we connect successive r and l Points in $O(n)$ time. For each edge, we test the circles for *re-association*. There may be $O(n)$ such circles associated with each edge and a total of $O(n)$ edges. Therefore, updating the CCI's takes $O(n^2)$ time.

Algorithm 1. Generating *TLC-tour* from *LC-tour*

Input: *LC-tour* T_{LC} and a path-gain threshold g_t
 1: Compute the CI's of all the nodes
 2: Compute the CCI's of all the edges
 3: **while** *path-gain* $> t_g$ **do**
 4:     derive tour T_{TLC} by connecting the r point of an edge with the l point of the next edge
 5:     **for** each edge e **do**
 6:         **for** each Circle c (of Node n) associated with Edge e **do**
 7:             *re-associate* Circle c (if necessary)
 8:         **end for**
 9:         update CCI of Edge e
10:     **end for**
11:     determine *path-gain*
12: **end while**
Output: *TLC-tour* T_{TLC}

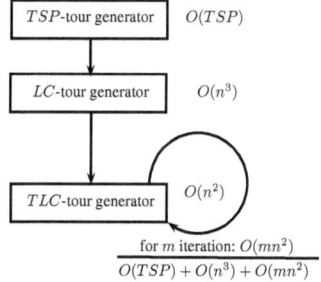

Fig. 7. Stages of computation along with the time complexity

If there are m iterations, then generating TLC-tour from LC-tour takes $O(mn^2)$ time. We illustrate the stages of computation along with the time complexity in Figure 7.

5 Experimental Results

We use $Castelia$[17] framework of $OMENT++$ simulator to distribute sensor nodes randomly. The sensor nodes also generated packet randomly. We use $Concord\ TSP\text{-}Solver$[18] to compute the optimal TSP-tour. We derive the LC-tour from the TSP-tour and the TLC-tour from the LC-tour using $path\text{-}gain$ threshold of 5%. The MDC tours continuously in TSP-tour, LC-tour and TLC-tour. During its travel, it collects the packets from the sensor nodes and deposits those to the sink node. We vary the value of TXR from $2m$ to $32m$. Low value of TXR indicates lower degree of connectivity and hence, sparse network. Similarly, higher value of TXR indicates dense network.

As shown in Figure 8(a), *average packet delivery latency* is always the lowest for TLC-tour and the highest for TSP-tour. The value is comparatively better in the case of the sparse WSN. In Figure 8(b), *throughput* for the entire run is shown. The value is always the highest for TLC-tour and the lowest for the TSP-tour.

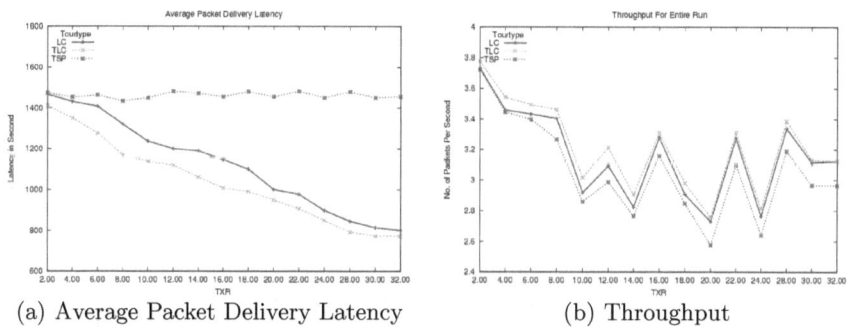

(a) Average Packet Delivery Latency   (b) Throughput

Fig. 8. Comparison among TSP-tour, LC-tour and TLC-tour

6   Conclusion

We have given a framework for shortening a given tour of the MDC. The resulting tour decreases packet delivery latency and increases throughput. The TLC-tour derived by *Linear Shortcut* of the LC-tour is highly suitable for real-time WSN in which high latency is undesirable.

In our future work, we shall consider more objectives besides minimizing latency, for example- facilitating multi-hop forwarding among the sensor nodes, load-balancing of the network traffic etc. We shall also extend the path-planning for a WSN with multiple MDC's and multiple sinks.

Acknowledgements. The authors would like to thank the department of Computer Science & Engineering, Bangladesh University of Engineering & Technology (BUET) for its generous support, laboratory, library and online resource facilities.

References

1. Akyildiz, I.F., Su, W., Sankarasubramaniam, Y., Cayirci, E.: Wireless sensor networks: a survey. Comput. Netw. 38(4), 393–422 (2002)
2. Rajagopalan, R., Varshney, P.K.: Data-aggregation techniques in sensor networks: a survey. IEEE Communications Surveys Tutorials 8(4), 48–63 (2006)
3. Di Francesco, M., Das, S.K., Anastasi, G.: Data collection in wireless sensor networks with mobile elements: A survey. ACM Transaction on Sensor Networks 8, 7:1–7:31 (2011)
4. Zhao, W., Ammar, M.H.: Message ferrying: Proactive routing in highly-partitioned wireless ad hoc networks. In: Proceedings of the The Ninth IEEE Workshop on Future Trends of Distributed Computing Systems, FTDCS 2003, pp. 308–314. IEEE Computer Society, Washington, DC (2003)
5. Shah, R.C., Roy, S., Jain, S., Brunette, W.: Data mules: Modeling a three-tier architecture for sparse sensor networks. In: IEEE SNPA Workshop, pp. 30–41 (2003)
6. Zhao, W., Ammar, M., Zegura, E.: A message ferrying approach for data delivery in sparse mobile ad hoc networks. In: Proceedings of the 5th ACM International Symposium on Mobile ad Hoc Networking and Computing, MobiHoc 2004, pp. 187–198. ACM, New York (2004)
7. Jun, H., Zhao, W., Ammar, M.H., Zegura, E.W., Lee, C.: Trading latency for energy in wireless ad hoc networks using message ferrying. In: Proceedings of the Third IEEE International Conference on Pervasive Computing and Communications Workshops, PERCOMW 2005, pp. 220–225. IEEE Computer Society, Washington, DC (2005)
8. Wang, Z.M., Basagni, S., Melachrinoudis, E., Petrioli, C.: Exploiting sink mobility for maximizing sensor networks lifetime. In: Proceedings of the 38th Annual Hawaii International Conference on System Sciences - Volume 09, HICSS 2005, p. 287.1. IEEE Computer Society, Washington, DC (2005)
9. Rao, J., Wu, T., Biswas, S.: Network-assisted sink navigation protocols for data harvesting in sensor networks. In: WCNC 2008, IEEE Wireless Communications & Networking Conference, Las Vegas, Nevada, USA, March 31-April 3. Conference Proceedings, pp. 2887–2892. IEEE (2008)

10. Conti, M., Pelusi, L., Passarella, A., Anastasi, G.: Mobile-relay Forwarding in Opportunistic Network. In: Adaptation and Cross Layer Design in Wireless Networks. CRC Press (2008)

11. Ma, M., Yang, Y.: Sencar: An energy-efficient data gathering mechanism for large-scale multihop sensor networks. IEEE Transaction on Parallel and Distributed System 18, 1476–1488 (2007)

12. Sugihara, R., Gupta, R.K.: Improving the data delivery latency in sensor networks with controlled mobility. In: 4th IEEE International Conference on Distributed Computing in Sensor Systems, pp. 386–399. Springer, Heidelberg (2008)

13. Sugihara, R., Gupta, R.K.: Path planning of data mules in sensor networks. ACM Transaction on Sensor Networks 8(1), 1:1–1:27 (2011)

14. Yuan, Y., Peng, Y.: Racetrack: an approximation algorithm for the mobile sink routing problem. In: Nikolaidis, I., Wu, K. (eds.) ADHOC-NOW 2010. LNCS, vol. 6288, pp. 135–148. Springer, Heidelberg (2010)

15. Bhadauria, D., Tekdas, O., Isler, V.: Robotic data mules for collecting data over sparse sensor fields. J. Field Robot. 28(3), 388–404 (2011)

16. Huang, P., Xiao, L., Soltani, S., Mutka, M., Xi, N.: The evolution of mac protocols in wireless sensor networks: A survey. IEEE Communications Surveys Tutorials PP(99), 1–20 (2012)

17. Castalia wsn simulator framwork for omnet++, http://www.castelia.org

18. Concorde tsp solver, http://www.tsp.gatech.edu/concorde.html

Towards Fault-Tolerant Chord P2P System: Analysis of Some Replication Strategies*

Rafał Kapelko

Institute of Mathematics and Computer Science
Faculty of Fundamental Problems of Technology,
Wrocław University of Technology, Poland
rafal.kapelko@pwr.wroc.pl

Abstract. In Peer-to-peer systems the process of unexpected nodes failures becomes critical. When some nodes depart from the network without uploading gathered information into the system, some data which are stored on the departed nodes will not be available. In our paper we address this issue for Chord-based P2P systems. We investigate two replication schemes: successor list and multiple Chord rings. We propose a novel analytical model to describe the process of unexpected nodes failures for investigated replication scheme. Let $\epsilon > 0$. Let d be replication degree in Chord with successor-list replication. We prove that for large n after failure of $n^{1-\frac{1}{d}-\epsilon}$ nodes from Chord with successor list replication no document is lost with high probability. We also prove that for k multiple chord rings after unexpected departures of $\frac{n^{1-\frac{1}{k}}}{n^{\frac{1}{2k}}}$ nodes, the system stays in safe configuration with high probability. As a practical illustration of our detailed theoretical analysis of replication schemes we discuss the recovery mechanism of partially lost information. We show how the system fault tolerance changes depending of different replica degree.

Keywords: peer-to-peer network, Chord, replication of documents, reliability.

1 Introduction

In recent years, Peer-to-Peer (P2P) networks (see e.g. [21], [17], [7] [15]) have been powerful environments for successful sharing of certain resources. Despite the advantages of the P2P systems, there exists a significant problem with fault tolerance. When some nodes depart from the system without uploading gathered information, some data which are stored on the departured nodes may be lost.

Among the structured P2P systems, Chord protocol introduced in [21] and developed in [14] is one of the most popular and we focus on it in this paper. In the Chord P2P system each of such unexpected node failures effects in loosing some data stored in the node (see [2], [11]).

The replication techniques are used for storing documents at many nodes in the system. The benefits of replication are that it can improve availability of documents

---

* Supported by grant nr 2012/S20027 of the Institute of Mathematics and Computer Science, Wrocław University of Technology, Poland.

Y. Ishikawa et al. (Eds.): APWeb 2013, LNCS 7808, pp. 686–696, 2013.

in case of failure [22]. Different strategies for replication in Chord P2P system such as successor-list, symmetric replication or multiple hash functions are discussed in many publications (see [5], [18], [10], [16], [3], [20], [19]). In the replication schemes successor-list replication is very popular and often applied in ring-based networks.

Our main contribution include:

- We propose a novel analytical model describing the process of unexpected nodes failures. Let $\epsilon > 0$. Let d be replication degree in Chord with successor-list replication. We prove that for large n after failure of $n^{1-\frac{1}{d}-\epsilon}$ nodes from Chord with successor list replication no document is lost with high probability (see Theorem 3). Our consideration dedicated to Chord P2P system is also applicable for other overlays based on ring such as Pastry [7] (with some modification), RP2P [1], Cassandra [12], CRP-overlay [13] and others.
- We also investigate the fault tolerance of overlay system called k multiple chord rings. We prove that after simultaneous unexpected departures of $\frac{n^{1-\frac{1}{k}}}{n^{\frac{1}{2k}}}$ nodes from k multiple chord rings no information disappears with high probability.
- As a practical application of our theoretical estimation we discuss the recovery mechanism of partially lost information in both replication schemes. We show how fault tolerance of Chord system changes depending of different replica degree. We also explain how to choose the proper number of replication to optimize those two replication strategies.

The remainder of this article is organised as follows. In Section 2 we describe the successor-list replication scheme in Chord P2P system and present analytical formulas describing the resistance to loss of document. Section 3 presents the overlay k − Chord. Then, in Section 4 we discuss the recovery mechanism of partially lost information. Finally, Section 5 summaries and concludes our work.

2   Successor-List Replication

In this section, we briefly describe the successor-list replication scheme in Chord P2P system [5]. For a more detailed description of Chord P2P system the reader is referred to [21]. Then, we analyze the process of unexpected departure.

In Chord P2P system nodes are arranged in a logical ring. The positions of nodes and *documents* are created by a hash operation, which results in random placement into the Chord ring. Let n denotes the number of nodes in Chord. Then, the identifier space of the nodes is defined as a set of integers $\{0, 1, \ldots, n-1\}$.

Let $(0, 1, \ldots, n-1)$ denotes Chord ring. The first successor of a node with identifier p is the first node found going in clockwise direction on the ring starting at p. The predecessor of a node with identifier p is the first node found going anti-clokwise direction on the ring starting at p. Each node p is responsible for storing *documents* between p's predecessor and p. To ensure the connectivity of the chord ring, each node knows the

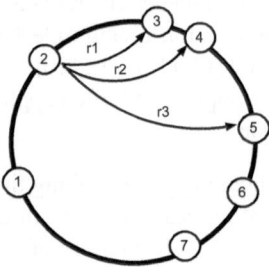

Fig. 1. Successor-list replication of degree 4

first s nodes which identifier is greater than the node's. These nodes are the node's successors. Additionally, each node maintains a table of fingers of size $O(\log(n))$, which point to other nodes in the ring. Hence the size of routing table is $O(\log(n) + s)$.

Let $d < s$. For a replication degree of d in successor-list replication a *document* is stored on the node p which is responsible for storing *document* and $d - 1$ immediate successor of p. Let the node p leaves the system in an unexpected way, i.e without uploading its *documents* into the system. After the departure there are still $(d - 1)$ replicas of *documents* of the node p.

For $p = 0, 1, \ldots, n - 1$ and $d > 1$ we define family of sets by the formula

$$U_{p,d} = \{p \mod n, \ (p+1) \mod n, \ \ldots (p+d-1) \mod n\}.$$

Definition 1. *Let n denotes the number of nodes in Chord. Let $A = \{n_1, \ldots, n_l\} \subseteq \{0, \ldots, n-1\}$ be a random subset of nodes from Chord with successor-list replication of degree d. We say that, the set A is **safe**, if*

$$\forall_{p \in \{0,1,\ldots,n-1\}} U_{p,d} \cap A \neq U_{p,d}.$$

*We say that, the set A is **unsafe**, if A is not safe.*

Remark 1. *Notice that, if the set A is safe, then no information disappears from Chord with successor-list replication of degree d after simultaneous unexpected departure of all nodes from A.*

Theorem 2. *Let $d \geq 2$. Let n be the number of nodes in Chord with successor-list replication of degree d and let A be the set of nodes. If $|A| \leq \frac{n^{1-\frac{1}{d}}}{n^{\frac{1}{2d}}}$ then $\Pr[A$ is safe$] \geq 1 - \frac{1}{\sqrt{n}}$.*

Proof. Firstly observe that for $|A| < d$ we have $\Pr[A$ is safe$] = 1$. Therefore, we may assume that $|A| \geq d$. Then

$$\Pr[A \text{ is unsafe}] = \Pr\left[\exists_{p \in \{0,1,\ldots,n-1\}} U_{p,d} \cap A = U_{p,d}\right] =$$

$$\Pr\left[\bigvee_{p=0}^{n-1} U_{p,d} \cap A = U_{p,d}\right] \le \sum_{p=0}^{n-1} \Pr\left[U_{p,d} \cap A = U_{p,d}\right] =$$

$$n\frac{\binom{n-d}{|A|-d}}{\binom{n}{|A|}} \le \frac{n|A|^d}{(n-(d-1))^d} \le \frac{1}{\sqrt{n}}\frac{1}{\left(1-\frac{d-1}{n}\right)^d} \le \frac{1}{\sqrt{n}}.$$

□

Theorem 3. *Let* $\epsilon > 0$. *Let* $d \ge 2$. *Let* n *be the number of nodes in Chord with successor-list replication of degree* d *and let* A *be the set of nodes. If* $|A| \le n^{1-\frac{1}{d}-\epsilon}$ *then*

$$\lim_{n \to \infty} \Pr[A \text{ is unsafe}] = 0$$

Proof. As in the proof of Theorem 2, observe that for $|A| < d$ we have $\Pr[A \text{ is safe}] = 1$. Therefore, we may assume that $|A| \ge d$. Then

$$\lim_{n \to \infty} \Pr[A \text{ is unsafe}] \le \lim_{n \to \infty} \left(n\frac{\binom{n-d}{n^{1-1/d-\epsilon}-d}}{\binom{n}{n^{1-1/d-\epsilon}}}\right) = 0.$$

Hence the proof is finished.

□

3 Multiple Chord Rings

The overlay system k − Chord was introduced and developed in [8], [9]. It the system of multiple chord rings. From the formal point of view k − Chord is the structure

$$k - \text{Chord} = \bigcup_{i=1}^{k} R_{i,n},$$

where $R_{i,n}$ is the i−th Chord ring. The nodes of the first Chord ring are expressed in the form $R_{1,n} = (0, 1, 2, \ldots, n-1)$ and on the i−th ring $(i = 2, \ldots, k)$ are generated by independent random permutation $\sigma_i \in S_n$ i.e., $R_{i,n} = (\sigma_i(0), \sigma_i(1), \ldots, \sigma_i(n-1))$. Each *document* has a unique position and is mapped into the same location on different Chord rings.

In k − Chord hybrid system each node maintains a k dimensional finger table and a successor list of a size s ($s > k$) on each Chord ring. Hence the routing table is of size $O(k(\log(n) + s))$. With k being constant and small (e.g, $k \le 5$.), the size of routing table is $O(\log(n) + s)$ which is the same like in Chord. Let us consider the event, when the node p leaves the system in an unexpected way without uploading storing documents into the system. In the classical Chord such event effects in loosing information stored on the node p. The situation changes in hybrid k − Chord. There are $(k-1)$ replicas of documents stored on the nodes $\sigma_i(p)$ on the i−th ring $(i = 2, \ldots, k)$. The probability of loosing information stored on the node p equals $\frac{1}{(n-1)^{k-1}}$.

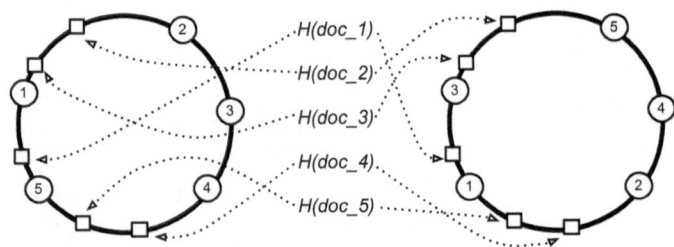

Fig. 2. An example of 2–Chord where 2–th ring is generated by permutation (3,5,4,2,1)

Definition 4. *Let* $A = \{n_1, \ldots, n_l\} \subseteq \{0, \ldots, n-1\}$ *be a random subset of nodes from the structure* k − Chord *with* n *nodes. We denote by* $K_{A,1}$ *the unions of intervals controlled by nodes from* A *in the first circle and we denote by* $K_{A,i}$ *the unions of intervals controlled by nodes from* A *in the* i−*th circle* $(i = 2, \ldots, k)$. *We say that the set* A *is **safe**, if* $\bigcap_{i=1}^{k} K_{A,i} = \emptyset$. *We say that the set* A *is **unsafe** if* A *is not safe.*

Let us remark that, if the set A is safe then no information disappears from the hybrid system after simultaneous unexpected departure of all nodes from A.

Theorem 2. *Let* $k \geq 2$. *Let* n *be the number of nodes in the structure* k − Chord. *let* A *be a random subset of nodes from the structure* k − Chord. *If* $|A| \leq \frac{n^{1-\frac{1}{k}}}{n^{\frac{1}{2k}}}$ *then* $\Pr[A \text{ is safe}] \geq 1 - \frac{1}{\sqrt{n}}$.

Notice that, the bounds in Theorem 2 do not depend on the number of documents put into the system.

Proof. Let $A = \{n_1, \ldots, n_l\} \subseteq \{0, \ldots, n-1\}$ be a random subset of nodes from the structure k − Chord with n nodes. Hence $|A| = l$. Let us recall that $K_{A,1} = \bigcup_{s=1}^{l} A_s$ are the unions of intervals controlled by nodes from A in the first circle, and $K_{A,i}$ are the unions of intervals controlled by nodes from A in the i−th circle $(i = 2, \ldots, k)$.

Notice that, the positions of nodes in the i−th copy of Chord are chosen independently from the position of nodes from the first copy. Let $\sigma_i(0), \sigma_i(1) \ldots \sigma_i(n-1)$ denotes the positions of nodes in the i−th copy of Chord, where $\sigma_i \in S_n$ are random and independent $(i = 2, \ldots, k)$. Let us consider an interval A_s controlled by a node n_s $(s = 1, \ldots, l)$.

The reason why an interval controlled by a node $\sigma_i(n_p)$ (p=1,…,l) may have a nonempty intersection with the set A_s is caused by the fact that $(\exists_p)\sigma_i(n_p) = n_s$. Therefore,

$$\Pr[A_s \cap K_{A,i} \neq \emptyset] = \Pr[(\exists_p)(\sigma_i(n_p) = n_s)] = \Pr[\bigcup_{p=1}^{l} \sigma_i(n_p) = n_s] = \frac{l}{n}.$$

Hence

$$\Pr[A \text{ is unsafe}] = \Pr\left[\bigcap_{i=1}^{k} K_{A,i} \neq \emptyset\right] \leq \sum_{s=1}^{l} \Pr\left[A_s \cap \bigcap_{i=2}^{k} K_{A,i} \neq \emptyset\right] \leq$$

$$\sum_{s=1}^{l} \Pr\left[\bigwedge_{i=2}^{k} A_s \cap K_{A,i} \neq \emptyset\right] = \sum_{s=1}^{l}\prod_{i=2}^{k} \Pr[A_s \cap K_{A,i} \neq \emptyset] \leq l\left(\frac{l}{n}\right)^{k-1}.$$

Finally notice that, for $l \leq \frac{n^{1-\frac{1}{k}}}{n^{\frac{1}{2k}}}$ we have $l\left(\frac{l}{n}\right)^{k-1} \leq \frac{1}{\sqrt{n}}$, so the Theorem is proved.

$\square$

4   The Recovery Process

In this section we discuss the recovery mechanism of partially lost information in the replication schemes: successor list and k − Chord. We show how fault tolerance of Chord system changes depending of different replica degree. Notice that the recovery mechanism based on Queueing Theory was introduced in [2] and applied for another modification of Chord. The notations defined in this chapter are listed in Table 2. From the Little's Law from Queueing Theory (see e.g. [4]) we get two equations

$$N = \mu \cdot T, \qquad N_r = u \cdot \mu \cdot T_r, \qquad (1)$$

Table 1. Time for recovery of partially lost information T_{recovery} as the function of the number of nodes $N \in [10^4, 10^6]$ for different replica degrees $D = 2, 3, 4, 5$ with fixed $T = 1800$ sec and $u = 5\%$

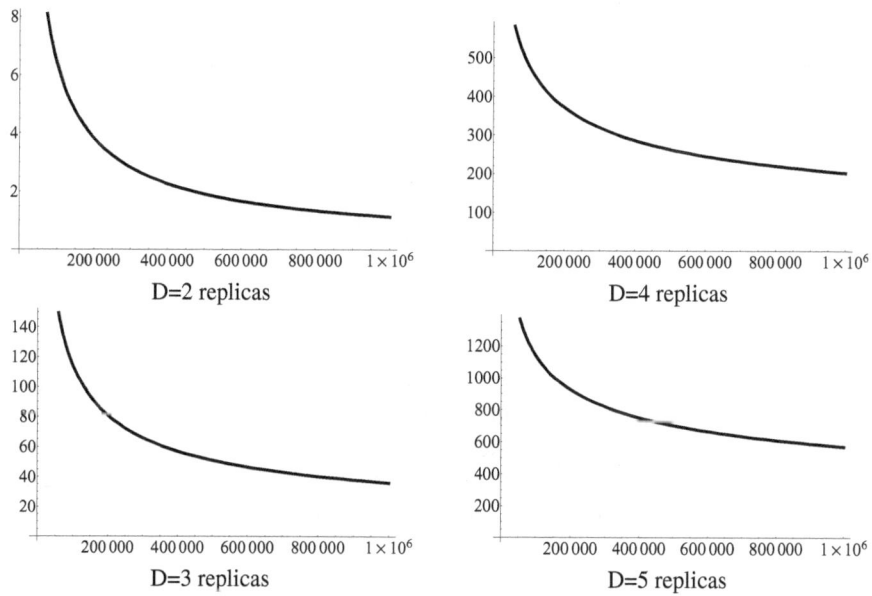

D=2 replicas

D=4 replicas

D=3 replicas

D=5 replicas

Table 2. Notations

Symbol	Meaning
N	the average number of nodes in the system
N_r	the medium number of nodes waiting for complementing
T	the average time a node spends in a system (during one session)
u	the proportion of unexpected departures among all departures
T_r	the average time needed for recovery of partially lost information
μ	the average number of departures from the system in a second
m	the average number of unexpected departures from the system in a second
$T_{c,d}$	the time in which nodes: $p, p+1, \ldots p+d-1$, check the node $p+d$
δ	the average number of documents in the system per one node
d_s	the size of the information item in the system
t_s	the transmission speed
T_{recovery}	time for recovery of partially lost information for different replica degrees

so

$$T_r = \frac{T}{u}\frac{N_r}{N}.$$

For two replication schemes discussed in this paper we have obtained the following safety bound

$$N_r = \frac{N^{1-\frac{1}{D}}}{N^{\frac{1}{2D}}}, \tag{2}$$

where D is a replication degree ($D = d$ for successor-list replication of degree d, $D = k$ for k $-$ Chord) . If we want to keep our system in a safe configuration with high probability greater than $1 - \frac{1}{\sqrt{N}}$, then the following inequality must hold

$$T_r \leq \frac{T}{u}\frac{N_r}{N}. \tag{3}$$

Therefore, time for recovery of partially lost information for different replica degrees is:

$$T_{\text{recovery}} = \frac{T}{u}\frac{N_r}{N}. \tag{4}$$

Let $T = 30$ minutes (see [6], [23]). Suppose that $u = 5\%$ nodes leave the system in unexpected way without uploading gathered information. In Table 1 we put four figures describing the function T_{recovery} of the number of nodes $N \in [10^4, 10^6]$ for different replica degrees $D = 2, 3, 4, 5$ with fixed $T = 1800$ sec. and $u = 5\%$. The system gains more time for recovery of partially lost information when replica degree increases.

The following table contains the time for recovery of partially lost information with fixed parameters $N = 10^5$, $T = 1800$ sec, $u = 5\%$, for different replica degree:

d	2	3	4	5	6	7	8
T_{recovery}	6 sec	113 sec	8 min	18.9 min	33.7 min	50.9 min	1h 9 min

---

Algorithm 1. Recovery process in Chord with successor-list replication of degree d

1: nodes: $p + 1, \ldots p + d - 1$, check the node $p + d$
2: **if** the node $(p + d)$ left the system in unexpected way **then**
3:     nodes: $p + 1, \ldots p + d - 1$, ask their successors for a portion of documents
4:     nodes: $p + 1, \ldots p + d - 1$, send some portion of documents to some of their successors
5: **end if**

4.1 Successor-List Replication

Let us consider the network with $N = 10^5$ average number of nodes in the system ($T = 30$ minutes, $u = 5\%$). From equation (1) we get $m = \frac{N}{T}u \approx 2.77$, so approximately 2.77 nodes unexpectedly leave the network in one second.

Assume that nodes: $p+1, \ldots p+d-1$ in the network check the node $p+d$ periodically with the period of $T_{c,d}$ seconds (see Algorithm 1). Assume that $T_{c,d} < 2$ sec.

During this time nodes: $p + 1, \ldots p + d - 1$, do not know that the node $p + d$ left the network. Then, there are $T_r - T_{c,d}$ seconds for information recovery (see inequality (3)). During this time nodes: $p + 1, \ldots p + d - 1$, ask their successors for a portion of documents in 0.2 sec. Then, nodes: $p + 1, \ldots p + d - 1$, wait for receiving necessary information. Nodes $p + 1, \ldots p + d - 1$, send some portion of documents to some of their successors. Therefore, the following inequality guarantees the system is in a safe configuration with high probability

$$T_{c,d} + 0.2 + 3\delta\frac{d_s}{t_s} \leq \frac{T}{u}N^{-\frac{3}{2d}}. \tag{5}$$

The inequality (5) can be used in a practical way. It shows how to choose the proper number of replication degree d to optimize the recovery process. For estimated parameters δ, d_s, t_s, T, u, N of the network and fixed $T_{c,d}$ the optimal replica degree is the minimal natural d ($d > 1$) for which the inequality (5) holds.

Let us assume, that the transmision speed $t_s = 100$ kB/s. As in [23] we consider two scenarios with size of documents, d_s of 1.57 kB and 1 MB, respectively.

The Table 3 contains the upper bounds for the average number of documents per one node for different replica degree, when the system stays in a safe configuration.

Table 3. Values of the upper bound of the average number of documents per one node for different replica degree and size of documents with fixed $T_{c,d} = 2$ sec., $T = 30$ minute, $u = 5\%$, $N = 10^5$, $t_s = 100$ kB/s

| | $d = 2$ | $d = 3$ | $d = 4$ | $d = 5$ |
	$T_{c,2} = 2$	$T_{c,3} = 2$	$T_{c,4} = 2$	$T_{c,5} = 2$
$d_s = 1,57$ kB	89	2370	10145	24123
$d_s = 1$ MB	0	3	15	37

Algorithm 2. Recovery process in k − Chord

1: nodes: $p + 1, \ldots p + k - 1$, check the node $p + k$
2: **if** the node $(p + k)$ left the system in unexpected way **then**
3:      nodes: $p + 1, \ldots p + k - 1$, localise in the remind $(k-1)$ Chord rings nodes responsible for storing lost documents
4:      nodes: $p + 1, \ldots p + k - 1$, ask localised nodes in the remind $(k - 1)$ Chord rings for receiving necessary documents
5:      nodes: $p + 1, \ldots p + k - 1$, send received documents to some of their successors
6: **end if**

4.2 k − Chord

In k − Chord network the recovery process of partialy lost information is more complicated than in successor-list replication.

Let us consider the network with $N = 10^5$ average number of nodes in the system ($T = 30$ minutes, $u = 5\%$). From equation (1) we get $m = \frac{N}{T} u \approx 2.77$, so approximately 2.77 nodes unexpectedly leave the network in one second.

Assume that nodes: $p + 1, \ldots p + k - 1$, in each chord ring check the node $p + k$ periodically with the period of $T_{c,k}$ seconds (see Algorithm 2). Assume that $T_{c,k} < 2$ sec.

During this time nodes: $p + 1, \ldots p + k - 1$, do not know that the node $p + k$ left the network. Then, there are $T_r - T_{c,k}$ seconds for information recovery (see inequality (3)). During this time nodes: $p + 1, \ldots p + k - 1$, localise in the remind $(k - 1)$ Chord rings nodes responsible for storing lost documents. The time for fulfilling this operation equals approximately $\frac{1}{2} \lg_2 N \cdot 0.2$ sec. Then, nodes: $p + 1, \ldots p + k - 1$, wait for receiving necessary information and the time is approximately $\delta \frac{d_s}{t_s}$. Nodes: $p + 1, \ldots p + k - 1$, send received documents to some of their successors. Therefore, the following inequality guarantees the system is in a safe configuration with high probability

$$T_{c,k} + \frac{1}{2} \lg_2 N \cdot 0.2 + 2\delta \frac{d_s}{t_s} \leq \frac{T}{u} N^{-\frac{3}{2k}}. \tag{6}$$

Similary, to successor-list replication inequality (6) can be used in a practical way. It shows how to choose the proper number of k Chord rings to optimize the recovery process. For estimated parameters $\delta, d_s, t_s, T, u, N$ of the network and fixed $T_{c,k}$, the optimal replica degree is the minimal natural k ($k > 1$) for which the inequality (6) holds.

Table 4. Values of the upper bound of the average number of documents per one node for different k − Chord and size of documents with fixed $T_{c,k} = 2$ sec., $T = 30$ minute, $u = 5\%$, $N = 10^5$, $t_s = 100$ kB/s

	$k = 2$ $T_{c,2} = 2$	$k = 3$ $T_{c,3} = 2$	$k = 4$ $T_{c,4} = 2$	$k = 5$ $T_{c,5} = 2$
$d_s = 1,57$ kB	80	3502	15165	36132
$d_s = 1$ MB	0	5	23	56

Let us assume that the transmition speed $t_s = 100$ kB/s. As in [23] we consider two scenarios with size of documents, d_s of 1.57 kB and 1 MB, respectively.

The Table 4 contains the upper bounds for the average number of documents per one node for different k − Chord, when the system stays in a safe configuration.

5   Conclusion

In this paper, we study two replication strategies for Chord P2P system: successor-list and multiple Chord rings. In both cases we obtain estimation describing the resistance to loss of documents. As a practical illustration of our estimated safety bounds we discuss the recovery mechanism of partially lost information. We show how fault tolerance of Chord system changes depending of different replica degree and explain how to choose the proper number of replication to optimize these two replication strategies. In future research we plan to experimentaly evaluate efficiency of proposed recovery algorithms. We also plan to extend our work to symmetric replication scheme.

References

1. Chen, S., Li, Y., Rao, K., Zhao, L., Li, T., Chen, S.: Building a scalable P2P network with small routing delay. In: Zhang, Y., Yu, G., Bertino, E., Xu, G. (eds.) APWeb 2008. LNCS, vol. 4976, pp. 456–467. Springer, Heidelberg (2008)
2. Cichoń, J., Jasiński, A., Kapelko, R., Zawada, M.: How to improve the reliability of chord? In: Meersman, R., Tari, Z., Herrero, P. (eds.) OTM-WS 2008. LNCS, vol. 5333, pp. 904–913. Springer, Heidelberg (2008)
3. Cichoń, J., Kapelko, R., Marchwicki, K.: Brief announcement: A note on replication of documents. In: Défago, X., Petit, F., Villain, V. (eds.) SSS 2011. LNCS, vol. 6976, pp. 439–440. Springer, Heidelberg (2011)
4. Cooper, R.B.: Introduction To Queueing Theory, 2nd edn. Elsevier North Holland, Inc. (1981)
5. Dabek, F., Kaashoek, M., Karger, D., Morris, R., Stoica, I.: Wide-area cooperative storgae with cfs. In: 18th ACM Symposium on Operating Systems Principles (SOSP), New York, USA, pp. 202–215 (2001)
6. Derek, L., Zhong, Y., Vivek, R., Loguinov, D.: On lifetime-based node failure and stochastic resilience of decentralized peer-to-peer networks. IEEE/ACM Transactions on Networking 5(15), 644–656 (2007)
7. Rowstron, A., Druschel, P.: Pastry: Scalable, Decentralized Object Location, and Routing for Large-Scale Peer-to-Peer Systems. In: Guerraoui, R. (ed.) Middleware 2001. LNCS, vol. 2218, pp. 329–350. Springer, Heidelberg (2001)
8. Flocchini, P., Nayak, A., Xie, M.: Hybrid-chord: A peer-to-peer system based on chord. In: Ghosh, R.K., Mohanty, H. (eds.) ICDCIT 2004. LNCS, vol. 3347, pp. 194–203. Springer, Heidelberg (2004)
9. Flocchini, P., Nayak, A., Xie, M.: Enhancing peer-to-peer systems through redundancy. IEEE Journal on Selected Areas in Communications 1(25), 15–24 (2007)
10. Ghodsi, A., Alima, L.O., Haridi, S.: Symmetric replication for structured peer-to-peer systems. In: Moro, G., Bergamaschi, S., Joseph, S., Morin, J.-H., Ouksel, A.M. (eds.) DBISP2P 2005 and DBISP2P 2006. LNCS, vol. 4125, pp. 74–85. Springer, Heidelberg (2007)

11. Park, G., Kim, S., Cho, Y., Kook, J., Hong, J.: Chordet: An Efficient and Transparent Replication for Improving Availability of Peer-to-Peer Networked Systems. In: 2010 ACM Symposium on Applied Computing (SAC), Sierre, Switzerland, pp. 221–225 (2010)
12. Lakshman, A., Malik, P.: Cassandra: a decentralized structured storage systems. ACM SIGOPS Operating Systems Review 2(44), 35–40 (2010)
13. Li, Z., Hwang, K.: Churn resilient protocol for massive data dissemination in p2p networks. IEEE Transactions on Parallel and Distributed Systems 22, 1342–1349 (2011)
14. Liben-Nowell, D., Balakrishnan, H., Karger, D.: Analysis of the Evolution of Peer-to-Peer Systems. In: ACM Conference on Principles of Distributed Computing, Monterey, California, USA, pp. 233–242 (2002)
15. Maymounkov, P., Mazières, D.: Kademlia: A Peer-to-Peer Information System Based on the XOR Metric. In: Druschel, P., Kaashoek, M.F., Rowstron, A. (eds.) IPTPS 2002. LNCS, vol. 2429, pp. 53–65. Springer, Heidelberg (2002)
16. Pitoura, T., Ntarmos, N., Triantafillou, P.: Replication, load balancing and efficient range query processing in dHTs. In: Ioannidis, Y., Scholl, M.H., Schmidt, J.W., Matthes, F., Hatzopoulos, M., Böhm, K., Kemper, A., Grust, T., Böhm, C. (eds.) EDBT 2006. LNCS, vol. 3896, pp. 131–148. Springer, Heidelberg (2006)
17. Ratnasamy, S., Francis, P., Handley, M., Karp, R., Shenker, S.: A Scalable Content-addressable Network. In: SIGCOMM 2001, San Diego, California, USA, pp. 161–172 (2001)
18. Rieche, S., Wehrle, K., Landsiedel, O., Götz, S., Petrak, L.: Reliability of data in structured peer-to-peer systes. In: HOT-P2P 2004, Volendam, The Netherland, pp. 108–113 (2004)
19. Kapelko, R.: Towards efficient replication of documents in chord: Case (r,s) erasure codes. In: Liu, B., Ma, M., Chang, J. (eds.) ICICA 2012. LNCS, vol. 7473, pp. 477–483. Springer, Heidelberg (2012)
20. Shafaat, T.M., Ahmad, B., Haridi, S.: ID-replication for structured peer-to-peer systems. In: Kaklamanis, C., Papatheodorou, T., Spirakis, P.G. (eds.) Euro-Par 2012. LNCS, vol. 7484, pp. 364–376. Springer, Heidelberg (2012)
21. Stoica, I., Morris, R., Karger, D., Kaashoek, M.F., Balakrishnan, H.: Chord: A Scalable Peer-to-Peer Lookup Service for Internet Applications. In: SIGCOMM 2001, San Diego, California, USA, pp. 149–160 (2001)
22. Vu, Q., Lupu, M., Ooi, B.: Peer-to-Peer Computing. Springer (2010)
23. Wang, C., Harfoush, K.: On the stability-scalability tradeoff of dht deployment. In: INFOCOM 2007, 26th IEEE International Conference on Computer Communications, Anchorage, Alaska, USA, pp. 2207–2215 (2007)

A MapReduce-Based Method
for Learning Bayesian Network from Massive Data

Qiyu Fang[1], Kun Yue[1,2,*], Xiaodong Fu[3], Hong Wu[1], and Weiyi Liu[1]

[1] Department of Computer Science and Engineering, School of Information Science and
Engineering, Yunnan University, 650091, Kunming, China
[2] Key Laboratory of Software Engineering of Yunnan Province, 650091, Kunming, China
[3] Faculty of Information Engineering and Automation, Kunming University of Science and
Technology, 650500, Kunming, China
kyue@ynu.edu.cn

Abstract. Bayesian network (BN) is the popular and important probabilistic
graphical model for representing and inferring uncertain knowledge. Learning
BN from massive data is the basis for uncertain-knowledge-centered inferences,
prediction and decision. The inherence of massive data makes BN learning be
adjusted to the large data volume and executed in parallel. In this paper, we
proposed a MapReduce-based approach for learning BN from massive data by
extending the traditional scoring & search algorithm. First, in the scoring
process, we developed map and reduce algorithms for obtaining the required
parameters in parallel. Second, in the search process, for each node we developed map and reduce algorithms for scoring all the candidate local structures in
parallel and selecting the local optimal structure with the highest score. Thus,
the local optimal structures of each node are merged to the global optimal one.
Experimental result indicates our proposed method is effective and efficient.

Keywords: Massive data, Bayesian network, Learning, Scoring and search,
MapReduce.

1 Introduction

With the rapid development of Internet and Web applications, more and more data are
generated and collected in scientific experiments, e-commerce and social network
applications [1]. In this case, computationally intensive analysis results in complex
and stringent performance demands that are not satisfied by any existing data management [2, 3]. Centered on massive data management and analysis, data intensive
computing may generate many queries, each involving access to supercomputer-class
computations on gigabytes or terabytes of data [4]. Thus, knowledge discovery from
massive data is naturally indispensable for the applications of data intensive paradigms, such as Web 2.0 and large-scale business intelligence. However, it is quite
challenging by incorporating the specialties of massive data [1].

---

* Corresponding author.

Y. Ishikawa et al. (Eds.): APWeb 2013, LNCS 7808, pp. 697–708, 2013.
© Springer-Verlag Berlin Heidelberg 2013

As one kind of important knowledge, probabilistic graphical models (PGMs) [5] comprise any model that uses the language of graphs to facilitate the representation and resolution of complex problems that use probability as representation of Statistical uncertainty knowledge. The most important and popular PGM is Bayesian network (BN) [5]. It is a directed acyclic graph (DAG) to represent a joint probability distribution of random variables and each node has a conditional probability table (CPT) that quantitatively reflects variable dependency. BN provides a framework for compact representation and efficient inferences about joint probability distribution of several interdependent parameters [6], and has been widely used in real applications.

Generally for BN construction, there are two kinds of methods: constructing from expert knowledge and learning from data [5]. The former relies on the knowledge of experts which is of subjectivity and one-sidedness frequently. However, the latter provides us a feasible way to obtain uncertain knowledge implied in data, frequently used to refine the initial expert knowledge. Learning BN from data has been paid much attention in machine learning and data mining [5, 6, 7, 8]. The inherence of massive data demands the task of BN learning to be adjusted to the large volume of sample data and executed in parallel. It is known that the tasks of BN learning include DAG and CPT learning [5], in which the former is the critical and challenging step. This makes us focus on the DAG learning from massive data in this paper.

Currently, dependency analysis and scoring & search are the methods for learning BN from data [6]. The typical dependency analysis methods focus on the dependence between data objects [6]. The typical scoring & search algorithms are K2 [8], CH [9], MMHC [10] and MDL [11]. These methods choose the candidate BN structure based on the scoring function quantitatively. In this paper, we adopt the scoring & search as the representative and initial exploration of BN learning from massive data, while the basic ideas can be universally suitable for the dependency analysis as well.

The crucial processes in K2 are structure scoring and structure search [8]. The structure search function is applied for each node's candidate structures, possible subgraphs made up of nodes likely to connect the current node, to compute their scores respectively, so the local optimal structure with the highest score can be obtained. Then, local optimal structures are merged to achieve the global optimal one. However, K2 will have to be extended w.r.t. the storage and retrieval of massive data for real applications. As the data volume grows, the execution time of K2 will be increased rapidly due to the expensive cost for computing the required parameters.

The scoring function [8] in K2 is as follows:

$$g(i, \pi_i) = \prod_{i=1}^{n} \prod_{j=1}^{q_i} \frac{(r_i - 1)!}{(N_{ij} + r_i - 1)!} \prod_{k=1}^{r_i} N_{ijk}! \tag{1}$$

where the parameters N_{ij} and N_{ijk} can be obtained by statistics from data. Actually, the values of N_{ij} and N_{ijk} of each record can be obtained in parallel, and then the values of these two parameters of all records can be summed to obtain the final values required in Equation (1). It can be observed that the parameters can be obtained separately (i.e., independently), and the scoring function for local structures can be evaluated separately as well. For each node, the local optimal structures could be chosen by scoring all candidate ones separately. So, we can merge all local optimal structures to obtain the global optimal one as the final result. The above observation and discussion illustrate that both the scoring and search process can be executed in parallel.

From the parallel point of view, some parallel methods for BN learning [12, 13, 14, 15] have been achieved. As a representative, the dynamic programming algorithm [14] is used to obtain the local and global optimal structures. However, these methods only focus on the parallelization during algorithm execution and the allocation for processor cooperation, while have not concerned the throughput computing to access massive data. Thus, it is desired to explore a BN learning method that can be executed in parallel and efficiently to access massive data.

Fortunately, MapReduce is a programming model for processing and generating large data sets [16]. It not only offers a parallel programming model, but also can conduct and process massive data. It has been used in data analysis and dependency-analysis-based BN learning [17, 18, 19, 20, 21]. Although MapReduce cannot reveal its advantages on conducting data of normal volume, it is suitable for learning BN from massive data. Using MapReduce to extend the current centralized K2 algorithm, we are to explore the parallel and data-intensive K2 algorithm. To make the MapReduce-based K2 fit the demand of massive data definitely is the purpose of our study. Hadoop is an open source implementation of MapReduce, and has been used extensively outside of Google by a number of organizations [22]. In this paper, we proposed a MapReduce-based K2 in which the scoring and search process can be executed in parallel on Hadoop. Generally, the contributions of this paper are as follows:

- We developed the architecture of BN learning system by extending the traditional K2 algorithm based on MapReduce.
- We introduced the scoring function based on MapReduce, which can be used to obtain all parameters in parallel. In the map process, the parameters in each record can be obtained by statistics. In the reduce process, the parameters obtained through the map process will be summed to obtain the requited ones.
- We proposed the search function based on MapReduce. In the map process, the candidate local structures are generate and scored in parallel. In the reduce process, all the local optimum structures are merged to generate the global optimal one. Thus, we obtained the BN structure from massive data by extending K2 algorithm.
- We implemented our proposed algorithms for learning BN from massive data. Experimental results show the feasibility of our method.

The rest of this paper is organized as follows: Section 2 introduces K2 algorithm and MapReduce. Section 3 extends the K2 algorithm based on MapReduce. Section 4 gives experimental results. Section 5 concludes and discusses future work.

2   Preliminaries

2.1   Scoring and Search for BN Learning

A BN is a DAG $G=(V, E)$, in which the following holds [5]:

(1) V is the set of random variables and makes up the nodes of the network.

(2) E is the set of directed edges connecting pairs of nodes. An arrow from node X to node Y means that X has a direct influence on Y (X, $Y \in V$ and $X \neq Y$).

(3) Each node has a CPT that quantifies the effects that the parents have on the node. The parents of node X are all those that have arrows pointing to it.

Let Z be a set of n discrete parameters, where a variable x_i in Z has r_i possible value assignments: $(v_{i1}, ..., v_{ij})$. Let D be a database with m records (i.e., rows), each of which contains a value assignment for each variable in Z. Let B_S denote a belief-network structure containing just the parameters in Z. Each variable x_i in B_S has a set of parents, which we represent with a list of r_i parameters. Let w_{ij} denote the jth unique instantial record of L_i relative to D. Suppose there are q_i such unique instantial records of L_i. Define N_{ijk} as the number of cases in D in which variable x_i has the value v_{ik} and L_i is instantiated as w_{ij}. Here let

$$N_{ij} = \sum_{k=1}^{r_i} N_{ijk} \tag{2}$$

For each node i and its parents π_i, we use Equation (1) to compute the score of the candidate structures of each node, and use $Pred(x_i)$ to denote the set of nodes preceding x_t. We consider the preceding parent of each node according to the nodes' sequence. In the scoring function $g(i, \pi_i)$, N_{ij} and N_{ijk} can be obtained by statistics counting from the given massive data in parallel, which will be discussed in Section 3. The following pseudo-code expresses the heuristic search algorithm, called K2 [8].

Algorithm 1. K2

```
Input:
1) A set of n nodes, a sequence of the nodes
2) An upper bound u on the number of parent nodes
3) A database D containing m records
Output: A printout of the parents for each node
Steps:
For i:=1 to n Do
    π:=∅
    P_old:=g(i, π) //by Equation (1)
    OKTo Proceed:=true
    while OKToProceed and |π|<u Do
Let z be a node in Pred(x_i)- π maximizing g(i, π∪{z_i})
P_new:=g(i, π∪{z_i})
If P_new>P_old Then
        P_old:=P_new
        π:= π∪{z}
    Else OKToProceed:=false
    End while
    output(`Node:', x_i, `Parents of this node:', π)
End for
```

For each node, we get a list of its parents. Actually, we can score the candidate local structures composed of the node and its parents in parallel. In the scoring of each local structure, the parameters in Equation (1) can be counted from data in parallel.

2.2 MapReduce Model

The computation takes or produces a set of input or output key/value pairs. Users of MapReduce library express the computation as two functions: Map and Reduce.

Map, written by users, takes an input pair and produces a set of intermediate key/value pairs. The MapReduce library groups all intermediate values associated with the same intermediate key together and passes them to the Reduce function.

Reduce, also written by users, accepts an intermediate key and a set of values for that key. It merges these values to form a possibly smaller set of values. The intermediate values are supplied to the user's reduce function via an iterator. This allows us to handle lists of values that are too large to fit in memory [15].

Fig. 1 (a) indicates that we send the input data to the Master node, by which the data are split into many maps sent to Slave node, which will return the result to Master node and reduce the map result to the final one for output. The NameNode JobTracker is allocated on Master node to control the whole task and DataNodes, and TaskTracker is allocated on Slave node to communicate with the NameNode JobTracker, shown in Fig.1 (b).

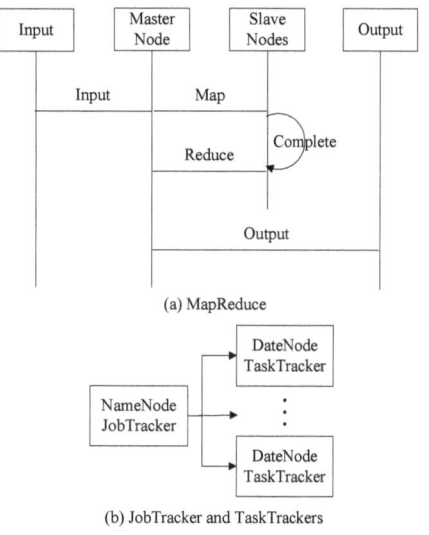

(a) MapReduce

(b) JobTracker and TaskTrackers

Fig. 1. MapReduce Process

3   MapReduce-Based Algorithm for BN Learning

The architecture of the MapReduce-based K2 algorithm is shown in Fig. 2:

- **Preprocessing.** The required parameters in Equation (1) are obtained by statistical counting from the given massive sample data in parallel based on MapReduce.
- **Searching.** For each node (i.e., a random variable) in a BN, the candidate local structures are scored in parallel by the map process, and the global optimal structure is obtained by the reduce process. The parameters obtained in preprocessing will be used for scoring each candidate structure, and the local optimal ones with the highest score of all nodes will be merged to achieve the global optimal one.

In this architecture, massive data will be split into m map processes dynamically, whose results will be collected to the final result by R (usually $R=1$) reduce processes.

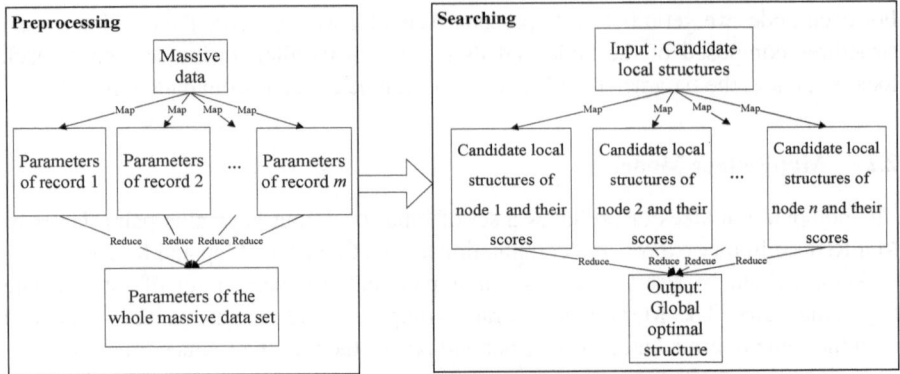

Fig. 2. Architecture for MapReduce-based K2

3.1 MapReduce-Based Scoring Function

In general, r_i in Equation (1) is a constant, which we do not compute in parallel. Assume there is a sample data set like Table 1, in which each column denotes a single variable and each record denotes a case upon the variables x_1, x_2 and x_3 that denote *smoking*, *cancer* and *Tuberculosis* respectively. For simplicity, 0 and 1 are used to denote *Absent* and *Present* respectively. In this example, r_i is 2.

Table 1. Sample data

Case	Variable values for each case		
	x_1	x_2	x_3
1	Present	Absent	Absent
2	Present	Present	Present
3	Absent	Absent	Present
4	Present	Present	Present
5	Absent	Absent	Absent
6	Absent	Present	Present
7	Present	Present	Present
8	Absent	Absent	Absent
9	Present	Present	Present
10	Absent	Absent	Absent

To obtain the parameters in Equation (1), we develop Algorithm 2 including the map process and reduce process respectively for accessing the sample data and summing the intermediate results. In each map process (Algorithm 2.1), the parameters N_{ij} and N_{ijk} of each record will be obtained as intermediate results.

Algorithm 2. MapReduce-based Scoring Function
```
Input:
1) m records upon n variables
2) record Lₓ (1≤x≤m) including values in {s₁, s₂, …, sₙ},
```

where s_i $(1 \leq i \leq n)$ is 1 or 0, denoting the ith variable is Present or Absent respectively

Output: <key/value> pairs of "N_{ij}" (or "N_{ijk}") and the value of N_{ij} (or N_{ijk}) $(1 \leq i \leq n; \ 1 \leq j \leq q_i; \ 1 \leq k \leq r)$

Algorithm 2.1. Map (String *key*, String *value*)
```
//key: key of data record, value: content of data record
Compute N_ij and N_ijk from L_x (1≤i≤n; 1≤j≤q_i; 1≤k≤r_i)
//by Equation (2)
If N_ijk =1 Then
  Generate key/value pairs <N_ijk, 1>
End If
If N_ij =1 Then
  Generate key/value pairs <N_ij, 1>
End If
If N_ijk =0 Then
  Generate key/value pairs <N_ijk, 1>
End If
If N_ij =0 Then
  Generate key/value pairs <N_ij, 1>
End If
```

For the first record in Table 1, we can obtain the <key/value> pair <$N(x_1=0, x_2=0), 0$> by Algorithm 2.1, that is the number of records satisfying $x_1=0$, $x_2=0$ is 0 and the value of N_{ij} is 0. Similarly, we can obtain for the pairs for other records.

In the reduce process (Algorithm 2.2), the intermediate results <N_{ij}, r_k> and <N_{ijk}, r_k> that have the same key will be collected and summer to obtain parameters <N_{ij}, $\sum r_k$> and < N_{ijk}, $\sum r_k$> required in Equation (1).

Algorithm 2.2. Reduce(String key, Iterator values)
```
//key: N_ij or N_ijk, values: in the Iterator and have the
same key in <key/value> pairs of <N_ij, r_k> or <N_ijk, r_k>
Result:=0
For each key/value pair Do
  value:=Interator.Next()
  result:=result+value
End For
Output<key, result>
```

For the data in Table 1, we can obtain <$N(x_1=0, x_2=0), 4$> by Algorithm 2.2.

3.2   MapReduce-Based Search Function

According to a sequence of the nodes *Seq* and the *Pred*(x_i) which represents the set of nodes that precede x_i in the predefined node sequence *Seq*, the candidate structures of node x_i can be obtained. For example, we assume *Pred*(x_1)={x_2, x_3}, *Pred*(x_2)={x_3}, *Pred*(x_3)={} and *Seq*={x_1, x_2, x_3}. For node x_1, there are 3 candidate structures ($x_1 \rightarrow$

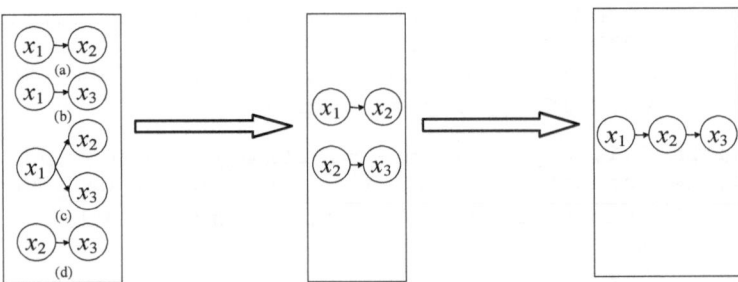

Fig. 3. Candidate structures **Fig. 4.** Local optimal structures **Fig. 5.** Global optimal structure
of x_1 and x_2 of x_1 and x_2

x_2), $(x_1 \rightarrow x_3)$ and $(x_1 \rightarrow x_2, x_1 \rightarrow x_3)$, shown in Fig. 3 (a), Fig. 3 (b) and Fig. 3 (c) respectively. For node x_2, the candidate structure $(x_2 \rightarrow x_3)$ is shown in Fig. 3 (d).

The candidate structures generated according to *Seq* and *Pred(xᵢ)* are formed as
<key/value> pairs $<x_i, S_i>$, where S_i is the set of l candidate structures of node x_i, and
$S_i = \{S_{i1}, S_{i2}, \dots, S_{il}\}$. For example, the candidate structure of x_2 will be $<x_2, x_2 \rightarrow x_3>$. In
each map process (Algorithm 3.1) of Algorithm 3, the candidate local structures will
be scored in parallel to obtain the local optimal one and split into M map processes..

Algorithm 3. MapReduce-based Search Function
Input: candidate local structures of each node
Output: the local optimal structure of each node

Algorithm 3.1. Map (String key, String value)
//key: node name x_i, value: a candidate structure of x_i
    $score_{ij}$:=g(i, S_{ij}); (1≤j≤l)//by Equation (1)
    Generate key/value pairs <i&S_{ij}, $score_{ij}$>
//key consists of node name x_i and structure S_{ij}, and value
is the score of S_{ij}

The intermediate results generated by map processes are formed as <key/value> pairs
$<x_i \& S_{ij}, score_{ij}>$. In each reduce process (Algorithm 3.2), the structure of node x_i with
the highest score will be chosen as the local optimal structure of x_i.

Algorithm 3.2. Reduce(String key, String value)
// key: node name x_i & structure S_{ij}, value: score of S_{ij}
max:=0;
For each S_{ij} of x_i Do
  If max<$score_{ij}$ Then
    max:=$score_{ij}$
    Content:=key.S_{ij}
    Node:=key.x_i
  End If
  Output <Node,Content>
End For

Now, we exemplify the structure search process on the data in Table 1. First, the scores of candidate structures $(x_1 \rightarrow x_2)$, $(x_1 \rightarrow x_3)$, $(x_1 \rightarrow x_2, x_1 \rightarrow x_3)$, shown in Fig. 3 (a), Fig. 3 (b) and Fig. 3 (c), can be computed in parallel by map processes. The scores of these three structures are 1.44×10^{-5}, 7.22×10^{-6} and 2.23×10^{-10} respectively. Thus, $(x_1 \rightarrow x_2)$ can be chosen as the optimal structure for x_1 with the highest score by the reduce process. By the same way, we can choose the only candidate structure $(x_2 \rightarrow x_3)$, shown in Fig. 3 (d), as the optimal structure for x_2 with the highest score 7.22×10^{-5}.

Second, by merging the local optimal structures $(x_1 \rightarrow x_2)$ and $(x_2 \rightarrow x_3)$ shown in Fig. 4, we can obtain the global optimal structure, shown in Fig. 5. According to the ideas in [8], the candidate local structures generated by Algorithm 1 are based on the sequence of *Seq*, which makes the merging result be the global optimal structure.

If the massive sample data have m records upon n variables. The time complexity of traditional K2 is $O(m*n^2)$. Assume the number of Slave nodes is large enough, and a map is executed in each Slave node, that is one Slave node exactly processes one record of data. If there are t Slave nodes, the time complexity of Algorithm 2 will be $O(m*n^2/t)$, whose order is much lower than $O(m*n^2)$. Comparing the time consummation of Algorithm 2 and that of Algorithm 3, the former fulfills the preprocessing and mainly determines the efficiency of the whole method. Thus, we conclude that the time complexity of our method is $O(m*n^2/t)$.

The BN's DAG can be obtained by Algorithm 2 and Algorithm 3. Then, we can easily compute the CPTs for each node in the DAG by the similar MapReduce-based ideas adopted in Algorithm 2 when accessing the massive data.

4   Experimental Results

4.1   Experiment Environment

Our experiment platform includes 4 machines. One machine is Master node and the others are Slave nodes. The CPU of each machine is Pentium(R) Dual-Core CPU E5700 @3.00GHz @3.01GHz with 2GB main memory. On each machine, the versions of Hadoop, Linux and Java are 0.20.2, Ubuntu 10.04, and JDK 1.6 respectively.

Chest-clinic [23] is a widely used BN in medical science, including 8 nodes, where each node corresponds to one kind of disease or its reason, and the arrows between nodes denote their causalities. We adopted the Chest-clinic test data and the corresponding BN in the experiments. The initial data set has 1000 records. To achieve the massive data, we generated the test data by multiple copies of the original sample set and the volume of the generated test data is larger than the initial data but the implied BN structure is the same. On average, every one million records of data have 15.26MB. We generated the test data from 15.62KB to 610.35MB.

4.2   Efficiency Comparison

To test the efficiency of Algorithm 2 and Algorithm 3, we considered two kinds of execution time, called total time and algorithm time and denoted as AT and TT respectively. Total time is the total execution time of the Hadoop system running our

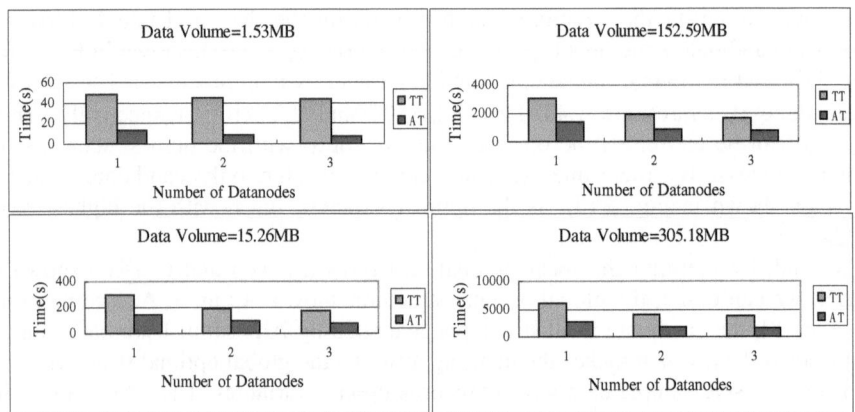

Fig. 6. Comparisons between total time and algorithm time on various Slave node numbers and data volume

algorithms, and algorithm time is the execution time of Algorithm 2 without including the time of initialization of Hadoop and I/O. First, we recorded TT and AT on different sized data with the increase of Slave nodes, shown in Fig. 6.

Then, we recorded the total time with the increase of data volume with 1, 2 and 3 Slave nodes respectively, shown in Fig. 7. It can be seen that the total time is increased slowly with the increase of data volume under 1, 2 or 3 Slave nodes. Meanwhile, for each size of the sample data, the smaller the data size, the less the total time will be. This means that the method for learning BN from sample data can be run efficiently on the Hadoop-based architecture. With the increase of machines in the cluster, BN can be constructed in less time.

Meanwhile, we recorded the algorithm time (including those of Algorithm 2 and Algorithm 3) with the increase of data volume with 1, 2 and 3 Slave nodes respectively, shown in Fig. 8. It can be seen that the algorithm time is increased with the increase of data volume under 1, 2 or 3 Slave nodes. Meanwhile, for each size of the sample data, the smaller the data size, the less the algorithm time will be. This means that the map and reduce steps can be executed efficiently in Hadoop, which verifies that our extension to the classical algorithm for BN learning is efficient and effective.

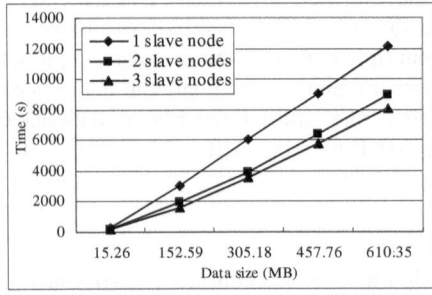 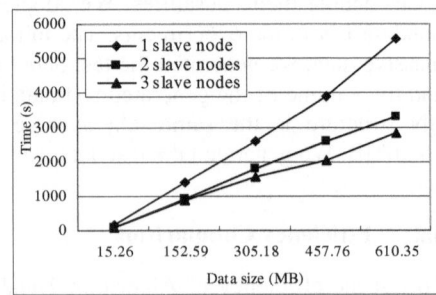

Fig. 7. Total time with the increase of data volume **Fig. 8.** Algorithm time with the increase of data volume

4.3   Structure Comparison

Power Constructor [23] is one of the current BN learning systems, by which we applied the chest-clinic sample data with 10000 records to construct BN, shown as Fig. 9. Also we used the same data to construct BN by our method, and we obtained the same structure as that obtained by Power Constructor. Thus, we can confirm that the BN structure learned by our algorithm is correct.

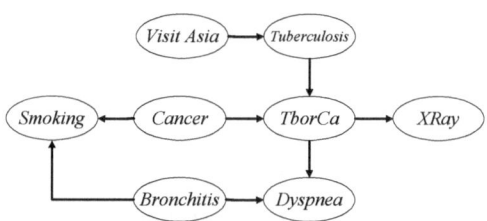

Fig. 9. Chest-clinic BN learned by the Power Constructor and our method

5   Conclusion and Future Work

In this paper, as the initial exploration of BN learning from massive data, we gave a MapReduce-based method for BN learning from massive data while focusing on the learning of BN's DAG structure. Our method is the extension but a more efficient version of the traditional scoring & search algorithm. The extended algorithm is suitable for data-intensive computing and can be adopted generically for representing and inferring uncertainties implied in massive data.

In the future, we will improve the environment for real massive data sets. Further, we can store the learned large-scaled BN in the distributed file system by looking upon it as a kind of informative data. Moreover, we can extend other BN learning algorithms by incorporating the ideas explored and proposed in this paper, so that the knowledge model can be obtained efficiently even oriented to massive data.

Acknowledgement. This paper was supported by the National Natural Science Foundation of China (61063009, 61163003, 61232002), the Ph. D Programs Foundation of Ministry of Education of China (20105301120001), the Yunnan Provincial Foundation for Leaders of Disciplines in Science and Technology (2012HB004), the Foundation for Key Program of Department of Education of Yunnan Province (2011Z015), and the Foundation of the Key Laboratory of Software Engineering of Yunnan Province (2012SE205).

References

1. Kouzes, R., Anderson, G., Elbert, S., Gorton, L., Gracio, D.: The changing paradigm of data–intensive computing. IEEE Computer 42(1), 26–34 (2009)
2. Agrawal, D., El Abbadi, A., Antony, S., Das, S.: Data Management Challenges in Cloud Computing Infrastructures. In: Kikuchi, S., Sachdeva, S., Bhalla, S. (eds.) DNIS 2010. LNCS, vol. 5999, pp. 1–10. Springer, Heidelberg (2010)
3. Borkar, V., Carey, M., Grover, R., Onose, N., Vernica, R.: Hyracks: A flexible and extensible foundation for data–intensive computing. In: Abiteboul, S., Böhm, K., Koch, C., Tan, K. (eds.) Proc. of ICDE 2011, pp. 1151–1162. IEEE Computer Society, Hannover (2011)
4. Deshpande, A., Sarawagi, S.: Probabilistic graphical models and their role in database. In: Koch, C., Gehrke, J., Garofalakis, M.N., et al. (eds.) VLDB 2007, pp. 1435–1436. ACM (2007)

5. Pearl, J.: Probabilistic reasoning in intelligent systems: network of plausible inference. Morgan Kaufmann, San Mates (1988)
6. Russel, S., Norvig, P.: Artificial intelligence-A modern approach. Pearson Education, Prentice Hall (2002)
7. Song, W., Yu, J.X., Cheng, H., Liu, H., He, J., Du, X.: Bayesian Network Structure Learning from Attribute Uncertain Data. In: Gao, H., Lim, L., Wang, W., Li, C., Chen, L. (eds.) WAIM 2012. LNCS, vol. 7418, pp. 314–321. Springer, Heidelberg (2012)
8. Cooper, G., Herskovits, E.: A Bayesian method for the induction of probabilistic networks from data. Machine Learning. Machine Learning 9(4), 309–347 (1992)
9. Heckerman, D., Geiger, D., Chickering, D.: Learning Bayesian networks: the combination of knowledge and statistic data. Machine Learning 20(3), 197–243 (1995)
10. Tsamardinos, I., Brown, L., Aliferis, C.: The max-min hill-climbing Bayesian network structure learning algorithm. Machine Learning 65(1), 31–78 (2006)
11. Suzuki, J.: Learning Bayesian belief networks based on the MDL principle: An efficient algorithm using the branch and bound technique. IEICE Trans. Information and Systems E82-D(2), 356–367 (1999)
12. Xiang, Y., Chu, T.: Parallel learning of belief networks in large and difficult domains. Data Mining Knowledge Discovery 3(3), 315–338 (1999)
13. Yu, K., Wang, H., Wu, X.: A parallel algorithm for learning Bayesian networks. In: Zhou, Z.-H., Li, H., Yang, Q. (eds.) PAKDD 2007. LNCS (LNAI), vol. 4426, pp. 1055–1063. Springer, Heidelberg (2007)
14. Yoshinori, T., Seiya, I., Satoru, M.: Parallel Algorithm for Learning Optimal Bayesian Network Structure. Journal of Machine Learning Research 12, 2437–2459 (2011)
15. Zhang, Q., Wang, S., Qin, B.: Cleaning Uncertain Streams by Parallelized Probabilistic Graphical Models. In: Chen, L., Tang, C., Yang, J., Gao, Y. (eds.) WAIM 2010. LNCS, vol. 6184, pp. 274–279. Springer, Heidelberg (2010)
16. Dean, J., Ghemawat, S.: MapReduce: a flexible data processing tool. Communications of the ACM 53(1), 72–77 (2010)
17. Chu, C., Kim, S., Lin, Y., Yu, Y., Bradski, G., Ng, A., Olukotun, K.: Map-Reduce for Machine Learning on Multicore. In: Schölkopf, B., Platt, J.C., Hoffman, T. (eds.) NIPS 2006, pp. 281–288. MIT Press, Vancouver (2006)
18. Low, L., Bickson, D., Gonzalez, J., Kyrola, A., Guestrin, G., Hellerstein, J.: Distributed GraphLab: a framework for machine learning and data mining in the cloud. PVLDB 5(8), 716–727 (2012)
19. Chen, W., Zong, L., Huang, W., Ou, G., Wang, Y., Yang, D.: An empirical study of massively parallel Bayesian networks learning for sentiment extraction from unstructured text. In: Du, X., Fan, W., Wang, J., Peng, Z., Sharaf, M.A. (eds.) APWeb 2011. LNCS, vol. 6612, pp. 424–435. Springer, Heidelberg (2011)
20. Bahmani, B., Kumar, R., Vassilvitskii, S.: Densest subgraph in streaming and MapReduce. PVLDB 5(5), 454–465 (2012)
21. Yuan, P., Sha, C., Wang, X., Yang, B., Zhou, A., Yang, S.: XML Structural Similarity Search Using MapReduce. In: Chen, L., Tang, C., Yang, J., Gao, Y. (eds.) WAIM 2010. LNCS, vol. 6184, pp. 169–181. Springer, Heidelberg (2010)
22. White, T.: Hadoop: The Definitive Guide. O'Reilly Media, Inc. (2009)
23. Cheng, J.: PowerConstructor system,
http://webdocs.cs.ualberta.ca/~jcheng/bnpc.htm

Practical Duplicate Bug Reports Detection in a Large Web-Based Development Community

Liang Feng[1], Leyi Song[1], Chaofeng Sha[2], and Xueqing Gong[1]

[1] Software Engineering Institute, East China Normal University
[2] School of Computer Science, Fudan University
{liang.feng,songleyi}@student.ecnu.edu.cn,
cfsha@fudan.edu.cn, xqgong@sei.ecnu.edu.cn

Abstract. Most of large web-based development communities require a bug tracking system to keep track of various bug reports. However, duplicate bug reports tend to result in waste of resources, and may cause potential conflicts. There have been two types of works focusing on this problem: relevant bug report retrieval[8][11][10][13] and duplicate bug report identification[5][12]. The former methods can achieve high accuracy (82%) in the top 10 results in some dataset, but they do not really reduce the workload of developers. The latter methods still need further improvement on the performance.

In this paper, we propose a practical duplicate bug reports detection method, which aims to help project team to reduce their workload by combining existing two categories of methods. We also propose some new features extracted from comments, user profiles and query feedback, which are useful for improving the detection performance. Experiments on real dataset show that our method improves the accuracy rate by 23% compared to state-of-the-art work in duplicate bug report identification, and improves the recall rate by up to 8% in relevant bug report retrieval.

Keywords: Bug Report, Duplicate Detection, Classification.

1 Introduction

For the development of large-scale software systems, project team members do not usually work together. They are distributed all over the world and communicate each other via various tools. In addition to e-mail, instant messaging softwares, web forums and wikis have become common as a communication medium for users. Moreover, Web-based software engineering tools are also widely used, such as bug tracking systems. Users and their various tools constitute a web-based development community.

A bug tracking system is designed to keep track of reported software bugs in a development community. Users can submit bug reports to the system which describe the software problems, and comments upon existing bug reports. Each bug report is triaged to a suitable developer who will in charge to address the related problems. Usually, there are thousands of developers in a large development

Y. Ishikawa et al. (Eds.): APWeb 2013, LNCS 7808, pp. 709–720, 2013.

community, and hundreds of bugs will be reported daily. Different QAs might report the same problem many times respectively. This causes an issue as duplicate bug reports. The phenomenon of duplicate bug reports is very common. If duplicate bug reports of the same problem are triaged to different developer, lots of efforts will be wasted and potential conflicts will be made.

There have been a number of works which focus on the duplicate bug reports detection issue in a bug tracking system. Existing methods [8][11][10][13][5][12][7] can be divided into two different categories: relevant bug report retrieval and duplicate bug report identification. Given a new submitted bug report, the former methods will return a sorted list of existing reports that are similar to it. However, the latter methods will identify that it is a duplicate report or not. As far as we know, the state-of-the-art search approach[7] can guarantee that the related bug reports will be found in the top-10 search results with 82% probability in some dataset if the new bug report is really duplicate. It is already a very good result. However, in actual development environment, it still requires users to spend lots of efforts to review the search results manually. From a practical point of view, duplicate bug reports identification can really reduce the developer's efforts. Yuan Tian etc.[10] reported their progress in the field of duplicate bug reports identification. The true positive rate was 24% while keeping the false positive rate of 9%. It still need to be improved in order to meet the needs of practical application. In this paper, we propose PDB, a practical duplicate bug reports detection method, which will help developers to reduce the efforts on processing bug reports by combining the relevant bug report retrieval and duplicate bug report identification.

For duplicate bug reports issue, any type of existing methods need to compute the similarity between reports. Different methods will choose different features to compute the similarity. such as textual similarity[8][11], surface features(timestamp, severity, product and component etc.)[10], execution trace[13] etc. Almost all of these features are extracted from the bug reports themselves. Existing works have tried almost all possible methods in the field of information retrieval and machine learning to address the duplicate bug reports issue. It is hard to improve the performance only depending on these features. We propose some new features extracted from comments and user profiles according to the observation on bug reports. In a development community, comments are necessary complement to bug reports. Furthermore, The reporter's knowledge and experience will affect the quality of the bug reports he submitted. Experiments show that these new features can effectively help us to further improve the duplicate bug reports detection performance.

The contributions of this paper include:

- We propose a practical duplicate bug reports detection method, which adopts three stages of classifiers to combine existing two categories of methods. It will really help developers to reduce the efforts on processing bug reports
- We propose some new features from comments, user profiles and query feedback to improve the effectiveness of duplicate bug report detection.

- Our extensive experiments on actual dataset shows PDB method outperforms the start-of-the-art approach on identification of duplicate bug reports, and makes sufficient improvement in duplicate bug reports retrieval compared to existing works.

The rest of this paper is organized as follows: section 2 gives a background knowledge on web-based bug tracking system, bug report, and duplicate bug report detection. Section 3 presents our practical solution to identify duplicate bug reports in a large web-based development community. The experimental study is shown in section 4. We discuss the related works in section 5. Section 6 concludes and describes some potential future works.

2   Background

2.1   Web-Based Bug Tracking System

Nowadays web-based development communities become grown up for its removing the geographical restrict, fostering communication and collaboration in group members. Various kinds of web-based development communities are widely used: Web forums provide place for users to submit and discuss the program questions. Distributed revision control system like GitHub facilitates sharing codes with other people and code management for multiple contributors.

Web-based bug tracking system also is one important kind of web-based development communities. It is designed to keep track of reported software bugs. It allows remote access of resources anytime, anywhere. It engages all the roles including QAs, developers and project managers on a single collaboration platform over the web. Users could review bugs and share comments and resources freely. All the knowledge are stored and centralized. Currently many web-based bug tracking systems such as Bugzilla and JIRA have been widely used in software projects, especially in large open source software projects like Firefox and Eclipse.

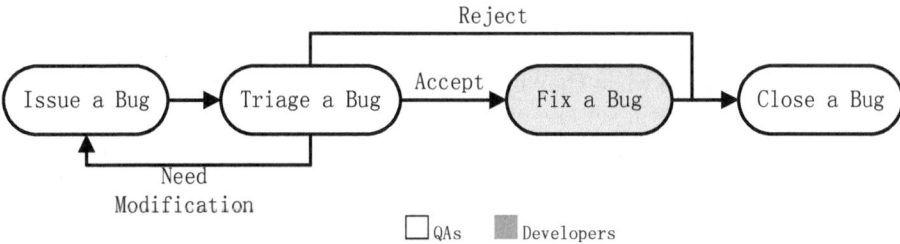

Fig. 1. General Workflow for Bug Processing

Figure 1 displays a general workflow for processing a bug in web-based bug tracking system. At first a reporter issues a bug for a special component. Then the bug is automatically sent to triager to check the quality of bug. If the triager judges that bug information is insufficient, the bug will be sent back to submitter to complete. If the bug is invalid, never fixed or duplicate, triage will set resolution and close bug. If triage have completed the entire checklist for a bug, bug will be marked as triaged and assigned to related developer. Next, bug is fixed by developer. After the bug is resolved, QAs verify the resolution of the bug, and close the bug when it is fixed.

2.2   Bug Report

A bug report is a structured file which is composed by multiple fields. These fields can be roughly categorized into three types: text data, metadata and attachments.

Text data include *summary, description* and *contents. Summary* is a one-sentence statement that presents problem; *Description* outlines problem in detail. It usually includes detailed steps to reproduce issue, the actual and expected result after performing the above steps; *Comment* is the opinion for this bug expressed by bug owner or others.

Metadata usually have several fields and generally contain following fields: *Product* and *component* fields indicate the artifact and module where the fault exist. *Assignee* is the person who is assigned to fix the bug. *Version* field shows the versions of product which the bug appears. *Priority* is the importance of the bug. *Reporter* is the person who submits the bug. *Create time* is when the bug is submitted. *Status* field indicates the processing state which the bug stays at. *Resolution* shows the result of bug, once it is resolved.

Attachments are usually added to bug reports in order to provide more data help to understand bug. They usually are screenshots, patches or logs.

2.3   Duplicate Bug Report Detection

As figure 1 shows, triage is a key procedure to assess the quality of bugs and prefilter some unnecessary bugs, which reduces nonessential effort, and improves the efficiency of project team. The checklist for bug triage usually contain several tasks. One main task is to mark duplicate bugs.

Bugs are duplicate when they have the same root cause. When labeling them duplicate, engineers will set one of them as master bug and others as duplicate ones. The master bug represents the whole group of duplicate ones with it.

Duplicate bugs occupy a large proportion in many projects, especially in large open source projects for its loose management. As [3] reported, issued from November 2006 to March 2008, about 38% of bugs are duplicate out of 60233 bugs in Firefox; about 49% of bugs are duplicate out of 19204 bugs in Thunderbird; Generally these duplicate bugs cause redundant work and increase workload for engineers. Meanwhile, as [1] pointed out that duplicate bug provides additional information about original bug, which is useful for developers to understand the bug. So identifying duplicate bugs and aggregating them will significantly improve the efficiency of bug processing.

However, it is not trivial to detect duplicate bugs manually. Several factors hinder the feasibility and efficiency of this activity in the large projects:

- Firstly, in general it costs much time to analyze a bug report and judge whether it is duplicate manually. The situation gets worse with the huge number of bug reports and limited manpower.
- Secondly, same error causes failure at different level of program. In some cases, duplicate bugs show completely different symptoms. Due to short of global knowledge, it is hard for triage to detect the relation of these bugs.
- Thirdly, a bug is not filed against appropriate component, which might mislead triagers.

Therefore, it need an automatical duplicate bug detection tool to help triagers. We propose PDB, a practical duplicate bug detection method. It is designed to complete two tasks and combine them: one is to make prediction whether one bug is duplicate, the other is to return top-k most possible duplicate bugs given a bug.

3   Practical Duplicate Bug Report Detection

In this section we describe the techniques in PDB in details; Overall, PDB adopts three stages of classification in duplicate bug detection. The first one is to predict how much contribution a comment makes to define a bug. The second one is to predict the possibility that two bugs are relevant and duplicate. The possibility is used to be score for ranking. The third one is to predict whether one bug is duplicate.

3.1   Modeling Comment Similarity

Comments in bugs are the points of view and discussions about the bugs issued by owner or others. Among these, a large proportion of comments are supplement to the original content, and help people better understand the appearing result and find root cause of bug. In some cases when the summary and description in a bug are vague and ambiguous, some key comments are critical to define to bug, and also contain important clues in finding duplicate bugs. So PDB puts comments into consideration when calculating the degree of relevance between bugs.

Comments in a bug report are issued by various roles with various kinds of intents. Some comments make some supplements to the description of bug phenomenon or possible rule causes, which can assist in identifying the relevance of bugs. But other comments just talks about bug fixing procedure or the results of bug processing, which deviate from the topic of main body for bug report and are not crucial in identifying the relevance of bugs. So it is intuitive to put comments weights differentially based on the type of comments, and set the former type of comments higher weight.

In order to measure the weight of comment, we divide all the comments into two classes: usefulness and less usefulness. The first class of comment is the one which describes bug phenomena, discusses root cause of bugs and solution of bug. The remaining of comments are categorized to the second class. We set comment weight as probability that the classifier put the content into usefulness class. For comment c_i in bug Y, we define the weight of this comment as follows:

$$W_Y(c_i) = p(usefulness|c_i) \tag{1}$$

SVM with RBF kernel is adopted as classifer. There are four types of features to classify the types of comments: average-idf, bug status, length of comment text, and unigrams of a term within comments.

Average-IDF: From our observation, those useful comments usually talks about the technical details of programs, and contain many professional vocabulary. These professional vocabularies generally have high idf value. On the contrary, the less useful comments usually are full of common words. It's much reasonable that these comments might appear in other bug reports, like "please fix as soon as possible".

Bug status: For a comment, the bug status indicates current state when the comment is submitted. Usually the intent of comment has strong relationship with the current bug status. E.g. when a bug stands on state of "NEW", QAs usually supplement some source and test environment of bug to facilitate triagers to understand bug.

Length of comment text: In generally, the length of comment text is also a useful measure for comment. We define the length of comment as the number of terms in comment. A comment with a large number of terms indicates more information and content. A comment with large information is more likely to be supplement to bug and be a useful one.

Unigrams of a term within comments: If a comment contains "root", "cause" and "likely", this comment is much likely to discuss phenomena or root cause of a bug. It is useful. On the other hand, comments containing words such as "close" and "verify" are likely to discuss the result of bug processing and be less useful. So we use vocabulary clues to determine type of comments.

Because summary and description field of a bug are representative of defining bug and tend not to contain unessential information, The weights of summary and description are set respectively with 2 and 1.

As discussed above, usually a comment in a bug report has its own intent and topic, the text content of comments in one report differs a lot. So it is common that it is only one comment has high similarity given a text content, and others comment in this bug are unmatched. We assume a new submitted bug have no comment. Given a query bug Q and a candidate bug Y, we propose *Weighted Comment similarity* $Ps(Q,Y)$ to calculate the similarity between summary and description in query bug and comments in candidate bug. Given a set of comments C in Bug Y and query text q, similarity function between them, $Score_Y(q, C)$, is the maximum score in similarity between text and comment. It is defined as follows:

$$Score_Y(q, C) = max\left\{Sim(q, c_i) \times W_Y(c_i) \mid c_i \in C\right\} \tag{2}$$

Where c_i denotes the ith comment in a bug report, c_i and query text q is modeled as vector space model, and $Sim(q, c_i)$ is cosine similarity between query text q and comment c_i .

Given query bug X, $text\_set(X)$ denotes the text set of bug X including summary and description. C denotes the set of comments in one bug Y, $Ps(Q, Y)$ is defined as follows:

$$Ps(Q, Y) = \frac{\sum_{q_i \in text\_set(Q)} W_Q(q_i) \times Score_Y(q_i, C)}{\sum_{q_i \in text\_set(Q)} W_Q(q_i)} \tag{3}$$

Where $W_Q(q_i)$ denotes weight of q_i in bug Q. It has been discussed above.

3.2   Duplicate Bug Reports Retrieval Model

PDB utilizes a discriminative model to calculate the possibility that two bug reports are duplicate. The possibility is used as score in ranking. Apart from weighted comment similarity and features proposed in [10], here we introduce several new features:

Topic model for document representation has widely used in machine learning and information retrieval. It makes excellent performance in many applications [14]. Topic model represents a document as a mixture of multiple topics. For all bugs report in a project, we learn LDA[2] topics from the corpus of past submitted bugs through gibbs sampling[4]. We put summary and description together as a bug document, and measure topic distance between two bug documents via both Jensen-Shannon and symmetric KL divergences[9] between two topic distributions. Finally, binary match/nonmatch of the most probable topic in each distribution is also feature. That is, 1 if the match of the most probable topic of two reports are the same; 0, otherwise.

Generally, the author of bug is much familiar with the bugs which are issued by him, and can roughly find the cause of bugs. Accordingly, same author is not likely to issue duplicate bugs. We use an indicator of the same author relationship as a feature, that is, 1 if the authors of two posts are the same, 0, otherwise.

Table 1 shows all the features in our algorithm.

3.3   Duplicate Bug Classifier Model

PDB leverages classifier to predict whether query bug is duplicate. The features in use consist of two aspects: user profile and query feedback.

User profile includes both demographic information, eg. name, age, etc, and his individual preference(technical background, interests and working styles).

Table 1. All the features used in bug retrieval model, CS is abbreviation for Cosine Similarity

# Feature name	# Feature name
1 CS : summary	10 Priority Similarity
2 CS : description	11 Version Similarity
3 CS : summary+description	12 Create date distance
4 CS : summary 2gram	13 Jensen-Shannon(LDA): summary+description
5 CS: description 2gram	14 Symmetric KL(LDA): summary+description
6 CS : summary+description 2gram	15 Maximum topic(LDA): summary+description
7 Product distance	16 Reporter distance
8 Component distance	17 CS: summary+description+comments
9 Severity Similarity	18 Weighted Comment similarity

The latter is hard to get, but valuable for predicting the resolution of submitting bugs. Intuitively novice users as well as expert or advanced users tend to submit bugs with different quality.

PDB takes some user profiles features into consideration. They include the total number of issuing bugs, the portion of issuing bugs with resolution as FIXED and the portion of issuing bugs with resolution as DUPLICATE. The total number of issuing bugs tries to identify the participation and familiarity of submitter in a project. Two portions shows submitter's habit of issuing bugs. These features are helpful to predict the resolution of a new query bug. For example, high total number of issuing bugs indicates that the submitter might have profound knowledge of project and program, and thus is less likely to submit duplicate bugs; A low portion of bugs with resolution as FIXED and a high portion of bugs with resolution as DUPLICATE mean the submitter might have not habit of searching relevant bugs before issuing a new bug, the resolution of the bug which this author issues has good chance of being duplicate.

The idea of query feedback is to judge the features of query from the returning sorted results by retrieval engine. Suppose that a user issues query bug Q to a duplicate bug retrieval system and a ranked list L of candidate bugs is returned. Intuitively, the more similar top one candidate bug is, the more probably query bug Q is duplicate bug. However, Due to convenience, submitters usually employ many same words and very similar writing style when filling in some relevant bug reports. It will lead to very high similarity score between these bugs which are submitted by same author, but not duplicate. In order to catch these situations, PDB consider several query feedback information. The features include the similarity score with first, fifth, tenth bug report, whether the first, fifth, tenth returning bug is submitted by same author, the portion of same author in top five returning bug reports, the portion of same author in top ten returning bug reports, the average similarity score for in top five returning bug reports and the average similarity score for in top ten returning bug reports. Similarity score of bugs is the probability that two bugs are duplicates, described in previous subsection.

4   Experiments Study

4.1   Dataset

MeeGo is a Linux-based free mobile operating system project. We crawled a dataset of bugs report from public MeeGo bugzilla[1] through open RPC/XML API. The dataset contains 22486 bugs, which are issued from March 2010 to August 2011. There are 1613 bugs with the resolution as DUPLICATE.

4.2   Evaluation and Discussion

The first experiment concerns at the performance of PDB's finding duplicate bug reports We compare PDB with the REP[10]. Several standard measures (MAP, Precision, Recall) are used in evaluation.

PDB adopts SVM classifer[2] with linear kernel. The number of topics in topic model is set to be 100. The ids of bugs for training set range from 1 to 5000, and the others are used for test. The query bug set contain both duplicate and master bugs. The result set are expected to contain all the duplicate bug reports.

Table 2 displays the performance on recall, precision and MAP respectively for these two approaches. PDB outperforms REP in all measures. PDB achieves higher recall rate by about 1%-8% for the resulting lists of top 1-20 bug reports. The gain gap grows as the size of list increases. In precision, PDB outperforms REF by 1%-8% in top 1-20 resulting bug reports. PDB improves MAP from 36% to 46%.

We analyze the contribution of the various features to the model by measuring the F-score through python script in LIBSVM. From all new features, weighted comment similarity has highest F-score, which means that comments in a bug play a major role in duplicate bug retrieval. The features related to topic model follow. Reporter distance is lowest-ranking.

Table 2. Performance results in Recall, Precision and MAP between two methods

Method	R@1	R@5	R@10	R@15	R@20	
PDB	0.2644	0.5089	0.6117	0.6623	0.6967	
REP	0.2569	0.4526	0.5307	0.5780	0.6095	

Method	P@1	P@5	P@10	P@15	P@20	MAP
PDB	0.3650	0.1411	0.0844	0.0607	0.0479	0.4677
REP	0.2846	0.1064	0.0637	0.0467	0.0371	0.3615

The second experiment is to evaluate the performance of identifying whether one bug is duplicate bug. We randomly picked up 534 duplicate and non-duplicate

---

[1] Available at http://bugs.meego.com
[2] Implemented by LIBSVM. Available at
http://www.csie.ntu.edu.tw/cjlin/libsvm/

bugs respectively, and then apply the Naive Bayes, Decision Tree[3] and SVM for classification. RBF kernel is used in SVM classifier. 10-fold cross validation is proceeded to measure accuracy, true positive rate and true negative rate. The parameters are grid searched to maximize accuracy.

We make comparison on the accuracy of three classifiers with PDB and state-of-the-art[12]. TIAN stands for [12] in the following. The best classifier refers to classifier that achieves the highest accuracy. Figure 2 shows the results. PDB outperforms TIAN consistently in all three classifiers. SVM and Decision Tree achieve close accuracy with both PDB and TIAN, but make increase about 10% on accuracy compared to the Naive Bayes.

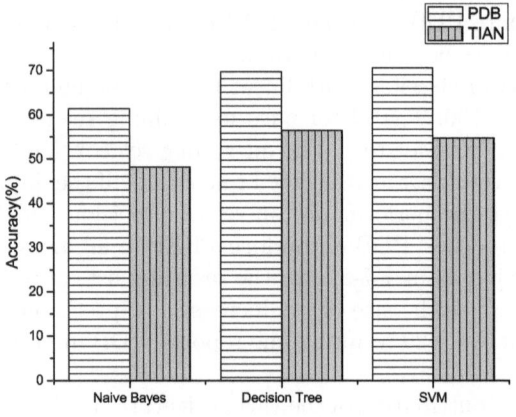

Fig. 2. Comparisons on the accuracies of three classifiers with PDB and TIAN methods

We define true positive(TP) rate as the fraction of duplicate bugs which are correctly classified, and true negative(TN) rate as the fraction of non-duplicate bug which are correctly classified. The higher are both these two rates, the better the classifier works.

Table 3 shows the comparison on the TP and TN rate for three classifiers with PDB and TIAN. In all three classifiers, PDB outperforms TIAN by 2%-31% on both TP rate and TN rate, except on TN rate in Decision Tree. Naive Bayes classifier shows best performance on TP rate, but get worst result on TN rate in three classifers. Similar with accuracy, SVM and Decision Tree achieve close performance on TP rate and TN rate overall. SVM outperforms Decision Tree on TP Rate and TN rate by a narrow margin. Therefore, Overall SVM do most stable and accurate work in three classifiers.

---

[3] Naive Bayes and Decision Tree are implemented by WEKA. Available at http://www.weka.net.nz

Table 3. Comparisons on the TP rate and TN Rate for three classifiers with PDB and TIAN methods

	PDB		TIAN	
Classifer	TP Rate	TN Rate	TP Rate	TN Rate
Naive Bayes	89.9%	33%	87.2%	12.2%
Decision tree	77.9%	61.6%	45.4%	66.7%
SVM	76.2%	68%	52%	57.2%

5   Related Works

In recent years, there have been a number of literatures on detecting duplicate bugs. These literatures can be roughly divided into two categories: one is to find most relevant bugs in bug report collection given a bug; The other aims to identify whether one bug is duplicate.

In the first category, Runeson et al.[8] work is the first one to address this problem. They make natural language process on the raw text data, use vector space model as computational framework, and set term weighting based on term frequency. Wang X et al.[13] combine text similarity and execution trace between bugs to rank bug reports. However, as Sun et al.[11] point out, the proportion of bugs contains execution information is fairly low in some projects like OpenOffice, Firefox. So is MeeGo dataset. Sun et al.[10] propose to train a support vector machine classifier to detect duplicate bugs with 54 text similarities feature including multiple text field. Sun et al.[10] propose REP, which combines BM25F with some meta-data like reported product, component, version. They make extensive experiment to show that REP outperforms the other works in effectiveness. Kaushik, N. et al.[6] make extensive comparison on different IR models. The experiment shows Log-Entropy based weighting scheme outperformed other models in effectiveness.

In second category of works, Jablert et al.[5] use surface features, textual semantics, and graph clustering to predict duplicate status. More recently, Tian, Y et al.[12] work based on [5]. They used maximum REP similarity, and add another category information, the product difference. Also they bring into relative similarity to determine the importance of the text similarity between bugs.

Our work aims to solve both two categories. In finding relevant bug report, we adopt discriminative model, and be first one to utilize comments to improve the effectiveness of duplicate bug detection. In identifying duplicate bug, we leverage user profile and query feedback to help predict query bug report resolution.

6   Conclusion

Detection of duplicate bugs is a necessary step to optimize resource allocation and improve working efficiency in the procedure of triage. In this work, we propose a practical duplicate bug reports detection method, which will help triagers to reduce their workload by combining existing two categories of methods. PDB adopts three stage to detect duplicate bug reports. In modeling the similarity of two bugs, PDB is the first one to utilize comments to enhance the effectiveness. In identifying duplicate bug, PDB makes use of user profiles and query feedback. We have evaluate our algorithm on MeeGo's bug reports which is issued in about one year. The experiment shows that PDB outperforms existing works [10] and state-of-the-art[12] in respective category of thread.

In future, we plan to investigate more bug reports from various software projects. Another idea is to utilize other text data like use cases and posts on online technical forums to enhance effectiveness of duplicate bug detection. Some of these are closely associated to and supplement to bug reports.

Acknowledgements. This work is partially supported by National Science Foundation of China under grant number 60803022.

References

1. Bettenburg, N., Premraj, R., Zimmermann, T., Kim, S.: Duplicate bug reports considered harmful... really? In: ICSM, pp. 337–345 (2008)
2. Blei, D.M., Ng, A.Y., Jordan, M.I.: Latent dirichlet allocation. Journal of Machine Learning Research 3, 993–1022 (2003)
3. Cavalcanti, Y.C., de Almeida, E.S., da Cunha, C.E.A., Lucrédio, D., de Lemos Meira, S.R.: An initial study on the bug report duplication problem. In: CSMR, pp. 264–267 (2010)
4. Griffiths, T.: Gibbs sampling in the generative model of latent dirichlet allocation. Unpublished note (2002), `http://citeseerx.ist.psu.edu/viewdoc/summary`
5. Jalbert, N., Weimer, W.: Automated duplicate detection for bug tracking systems. In: DSN, pp. 52–61 (2008)
6. Kaushik, N., Tahvildari, L.: A comparative study of the performance of ir models on duplicate bug detection. In: CSMR, pp. 159–168 (2012)
7. Nguyen, A.T., Nguyen, T.T., Nguyen, T.N., Lo, D., Sun, C.: Duplicate bug report detection with a combination of information retrieval and topic modeling. In: ASE, pp. 70–79 (2012)
8. Runeson, P., Alexandersson, M., Nyholm, O.: Detection of duplicate defect reports using natural language processing. In: ICSE, pp. 499–510 (2007)
9. Steyvers, M., Griffiths, T.: Probabilistic topic models. Handbook of Latent Semantic Analysis 427(7), 424–440 (2007)
10. Sun, C., Lo, D., Khoo, S.-C., Jiang, J.: Towards more accurate retrieval of duplicate bug reports. In: ASE, pp. 253–262 (2011)
11. Sun, C., Lo, D., Wang, X., Jiang, J., Khoo, S.-C.: A discriminative model approach for accurate duplicate bug report retrieval. In: ICSE (1), pp. 45–54 (2010)
12. Tian, Y., Sun, C., Lo, D.: Improved duplicate bug report identification. In: CSMR, pp. 385–390 (2012)
13. Wang, X., Zhang, L., Xie, T., Anvik, J., Sun, J.: An approach to detecting duplicate bug reports using natural language and execution information. In: ICSE, pp. 461–470 (2008)
14. Wei, X., Croft, W.B.: Lda-based document models for ad-hoc retrieval. In: SIGIR, pp. 178–185 (2006)

Selecting a Diversified Set of Reviews

Wenzhe Yu[1], Rong Zhang[1], Xiaofeng He[1], and Chaofeng Sha[2]

[1] East China Normal University, Shanghai, China
[2] Fudan University, Shanghai, China
51121500035@ecnu.cn,
{rzhang,xfhe}@sei.ecnu.edu.cn,
cfsha@fudan.edu.cn

Abstract. Online product reviews provide helpful information for user decision-making. However, since user-generated reviews proliferate in recent years, it is critical to deal with the information overload in e-commerce sites. In this paper, we propose an approach to select a small set of representative reviews for each product, which shall consider both the attribute coverage and opinion diversity under the requirement of providing high quality reviews. First, we assign weights to each attribute, which measure the attribute importance and help realize useful review selection; second, we cluster reviews into different groups representing different concerns which lead to better diversification results especially for selecting smaller sets of reviews; finally, we perform a set of experiments on real datasets to verify our ideas.

1 Introduction

Online user reviews have great effects upon decision-making process of users. Customers would like to read reviews to get a full picture of an item (i.e., a product or service) when they are selecting items. For example, according to the reviews of a camera, a customer can have knowledge of the size, color and convenience of usage though they have never bought it.

However, since user-generated reviews proliferate in recent years, e-commerce sites are facing challenges of information overload. In eBay.com, there are typically several hundreds of reviews for popular products such as the latest model of iPhone. The massive volume of reviews prevent users from catching the clear view of products. On one hand, it is not easy for users to read all of the reviews. Users may be unable to go through all the reviews for each item, for it is too time-consuming. On the other hand, an increasing number of users choose to shop online via mobile phones, such as reserving a table at a restaurant or picking a movie to watch when they are outside. Those users have to make their purchase decision in a short time, so they can only read a small fraction of reviews for each item. In addition, due to the limited screen size of mobile phones, it is inconvenient for users to scan more reviews.

To address the problem of information overload, e-commerce sites have adopted several kinds of review ranking and selection methods. Review ranking ranks reviews according to their helpfulness votes, so as to provide top-k reviews to users.

Y. Ishikawa et al. (Eds.): APWeb 2013, LNCS 7808, pp. 721–733, 2013.
© Springer-Verlag Berlin Heidelberg 2013

Helpfulness votes are evaluated by users to those reviews that are helpful to them. And there are also a number of researches on automatically estimating the quality of reviews [8][11][15][6]. However, there are two drawbacks in these review ranking work. First, the resulting top-k reviews of an item may contain redundant information while some important attributes may not be covered. For example, top-k reviews of a mobile phone may comment on display, camera and battery life, but mention nothing about whether it is easy to use or carry. Second, since previous experiments [3] showed that users tend to consider helpful the reviews that follow the mainstream, the resulting top-k reviews may lack diversity of opinions. By these two observations, review selection based on attributes coverage [14] is proposed. It prefers to select reviews covering as many attributes as possible. But it is found that such kind of method does not reflect the original customer opinion distributions and may lead to unfair picture to users. Then [9] comes to solve the problem by selecting a set of diversified reviews that keep the proportion of positive and negative opinions. But we have found that this work does not perform well especially for selecting a smaller set of reviews, which is caused by the overlooking of attribute importance. For example, for a T-shirt, there are a lot of reviews complaining about its delivery speed while a smaller number of reviews complaining on the product quality. In such a case, it may be better to show reviews for clothes quality.

Hence, to improve the overall value of the top-k reviews, we view the top-k reviews as a review set rather than simple aggregation of reviews. We propose an approach to select a small set of reviews that cover important attributes with high quality, while at the same time have better diversified attributes and opinions by clustering. So our contributions in this paper are as follows:

1. We propose to evaluate attribute importance by calculating weights to them. We suppose to return users reviews covering attributes to their concerns.

2. In order to improve the diversity of the top-k results, we propose to cluster reviews to different topic groups. We design an algorithm to select diversified reviews from different groups which can help improve diversification results.

3. We perform experiments on real data crawled from e-commerce site to evaluate our algorithm.

The rest of the paper is organized as follows. Section 2 introduces the related work. Section 3 defines the problem. Section 4 gives the algorithms to the problem. Section 5 presents the experimental evaluation. Section 6 reaches the conclusion.

2   Related Work

Our work is related to the existing work that addresses the problem of review selection, review assessment and ranking and review summarization. A sequence of recent work has focused on these problems.

Review Selection: The review selection problem is to select a set of reviews from a review collection. Lappas and Gunopulos [10] proposed to select a set of reviews that represent the majority opinions on attributes. The drawback of this approach is that it reduces the diversity of opinions, regardless of the fact that users tend to make purchase decision after viewing different opinions on an item. Tsaparas et al. [14] intended to select a set of reviews that contain at least one positive and one negative opinion on each attribute. This method fails to reflect the distribution of opinions in the entire review collection, thus misleading users. Lappas et al. [9] proposed the selection of a set of reviews that capture the proportion of opinions in the entire review collection. The shortcoming of this approach falls in the lack of consideration on the quality of reviews. Furthermore, the three methods view all the attributes as the same, which may lead to the overload of attributes.

Review Assessment and Ranking: The review assessment and ranking problem is to rank reviews according to their estimated quality. Kim et al. [8] trained a SVM regression system on a variety of features to learn a helpfulness function and applied it to automatically rank reviews. They also analyzed the importance of different feature classes to capture review helpfulness and found that the most useful features were the length, unigrams and product rating of a review. Hong et al. [6] used user preference features to train a binary classifier and a SVM ranking system. The classifier divides reviews into helpful and helpless reviews, and the ranking system ranks reviews based on their helpfulness. Their evaluation showed that the user preference features improved the classification and ranking of reviews, and jointly using textual features proposed by Kim et al. [8] can achieve further improvement. However, these methods suffer from low coverage of attributes as the resulting top-k reviews may contain redundant information. It is in that they estimated the review quality separately and did not consider the overall quality of the top-k reviews.

Review Summarization: The review summarization problem is to extract attributes and opinions from the review collection and summarize the overall opinion on each attribute. Hu and Liu [7] proposed a three-step approach: first extract attributes from reviews, then identify opinion sentences and whether they are positive or negative, and finally summarize the number of positive and negative opinion sentences for each attribute. Yu et al. [12] intended to predict rating for each attribute from the overall rating, and extract representative phrases. Although these methods extract useful information from the review collection, they fail to provide the context of reviews from which users can view the opinions in a more comprehensive and objective way.

3   Problem Definition

In this section, we formally define our problem and the properties of our problem.

Given an item $i \in I$, where I denotes a set of items in a certain category, let $R = \{r_1, r_2, \ldots, r_n\}$ be a set of reviews commenting on i which cover a set of attributes $A = \{a_1, a_2, \ldots, a_m\}$ and express a set of opinions $O = \{o_1, o_2, \ldots, o_z\}$. We assume that reviewers may express two types of opinions on each attribute: positive or negative opinion. Hence, the size of O is $z = 2m$. We also say that a review r covers a subset of attributes $A_r \subseteq A$ and expresses a subset of opinions $O_r \subseteq O$.

Our task is to select a small set of reviews that have high quality, while at the same time cover important attributes and have different opinions. We now formalize our problem as shown in Problem 1.

Problem 1 (Diversified Review Selection Problem). Given a set of reviews R of an item i that cover a set of attributes A and express a set of opinions O, find a subset of reviews $S \subseteq R$ of size $|S| \leq k$ that maximize the overall value of S:

$$F(S) = (1 - \lambda) Q(S) + \lambda D(S) \tag{1}$$

where $Q(S)$ measures the quality of S, $D(S)$ rewards the diversity of S, k is an integer number, and $\lambda \in [0, 1]$ is a diversity factor. The definition of $Q(S)$ and $D(S)$ are given in the rest of the section.

3.1   The Quality of Review Set

Since the overall quality of a review set is impacted by the quality of every single review in the set, we regard the quality of a review set as the average quality of reviews in the set. Therefore, the review set quality function $Q(S)$ is defined as follows:

$$Q(S) = \frac{\sum_{r \in S} q(r)}{|S|} \tag{2}$$

where $q(r) \in [0, 1]$ measures the quality of review r.

Assessment of Review Quality. Motivated by the helpfulness vote mechanism adopted in e-commerce sites, we view the quality of a review as its proportion of helpfulness votes. Therefore, the review quality function $q(r)$ is defined as follows:

$$q(r) = \frac{q^+(r)}{q^+(r) + q^-(r)} \tag{3}$$

where $q^+(r)$ (or $q^-(r)$) represents the number of users that consider helpful (or helpless) the review r.

However, some reviews such as newly-written reviews have few votes, which cannot properly estimate the quality of reviews, so instead of computing $q(r)$ directly, we use SVM regression to learn the function $q(r)$. It takes the following two steps to finish the task.

Step 1: define features impacting review quality. We consider two types of features: textual features [8] and user preference features [6], as shown in [6] that the combination of the two feature classes can improve the assessment. The textual features include the length and unigrams of a review, while the user preference features include the coverage of important attributes and the divergence from mainstream viewpoint of a review. The user preference features are defined as follows.

Attribute Coverage: Covering as many attributes as possible is not enough, users pay more attention to those key attributes, so we consider the coverage of important attributes. Let $A_r \subseteq A$ be a set of attributes covered in review r, the coverage of important attributes $cov(r)$ is defined as follows:

$$cov(r) = \frac{\sum_{a \in A_r} w(a)}{\sum_{a' \in A} w(a')} \qquad (4)$$

where $w(a) \in (0,1)$ represents the weight of attribute a. We will introduce the attribute weight later.

Mainstream Divergence: Users tend to consider helpful the reviews that follow the mainstream, so we measure the gap between the viewpoint of a review and the mainstream viewpoint. In e-commerce sites, users are required to give an item an overall rating ranging from one to five along with their reviews, so we can say that the overall rating expresses the same viewpoint as the review, and the average rating reflects the mainstream viewpoint. Therefore, the divergence from mainstream viewpoint $div(r)$ is defined as follows:

$$div(r) = \frac{\left| v(r) - \frac{\sum_{r' \in R} v(r')}{|R|} \right|}{\max_{r' \in R} v(r')} \qquad (5)$$

where $v(r) \in \{1, 2, \ldots, 5\}$ is the viewpoint of review r, $\frac{\sum_{r' \in R} v(r')}{|R|}$ is the mainstream viewpoint, and $\max_{r' \in R} v(r')$ is the maximum rating.

Step 2: construct training data for SVM regression. We randomly choose reviews with a certain amount of votes and estimate their quality according to $q(r)$ so as to form a labeled dataset of $\{r, q(r)\}$ pairs. We employ the labeled dataset to train our regression model and use the model to estimate review quality.

Assignment of Attribute Weight. Users read reviews to acquire information on attributes. However, the review collection of an item may cover dozens of attributes, while users are only interested in part of them. In order to provide key information to users, we measure the importance of attributes and assign relevant weights.

According to user behaviors of review-writing, we regard attributes that are frequently referred in the reviews of an item and items in the same category as important attributes. Here, attributes that appear in the reviews of most items in a category are taken into consideration since they are common attributes for a category. When users are selecting items of this category, they are more likely to refer to those attributes first. Therefore, let $R_a \subseteq R$ be a set of reviews commenting on an item $i \in I$ that cover attribute a, where I denotes a set of items in a certain category, and $I_a \subseteq I$ be a set of items that have reviews covering a, the attribute weight $w(a)$ is defined as follows:

$$w\left(a\right) = \frac{|R_a|}{\max_{a' \in A} |R_{a'}|} \cdot \frac{|I_a|}{\max_{a' \in A} |I_{a'}|} \qquad (6)$$

where $\frac{|R_a|}{\max_{a' \in A} |R_{a'}|}$ measures the importance of a for an item, and $\frac{|I_a|}{\max_{a' \in A} |I_{a'}|}$ measures the importance of a for a category.

The definition of $w\left(a\right)$ is similar to that of TF-IDF, as an attribute can be viewed as a term, and the review collection of an item as a document. Hence, the importance of an attribute for an item is equal to TF, while the importance of an attribute for a category is equal to DF instead of IDF since we value attributes that are frequently referred in the reviews of items in the same category.

3.2   The Diversity of Review Set

Since each time we select one review to add into the review set, we regard the diversity of a review set as the sum of the diversity of each review in the set when it is selected. Therefore, the review set diversity function $D\left(S\right)$ is defined as follows:

$$D\left(S\right) = \sum_{r \in S} d\left(r\right) \qquad (7)$$

where $d\left(r\right)$ rewards the diversity of review r.

Diversification of Opinions. To diversify opinions in a review set, we take the following three-step method.

In **step 1**, we cluster reviews with similar concerns. We assume that there are l clusters, and randomly select l reviews to each initialize a cluster. Since there may exist opinion sparsity in the review collection, we represent a review as a vector of TF-IDFs for the unigrams in the review. We compute the cosine distance between two review vectors as the similarity between them. Each time a review is added into a cluster based on the similarity between the review vector and the cluster centroid vector, and the cluster centroid vector will be updated accordingly.

In **step 2**, we decide the number of reviews selected from each cluster. We compute the proportion of each cluster in the review collection as the proportion of reviews from different clusters in the selected review set. Therefore, the proportion of reviews from cluster C_i in the selected review set is defined as follows:

$$p\left(C_i\right) = \frac{|C_i|}{|R|} \qquad (8)$$

Given an integer k as the size of the selected review set, the number of reviews selected from each cluster is defined as follows:

$$n\left(C_i\right) = \lfloor k \cdot p\left(C_i\right) + 1/2 \rfloor \qquad (9)$$

In **step 3**, we use MMR [1] to select reviews from different clusters. That is, in addition to the assumption that review r has not been added into S, it is in cluster C_i, and the number of reviews selected from C_i is less than $n\left(C_i\right)$, select r if it is similar with C_i, while at the same time not similar with S. Therefore, the review diversity function $d\left(r\right)$ is defined as follows:

$$d\left(r\right) = \lambda' sim\left(r, C_i\right) - \left(1 - \lambda'\right) sim\left(r, S\right) \tag{10}$$

where $\lambda' \in [0, 1]$ is a trade-off coefficient.

Here, let $O_r \subseteq O$ (or $O_{R'} \subseteq O$) be the set of opinions expressed by review r (or review set R'), the review similarity $sim\left(r, R'\right)$ is defined as follows:

$$sim\left(r, R'\right) = \frac{\sum_{o \in O_r \cap O_{R'}} w\left(a_o\right)}{\sum_{o' \in O_{R'}} w\left(a_{o'}\right)} \tag{11}$$

where a_o represents the attribute corresponding to opinion o.

4   Review Selection Algorithm

Algorithm 1. Diversified Review Selection Algorithm

Input: a set of reviews $R = \{r_1, r_2, \ldots, r_n\}$;
a set of attributes $A = \{a_1, a_2, \ldots, a_m\}$; a set of opinions
$O = \{o_1, o_2, \ldots, o_z\}$; an integer k
Output: a subset of reviews $S \subseteq R$ of size $|S| \leq k$
Review Clustering
1: randomly select $\{r_1, r_2, \ldots, r_l\} \subseteq R$
2: for $i = 1, \ldots, l$ do
3:    $C_i = \{r_i\}$
4: end for
5: for all $r \in R \backslash \{r_1, r_2, \ldots, r_l\}$ do
6:    cluster r
7: end for
8: for $i = 1, \ldots, l$ do
9:    compute $p\left(C_i\right), n\left(C_i\right)$
10: end for
Review Selection
1: $S = \emptyset$
2: for $i = 1, \ldots, k$ do
3:    $r = argmax_{r \in R \backslash S, r \in C_j, n\left(C_j\right) > 0} F\left(S \cup \{r\}\right)$
4:    $S = S \cup \{r\}$
5:    $n\left(C_j\right) = n\left(C_j\right) - 1$
6: end for
7: return S

In this section, we give the algorithms to our problem including the algorithms for opinion extraction and review selection. Before we select reviews, we apply opinion extraction algorithm that extracts attributes and opinions from the original review collection to generate input for review selection algorithm.

4.1 Opinion Extraction

There are a number of researches [7][5][12] on extracting attributes and mining opinions from reviews. Our method is lexicon-based as well as rule-based to handle context-dependent opinion words. Given a review, it takes the following three steps.

First, extract opinion phrases. We define an opinion phrase $\{a, o\}$ as a pair of attribute a and opinion word o. We refer to the approach proposed in [13]. They employ the grammatical relations produced by Stanford Dependency Parser [4] to define a set of dependency patterns, and generate opinion phrases according to the patterns.

Second, identify opinions expressed by opinion words in opinion phrases. There are two types of opinions: positive or negative opinion. Opinions are decided by opinion word lists [7], including a list of positive words and a list of negative words. If opinions cannot be determined according to the opinion word lists, we apply intra-sentence rule and inter-sentence rule in [5].

Intra-Sentence Rule: given an opinion word o and the sentence s where o is, identify its opinion by the other opinion words in s. This rule is based on the fact that if there are conjunctions such as "and" and "but" near o in s, we can decide its opinion by the opinion word o' on the other side of the conjunction.

Inter-Sentence Rule: given an opinion word o and the sentence s where o is, if s (or the next sentence to s) begins with a word like "but" or "however", it means o has different opinion from the opinion word in the last sentence (or the next sentence), otherwise they have the same opinion.

Finally, summarize opinions of attributes. Since an attribute may appear several times in a review, we aggregate all the opinions to produce an overall opinion. Given an attribute, if the number of positive opinions is larger than that of negative opinions, the attribute gains a positive opinion, or otherwise it gains a negative opinion.

4.2 Review Selection

We use a greedy algorithm to implement review selection as shown in Algorithm 1. It first clusters reviews according to their concerns, and then selects reviews proportionally from different clusters to maximize the overall value of the selected review set.

5 Experimental Evaluation

In this section, we evaluate our algorithm for review selection. We set the size of the selected review set as 5. We apply the diversity factor with $\lambda \in \{0, 0.1, \ldots, 0.9\}$ to observe the effects on the quality and diversity of the selected review set with the increase on λ.

5.1 Datasets

In our experiments, we use data crawled from e-commerce site eBay.com. This dataset includes reviews from three categories: tablets and e-book readers, portable audio and digital cameras. There are 4789 items and 110480 reviews in all. After pruning items having less than 20 reviews, there are 121 items on tablets and e-book readers with 9555 reviews, 286 items on portable audio with 38799 reviews, and 583 items on digital cameras with 42783 reviews.

5.2 Experiment Setup

In assessment of review quality, we select reviews having at least 5 votes as labeled data and use LIBSVM [2] to train regression model. To measure our algorithm for quality assessment, we select 60% of the labeled data as training data and the rest as test data to predict review quality. We adopt Pearson correlation coefficient as the metric to the correlation between the estimated quality and the labeled quality, so as to find whether the quality function is well learned. According to the evaluation in [8], the resulting coefficient is around 0.5, so we set 0.5 as the borderline between well-learned quality functions and poorly-learned quality functions. Our algorithm results in a coefficient of 0.6405749, which is a little higher than 0.5. In diversification of opinions, since the size of the selected review set is small, we set the number of clusters as 3 to ensure the selection of reviews from each cluster. We set the trade-off coefficient λ' as 0.7.

5.3 Result Analysis

We are interested in how quality behaves when raising diversity factor λ from 0.1 up to 0.9, we hypothesize that the quality will reduce with the increase on λ. On the other hand, we will focus on whether attributes and opinions are diversified through our algorithm. In addition, we will compare our algorithm with other algorithms for review selection in both the quality and diversity of the selected review set.

Review Set Quality Analysis. We apply our algorithm on all three categories of items and compute the average quality of all the selected review sets. As shown in Fig. 1, the quality goes down smoothly when raising λ by 10% each time, which agrees with our hypothesis, and reveals that the diversification of review set have detrimental effects on the quality.

Review Set Diversity Analysis. To decide whether attributes and opinions are diversified through our algorithm, we compute the average opinion coverage and the average cluster coverage of all the selected review sets.

The opinion coverage includes opinion coverage and weighted opinion coverage. Let $O_S \subseteq O$ be the set of opinions expressed by the selected review set S, the opinion coverage is defined as: $cov_{op}(S) = \frac{|O_S|}{|O|}$. The weighted opinion

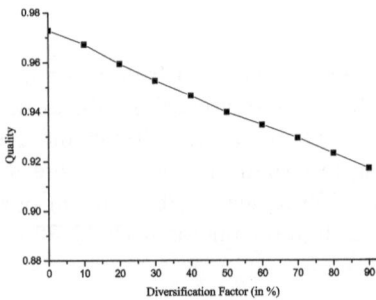

Fig. 1. Average Quality for Increasing λ

coverage is defined as: $cov_{wop}(S) = \frac{\sum_{o \in O_S \cap O} w(a_o)}{\sum_{o' \in O} w(a_{o'})}$, where a_o represents the attribute corresponding to opinion o. The opinion coverage measures the diversity of opinions while the weighted opinion coverage measures the coverage of important attributes.

The cluster coverage includes cluster coverage and weighted cluster coverage, the definitions of which are similar with that of cov_{op} and cov_{wop}, except that O in the denominator is replaced by $O_{C_i} \subseteq O$, the set of opinions expressed by cluster C_i, and after computing the cluster coverage for each cluster, use the average cluster coverage as the result.

As depicted in Fig. 2 and Fig. 3, the opinion coverage grows rapidly at first, peaking at 31.5% when λ is 0.3, while the weighted opinion rises up to around 70% and remain so afterwards. Hence, with the introduction of diversity factor λ, the opinion coverage improves.

In Fig. 4 and Fig. 5, the trend of cluster coverage is in line with that of opinion coverage, while the weighted cluster coverage goes up faster than the weighted opinion coverage in the first step. Hence, the diversity factor λ enhances the cluster coverage as well.

In conclusion, the attributes and opinions of the selected review set are diversified through our algorithm.

Comparison of Algorithms. We compare our algorithm with the characteristic method proposed in [9] that selects a set of reviews that capture the proportion of opinions in the review collection of an item. We implement the characteristic method with a greedy algorithm.

As presented in Review Set Diversity Analysis, the diversity of the selected review set reaches its peak when λ is 0.3, so we set λ as 0.3, and compare the average quality **QLTY**, the average (weighted) opinion coverage **(W)OP-COV** and the average (weighted) cluster coverage **(W)C-COV** of all the selected review sets produced by the two algorithms. We will also consider the average error of the proportion of opinions **PROP-ERR** defined in [9].

In Table 1, our algorithm outperforms the characteristic algorithm in all aspects except **PROP-ERR**. Since the characteristic algorithm aims at reflecting the distribution of opinions in the review collection of an item, it is as expected

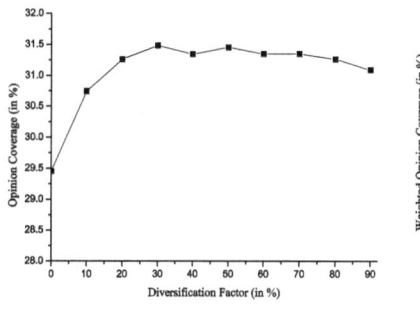

Fig. 2. Average Opinion Coverage for Increasing λ

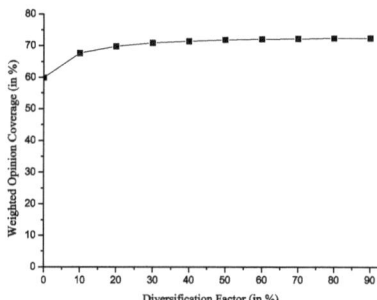

Fig. 3. Average Weighted Opinion Coverage for Increasing λ

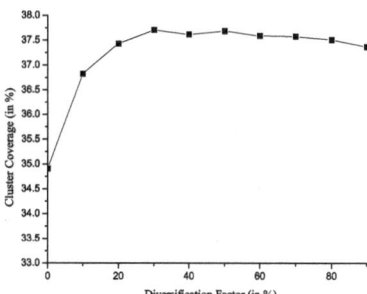

Fig. 4. Average Cluster Coverage for Increasing λ

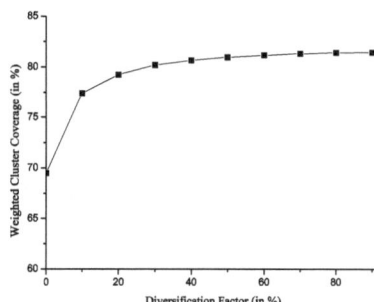

Fig. 5. Average Weighted Cluster Coverage for Increasing λ

to perform better than our algorithm in **PROP-ERR**. However, since the size of the selected review set is small and the characteristic algorithm diversifies the selected review set by maintaining the proportion of opinions in the original review collection, some rarely mentioned attributes may not be covered, resulting in the low coverage of opinions and important attributes as well.

In our work, we define important attributes as those attributes that are frequently referred in the reviews of an item and items in the same category. Some attributes may only appear in a small fraction of reviews of an item, but they are likely to be important for the category where the item belongs to. For example, the reviews selected by our algorithm for **Canon EOS 1D 4.2 MP Digital SLR Camera** cover not only those common attributes such as:

picture quality: "...It delivers excellent image quality, with great colors and detail..."

shooting: "...I also like the 6.5fps shooting, which is handy for my wildlife shooting..."

price: "...This is one of the best cameras out there for the price in my opinion..."

but also attributes that are not frequently mentioned in the reviews of this item while in fact are of great importance to the category of cameras, such as:

screen and interface: "...What I love in it? Large viewfinder, great ISOs range, nice speed, large screen and comfortable interface..."

menu system: "...The menu system is also very intuitive and easy to learn and the my menu feature is good for me about 90% of the time and keeps me from having to go through the whole menu interface..."

Users may concern about screen, interface and menu system when selecting cameras as they are key to user experience. Though they are not frequently referred in the reviews of this item, they are definitely important for the category of cameras.

Table 1. Comparison between Characteristic Algorithm and Diversified Algorithm

	Characteristic	Diversified
QLTY	0.86128730	0.95268065
OP-COV	0.10075375	0.31482820
WOP-COV	0.46727788	0.70877750
C-COV	0.17052619	0.37709308
WC-COV	0.60632420	0.80178090
PROP-ERR	0.21768339	1.30870930

6   Conclusion

We have proposed an approach to select a small set of reviews from the review collection of an item that have high quality, while at the same time have diversified attributes and opinions. We first train a regression model to estimate review quality. Then, to cover important attributes, we assign weights to attributes according to their importance, and to cover different opinions, we cluster reviews according to their opinions and select reviews proportionally from different clusters. Finally, we use a diversity factor to balance the quality and diversity of the selected review set, and implement our method with a greedy algorithm to maximize the overall value of the selected review set. The experimental results show that our algorithm helps diversify the selected review set.

Acknowledgements. The work is partially supported by the Key Program of National Natural Science Foundation of China (Grant No. 61232002), National High Technology Research and Development Program 863 (Grant No. 2012AA011003), National Natural Science Foundation of China (Grant No. 61103039, No.61021004, No.60903014).

References

1. Carbonell, J., Goldstein, J.: The use of mmr, diversity-based reranking for reordering documents and producing summaries. In: Proceedings of the 21st Annual International ACM SIGIR Conference on Research and Development in Information Retrieval, pp. 335–336. ACM (1998)

2. Chang, C.C., Lin, C.J.: Libsvm: a library for support vector machines. ACM Transactions on Intelligent Systems and Technology (TIST) 2(3), 27 (2011)
3. Danescu-Niculescu-Mizil, C., Kossinets, G., Kleinberg, J., Lee, L.: How opinions are received by online communities: a case study on amazon. com helpfulness votes. In: Proceedings of the 18th International Conference on World Wide Web, pp. 141–150. ACM (2009)
4. De Marneffe, M., MacCartney, B., Manning, C.: Generating typed dependency parses from phrase structure parses. In: Proceedings of LREC, vol. 6, pp. 449–454 (2006)
5. Ding, X., Liu, B., Yu, P.S.: A holistic lexicon-based approach to opinion mining. In: Proceedings of the International Conference on Web Search and Web Data Mining, pp. 231–240. ACM (2008)
6. Hong, Y., Lu, J., Yao, J., Zhu, Q., Zhou, G.: What reviews are satisfactory: novel features for automatic helpfulness voting. In: Proceedings of the 35th International ACM SIGIR Conference on Research and Development in Information Retrieval, pp. 495–504. ACM (2012)
7. Hu, M., Liu, B.: Mining and summarizing customer reviews. In: Proceedings of the Tenth ACM SIGKDD International Conference on Knowledge Discovery and Data Mining, pp. 168–177. ACM (2004)
8. Kim, S.M., Pantel, P., Chklovski, T., Pennacchiotti, M.: Automatically assessing review helpfulness. In: Proceedings of the 2006 Conference on Empirical Methods in Natural Language Processing, pp. 423–430 (2006)
9. Lappas, T., Crovella, M., Terzi, E.: Selecting a characteristic set of reviews. In: Proceedings of the 18th ACM SIGKDD International Conference on Knowledge Discovery and Data Mining, pp. 832–840. ACM (2012)
10. Lappas, T., Gunopulos, D.: Efficient confident search in large review corpora. Machine Learning and Knowledge Discovery in Databases, 195–210 (2010)
11. Liu, J., Cao, Y., Lin, C., Huang, Y., Zhou, M.: Low-quality product review detection in opinion summarization. In: Proceedings of the 2007 Joint Conference on Empirical Methods in Natural Language Processing and Computational Natural Language Learning (EMNLP-CoNLL), pp. 334–342 (2007)
12. Lu, Y., Zhai, C.X., Sundaresan, N.: Rated aspect summarization of short comments. In: Proceedings of the 18th International Conference on World Wide Web, pp. 131–140. ACM (2009)
13. Moghaddam, S., Ester, M.: On the design of lda models for aspect-based opinion mining (2012)
14. Tsaparas, P., Ntoulas, A., Terzi, E.: Selecting a comprehensive set of reviews. In: Proceedings of the 17th ACM SIGKDD International Conference on Knowledge Discovery and Data Mining, pp. 168–176. ACM (2011)
15. Tsur, O., Rappoport, A.: Revrank: A fully unsupervised algorithm for selecting the most helpful book reviews. In: International AAAI Conference on Weblogs and Social Media (2009)

Detecting Community Structures
in Microblogs from Behavioral Interactions

Ping Zhang[1], Kun Yue[1,*], Jin Li[2], Xiaodong Fu[3], and Weiyi Liu[1]

[1] Department of Computer Science and Engineering, School of Information Science
and Engineering, Yunnan University, Kunming, China
[2] Department of Software Engineering, School of Software,
Yunnan University, Kunming, China
[3] Faculty of Information Engineering and Automation, Kunming University of Science
and Technology, 650500, Kunming, China
kyue@ynu.edu.cn

Abstract. Microblogs have created a fast growing social network on the Internet and community detection is one of the subjects of great interest microblogs network analysis. The "follow" relationships highlighted in many existing methods are not sufficient to capture the actual relationship strength between users, which leads to imprecise communities of microblog users. In this paper, we presented the graph-based method for modeling microblog users and correspondingly proposed a new metric for obtaining the relationship quantitatively based on interaction activities. We then gave a hierarchical algorithm for community detection by incorporating the quantitative relationship strength and modularity density criterion. Experimental results showed and verified the effectiveness, applicability and efficiency of our method.

Keywords: Microblog, Community detection, Relationship strength, Graph model, Modularity density.

1 Introduction

Microblogging is a relatively new phenomenon defined as "a form of blogging that lets you write brief text updates, usually less than 200 characters, about our life on the go and send them to friends and interested observers via text messaging, instant messaging, email or the web" [1]. Provided by several services including Twitter, Sina, Tencent, etc., microblog has become one of the most popular applications in the real world. The microblog services with user-generated contents have made a staggering amount of available online information. In 2010, Twitter generated an estimated 55 million tweets a day [2].

The analysis and the corresponding knowledge discovery of microblog data have received more and more attention from government, business, sociology etc. with the increased volume of microblogging data generated by more and more users.

---

* Corresponding author.

Y. Ishikawa et al. (Eds.): APWeb 2013, LNCS 7808, pp. 734–745, 2013.

For example, many companies adopt microblogging social media services as ideal places to obtain timely opinions from customers and potential customers directly. Among the various perspectives of knowledge discover in microblog related applications, the analysis of the so-called community structure of online social networks has acquired much attention during the latest years. For instance, Guan et al. [3] found that it is likely to stimulate new purchases effectively by targeting a product on potential users of a community that has a significant number of users who have already purchased the product.

Naturally, in a specific physical domain or logical range, the users with communications of microblog messages can be described by a network, from which their associations can be represented globally. To identify the community structures in microblogs could help us understand and exploit these networks more effectively, centered on which various studies on community discovery or detection were conducted [4, 5]. For this purpose, many scientists cluster users by the "follow" relationships [1, 6] based on the hypothesis that microblog users are likely to follow (or be followed by) other similar ones if their relationships of message communication could be reflected in a community. Intuitively, by comparing the lists of followers of the users in small communities, large communities can be generated by merging the small ones that share a certain number of users according to a pre-defined threshold. However, the "follow" relationships actually do not reflect the strength of real relationships from weak to strong. Thus, these methods [1, 6] might group the users into a community mistakenly, since the close or remote relationships not only depend on the number of associations, but also on the strength of these associations.

Consequently, some sociologists [7, 8] proposed the method for measuring the relationship strength based on communication reciprocity and interaction frequency. By this way, the strength was predicted by analyzing the behavioral interaction data [9, 10], but just two levels of strength, namely weak and strong, were considered. Therefore, from the community discovery point of view, it is desired to develop a strength metric to describe the relationships both qualitatively and quantitatively.

As stated above, the relationships among microblog users can be represented by a weighted network composed of vertices and weighted edges, on which the relationship strengths can be looked upon as the weight. The property of community structure is as follows: network nodes are joined together in tightly knit groups with strong relationship strength, while there are only looser connections with weak relationship strength between them. Starting from this observation, many clustering algorithms in social network analysis, such as GN, CNM, have been applied for community discovery on weighted networks [11, 12]. But only the coarse structure of the communities can be discovered by these algorithms. As the improvement, Li [13] introduced the concept of modularity density criterion and proved the k-means algorithm which can optimize the criterion to obtain finer results. Actually, in the clustering algorithms like k-means, the number of clusters is to be specified in advance, which cannot be made for the microblog network since the actual topology and the scale of the concerned microblog users cannot be known clearly.

Therefore, inspired by the basic idea of clustering algorithms and starting from the relationship strengths of microblog users, we proposed the precise method for hierarchical community discovery by statistical analysis on the behavioral interactions while avoiding predefined parameters. More specifically, the main contributions of this paper can be summarized as follows:

- To measure the actual relationships between users by a finer manner, we presented the graph-based method for modeling the microblog users. Then, we proposed a new metric based on the interaction activities to obtain the relationship strength quantitatively, as the generalization of current metric including just weak and strong states.
- To discover the communities precisely, we proposed the concept of subgraph proximity and gave a hierarchical algorithm by incorporating the quantitative relationship strength and the modularity density criterion. By this algorithm, we can discover the communities precisely with high resolution ratio.
- To evaluate the feasibility of our idea given in this paper, we implemented the algorithm and made preliminary experiments. By applying our algorithm to analyzing some classic benchmark data sets, we tested and verified the effectiveness, applicability and efficiency of our algorithm.

The remainder of this paper is organized as follows: In Section 2, we introduce the weighted graph model for simulating user relationships. In Section 3, we give the method for community identification. In Section 4, we show experimental results and performance analysis. In Section 5, we conclude and discuss future work.

2   Weighted-Graph Based Modeling of User Relationships

Suppose that user i calls user j for w_{ij} times and user j calls user i for w_{ji} times. What is the degree of relationship strength between them? In this section, we discuss the metric to quantify the relationship strength between a given pair of users.

Microblog has two special tweet types for communication among users, called replies and retweets. In the network of microblog users, the edges correspond to the directed interactions of replies or retweets among users.

A "reply" is a tweet sent in response to an earlier tweet and it includes the earlier tweet ID. This type of tweet typically includes "@account_name" followed by the new message. All of the users following either the replying user or the replied-to user can read this tweet. When a tweet is "retweeted" (abbreviated as RT) by a user, it is broadcast to the user's followers. For an official RT with twitter's retweet command, the tweet is simply reposted without changing the original message. An unofficial RT is created by retweet users using the "RT @account_name_original message" notation, possibly followed by additional text.

It is well known that the more reciprocal and frequent of interaction of a pair, the closer their relation will be. Therefore, we can simply define the relationship strength as follows:

$$F_{ij} = Rec_{ij} * Fre_{ij} \qquad (1)$$

where F_{ij}, Rec_{ij}, Fre_{ij} represent the relationship, reciprocity and frequency strength between i and j respectively. It is worth noting that F_{ij} is computed by adopting the weights in the directed micoblog behavioral network. The most intuitionistic metric to quantify the F_{ij} can be computed as follows:

$$F_{ij} = Rec_{ij} * Fre_{ij} = \left(\frac{\min(w_{ij}, w_{ji})}{\max(w_{ij}, w_{ji})}\right)(w_{ij} + w_{ji}) \tag{2}$$

However, Eq. (2) is not suitable for measuring the relationship strength. For example, the values of Eq. (2) for different pairs (1, 10) and (1, 1000) are equal. Actually in the microblog applications, the information on some nodes is always transmitted enormously by others nodes, but these nodes themselves transmit little from other nodes. This appears frequently between the nodes corresponding to stars and their fans. Although the sum of interactions is very high, stars and their fans are generally not very close from each other. From this standpoint, when Rec_{ij} is large enough, the value of F_{ij} should mainly depend on Rec_{ij}. Thus, the relationship between 1 and 10 should be higher than that between 1 and 1000.

To make a distinction of the relationships as illustrated above, the increase of Fre_{ij} should tend to flat with the increase of total interactions. So, we can compute the frequency strength as $Fre_{ij}^* = \sqrt{w_{ij} w_{ji}}$, and Eq. (2) can be improved as follows:

$$F_{ij} = Rec_{ij} * Fre_{ij}^* = \frac{\min(w_{ij}, w_{ji})}{\max(w_{ij}, w_{ji})}\sqrt{w_{ij} w_{ji}} \tag{3}$$

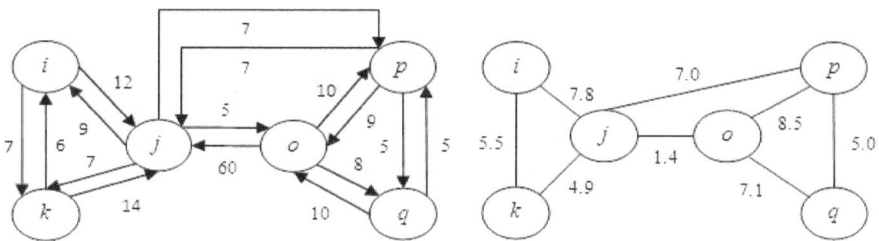

Fig. 1. Microblog network   **Fig. 2.** Quantitative microblog network

By using Eq. (3), we can transform the directed weighted network to an undirected one while deriving the general relationship strength between each pair of users. For example, we can transform the directed weighted network shown in Fig. 1 into the undirected one shown in Fig. 2. In Section 3, we will present our method for community detection based on the undirected network.

3   Community Detection in Microblog Networks

A widely used quantitative measure for evaluating the partition of a network is called modularity (known as Q), which was introduced in [14]. Recently, Fortunato et al. [15]

claimed that modularity contains an intrinsic scale that depends on the total size of links in the network, and stated that Q-based method only can detect the coarse structure of the communities. To improve the situation, Li et al. [13] introduced modularity density (D) which takes both edge and node into account, and was proved to be able to recognize the various sizes communities in the undirected network. In our algorithm, we exactly take modularity density as the quantitative measure of partition. First, we introduce some related concepts.

Definition 1. [13] The density of subgraphs is defined as:

$$d(G_i) = \frac{L(V_i, V_i) - L(V_i, \overline{V_i})}{|V_i|} \tag{4}$$

where V is the vertex set and E is the edge set of a network $G = (V, E)$, and A is the adjacent matrix of G. Given the partition of a network $G_1 = (V_1, E_1), \dots, G_m = (V_m, E_m)$, where V_i and E_i are the node set and edge set of G_i respectively, for $i = 1, \dots, m$ and m is not larger than n. $d(G_i)$ is the density of subgraph G_i. $|V_i|$ is the number of vertexes in V_i, $L(V_i, V_i) = \sum_{i \in V_i, j \in V_i} A_{ij}$ (representing the total number of edges that fall in the subgraph G_i), and $L(V_i, \overline{V_i}) = \sum_{i \in V_i, j \in \overline{V_i}} A_{ij}$, where $\overline{V_i} = V - V_i$.

Given a valid partition of a network G, the summation of the density value of all communities, computed by Eq. (4), should be as large as possible. Li [13] also gave the definition of modularity density.

Definition 2. [13] Modularity density of G is defined as

$$D = \sum_{i=1}^{m} d(G_i) = \sum_{i=1}^{m} \frac{L(V_i, V_i) - L(V_i, \overline{V_i})}{|V_i|} \tag{5}$$

This measure provides a way to determine whether a certain microscopic description of the graph is accurate in terms of communities. The larger the value of D, the more accurate a partition will be. So the problem of community detection can be looked upon as the problem of finding a partition of a network such that its modularity density D is maximized.

Naturally, this can be done by optimizing the quantitative measure or adopting the quantitative measure into a certain clustering algorithm. The FN algorithm [14] used the optimized idea to find communities, which is based on the greedy algorithm to choose a pair of subgraphs that can maximum the increase of Q with respect to the previous configuration. However, this optimized idea does not suit for the modularity density in Definition 2. The problem is that the subgraph containing a single node has arbitrary density, so we cannot compute which pairs can maximum ΔD (denoted as the increase of D). To obtain the communities, instead of optimizing D directly, we suppose that D's value of all single node subgraphs is 0, and define the subgraph proximity for merging nodes. Our basic idea is as follows: two subgraphs will be merged if the combination of them could decrease the proximity by a maximal degree and make $\Delta D>0$, specified in Definition 3.

Definition 3. The subgraph proximity between V_i and V_j is defined as

$$pro_{ij} = \frac{2L(V_i,V_j)}{L(V_i,\overline{V_i})+L(V_j,\overline{V_j})} \tag{6}$$

The greater the value of pro_{ij}, the closer the relationship between G_i and G_j. If there is no connection between G_i and G_j, then pro_{ij} equals 0, and if G_i and G_j are only connected by each other, then pro_{ij} equals 1.

In some network, the idea of joining the pair of subgraphs with the maximal proximity may lead to the asymmetry growth: large subgraphs will be more likely to be merged than small ones. In extreme cases, the whole network may be clustered as just one community. The reason is that large subgraphs have more nodes, even if they are linked with others by a very small rate, and the number of edges between them is also more than those in small ones. For example, w.r.t. the two communities in Fig. 2, the result of subgraph merging is shown in Fig. 3, where all the nodes will be clustered together as one community in each time when D is increased. To avoid this situation, we choose subgraphs randomly during the merging process. According to this idea, the large communities will have some possibilities that will not be chosen to merge. Fig. 4 illustrates the random choose, from which the correct result can be obtained.

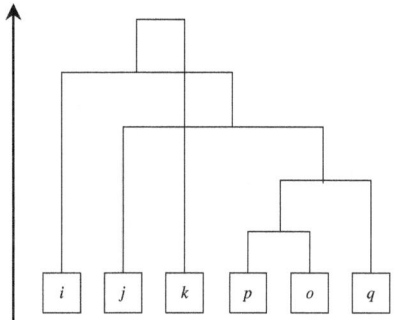

Fig. 3. General combination of pairs

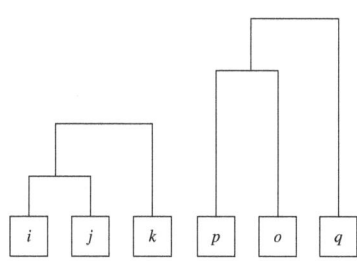

Fig. 4. Combination of random choose

```
Algorithm HCD: CommuDis (G)
Input: the relationship network G
Local variables:
  C: adjacent matrix of a subgraph of G. The value of
each cell C_ij is the number of edges connecting vertices
in subgraph i to those in subgraph j.
  V: node set of a subgraph. V_i records the nodes in sub-
graph i. Initialize each node as a subgraph: V[i] ← node
i.
  N: the number of nodes in G
  F: termination of array. The value of F_i is the index of
V_i which is not a community. Initialization, all the in-
dexes of the subgraphs in V are added into F. When F is
empty, the algorithm could be stopped.
for i←0 to N-1 do
  while true do
    if F.length = 0 then
        return V
    end if
    q•random(0, F.length-1)
    m←F_q; V_k←ϕ
    for i←0 to V.length-1 do
        if i•m and C_im •0 then
            compute P_im and ΔD if merging them
            V_k←the subgraph V_i that maximizes P_im when ΔD>0
        end if
    end for
    if V_k = ϕ then  //V_m is already a community, delete the
                      //mth element from F
        continue   // select m again
    else
        delete the kth element from F
        V_m←V_m + V_k // combine the node set
        delete the kth element from V
        C[m]←C[k] + C[m]
        //update the adjacent matrix of a subgraph
        delete the kth column and row from C
    end if
  end while
end for
return V
```

Algorithm HCD starts from a state in which each vertex is the sole member of one of n subgraphs. At each step for the randomly chosen subgraph m, joining subgraph i with the greatest pro_{mi}, which makes modularity density D increased. The value of D

will be maximized, if all the nodes are partitioned correctly. Randomly choosing a subgraph in F will take $O(1)$ time. Since the joining of a pair of communities between which there are no edges at all can never result in an increase in D, so computing pro_{mi} and ΔD only depends on k (the mean of the node degree of the whole network) in the worst case. Updating the values of V, F and C totally takes $O(n)$ time. Thus, each step of the algorithm takes $O(k+n)$ time in the worst case. There is a maximum of $n-1$ join operations necessary to construct the complete dendrogram and hence the entire algorithm runs in $O(kn+n^2)$ time on a dense graph or $O(n^2)$ time on a sparse one..

4   Experimental Results

To evaluate the performance of the proposed method, we made experiments on resolution ratio, precision and efficiency. The resolution ratio was tested to represent the granularity of discovered communities, which indicates the range of communities that can be identified. The precision was used to indicate the number of users included in the communities that are discovered correctly. The efficiency was tested by comparing the execution time of our algorithm with those of other 2 existing ones. All experiments were run on a PC with 3.0 GHz processor, 2.0 GB memory, running Windows XP Professional. The codes were written in JAVA.

4.1   Resolution Ratio

Due to the small network can be clearer to illustrate, we tested the resolution ratio of our result on well-studied small networks. The Zachary's karate club network [16] is one of the classic studies in social network analysis. It consists of 34 members of a karate club as nodes and 78 edges representing friendship between members of the club which was observed over a period of two years. Due to a disagreement between the club's administrator and the club's instructor, the club can be split into 2 small ones. Fig. 5 shows the Karate club network, where the shapes of the vertices represent 2 groups: the instructor and administrator factions respectively.

Table 1 shows that the results of the different algorithms are reasonable from the topology of the network. However, only two coarse structures were found by the famous algorithm based on Q proposed by Clauset, Newman and Moore (CNM) [17]. The Community 1# and 2# were split into 4 subgraphs by the density-based k-means algorithm [13]. The Community 1# was split into 2 subgraphs by Algorithm HCD we proposed in this paper. This means that the resolution ratio of the result of Algorithm HCD is better than that of the CNM algorithm and not far from the density-based k-means algorithm. Furthermore, we note that the underlined node 10 cannot be assigned properly, since it should be assigned to Community 1# or Community 2# directly from the topology of the network. By the density-based k-means algorithm, 4 factions can be detected perfectly, since the number of groups is known as human knowledge in advance.

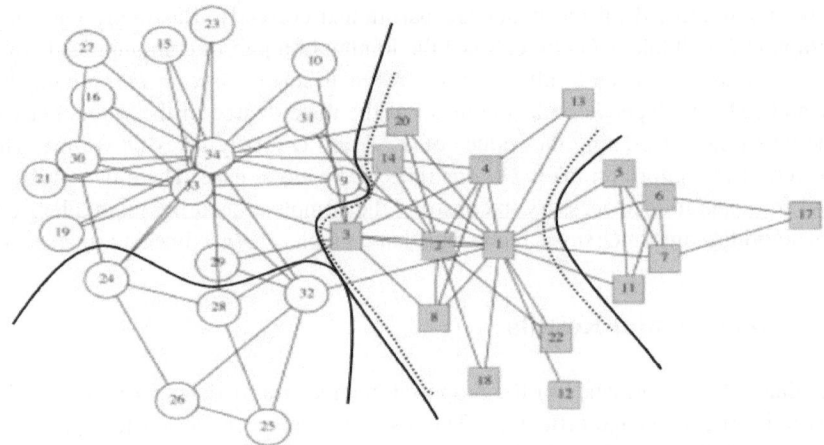

Fig. 5. Karate club network. Square nodes and circle nodes represent the instructor's faction and the administrator's faction, respectively. The dotted line represents our partition, and solid line represents the partition by the method in [13]. Shaded and open nodes represent the partition by CNM [17].

Table 1. Community structure in the Karate club network

	Community 1#	Community 2#	Community 3#	Community 4#
CNM algorithm	3, 10, 4, 13, 8, 14, 2, 22, 18, 20, 1, 7, 6, 17, 5, 11, 12	34, 30, 27, 24, 28, 32, 25, 26, 29, 33, 9, 31, 23, 19, 21, 16 15		
k-means algorithm based on density	18, 20, 22, 2, 12, 1, 3, 8, 13, 14, 4	29, 30, 31, 9, 34, 27, 15, 16, 19, 21, 33, 23, 10	28,32,26,25,24	11, 5, 6, 17, 7
Algorithm HCD	18, 20, 22, 2, 12, 1, 10, 3, 8, 13, 14, 4	32, 29, 26, 28, 25, 30, 24, 31, 9, 34, 27, 15, 16, 19, 21, 33, 23	11, 5, 6, 17, 7	

4.2   Precision

We tested the precision of our algorithm on different generated networks with the community structure from apparent to fuzzy. Our method was applied to the bench-mark computer-generated networks (denoted FLR) proposed by Lancichinetti et al. [18]. The networks have a heterogeneous distribution of node degree, edge weight and community size, which are exactly the features of real microblog networks [1, 19]. Thus, it is generally accepted as a proxy of a real network with community structure and appropriate to evaluate the performance of the community detection algo-rithms. FLR allows the users to specify distributions for the community sizes, the vertex degrees and the edge weights, with exponents γ, β and *muw*. Users can also adjust the average ratio μ to alter the community structure from apparent to fuzzy.

To evaluate the precision of the community identification, we used the normalized mutual information (NMI) [20] to quality the precision of Algorithm HCD. Fig. 6 shows the variation of the normal mutual information obtained by our method on the benchmark networks, with the mixing parameter μ from 0.1 to 0.6, $k = 20$, $\gamma = 2.5$, $\beta = 1.5$ and $muw = 0.1$. The 2 curves in Fig. 6 correspond to the 2 different scales of networks which have 1000 nodes and 10000 nodes respectively. With the increase of μ, the community structure in the networks becomes more fuzzy and harder to identify.

Comparing the results obtained by Algorithm HCD and CNM, we note that Algorithm HCD is more precise than CNM, and the precision of Algorithm HCD declines slower than that of CNM. Moreover, as the number of nodes increasing, from 1000 to 10000, the curves of CNM go down more quickly with respect to larger networks. On the contrary, the rangeability of the curves of Algorithm HCD is smaller when the size of network gets larger.

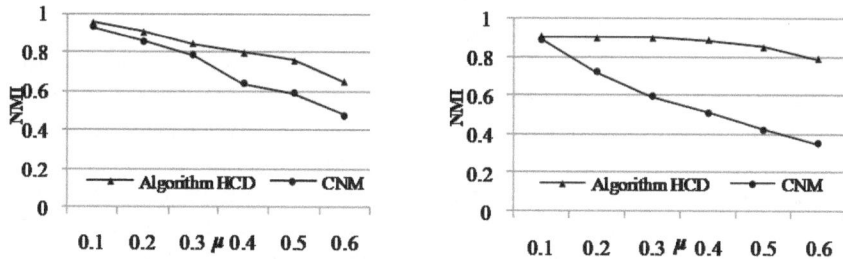

Fig. 6. Comparison of the normalized mutual information between Algorithm HCD and the CNM algorithm [17] on the benchmark computer-generated network

4.3   Efficiency

We tested the efficiency of our algorithm on the generated networks with different scales by comparing the execution time of Algorithm HCD proposed in this paper and those of FN and CNM as the improvement of FN. The comparisons were made with the increase of nodes in the generated networks and shown in Fig. 7. It can be seen

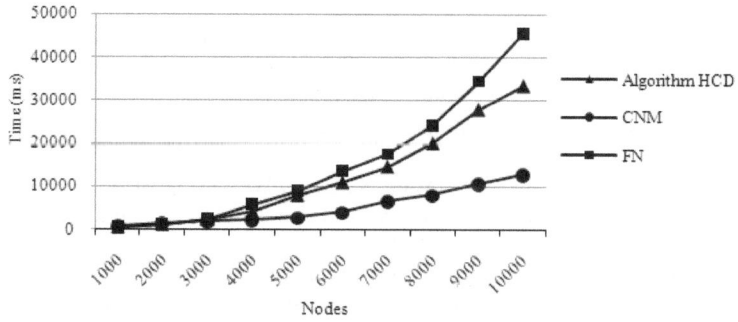

Fig. 7. Execution time of Algorithm HCD, CNM and FN

that the execution time of Algorithm HCD(i.e., CommuDis) is larger than that of CNM but smaller than that of FN when the network gets larger and larger.

To sum up, we can conclude from the above experimental results that our algorithm for community detection can achieve more precise results than CNM, but is less efficient than CNM although more efficient than FN. This makes us improve our algorithm to achieve better scalability inspired by some ideas of the improvement from FN to CNM, which is our future work.

5   Conclusions and Future Work

In this paper, we used the weighted graph to model microblog users and their relationships by user interaction activities, and gave a hierarchical algorithm which introduced the modularity density D successfully to discover the communities precisely. From a certain extent, experimental results showed that our method is effective and efficient for community detection in the microblog networks.

The methods proposed in this paper also leave open some other research issues. For example, different types of behavioral interactions may have different effects on relationships, but our method does not make a distinction between them. According to the fact that different interactions correspond to different weights, it may be possible to make better predictions on relationships. Actually, the scales of real microblog networks are frequent large, so the efficiency and scalability of our method is desired to be improved. In addition, the interaction in microblog occurs over time gradually, the features we used in this work have not concerned the dynamical aspects of data. These issues are exactly our future work.

Acknowledgement. This work was supported by the National Natural Science Foundation of China (Nos. 61063009, 61163003, 61232002), the Ph. D Programs Foundation of Ministry of Education of China (No. 20105301120001), the Natural Science Foundation of Yunnan province (Nos. 2011FZ013, 2011FB020), and the Yunnan Provincial Foundation for Leaders of Disciplines in Science and Technology (No. 2012HB004).

References

1. Song, X., Finin, T., Tseng, B.: Why we twitter: understanding microblogging usage and communities. In: Zhang, H., Spiliopoulou, M., Mobasher, B., Lee Giles, C., McCallum, A., Nasraoui, O., Srivastava, J., Yen, J. (eds.) Proc. of the 9th WebKDD and 1st SNA-KDD 2007 Workshop on Web Mining and Social Network Analysis, pp. 56–65. ACM Press (2007)
2. Kavanaugh, A., Fox, E., Sheetz, S., Yang, S., Li, L., Whalen, T., Shoemaker, D., Natsev, P., Xie, L.: Social Media Use by Government: From the Routine to the Critical. Government Information Quarterly 29(4), 480–491 (2012)
3. Guan, Z., Wu, J., Zhang, Q., Singh, A., Yan, X.: Assessing and Ranking Structural Correlations in Graphs. In: Sellis, T., Miller, R., Kementsietsidis, A., Velegrakis, Y. (eds.) Proc. of the 2011 ACM SIGMOD International Conference on Management of Data, pp. 937–948. ACM Press (2011)

4. Ferrara, E.: A Large-scale Community Structure Analysis in Facebook. EPJ Data Science 1(9) (2012), doi:10.1140/epjds9

5. Liao, Y., Moshtaghi, M., Han, B., Karunasekera, S., Kotagiri, R., Baldwin, T., Harwood, A., Pattison, P.: Mining Micro-Blogs: Opportunities and Challenges. Computational Social Networks, Part 1, 129–159 (2013)

6. Enoki, M., Ikawa, Y., Rudy, R.: User Community Reconstruction using Sampled Micro-blogging Data. In: Mille, A., Gandon, F., Misselis, J., Rabinovich, M., Staab, S. (eds.) Proc. of the 21st International Conference Companion on World Wide Web, Workshop, pp. 657–660. ACM Press (2012)

7. Burt, R.: Structural Holes: The Social Structure of Competition. Harvard University (1995)

8. Marsden, P.: Core Discussion Networks of Americans. American Sociological Review 52(1), 122–131 (1981)

9. Gilbert, E., Karahalios, K.: Predicting tie strength with social media. In: Dan, O., Ken, H. (eds.) Proc. of the SIGCHI Conference on Human Factors in Computing Systems, pp. 211–220. ACM Press (2009)

10. Kahanda, I., Neville, J.: Using transactional information to predict link strength in online social networks. In: Cohen, W., Nicolov, N., Glance, N. (eds.) Proc. of the Third International AAAI Conference on Weblogs and Social Media, pp. 74–81. AAAI Press (2009)

11. Coscia, M., Giannotti, F., Pedreschi, D.: A classification for Community Discovery Methods in Complex Networks. Statistical Analysis and Data Mining 4(5), 512–546 (2011)

12. Huang, L., Li, R., Li, Y., Gu, X., Wen, K., Xu, Z.: ℓ1-Graph Based Community Detection in Online Social Networks. In: Sheng, Q.Z., Wang, G., Jensen, C.S., Xu, G. (eds.) APWeb 2012. LNCS, vol. 7235, pp. 644–651. Springer, Heidelberg (2012)

13. Li, Z., Zhang, S., Wang, R., Zhang, X., Chen, L.: Quantitative Function for Community Detection. Physical Review E 77(3) (2008), doi:10.1103/PhysRevE.77.036109

14. Newman, M.: Fast Algorithm for Detecting Community Structure in Networks. Physical Review E 69(6) (2004), doi:10.1103/PhysRevE.69.066133

15. Fortunato, S., Barthélemy, M.: Resolution limit in community detection. The National Academy of Sciences of the United State of America, PNAS 104(1), 36–41 (2007)

16. Zachary, W.: An information flow model for conflict and fission in small groups. Anthropological Research 33(4), 452–473 (1977)

17. Clauset, A., Newman, M., Moore, C.: Finding community structure in very large networks. Physical Review E 70(6) (2004), doi:10.1103/PhysRevE.70.066111

18. Lancichinetti, A., Fortunato, S., Radicchi, F.: Benchmark graphs for testing community detection algorithms. Physical Review E 78(4) (2008), doi:10.1103/PhysRevE.78.0461-10

19. Kwak, H., Lee, C., Park, H., Moon, S.: What is twitter, a social network or a news media? In: Proc. of the 19th International Conference on World Wide Web 2010, pp. 591–600. ACM Press (2010)

20. Danon, L., Díaz-Guilera, A., Duch, J., Arenas, A.: Comparing community structure identification. J Statistical Mechanics 2005(9) (2005), doi:10.1088/1742-5468/2005/09/P09008

Towards a Novel and Timely Search and Discovery System Using the Real-Time Social Web*

Owen Phelan, Kevin McCarthy, and Barry Smyth

CLARITY: Centre for Sensor Web Technologies
School of Computer Science & Informatics, University College Dublin
`firstname.lastname@ucd.ie`

Abstract. The world of web search is changing. Mainstream search engines like Google and Bing are adding social signals to conventional query-based services while social networks like Twitter and Facebook are adding query-based search to sharing-based services. Our search and discovery system, Yokie, harnesses the wisdom of the crowd of communities of Twitter users to create indexes of proto-content (or recently shared content) that is typically not yet indexed by mainstream search engines. The system includes an architecture [13] for a range of contextual queries and ranking strategies beyond standard relevance. In this paper, we focus on evaluating Yokie's ability to retrieve timely, relevant and exclusive results with which users interacted and found useful, compared to other standard web services.

1 Introduction

Yokie is a novel search and discovery service that harnesses the social content streams of curated sets of users on Twitter. The system aims to provide timely, novel, useful and relevant results by streaming content from Twitter. It also allows for partitioning of the Twitter social graph into topical or arbitrary groups of users (called *Search Parties*). While query-based search is at the core of both Yokie and traditional search services, our system is not designed or conceived to compete with or replace the latter. Yokie gathers its content from entirely different sources for a start, and this content is indexed in response to real-time search needs. As a result, the Yokie use-case of returning high-quality, real-time, breaking or trending content is very different from archival search approaches of traditional search engines.

Our work is motivated by the fact that the nature of online content is changing. The conventional web of links and authored content is giving ground to a web of user-generated content, opinions and ratings. At the same time, users who have previously relied on traditional search as their basic discovery service are finding more content through sharing on social networks. The advent of the social web has seen an important shift in the role of users from passive consumers of content to active contributors of content creation and dissemination.

---

* Supported by Science Foundation Ireland Grant No. 07/CE/11147 CLARITY CSET.

Y. Ishikawa et al. (Eds.): APWeb 2013, LNCS 7808, pp. 746–757, 2013.

New technologies and services have emerged that allow users to directly manipulate and contribute diverse web content [7]. The Twitter microblogging service has, like the web as a whole, grown significantly in recent years [6][12].

While emphasis is placed on simple 140-character messaging, tweets can contain a wealth of contextual metadata relating to the time, location content and the user profile. Both Twitter and Facebook have become vital link-sharing networks [13][4]— our previous work has shown that approximately 22% of Twitter tweets contain a hyperlink string [13]. This prevalence of link sharing on social networks motivates an alternative approach to sourcing fresh, relevant and human-selected search engine content, instead of crawling the entire web-sphere. Also, by focussing on sourcing content from humans rather than the entire web, we can subdivide the landscape of content sharers into useful groups. This has numerous benefits, including the opportunity for users to curate their own adhoc lists, algorithmically group topical producers, and to aid in eliminating spam contamination.

A proof-of-concept prototype of Yokie was described and presented at [13], and a full system demonstration was presented at [14]. We have since added several user-feedback features to gain insight into the system's usefulness, and propose a per-item global interestingness score; both of which are described below. In this paper, we are particularly interested in evaluating the performance of the system at providing timely, relevant and useful content. This is done by a live user trial, and an attempt at capturing comparative data with other established online systems.

1.1   System Background and Related Work

Information Retrieval and Search systems are rich and well-established research landscapes. The technique of Web Crawling [1] has been widely lauded and adopted, however it is computationally expensive and time consuming because it attempts to operate over the entire web. As such, some of the major research themes and challenges include introducing social signals for better relevance, and obtaining better content freshness in their indexes [4]. More recently, efforts have been made in applying social signals to search and content recommendation results [10][4]. For example, Dong et al. successfully exploit features of *recent* Twitter content to improve the performance of recency-sensitive queries of a standard portal search engine such as Yahoo [4]. Alternatively, recent work in retrieval has proven temporally-sensitive microblogs can be ranked and retrieved with reasonable precision and recall, while ignoring profile data altogether [9].

Yokie's content is entirely mined from social streams. Increasingly, web users' online time is spent mostly on social network activity [7]. Many of these services allow relatively open access to user content streams. It is no surprise then that many instances of recent literature includes analyses of Twitter's real-time data and social graph, largely with a view to developing an understanding of why and how people are using such services; see for example [4][5][7][12]. The free and abundant access to the real-time web has also given rise to applications in other related fields beyond search. Phelan et al. describe useful recommendation techniques for harnessing recently shared content from Twitter for a news discovery and recommendation service [12]. Chen et al. and Bernstein et al. have explored using the content of messages to recommend topical content [3]. Garcia et. al. successfully apply microblog data to product recommendation [5].

A main focus of applying social network properties to search systems should be on the users and contributors. Indeed, a core component of Yokie is the selection of *groups* of users for sourcing content (explained further below). Krishnamurthy et al. demonstrate that the user base of Twitter can be split into definitive classes based on behaviours and geographical dispersion [6]. They highlight the process of producing and consuming content based on retweet actions, where users source and disseminate information through the network. Several services have emerged that take advantage of the open access to profile data with a view to providing a score of a user's online activities (provided the user has given them permission). One such service, Klout (http://www.klout.com), aims to provide an engagement and influence score based on an aggregation of a user's activities across a broad set of online social networks. Other work has focussed on applying user-level reputation systems to social search contexts [10] and recommender systems [15].

Grouping users can have numerous benefits including the opportunity for users to curate or generate their own preferred lists of personal or topical sources from which to receive content. This would, in turn, allow for vertical searching across multiple domains [2]. This may be favourable compared to a catch-all attempt to gather all shared content. Previous work by [6][18] postulate a large and active subgraph of social networks like Twitter exist that exclusively publish spam. As such, any steps towards algorithmic subdivision of the social graph may need to consider the potential of irrelevant or spam-laden accounts.

1.2   The Yokie System

To better understand the system and our goals, we must briefly review the system architecture. Yokie brings together a number of key ideas to provide an alternative discovery service that emphasizes real-time, trending information rather than more static and archival content of traditional web search engines. The end goal is a service that allows users to query to find content from specified groups of users, with a variety of contextual querying and ranking functions.

"Search Parties" and Content Acquisition. A key aspect of the system is the curated list of content sources, what we've termed a *Search Party*. Yokie sources its information from topical communities of Twitter users. These communities are called *Search Parties*, and are intended to be either manually curated or automatically created. Links each member share are indexed in Yokie using the text of the Tweet with which it is shared. The system uses Twitter's API to acquire content as defined by the search parties. These messages are then stored and indexed using the service described in the following subsection. This component also carries out real-time language classification and finds other messages that contain the same URL so the system can calculate item popularity, and extracts tags so as to allow collaborative tagging.

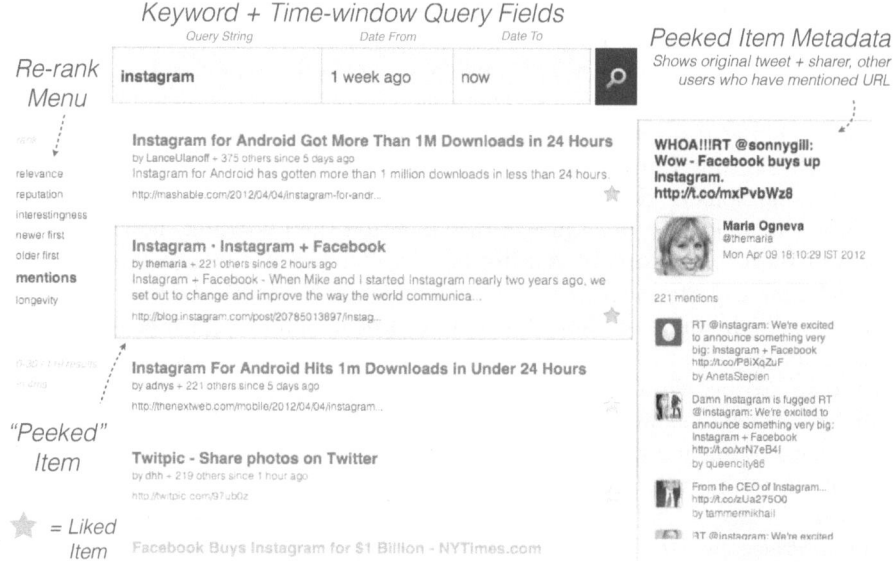

Fig. 1. Yokie's UI includes functions for viewing extra explanatory metadata related to the item, and unary relevance ratings. This example resultlist is based on "links labelled Instagram in the past week by high-value technology people, ranked by popularity."

Storage and Indexing. This subsystem is responsible for extracting metadata regarding the tweets, for instance timestamp data, hashtags (#election2012, etc.), user profile information, location, etc. as well as the message content itself. The main content, the urlID of the URL mentioned in the message, and the timestamp is pushed to an indexer for storage and querying. In our current implementation we use Apache Solr (http://lucene.apache.org/solr/) for this. We also store the remaining extracted metadata in a MongoDB (http://mongodb.org) NoSQL database for easy retrieval and fast MapReduce functionality.

UI and Querying. The system is presented with a query interface, currently comprising of a query string field, and two temporal fields, *date from* and *date to*. The system takes as input a query string with an associated time window, which can be either a natural language query (eg. "1 day ago", "now", "last week", etc) or a fixed date ("3 March 2013"), and then parses these into UNIX timestamps. The querying subsystem represents user queries, Q, based on the notation $\mathbf{Q} = (\mathbf{QueryString}, \mathbf{SearchParty}, T_{max}, T_{min})$. The user can *click through* to visit a webpage and *peek* at an item, and drill-down on results to explore related content such as the original tweet that the URL was shared with, the time and date it was shared, and the related Tweet mentions (if any). Each item UI element contains a unary rating *like*, which users select when they believe an item is relevant to the query.

User and Item-Based Result Ranking. Users can rank using typical **Relevance**, which is vanilla Term Frequency Inverse Document Frequency scoring (TFxIDF) [16]. However, a set of ranking strategies beyond standard relevance have been devised using the added contextual metadata of the microblogs. These include ranking using temporal features such as item age (**Newer-First** and **Older-First**) and **Longevity** (the size of the time window between the first and last mentions of the given hyperlink). Popular items can be ranked as we capture a number of sample tweets that mention the hyperlink (labelled **Mentions** in the UI). **Potential Audience** is a user-based sum of follower-counts for each search-party member who have mentioned the link. The premise of this strategy is high-volume in-links for users can be a sign of high user reputation. It is easily possible to derive a range of user-based scores as we are specifically "listening" to people for content, as opposed to relying on a document graph that ignores content publishers.

We have recently implemented two new ranking strategies, both of which are part of our evaluation. **Klout** is a service that provides users of social networks an *influence score* based on user reach, engagement and their ability to drive other interactions. Using the Klout API, we can gather scores for each user (once Klout has a score computed for them). It is possible to rank content based on the *original publishers/sharers Klout score*. **Interestingness**, is an attempt to measure how interesting an item is based on activity of the item on the social network, and in other Yokie query sessions. Since we are privy to a multitude of content, context and user interaction data for each item, and much of this data is dependent on the conditions of the query, we can create a novel interestingness algorithm. Given the Query Q tuple, Interestingness is defined as:

$$Int(U_i, Q) = \left(\frac{Pop(U_i, Q)}{Lng(U_i, Q)} \right) \cdot \left(\frac{|Clk_{\forall U_i}|}{|Pk_{\forall U_i}|} \right) \cdot |Lk_{\forall U_i}| \tag{1}$$

$Pop(U_i, Q)$ and $Lng(U_i, Q)$ are the popularity and longevity of the current item U_i, given the parameters of the query tuple, (meaning their value is dependent on the query T_{max} and T_{min} values). $|Clk_{\forall U_i}|$, $|Pk_{\forall U_i}|$ and $|Lk_{\forall U_i}|$ represent the total number of clicks, peeks and likes for the item, irrespective of the parameters of Q. These values have a default value of 1 so as to avoid null values for interestingness of items with no user engagement. This algorithm utilizes the contextual features of the item, benefits from prior user engagement with the item, and ignores the profile data of the user. This allows the item's score to be potentially independent of the original sharers' explicit profile data.

2   Evaluating Yokie

In the following evaluation of Yokie we consider two main objectives. First, we explore how users engage and interact with Yokie, paying particular attention to their engagement with the different ranking strategies and their subsequent click-through rates. Second, we consider the potential for Yokie to succeed where mainstream search engines fall short — as an exclusive source of timely or breaking content — by comparing Yokie usage to Bing Search and Bing News. In order to explore these evaluation objectives, we have conducted a live user trial of Yokie, to explore levels of engagement and interactivity. We use the data generated during this trial (queries and interaction data) as the basis for a comparative evaluation between Yokie, Bing Search and Bing News, focussing on the ability of the former to identify novel and timely content. Broadly, we have three main objectives we wish to evaluate from these two trials: **Objective 1** - Users *engage and interact with* user-based and contextually-ranked results. **Objective 2(A)** - Yokie locates *exclusive and timely results* compared with a mainstream Search Engine. **Objective 2(B)** - Yokie Users *engage and interact with* timely and exclusive content, compared with a mainstream Search Engine.

2.1   Open User Trial

We curated a Search Party of 1000 Twitter users whose profiles were designated as being *Technology* oriented. These were gathered by mining for names listed in 10 "Technology" labelled lists of users on the Listorious (http://www.listorious.com) Twitter list website. To garner high-quality users, each unique user found in these lists was assigned a Klout score that we used to determine the top 1000 users.

The search index was initially seeded by retroactively mining links from these user profiles and during the course of the evaluation these profiles continued to be mined for current links. As mentioned previously, the user interface allows users to perform three kinds of item-based interactions — namely click through (visit the webpage), "peek" for more metadata, and "like" using the star-shaped button beside each item. Users were also encouraged to re-rank results using the re-ranking UI. While queries retrieved an average of 50 results, users were presented with a maximum of 30 results at any given time. There was a possibility of being presented with different results when re-ranked, however we did not consider any particular ranking strategy as a benchmark to compare with others. Rather, we preferred to gain insight into the performance of the complete system. Each time the user began a new search session, a random ranking strategy was chosen so as to eliminate bias towards any individual options.

Summary Statistics. During the course of the two month trial, a total of 365,735 unique hyperlinks were mined from the Twitter profiles of our 1000 technology influencers. The user study attracted 125 unique users who submitted 1100 queries across 327 sessions. These queries led to the retrieval of 44,856 URLs, and in 95% of the cases, had a temporal window T_{min} set to "now" — indicating users' interest in up-to-the-minute results. Users interacted with 1061 items — 203 of which were click-throughs, and 858 were exploratory "peeks". While we encouraged users to click on the "Like" button for items, we found it difficult to elicit meaningful volumes of user feedback during the trial. As such, we will focus on user engagement of items based on Click-through and Peek actions. Further dataset and usage statistics are discussed below.

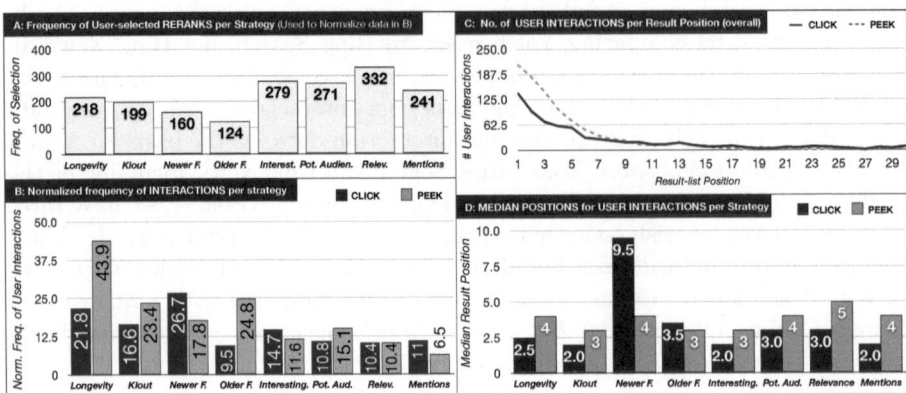

Fig. 2. Chart (A) shows frequency of user selected ranking strategies. Charts (B) shows normalized User Interaction Data for Clicks and Peeks. Charts (C) and (D) show Click Position Data for User Interactions.

Objective 1 — Users Engaged and Interacted with User-Based and Contextually-Ranked Results. Unlike more routine IR and Web Search evaluations which tend to focus on exclusivity of relevance-based rankings of results, in this work our aim is to explore the utility of different ranking strategies. As shown in Figure 2, users tended to actively test a variety of different strategies with a slight preference for strategies such as *Interestingness*, *Relevance* and *Mentions*. We use this data as a basis for normalizing per strategy user engagement of items.

Interactions vs. Ranking Strategies. While *Relevance* was the most frequently selected strategy by users, it performed poorly in terms of normalized user interactions. As presented in Figure 2 (B) users performed 2.6 times more click-throughs on items ranked by *Newer First*, and over twice as many items ranked by *Longevity*. Post-trial, we discovered that over 90% of the queries performed had a T_{max} of "now" (i.e. the query time). As such, it is conceivable the items selected in Newer First results lists were more topical or new to the user, or contained breaking news. Nevertheless, this is an interesting trend towards a preference for novel context-sensitive ranking. While "Peek" engagement of *Relevance* ranked items remains low, the trends are starkly different compared to the corresponding frequencies of selected ranking strategies. Users peeked at more items ranked by *Longevity and Klout* scores.

Result Positioning. Figure 2(C) represents median positions for Clicks and Peeks across all strategies. Both interaction methods follow a standard F-pattern of user interactivity as observed in other search result evaluations [11][8][17]. This chart presents a clear peak of user interaction for items on average above the 7th element in the Yokie result-lists. It also illustrates higher preference for users to perform exploratory "peek" actions on items instead of visiting the hyperlinks themselves. Finally, Figure 2(D) presents the median positions per strategy for each interaction method. Here, we can interpret the positioning of items lower in the list as the system performing well at capturing interactions in the first few result items. In this case, Potential Audience, Klout, Interestingness and Mentions gained user click-throughs with a median position of the second item in the list, while the rest slightly higher. While it can be assumed that a better-performing engine would present higher-frequency interactions at the earliest position in the list, we can also argue the benefits of users exploring further beyond the top-n results. Recency-based (Newer-first) results garnered a higher median positioning — users explored more *recent items* lower in the list.

Discussion. There are a number of take-home messages that arise from these results. First, users do appear to be motivated to try a variety of different ranking approaches. There is a tendency to prefer those such as *Relevance* and *Interestingness*, at least with relation to their frequency of selection, perhaps because they align well with conventional Search Engine ranking. However, based on result engagement, we can see our novel strategies such as *Interestingness, Klout* and temporal strategies seem to outperform others.

2.2   Comparative Study with Bing

For every Yokie query performed during the open user trial, we also gathered search results from Bing Web and Bing News using a vector of the same *keyword queries* users submitted to Yokie (Bing's API's don't support contextual features of the Yokie query tuple). For every Yokie query, we continued to collect Bing's responses every 30 minutes so that we could track how these responses evolved over the evaluation period. In each case, Bing Web and News returned a maximum of 50 results, and each result item was stored based on its query, page title, timestamp of indexing by Bing, item positioning and description. Useful pieces of metadata returned by the API included a corresponding timestamp that represented the moment Bing captured or indexed the item, and the result-list position of an item with respect to that query in Yokie. These two features, along with the ability to compare and contrast overlapping content with Yokie allow us to do comparative studies to see if Yokie has gathered exclusive content, and whether it can present timely results higher up its result lists. Our assumption that the Twitter-sourced content would be newer and easier to index in a more timely fashion seems obvious, however little has been done to test it before.

Objective 2 (A) — Yokie Locates Exclusive and Timely Results Compared with a Mainstream Search Engine. Figure 3 illustrates dataset size and item overlap between the content captured from Bing Web, Yokie and Bing News. In the first panel, Figure 3(A), the comparison is based on the entire datasets of unique items captured by Bing Web and News using Yokie's user trial queries, compared to the URL index of Yokie. In the second panel, Figure 3(B), represents the total coverage of user-interacted items compared to the Bing datasets. In each case, we pay particular attention to whether overlapping items appeared earlier in Yokie or Bing Web/Bing News, so we present a breakdown

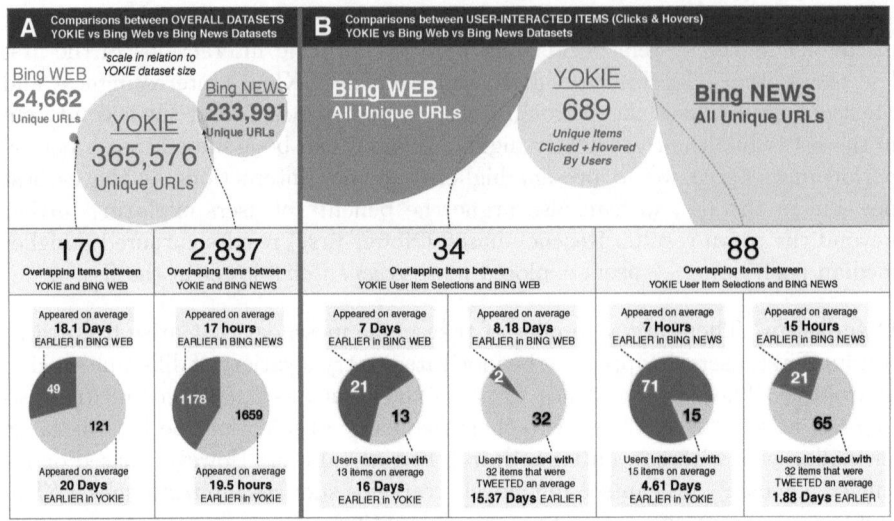

Fig. 3. Yokie dataset comparison with Bing Web Search and Bing News Search. Figure A shows a breakdown of item overlaps in the Bing Web / Yokie / Bing News datasets. Figure B shows the overlaps of user-engaged Yokie items and the total Bing datasets, and also highlight items that appeared earlier in respective datasets.

of the items which appeared earliest in either Yokie or Bing, and the average time of how *early* it was in the corresponding dataset.

Overall Dataset Coverage and Timeliness. As illustrated in Figure 3(A), we found only 170 Yokie results in the entire Bing Web result list — representing less than 0.1% of the total Bing Web items for the same queries. Of those 170 items, 121 (70%) appeared on average 20 days before they appeared in Bing Web. The remaining 49 items appeared earlier in Bing Web by an average of 18.1 days. When comparing the Yokie set with the arguably more timely of the comparative systems, Bing News, we see a much higher overlap set with 2837 items — almost 1% of the total News dataset. In this case, Yokie garners a considerably higher rate of coverage — 28% more of a share of those overlapping items earlier than Bing News. We believe that while gaining earlier items can be important, an overall broader coverage of items trumps Bing News' gaining items on average 2 hours earlier across a smaller share.

Clearly Yokie is maintaining a very different type of information index compared to Bing (Web and News). It is less comprehensive than Bing Web, but it focusses on trending URLs and produces result-sets that are entirely different from those produced by Bing. The results returned by Yokie are exclusive to Yokie. If they have been indexed by Bing they are rarely retrieved, or swamped by more conventional results. If the results do come to be recognised by Bing, this usually happens after a significant latency period which can be measured in days for Bing Search and tens of hours for Bing News.

Item Exclusivity and Position Performance. Our next comparison is aimed at exploring how well Yokie performs at ranking and retrieving exclusive items that didn't appear in Bing. During the user trial, Yokie retrieved 44,856 of it's 365,735 items from the 1,100 user-performed queries (not illustrated above).

If we compare this set of items with both Bing sets, we see negligible overlaps — 109 items appeared in any Bing Web result list, whereas 1,292 of them appeared in Bing News result lists. Therefore, Yokie's results contained 99.8% exclusive items compared to Bing Web, and 97.2% compared to the Bing News set (also of note is Bing News' 10 times larger set compared to Bing Web).

A more useful metric is comparing the Bing itemsets with items with which users interacted. To return your attention to Figure 3(B), we analyze a similar breakdown of overlaps as before, except here we compare them with overlapping items that are dated from when the user *interacted* with them. Also depicted is another interesting trend — of the 689 unique items, only a total of 122 overlapping items. Combined, this says that of all the items with which users interacted, 82% were *exclusively* present in Yokie's index.

Objective 2 (B) — Yokie Users Interacted with Timely and Exclusive Content. The previous objective was limited to analysing the overall composition of the datasets. Here, we examine the items explicitly selected by users from Yokie result lists, as these represent actual indications of user interest in the content. We separate each overlapping item based on whether its timestamp was earlier in Yokie or Bing. The timestamp can either represent the moment the user interacted with it, or the moment when the original item was published on Twitter (since Yokie live-streams content, conceivably its indexing time also).

Time of User Interaction vs. Bing Indexing. In Figure 3(B), we see that during the trial, 689 results were interacted with by Yokie users. Once again, the vast majority were exclusive to Yokie; only 34 items found by Bing web, and 89 by Bing News. In terms of timeliness, we also see Yokie performing comparatively well as opposed to Bing Web — 13 items were interacted with by users an average of 16 days earlier than they even appeared in the Bing Web index. Compare this to Bing Web's larger share of 21 items which appear on average only a week earlier in the Bing Web index than in Yokie's. Yokie is also seen to fall behind in terms of user engagement and Bing News publication. While Yokie users had clicked on items that were on average almost 5 days newer than their appearance in Bing News, this was only represented by 15 items. Bing News had a considerably broader coverage with 71 items with which Yokie users interacted, however these were in its indexes only 7 hours before Yokie. Overall, comparing the time of item interactions with Bing index time shows Yokie under-performing in terms of coverage of items with which its users interacted, regardless of its good performance gaining content earlier than Bing.

Interacted Items' Tweet Time vs. Bing Indexing. Here, we describe a similar comparison outlined earlier — but here we compare like-for-like analyses between the moment users *interacted with* an item, and its index time in Bing

Web and News. As described in Figure 3(B), Yokie users interacted with a total of 32 items that appeared on average 15.3 days earlier than they appeared in Bing Web. They also interacted with the majority of the news overlaps on average 1.88 days earlier than they even appeared in Bing News.

Discussion. One could sensibly assume that Bing Web and Bing News would garner higher coverage and timely content respectively compared to our system. However, the vast majority of Yokie results were *exclusive* to the system, apart from the 170 web items and 2837 news items that are overlapping. We can use this overlapping data to further support the hypothesis that Yokie is able to identify these results earlier than Bing — with over 70% of Bing Web/Yokie overlaps appearing almost 20 days earlier in Yokie, and almost 60% of the Bing News/Yokie coverage appearing nearly a full day beforehand in Yokie.

Once again, the data supports the argument for the potential of a real-time streaming technique for indexing web content. While Bing had more coverage of earlier items amongst user-interacted results, Yokie's interacted content was not only vastly exclusive, but also in many cases more timely than Bing. This is a surprising result considering Bing Web represents a capture of the entire web, and Bing News represents a real-time capture of topical breaking news.

3   Trial Discussion, Limitations and Conclusions

Our user trial and comparative evaluations have highlighted Yokie's potential for timely, exclusive and relevant content compared with popular web and news search services. Our approaches to contextual and user-based ranking strategies also performed well at providing relevant content with which users interacted. In particular, the temporally sensitive rankings that showed newer content and also content with high longevity garnered significant interest from users. Our second trial illustrates the shortcomings of both archival Search Engines and, surprisingly, some reputable online news search services at providing timely content.

This work is not without its limitations. In building Yokie we have designed a complex and novel information discovery service with many moving parts. This presented something of an evaluation challenge. Unlike traditional IR systems, there were no benchmark datasets (e.g. TREC [9], etc). Our approach has been to attempt a live user trial. Obviously in an ideal world one would welcome the opportunity to engage a large cohort of users — perhaps thousands or tens of thousands — in such an evaluation. Unfortunately this was not possible with the resources available, and so our trial is limited to a smaller cohort of just 125 users. Nevertheless these users did help to clarify our primary evaluation objectives and their usage provided a consistent picture of their common engagement patterns to show how they interacted with different ranking strategies and gained benefit from the results they provided. Moreover we bolstered this live evaluation with an additional study to compare Yokie with more conventional search and news discovery services by harnessing the activity of our live users as the basis for search queries. Our focus on Bing was purely pragmatic as it's API's Terms and

Conditions provided us with access to the data we needed, whereas alternatives such as Google do not.

Yokie's architecture and content source enables us to perform a broad range of future evaluations. Indeed, a multitude of ranking and retrieval techniques remain to be explored. Emphasis will be placed on trials in which users may curate their own Search Parties, and scope exists for these Search Parties to be collaboratively built. While the findings in this paper validate the system's merits and novelties, significantly more work can be done with the general retrieval aspects, such as the reasoning behind the comparatively lower positioning of user interesting items compared to Bing, comparative studies between the user-selected rankings and randomly assigned ranking strategies, and so on.

References

1. Brin, S., Page, L.: The anatomy of a large-scale hypertextual web search engine. Comput. Netw. ISDN Syst. 30, 107–117 (1998)
2. Chau, M.: Comparison of three vertical search spiders. IEEE Computer 36 (2003)
3. Chen, J., Geyer, W., Dugan, C., Muller, M., Guy, I.: Short and tweet: experiments on recommending content from information streams. In: CHI 2010. ACM (2010)
4. Dong, A., Zhang, R., Kolari, P., Bai, J., Diaz, F., Chang, Y., Zheng, Z., Zha, H.: Time is of the essence: improving recency ranking using twitter data. In: WWW 2010, ACM, New York (2010)
5. Esparza, S.G., O'Mahony, M.P., Smyth, B.: On the real-time web as a source of recommendation knowledge. In: RecSys 2010, September 26-30. ACM (2010)
6. Krishnamurthy, B., Gill, P., Arlitt, M.: A few chirps about twitter. In: WOSP 2008, NY, USA. ACM (2008)
7. Kwak, H., Lee, C., Park, H., Moon, S.: What is twitter, a social network or a news media? In: WWW 2010, pp. 591–600 (2010)
8. Lewandowski, D.: A framework for evaluating the retrieval effectiveness of search engines. In: Next Generation Search Engines (2012)
9. Macdonald, C., Ounis, I., Lin, J., Soboroff, I.: Microblog track. In: TREC (2011)
10. McNally, K., O'Mahony, M.P., Smyth, B., Coyle, M., Briggs, P.: Towards a reputation-based model of social web search. In: IUI 2010. ACM (2010)
11. Nielsen, J., Pernice, K.: Eye tracking research into web usability, http://www.useit.com/eyetracking/ (accessed May 2012)
12. Phelan, O., McCarthy, K., Bennett, M., Smyth, B.: Terms of a feather: Content-based news recommendation and discovery using twitter. In: Clough, P., Foley, C., Gurrin, C., Jones, G.J.F., Kraaij, W., Lee, H., Mudoch, V. (eds.) ECIR 2011. LNCS, vol. 6611, pp. 448–459. Springer, Heidelberg (2011)
13. Phelan, O., McCarthy, K., Smyth, B.: Yokie - a curated, real-time search and discovery system using twitter. In: 2nd RSWEB Workshop, RecSys 2011 (2011)
14. Phelan, O., McCarthy, K., Smyth, B.: Yokie: explorations in curated real-time search & discovery using twitter. In: ACM RecSys 2012. ACM (2012)
15. Resnick, P., Kuwabara, K., Zeckhauser, R., Friedman, E.: Reputation systems. Commun. ACM 43, 45–48 (2000)
16. Sebastiani, F.: Machine learning in automated text categorization. ACM Comput. Surv. 34, 1–47 (2002)
17. Smyth, B., Balfe, E., Boydell, O., Bradley, K., Briggs, P., Coyle, M., Freyne, J.: A live-user evaluation of collaborative web search. In: IJCAI (2005)
18. Yardi, S., Romero, D., Schoenebeck, G., Boyd, D.: Detecting spam in a Twitter network. First Monday 15(1) (January 2010)

GWMF: Gradient Weighted Matrix Factorisation for Recommender Systems

Nipa Chowdhury and Xiongcai Cai

School of Computer Science and Engineering,
The University of New South Wales, Sydney, NSW 2052, Australia
{nipac,xcai}@cse.unsw.edu.au

Abstract. In this paper, we developed a new algorithm, Gradient Weighted Matrix Factorisation (GWMF), for matrix factorisation. GWMF uses weights to focus the approximation in the matrix factorisation to the higher approxmation residual. Therefore, it improves the matrix factorisation accuracy and increases the speed of convergence. We also introduce a regularisation parameter to control overfitting. We applied our algorithm to a movie recommendation problem and GWMF performs better than ordinal gradient descent-based matrix factorisation (GMF) on Movielens dataset. GWMF converges faster than GMF and it guarantees lower root mean square error (RMSE) at earlier iterations on both training and testing.

Keywords: Weighted gradient descent, Matrix factorisation, Recommendation.

1   Introduction

The population of the Internet is increasing day by day. Now 50 billion webpages are available on the Internet and the size of the Internet is about to double in every 5 years[1, 2]. This gigantic size of the Internet creates some problems too. When people search in the Internet, they are now experiencing overwhelmed with lots of information. A little portion of the information is relevant to the person's interest. So a recommender system is needed to guide the user when searching such a vast space. Recommender systems suggest items or persons to a user based on the user's taste so that the user can quickly reach to the point of his interest without going through irrelevant information. Furthermore, a recommendation list itself may be large enough and usually the user considers only the top few items. So it is a challenging problem in recommender systems to rank top items accurately.

Recommender systems are mainly divided into two categories: content-based filtering and collaborative filtering [3]. In recommender systems, an item is recommended to an active user based on 1) similarity of the item with other items in which the active user has already shown some preferences, or 2) similarity of users to the active user, whom active user has similar preference on some set of items. In content-based filtering, the similarity is based on the item or user profile. For example, in a book recommendation, a book is recommended to a user based on the books profile

Y. Ishikawa et al. (Eds.): APWeb 2013, LNCS 7808, pp. 758–769, 2013.

similarity (book type, writer and so on) with other books that were already rated highly by that user [3-5]. Rather than computing similarity from profile, collaborative filtering uses users choice on items to derive similarity. Collaborative filtering (CF) is widely used in social network and also in online web [3-5]. As the size of the Internet is increasing day by day, scalability becomes an important issue and matrix factorisation is the ultimate solution to handle this scalability problem [6].

In matrix factorisation (MF), data are placed into a matrix, which indicates user interaction over item. For example, $R \in R^{N \times M}$ is the rating matrix represents preferences on N users over M items. The major purpose of matrix factorisation in this context is to obtain some form of lower-rank approximation to R for understanding how the users relate to the items [7]. User preferences over item are modelled by user's latent factors and item preferences by user are modelled by item's latent factors. These factor matrices are learned from observed preferences in such a way that their scalar product equals to original rating matrix [8]. For example, the target matrix $R \in R^{N \times M}$ is approximated by two low rank matrix $P \in R^{N \times K}$ and $Q \in R^{M \times K}$

$$R \approx P.Q^t \tag{1}$$

Here K represents the dimensionality of the approximation. After learning, these latent factors are used to predict unobserved ratings. Gradient descent algorithm [8] is widely used for learning these latent factors.

In this paper, we present a new algorithm for learning latent factors. Our main contributions are as follows. Firstly, we develop a new gradient descent based matrix factorisation algorithm, namely GWMF, by using weights to update factor matrix in each steps to learn latent factors. We also introduce a regularisation parameter to reduce overfitting. Result shows our algorithm achieves converge faster with higher accuracy than the ordinal gradient descent based matrix factorisation algorithm.

2   Literature Survey

For its scalability and efficiency, matrix factorisation has become most popular method in recommendation. A lot of research was done for further improvement of accuracy and reducing computational time.

In [6] a probabilistic matrix frame (PMF) is proposed to approximate the user feature vector and movie feature vector from user movie rating matrix. The work addresses two crucial issues of recommender system: i) complexity in terms of observed rating and ii) sparsity of ratings. With a cost function by a probabilistic linear model with Gaussian noise, this method achieves 7% performance improvement over Netflix own system [17]. Authors in [9] identify drawback of their previous PMF models [6] as PMF is computationally complicated. To solve the problem, they propose Bayesian Probabilistic Matrix Factorisation (BPMF) where user and movie features have no zero mean which need to be modeled by some hyper-parameters. Therefore, the predictive distribution of the rating is approximated by Monte Carlo algorithm. When the number of features is large, BPMF outperforms PMF in terms of time complexity.

A hybrid approach is used in [10] to learn incomplete rating from a rating matrix. Here they enforce non-negativity constrain to model matrix factorisation. At first, Expectation Maximization algorithm is performed several iterations to select initial values of user and item matrix. These initial values are used by Weighted Non-negative Matrix Factorisations (WNMF). Here they use weights as indicator matrix. For example, a weight equals to one if rating is observable or zero if unobservable. Authors in [11] extend WNMF in [10] to Graph Regularized Weighted Nonnegative Matrix Factorisation (GWNMF) to approximate low rank matrices of user and item. They use both user and item's internal information, external information and also social trust information. Experimental result shows that these internal and external information increase recommendation accuracy.

Clustering is applied with Orthogonal Non-negative Matrix Factorisation (ONMTF) in [12]. ONMTF is applied to decompose a user-item rating matrix into a user matrix (U) and an item matrix (SVT). By using U and SVT, user/item clusters and their centroids are identified. To predict unknown rating of a user to an item, at first similarity between the test user and item is calculated and similar user/item clusters are identified. Within the identified clusters, K most similar neighbours are found. The linear combination model based on user-based approach, item-based approach and ONMTF is applied to calculate the test user's rating on the item. Although experimental result does not show any improvement, their method requires less computational cost.

An ensemble of collaborative filters is applied in [13] to predict ratings on Netflix dataset. As MF algorithm is sensitive to local minima, the author applies different MF algorithms with different initialisation parameters several times. And then the mean of their prediction is used to compute final ratings.

Authors in [14] develop non-negative matrix factorisation algorithm where they put additional constraint that all elements of the factor matrices must be non-negative. They develop two algorithms by using multiplicative update steps, one attempts to minimise the least squares error and the other to minimises the Kullback-Leibler divergence. Both of the algorithms guarantee locally optimal solutions and are of easy implementations.

To the best of our knowledge, there is no research done by using user or item weights with latent factor vectors to approximate ratings. Our algorithm, GWMF gives larger weight values to an item and user with high approximation error, i.e. latent factors not approximated the rating between the user and item well in the current iteration, so that in next gradient step the algorithm can focus on those items and users and give more update on them. We also introduce a regularisation parameter to reduce overfitting. Result shows that GWMF guaranteed lower training and test error and it achieves minimum RMSE faster than gradient descent.

The rest of the paper is arranged as follows: in next section we present gradient descent-based matrix factorisation algorithm (GMF) and our gradient weighted matrix factorisation algorithm (GWMF). Experimental settings and result of GWMF vs. GMF is presented in section 4. Finally we draw conclusions and future work.

3   Algorithm

3.1   Gradient Descent Algorithm for Matrix Factorisation

Given a rating matrix R and mean square error (MSE) as optimisation function, Gradient descent based matrix factorisation aims to find two low rank matrix P and Q, which minimise the following objective.

$$GMF = \sum_{i=1}^{N} \sum_{j=1}^{M} \left(R - P.Q^t \right)_{ij}^2$$

(2)

The algorithm is presented in Algorithm 1.

Algorithm 1: GMF

Input: Rating Matrix R, learning rate α, regularisation parameter β, no. of iterations steps, dimension of low rank matrix K.
Output: Factor matrix P and Q.

1 Initialise P and Q randomly.
2 for s=1:steps do
3   for i=1:N do
4   for j=1:M do
5   if R(i,j)>0
5.1   er←R(i,j)-(P(i,:)*Q(:,j));
5.2   P(i,:) ←P(i,:)+(α*er*2*Q(:,j)'-(α*β*P(i,:)));
5.3   Q(:,j) ←Q(:,j)+(α*er*2*P(i,:)'-(α*β*Q(:,j)));
         end for
      end for
   end for
6 return P,Q

Algorithm 1. Gradient Descent Algorithm for Matrix Factorisation

As shown in Algorihtm 1, GMF treats each element equally, i.e. assign equal learning rate for each element, in Q when update P and each equally in P when update Q.

3.2   Gradient Weighted Algorithm for Matrix Factorisation

In GMF, a rating $R(i,j)$ is approximated by the product of the corresponding factor vectors $P(i,:)$ and $Q(:,j)$. The approximation can be formulated as a learning problem for each row $R(i,:)$:

$$R(i,:) = P(i,:) * Q$$

When we fix Q and solve $P(i,:)$, in the above equation, each rating $R(i,j)$ is a target of an instance, each corresponding column $Q(:,j)$ is the feature value vector of that instance, and $P(i,:)$ is the linear regression model parameter. Similarly, for each column $R(:,j)$, we have:

$$R(:,j) = P*Q(:,j)$$

where we can fix matrix P and solve the parameter vector $Q(:, j)$ for a linear regression model. In such a configuration, the matrix factorlisation is formulated as a set of linear regression models: one for each row of P and one for each column of Q.

One of the common assumptions, underlying most linear and nonlinear least squares regression, is that each data instance provides equally precise information about the deterministic part of the total process variation. More specific, the standard deviation of the error term is constant over all values of the predictor or explanatory variables. However, this assumption does not hold, even approximately, in every modelling application. In such situations, weighted least squares can be used to maximise the efficiency of parameter estimation [18].

To this end, giving each instance its proper amount of influence over the parameter estimates may improve the performance of the approximation since a method that treats all of the data instance equally would give less precisely measured instance more influence than they should have and would give highly precise points too little influence.

Therefore, in GWMF, we use weights to optimise approximation of the target matrix. The idea behind this is, if a user or item is not approximated well in previous iterations, the algorithm focuses on reducing its errors and thus which contributes more to the update than others. In GWMF we give more weight to instances that are not approximated well by previous iterations. We use two weight matrices 1) user weight $WP \in R^{N \times M}$ that indicates user approximation over item and item weight $WQ \in R^{N \times M}$ that indicates item approximation by user in current iterations. At initialisation all user weight values for a rated item and all rated item weight values for a user are equal (for example as shown in Fig. 1(a)). Then at each gradient steps this weight values keep tracks how error of rating matrix to factor matrix i.e. $(R-PQ)$ is changed for each observed user and item. We also enlarge this effect by exponential cost function for weight updates, which results in large weight values for instances which have higher error. Weight update formula of WP is given below. WQ follows the same update formula. Furthermore, we also normalise the weight values by maximum normalisation.

$$WP = e^{|(R-PQ)|}$$

Therefore, the algorithm converges to the training objective quickly. But in real world, there may be some outliers or noise in datasets. So this weight update formula gives more weights to those instances constantly and thus could make the algorithm lose generalisation ability, i.e. overfitting. In this concern, we introduce a parameter called *rbeta* inspired by [15] which takes $(R-rbeta*PQ)$ for weight update. So the weight update formula becomes:

$$WP = e^{|(R-rbeta\,PQ)|} \tag{3}$$

From Fig. 1 we see how weight values of *WP* and *WQ* changes according to equation 3 in each gradient steps. After the 1st iteration (Fig. 1(b)) the weight values of each user-item vary by a large amount. So more gradient update is performed to user-items that have high weight values than others. When the algorithm converges, user weight values for an item and item weight values for a user become equal. At 1000 iterations (Fig. 1(c)) and 2000 iterations (Fig.1(d)) training errors become very small, i.e. similar, and weight values of user item become 1. The GWMF algorithm is given in Algorithm 2.

Algorithm 2: GWMF

Input: Rating Matrix R, learning rate α, regularisation parameter β, no. of iterations steps, dimension of low rank matrix K.
Output: Factor matrix P and Q.

   1. Initialise P and Q randomly. Initialise WP=1/M and WQ=1/N.
   2. for s=1:steps do
   3   for i=1:N do
   4   for j=1:M do
   5   if R(i,j)>0
   5.1   er←R(i,j)-(P(i,:)*Q(:,j));
   5.2   P(i,:)←P(i,:)+(α*er*2*WP(i,j)*Q(:,j)'-(α*β*WQ(i,j)*P(i,:)));
   5.3   Q(:,j)←Q(:,j)+(α*er*2*WQ(i,j)*P(i,:)'-(α*β*WP(i,j)*Q(:,j)));
           end for
        end for
      end for
   6 return P,Q

Algorithm 2. Gradient Weighted Algorithm for Matrix Factorisation

4   Results

To evaluate performance of GWMF, we use two datasets: (i) Small Dataset that consists of first 100-by-50(user-by-item) rating for training from Movielens u1.base dataset[16]. Testing is done by those users-items rating from u1.test dataset. (ii) The Movielens Dataset: which consists of 100,000 ratings from 1000 users on 1700 movies. We perform experiments separately on each five Movielens dataset and average the results. We compare the proposed GWMF with the standard GMF algorithm. For fair comparison, we need to make learning rates of these two algorithms equal. To achieve this, we set learning rates of two algorithms in two different ways. For the smaller dataset, we divide learning rate of GWMF by initialisation weight values. Here initialisation weight values means if a user rated 5 items, all items receive equal weight values 1/5=0.2. This is because in GWMF we multiplied weight with errors so it decreases the learning rate. To make it the same as GMF, we divide learning rate of GWMF again by initialisation weight. Here initialisation weight values remain the same for all iterations whereas other weight values (i.e multiplied *WP* and *WQ*) change according to GWMF. In the settings for large Movielens dataset, we keep learning rate of GWMF unchanged but multiplied

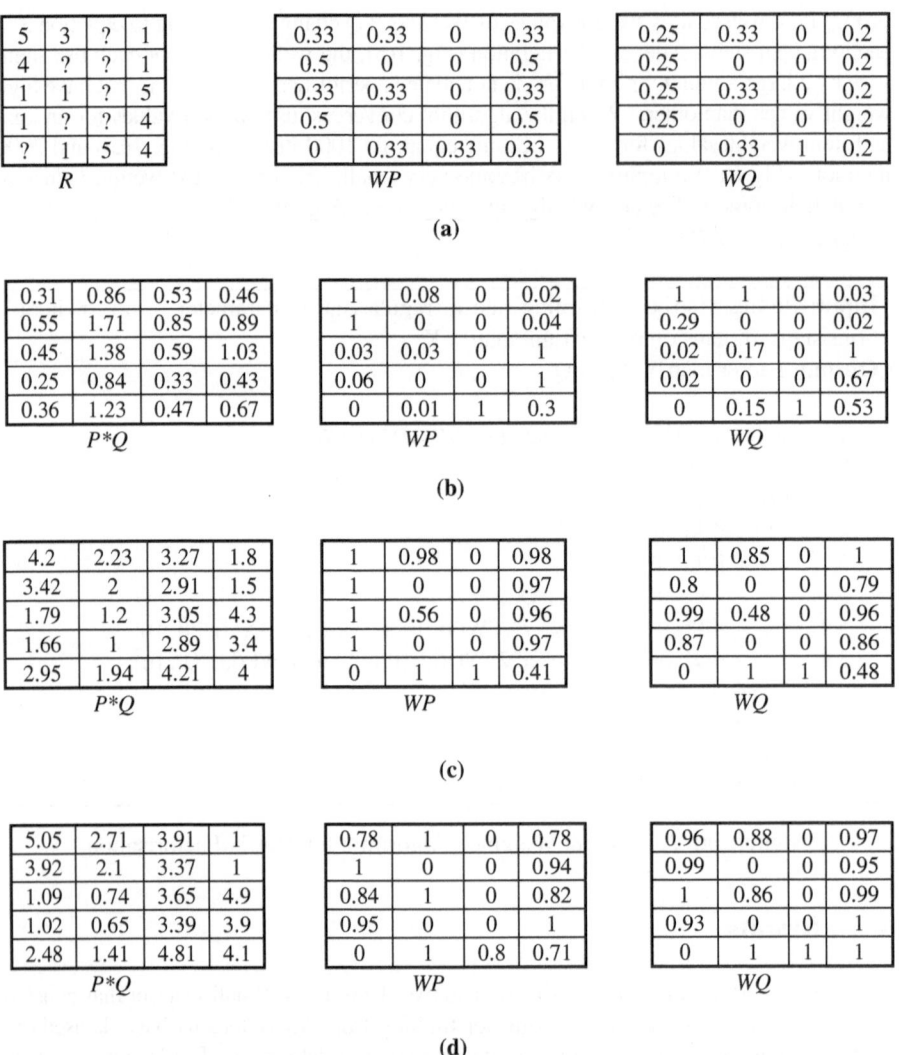

Fig. 1. Rating matrix R and initialization weight values of WP and WQ is shown in (a). PQ, WP, WQ is shown after the 1st (b), after 1000 (c) and after 2000 (d) iterations.

weight value with learning rate of GMF. So this rate is also decreased and learning rate of GWMF and GMF are comparable.

RMSE is used to measure performance. For the small dataset, as the size of our training set is smaller and thus more sparse, rating prediction is more challenging than that of the original Movielens dataset.

To see how weights work on matrix factorisation, we perform two experiments on small dataset. In one experiment we use one weight matrix to examine how instances are approximated in previous iterations and give more weight to instances that are not approximated well. In other experiment, we use two weight matrices: the user weight

matrix *WP* which indicates how items are approximated for a user in previous iterations and the item weight matrix *WQ* which indicates how users are approximated
for an item. The idea behind this experiment is: the user factor (item factor) is multiplied by rated item factors (user factor) in matrix factorisation. So these two weight
matrices give an idea about how well users and items are approximated in previous
iterations. RMSE of two weight matrices vs. one weight matrix on smaller version of
Movielens dataset is given in Table 1. In Fig. 2, RMSE on test dataset is shown.

Table 1. RMSE of two weight matrix vs. one weight matrix in training and testing

Iterations	MF with one weight matrix		MF with two weight matrices	
	Train	Test	Train	Test
10	1.0155	1.05	1.04	1.0772
20	0.8918	0.9676	0.9106	0.9879
50	0.7082	0.8765	0.7216	0.8892
100	0.5306	0.8245	0.5549	0.8403
120	0.4762	0.8142	0.5044	0.8306
200	0.3258	0.797	0.3563	0.8117
250	0.2681	0.7934	0.2967	0.8074
500	0.125	0.7843	0.1499	0.7959
1000	0.0473	0.7715	0.0562	0.7777
2000	0.0326	0.757	0.033	0.7583
5000	0.0279	0.7515	0.0282	0.7481
10000	0.026	0.7632	0.0262	0.7592
20000	0.0251	0.781	0.0252	0.7773

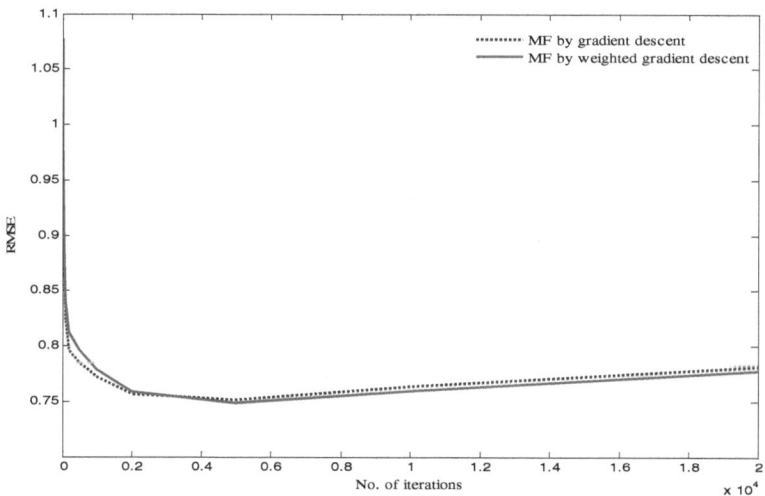

Fig. 2. RMSE of two weight matrices vs. one weight matrix on the small dataset

From Fig. 2 we see that MF by two weight matrices performs better that one weight matrix. In this concern we use two weight matrices for the rest of our experiment. Next we present performance of user parameter *rbeta* in our algorithm. When *rbeta*=1, it means that it takes *PQ*'s opinion completely for next iterations and when *rbeta*=0 it results unique weight update for all iterations. For this small dataset, we achieve minimum RMSE for *rbeta*=0.5. Here in Fig. 3 we present performance of various *rbeta* values on GWMF with two weight matrices.

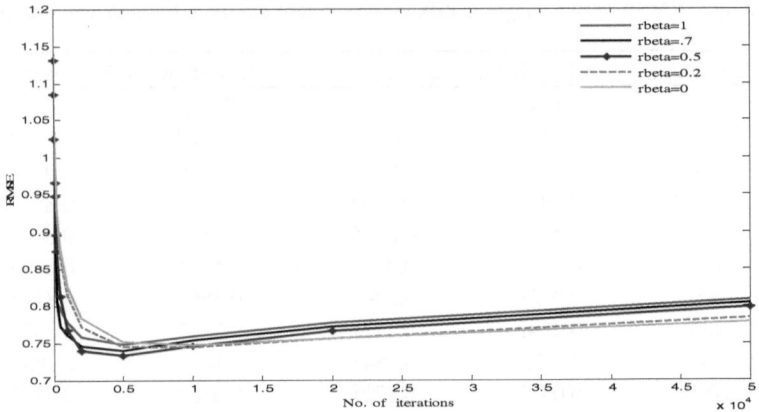

Fig. 3. RMSE of various *rbeta* values on GWMF with two weight matrices on small dataset

We also compare our algorithm to gradient descent and the result is presented in Fig. 4 and 5 on the smaller dataset and Fig. 6 and 7 for the Movielens dataset. We set *rbeta*=1 for both dataset and learning rate=.0002 for smaller dataset and 0.01 for Movielens dataset.

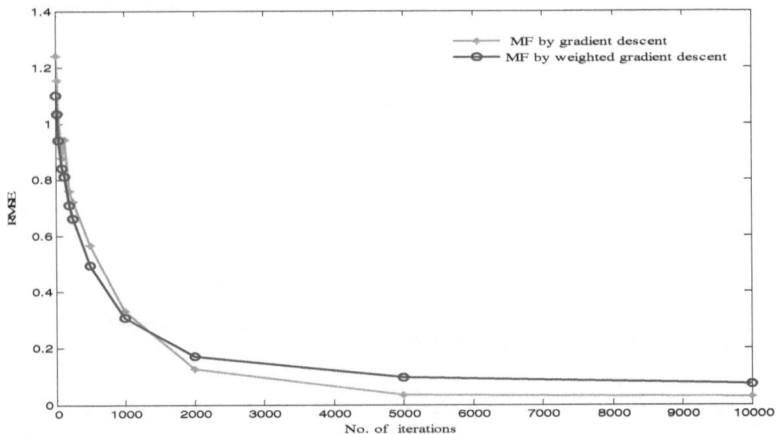

Fig. 4. MF by gradient descent vs. MF by weighted gradient descent on the training set (Small).

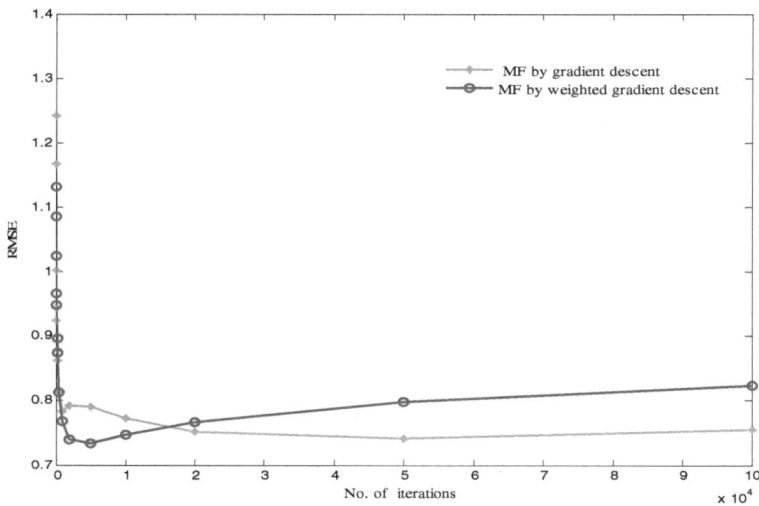

Fig. 5. MF by gradient descent vs. MF by weighted gradient descent on the test set (Small).

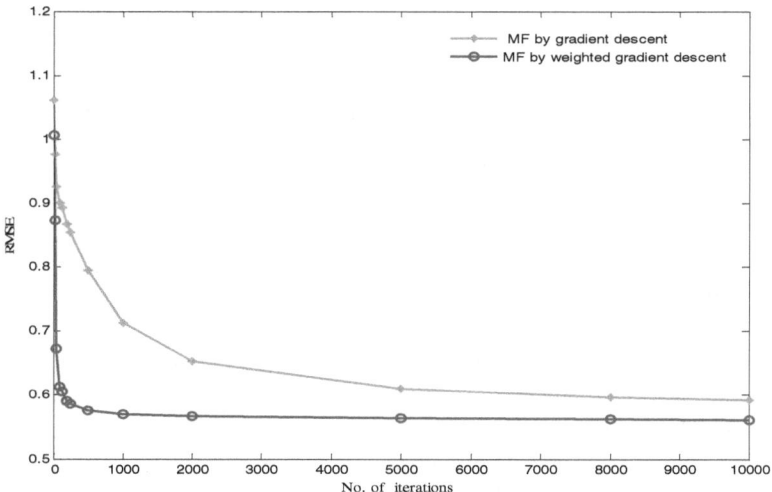

Fig. 6. MF by gradient descent vs. MF by weighted gradient descent on the training set (Movielens)

From Fig. 4 and 5, we see that GWMF achieves lower RMSE on earlier iterations than GMF but higher RMSE than GMF at the rest of the training phase. But in testing (Fig. 5), GWMF achieves lower RMSE (0.7334) which is 10 times earlier than GMF.

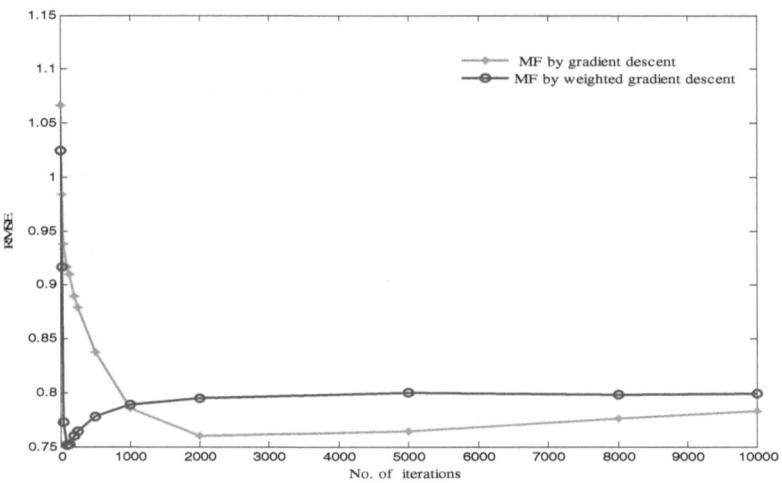

Fig. 7. MF by gradient descent vs. MF by weighted gradient descent on the test set (Movielens)

For earlier iterations, it shows GWMF achieves lower RMSE and in 5000 iterations GWMF achieves its lowest RMSE. However, GMF reaches its lowest RMSE at 50000 iterations that is still larger than the lowest RMSE of GWMF.

In Movielens dataset, GWMF also outperforms GMF (Fig. 6 and 7). On test set GWMF achieves the lowest RMSE in 100 iterations (0.75) whereas GMF achieves the lowest RMSE at 2000 iterations (0.76) that is also higher than GWMF. So our GWMF performs better than GMF in terms of accuracy and computational cost in both Movielens dataset.

5   Conclusion

Although Gradient descent is the widely used techniques in matrix factorisation, it has slow convergence. Researchers use multiplicative update steps or other techniques to resolve this problem. However, to the best of our knowledge, there is no research was done to optimise matrix factorisation by weighted gradient descent. GWMF is the first work in this direction, where we use weights directly in gradients update steps. As gradient descent moves forward to objective function in each step, putting extra weights is very challenging. There is a risk to destroy gradient descents convergence property. But using an appropriate cost function and regularisation parameter, we achieve this and our results are very encouraging. In future, we plan to apply this algorithm on larger datasets and other problems where gradient descent algorithm is also applied.

Acknowledgement. This project is funded by the Smart Services CRC under the Australian Government's Cooperative Research Centre program.

References

1. The size of the World Wide Web, `http://www.worldwidewebsize.com/`
2. The Size of Internet to Double Every 5 Years,
   `http://www.labnol.org/internet/`
   `internet-size-to-double-every-5-years/6569/`
3. Adomavicius, G., Tuzhilin, A.: Toward the Next Generation of Recommender Systems: A Survey of the State-of-the-Art and Possible Extensions. In: IEEE Transaction of Knowledge and Data Engineering, pp. 734–749 (2005)
4. Pazzani, M., Billsus, D.: Learning and Revising User Profiles: The identification of Interesting web sites. J. Mac. Lea. 27, 313–331 (1997)
5. Jannach, D., Friedrich, G.: Tutorial: Recommender Systems. In: Joint Conference on Artificial Intelligence, Barcelona (2011)
6. Salakhutdinov, R., Mnih, A.: Probabilistic Matrix Factorization. In: Neural Information Processing Systems (2008)
7. Hubert, L., Meulman, J., Heiser, W.: Two Purposes for Matrix Factorization: A Historical Appraisal. Society for Industrial and Applied Mathematics 42, 68–82 (2000)
8. Koren, Y., Bell, R.M., Volinsky, C.: Matrix Factorization Techniques for Recommender Systems. IEEE Computer 42(8), 30–37 (2009)
9. Salakhutdinov, R., Mnih, A.: Bayesian Probabilistic Matrix Factorization using Markov Chain Monte Carlo. In: 25th International Conference on Machine Learning (ICML), Helsinki (2008)
10. Zhang, S., Wang, W., Ford, J., Makedon, F.: Learning from Incomplete Ratings Using Non-negative Matrix Factorization. In: Proceedings of the SIAM International Conference on Data Mining, SDM (2006)
11. Gu, A., Zhou, J., Ding, C.: Collaborative Filtering: Weighted Nonnegative Matrix Factorization Incorporating User and Item Graphs. In: 10th SIAM International Conference on Data Mining (SDM), USA (2010)
12. Chen, G., Wang, F., Zhang, C.: Collaborative Filtering Using Orthogonal Nonnegative Matrix Trifactorization. J. of Inf. Pro. and Man. 45, 368–379 (2009)
13. Wu, M.: Collaborative Filtering via Ensembles of Matrix Factorizations. In: Proceedings of KDD Cup and Workshop, pp. 43–47 (2007)
14. Lee, D.D., Seung, H.S.: Algorithms for Non-Negative Matrix Factorization. In: Advances in Neural Information Processing, pp. 556–562 (2000)
15. Jin, R., Liu, Y., Si, L., Carbonell, J., Hauptmann, A.G.: A New Boosting Algorithm Using Input-Dependent Regularizer. In: Proceedings of the 20th International Conference on Machine Learning, Washington (2003)
16. MovieLens Data Sets, `http://www.grouplens.org/node/73`
17. Netflix Prize, `http://www.netflixprize.com/`
18. Gentle, J.: 6.8.1 Solutions that Minimize Other Norms of the Residuals. In: Matrix Algebra. Springer, New York (2007)

Collaborative Ranking
with Ranking-Based Neighborhood

Chaosheng Fan and Zuoquan Lin

School of Mathematical Sciences, Peking University, Beijing 100871, China
{fcs,lz}@pku.edu.cn

Abstract. Recommendation system is a very important tool to help users to find what they are interested in on the web. In many commercial recommendation systems, only the top-K items are shown to users, and recommendation becomes a ranking task rather than a classical rating prediction task. In this paper, we propose a new collaborative ranking algorithm based on learning to rank framework in information retrieval. For a given user-item pair (u, i), we use Kendall Rank Coefficient as similarity metric to choose neighborhood for user u and use the ranking statistical information of item i from user u's neighborhood as the feature representation of pair (u, i). We apply LambdaRank to learn the ranking model and experimentally demonstrate the effectiveness of our method by comparing its performance with several collaborative ranking approaches. Moreover, we can address scenarios where users' feedbacks are non-numerical scores.

Keywords: collaborative ranking, collaborative filtering, recommendation system.

1 Introduction

Electronic retailers and content providers offer a huge selection of products for modern consumers. Matching consumers with products they like is very important for user's satisfaction on the web. Recommendation systems, which provide personalized recommendation for users, become a very important technology to help users to filter useless information. Through automatically identifying relevant information by learning users' tastes from their behavior data in the system, recommendation systems greatly improve users' experience on the web. Internet leaders like Amazon, Google, Netflix and Yahoo increasingly adopt different kinds of recommendation systems, which generate huge profits for them.

Recommendation systems are usually divided into two classes: content-based filtering and collaborative filtering. Content-based filtering approaches use the content information of the users and items (products) to make recommendations, while collaborative filtering approaches just collect users' ratings on items to make recommendations. Recommendation has been treated as a rating prediction task for a long time. Two successful approaches to rating prediction task are neighborhood model and latent factor model. Neighborhood models prediction a target user's rating on an item by (weighted) averaging the item's ratings

Y. Ishikawa et al. (Eds.): APWeb 2013, LNCS 7808, pp. 770–781, 2013.

of her/his k-nearest neighbors, who have the similar rating pattern with the target user. Latent factor models transform both items and users to the same latent factor space and predict a user's rating on an item by multiplying the user's latent feature with the item's latent feature. The combination of neighborhood model and latent factor model successfully completes the rating prediction task [1].

However, in many recommendation systems only the top-K items are shown to users. Recommendation is a ranking problem in the top-K recommendation situation. Ranking is more about predicting items' relative orders rather than predicting rating on items and it is broadly researched in information retrieval. The problem of ranking documents for given queries is called Learning To Rank(LTR) [2] in information retrieval. If we treat users in recommendation systems as queries and items as documents, then we can use LTR algorithms to solve the recommendation problem. A key problem in using LTR models for recommendation is the lack of features. In information retrieval, explicit features are extracted from $(query, document)$ pairs. Generally, three kinds of features can be used: query features, document features, and query-document-dependent features. In recommendation system, users' profiles and items' profiles are not easy to be represented as explicit features. Extracting efficient features for recommendation systems is emerged and also very important for learning a good ranking model.

Some work tries to extract features for user-item pairs [3,4] etc. In [3], the authors extract features for a given (u, i) pair from user u's k-nearest neighbors who rated item i. They use rating-based user-user similarity metric to find neighbors for a target user. However, in some e-commerce systems users' preferences to items are perceived by tracking users' actions. A user can search and browse an item page, bookmark an item, put an item to the shopping cart and purchase an item. Different action indicates different preference to the item. For example, if a user u purchased item i and bookmarked item j, we can assume that user u prefers item i to item j. Mapping the user actions into a numerical scale is not natural and trivial. So it is hard to accurately compute the user-user similarity in e-commerce systems where users' feedbacks are non-numerical scores.

In this paper, we argue that using ranking-based neighbors to extract features for user-item pairs is more suitable than using rating-based neighbors, because we are facing and completing a ranking task. Choosing neighbors through a ranking-based user-user similarity metric is natural and much closer to the essence of the ranking. Neighbors with the similar preference over items with the target user will give more accurate ranking information on items than rating-based neighbors. Moreover, We can also easily use ranking-based user-user similarity metric to find neighbors for a target user even that the users' feedbacks are no-numerical scores. Our contributions in the paper can be summarized as:

1. We use ranking-based neighbors to extract features for user-item pairs and apply learning to rank algorithm LambdaRank for making recommendations. Based on the characteristic of the features we used, we significantly improve top-K recommendation performance.

2. The feature extraction method we used can be applied to recommendation systems where users' feedbacks are non-numerical scores.

The paper is organized as follows: In Section 2 we discuss notations and related works. The algorithm framework is presented in Section 3. Our experimental steps and experimental results are discussed in Section 4. In Section 5, we summarize our work and discuss the future work.

2   Notations and Related Work

In this section, we briefly discuss the notations used in the paper and some related topics.

2.1   Notations

In recommendation system, we are given a set of m users $U = \{u_1, u_2, ..., u_m\}$ and a set of n items $I = \{i_1, i_2, ..., i_n\}$. We also use u, v to represent any user and use i, j to represent any item for convenience. The users' ratings on the items are represented by an $m \times n$ matrix R where entry $r_{u,i}$ represents user u's true rating on item i. $\hat{r}_{u,i}$ represents user u's prediction rating on item i. R_u represents all the ratings of user u.

2.2   Collaborative Filtering

There are two kinds of collaborative filtering algorithms: user-based and item-based [5]. Given a user-item pair (u, i), user-based neighborhood model estimates the rating on item i based on the ratings by a set of neighbors who have the similar rating pattern with u. Possible choices for user-user similarity $s_{u,v}$ include the Vector Space Similarity (VSS), Adjusted Cosine Similarity (ACS) and Pearson Correlation Coefficient (PCC). For example, the VSS similarity represents a user as a vector in a high dimensional vector space based on her/his ratings. The cosine of the angle between two users' ratings is used to measure their similarity:

$$s_{u,v} = \frac{\sum\limits_{i \in R_u \cap R_v} r_{ui} r_{vi}}{\sqrt{\sum\limits_{i \in R_u \cap R_v} r_{ui}^2} \sqrt{\sum\limits_{i \in R_u \cap R_v} r_{vi}^2}} \tag{1}$$

Since different users and items have different bias, PCC and ACC are proposed to relieve the bias problem. An alternative form of the neighborhood-based approach is the item-based model. Item-item similarity can also be computed like user-user similarity. Given a user-item pair (u, i), item-based neighborhood model estimates the rating on item i based on the k most similarity items that have been rated by u. PCC, ACC and VSS are rating-based similarity metric. The similarity is computed purely based on rating vector, and it ignores the explicit preference over items. In [6], the similarity between users is computed by

Kendall Rank Coefficient [7] and a ranking neighborhood-based approach is used to make recommendations. The major problem with the neighborhood-based approaches is that the rating matrix R is highly sparse, making it very difficult to find similar neighbors reliably. Matrix Factorization approach is better than neighborhood-based approaches at minimizing rating prediction error. Matrix Factorization approximates the observed rating matrix R as the inner product of two low rank matrices. The prediction rating on item i by user u is equal to the inner product of the corresponding user and item features. Classical matrix factorization approach is more about rating prediction, but in many recommendation systems only the top-K items are shown to user and ranking becomes more important than rating prediction.

2.3   Collaborative Ranking

We will introduce ranking algorithms in recommendation system in this part. In [8], by considering the missing entries as zeros, the authors perform conventional SVD on sparse matrix R and make top-K recommendation based on the value of the test items in the reconstruct matrix. Overfitting on training dataset is avoided by only keeping high singular values. The algorithm shows good performance (precision/recall) than SVD++ [1], which is a famous model based on matrix factorization.

Some collaborative ranking algorithms are proposed based on LTR and matrix factorization. Ordrec [9] is an ordinal rank algorithm based on ordinal regression and SVD++. It models user's rating via SVD++ and aims at minimizing an ordinal regression loss. It brings the advantage to estimate the confidence level in each individual prediction and can also handle user's non-numerical feedbacks. In [10], the authors present a generic optimization criterion (Bpropt) for personalized ranking that is the maximum posterior estimator derived from a Bayesian analysis of the ranking problem. Bpropt optimizes the measure of the Area Under the ROC Curve (AUC) based on matrix factorization and adaptive KNN. It uses stochastic gradient descent with bootstrap sampling to update parameters and converges very fast. PMF-co [4] uses pairwise learning to rank algorithm to solve recommendation problem, both user-item features and the weights of the ranking function are optimized during learning. ListRankMF [11] aims at minimizing the cross entropy between the predict item permutation probability and true item permutation probability. In [6], the authors propose a probabilistic latent preference analysis model for ranking prediction by directly modeling user preferences by a mixture distribution based on Bradley-Terry model. Some collaborative ranking work tries to directly optimize evaluation measures. TFMAP [12] uses tensor factorization to model implicit feedback data with contextual information, and directly maximizes mean average precision under a given context. In CLiMF [13] , the model parameters are learned by directly maximizing the Mean Reciprocal Rank and in CofiRank [14], the authors fit a maximum margin matrix factorization model to minimize the upper bound of NDCG. In [3], the authors extract features for a given (u, i) pair from user u's k-nearest neighbors who have rated item i. They use rating-based user-user similarity metric (e.g.

Vector Space Similarity) to find neighbors for a target user. Their work has some drawbacks. First, they can not well define user-user similarity when user's feedbacks are not numerical scores. Second, they choose neighbors by rating-based user-user similarity metric, which is not corresponding to the ultimate goal of ranking. Our work is different from their work. We choose neighbors with similar preference pattern rather than rating pattern. The problem with rating oriented approaches is that the focus has been placed on approximating the ratings rather than the rankings, but ranking is a more important goal for modern recommender systems.

3   Ranking-Based Collaborative Ranking Framework

In this section, we will discuss our ranking-based collaborative ranking framework. Since users and items are not easy to be represented in term of explicit features, we should design new method to extract efficient feature for ranking. We extract features from neighbors of a target user to capture the local information from the neighborhood.

3.1   Neighborhood Models

In some movie recommendation systems, users rate movies with non-numerical values, e.g., $A, B, ..., F$. There is no direct resemblance this kind of feedback to numerical or binary values. Furthermore, we argue that even when the feedback is related to absolute numbers, taking the scores as numerical may not reflect the users' intentions well. Different users tend to have different internal scales. For example, taking star ratings as numeric will put the same distance between 3 stars and both 4 stars and 2 stars. However, one user can take 3 stars as similar to 4 stars, while another user strongly relates 3 stars to low quality, being similar to 1-2 stars. Computing user-user similarity is difficult in this scene. Moreover, the rating-oriented neighborhood models focus on rating prediction rather than ranking. Since we should complete a collaborative ranking task, we prefer a ranking-based user-user similarity metric to choose neighbors for users. In the ranking-oriented approach, two users with similar preferences over the items are more close. The preference is reflected by their ranking of the items. We modify the Kendall Tau Correlation Coefficient [6,7] to qualify the user-user similarity. We define the user-user similarity as:

$$s_{u,v} = 1 - \frac{N_{uv}}{M_{uv}} \tag{2}$$

where

$$M_{uv} = \frac{|R_u \cap R_v| \cdot (|R_u \cap R_v| - 1)}{2}$$

$$N_{uv} = \sum_{i,j \in I_u \cap I_v} I^-(r_{u,i} - r_{u,j}, r_{v,i} - r_{v,j})$$

$$I^-(x, y) = \begin{cases} 1 \text{ , if } x \cdot y \leq 0 \wedge (x \neq 0 \vee y \neq 0) \\ 0 \text{ , otherwise} \end{cases}$$

The similarity is defined as the ratio between the number of concordant pairs $M_{uv} - N_{uv}$ and the number of all possible pairs M_{uv}. It is negatively correlated with the number of dis-concordant pairs N_{uv} and its value is in the range of $[0, 1]$. We can compute the user-user similarity even that the users' feedbacks are non-numerical scores from the definition. From the ranking-based user-user similarity definition, we can see that if two users assign on four items with ratings $\{1, 2, 3, 4\}$ and $\{4, 3, 2, 1\}$ respectively, their similarity is equal to 0, so they have totally different ranking-based preference. While the user-user cosine similarity between the two users is 0.77, they will be treated as very similar users. We can see that ranking-based user-user similarity is very different from rating-based user-user similarity. Choose neighbors from ranking-based neighborhood is more close to the essence of ranking. Given a target user u, we should carefully choose neighbors for her/him, and those users who have high rating-based similarity and low ranking-based similarity with the target user should not be put into u's neighborhood. The ranking-based neighbors can give more significant and accurate ranking information about items and contribute more information for feature extraction.

3.2   User-Item Feature Extraction

In this section, we use the ranking-based neighbors to extract the features for user-item pairs. A feature vector $\gamma(u, i)$ for a user-item pair (u, i) is extracted based on the K nearest ranking-based neighbors $K(u, i) = \{u_k\}_{k=1}^{K}$ of user u who have rated i. The features are descriptive statistics information of the ranking of item i. Every neighbor u_k of user u has her/his opinion on the ranking of item i. For every item j ($j \neq i$) in u_k's profiles, u_k has her/his preference over item i and item j. We aggregate K neighbors' opinions and extract statistics information from their opinions as features. We summarize the preference for i into three $K \times 1$ matrices $\mathbf{W}_{ui}$, $\mathbf{T}_{ui}$ and $\mathbf{L}_{ui}$:

$$\mathbf{W}_{ui}(k) = \frac{1}{|Ru_k| - 1} \sum_{i' \in Ru_k \backslash i} I[r_{u_k, i} > r_{u_k, i'}]$$

$$\mathbf{L}_{ui}(k) = \frac{1}{|Ru_k| - 1} \sum_{i' \in Ru_k \backslash i} I[r_{u_k, i} < r_{u_k, i'}]$$

$$\mathbf{T}_{ui}(k) = \frac{1}{|Ru_k| - 1} \sum_{i' \in Ru_k \backslash i} I[r_{u_k, i} = r_{u_k, i'}]$$

The three matrices describe the relative preference for i across all the neighbors of u. Then we summarize all matrices with a set of simple descriptive statistics.

$$\gamma(\mathbf{W}_{ui}) = [\mu(\mathbf{W}_{ui}), \sigma(\mathbf{W}_{ui}), \max(\mathbf{W}_{ui}), \min(\mathbf{W}_{ui}), \frac{1}{k} \sum_k I[\mathbf{W}_{ui}(k) \neq 0]$$

$$\gamma(\mathrm{L}_{ui}) = [\mu(\mathrm{L}_{ui}), \sigma(\mathrm{L}_{ui}), \max(\mathrm{L}_{ui}), \min(\mathrm{L}_{ui}), \frac{1}{k}\sum_k I[\mathrm{L}_{ui}(k) \neq 0]$$

$$\gamma(\mathrm{T}_{ui}) = [\mu(\mathrm{T}_{ui}), \sigma(\mathrm{T}_{ui}), \max(\mathrm{T}_{ui}), \min(\mathrm{T}_{ui}), \frac{1}{k}\sum_k I[\mathrm{T}_{ui}(k) \neq 0]$$

where μ is mean function, σ is the deviation function, max and min are the maximize and minimize function. The last term counts the number of neighbors who express any preference towards i. We convert neighbors' ratings on i to preference descriptive features in this way. After concatenating the three statistics features we have the last feature representation for (u, i):

$$\gamma(u, i) = [\gamma(\mathrm{W}_{ui}), \gamma(\mathrm{L}_{ui}), \gamma(\mathrm{T}_{ui})] \tag{3}$$

The features are descriptive statistics information for the ranking of item i. Since these feature are statistics of neighbors' opinions on the ranking of items and our experiment show that they are very robust. It is obvious that ranking-based neighbors will give more significant and accurate ranking information than rating-based neighbors in this feature extraction setting. After we have extracted features for every user-item pairs, we can use any learning to rank algorithm to learn a ranking model and make recommendation. We will introduce the procedure in the next part.

3.3   Learn and Prediction

We use a simple linear function as our prediction function. The output of the prediction function indicates the relative position of items. The prediction function is:

$$r(u, i) = \boldsymbol{w} \cdot \gamma(u, i) + w_0 + I[K(u, i) = \emptyset]b_0 \tag{4}$$

where $\boldsymbol{w}$ and w_0 is the ranking model parameters to be learned (w_0 is the offset term). $I[K(u, i) = \emptyset] = 1$ if there are no neighbors who rate item i (corresponding to $\gamma(u, i) = 0$). The bias term b_0 provides a base score for i if i does not have enough ratings in the training dataset and we can't find neighbors who have rated item i. We can apply any LTR algorithm to learn the ranking model, and we use LambdaRank in our experiment because we evaluated our rank quality by NDCG. We omit the details of LambdaRank due to the lack of space and refer the reader to [15] for details description. Given a user u, we predict every item i in user u's test dataset by:

$$\hat{r}(u, i) = \boldsymbol{w}^* \cdot \gamma(u, i) + w_0^* + I[K(u, i) = \emptyset]b_0^* \tag{5}$$

where $\boldsymbol{w}^*$ and w_0^* is the learned ranking model. Only ratings in the training dataset are used to select the neighbors, and make predictions for the items in the validating and testing dataset. After all unrated items' prediction ratings are computed, we can order them based on the prediction rating value and make top-K recommendation. We can see that the rating is not on a scale of 1 to 5 anymore and it represents the relative position of the items. Moreover, based on the ranking-based neighbor's statistical feature and the personalized ranking function we used, we can see that if $\hat{r}_{u,i} > \hat{r}_{u,j}$ and $\hat{r}_{u,j} > \hat{r}_{u,k}$, we will have $\hat{r}_{u,i} > \hat{r}_{u,k}$. The prediction function keeps the symmetric attribute on items, so we don't need use a greedy algorithm to infer the final ranking list like [6].

4   Experiments

4.1   Evaluation Measure

Since recommendation systems often show several items to the user, a good ranking mechanism should put relevant impressions in the very top of a ranking list extremely well. So we choose Normalized Discount Cumulative Gain (NDCG) [16] metric in information retrieval as the evaluation measure, because it is very sensitive to the ratings of the highest ranked items, making it highly desirable for measuring ranking quality in recommendation system. A ranking of the test items can be represented as a permutation $\pi : \{1, 2, ..., N\} \rightarrow \{1, 2, ..., N\}$ where $\pi(i) = p$ denotes that the ranking of the item i is p and $i = \pi^{-1}(p)$ is the item index ranked at p-th position. Items in test dataset are sorted based on their prediction ratings. The NDCG value for a given user u is computed by:

$$\text{NDCG}(u, \pi)@\text{K} = \frac{1}{\text{MDCG}(u)} \sum_{p=1}^{K} \frac{2^{r_{u,\pi^{-1}(p)}} - 1}{\log(p+1)} \tag{6}$$

We assume that π^* is a permutation over the test items that are sorted on the true ratings and MDCG@K is DCG@K value corresponding to permutation π^*. We can see that a perfect rank's NDCG value is equal to one. We set k to be 10 in all our experiments like [11,14,4][1]. We report the macro-average NDCG@K across all users in our experiment results.

4.2   Datasets

We use two well known movie ratings datasets in our experiment: Movielens100K and Movielens1M [2]. Most of the state-of-the-art ranking-based collaborative ranking algorithms also use them. The Movielens100K dataset consists of about 100K ratings made by 943 users on 1682 movies. The Movielens1M dataset consists of about 1M ratings made by 6040 users on 3706 movies. Since users in the two datasets rate at least 20 rating, the cold start users won't appear. The ratings for Movielens100K and Movielens1M are on a scale from 1 to 5. Table 1 gives statistics information associated with the two datasets.

We divide the whole dataset into training dataset and Testing dataset. We randomly sample a fixed number N of ratings for every user for training test on the rest data. We sample 10 ratings from training dataset for validating and

---

[1] NDCG values at other cutoffs give us different numerical values but qualitative almost the same results.

[2] http://www.grouplens.org/node/73

Table 1. Datasets Statistics

Dataset	Users	Items	Train 20	Train 30	Train 40	Train 50
Movielens100K	943	1682	744/1682	645/1681	568/1681	497/1677
Movielens1M	6040	3706	5289/3701	4737/3694	4297/3689	3938/3677

after we found the best hyper parameters, we put validating dataset into training dataset and retrain the model, then we report the rank performance on testing dataset. In our experiment, we set N as 20, 30, 40 and 50, so users with less than 30, 40, 50 and 60 ratings are removed to ensure that we can compute NDCG@10 on the testing dataset. The number of users and items after filtering for different N are reported in Table 1.

4.3 Parameters and Baselines

We train our algorithm (RankWLT) using user-wise stochastic gradient descent. We update parameter with learning rate in $\{1, 0.1, 0.01, 0.001\}$. In our experiment, the test performance is the best when learning rate is 0.01. Our neighborhood size is chosen in range $[10, 500]$. We filter neighbors who have very low similarity with the target user with threshold $t = 0.05$, so these users won't appear in the target user's neighborhood. When the neighborhood size becomes too large, no performance gain are observed.

We compare our algorithm with Probability Matrix Factorization (PMF) and PureSVD [8]. Since TOPK-ListMLE is state-of-the-art learning to rank algorithm, so we also extend TOPK-ListMLE to recommendation systems based on the framework proposed by [11]. For PureSVD algorithm, the dimension of the singular value matrix Σ is 10. The user and item latent dimension size D in TOPK-ListMLE and PMF are chosen from $\{5, 10, 15, 20, 50\}$ by cross validation and the regularization term for latent factor are chosen from $\{1, 0.1, 0.01, 0.001\}$. We report the best rank performance of TOPK-ListMLE and the rank performance of PMF aiming at minimizing RMSE. We also report the best rank performance for PMF aiming at maximizing NDCG@10 (We call it RankPMF model). The user latent factor and item latent factor in TOPK-ListMLE, PMF and RankPMF are initialized as a random positive value in range $[0, 0.1]$ (we find that initializing the latent factors with positive values converges faster than initializing them with random values).

4.4 Experiment Results

The experiment results are listed in Table 2 and Table 3.

We have some intuitions from the experiment results:

1. NDCG@10 increases when the sample number N increases in most of the models, because the training dataset size increases and we have more information to learn users' preference. For example, NDCG@10 of PMF increases from 0.681 to 0.708 on Movielens100K.

Table 2. NDCG@10 on Movielens100K

Algorithm	Train 20	Train 30	Train 40	Train 50
PMF	0.681	0.691	0.696	0.708
RankPMF	0.706	**0.705**	0.707	0.712
PureSVD	0.646	0.649	0.669	0.662
TOPK-ListMLE	**0.710**	0.704	0.711	0.709
RankWLT	0.685	0.701	**0.716**	**0.720**

Table 3. NDCG@10 on Movielens1M

Algorithm	Train 20	Train 30	Train 40	Train 50
PMF	0.729	0.733	0.731	0.737
RankPMF	0.747	0.744	0.745	0.748
PureSVD	0.699	0.704	0.708	0.710
TOPK-ListMLE	0.740	0.745	0.745	0.746
RankWLT	**0.749**	**0.754**	**0.753**	**0.756**

2. Surprisingly, PureSVD performs non-competitively compared to other methods on NDCG@10. This is very different from the conclusion in [8]. We think that this is mainly due to the nature of their evaluation metric (precision/recall). They evaluate precision/recall performance on the test dataset which is constructed by high ratings (above 5 point) and a large number of items sampled at random. We prefer our evaluation using NDCG which avoids random sampling and has very little bias.

3. Since improvements in RMSE (Root Mean Square Error) often don't translate into accuracy improvements, model (PMF) which purses of minimizing prediction error behaves not well in completing ranking task. It is clear that PMF behaves worse than other ranking models (except PureSVD) and RankPMF behaves better than PMF.

4. The results demonstrate the robust of the statistical features. Our algorithm generalizes well when both few and many ratings are available.

5. We have better NDCG@10 than other approaches. We have many informative ranking-based neighbors, and these neighbors have the same preference with the target user, so they contain accuracy preference information.

5   Conclusion and Future Work

In this paper, we propose a new collaborative ranking approach. We use the statistical information of the rankings of item from user's nearest neighbors as the feature representation for user-item pairs. We use Kendall Rank Coefficient as similar metric to choose neighbors and experimentally demonstrate the effectiveness of our method. Interestingly, we can address scenarios where assigning numerical scores to different types of users' feedbacks would not be easy.

There are some future work based on this work. Since classical collaborative filtering algorithm ignores the social connections between users, but in real-world user often asks her/his friends for recommendations and user's taste is influenced by friends' taste on the social network. Incorporating social effects in our collaborative ranking framework is our next work. Different users in recommendation systems may vary largely. For example, the influence of neighbors on different users are very different. Considering the large difference between users, it is not the best choice to use a single ranking function to deal with all users. So we also plan to investigate how to design personalized collaborative ranking algorithm in the future.

Acknowledgement. We thank the anonymous reviewers for helpful comments. This work is supported by the program of the National Natural Science Foundation of China (NSFC) under grant 60973003.

References

1. Koren, Y.: Factor in the neighbors: Scalable and accurate collaborative filtering. ACM Trans. Knowl. Discov. Data 4(1), 1:1–1:24 (2010)
2. Liu, T.Y.: Learning to rank for information retrieval, Berlin, German, vol. 3(3), pp. 225–331. Springer (2011)
3. Volkovs, M.N., Zemel, R.S.: Collaborative ranking with 17 parameters. In: Proceedings of the Twenty-sixth Annual Conference on Neural Information Processing Systems, NIPS 2012, Lake Tahoe, Nevada, United States, December 3-6, MIT Press (2013)
4. Balakrishnan, S., Chopra, S.: Collaborative ranking. In: Proceedings of the Fifth ACM International Conference on Web Search and Data Mining, WSDM 2012, pp. 143–152. ACM, New York (2012)
5. Deshpande, M., Karypis, G.: Item-based top-n recommendation algorithms. ACM Trans. Inf. Syst. 22(1), 143–177 (2004)
6. Liu, N.N., Yang, Q.: Eigenrank: a ranking-oriented approach to collaborative filtering. In: Proceedings of the 31st Annual International ACM SIGIR Conference on Research and Development in Information Retrieval, SIGIR 2008, pp. 83–90. ACM, New York (2008)
7. Marden, J.I.: Analyzing and modeling rank data. Chapman & Hall Press, New York (1995)
8. Cremonesi, P., Koren, Y., Turrin, R.: Performance of recommender algorithms on top-n recommendation tasks. In: Proceedings of the Fourth ACM Conference on Recommender Her Systems, RecSys 2010, pp. 39–46. ACM, New York (2010)
9. Koren, Y., Sill, J.: Ordrec: an ordinal model for predicting personalized item rating distributions. In: Proceedings of the Fifth ACM Conference on Recommender Systems, RecSys 2011, pp. 117–124. ACM, New York (2011)
10. Rendle, S., Freudenthaler, C., Gantner, Z., Schmidt-Thieme, L.: Bpr: Bayesian personalized ranking from implicit feedback. In: Proceedings of the Twenty-Fifth Conference on Uncertainty in Artificial Intelligence, UAI 2009, Arlington, Virginia, United States, pp. 452–461. AUAI Press (2009)
11. Shi, Y., Larson, M., Hanjalic, A.: List-wise learning to rank with matrix factorization for collaborative filtering. In: Proceedings of the Fourth ACM Conference on Recommender Systems, RecSys 2010, pp. 269–272. ACM, New York (2010)

12. Shi, Y., Karatzoglou, A., Baltrunas, L., Larson, M., Hanjalic, A., Oliver, N.: Tfmap: optimizing map for top-n context-aware recommendation. In: Proceedings of the 35th International ACM SIGIR Conference on Research and Development in Information Retrieval, SIGIR 2012, pp. 155–164. ACM, New York (2012)
13. Shi, Y., Karatzoglou, A., Baltrunas, L., Larson, M., Oliver, N., Hanjalic, A.: Climf: learning to maximize reciprocal rank with collaborative less-is-more filtering. In: Proceedings of the Sixth ACM Conference on Recommender Systems, RecSys 2012, pp. 139–146. ACM, New York (2012)
14. Weimer, M., Karatzoglou, A., Bruch, M.: Maximum margin matrix factorization for code recommendation. In: Proceedings of the Third ACM Conference on Recommender Systems, RecSys 2009, pp. 309–312. ACM, New York (2009)
15. Burges, C.J.C., Ragno, R., Le, Q.V.: Learning to rank with nonsmooth cost functions. In: Proceedings of the Twentieth Annual Conference on Neural Information Processing Systems, NIPS 2006, Vancouver, British Columbia, Canada, December 4-7, pp. 193–200. MIT Press (2007)
16. Järvelin, K., Kekäläinen, J.: Cumulated gain-based evaluation of ir techniques. ACM Trans. Inf. Syst. 20(4), 422–446 (2002)

Probabilistic Top-k Dominating Query over Sliding Windows

Xing Feng[1], Xiang Zhao[2], Yunjun Gao[1], and Ying Zhang[2]

[1] Zhejiang University, China
{3090103362,gaoyj}@zju.edu.cn
[2] The University of New South Wales, Australia
{xzhao,yingz}@cse.unsw.edu.au

Abstract. Probabilistic queries on uncertain data have been intensively investigated lately. Top-k dominating query is important in many applications, e.g., decision making, as it offers choices which are better than the most of others. In this paper, we study the problem of probabilistic top-k dominating query over sliding windows. An efficient algorithm is developed to compute the exact solution. Extensive experiments are conducted to demonstrate the efficiency and effectiveness of the algorithm.

1 Introduction

Spatial database has drawn attention in recent decades. Beckmann *et al.* [1] design the R-tree as an index for efficient computation. Roussopoulos *et al.* [8] study nearest neighbor queries, and propose three effective pruning rules to speed up the computation. They also the extend nearest neighbor queries to k nearest neighbor queries, which return top-k preferred objects. Börzsönyi *et al.* [4] is among the first to study the skyline operator, and propose an SQL syntax for the skyline query.

Top-k dominating query is first studied by Yiu *et al.* [10], which retrieves the top-k objects that are better than the largest number of objects in a dataset. This is quite different from the skyline query in [4] that retrieves objects which are not worse than other objects.

Uncertain data analysis is of importance in many emerging applications, such as sensor network, trend prediction and moving objects management. There has been a lot of works focusing on uncertain data management, [5,11] to cite a few.

In this work, we study probabilistic top-k dominating query over sliding windows. We employ the data model with sliding windows as used in [9,13]. In sliding windows, data is treated as a stream, and only recent n objects are considered.

There are some closely related work, seen in [7,12]. However, they study objects with multiple instances. Besides, their query is to get top-k dominating objects from the total data set. While our paper studies objects from append-only data stream. And our main concern is to maintain top-k dominating objects from recent N objects where N is the window size.

Probabilistic top-k dominating query is desirable in various real-life applications. Table 1 evidences a scenario of house rental, where answering such type of query can be

Y. Ishikawa et al. (Eds.): APWeb 2013, LNCS 7808, pp. 782–793, 2013.
© Springer-Verlag Berlin Heidelberg 2013

beneficial. Lessees are interested in knowing the top-1 house from all recent 4 advertisements. Each advertisement is associated with its house ID, post time, price, distances to supermarket and trustability. Trustability is derived from Lessee's feedback on the lessors product quality; this "trustability" value can also be regarded as the probability that the house is the same as what it claims.

We assume lower price and closer to supermarket are preferred and lessees do not care about other attributes. We also assume that lessees want a house that are better than the most of others. A top-k dominating ($k = 1$) query would retrieve $\{H1\}$ as the result for the first 4 objects $\{H1, H2, H3, H4\}$ because $H1$ is better than $H2, H3, H4$. $\{H4\}$ is the result for the next sliding window $\{H2, H3, H4, H5\}$. However, we may also notice that $H4$ is of low trustability. It is more reasonable to take the trustability into consideration. We model such on-line selection problem as probabilistic top-k dominating query against sliding windows by regarding on-line advertisements as an append-only data stream. This query will be formally defined in Section 2. Hence, the probabilistic top-k dominating query over sliding windows provides a more reasonable solution under the given scenario.

Table 1. Example of house rental

House ID	Post time	Rent per week	Distance to supermarket	Trustability
$H1$	100 days ago	$167	100m	0.80
$H2$	97 days ago	$205	340m	0.90
$H3$	63 days ago	$206	820m	0.85
$H4$	55 days ago	$185	110m	0.50
$H5$	50 days ago	$230	700m	1.00
$H6$	43 days ago	$200	760m	0.90

In summary, we make the following contributions in this paper:

- We identify the problem of probabilistic top-k dominating query over sliding windows. To the best of our knowledge, there is no prior work regarding this.
- We develop novel algorithms to continuously compute the top-k dominating objects, and analyze the complexity.
- We conduct extensive experiments to demonstrate the correctness and efficiency of our algorithms.

Organization.   The rest of the paper is organized as follows. In Section 2, we present background information and formally define the problem. Sections 3 and 4 present the techniques for processing top-k dominating query over sliding windows. Experimental results are provided and analyzed in Section 5. Section 6 concludes the paper.

2   Preliminary

This section introduces the background information, and defines the problem.

2.1 Background

When comparing two objects, if one object p is not worse than another q in all aspects and p is better than q in at least one aspect, we say p *dominates* q. Formally, we have Definition 1.

Definition 1 (Dominate). *Consider two distinct d-dimensional objects p and q, and $p[i]$ denotes the i-th value of p. p dominates q (denoted by $p \prec q$) iff $p[i] \leq q[i]$ holds for $1 \leq i \leq d$, and there exists j, such that $p[j] < q[j]$.*

Based on Definition 1, [10] develops a score function to count the number of objects dominated by an object as

$$\mu(p) = |\{p' \in \mathcal{D} | p \prec p'\}|. \tag{1}$$

Consequently, top-k dominating query in a data set is to retrieve the top-k objects with maximal $\mu(p)$ in the data set.

Given a sequence data stream DS of uncertain data objects, a possible world W is a subsequence of DS; i.e., each object α from DS can be either in W or not. For example, given a data stream with 3 objects, $\{a, b, c\}$, there are totally 8 possible worlds — $\{\emptyset, \{a\}, \{b\}, \{c\}, \{a, b\}, \{a, c\}, \{b, c\}, \{a, b, c\}\}$. The probability of W to appear is the product of probabilities of all objects which appear in W and 1 − the probabilities of all objects which do not appear in W; i.e., $P(W) = \prod_{\alpha \in W} P(\alpha) \times \prod_{\alpha \notin W}(1 - P(\alpha))$.

We compute the μ function in Equation (1) for every object in the certain possible world W. For a top-k dominating query over sliding window, probability of each qualified possible world W is accumulated in order to get the overall probability.

2.2 Problem Definition

In many applications, a data stream is append-only; that is, there is no deletion of data element involved. In this paper, we study the top-k dominating problem by employing the append-only data stream model. Given an uncertain data stream DS consisting of n uncertain objects, sliding window size W, probability threshold q, we use l to measure the dominating ability of each object in a sliding window. That is, an object dominates at least l objects in possible worlds, of which the total probability is over q in a sliding window of size W. Thus, top-k dominating over sliding windows maintains the top-k objects which are ranked according to l over sliding windows with regards to threshold q.

Example 1. Consider the running example in Table 1, where we need the top-1 house over a sliding window of size $N = 4$, and probability threshold $q = 0.6$. We have in total 16 possible worlds in each window. When a window contains $\{H3, H4, H5, H6\}$, it is a little different. There are 8 possible worlds rather 16 possible worlds, since $P(H5) = 1$. First, we compute the dominating number of each object in every possible world (as shown in Table 2). When computing l value, we sum up the probability of each possible world according to the number of objects it dominated in descending order, till the probability exceeds the threshold q. Thus, $H6$ is $76.5\%(38.25\% + 38.25\%)$, and $H4$ gets $50\%(38.25\% + 6.75\% + 4.25\% + 0.75\%)$. $H3$ and $H5$ are both 0, since they do not dominate others. Therefore, we have $H3.l = 0$, $H4.l = 0$, $H5.l = 0$, and $H6.l = 1$; and hence, $H6$ is the top-1 object in the probabilistic top-1 dominating query with probability threshold $= 0.6$ over the sliding window $\{H3, H4, H5, H6\}$.

Table 2. Dominating number in possible world

Probability	Possible world	$H3$	$H4$	$H5$	$H6$
0.75%	$H5$	0	0	0	0
4.25%	$H3, H5$	0	0	0	0
0.75%	$H4, H5$	0	1	0	0
6.75%	$H5, H6$	0	0	0	0
4.25%	$H3, H4, H5$	0	2	0	0
38.25%	$H3, H5, H6$	0	0	0	1
6.75%	$H4, H5, H6$	0	2	0	0
38.25%	$H3, H4, H5, H6$	0	3	0	1

2.3 Dominance Relationships

We implement our techniques based on the R-tree. Thus, we define the following relationships between a pair of entries E and E' in the R-tree. $E.min$ denotes the lower-
left corner of the minimum bounding box (MBB) of the objects contained by E, and
$E.max$ denotes the upper-right corner of MBB. Note if an entry contains only one
object α, $E.min = E.max = \alpha$.

An entry E is said to *fully dominate* another E', iff $E.max \prec E'.min$ or $E.max =
E'.min$, denoted as $E \prec E'$. An entry E *partially dominates* E', iff $E.max \not\prec
E'.min, E.min \prec E'.max$, denoted as $E \prec_{partial} E'$. An entry E does not domi-
nate E', iff $E.min \not\prec E'.max$, denoted as $E \not\prec E'$. As depicted in Figure 1, $E2$ fully
dominates $E3$, E partially dominates $E2$, and $E1$ does not dominate E.

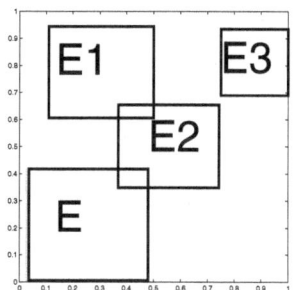

Fig. 1. Dominance relationship

3   Dominating Score Calculation

As aforementioned, we may calculate the dominating score by the finding maximal l
whose accumulating possibility of all the qualified possible worlds is over q. However,
this can be rather time-consuming due to the exponential number of possible worlds;
i.e., there are 2^N possible worlds, where N is the number of all objects.

We observe that the complexity can be reduced by viewing the problem from another perspective. Instead of enumerating possible worlds, we retrieve the largest l in the condition, where at least l objects exist of M objects, where M is the number of all dominated objects.

Specifically, let $m_{i,j}$ denote the probability that exactly i objects out of j objects exist, and $P(i)$ denote the probability that the i-th object exists. We have the following relations.

$$m_{i,j} = \begin{cases} m_{0,j-1} \times (1 - (P(j))) & i = 0 \\ m_{i-1,j-1} \times P(i) & i = j \\ m_{i-1,j-1} \times P(j) + m_{i,j-1} \times (1 - P(j)) & otherwise \end{cases} \qquad (2)$$

Equation (2), known as Poisson-Binomial Recurrence [6], is widely used in uncertain database analysis [2,3]. Applying Equation (2), we propose a dynamic programming algorithm that efficiently computes l value of a certain object, encapsulated in Algorithm 1.

Algorithm 1 takes Ω and α as inputs, where α is the object of which we need to compute l value, and Ω contains all the objects which are dominated by α. A matrix is used to store the intermediate result, initialized in Line 2. Then, we use Equation (2) to fill in the cells of the matrix (Lines $4 - 13$). Finally, we remove $m_{i,size}$ from f at a time till f is less then q (Lines 15 - 17), and $i - 1$ is returned as the result.

Algorithm 1. calDS(Ω, α)

1   **if** $P(\alpha) < q$ **then return 0;**
2   $m_{0,0} = 1; i = 1;$
3   $size = |\Omega|; f = P(\alpha);$
4   **while** $j \leq size$ **do**
5      $m_{0,j} = m_{0,j-1} \times (1 - P(j));$
6      $j = j + 1;$
7   **while** $i \leq size$ **do**
8      $m_{i,i} = m_{i-1,i-1} \times P(i);$
9      $j = 1;$
10   **while** $j < i$ **do**
11        $m_{i,j} = m_{i-1,j-1} \times P(j) + m_{i,j-1} \times (1 - P(j));$
12        $j = j + 1;$
13     $i = i + 1;$
14   $i = 0;$
15   **while** $f > q$ **do**
16      $f = f - m_{i,size} \times P(\alpha);$
17      $i = i + 1;$
18   **return** $i - 1;$

Correctness and Complexity Analysis. It is immediate that Algorithm 1 correctly computes the l value. The most time-consuming part is the loop in Lines $7 - 13$, and thus, the complexity of algorithm is $O(|\Omega|^2)$.

Example 2. Consider the running example in 2.2. We only compute $H4$ and $H6$, since Ω of $H3$ or $H5$ is empty. For $H4$, $P(H4) = 0.5 < 0.6$, and hence , $H4.l = 0$. For $H6$, $P(H6) = 0.9 > 0.6$, $m_{0,1} = 0.15$, $m_{1,1} = 0.85$. As $0.9 - 0.15 \times P(H6) = 0.765 > 0.6$, $0.765 - 0.85 \times P(H6) = 0 < 0.6$, the $H6.l$ is 1. Therefore, top-1 object among $\{H3, H4, H5, H6\}$ is $H6$.

4   Updating Technique

A naive solution to retrieving top-k dominating objects over sliding windows is to visit each object every time the window slides. When visiting an object, it also gets all the objects it dominates by traversing R-tree and then computes its l value using Algorithm 1. Eventually, it chooses k objects with maximal l values. We observe that this tedious process can be accelerated. Given a probability threshold q and a sliding window with size N, Algorithm 2 shows how we process every arriving object.

Algorithm 2. Probabilistic top-k dominating over Sliding Windows

---

1 **while** a new object α_{new} arrives **do**
2   **if** $|\mathfrak{S}| \leq N$ **then**
3          $insert(\alpha_{new})$;
4   **else**
5          $remove(\alpha_{old})$;
6          $insert(\alpha_{new})$;
7   collect top-k dominating objects;

When a new object α_{new} comes, if there are exactly N objects in the window, we remove the oldest object α_{old}. Then, we use function $insert(\alpha_{new})$ to update the l value of objects. Finally we collect the top-k dominating objects using min-heap.

In the following subsections, we present novel algorithms to efficiently execute Algorithm 2. We first introduce the data structure, then provide our algorithm to handle the situation when a new object is inserted ($insert(\alpha_{new})$), followed by an algorithm to deal with a removed old object ($remove(\alpha_{old})$).

4.1   Aggregate R-Trees and Heap

We store all the N objects in an aggregate R-tree. We store P, l and $\lambda_{\text{top-k dominating}}$ with each object as aggregate information. Specifically, P is the probability of the object, l is the measure of dominating ability regarding probabilistic threshold q. Additionally, $\lambda_{\text{top-k dominating}}$ is the last column of matrices that we use to compute l by Algorithm 1. Note that each column only depends on the previous column. In another word, when a new object α_{new} comes, keeping last column of the matrix is enough to update the l value.

By storing these aggregate information, we do not have to recalculate the l value of all objects when a new object comes. We only need to update $\lambda_{\text{top-k dominating}}$ and l value every time a new object comes.

In addition, we maintain the top-k dominating objects in a min-heap; i.e., we insert the first k objects into a min-heap. For each following object having larger l than the minimum object in the heap, we first dequeue the minimal object, and then insert the new object into the min-heap.

4.2   Insert

When a new object α_{new} comes, we need to consider all the objects dominating it (denote as $S1$) and being dominated by it (denote as $S2$). To this end, we process the following tasks: 1) compute $S1$ and $S2$ by traversing the aggregate R-tree, 2) update l of each object in $S1$; and 3) compute l of the new object.

Algorithm 3 describes the steps we take to handle an insertion. R is the aggregate R-tree we used to store all objects in current sliding window. $S3$ is used to collect all the entries which partially dominate α_{new}, $S4$ is used to collect all the entries which are partially dominated by α_{new}, and $S34$ is used to collect the entries which partially dominate α_{new} and are partially dominated by α_{new}. First, we classify all the entries to different sets (Lines 1 - 6). Then, we refine the result and get all the entries of interest, namely $S1$ and $S2$, using **probe1**, **probe2** and **probe3** (Lines 7 - 9). So far, we have finished task 1). Then, task 2) is done by updating all the entries in $S1$ using **update** (Line 10). Afterwards, we achieve task 3) by computing l of α_{new} (Line 11).

Algorithm 3. insert(α_{new})

---

1 **for each** $E \in R.root$ **do**
2   if$E \prec \alpha_{new}$ **then add** E **to** $S1$;
3   **else if**$\alpha_{new} \prec E$ **then add** E **to** $S2$;
4   **else if**$E \prec_{partial} \alpha_{new} \& \alpha_{new} \nprec E$ **then add** E **to** $S3$;
5   **else if**$E \prec_{partial} \alpha_{new} \& \alpha_{new} \prec_{partial} E$ **then add** E **to** $S34$;
6   **else if**$\alpha_{new} \prec_{partial} E \& E \nprec \alpha_{new}$ **then add** E **to** $S4$;
7 if$S3 \neq \emptyset$ **then probe1**($S3$);
8 if$S4 \neq \emptyset$ **then probe2**($S4$);
9 if$S34 \neq \emptyset$ **then probe3**($S34$);
10 if$S2 \neq \emptyset$ **then update1**($S1$);
11 if$S1 \neq \emptyset$ **then** $\alpha_{new}.l =$ **calDS**($S2$, α_{new});

Algorithm 4 shows how we refine $S3$. Note that all the entries in $S3$ cannot be dominated by α_{new}. We add entries dominating α_{new} to $S1$ (Line 4 in Algorithm 4), and leave entries which partially dominate α_{new} in $S3$ (Line 5 in Algorithm 4).

Algorithm 4. probe1($S3$)

---

1 **while** $S3 \neq \emptyset$ **do**
2      $E =$ **Dequeue**($S3$);
3   **for each** child E' of E **do**
4   **if** $E' \prec \alpha_{new}$ **then add** E' **to** $S1$;
5   **else if** $E' \prec_{partial} \alpha_{new}$ **then add** E' **to** $S3$;

The ideas behind Algorithm 4 and Algorithm 5 are similar. We add entries which are dominated by α_{new} to $S2$ (Line 4 in Algorithm 5), and leave entries which are partially dominated by α_{new} in $S4$ (Line 5 in Algorithm 5).

Algorithm 5. probe2($S4$)

1 **while** $S4 \neq \emptyset$ **do**
2      $E = \mathsf{Dequeue}(S4)$;
3   **for each** child E' of E **do**
4   **if** $\alpha_{new} \prec E'$ **then** add E' to $S2$;
5   **else if** $\alpha_{new} \prec_{partial} E'$ **then** add E' to $S4$;

$S34$ is special in that it may contain entries which dominate α_{new} and entries which are dominated by α_{new}. Thus, we first partition them (Lines 1 - 8), and apply probe1 and probe2 thereafter.

Algorithm 6. probe3($S34$)

1 **while** $S34 \neq \emptyset$ **do**
2      $E = \mathsf{Dequeue}(S34)$;
3   **for each** child E' of E **do**
4   **if** $\alpha_{new} \prec E'$ **then** add E' to $S1$;
5   **else if** $E' \prec \alpha_{new}$ **then** add E' to $S2$;
6   **else if** $\alpha_{new} \prec_{partial} E' \& E' \prec_{partial} \alpha_{new}$ **then** add E' to $S34$;
7   **else if** $E' \prec_{partial} \alpha_{new} \& \alpha_{new} \nprec E'$ **then** add E' to $S3$;
8   **else if** $\alpha_{new} \prec_{partial} E' \& E' \nprec \alpha_{new}$ **then** add E' to $S4$;

9 **if** $S3 \neq \emptyset$ **then** **probe1**($S3$);
10 **if** $S4 \neq \emptyset$ **then** **probe2**($S4$);

Algorithm 7 describes the update of $S1$. For each object at leaf node, we update its l and $\lambda_{\text{top-k dominating}}$ by using α_{new}. As a consequence, the time complexity of the update is $O(l)$.

Algorithm 7. update1($S1$)

1 **while** $S1 \neq \emptyset$ **do**
2      $E = \mathsf{Dequeue}(S1)$;
3   **for each** child E' of E **do**
4   **if** E' is at leaf level **then**
5   Read E';
6   **for each** child E'' of E' **do**
7               $\mathsf{updateLambda}(E'', \alpha_{new})$;
8               $\mathsf{updateL}(E'')$;

9   **else**
10   **for each** child E'' of E' **do**
11   add E'' to $S1$;

We analyse the worst case complexity of Algorithm 3; i.e., when every object is dominated by all the previous objects. Let N denote the window size. In this case, $S1$ contains all N objects. Immediately, the most time consuming part is in Line 11, and thus, the time complexity is given by $O(N^2)$.

Since Algorithm 3 is complicated, we give an example below.

Example 3. Regarding the example in Table 1. We assume windows size $N = 4$, probabilistic threshold $q = 0.6$, and there are exactly $H1, H2, H3$ in current window. $H4$ comes as α_{new}. Before $H4$ comes, we use Algorithm 1 to compute the corresponding l values of $H1, H2, H3$. As an example, the first three columns in Table 3 shows how to compute the l value of $H1$. And we keep $\{0.015, 0.22, 0.765\}$ as $\lambda_{\text{top-}k \text{ dominating}}$ of $H1$. The aggregate information of other objects are listed in the first three rows of Table 4. When $H4$ comes, according to Algorithm 3, we first collect all objects dominating $H4$ in $S1$ and all objects dominated by $H4$ in $S2$. Here, we have $S1 = \{H1\}, S2 = \{H2, H3\}$. Then, we update l and $\lambda_{\text{top-}k \text{ dominating}}$ of $H1$. Using Equation (2), we get the 4-th column in Table 3. Because $(0.3825 + 0.4925) \times P(H1) > q = 0.6$, the l value still equals 2. And $\lambda_{\text{top-}k \text{ dominating}}$ is updated as $\{0.0075, 0.1175, 0.4925, 0.3825\}$. Finally, we compute l and $\lambda_{\text{top-}k \text{ dominating}}$ for $H4$. Because $P(H4) < q = 0.6$, we have $l = 0$ and $\lambda_{\text{top-}k \text{ dominating}} = \emptyset$. Therefore, the top-1 dominating query for window $\{H1, H2, H3, H4\}$ is $H1$.

Table 3. The matrix of $H1$

$H1(0.80)$	$H2(0.90)$	$H3(0.85)$	$H4(0.5)$
1	0.1	0.015	0.0075
	0.9	0.22	0.1175
		0.765	0.4925
			0.3825

Table 4. Aggregate information

House ID	l	$\lambda_{\text{top-}k \text{ dominating}}$
$H1$	2	$\{0.015, 0.22, 0.765\}$
$H2$	1	$\{0.15, 0.85\}$
$H3$	0	$\emptyset$
$H1$(updated)	2	$\{0.0075, 0.1175, 0.4925, 0.3825\}$
$H4$	0	$\emptyset$

4.3 Expire

When α_{old} expires, the l value of each object dominated by α_{old} will not change. Therefore, we only need to collect all the objects dominating α_{old}, re-calculate their l values.

Algorithm 8. remove(α_{old})

---

1 **for each** $E \in R.root$ **do**
2 $\quad$ **if** $E \prec \alpha_{old}$ **then add** E to $S1$;
3 $\quad$ **else if** $E \prec_{partial} \alpha_{old}$ **then add** E to $S3$;
4 **while** $S3 \neq \emptyset$ **do**
5 $\quad$ $E = \mathsf{Dequeue}(S3)$;
6 $\quad$ **for each** child E' of E **do**
7 $\quad\quad$ **if** $E' \prec \alpha_{old}$ **then add** E' to $S1$;
8 $\quad\quad$ **else if** $E' \prec_{partial} \alpha_{old}$ **then add** E' to $S3$;
9 $\mathsf{update2}(S1)$;
10 delete α_{old} from R;

The update process of $S1$ is shown in Algorithm 9. For each object at leaf node, we re-calculate its l and $\lambda_{\text{top-k dominating}}$ by traversing the R-tree to collect all the objects which are dominated by the object in Ω.

Algorithm 9. update2($S1$)

---

1 **while** $S1 \neq \emptyset$ **do**
2      $E = \text{Dequeue}(S1)$;
3   **for each** child E' of E **do**
4   **if** E' is at leaf level **then**
5   read E';
6   **for each** child E'' of E' **do**
7   traverse R to collect all objects dominated by E'' in Ω;
8                  $E''.l = \text{calDS}(\Omega, E'')$;
9   **else**
10   **for each** child E'' of E' **do**
11   add E'' to $S1$;

Similarly, we provide a worst case complexity analysis of Algorithm 8. When all objects dominate all previous objects, we need to re-calculate l value of all remaining objects. Therefore, the time complexity is given by $O(N^3)$, where N is the window size.

Correctness.   The correctness of our techniques is guaranteed by computing the exact l value of each object in current sliding window. Besides, we get the top-k dominating objects using a min-heap. This is also correct because the k objects are guaranteed to have larger l values than these not in the min-heap.

5 Experiments

This section reports the experimental studies.

5.1 Setup

We ran all experiments on a MacBook Pro with Mac OSX 10.8.1, 2.26GHz Intel Core 2 Duo CPU and 4G 1333 MHz RAM. We evaluated the efficiency of our algorithm against sliding window size, dimensionality, and probabilistic threshold, respectively. The default values of the parameters are listed in Table 5.

Parameters are varied as follows:

- sliding window size: 2k, 4k, 6k, 8k, 10k
- dimensionality: 2, 3, 5
- probabilistic threshold: 0.1, 0.3, 0.5, 0.7, 0.9

We ran all experiments with 100 windows, and repeated all experiments 10 times to get an average execution time. The execution time in each following figure is the running time for processing 100 windows.

Table 5. Experimental parameters

Notation	Definition(Default Values)
N	sliding window size (4K)
d	dimensionality of the data set (3)
D	data set (Anti)
D_p	probabilistic distribution of appearance (uniform)
P_μ	expected appearance probability (0.5)
q	probabilistic threshold (0.3)

5.2 Evaluation

Figures 2, 3 and 4 show the results when window size is varied. In Figure 5, we vary the threshold probability to see the running time of baseline algorithm. In Figure 6, we vary the threshold probability to see the running time of the efficient algorithm. From the figures, we see that the efficient algorithm outperforms the baseline algorithm in all cases.

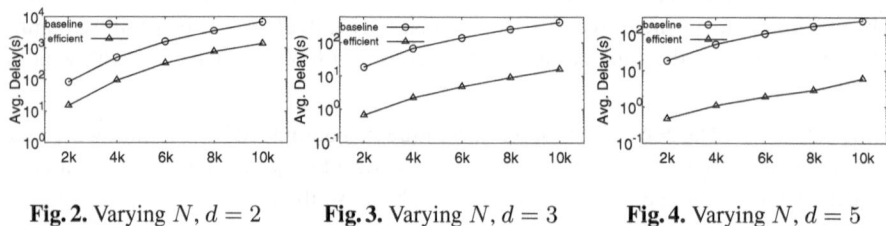

Fig. 2. Varying N, $d = 2$ **Fig. 3.** Varying N, $d = 3$ **Fig. 4.** Varying N, $d = 5$

As shown in Figures 2, 3 and 4, when window size increases linearly, the running time increases in a super-linear manner for of the both efficient and baseline algorithms. Moreover, we conclude, by comparing Figures 2, 3 and 4, that the running time decreases greatly with the increase of dimensionality. This is because the most time-consuming part of the baseline algorithm is to compute l; i.e., collect all objects and apply Algorithm 1 to compute l. With higher dimensionality, there is a smaller Ω for each object. Similar conclusion can be drawn for the efficient algorithm.

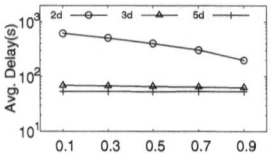   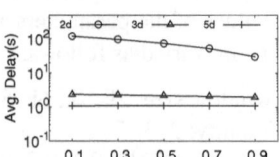

Fig. 5. Baseline, Varying q **Fig. 6.** Efficient, Varying q

From Figures 5 and 6, when threshold grows, the running time decreases. With larger threshold, there are more objects being pruned, i.e. objects whose probability is lower than threshold. In addition, this effect is more obvious with low dimensionality, as more objects to be pruned by the algorithm in the lower dimensional space.

6   Conclusion

In this paper, we have investigated the problem of top-k dominating query in the context of sliding window on uncertain data. We first model the probability threshold based top-k dominating problem, and then present a framework to handle the problem. Efficient techniques have been presented to process such query continuously. Extensive experiments demonstrate the effectiveness and efficiency of our techniques.

References

1. Beckmann, N., Kriegel, H.-P., Schneider, R., Seeger, B.: The r*-tree: An efficient and robust access method for points and rectangles. In: SIGMOD Conference, pp. 322–331 (1990)
2. Bernecker, T., Kriegel, H.-P., Mamoulis, N., Renz, M., Züfle, A.: Scalable probabilistic similarity ranking in uncertain databases. IEEE Trans. Knowl. Data Eng. 22(9), 1234–1246 (2010)
3. Bernecker, T., Kriegel, H.-P., Mamoulis, N., Renz, M., Zuefle, A.: Continuous inverse ranking queries in uncertain streams. In: Bayard Cushing, J., French, J., Bowers, S. (eds.) SSDBM 2011. LNCS, vol. 6809, pp. 37–54. Springer, Heidelberg (2011)
4. Börzsönyi, S., Kossmann, D., Stocker, K.: The skyline operator. In: ICDE, pp. 421–430 (2001)
5. Cormode, G., Garofalakis, M.N.: Sketching probabilistic data streams. In: Chan, C.Y., Ooi, B.C., Zhou, A. (eds.) SIGMOD Conference, pp. 281–292. ACM (2007)
6. Lange, K.: Numerical analysis for statisticians. Statistics and Computing (1999)
7. Lian, X., Chen, L.: Top-k dominating queries in uncertain databases. In: Kersten, M.L., Novikov, B., Teubner, J., Polutin, V., Manegold, S. (eds.) EDBT. ACM International Conference Proceeding Series, vol. 360, pp. 660–671. ACM (2009)
8. Roussopoulos, N., Kelley, S., Vincent, F.: Nearest neighbor queries. In: SIGMOD Conference, pp. 71–79 (1995)
9. Shen, Z., Cheema, M.A., Lin, X., Zhang, W., Wang, H.: Efficiently monitoring top-k pairs over sliding windows. In: Kementsietsidis, A., Salles, M.A.V. (eds.) ICDE, pp. 798–809. IEEE Computer Society (2012)
10. Yiu, M.L., Mamoulis, N.: Efficient processing of top-k dominating queries on multidimensional data. In: VLDB, pp. 483–494 (2007)
11. Zhang, Q., Li, F., Yi, K.: Finding frequent items in probabilistic data. In: SIGMOD Conference, pp. 819–832 (2008)
12. Zhang, W., Lin, X., Zhang, Y., Pei, J., Wang, W.: Threshold-based probabilistic top-k dominating queries. VLDB J. 19(2), 283–305 (2010)
13. Zhang, W., Lin, X., Zhang, Y., Wang, W., Yu, J.X.: Probabilistic skyline operator over sliding windows. In: Ioannidis, Y.E., Lee, D.L., Ng, R.T. (eds.) ICDE, pp. 1060–1071. IEEE (2009)

Distributed Range Querying Moving Objects in Network-Centric Warfare

Bin Ge, Chong Zhang, Da-quan Tang, and Wei-dong Xiao

Science and Technology on Information Systems Engineering Lab,
National Univ. of Defense Technology, Changsha 410073, China
gebin1978@gmail.com

Abstract. Sensing and analyzing moving objects is an important task for commanders to make decisions in Network-centric Warfare(NCW). Spatio-temporal queries are essential operations to understand moving objects situation. This paper proposed a new method for addressing distributed range querying moving objects problem, which was not considered in related studies. Firstly, the paper designed an index structure called MOI which utilized hilbert curve to mapping moving objects into 1-dimensional space and then used DHT to index. Then based on locality-preserved characteristic of hilbert curve, the paper devised a high-efficient spatio-temporal range query algorithm, which greatly reduces routing messages. Experimental results reveal that MOI outperforms other related algorithms and exhibits a low index maintaining cost.

Keywords: P2P, Distributed, MOI, Hilbert curve, Spatio-temporal Indexing.

1 Introduction

Moving objects(MO) data is a kind of important information for analysis, planning and fire striking in NCW, especially in the future digital battlefield. Awaring battlefield situation by sharing moving objects data, it is possible to accomplish tasks such as effective alert action, battlefield environment protection and weapon unit precision strikes, which is of great importance on improvement of battlefield transparency and joint operations ability and implementation of high efficient command[1,2]. Accumulating battlefield information, forming large battlefield spatio-temporal data set, new requirements for spatio-temporal index technology in collaborative, efficiency, agility and reliability have been put forward.

Based on Peer-to-Peer computing (P2P) technology, this paper presented a distributed spatio-temporal index structure MOI (Moving Objects Index). Utilizing DHT to organize node topology structure, adapting Hilbert curve to map spatio-temporal objects to one dimensional space and utilizing locality preservation property of the curve, the index structure designs efficient query algorithm, with which the number of routing messages has been greatly reduced and performance of spatio-temporal query has been improved, ensuring efficient spatio-temporal information sharing in battlefield.

Y. Ishikawa et al. (Eds.): APWeb 2013, LNCS 7808, pp. 794–803, 2013.
© Springer-Verlag Berlin Heidelberg 2013

2   Problem Description

Moving objects constantly change their position and shape with time passing by. This paper considers that spatio-temporal data is a kind of data related to the position detected by various detection methods in the earth reference space, expressing the gradual changes over time of entity's and process's state of being (position, shape, size, etc.) and property.

In the battlefield, there exists all kinds of objects whose properties evolves over time, such as position changing individual, tanks, and shape changing cloud, smoke, fire, etc. In this paper, the use of thought about traditional multidimensional index (such as R-trees) abstracts time-space objects in the battlefield into minimum bounding rectangle (MBR). The corresponding coordinates of MBR (i.e., two point coordinates in left bottom corner and right-top-corner) change with the position and shape of time-space object.

Besides moving objects, other command unit exists in battlefield. With positioning technology, the position and shape of objects can be transmitted to these units. Moving objects (such as individual, tanks and other autonomy type objects) can register nearby to report to corresponding units, whereas the units can collect space information to local of the objects by continuous reconnaissance and detection.

Through the description above, the problem can be defined as: in the battlefield, command unit can be abstracted as network nodes (i.e., peer), each node gets physical connection through the ground link, the investigation data is position and shape changing over time rectangular data (i.e., MBR), in the distributed environment, it is required to set up a kind of spatio-temporal index mechanism, i.e., set up a network, each node maintains part of the global index, each node, through the mutual communication, completes the whole index operation as query, insert, update, and etc, as shown in figure 1. This paper here assumes that each moving object has a global object ID.

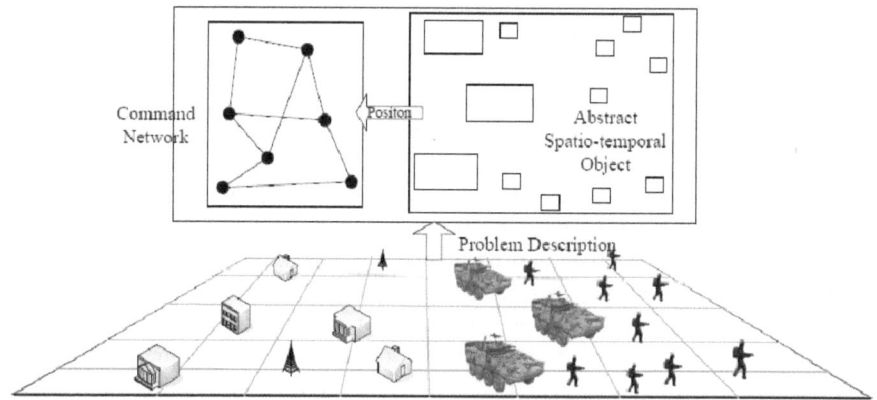

Fig. 1. Problem Description

3   MOI: Distributed Spatio-temporal Index Based on DHT

Using consistent hash, Chord[3] maps resources to one dimensional space, which destroys the locality of resources. Namely resources nearby in the original space through mapping are no longer neighbors, which cannot meet the needs of queries in space range. So this paper maps the time-space information to one dimension information using Hilbert curve[4].

3.1   Index Structure

MOI is actually a kind of overlay network. Based on DHT technology, each node has a node identifier, and is responsible for part of key areas. Moving objects belonging to this key area through mapping, their spatio-temporal information is maintained by nodes responsible for. The detail description is as below:

1) This paper views the moving objects to be considered as extent objects, and present its space state with Minimal Bounding Rectangle (MBR). This paper uses MBR to represent moving objects' space state below. This paper assumes the spatio-temporal information format of moving objects obtained by nodes as (Oid, X_{min}, X_{max}, Y_{min}, Y_{max}, t), in which Oid means the object's global ID, (X_{min}, X_{max}, Y_{min}, Y_{max}) means the coordinates of MBR, and t means MBR timestamp.

2) This paper sets the whole space to be considered as square in two dimensions with the side length equals to value E, and uses D order Hilbert curve. The whole space is divided into 2^{2D} cells of the same size. Through the construction of Hilbert curve, the mapping from the set P of all the points in the whole space to the set V of Hilbert value exists, among which $P = \{(x, y) | 0 \leq x, y \leq E\}$ and $V = \{v | 0 \leq v \leq 2^{2D} - 1, v \in \mathbb{Z}\}$.

3) This paper uses Chord as overlay network structure, with $[0, 2^m - 1]$ as its key space. We set $m=2D$, namely ensure the consistency of key space and the set of Hilbert value.

4) The sizes of corresponding MBR for every MO may be different. One MBR may overlap several cells while one cell may contain several MBR. According to the Hilbert curve, all points in each cell have unique counterpart Hilbert value, which can be regarded as each cell has an unique corresponding Hilbert value, so that any one of the MBR can be mapping for one or several Hilbert value based on Hilbert curve. When an MBR maps for several Hilbert value, because of the locality preservation property of the Hilbert curve, in these Hilbert values, some may be continuous, which can be regarded that the MBR will be mapped to a series of 1 dimensional region. This paper calls one dimension region for segments.

5) Each segment has a starting value and an end value, and according to the Chord's position in key space, MBR will be stored to the corresponding node, and the specific process will be described in section 3.2. According to its different size, each MBR can be stored to one or more nodes, which ensures the spatio-temporal query rate. Each node establish HR-trees[5] index in local according to the spatio-temporal information (X_{min}, X_{max}, Y_{min}, Y_{max}, t) of stored MBR.

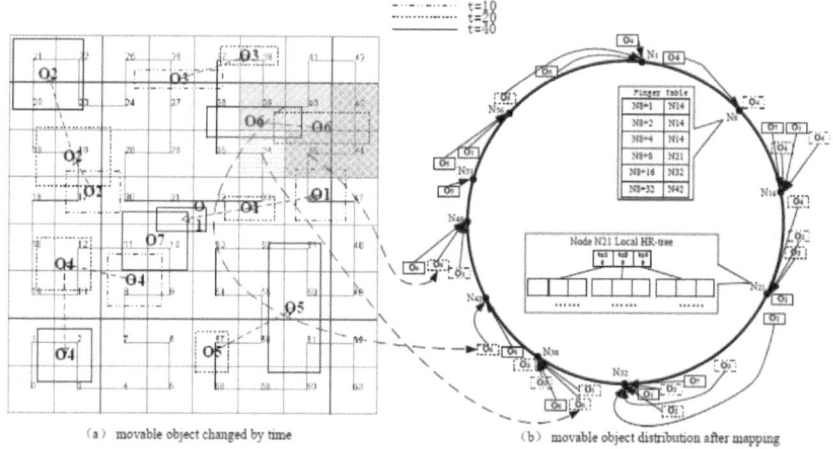

Fig. 2. MOI Structure

Figure 2(a) reveals the space state of $STO_1 \sim STO_7$ (for simplicity, we use $O_1 \sim O_7$ in the figure) respectively at $t=10$, $t=20$ and $t=40$ (namely MBR). Adopting 3 order Hilbert curve, the whole space is divided into 64 cells and numbered according to Hilbert curve. Figure 2(b) shows the distribution of MO in the MOI index. If the segments of O_6 through mapping are [43,45],[39,40] and 34 respectively (figure 2 (a) distinguish different segment with different shadow patterns), according to the segment value and Chord resource placement strategy, referring data maintenance algorithm for detail, when t=20, O_6 distributed to node N_{38}, N_{42}, and N_{48}. Each node establishes HR-tree index locally, figure 2(b) samples the HR-tree index of node N_{21} and Finger Table of node N_8.

3.2   Spatio-temporal Query Algorithm

Two types of spatio-temporal queries are discussed: time point query and time interval query. Time point query is that at time T_Q, all the objects in the space range (Q_{xmin}, Q_{xmax}, Q_{ymin}, Q_{ymax}) return as result. Time interval query is that in the period (T_S, T_E), all the objects in the space range (Q_{xmin}, Q_{xmax}, Q_{ymin}, Q_{ymax}) return as result. They can be presented respectively as ((Q_{xmin}, Q_{xmax}, Q_{ymin}, Q_{ymax}), T_Q) and ((Q_{xmin}, Q_{xmax}, Q_{ymin}, Q_{ymax}), (T_S, T_E)).

Spatio-temporal query algorithm described as below:

1) In accordance with Hilbert curve mapping rules that any coordinate in each cell maps for the cell number Hilbert curve running through, the space range (Q_{xmin}, Q_{xmax}, Q_{ymin}, Q_{ymax}) can be converted to a series of segments, one of which is a continuous Hilbert value region. Each segment has the structure of (H_{enter}, H_{exit}). H_{enter} is a Hilbert value as Hilbert curve into the segment, H_{exit} is a Hilbert value as Hilbert curve out the segment, $H_{enter} \leq H_{exit}$;

2) For each segment, according to the Chord routing rules that stated in [3] as find_successor() function, node P routes the spatio-temporal query conditions and it to key's successor node S;

3) When receiving spatio-temporal query request, node S carries on the judgment: if node R, the predecessor node of node S, had identifier value equal to or greater than the value of the received segment, the query request would be forward to node R;

4) Using of local HR-tree, node S conducts time point queries or time interval queries, then routes the result to node P.

All nodes which is the equivalent of node S (successor node corresponding to H_{exit} value of each segment) and node R (predecessor node meet the space overlapping conditions through judgment by the equivalent node of node S) will go to step 3) and 4).

The example of spatio-temporal query algorithm is shown in figure 3. According to the four segments, query conditions will route to node N_{38}, N_{48} and N_{56} respectively. After comparison, node N_{56} will forward query conditions to node N_{51}. Each node searches through time point queries or time interval queries in the local HR-tree and consequently returns the result to the original inquiring node.

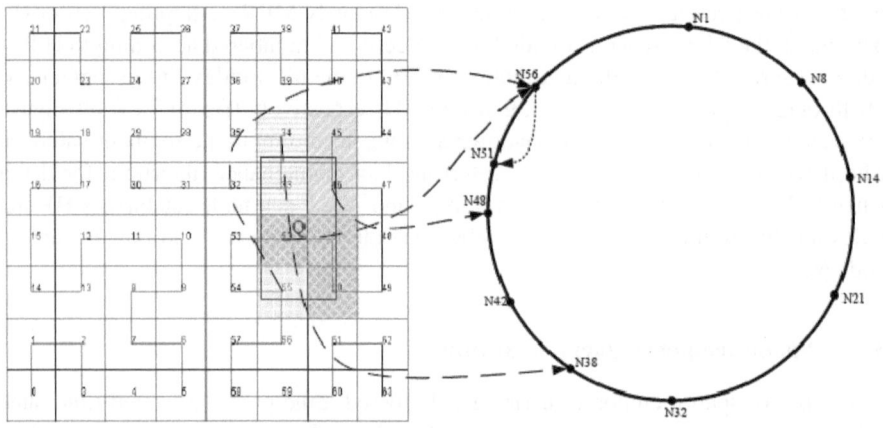

Fig. 3. Spatio-temporal Query

4   Construction and Maintenance of Index

4.1   Construction

In the battlefield, when a new command unit added, namely supposing that node Q joins the network, implement steps as follows:

1) Node Q accesses to node identifier through connecting to match-making server and other ways. Node Q then arbitrarily connects a node M, sending node identifier. On the basis of Chord routing rules, namely the find_successor() function in [3], node

M forward the message. The join request of node Q is routed to the successor node N of the key where node Q's identifier is.

2) Node N and node Q are respectively responsible for updating the key space themselves (in agreement with chord protocol's node adding algorithm).According to the relationships between Hilbert mapping rules and key space itself, node N shifts MO in the scope of node Q to node Q, by the way updating local HR-tree index.

3) Node Q receives the MO, and sets up the local HR-tree index.

4.2   Maintenance

Index maintenance is that when a node leaving or failing and data on node changing, relevant nodes need to update the index, this kind of situation on the battlefield environment often occur.

Leaving and Failing of Node

Node leaving is that before exiting system, the node will inform its successor node or predecessor node that "I will leave system", while node failing is that because of the network connection instability or hardware failure, the line between nodes breaks, the successor node or predecessor node do not know its exiting. Although in the two situations, node leaves system in different ways, the steps it takes are similar. The specific process is as below:

1) Node L sends "I will leave system" message and local MO to the successor node K, node K updating local HR - trees, followed by the exiting of node L;

2) The predecessor node of node J keeps detecting the state of node L, once found node L leaving the system, according to its own successor list (the same with Chord), it would contact node K and save node K as successor node;

3) Node K saves node J for predecessor node, meanwhile, the index update completes.

For node failure, we adopt Heartbeat mechanism that periodically tests successor nodes and predecessor node's state, to assess whether there is a node failure, and then, take steps 2)~3), completing index update.

Data Change

In the battlefield, position and shape of all kinds of MO change continuously as time goes on, new situation, new object appear constantly in the battle space while some objects possibly will quit or disappear (every MO has the only object ID), these are corresponding to the system operations including update, insert and delete. These operations are similar to the spatio-temporal query, specifically, when a node in the system obtains a new object O (Oid, X_{min}, X_{max}, Y_{min}, Y_{max}, t) to insert, take steps as follows:

1) Based on Hilbert mapping rules, the space range of object O (X_{min}, X_{max}, Y_{min}, Y_{max}) is transformed into a series of segments (H_{enter}, H_{exit});

2) For each segment, according to Chord routing rules, node I routes object O and this segment to key H_{exit}'s successor node G;

3) After receiving object insertion request, node G start to assess: if identifier value of node G's predecessor node H is greater than or equal to the value H_{enter} from the received segment, forward this request to node H;

4) Using HR-tree insertion algorithm, node G maintains the spatio-temporal data and returns the result of successful insertion operation to node I.

All nodes equal to node G (successor node corresponding to each segment's H_{exit} value) and node H (predecessor node in accordance with space overlap condition judged by the equivalent node of node G) take steps 3)~4).

Both delete and update operations of MO are similar to insert operation, the difference is, the maintenance of local data need be shifted into HR-tree's deletion and update algorithm. Due to hardly space, no more statements is listed here.

5   Performance Evaluation

In order to evaluate performance of the proposed method in this paper, we use open source simulation platform PeerSim[6] to conduct the simulation experiment. All experiments are carried out on computers with Intel Core2 Quad 2.5 GHz and 2 GB main memory. We set the whole space to be considered as 1×1 unit rectangular, and set the length of time as 1 unit time region; Owing to the side length of the rectangle in experiment is far less than that of the rectangle to be considered, we take 7 order Hilbert curve for default.

5.1   Data Set Description

In order to simulate the battle space objects, we implement the GSTD[7] spatio-temporal data generation method, which is widely used in spatio-temporal index experiment. Based on GSTD method, two data sets are used: analog data as the base state data (hereinafter referred as data set A) and real data as base state data (hereinafter referred as data set B).

In data set A, the center of base state data follows Gaussian distribution with μ =0.5 (expectation) and σ =0.1 (variance), the side length presents 1000 rectangular data which follows（0,0.1）uniform distribution; time sampling point numbers for 100, and follows uniform distribution. The variation of the center (Δcenter) follows (0.2, 0.2) uniform distribution and the variation of side length (Δextent) follows (0.05, 0.05) uniform distribution. With time goes on, the overall change of data is that they spread from the center to all around.

In data set B, we adopt TIGER®[8] data set as the ground state data. The detailed generate process is: choose all the ways in Maryland Montgomery county of TIGER ® data set, with each road composed by limited lines. According to the starting point and end point of the line segment, rectangular is generated, which is provided for ground state data. Time sampling point numbers for 100, and follows uniform distribution. The variation of the center (Δcenter) follows (0, 0.4) uniform distribution, the side length variation (Δextent) follows (-0.05, 0.05) uniform distribution. As time goes on, the overall change of data is that they gradually move to northeast of the space.

5.2   Query Performance Evaluation

We evaluate the query performance by the number of messages when querying. Message number refers to the total message number sent by all the nodes from the network for one time of spatio-temporal query. For each message number statistics, we randomly generate 300 queries and take the average number of messages as the evaluation result. Compare MOI with Distributed Quadtree[9], P2P Meta-index [10], and Service Zone[11], we conduct the query performance comparison. The size of message number is tested under different network scales and different selection rate respectively. For data set A and data set B, the experiment results are similar, here we only take the result for data set A.

Set the query's space selection rate as 0.01. The number of nodes in the network rises from 250 to 10000. And then, count the corresponding average message number. As shown in figure 4(a), the result of the experiment reveals that with the increase of the network scale, the query performance of the method proposed by this paper is superior to the other three kinds of methods. This is because the space division method adopted by this paper using Hilbert curve with local preservation property, which maps the query into continuous one dimensional queries. And the query algorithm proposed by this paper considers checking the predecessor node whether it intersects with query region by node itself, which greatly reduces the number of messages and optimizes the query performance.

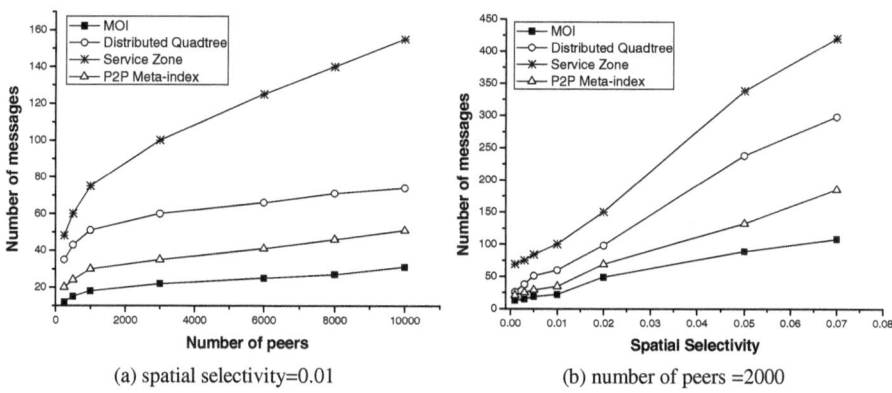

(a) spatial selectivity=0.01   (b) number of peers =2000

Fig. 4. Query Performance Comparison

Set the number of node as 2000. The query space selection rate changes from 0.001 to 0.07. And then, count the corresponding average message number. As shown in figure 4(b), the result of the experiment reveals that with the increase of the space selection rate, the Service Zone methods and Distributed Quadtree method very rapid increase in the number of messages. And the proposed MOI method produced little change in the number of messages, which is lower than the P2P Meta-index method. This is because the Service Zone method uses CAN's[12] thought to divide the space binary tree, finding low efficiency. Distributed Quadtree method uses consistent hash, where query is decomposed into a large number of one-dimensional intervals, increasing traffic, resulting in query performance degradation. P2P Meta-index method with the crawlers, need to continue to contact each server, increasing the communication overhead.

5.3   The Evaluation of Index Maintenance Cost

Based on DHT, the processing performance for node join, exiting and the failure is similar to DHT, so no more narrative hare. In this section, the experimental evaluation mainly bases on data insertion cost. The number of messages caused by spatial and temporal data insertion for data set A and data set B were tested under different network size. The space default selection rate is 0.01, where spatial selectivity refers to the size of the insertion of the MBR, the results shown in Figure 5.With the increase in the number of nodes, for two different sets of data, the cost of maintaining indexes showed a logarithmic growth. It illustrates that the index maintenance performance of MOI is able to adapt to the changes in the size of network.

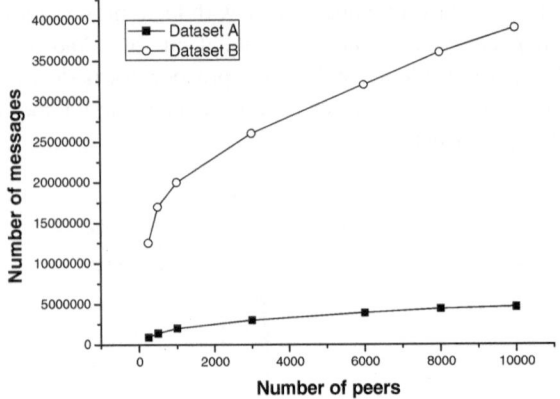

Fig. 5. Index maintenance cost, spatial selectivity = 0.01

6   Conclusion

In future digital battlefield, the effective index of the temporal and spatial objects will greatly help to improve the battlefield space-time query performance, which is of great significance. For the needs of fast access to spatial and temporal information, faced with the distributed battlefield environment, based on peer-to-peer computing, this article proposes distributed spatio-temporal indexing mechanism MOI. Make use of Hilbert curve combined the Chord to design spatio-temporal query algorithms, indexing and maintenance algorithm. We test query performance on two different sets of data, cost of the maintenance. The experimental results show that the local Hilbert curve's characteristics greatly improve the efficiency of the spatio-temporal query, meanwhile the index maintenance performance is acceptable, so as to provide real-time rapid access to the distributed spatio-temporal information in the future digital battlefield with technical support.

Acknowledgement. This work is supported by the project of National Science Foundation of China (No.71103203).

References

1. Ilarri, S., Mena, E., Illarramendi, A.: Location-Dependent Query Processing: Where We Are and Where We Are Heading. ACM Computing Surveys 42(3), 1–73 (2010)
2. Wolfson, O., Xu, B., Chamberlain, S., Jiang, L.: Moving Objects Databases: Issues and Solutions. In: Proceedings of the 10th International Conference on Scientific and Statistical Database Management (SSDM), pp. 111–122 (1998)
3. Stoica, I., Morris, R., Karger, D., Kaashoek, M.F., et al.: Chord: A Scalable Peer-to-Peer Lookup Service for Internet Applications. In: Proceedings of the ACM SIGCOMM, pp. 149–160 (2001)
4. Faloutsos, C., Roseman, S.: Fractals for Secondary Key Retrieval. In: Proceedings of the 8th ACM SIGACT–SIGMOD–SIGART Symposium on Principles of Database Systems, pp. 247–252 (1989)
5. Nascimento, M., Silva, J.: Towards Historical R-trees. In: Proceedings of ACM Symposium on Applied Computing (ACM-SAC), pp. 235–240 (1998)
6. PeerSim Simulator Project[EB/OL] (2005), http://peersim.sourceforge.net
7. Theodoridis, Y., Silva, J.R.O., Nascimento, M.A.: On the Generation of Spatiotemporal Datasets. In: Proceedings of the 6th International Symposium on Spatial Databases (SSD), pp. 147–164 (1999)
8. Tiger Datasets[EB/OL], http://www.census.gov/geo/www/tiger/
9. Tanin, E., Harwood, A., Samet, H.: Using a Distributed Quadtree Index in Peer-to-Peer Networks. VLDB Journal 16(2), 165–178 (2007)
10. Hernández, C., Rodríguez, M.A., Marín, M.: A P2P Meta-Index for Spatio-Temporal Moving Object Databases. In: Proceedings of the 13th International Conference on Database Systems for Advanced Applications, pp. 653–660 (2008)
11. Wang, H., Zimmermann, R., Ku, W.-S.: Distributed continuous range query processing on moving objects. In: Bressan, S., Küng, J., Wagner, R. (eds.) DEXA 2006. LNCS, vol. 4080, pp. 655–665. Springer, Heidelberg (2006)
12. Ratnasamy, S., Francis, P., Handley, M., Karp, R., Shenker, S.: A Scalable Content-Addressable Network. In: Proceedings of the ACM SIGCOMM, pp. 161–172 (2001)

An Efficient Approach on Answering Top-k Queries with Grid Dominant Graph Index

Aiping Li, Jinghu Xu, Liang Gan, Bin Zhou, and Yan Jia

School of Computer Science, National University of Defense Technology
410073 Changsha, China
apli1974@gmail.com, peaceful_lake@163.com,
71628959@qq.com, binzhou@nudt.edu.cn, jiayanjy@vip.sina.com
http://www.springer.com/lncs

Abstract. The *top-k* queries are well used in processing spatial data and dimensional data stream to recognize important objects. To address the problem of *top-k* queries with preferable function, a grid dominant graph (GDG) index based on reserve data points set (RDPS) is presented. Because of the correlation between *top-k* and skyline computation, when the count of one point's RDPS is bigger than or equals to the value of k, it is for sure to be pruned the point's dominant set. This approach in used to prune a lot of free points of *top-k* queries. GDG used grid index to calculate all RDPS approximately and efficiently. When construction GDG, grid index figures out *k-max* calculating region to prune free points that all *top-k* function would not visit, which decreases the amount of index dramatically. When travelling GDG, grid index figures out *k-max* search region by *top-k* function which avoids travelling those free points of ad-hoc *top-k* function. Because GDG uses grid index to rank those data points in the same layer approximately by *k-max* search region by *top-k* function, this index structure make less visited points than traditional dominant graph (DG) structure. What's more, grid index needs less additional computation and storage that make the GDG index more adaptive for *top-k* queries. Analytical and experimental evidences display that GDG index approach performs better on both storage of index and efficiency of queries.

Keywords: *top-k* query, database, index, grid index, reverse dominant point set.

1 Introduction

Given a record set D and a query function F, a *top-k* preference query (*top-k* query for short) returns k records from D, whose values of function F on their attributes are the highest. The *top-k* query is crucial to many multi-criteria decision making applications. For example, autonomous robot explore a given spatial region under dangerous environment and returns the search results as a sequence of objects ranked by a scoring function. As another example, in self-determinded soccer robet game, mobile robotic system list an ordered set of

Y. Ishikawa et al. (Eds.): APWeb 2013, LNCS 7808, pp. 804–814, 2013.

objects based on how aggressive the teammate object is or how weakly the rival objects is matching the score function used in the game.

Existing work on *top-k* query processing is to avoid searching all the objects in the underlying dataset, and limiting the number of random accesses to the spatial data. A key technique to evaluate *top-k* query is building index on the dataset. They can be divided into four categories: Sorted list-based, Layer-based, View-based and Sketch-based.

1. *Sortedlists-based*. The methods in this category sort the records in each dimension and calculate the rank by parallel scanning each list until the *top-k* results are returned, such as TA[6] and SA[2].

2. *Layer-based*. The algorithms organize all records into consecutive layers, such as DG[9],Onion[3] and AppRI[7]. The organization strategy is based on the common property among the records, such as the same convex hull layer in Onion[3]. Any *top-k* query can be answered by up to *k* layers of records.

3. *View-based*. Materialized views are used to answer a *top-k* query, such as PREFER[5] and LPTA[4].

4. *Sketch-based*. Data stream sketch-based method is used to calculate *top-k* query, such as RankCube[8].

In this paper we proposed a hybrid index structure gridded dominant graph (GDG) method to improve *top-k* query efficiency, which integrates grid index into dominant graph (DG) index. The rest of this paper is organized as follows. The next section discusses related work. Section 3 describes the problem description and definition. Section 4 gives the index structure and algorithm. Section 5 present and analyze the experiments, and Section 6 concludes.

2   Related Works

There are considerable amount of query answering algorithms proposed for *top-k* queries in the literatures. The threshold algorithm (TA) [6] is one of the most common algorithms. TA first sorts the values in each attribute and then scans the sorted lists in parallel. A threshold value is used to prune the tuples in the rest of the lists if they cannot have better scores than the threshold. A lot of follow-on methods have built on the TA algorithm as it addresses a common category of *top-k* queries, where the scoring function is a monotone function composed by using various attributes associated with the objects in the input dataset.

The layer-based method divided all the tuples of the dataset into several layers; any *top-k* query can be answered by up to *k* layers of records. There are some layered methods, such as DG[9],Onion[3] and AppRI[7]. The Onion method takes convex hull as origination rule. For given preference liner function, the result only exists in the convex hull layer. The process of Onion is calculating the convex hull of the dataset. The first convex hull is calculated and then the next is calculated on the rest dataset, and so on, until no dataset is left. The layer rule in AppRI method is the tuple t is put into layer *l*, if and only if it satisfy two conditions: a) For any liner function, *t* is not in the result of *top-(l-1)*; b) Exists at least one function satisfy *t* belongs the result of *top-l*. The layer rule in DG is each layer is the former Skyline. The first skyline is calculated, then the second skyline is computed on the rest dataset, till no dataset is left. The most important property in DG is the necessary condition that a record can be in *top-k* answers is that all its parents in DG have been in *top-(k-1)* answers. Benefiting from the property, the search space (that is the number of retrieved records from the record set to answer a *top-k* query) of a *top-k* query is reduced greatly. The difference between DG and the other two methods is there is dominant relationship in DG, so it is not necessary to access and calculate all the former record set in front of k layer.

The DG method based on *k-skyband* is to keep adding the candidate points. Our method is to prune the free points based on Reverse Dominant Point Set. In other words, if exists k-1 points dominate the point e, all the points dominated by e can be pruned.

3   Problem Definitions and Theorem

We begin by specifying the notation that will be used in the rest of this paper. We use $D = \{O_1, O_2, \cdots, O_N\}$ to denote a set of objects, the state of each object is denoted by $< id, x_1, x_2, \cdots, x_d >$, where id is the identification of the object, $x_i (i = 1, \cdots, d)$ is the ith property of the object. All the properties composed d dimension space; each state of an object is a point the space.

Definition 1. (Dominate [1]). *Given two objects O_1 and O_2 in a d dimension space, we say that O_1 dominates O_2 if and only if they satisfy the following two conditions: 1) in any dimension i, the value of O_2 is larger than or equal to O_1, i.e.$O_2.x_i \geq O_1.x_i$; 2) there must exist at least one dimension j, $O_2.x_j > O_1.x_j$.*

Definition 2. (Aggregate Monotone Function [9]). *An aggregation function f is a monotone if $f(x_1, \cdots, x_n) \leq f(y_1, \cdots, y_n)$, whenever $x_i \leq y_i$, for every i.*

Both DG and Grid index implement aggregate monotone function *top-k* preference query through order the records in advance.

Definition 3. (Free). *If one point or Cell can never be the top-k result, which needn't to be calculated, then we call it is free.*

Definition 4. (Reverse Dominant Point Set). *In data points set P, RDPS = $\{e\} \cup \{t|Dominate(t, e), e, t \in P\}$, is the RDPS of e, denoted by RDPS(e).*

The number of Reverse Dominant Point Set is denoted by $RDPSCount(e)$. The free points in the *top-k* query results can be pruned by $RDPS$, as shown in the next lemma.

Lemma 1. *Let $e \in P$, $m = RDPSCount(e)$, if $m = k$, then for any monotone top-k preference query function f, the points dominated by e are free.*

The proof is omitted.

Definition 5. (Determine Point). *Let* $e, e' \in P$, e' *dominate* e, $m = RDSPCount(e)$, $m' = RDSPCount(e')$, *if* $m' < k$ *and* $m \geq k$, *we call* e *is the Determine Point of top-k query.*

Definition 6. (Dominant Graph Index[9]). *Given a set D of objects in a d dimension space, D has k nonempty maximal layers $L_i, i = 1, \cdots, n$. The objects O in ith maximal layer and records O' in $(i+1)$th layer form a bipartite graph $g_i, i = 1, \cdots, (m-1)$. There is a directed edge from O to O' in g_i if and only if record O dominates O'. We call the directed edge as "parent-children relationship". All bipartite graphs g_i are joined to obtain Dominate Graph (DG for short). The maximal layer L_i is called ith layer of DG.*

To build a DG offline, any skyline algorithm can be used to find each layer of DG. Then, the "parent-children relationship" between records in the ith layer and that in the $(i+1)$th layer. Essentially, DG is a lattice structure, which uses "dominating relationship" as "partial order," as defined in Ref[9].

As a popular index in data stream management, grid index is very efficient. The extent of each cell on every dimension is δ_1 and δ_2. Assuming a 2-dimensional space, the cell $C_{i,j}$ at column i and row j contains all the tuples whose $i \times \delta_1 < O.x_1 \leq (i+1) \times \delta_1$ and $j \times \delta_2 < O.x_2 \leq (j+1) \times \delta_2$, denoted by $IN(O, C_{i,j})$. Then for each object O, the cell $C_{i,j}$ it belongs to can be easily computed by $i = \lfloor O.x_1/\delta_1 \rfloor$ and $j = \lfloor O.x_2/\delta_2 \rfloor$. The right-top point of $C_{i,j}$ is denoted $C_{i,j}.Max$, while the left-bottom point of $C_{i,j}$ is denoted $C_{i,j}.Min$.

Definition 7. (Cell Dominate). *Let* $C_a, C_b \in Cell$, *if* $\forall e \in C_a, \forall s \in C_b$, $Dominate(e, s)$ *is right, then call C_a Dominate C_b, denoted by (C_a, C_b). We call C_a is parent of C_b, and C_b is child of C_a. The Cells which are neither parent nor child of C_a are called siblings of C_b.*

We can determine the dominant relationship between two Cells via $C_a.Min$, $C_a.Max$, $C_b.Min$ and $C_b.Max$. Building the DG index is expensive as it requires a full scan of the entire dataset. When k is known, we can only build the k layer skyline index, and then get the results. Due to the property of RDSP, we can prune more records while build the k layer index.

Definition 8. (k-Max Calculating Region, k-MCR.) *If there is Cell C_i, the sum of the points in the parent Cell of C_i is m, the sum of the points in the siblings Cell of C_i is l, and the sum of points in C_i is n, if $m < k$ and $m + n + l \geq k$, then C_i is Determine Cell. The Union region of C_i parent and sibling Cells is k-Calculating Region, k-CR. All the child Cells of C_i are k-Free Region. The Intersection Region of all the k-CR are the k-Max Calculating Region, k-MCR.*

Theorem 1. *For any monotone top-k preference query function f, its results are in the k-MCR.*

Proof is omitted.

Corollary 1. *If $n < k$, then top-n query can be calculated from k-MCR.*

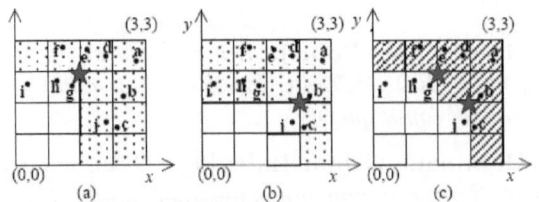

Fig. 1. Schematic diagram of computing k-MCA

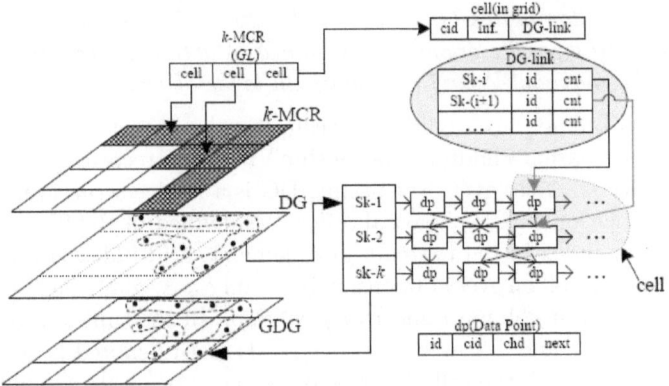

Fig. 2. Architecture of GDG index

Let $C_a, C_b \in CELL$, if $\forall t \in C_a, \forall s \in C_b, f(t) > f(s)$, then a dominates b by function f, denoted by $f - Dom(C_a, C_b)$.

Property 1. $C_a, C_b \in CELL$, if $Dominate(C_a, C_b)$, then $f - Dom(C_a, C_b)$.

Property 2. $C_a, C_b \in CELL$, for given function f, $iffminscore(a) \geq fmaxscore(b)$, then $f - Dom(a, b)$.

Where $fminscore(C_i)$, $fmaxscore(C_i)$ is the minimum and maximum f value of Cell C_i.

Example 1. Figure. 1 is 4×4 grid. The a k points are data points, we execute top-2 query, where a, d, e, f are skyline points, b, g, h are sk-2 points, c, k, i are sk-3 points. $C_{3,3}$ is the older brother of $C_{2,3}$. The total number of $C_{3,3}$ and $C_{2,3}$ is 3, for $3 > 2$, then $C_{2,3}$ is determine cell. The shadowed region in Figure. 1(a) is 2-CR of $C_{2,3}$. The non-shadowed region isn't the calculated region. Similarly, $C_{3,2}$ is the determine cell in Figure. 1(b). The shadowed region in Figure. 1(b) is 2-CR of $C_{2,3}$. The non-shadowed region isn't the calculated region. The shadowed region in Figure. 1(c) is 2-CR of intersection of $C_{2,3}$ and $C_{3,2}$, i.e. 2-MCA.

Definition 9. (Max Search Region by f, k-$MSRf$). For top-k function f, if there is Cell C_i, the sum of the f-parent Cell of C_i is m, the sum of f-siblings Cell of C_i is l, and the sum of points in C_i is n, if $m < k$ and $m+n+l \geq k$, then C_i is f-Determine Cell. The Union region of C_i f-parent and f-sibling Cells is the k-Max Search Region by f, k-$MSRf$.

That is, when f is given, we can improve the efficiency more by pruned the redundant Cells in the k-MCR.

Theorem 2. *For any monotone top-k preference query function f , its results are in the k-$MSRf$.*

Proof is omitted.

4   GDG Index

4.1   Grid Index Structure

The GDG index is based on grid index and DG index, as shown in Figure. 2.

There are two parts in the GDG index, which are DG and k-MCR. We maintain the layer and dominate relationship between the data point dp. Each data point dp is composed of id, cid (cell id), the pointer set of its children chd, and the *next* pointer to the next data in the same layer sk-i. The sk-i($i = 1, \cdots, k$) (skyline-i) , represents the ith skyline of the DG index. The same sk-i and the data points dp in the same cell are stored in the continuous memory via link structure. There is one GL link table in the k-MCR, each node of the link table is the cell in the grid. The basic information Inf and link information with DG DG-$link$ is stored in the cell.

Theorem 3. *GCG index correctly finds top-k answers.*

Proof is omitted.

4.2   Initial Computation Module

We calculated the k-MSRf through the initial computation module(CM), and can get the approximate order relationships among the grid cells. Then we can obtain the free cells in the top-k queries about function f, which can be pruned. It is shown in algorithm 1.

As shown in algorithm 1, Lines 1 to 2 calculate the maximum and minimum point of cell C from the k-MCR points. Lines 5 to 11 calculate the k-MCRf of the cells. Because the results of the data sets function f are mapped into one dimensional space, it is simple to calculate the f-relationship between the data points. For lines 8 to 9 read the ordered cells, we can judge cell C is f determine cell without judge the father of cell C.

4.3   GDG Query Module

After calculated k-MCRf, we can get the result sets of top-k query must be located in the data points in the region of the k-MCRf. The data points outside the region are free, can not appear in the result set.

As shown in algorithm 2, the top-k query algorithm is similar with DG, with the $kMSRf$ points are pruned. Line 1 read the data in the first layer $Sk-1$ of DG, pruned the points without in the $kMSRf$ region, and put the results into

---

Algorithm 1. Algorithm of Initial Computing $k - MSRf$

Input: Data Set P;
Output: $kMCRf$ of P;
 1: **for** each cell C in k-MCR **do**
 2:    Calculate f-range $[[fminscore(C), fmaxscore(C)]]$ of C;
 3: **end for**
 4: Sort cells in k-MCR by f-range;
 5: $kMCRf = kMCR$;
 6: **while** $kMCR$ is not null **do**
 7:    Read first cell C in kMCR;
 8:    kMCR= kMCR - C;
 9:    $t = fFatherDPCount(C) + fElderBrotherDPCount(C)$;
10:    **if** $t \geq k$ **then**
11:      $kMCRf = kMCRf - C.fchildren$;
12:      $kMCR = kMCR - C.fchildren$;
13:    **end if**
14: **end while**

---

the CL list. The data in CL are ordered by the f value, and the count of the data in CL is $k - n$, where n is the number of result set(RS). Line 1 put the data have largest R value into RS. Lines 2 to 21 calculated the rest of the RS one by one. The proof of algorithm 2's correctness is omitted.

Algorithm 2. k-MCR Traveler Algorithm

Input: A GDG of the database D, top-k query function f and kMSRf;
Output: the top-k answers;
 1: Move the first largest record R in CL into result set RS;
 2: Set $n=1$;
 3: **while** kMSRf is not NULL and $RS.size() < k$ **do**
 4:    **for** each child C of R **do**
 5:      **if** C has at least one parent that is not in RS or C has been computed using query function f before or not in kMSRf **then**
 6:        Continue;
 7:      **else**
 8:        Compute C by query function f and insert it into CL;
 9:      **end if**
10:      **if** $CL.size() > k - n$ **then**
11:        Only keep the first $k - n$ records in CL;
12:      **end if**
13:      **for** each cell Ce in kMSRf **do**
14:        **if** $Ce.Max < min(CL)$ **then**
15:          Delete cell Ce from kMSRf;
16:        **end if**
17:      **end for**
18:    **end for**
19:    Move the first largest record R in CL into the answer set RS.
20:    Set $n=n + 1$;
21: **end while**

4.4   Space Complexity and Time Complexity

We firstly analysis the space complexity of the algorithm. Suppose the dimension of data set is d, the m is the number of intervals in index. We can get the total number of cells is $M = m^d$. Each data point needs d memory units. The ratio of k-MCR in total cells is P, with two points in this type cells, one is total numers, the f-value of this cell. So the number of total memory units is $m^d * P * (2d + 4)$. Let the ratio of k-MSRf in total cells is P', where cell's f-value is mapped to one cell ID, is $m^d * P'$. The number of k-MSRf access points in GDG is $s * d$. So, the space complexity of our method is $O(m^d)$, where m is the number of intervals in index, d is the dimension.

Then we analysis the time complexity. Suppose the distribution of the data is uniform distribution, then the number of points in each cell is about N/m^d. During the process of building index k-MCR, the complexity of k-MCR equals scan the data set time plus find the determine point time. Let N is the sum of the data points in the window, then scan the data set time is $O(N)$. Find the determine point time is find the *cell.Min* point. The worst case is find the whole cell set, the time is $O(M)$, where M is the total number of cells in the Grid Index. If N is large, and M is small, then the complexity of k-MCR is $O(N)$, where N is the window size. Similarly, we can get the complexity of k-MSRf is $O(M * P + log(M * P'))$, approximately equal $O(M)$, where M is the total number of cells in the Grid Index.

5   Experimental Result

All experiments were carried out on a PC with Intel Core2 6400 dual CPU running at 2.13GHz and 2GB RAM. The operating system is Fedora with the Linux kernel version 2.6.35.6. All algorithms were implemented in C++ and compiled by GCC 4.5.1 with -O3 flag. Our experiments are conducted on both real and synthetic datasets.

Synthetic datasets are generated as follows. We first use the methodologies in [10] to generate 1 million data elements with the dimensionality from 2 to 5 and the spatial location of data elements follow *independent* distribution. Synthetic datasets are represented with U_i, which have 1000K tuples, the granularity is 8000, with normal distribution between [0,7999]. The real data dataset is from http://archive.ics.uci.edu/ml/machine-learning-databases/ covtype/, 54 dimensions, 10 dims are numeric. We choose 3 numeric variable as dimension, while the granularity is 1989, 5787 and 5827 respectively.

We first compare the space and time cost during building the index between DG and GDG.

The method we compared with is RankCube[4] and DG[9], the dataset is offline. Our methods are GDG40 and GDG80, where each dimension is divided into 40 intervals and 80 intervals. As synthetic data (3 dimension) in Figure. 3(a),

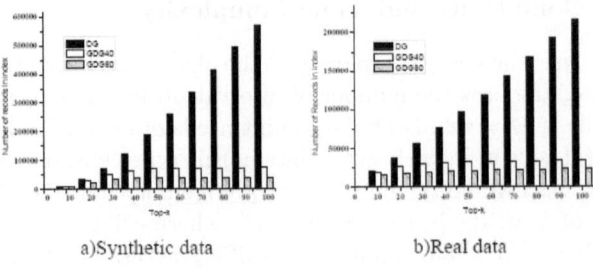

a)Synthetic data   b)Real data

Fig. 3. Reduce the size of index

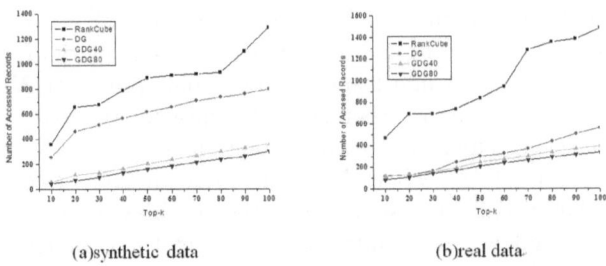

(a)synthetic data   (b)real data

Fig. 4. Reduce the number of access data

when answer top-100 queries, the DG index is 10K 572K tuples, the maximum
ratio is 57.2%. While the GDG40 index is 9K 74K tuples, the maximum ratio is
7.4%, and the GDG40 index is 8K 39K tuples, the maximum ratio is 3.9%. As
real data in Figure. 3(b)(580 tuples), when answer top-100 queries, the DG index
is 216K tuples, the ratio is 37%. While the GDG40 index is 19K 34K tuples, the
maximum ratio is 6%, and the GDG80 index is 6K 24K tuples, the maximum
ratio is 4%. So, our methods can reduce the size of index obviously.

Secondly, the comparisons of the number of access data, query response time
among our method and RankCube[4] , DG[9] is completed.

As shown in Figure. 4 and Figure. 5, for each function f has different number
of access data and query response time, we choose 10 functions, then take average
value as the result. As shown in Figure. 4, our methods have less number of access
data than RankCube[4] and DG[9], where GDG80 is the best.

As shown in Figure. 5, our methods have better performance than the existing RankCubee[4] and DG[9] method, both using synthetic data and real data
environments.

We can draw conclusions from the experimental results as follows. Firstly,
some data sets accessed in DG method are free, which are pruned in GDG
method. So GDG method can decrease the number of access points and response
time. Secondly, the RankCube method access the data set unordered and have

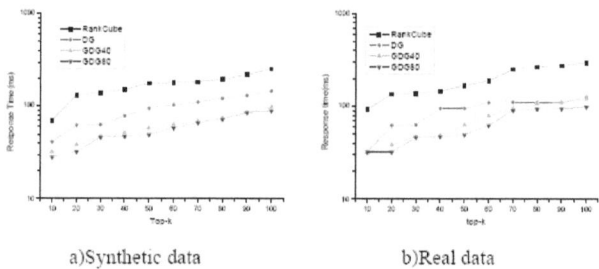

a)Synthetic data   b)Real data

Fig. 5. Query time response Comparison

no hierarchical relationships, but GDG method did. So the number of access points and response time of GDG method is better. Finally, Compared GDG80 and GDG40, the former one has more intervals, the result is more precise, so the the number of access points is better. But the time of precalculated kMSRf is increased a little.

6   Conclusions

The *top-k* query is well used in spatial dara or searching multi-dimensional data stream to recognize important objects. A grid-index based on reserve data points set ($RDPS$) is presented to calculate multi-dimensional preference *top-k*. Because of the correlation between *top-k* and *skyline* computation, when the count of one point's $RDPS$ is bigger than or equals to the value of k, it is for sure to be pruned the point's dominant set. $RDPS$ can be calculated with grid-index algorithm quickly. Combining grid-index and dominant graph index, we get gridded dominant graph (GDG) index. Analytical and experimental evidences display that GDG index approach performs better on both storage of index and efficiency of queries.

The GDG index we proposed in this paper can clearly reduce the index size, the random access of the dataset and increase the performance of *top-k* queries, which can satisfy the critical performance demand in robots environments.

In our future work, we will study how apply $RDSP$ in other layered index, such as Onion and AppRI, to reduce random access number of datasets. Morever, we will use our methods to analysis online data streams, which can be used in the mobile robots exploring in the computational complicated environments.

Acknowledgments. The authors acknowledge the financial support by State Key Development Program of Basic Research of China (No. 2013CB329601), National Key Technology R&D Program (No. 2012BAH38B-04), "863" program (No. 2010AA -012505, 2011AA010702) and NSFC(No. 60933005). The author is grateful to the anonymous referee for a careful checking of the details and for helpful comments that improved this paper.

References

1. Börzsönyi, S., Bailey, M., Kossmann, D., Stocker, K.: The Skyline Operator. In: Proc. 17th International Conference on Data Engineering, pp. 421–430 (2001)
2. Chang, K.C., He, B., Li, C., Patel, M., Zhang, Z.: Structured databases on the web: observations and implications. SIGMOD Rec., 61–70 (2004)
3. Chang, Y.C., Bergman, L., Castelli, V., Li, C.S., Lo, M.L., Smith, J.R.: The onion technique: indexing for linear optimization queries. SIGMOD Rec., 391–402 (2000)
4. Das, G., Gunopulos, D., Koudas, N., Tsirogiannis, D.: Answering top-k queries using views. In: Proc. 32nd International Conference on Very Large Data Bases, pp. 451–462 (2006)
5. Hristidis, V., Koudas, N., Papakonstantinou, Y.: PREFER: a system for the efficient execution of multi-parametric ranked queries. In: Proc. 2001 ACM SIGMOD International Conference on Management of Data, pp. 259–270 (2009)
6. Nepal, S., Ramakrishna, M.V.: Query Processing Issues in Image(Multimedia) Databases. In: Proc. 15th International Conference on Data Engineering, pp. 22–29 (1999)
7. Dong, X., Chen, C., Han, J.W.: Towards robust indexing for ranked queries. In: Proc. 32nd International Conference on Very Large Data Bases, pp. 235–246 (2006)
8. Dong, X., Han, J.W., Cheng, H., Li, X.L.: Answering top-k queries with multi-dimensional selections: the ranking cube approach. In: Proc. 32nd International Conference on Very Large Data Bases, pp. 463–474 (2006)
9. Zou, L., Chen, L.: Dominant Graph: An Efficient Indexing Structure to Answer Top-K Queries. In: Proc. 24th International Conference on Data Engineering, pp. 536–545 (2008)
10. Li, A.P., Han, Y., Zhou, B., Han, W., Jia, Y.: Detecting Hidden Anomalies Using Sketch for High-speed Network Data Stream Monitoring. Appl. Math. Inf. Sci. 6(3), 759–765 (2012)

A Survey on Clustering Techniques for Situation Awareness[*]

Stefan Mitsch[2], Andreas Müller[1], Werner Retschitzegger[1], Andrea Salfinger[1],
and Wieland Schwinger[1]

[1] Johannes Kepler University Linz, Altenbergerstr. 69, 4040 Linz, Austria
[2] Computer Science Dept., Carnegie Mellon University, Pittsburgh, PA-15213, USA

Abstract. Situation awareness (SAW) systems aim at supporting assessment of critical situations as, e.g., needed in traffic control centers, in order to reduce the massive information overload. When assessing situations in such control centers, SAW systems have to cope with a large number of heterogeneous but interrelated real-world objects stemming from various sources, which evolve over time and space. These specific requirements harden the selection of adequate data mining techniques, such as clustering, complementing situation assessment through a data-driven approach by facilitating configuration of the critical situations to be monitored. Thus, this paper aims at presenting a survey on clustering approaches suitable for SAW systems. As a prerequisite for a systematic comparison, criteria are derived reflecting the specific requirements of SAW systems and clustering techniques. These criteria are employed in order to evaluate a carefully selected set of clustering approaches, summarizing the approaches' strengths and shortcomings.

1 Introduction

Situation Awareness (SAW). SAW systems are increasingly used in large control centers for air or road traffic management in order to reduce the information overload of operators induced by various data sources. This is done by automatically assessing critical situations occurring in the environment under control (e. g., an accident causing a traffic jam) [3].

Data Mining (DM) for SAW. The definition of relevant critical situations is provided in current SAW systems (e.g., [3,26]) explicitly by domain experts during a *configuration phase*, representing a time-consuming task. This effort of providing explicit knowledge could be complemented and eased by a data-driven approach in terms of DM techniques making direct use of the observed data, through detecting "interesting" or uncommon relationships, which the domain experts might not be explicitly aware of (i.e., intrinsic knowledge). Furthermore, during *runtime* of the SAW system ongoing changes and additional uncommon

---

[*] This work has been funded by the Austrian Federal Ministry of Transport, Innovation and Technology (BMVIT) under grant FFG FIT-IT 829589, FFG BRIDGE 838526 and FFG Basisprogramm 838181.

Y. Ishikawa et al. (Eds.): APWeb 2013, LNCS 7808, pp. 815–826, 2013.

relationships can be spotted that have not even been explicitly defined so far. Especially *clustering techniques* are considered to be beneficial, since they require neither a-priori user-created test data sets nor other background knowledge and also allow anomaly detection, such as uncommon relations. If we consider a road traffic SAW system, clustering might for example be used to detect (ST) hotspots on a highway, i. e., road segments and time windows where atypically many accidents occur, or to reveal current major traffic flows in an urban area.

Specific Requirements of SAW. The nature of SAW systems, however, poses specific requirements on the applicability of existing clustering techniques. SAW systems have to cope with a large number of heterogeneous but interrelated real-world objects stemming from various sources, which evolve over time and space, being quantitative or qualitative in nature (as, e.g., demonstrated in [3]). The focus of this paper is therefore on evaluating existing spatio-temporal (ST) clustering techniques with respect to their ability to complement the specification of critical situations in SAW systems.

Contributions. In this paper, we first systematically examine the requirements on clustering techniques to be applicable in the SAW domain. Then, we survey several carefully selected approaches according to our criteria stemming from the fields of SAW and ST clustering, and compare them in our lessons learned, highlighting their advantages and shortcomings.

Structure of the Paper. In the following section, we compare our survey with related work (cf. Section 2). Then, we introduce our evaluation criteria (cf. Section 3), before we present lessons learned (cf. Section 4). Due to space limitations, the in-depth criteria-driven evaluation of each approach surveyed can be found online[1] only.

2   Related Work

Despite a plethora of work exists in the field of data mining, to the best of our knowledge no other survey has dealt with the topic of clustering ST data for SAW applications so far. However, there exist surveys about ST clustering (e. g., [21]) and ST data mining in general (e. g., [19], [29]), surveys about spatial (e. g., [11]) and temporal (e. g., [34]) clustering and various surveys on clustering in other domains (e. g., [15], [36]). Furthermore the field of stream data mining has received a lot of attention over the last few years and as a result several surveys on stream data clustering (e. g., [20]) and on stream data mining (e. g., [8], [13]) in general have been conducted.

In the following paragraphs these mentioned surveys are briefly discussed with respect to the contributions of our survey.

ST Clustering. Kisilevich et al. [21] propose a classification of ST data and focus their survey of clustering techniques on *trajectories*, being the most complex setting in their classification. The main part of the survey is a discussion of several groups of clustering approaches including different example algorithms for

---

[1] http://csi.situation-awareness.net/stc-survey

each group. Finally, they present examples from different application domains where clustering of trajectory data is an issue, like studying movement behavior, cellular networks or environmental studies. In contrast to our survey, Kisilevich et al. focus on presenting various groups of ST data mining approaches, rather than systematically analyzing different techniques backed by a catalog of evaluation criteria. Furthermore, we considered the applicability of the techniques to our domain on basis of SAW-specific criteria, while they conducted their analysis in a more general way. Nevertheless, their work represented a valuable starting point for our survey, from which we especially adopted their classification of ST data, like trajectories or ST events.

Spatial, Temporal and General Clustering. Several surveys on spatial clustering (e.g., [11]) and temporal clustering (e.g., [34]) have been conducted for different application domains. Nevertheless, none of the surveyed approaches specifically deals with the unique characteristics that arise from the domain of SAW respectively from ST data in general (cf. Section 3.1), but take only either spatial- or temporal characteristics into account.

Finally, there exist numerous and comprehensive surveys on non-ST clustering (e.g., [15], [36]). These approaches cannot simply be used to work with ST data, but have to be adopted manually to deal with its particular nature.

ST Data Mining. Kalyani et al. [19] describe the peculiarities of ST data models and the resulting increase of complexity for the data mining algorithms. In their survey they outline different data mining tasks compared to their spatial counterparts and motivate the need for dedicated ST data mining techniques. Geetha et al. [29] present a short overview of challenges in spatial, temporal and ST data mining. However, none of the above focuses on concrete data mining techniques, but rather gives an overview of the field of ST data mining.

Stream Data Mining. In their survey on clustering of time series data streams, Kavitha et al. [20] review the concepts of time series and provide an overview of available clustering algorithms for streaming data. More general surveys of stream data mining were conducted by Gaber et al. [8] and Ikonomovska et al. [13]. Both review the theoretical foundations of stream data analysis and give a rough overview of algorithms for the various stream data mining tasks and applications. Furthermore, they mention that stream data clustering is a major task in stream data mining and lots of algorithms have been adapted to work on streaming data (e.g., CluStream [1], DenStream [7], ClusTree [22]).

However, in contrast to our SAW systems which store a history of the data received, thus allowing an arbitrary number of read accesses, such stream data mining approaches do not store the huge amount of data they process. Even though the elements of a data stream are temporally ordered, they might arrive in a time-varying and unpredictable fashion and do not necessarily contain a timestamp, which does not allow for the identification of temporal patterns (especially cyclic ones). Furthermore, most stream data mining approaches surveyed by these authors do not deal with spatial data at all and hence are not applicable here.

Consequentially, we conduct this survey on ST clustering to investigate the applicability of various clustering techniques for the field of SAW.

3   Evaluation Criteria

In this section, we derive a systematic criteria catalog, (methodologically adhering to some of our previous surveys, e.g., [35]), which we use for evaluating selected ST clustering techniques. The criteria catalog consists of two sets of criteria, as depicted in Fig. 1, comprising SAW-specific criteria (cf. Section 3.1) and clustering-specific criteria (cf. Section 3.2), which are detailed in the following. Each criterion is assigned an abbreviation for reference during evaluation.

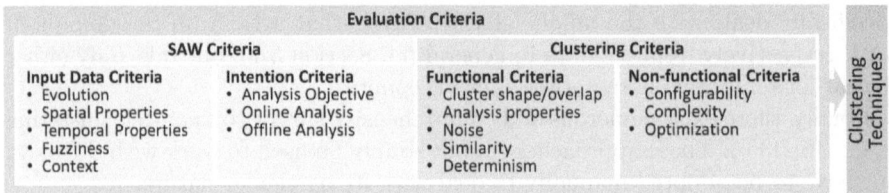

Fig. 1. The evaluation criteria at a glance

3.1   SAW-Specific Criteria

The rationale behind the following SAW-specific criteria, which are illustrated by examples from the domain of road traffic control, is based on considering the kind of input data a clustering technique has to cope with ("What do we have") and further on reflecting on the intent of the analysis ("What do we want").

Spatial, Temporal and Other Input Data Properties. As already mentioned, SAW systems employed in control centers need to monitor a large number of interrelated objects anchored in space and time, both either with an extent or without. In particular the *temporal properties* (TP) range from instants to intervals, whereas the *spatial properties* (SP) comprise points, one-dimensional intervals (e. g., lines) or two-dimensional intervals (e. g., regions). Non-ST properties can be of qualitative (e.g., freezing temperature) or quantitative (e.g., $0\,°C$) nature. Regarding SP and TP, however, we assume quantitative data, as typically required by SAW applications to monitor the ST environment.

Heterogeneity of Input Data. Another SAW-specific requirement is that objects constituting a certain critical situation often exhibit a mixture of different ST properties. This can be exemplified by a situation involving three objects, (i) a bus, sending location points at time instants via GPS to the control center (i. e., a trajectory), (ii) a traffic jam, comprising a spatial and temporal interval, both evolving over time (i. e., growing, shrinking, moving) and (iii) a fog area being spatially extended over a certain region and characterized by a temporal interval (e. g., predicted for two hours). The more different ST properties data a clustering technique supports, i.e., allows for *heterogeneous input data* (HID), the better it is applicable to the SAW domain.

Evolution of Input Data. As indicated in the examples above, the temporal dimension furthermore captures the potential *evolution* of the observed objects (E) with respect to their spatial (e.g., location or length) or non-spatial (e.g., temperature) properties. An evolution along the spatial dimension corresponds to a *mobile*, i.e., moving, object, whereas an object solely evolving with respect to the non-spatial properties represents an *immobile*, i.e., static, object.

Context of Input Data. The input data used in SAW systems often comprises objects which are bound to a certain *context* (CID), enforcing constraints on the interpretation of the input data (e.g., in road traffic, the majority of objects cannot move around in space freely, but is bound to the underlying road network), or further requirements on the input data (e.g., necessity of velocity information). We evaluate if and what kind of context information is required.

Fuzzy Input Data. SAW systems often have to cope with *fuzzy input data* (FID), for example incidents reported by humans who only have a partial overview of the situation. Fuzzy input data might address the spatial properties (SP) (e.g., uncertainty about the exact location of an object), the temporal properties (TP) (e.g., an accident has occurred within the last half an hour), or the non-ST properties (e.g., it cannot be defined exactly what has happened).

Besides the criteria mentioned above, which detail the nature of the input data, the following further deal with the goal of the analysis.

Intention. This criterion (I) reflects the *objective of the analysis*, i.e., the kind of implicit knowledge that should be extracted by the clustering technique. Possible intentions are clustering of *events* or *regions* with similar ST characteristics, clustering of *trajectories*, or the detection of *moving clusters*.

Online or Offline Analysis. The requirements imposed on the clustering techniques differ with respect to the phase they should be employed. During the configuration phase of a SAW system, we have to perform an analysis on a complete, historical data set, i.e., *offline analysis*. However, since an SAW system typically operates on a real-time environment, we also want to perform an analysis at runtime, i.e., *online analysis*. Since clustering techniques devoted to online analysis often rely on optimizations and approximations in order to deliver fast results (e.g., compute only locally optimal clusters), these approaches are less suited for configuration tasks where computation time is not a major issue, whereas an exact result is preferred. Thus, this criterion (OOA) reflects whether the clustering technique is only suited for the offline configuration phase, or if it is also applicable to or favored for runtime analysis.

3.2 Criteria Imposed by Clustering Techniques

Whereas in the previous section, the criteria have been derived from the nature of data and the intention of the analysis stemming from the SAW domain, the present section focuses on assessing *how* these goals can be achieved with dedicated ST clustering techniques.

We first give a short description of the main contribution (MC) and distinguish the clustering techniques according to their *algorithm class* (AC), describing the method how the clusters are obtained. Following [12] this can be, due to *partitioning* (i.e., the data space is partitioned as a whole into several clusters), *hierarchical clustering* (i. e., clusters are created by merging close data points bottom-up or splitting clusters top-down), *density-based clustering* (i. e., clusters are defined as exceeding a certain density of data points over a defined region), and *grid-based clustering* (i. e., the data space is overlaid by a grid, grid cells containing similar structures are merged).

The remaining criteria are structured into functional and non-functional ones.

Functional Criteria. The distinct algorithmic methods yield different *cluster shapes* (CS), distinguishing *spherical, rectangular* or *arbitrarily* shaped clusters. Besides, it is investigated if a technique can handle *clusters which overlap* (CO). For *analysis properties*, we consider *spatial analysis properties* (SA), i.e., what kind of spatial data is internally processed by the algorithm, *temporal analysis properties* (TA), i.e., time instants or intervals, *temporal patterns* (TAP), i.e., linear or cyclic time patterns, and the *focus* of the analysis (F), i.e., if spatial and temporal aspects are handled equally or if one is favored without ignoring the other. We evaluate if the techniques support dedicated handling of *noise* (N) within the data set and furthermore investigate the employed *similarity measures* (SM) and the *determinism* of the produced results (D).

Non-functional Criteria. These criteria comprise the *configurability* (C) of the approaches, reflecting to which degree the clustering technique can be tweaked, and the computational *complexity* (CC). Furthermore, we examine whether and which *optimization strategies* (O) are provided, which would be beneficial in reducing runtime, however might also affect the quality of the obtained clustering result.

4   Survey of ST Clustering Techniques

In the following we present our selection of techniques, and group and analyze them according to our criteria.

4.1   Rationale behind the Selection of Techniques

We carefully selected ST clustering approaches ranging from very recent ones on the one hand, to several more mature ones on the other hand to provide a broad overview of the field of ST clustering. We did not include *visually aided approaches* (e. g., Andrienko et. al [2]) in our survey as they contradict our approach to complement the configuration of SAW systems in a more data-driven way because a purely user-guided clustering approach again has to be steered by a domain expert.

In the following, we structure our discussion into three groups according to the *kind of evolution* the techniques consider. The rationale behind these categories is that algorithms from the different groups share several properties, which is reflected in the evaluation criteria tables.

		Input Data									Intention (I)					
		Evolution (E)		Properties (HID)							No Evolution				Analysis (OOA)	
		Spatial		Spatial (SP)			Temporal (TP)		Fuzziness (FID)	Context (CID)						
		Immobile	Mobile	Point	Line	Region	Instant	Interval			Cluster ST-Events	Cluster ST-Regions	Cluster Trajectories	Moving Clusters	Online	Offline
No Evo.	Wang, 2006	✓	-	✓	-	-	✓	-	-		✓	-	-	-	-	✓
	Birant, 2006	✓	-	✓	-	-	✓	-	-		-	✓	-	-	-	✓
	Tai, 2007	✓	(✓)	✓	-	-	✓	-	-		✓	-	-	-	✓	✓
Object Evo.	Gaffney, 1999	-	✓	✓	-	-	✓	-	-		-	-	✓	-	-	✓
	Nanni, 2006	-	✓	✓	-	-	✓	-	-		-	-	✓	-	-	✓
	Li, 2010	-	✓	✓	-	-	✓	-	-		-	-	✓	-	✓	✓
	Lu, 2011	-	✓	✓	-	-	✓	-	-		-	-	✓	-	-	✓
	Gariel, 2011	-	✓	✓	-	-	✓	-	-		-	-	✓	-	✓	✓
Cluster Evolution	Kalnis, 2005	-	✓	✓	-	-	✓	-	-		-	-	-	✓	✓	✓
	Iyengar, 2004	✓	-	✓	-	-	✓	-	-		✓	-	-	-	-	✓
	Neill, 2005	✓	-	✓	-	-	✓	-	-		✓	-	-	-	-	✓
	Li, 2004	-	✓	✓	-	-	✓	-	-	requires velocity	-	-	-	✓	-	✓
	Jensen, 2007	-	✓	✓	-	-	✓	-	✓	requires velocity	-	-	-	✓	✓	-
	Chen, 2007	-	✓	✓	-	-	✓	-	-	spatial network & velocity	-	-	-	✓	✓	-
	Rosswog, 2008	-	✓	✓	-	-	✓	-	-		-	-	-	✓	-	✓
	Jeung, 2008	-	✓	✓	-	-	✓	-	-		-	-	✓	-	-	✓
	Rosswog, 2012	-	✓	✓	-	-	✓	-	-		-	-	-	✓	-	✓
	Zheng, 2013	-	✓	✓	-	-	✓	-	-		-	-	✓	✓	(✓)	✓

Fig. 2. Situation Awareness Criteria Table

The first group comprises techniques that *do not entail any evolution* (NE), like clustering ST events (i. e., grouping events in close or similar regions) or ST groups (i. e., finding regions sharing similar physical properties over time). Techniques for trajectory clustering (i. e., grouping similar trajectories) deal with the *evolution of objects* that move along these trajectories (OE). And finally, the area of moving-object clustering (i. e., discovery of groups of objects moving together during the same time period, so called moving clusters) and the detection of spatio-temporal trends (e.g., disease outbreaks) deal with the *evolution of clusters* over time (CE).

4.2   Lessons Learned

Our evaluation of ST clustering approaches for the field of SAW has revealed interesting peculiarities of current clustering techniques. In the following, we explain our findings grouped according to our evaluation criteria (cf. also Figure 2, 3 and 4). Note that a tick in parentheses means that the criteria is only partly fulfilled by the approach.

Evolution Mostly Supported (E). Algorithms from the NE group work with objects in fixed locations and thus do not support spatial evolution at all. All other techniques allow for moving objects.

Spatial and Temporal Extent Not Handled (SP, TP). None of the surveyed techniques can handle anything but spatial points or temporal instances —

lines, rectangles or temporal intervals are not dealt with. Hence, currently none of the approaches is able to directly deal with the heterogeneity (HID) criterion.

Fuzziness of Data Is Not an Issue (FID). All of the approaches treat objects as facts and do not consider any uncertainties, except Jensen et al. [16] who consider that objects *might disappear* without notifying the server and reduce the confidence in the assumed object movement as time passes.

Almost NO Context Knowledge Supported (CID). Most of the algorithms work without any knowledge of context and do not allow any further information than a data set and parameter values. Exceptions are techniques that cluster the objects backed by a network graph (e. g., [5]), or techniques that make use of velocity information for moving objects (e. g., [16], [23]).

Techniques Mainly Focus on Offline Clustering (OOA). While the surveyed techniques mainly aim at offline clustering of closed data-sets, only few exceptions (from each group) allow online clustering of ever-changing data-sets (e. g., [32], [10], [16]).

Majority of Algorithms Is Density-Based (AC). Our survey comprises clustering techniques from all classes of algorithms, whereby OE techniques are usually density based, while CE techniques cover all algorithm classes.

Predominance of Arbitrarily Shaped Clusters (CS). OE and NE techniques mostly result in arbitrary clusters, whereas clusters produced by CE techniques are often restricted to spherical or rectangular shape (e. g., [14], [30]).

Cluster Overlap in CE Techniques (CO). Overlapping clusters are only considered by few of the techniques mostly from the CE group (e. g., Kalnis et al. [18]). These approaches are able to detect different clusters moving through each other and can keep them separated until they split again.

Spatial Analysis Properties Highly Dependent on Intent of Analysis (SA). While NE and CE techniques are focused on clustering data points, OE techniques mostly deal with clustering of lines, since a trajectory can be interpolated to a line. Li et al. [23] first group the objects into micro-clusters (i. e., small regions) and then combine these regions to complete clusters. However, as they cluster the micro-clusters center points, they also deal with points only.

Only Temporal Instants (TA) and Linear Patterns (TAP). Linear patterns for time instants are predominant in the surveyed approaches, while other temporal properties like cyclic time patterns or intervals are unhandled in the majority of cases, especially by CE techniques. Only Birant et al. [4] include cyclic time patterns, and Nanni et al. [27] consider time intervals.

Focus on ST data (F). Most of the techniques are focused on handling the specific nature of ST data. However, there are several spatial-data dominant techniques originally stemming from the field of spatial clustering, enriched with the ability of dealing with temporal data aspects (e. g., [4], [32]). Only Nanni et al. [27] presented a temporal-data dominant variation of their algorithm.

Sparse Explicit Noise Handling (N). As most of the proposed techniques extend well-known clustering algorithms to work with ST data, the handling

		Main Contribution (MC)	Partitioning	Hierarchical	Density-based	Grid-based	Configurability (C)	Computational Complexity (CC)	Optimization strategies (O)
No Evo.	Wang, 2006	DBSCAN-based / grid based ST-clustering	-	-	✓	✓	•calculated with k-dist graph	O(n) / O(n²)	fast vs. precise
	Birant, 2006	DBSCAN-based ST-clustering	-	-	✓	-	•min # points •density •similarities	O(n*log n)	improved R-Tree, filters to reduce search space
	Tai, 2007	incremental ST-clustering of dual data	✓	-	-	-	•mostly determined automatically	not specified	NeiGraph; fast vs. precise
Object Evo.	Gaffney, 1999	clustering trajectories through regression	✓	-	-	-	•# clusters	O(N K l) (K...# clust., l...# iter.)	-
	Nanni, 2006	clustering trajectories with focus on time	-	-	✓	-	•min # points •time window •similarities	O(n*log n) / O(n_iter*n*log n)	M-Tree
	Li, 2010	incremental clustering trajectories with micro clusters	-	-	✓	-	•distance threshold •parameters for macro-clustering	not specified	micro-clusters, kinetic heap
	Lu, 2011	clustering trajectories through semantic regions	-	-	✓	-	•density threshold •distance radius •min # points	not specified	-
	Gariel, 2011	detection of aircraft waypoints and nominal trajectories, aircraft anomaly detection	-	-	✓	-	•DBSCAN parameters •# PCs •# sample points	not specified	-
Cluster Evolution	Kalnis, 2005	detecting moving clusters	-	-	✓	-	•DBSCAN parameters •integrity threshold Θ		pruning redundant cluster comb./approximations
	Iyengar, 2004	detecting evolving space-time clusters	-	-	-	✓	•max. population candidate size •nr. of search iterations	not specified	-
	Neill, 2005	detecting evolving space-time clusters	-	-	-	✓	•temporal window size •level of data aggregation	O(N_S T+N⁴T)	-
	Li, 2004	clustering moving objects into micro-clusters to enable generic clustering	-	✓	-	-	•split threshold	O((N+U)*log(N+U)* log(N)+N*log³(N))	heuristics
	Jensen, 2007	continuous moving-object clustering	-	✓	-	-	•max. update time/cluster capacity •time window size	not specified	disk-based data structure, event queue, hash tab.
	Chen, 2007	clustering moving-objects on a network	✓	-	✓	-	•max distances	not specified	cluster blocks
	Rosswog, 2008	detecting and tracking spatio-temporal clusters with adaptive history filtering	✓	-	-	-	•order of the filter, filter type (flat, adaptive, adaptive+stability)	not specified	results improve as history grows
	Jeung, 2008	clustering of convoys, i.e., groups of objects that travel together for at least a min. duration of time	-	-	✓	-	•min # of cluster objs. •distance value e •lifetime k, length of time partition λ •tolerance of trajectory simplification δ	not specified	clustering on simplified trajectories for detection of convoy candidates
	Rosswog, 2012	DBSCAN-based ST-clustering backed by a SVM	-	-	✓	-	•parameterized kernel •DBSCAN •training data set required	Relationship graph: O(n*log n)	R-Tree
	Zheng, 2013	clustering of gatherings	-	-	✓	-	•DBSCAN parameters •variation threshold δ •lifetime threshold of crowd/participator •support threshold of crowd/gathering	not specified	diff. spatial indexing techniques, pruning

Fig. 3. Clustering Criteria Table (part 1)

of noise highly depends on the abilities of the underlying base algorithm. An example for an often used algorithm is DBSCAN [6], which is generally tolerant to noise, although it is prone to errors if clusters of different densities exist. Only Birant et al. [4] and Rosswog et al. [30] explicitly address the topic of noise and inconsistencies by proposing an additional density parameter or a stability filter. As detection of noise might be used to find ST outliers, approaches that deal with anomalies might be applicable in different ways within an SAW system.

Euclidean Distance as Similarity Measure (SM). Regarding similarity measures, most authors make use of an adjusted version of the Euclidean distance (ED) (e.g., weighted ED, squared ED). Obvious exceptions are Chen et al. [5] who operate on a network graph and thus use a network distance and some special approaches that do not require a similarity measure at all (e.g., [33]). A border case is the similarity measure used by Tai et al. [32], who define their own cost function looking similar to the ED, considering only (what they call) optimization attributes, thus excluding spatial components.

Mostly Deterministic Techniques (D). Only few of the techniques use heuristic approaches to deliver fast, but more inexact results (e.g., Iyengar et al. [14] who propose a heuristic cluster search because of the huge search space).

Various Levels of Configurability (C). The number of parameters ranges from none or automatically approximated to up to seven parameters. For instance,

		Cluster		Analysis Properties										Noise (N)	Similarity Measures (SM)	Determinism (D)	
				Spatial (SA)			Pattern (TAP)		Temporal (TA)		Focus (F)						
		Shape (CS)	Overlap (CO)	Point	Line	Region	Linear	Cyclic	Instant	interval	spatial	temporal	spatio-temp.			deterministic	heuristic
No Evo.	Wang, 2006	A	-	✓	-	-	✓	-	✓	-	-	-	✓	-	n.a. / arbitrary	✓	-
	Birant, 2006	A	-	✓	-	-	✓	✓	✓	-	✓	-	-	density parameter	arbitrary (ED used for evaluation)	✓	-
	Tai, 2007	A	-	✓	-	-	✓	-	✓	-	✓	-	-	-	cost function	✓	-
Object Evo.	Gaffney, 1999	A	-	-	✓	-	✓	-	✓	-	-	-	✓	-	Membership probabilities	-	✓
	Nanni, 2006	A	-	-	✓	-	✓	-	-	✓	-	✓	✓	-	average ED between objects	✓	-
	Li, 2010	A	-	-	✓	-	✓	-	✓	-	-	-	✓	micro-cluster extent	composed three-part distance	✓	-
	Lu, 2011	A	-	-	✓	-	✓	-	✓	-	-	-	✓	-	weighted squared ED	✓	-
	Gariel, 2011	A	✓	-	✓	-	✓	-	✓	-	-	-	✓	projecting trajectories to PCs	dist. betw. projections on PCs	✓	-
Cluster Evolution	Kalnis, 2005	A	-	✓	-	-	✓	-	✓	-	✓	-	-	-	spatial distance measure (e.g., ED)	✓	-
	Iyengar, 2004	R	-	✓	-	-	✓	-	✓	-	-	-	✓	-	n.a.	-	✓
	Neill, 2005	R	-	-	-	✓	✓	✓	-	✓	-	-	✓	-	n.a.	-	✓
	Li, 2004	S	-	✓	-	-	✓	-	✓	-	-	-	✓	-	ED (exchangeable)	✓	-
	Jensen, 2007	S	-	✓	-	-	✓	-	✓	-	-	-	✓	-	weighted squared ED over time period	✓	-
	Chen, 2007	A	-	✓	-	-	✓	-	✓	-	✓	-	-	-	network distance	✓	-
	Rosswog, 2008	S	✓	✓	-	-	✓	-	✓	-	✓	-	-	-	filtered ED squared over time period	-	(✓)
	Jeung, 2008	A	-	-	✓	-	✓	-	✓	-	-	-	✓	-	specialized measure of ED between simplified trajectories	✓	-
	Rosswog, 2012	A	✓	✓	-	-	✓	-	✓	-	-	-	✓	SVM	n.a.	✓	-
	Zheng, 2013	A	-	-	✓	-	✓	-	-	✓	-	-	✓	-	Hausdorff distance	✓	-

S... spherical R... rectangular A... arbitrary

Fig. 4. Clustering Criteria Table (part 2)

DBSCAN-based algorithms usually require a minimum number of points per cluster and a desired cluster radius, while grid based clustering techniques require a grid-cell border length. We highlight the approach by Wang et al. [33], where parameters can be chosen arbitrarily, but might be approximated by an optional technique, thus offering a hybrid approach of parameter settings.

Computational Complexity Rarely Included (CC). Most authors did not provide the computational complexity of their approaches. Noticeable is the grid based algorithm proposed by Wang et al. [33] that has the advantage of a linear runtime complexity, as a single iteration of the grid is sufficient to discover ST clusters.

Few Optimization Strategies (O). Only some of the authors suggest usage of special optimization strategies, like indexes or pruning. Some authors suggest different variants of their algorithms, offering a trade-off between execution efficiency and quality of results. For example Tai et al. [32] suggest an exact technique for cluster discovery, and a second variant that is more efficient in execution, but might not yield the best results.

4.3 Conclusion

In this paper we focused on ST clustering approaches in the domain of SAW. We proposed evaluation criteria stemming from the field of SAW on the one hand and from the field of ST data mining on the other hand, to evaluate the approaches with respect to their applicability to the field of SAW.

Summing up, we state that none of the surveyed approaches fulfills all the criteria stemming from the special nature of data in SAW systems. Spatial and temporal extent (SP) as well as cyclic time patterns (TAP) are not the focus of ST clustering techniques. Also fuzzy input data (FID) is not handled in the predominant number of cases and only few online approaches (OOA) exist. As long as no appropriate techniques are available, we suggest transformation of the input data to enable application of the clustering techniques reviewed in this survey (e. g., only the starting points of traffic jams could be used for clustering, in order to apply techniques operating on point input data only and thus discarding the information about their extent).

References

1. Aggarwal, C.C., et al.: A framework for clustering evolving data streams. In: Proc. of 29th Int. Conf. on Very Large Data Bases, pp. 81–92. VLDB Endowment (2003)
2. Andrienko, G., Andrienko, N.: Interactive cluster analysis of diverse types of spatio-temporal data. ACM SIGKDD Explorations Newsletter 11(2), 19–28 (2009)
3. Baumgartner, N., et al.: Beaware!—situation awareness, the ontology-driven way. Int. Journal of Data and Knowledge Engineering 69(11), 1181–1193 (2010)
4. Birant, D., Kut, A.: St-dbscan: An algorithm for clustering spatial and temporal data. Data & Knowledge Engineering 60(1), 208–221 (2007)
5. Chen, J., Lai, C., Meng, X., Xu, J., Hu, H.: Clustering moving objects in spatial networks. In: Kotagiri, R., Radha Krishna, P., Mohania, M., Nantajeewarawat, E. (eds.) DASFAA 2007. LNCS, vol. 4443, pp. 611–623. Springer, Heidelberg (2007)
6. Ester, M., et al.: A density-based algorithm for discovering clusters in large spatial databases with noise. In: Proc. of 2nd Int. Conf. on Knowledge Discovery and Data Mining (KDD), pp. 226–231. AAAI Press (1996)
7. Cao, F., et al.: Density-based clustering over an evolving data stream with noise. In: SIAM Conf. on Data Mining, pp. 328–339 (2006)
8. Gaber, M.M., et al.: Mining data streams: a review. ACM SIGMOD Record 34(2), 18–26 (2005)
9. Gaffney, S., Smyth, P.: Trajectory clustering with mixtures of regression models. In: Proc. of the 5th ACM SIGKDD Int. Conf. on Knowledge Discovery and Data Mining, KDD 1999, pp. 63–72. ACM (1999)
10. Gariel, M., et al.: Trajectory clustering and an application to airspace monitoring. Trans. Intell. Transport. Sys. 12(4), 1511–1524 (2011)
11. Han, J., et al.: Spatial clustering methods in data mining: A survey. In: Geographic Data Mining and Knowledge Discovery, pp. 1–29 (2011)
12. Han, J., et al.: Data Mining Concepts and Techniques, 3rd edn. Morgan Kaufmann (2011)
13. Ikonomovska, E., et al.: A survey of stream data mining. In: Proc. of 8th National Conf. with Int. Participation, ETAI (2007)
14. Iyengar, V.S.: On detecting space-time clusters. In: Proc. of 2nd Int. Conf. on Knowledge Discovery and Data Mining (KDD), pp. 587–592. AAAI Press (1996)
15. Jain, A.K., et al.: Data clustering: a review. ACM Computing Surveys 31(3), 264–323 (1999)
16. Jensen, C., et al.: Continuous clustering of moving objects. IEEE Transactions on Knowledge and Data Engineering 19(9), 1161–1174 (2007)
17. Jeung, H., et al.: Discovery of convoys in trajectory databases. Proc. VLDB Endow. 1(1), 1068–1080 (2008)

18. Kalnis, P., Mamoulis, N., Bakiras, S.: On discovering moving clusters in spatio-temporal data. In: Medeiros, C.B., Egenhofer, M., Bertino, E. (eds.) SSTD 2005. LNCS, vol. 3633, pp. 364–381. Springer, Heidelberg (2005)
19. Kalyani, D., Chaturvedi, S.K.: A survey on spatio-temporal data mining. Int. Journal of Computer Science and Network (IJCSN) 1(4) (2012)
20. Kavitha, V., Punithavalli, M.: Clustering time series data stream - a literature survey. Int. Journal of Computer Science and Inf. Sec. (IJCSIS) 8(1) (2010)
21. Kisilevich, S., et al.: Spatio-temporal clustering: a survey. In: Data Mining and Knowledge Discovery Handbook, pp. 1–22 (2010)
22. Kranen, P., et al.: The clustree: Indexing micro-clusters for anytime stream mining. Knowledge and Information Systems Journal 29(2), 249–272 (2011)
23. Li, Y., et al.: Clustering moving objects. In: Proc. of the 10th ACM Int. Conf. on Knowledge Discovery and Data Mining (KDD), pp. 617–622. ACM (2004)
24. Li, Z., Lee, J.-G., Li, X., Han, J.: Incremental clustering for trajectories. In: Kitagawa, H., Ishikawa, Y., Li, Q., Watanabe, C. (eds.) DASFAA 2010. LNCS, vol. 5982, pp. 32–46. Springer, Heidelberg (2010)
25. Lu, C.-T., Lei, P.-R., Peng, W.-C., Su, I.-J.: A framework of mining semantic regions from trajectories. In: Yu, J.X., Kim, M.H., Unland, R. (eds.) DASFAA 2011, Part I. LNCS, vol. 6587, pp. 193–207. Springer, Heidelberg (2011)
26. Matheus, C., et al.: Sawa: An assistant for higher-level fusion and situation awareness. In: Proc. of SPIE Conf. on Multisensor, Multisource Information Fusion. Architectures, Algorithms, and Applications, pp. 75–85 (2005)
27. Nanni, M., Pedreschi, D.: Time-focused clustering of trajectories of moving objects. Journal of Intelligent Information Systems 27(3), 267–289 (2006)
28. Neill, D.B., et al.: Detection of emerging space-time clusters. In: Proc. of the 11th ACM SIGKDD Int. Conf. on Knowledge Discovery in Data Mining, KDD 2005, pp. 218–227. ACM (2005)
29. Geetha, R., et al.: A survey of spatial, temporal and spatio-temporal data mining. Journal of Computer Applications 1(4), 31–33 (2008)
30. Rosswog, J., Ghose, K.: Detecting and tracking spatio-temporal clusters with adaptive history filtering. In: Proc. of 8th IEEE Int. Conf. on Data Mining, Workshops (ICDMW), pp. 448–457 (2008)
31. Rosswog, J., Ghose, K.: Detecting and tracking coordinated groups in dense, systematically moving, crowds. In: Proc. of the 12th SIAM Int. Conf. on Data Mining, pp. 1–11. SIAM/Omnipress (2012)
32. Tai, C.-H., Dai, B.-R., Chen, M.-S.: Incremental clustering in geography and optimization spaces. In: Zhou, Z.-H., Li, H., Yang, Q. (eds.) PAKDD 2007. LNCS (LNAI), vol. 4426, pp. 272–283. Springer, Heidelberg (2007)
33. Wang, M., Wang, A., Li, A.: Mining spatial-temporal clusters from geo-databases. In: Li, X., Zaïane, O.R., Li, Z. (eds.) ADMA 2006. LNCS (LNAI), vol. 4093, pp. 263–270. Springer, Heidelberg (2006)
34. Warren Liao, T.: Clustering of time series data-a survey. Pattern Recogn. 38(11), 1857–1874 (2005)
35. Wimmer, M., et al.: A survey on uml-based aspect-oriented design modeling. ACM Computing Surveys 43(4), 28:1–28:33 (2011)
36. Xu, R., Wunsch II, D.C.: Survey of clustering algorithms. IEEE Transactions on Neural Networks 16(3), 645–678 (2005)
37. Zheng, K., et al.: On discovery of gathering patterns from trajectories. In: IEEE Int. Conf. on Data Engineering, ICDE (2013)

Parallel k-Skyband Computation on Multicore Architecture

Xing Feng[1], Yunjun Gao[1], Tao Jiang[2], Lu Chen[1], Xiaoye Miao[1], and Qing Liu[1]

[1] Zhejiang University, China
{3090103362,gaoyj,chenl,miaoxy,liuq}@zju.edu.cn
[2] Jiaxing University, China
jiangtao_guido@yahoo.com.cn

Abstract. Given a set of elements $\mathcal{D}$ in a d-dimensional space, a k-skyband query returns all elements which are worse than at most k other elements. The k-skyband query is a fundamental analysis query. It can offer minimal candidate set for other queries such as top-k ranking query where the ranking functions are monotonic. With the development of multicore processors, it has been a trend that more algorithms focus on parallel execution to exploit the computation resources. In this paper, rather than use traditional methods to compute k-skyband, we employ parallel techniques to get the result efficiently based on multicore architecture. Extensive experiments show the efficiency and effectiveness of our proposed algorithm.

1 Introduction

In recent decades, more manufacturers are producing multicore architecture processors due to the frequency wall. As a consequence, even personal computer has 4 cores, 8 cores or more. It is meaningful to design algorithms to leverage this feature, especially for problems which are computationally expensive. k-skyband query is a typical problem of such kind that requires extensive computation. This paper investigates the problem, and designs a parallel solution using multicore processors.

Given a data set $\mathcal{D}$ of elements in a d-dimensional numerical space, we say an element p dominates another element q, if p is not worse than q in every dimension and is better than q in at least one dimension. Without lose of generality, in this paper, we assume lower value is preferred. Given a data set $\mathcal{D}$, a k-skyband query retrieves the set of elements which are dominated by at most k elements. The concept of k-skyband is closely related to another two concepts, *skyline* and *top-k ranking* queries. Given a data set $\mathcal{D}$, skyline query retrieves the elements of a data set that are not worse than others. Clearly, skyline query is a special case of k-skyband where $k = 0$. Given a data set $\mathcal{D}$ and a ranking function f, top-k ranking query retrieves the k elements with highest score according to f. Note k-skyband query keeps a minimum candidate set for top-k query. Thus, keeping k-skyband of a data set is sufficient to answer all top-k queries. Below, we give an example of k-skyband.

Example 1. Consider the 13 2-dimensional elements in a data set as shown in Figure 1. k-skyband query with $k = 0$ (or skyline) returns the set $\{a, i, k\}$. 1-skyband

Y. Ishikawa et al. (Eds.): APWeb 2013, LNCS 7808, pp. 827–837, 2013.

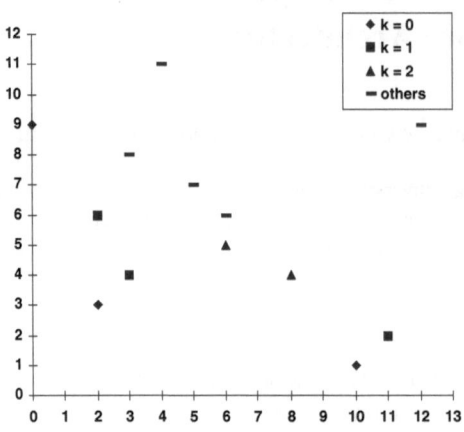

Fig. 1. A data set with 13 points

Table 1. A data set with 13 points

ID	Coordinate
a	$(10, 1)$
b	$(11, 2)$
c	$(8, 4)$
d	$(6, 6)$
e	$(12, 9)$
f	$(5, 7)$
g	$(6, 5)$
h	$(3, 4)$
i	$(2, 3)$
k	$(0, 9)$
l	$(4, 11)$
m	$(2, 6)$
n	$(3, 8)$

consists $\{a, i, k, b, h, m\}$ which are worse than at most 1 other element. 2-skyband consists $\{a, i, k, b, h, m, c, g\}$ which are worse than at most 2 other elements.

To the best of our knowledge, this paper is among the first to study the problem of parallel k-skyband computation on multicore architecture. Our main contributions are summarized as follows.

- We design a parallel algorithm to efficiently compute k-skyband query and analyse its time complexity.
- We conduct extensive experimental evaluation to verify the effectiveness and efficiency of our proposed algorithm.

The rest of the paper is organized as follows. In Section 2, we formally define the k-skyband problem and provide some necessary background information about our parallel techniques. In Section 3, we present the BBSkyband algorithm for k-skyband computation, which is used as the baseline algorithm. Section 4 presents the parallel version of BBSkyband algorithm (Parallel BBSkyband). In Section 5, we examine the efficiency of our algorithm. Related work is summarized in Section 6. This is followed by conclusions.

2   Background Information

We present the problem definition and introduce a parallel library, OpenMP, in this section.

2.1   Problem Definition

Definition 1 (Dominate). *Consider two distinct d-dimensional elements p and q, and p[i] denotes the i-th value of p. p dominates q (denoted by $p \prec q$) iff $p[i] \leq q[i]$ holds for $1 \leq i \leq d$, and there exists j, such that $p[j] < q[j]$.*

Definition 2 (k-**skyband query**). *Given a data set $\mathcal{D}$ of elements in a d-dimensional numerical space, the k-skyband query retrieves the subset of $\mathcal{D}$ which are dominated by at most k elements in $\mathcal{D}$.*

Problem Statement. In this paper, we study the problem of efficiently retrieving k-skyband elements from a data set using parallel techniques.

2.2 OpenMP

In this paper, we use OpenMP [5] to parallelize the k-skyband computation. OpenMP (Open Multiprocessing) is an API that supports multi-platform shared memory multi-processing programming in C, C++, and Fortran, on most processor architectures and operating systems. It consists of a set of compiler directives, library routines, and environment variables that influence run-time behavior.

OpenMP is an implementation of multithreading, a method of parallelizing whereby a master thread (a series of instructions executed consecutively) forks a specified number of slave threads and a task is divided among them. The threads then run concurrently, with the runtime environment allocating threads to different processors.

There are many details about the usage of OpenMP. Here, we only present the two examples we used in our implementation. More information can be found on their website[1].

Example 2.
  $\sharp$ pragma omp parallel for default (shared) private (i)
  $for(i = 0; i < size; i + +)$
    $output[i] = f(input[i]);$

The first line of Example 2 instructs the compiler to distribute loop iterations within the team of threads that encounters this work-sharing construct. Besides, "default (shared) private (i)" means all the variables in the loop region be shared across all threads except i. In the following sections, for a concise presentation, we use "parallel" instead of "$\sharp$ pragma omp parallel for default (shared) private (i)".

Example 3.
  $omp\_set\_num\_threads(\text{NUMBER\_OF\_PROCESSORS});$

The API in Example 3 instructs the compiler to set the number of threads in subsequent parallel regions to be NUMBER_OF_PROCESSORS. This can be useful to see how the running time changes against varying number of threads.

3 Baseline Algorithm

Papadias *et al.* develop a branch-and-bound algorithm (BBS) to get the skyline with the minimal I/O cost [11]. They also point that their algorithms can be used to compute k-skyband with some modifications.

First, we demonstrate the main idea behind BBS.

---
[1] http://openmp.org

Theorem 1. *An element p cannot dominate another element q, if the sum of q's all dimension is smaller than that of p. That is,* $\sum_{i=1}^{d} p[i] < \sum_{i=1}^{d} q[i] \Rightarrow q \not\prec p$.

This theorem is intuitive and can be proved using Definition 1.

Then, we modify the BBS algorithm to compute k-skyband. Algorithm 1 shows the details of an algorithm adapted from BBS algorithm. H is a minheap. S is a buffer used to collect all the elements of k-skyband. k denotes the k-skyband width. R is the root of the data set's R-tree index. All elements that cannot be in S are discarded in Line 6 (dominated by more than k elements). All elements that should be in S are inserted into S in Line 12. All entries that may contribute to final result are inserted into the heap in Line 10.

Algorithm 1. BBSkyband

---

1 $S = \emptyset$;
2 insert all entries of the root R in the heap H;
3 **while** H not empty **do**
4   remove top entry e;
5   **if** e is dominated by more than k elements in S **then**
6   discard e;
7   **else if** e is an intermediate entry **then**
8   **for each** child e_i of e **do**
9   **if** e_i is dominated by at most k elements in S **then**
10   insert e_i into H;
11   **else**
12   insert e into S

Correctness and Complexity Analysis. We prove the correctness in two steps. First, all the elements in S must be in k-skyband. This is guaranteed by Theorem 1. Second, all the elements that are discarded cannot be in k-skyband. This is intuitive because these elements are already dominated by $k + 1$ elements. We analyse the worst case complexity of Algorithm 1; i.e., when all the elements are k-skyband elements. Let N denote the size of the data set. In this case, the most time-consuming parts are Line 5 and Line 9 of Algorithm 1; and thus, the time complexity is given by $O(N^2)$.

Although the worst case time complexity is $O(N^2)$, this algorithm is still efficient because it discards the non-k-skyband elements at a level as high as possible. Below is an example for Algorithm 1.

Example 4. Consider the running example in Section 1. Suppose we are going to find 2-skyband and the R-tree is depicted in Figure 2. Applying the BBSkyband (Algorithm 1), we have the following steps in Table 2. We omit some steps in which the top entry is an element. In the initial step, we insert all the entries of root R, e_6 and e_7, into H. Then we remove top entry, e_7, from H. Note the distance of e_7 is $0 + 3 = 3$ which is smaller than that of e_6, $5 + 1 = 6$. Because no element in S dominates e_7, we expand e_7. After expanding e_7, we have $\{e_3, e_6, e_5, e_4\}$ in H. Then we remove and expand e_3. After that, we have $\{i, e_6, h, e_5, e_4, g\}$ in H. Then, we remove i. Because i is at leaf level, we

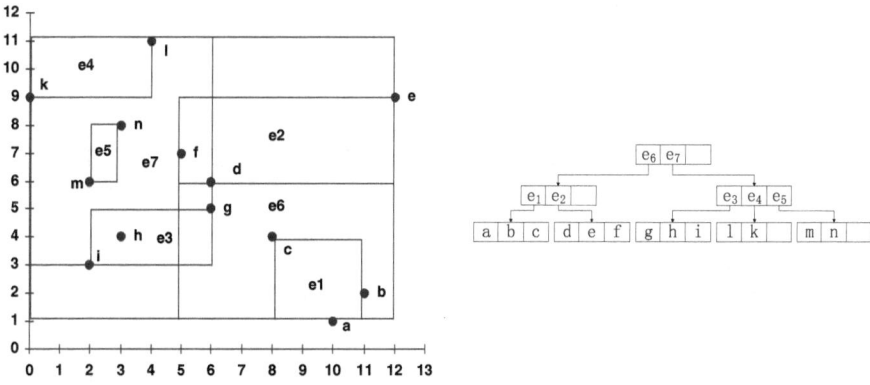

Fig. 2. The R-Tree of Figure 1

Table 2. BBSkyband Steps

Action	Heap content after action	S after action
Initial	$< e_7, 3 >, < e_6, 6 >$	$\emptyset$
Expanding e_7	$< e_3, 5 >, < e_6, 6 >, < e_5, 8 >, < e_4, 9 >$	$\emptyset$
Expanding e_3	$< i, 5 >, < e_6, 6 >, < h, 7 >, < e_5, 8 >,$ $< e_4, 9 >, < g, 11 >$	$\emptyset$
Expanding e_6	$< h, 7 >, < e_5, 8 >, < e_1, 9 >, < e_4, 9 >,$ $< g, 11 >, < e_2, 11 >$	i
Expanding e_5	$< m, 8 >, < e_1, 9 >, < e_4, 9 >, < g, 11 >,$ $< n, 11 >, < e_2, 11 >$	i, h
Expanding e_1	$< e_4, 9 >, < a, 11 >, < g, 11 >, < n, 11 >,$ $< e_2, 11 >, < c, 12 >, < b, 13 >$	i, h, m
Expanding e_4	$< k, 9 >, < a, 11 >, < g, 11 >, < n, 11 >,$ $< e_2, 11 >, < c, 12 >, < b, 13 >$	i, h, m
Expanding e_2	$< c, 12 >, < b, 13 >$	i, h, m, k, a, g
	$\emptyset$	i, h, m, k, a, g, c, b

insert i into S instead of expanding. This process goes until there is no element or entry left in H. Finally, S contains the 2-skyband, that is, $\{i, h, m, k, a, g, c, b\}$.

4   Parallel BBSkyband

In this section, we are going to parallelize the baseline algorithm in Section 3.

By analysing the algorithm, we find the most computation consuming parts are Line 5 and Line 9 in Algorithm 1 because the element is compared with all the elements in S.

A naive solution is to parallelize these two lines by using the parallel command in Example 2. However, for example, when executing Line 5, a thread has found more than k dominating elements and break out, how could other threads be notified? To solve that, we either have to let a thread waits for all other comparing threads terminate, or use communication method between threads. Both of the two ways decrease the benefit we get from parallelization.

Algorithm 2. Parallel BBSkyband

---

1 $S = \emptyset; S' = \emptyset;$
2 insert all entries of the root R in the heap H;
3 **while** H or S' is not empty **do**
4   **if** there is no intermediate S' and H is not empty **then**
5   e = remove top entry of H;
6   **for each** $e' \in S'$ and $e' \prec e$ **do**
7              $e_k++;$
8   **if** $e_k \leq k$ **then**
9                  $S' = S' \cup \{e\};$
10   **else**
11   parallel **for each** e' of S' **do**
12   **for each** $e_i \in S, e_i \prec e'$ **do**
13                  $e'_k++;$
14   **if** $e'_k > k$ **then**
15   break;
16   **for each** e' of S' **do**
17   **if** $e'_k \leq k$ **then**
18   **if** e' is an intermediate node of R **then**
19   parallel **for each** child c_j of e' **do**
20   **for each** $e \in S$ and $e \prec c_j$ **do**
21                              $c_{jk}++;$
22   **if** $c_{jk} > k$ **then**
23   break;
24   **for each** child c_j of e' **do**
25   **if** $c_{jk} \leq k$ **then**
26   insert c_j into heap;
27   **else**
28                      $S = S \cup \{e'\};$
29          $S' = \emptyset;$
30 return S;

Algorithm 2 shows the details about our algorithm to parallelize BBSkyband. S is used to store all the k-skyband elements we have found. S' is a buffer to store candidate k-skyband elements. k is the width of the k-skyband. e_k, e'_k and c_{jk} are used to record the number of elements dominating e_k, e'_k and c_{jk} respectively. Our solution is to use a buffer S' to store all candidate entries which are not dominated by more than k other elements in this buffer (Lines $4 - 9$). Then, we compare these candidate entries with elements that are already in k-skyband, i.e. elements in S (Lines $11-15$). This operation can be run concurrently. After that, if the entry is an intermediate node, we compare each child node of the entry with k-skyband elements in S concurrently (Lines $19-23$); If the element is a leaf node, we insert it into S (Line 28). By doing this, we parallelize the most time-consuming parts, Line 5 and Line 9 in Algorithm 1.

Correctness and Complexity Analysis.   The proof of correctness is quite similar to that of Algorithm 1. All the elements in S must be in k-skyband and all the elements that are discarded cannot be in k-skyband. Next, We analyse the worst case complexity of Algorithm 2; Similarly, we assume all the elements are k-skyband elements. Let N denote the size of the data set. In this case, the most time-consuming parts are Line 12 and Line 20 of Algorithm 2. And thus, the time complexity is given by $O(N^2/c)$ where c is the number of cores.

Example 5. Consider the running example in Section 3. Suppose we are going to find 2-skyband using *parallel BBSkyband* (Algorithm 2). Table 3 shows the steps of the algorithm. We omit some steps where there is no intermediate entry in S'. In the initial step, we insert all the entries of root R, e_6 and e_7, into minheap H. Then we remove top entry e_7. Because no element in S' dominates e_7, e_7 is stored in S'. In the next loop, because there is an intermediate entry (e_7) in S', we compare e_7 with every element in S and record the number of elements that dominate e_7 (Here is 0). Note that, if there are other elements in S', we can compare them with elements in S concurrently. Then, we expand e_7 (like Lines $19 - 23$ in Algorithm 2). Note that, this process can be done concurrently. After expanding e_7, we add entries that are not dominated by more than k elements to H (like Lines $24 - 26$ in Algorithm 2). Here, we have $\{e_3, e_6, e_5, e_4\}$ in H. Then we remove and expand e_3. After that, we have $\{i, e_6, h, e_5, e_4, g\}$ in H. Note that, i will be moved to S'. Similar process goes on until there is no element or entry left in H and S'. Finally, S contains the 2-skyband, that is, $\{i, h, m, a, k, g, b, c\}$.

Table 3. Parallel BBSkyband Steps

Action	Heap content after action	S' after action	S after action
Initial	$< e_7, 3 >, < e_6, 6 >$	$\emptyset$	$\emptyset$
Expanding e_7	$< e_3, 5 >, < e_6, 6 >, < e_5, 8 >, < e_4, 9 >$	$\emptyset$	$\emptyset$
Expanding e_3	$< e_6, 6 >, < h, 7 >, < e_5, 8 >, < e_4, 9 >,$ $< g, 11 >$	i	$\emptyset$
Expanding e_6	$< e_5, 8 >, < e_1, 9 >, < e_4, 9 >, < g, 11 >,$ $< e_2, 11 >$	h	i
Expanding e_5	$< e_1, 9 >, < e_4, 9 >, < g, 11 >, < n, 11 >,$ $< e_2, 11 >$	m	i, h
Expanding e_1	$< e_4, 9 >, < a, 11 >, < g, 11 >, < n, 11 >,$ $< e_2, 11 >, < c, 12 >, < b, 13 >$	$\emptyset$	i, h, m
Expanding e_4	$< e_2, 11 >, < c, 12 >, < b, 13 >$	k, a, g, n	i, h, m
Expanding e_2	$\emptyset$	c, b	i, h, m, k, a, g
	$\emptyset$	$\emptyset$	i, h, m, k, a, g, c, b

5   Experiment

This section reports the experimental studies.

5.1   Setup

We ran all experiments on a DELL OPTIPLEX 790 with Windows 7, two quad-Intel Core CPU and 2G 1333 MHz RAM. We evaluated the efficiency of our algorithm against data set distribution, dimensionality, data set size, number of cores and k-skyband width, respectively. Our experiments are conducted on synthetic data sets.

The synthetic data sets are generated as follows. We use the methodologies in [12] to generate 1 million data elements with dimensionality from 2 to 5 and the spatial location of data elements follow two kinds of distributions, *independent* and *anti-correlated*.

The default values of the parameters are listed in Table 4. *Data set distribution* and *number of cores* do not have default values because we vary all the values of them in all experiments.

Table 4. Experimental parameters

Notation	Definition (Default Values)
D	data set distribution
d	dimensionality of the data set (4)
N	data set size (256K)
c	number of cores
k	k-skyband width (4)

Parameters are varied as follows:

- data set distribution: Independent, Anti-correlated.
- dimensionality: $2, 3, 4, 5$.
- data set size: 64K, 128K, 256K, 512K, 1024K.
- number of cores: $1, 2, 4, 8$.
- k-skyband width: $2, 4, 6, 8$.

5.2   Evaluation

We evaluate our algorithm on both Independent and Anti-correlated data set. Besides, as demonstrated in Example 3, we use the command in Example 3 to set different number of threads which can also be regarded as different cores. Particularly, when $c = 1$, we use BBSkyband as baseline algorithm rather than Parallel BBSkyband.

The first set of experiments is reported in Figure 3 where dimensionality varies. The performance of our algorithm decreases as the dimensionality increases. This is because when dimensionality increases, the cost of comparing two elements increases (Note the comparing cost is $O(d)$ where d is the dimensionality). Besides, there are usually more elements in a k-skyband due to higher dimensionality. Because with higher dimensionality, an element has to be better than another one in more dimensions. Note having more elements in k-skyband also increases the cost for operations such as Line 12 and Line 20 in Algorithm 2.

Figure 4 evaluates the system scalability towards the data set size. The performance of our algorithm decreases as the data set size increases. When data set size increases, there is a larger R-tree as index. The operation on R-tree takes more time. What is more, larger data set usually results in more elements in a k-skyband.

Figure 5 evaluates the impact of k-skyband width. As expected, Figure 5 shows that processing cost increases when k increases. k is the width of k-skyband. With larger k, we have to do more dominating test and we usually have more elements in a k-skyband.

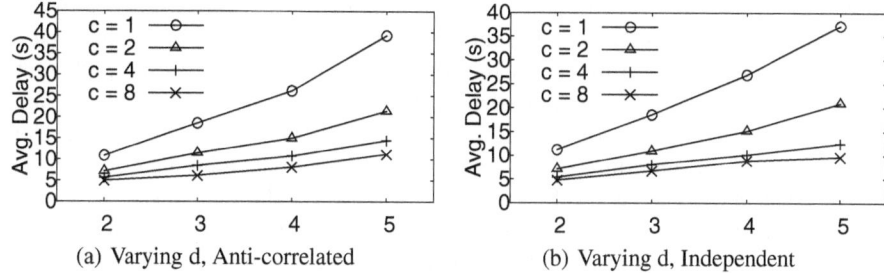

(a) Varying d, Anti-correlated   (b) Varying d, Independent

Fig. 3. Avg. Delay vs d

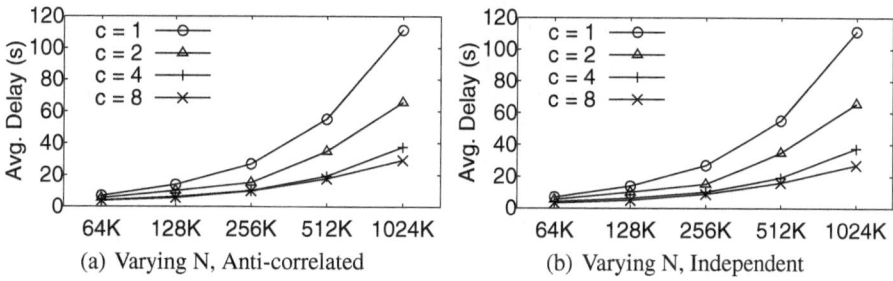

(a) Varying N, Anti-correlated   (b) Varying N, Independent

Fig. 4. Avg. Delay vs N

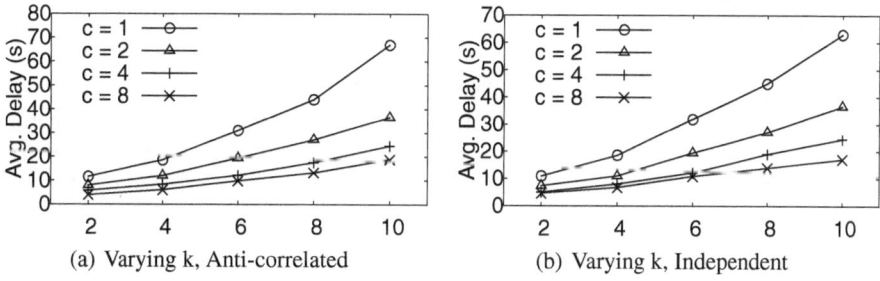

(a) Varying k, Anti-correlated   (b) Varying k, Independent

Fig. 5. Avg. Delay vs k

In all figures, there is an obvious improvement when the number of cores rises. However, it is hard to get a speedup that strictly equals the number of cores. This is because there are other costs that are not parallelized in our algorithm (I/O operation, collecting candidate elements, insertion into heap, etc.).

6 Related Work

Spatial database has draw many attention in recent decades. Beckmann *et al.* [2] design the R-tree as an index for efficient computation. Roussopoulos *et al.* [12] study Nearest Neighbor Queries and propose three efficient pruning rules. They also extend Nearest Neighbor Queries to K Nearest Neighbor Queries which return top-K preferred objects. Börzsönyi *et al.* [3] first study the skyline operator and propose an SQL syntax for the skyline query. After that, the skyline query is widely studied by a lot of papers, like Chomicki *et al.* [4], Godfrey *et al.* [7] and Tan *et al.* [13]. Papadias *et al.* develop a branch-and-bound algorithm (BBS) to get the skyline with the minimal I/O cost [11].

The concept of k-skyband is first proposed in [11]. [10] shows that it suffices to keep the k-Skyband of the data set to provide the top-k ranked queries for the monotone preference function f. Reverse k-skyband query is recently studied in [9].

With the development of multi-core processors, a recent trend is to study the parallel skyline computation. Most of them are divide and conquer algorithms. Some papers focus on distributed skyline computation. That is, their algorithm runs on server cluster with several servers, such as [1]. Other papers focus on parallize skyline computation on different cores instead of servers. This is different since all the threads share the same memory and there is much less communication cost. Park *et al.* design a parallel algorithm on multicore architecture with an easy idea. But it distributes the most computation-consuming part to every core. Our work is based on their algorithm. Vlachou *et al.* [14] employ the technique which partitions the data set according to the weight of each axis. Moreover, Köhler *et al.* [8] use a similar idea but a faster way to partition the data set. Besides, Gao *et al.* [6] design a parallel algorithm for skyline queries in multi-disk environment.

7 Conclusion

In this paper, we investigate the problem of k-skyband query. We give a modified Algorithm (BBSkyband) from the traditional algorithm and parallelize it, namely, Parallel BBSkyband algorithm. The experimental results demonstrate that a simple design of the parallel algorithm provides a satisfactory runtime performance. We believe that more efficient algorithm can be further developed based on our presented method.

Acknowledgements..   This work was supported in part by NSFC 61003049, the Natural Science Foundation of Zhejiang Province of China under Grant LY12F02047, the Fundamental Research Funds for the Central Universities under Grant 2012QNA5018, the Key Project of Zhejiang University Excellent Young Teacher Fund (Zijin Plan), and the Bureau of Jiaxing City Science and Technology Project under Grant 2011AY1005.

References

1. Afrati, F.N., Koutris, P., Suciu, D., Ullman, J.D.: Parallel skyline queries. In: Deutsch, A. (ed.) ICDT, pp. 274–284. ACM (2012)
2. Beckmann, N., Kriegel, H.-P., Schneider, R., Seeger, B.: The r*-tree: An efficient and robust access method for points and rectangles. In: SIGMOD Conference, pp. 322–331 (1990)
3. Börzsönyi, S., Kossmann, D., Stocker, K.: The skyline operator. In: ICDE, pp. 421–430 (2001)
4. Chomicki, J., Godfrey, P., Gryz, J., Liang, D.: Skyline with presorting. In: ICDE, pp. 717–719 (2003)
5. Clark, D.: Openmp: a parallel standard for the masses. IEEE Concurrency 6(1), 10–12 (1998)
6. Gao, Y., Chen, G., Chen, L., Chen, C.: Parallelizing progressive computation for skyline queries in multi-disk environment. In: Bressan, S., Küng, J., Wagner, R. (eds.) DEXA 2006. LNCS, vol. 4080, pp. 697–706. Springer, Heidelberg (2006)
7. Godfrey, P., Shipley, R., Gryz, J.: Maximal vector computation in large data sets. In: VLDB, pp. 229–240 (2005)
8. Köhler, H., Yang, J., Zhou, X.: Efficient parallel skyline processing using hyperplane projections. In: SIGMOD Conference, pp. 85–96 (2011)
9. Liu, Q., Gao, Y., Chen, G., Li, Q., Jiang, T.: On efficient reverse *k*-skyband query processing. In: Lee, S.-g., Peng, Z., Zhou, X., Moon, Y.-S., Unland, R., Yoo, J. (eds.) DASFAA 2012, Part I. LNCS, vol. 7238, pp. 544–559. Springer, Heidelberg (2012)
10. Mouratidis, K., Bakiras, S., Papadias, D.: Continuous monitoring of top-k queries over sliding windows. In: Chaudhuri, S., Hristidis, V., Polyzotis, N. (eds.) SIGMOD Conference, pp. 635–646. ACM (2006)
11. Papadias, D., Tao, Y., Fu, G., Seeger, B.: Progressive skyline computation in database systems. ACM Trans. Database Syst. 30(1), 41–82 (2005)
12. Roussopoulos, N., Kelley, S., Vincent, F.: Nearest neighbor queries. In: SIGMOD Conference, pp. 71–79 (1995)
13. Tan, K.-L., Eng, P.-K., Ooi, B.C.: Efficient progressive skyline computation. In: VLDB, pp. 301–310 (2001)
14. Vlachou, A., Doulkeridis, C., Kotidis, Y.: Angle-based space partitioning for efficient parallel skyline computation. In: SIGMOD Conference, pp. 227–238 (2008)

Moving Distance Simulation for Electric Vehicle Sharing Systems for Jeju City Area*

Junghoon Lee and Gyung-Leen Park

Dept. of Computer Science and Statistics,
Jeju National University, 690-756, Jeju Do, Republic of Korea
{jhlee,glpark}@jejunu.ac.kr

Abstract. This paper measures the moving distance for relocation strategies in electric vehicle sharing systems. The moving distance denotes the distance a customer actually moves to pick up a vehicle and indicates the service quality for carsharing. For even, utilization-based, and morning-focused relocations schemes, the experiments are conducted using taxi trip records in Jeju City area. The measurement results show that the morning-focused scheme generally outperforms the others, relocating vehicles according to the demand just in the morning time. For a large number of vehicles, the moving distance and the serviceability is not affected by static relocation schemes, revealing the essentiality of proactive relocation.

Keywords: Electric vehicle, carsharing service, relocation strategy, moving distance, analysis framework.

1 Introduction

Considering power efficiency and low carbon emissions, electric vehicles, or EVs, are hopefully expected to penetrate into our daily lives [1]. However, private ownership is not yet affordable to customers due to their high cost. Hence, EV sharing is considered to be a good intermediary business model to accelerate their deployment. EV sharing has many operation models according to the manageability, user convenience, serviceability, and others. In one-way rental, a customer picks up an EV at a station and return to a different station. Even if this model provides great flexibility to customers, it faces uneven EV distribution. Some stations do not have an EV, so a pick-up request issued at those stations cannot be admitted. This problem can be solved only by an appropriate vehicle relocation strategy. Obviously, there is trade-off between relocation frequency and relocation cost.

* This research was supported by the MKE, Republic of Korea, under IT/SW Creative research program supervised by the NIPA (NIPA-2012-(H0502-12-1002). This research was also financially supported in part by the Ministry of Knowledge Economy (MKE), Korea Institute for Advancement of Technology (KIAT) through the Inter-ER Cooperation Projects.

Y. Ishikawa et al. (Eds.): APWeb 2013, LNCS 7808, pp. 838–841, 2013.

Practically, relocation is highly likely to be carried out during non-operation hours to prevent service discontinuity. Then, the relocation problem is reduced down to decide the number of EVs for each station for the next operation hours. To begin with, service ratio, namely, the probability that an EV is available when a sharing request arrives, is an important performance metric. Besides, the moving distance, which denotes the distance a customer moves to take an EV, can indicates the fitness of sharing station, the efficiency of relocation policy, and user-side service convenience. In this regard, this paper measures the moving distance for three relocation policies, exploiting our analysis framework, which takes trip records collected from the Jeju Taxi Telematics System [2].

There have been some related researches on EV relocation schemes and simulations methods. First, [3] has analyzed the performance of multiple-station shared-use vehicle systems using diverse simulation tools. For the better knowledge of operation characteristics, UCR (University of California at Riverside) IntelliShare system began its operation with tens of electric vehicles. Both reservation-based and on-demand sharing is provided to customers. Vehicle relocation can be performed through towing and ridesharing. In ridesharing, multiple drivers take separate vehicles and then return by sharing a ride in a single vehicle. In order to reduce relocation complexity, UCR IntelliShare has introduced techniques of trip joining and trip splitting. In trip splitting, the coordinator recommends a passenger team having more than one person to take separate vehicles when the number of vehicles at the destination station goes below the lower bound. On the contrary, trip joining merges multiple trips into one if possible.

[4] begins with the two operation-based relocation strategies, namely, the shortest time and inventory balancing techniques, both of which take advantage of vehicle-to-operation center communication channels. The first one relocates vehicles to or from a neighbor station in the shortest possible time, putting emphasis on service levels. The second scheme moves vehicles from a station having surplus to a station with vehicle shortage to enhance cost efficiency. Then, a time-stepping simulation model is developed, exploiting a tremendous trip records obtained from Honda Diracc system in Singapore. Tracing a series of dynamic events on pick-up and return patterns, staff status, and job status, the number of vehicles in each station is checked against the predefined threshold values for respective stations. The simulation yields performance indicators of zero-vehicle time, full-port time, and the number of relocations.

2 Paramter Setting

The EV relocation policy must consider such factors as relocation time, the number of service men, and relocation goal. Practically speaking, relocation during the operation hours may lead to service discontinuity, so nonoperation hours are appropriate for relocation and other management activities. This paper considers 3 strategies and evaluates their performance using the sharing system analysis framework implemented in our previous work [2]. First, even relocation strategy makes the number of EVs equal for each station. Second, utilization-based relocation scheme distributes EVs according to the pick-up ratio of each station during the whole observation interval. Next, morning-focused relocation strategy just focuses on the pick-ups during the morning time, namely, just the first a few hours in the operation time.

3   Experiment Result

Figure 1 measures the moving distance according to the number of EVs. It is the distance the customer actually moves to the sharing station. The moving distance decreases according to the increase of the number of EVs, as more requests can be served. The moving distance for utilization-based scheme is the worst. The difference between morning-focused and even relocation schemes reaches 2.5 m. Here again, the morning-focuses scheme shows slightly shorter moving distance than the even relocation. The gap gets smaller when the access distance is 500 m, while both even and morning-focused schemes show almost the same distance, as shown in Figure 1(b).

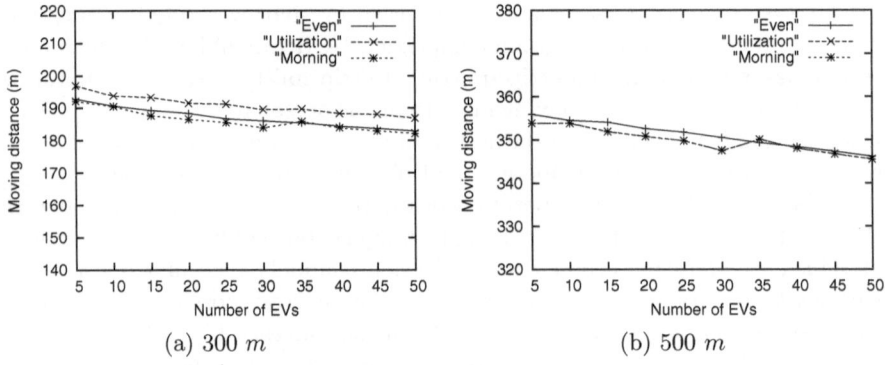

(a) 300 m (b) 500 m

Fig. 1. Effect of number of EVs

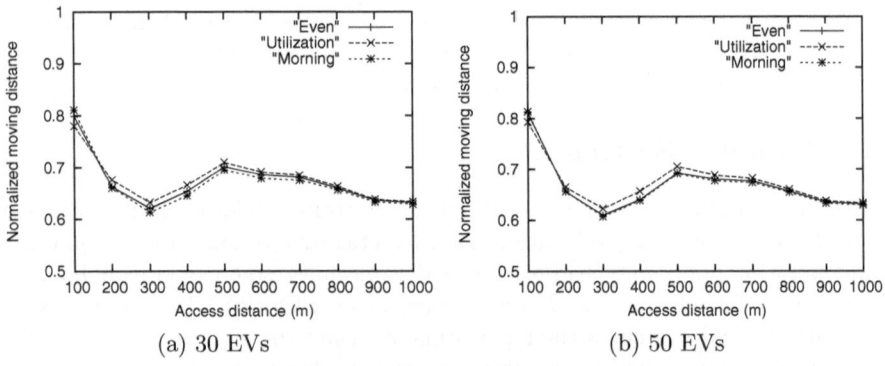

(a) 30 EVs   (b) 50 EVs

Fig. 2. Access distance effect analysis

Next, Figure 2 measures the moving distance according to the access distance. As the moving distance is sure to increase according to the increase of the access distance, the moving distance is normalized to 1.0. It denotes how far requests happen away from sharing stations. Figure 2(a) and Figure 2(b) have almost the same shape, indicating independence to the number of EVs. When the access distance is 100 m, the number of trip records is not sufficient, so this point is not consistent with other access distance range. Anyway, when the access distance is around 500 m, the sharing requests from wider range can be served. Again, the utilization-based scheme is the worst out of the three.

4   Conclusions

This paper has measured the moving distance for three practical relocation strategies in electric vehicle sharing systems. Combined with the serviceability, the moving distance indicates the easiness to access the sharing system and thus the service quality. For even, utilization-based, and morning-focused relocations schemes, the experiments are conducted using taxi trip records in Jeju City area. The measurement results show that the morning-focused scheme generally outperforms the others, making customers conveniently access the EV sharing service, discovering the necessity of a proactive relocation scheme based on the future demand forecast.

References

1. Lee, J., Kim, H.-J., Park, G.-L., Kwak, H.-Y., Lee, M.Y.: Analysis Framework for Electric Vehicle Sharing Systems Using Vehicle Movement Data Stream. In: Wang, H., Zou, L., Huang, G., He, J., Pang, C., Zhang, H.L., Zhao, D., Yi, Z. (eds.) APWeb 2012 Workshops. LNCS, vol. 7234, pp. 89–94. Springer, Heidelberg (2012)
2. Lee, J., Kim, H., Park, G., Kang, M.: Energy Consumption Scheduler for Demand Response Systems in the Smart Grid. Journal of Information Science and Engineering, 955–969 (2012)
3. Kek, A., Cheu, R., Meng, Q., Fung, C.: A Decision Support System for Vehicle Relocation Operations in Carsharing Systems. Transportation Research Part E, 149–158 (2009)
4. Barth, M., Todd, M., Xue, L.: User-based Vehicle Relocation Techniques for Multiple-Station Shared-Use Vehicle Systems. Transportation Research Record 1887, 137–144 (2004)

Author Index